SELECTED PAPERS OF

LEV P GOR'KOV

SELECTED PAPERS OF
LEV P GOR'KOV

Lev P Gor'kov
National High Magnetic Field Laboratory, Florida State University, USA

World Scientific
NEW JERSEY • LONDON • SINGAPORE • BEIJING • SHANGHAI • HONG KONG • TAIPEI • CHENNAI

Published by

World Scientific Publishing Co. Pte. Ltd.
5 Toh Tuck Link, Singapore 596224
USA office: 27 Warren Street, Suite 401-402, Hackensack, NJ 07601
UK office: 57 Shelton Street, Covent Garden, London WC2H 9HE

Library of Congress Cataloging-in-Publication Data
Gor'kov, L. P. (Lev Petrovich), 1929– author.
[Papers. Selections]
Selected papers of Lev P. Gor'kov / Lev P. Gor'kov, Florida State University, USA.
pages cm
Includes bibliographical references.
ISBN 978-9814366953 (hardcover : alk. paper)
1. Superconductivity. 2. Superconductors. 3. Quantum field theory. I. Title.
QC611.92.G67 2014
530--dc23

2014011414

British Library Cataloguing-in-Publication Data
A catalogue record for this book is available from the British Library.

Typeset by Stallion Press
Email: enquiries@stallionpress.com

Printed in Singapore

Preface

In 2010, at a Symposium organized by the National High Magnetic Field Laboratory (NHMFL) to mark my 50-plus years in theoretical physics, two of my colleagues, Nick Bonesteel and Vlad Dobrosavljevic, asked if I have ever considered publishing my collected papers. I thought then that in any event it could be very rewarding to keep such a volume in hand.

I have seen such books before. Most often several people are asked to write introductions to or give a summary of the honoree's achievements. However, upon chance I have worked in a number of fields and in a vastly different scientific environment. Besides, I did not like the idea of troubling friends with that kind of request.

So, I decided to write something myself. I thought that a few reminiscences from my Russian past would not be barren of interest even for modern readers as not much is known in the West about life of scientists in the Soviet Russia. However, there were many great names in Science, even in the Soviet period, in the fields of physics and mathematics. The Soviet system of education and research was very effective and made several significant advancements in Science and Technology possible in Russia after World War II and in very unfavorable conditions, too. It was not my intention, of course, to draw any systematic picture, just a few separate strokes. My making as a scientist was with the famous Landau School; for the Russian part of my career I always worked at the leading research Institutes of the Academy of Science of the former Soviet Union and I taught at the elite Moscow Institute for Physics and Technology.

Such as it was my plan then, I kept procrastinating as each year something or other would distract me from the project again and again. Thus, instead of being published as expected in late 2011, the book appears sometime in 2014.

A few words concerning the publications selected for this volume. Many of my papers have first appeared in Russian journals. Translated into English, they were republished in Sov. Phys.: JETP, Doklady, and JETP Letters. It so happened that for the period (1955–1965) JETP and Doklady are not available online, but namely these years were times of the remarkable advances, especially in the physics of condensed matter and statistical physics. So, papers of that period were all copied and are all reproduced in the present volume. I also included some more recent publications that seem to deserve to be read in their original form, although most of these papers are currently available online. Where it is relevant, I provide a summary of the results and remarks on the ideas or technicalities that seem important to me.

I am glad that I returned to the project at last. It gave me an opportunity to reflect on my rather long career in physics somewhat from the outside. There could be no regrets, of course, still, it was curious to look anew on what has been done and realize how much could be done instead or done differently. I thank Vlad and Nick for their advice.

The book could never have appeared without the most active help from Mr. Arshad Javed, Grants Compliance Analyst at Condensed Matter Science, NHMFL. It was he who took the pains of digging out and bringing the old journals buried in archives of the Florida State University Library and having them scanned. That last part of the job was terribly physical; one should see these huge and heavy volumes of old Sov. Phys. JETP, JETP Letters and Doklady of six issues each tightly bound together!

Lev P. Gor'kov
Tallahassee, Florida

Contents

Chapter 1

Introduction

1.1 Several words about me

I have finished a standard ten-grade school certification in Moscow in the spring of 1947. At that time, compulsory in the Soviet Union, were only the seven grades of school education. Everyone who planned to go to Moscow University or to a technical university belonging to the Ministry of the Highest Education of the USSR had to continue school for another three years and after that, pass the state examination to receive the ten-grade school certificate. Competition for enrollment in the most prestigious universities was strong. The applicants, except those who had graduated with honors, would need to pass basically the same exams anew and get good scores.

I was thinking initially about joining the Mathematics Department at Moscow State University, although I could hardly say that it would be a well-informed choice. So, the advertisement of a new school, the Department of Physics and Technology, to be organized within the Moscow State University happened very timely.

The prospectus of the new Faculty proclaimed a fully new process of education. Most impressively for the imagination of a prospective student was that from the very first day, apart from attending general lectures on the fundamentals of physics, mathematics, chemistry *et cetera*, each and every student would be assigned to one of the scientific facilities, mainly at Institutes of the Soviet Academy of Science, to take specialized courses there and gradually be involved in research. To any ambitious young fellow it was, of course, a challenge.

Each applicant had to pass the entrance exams on literature, physics and mathematics anew. But the most significant stage was the oral interview. As a student graduated with honors I was exempt from the entrance exams, but the interview was a serious matter. In my case, standing at the blackboard for about one hour, I solved problems in physics and, more importantly in mathematics offered by (as I was told later) the team of one Academician (i.e., a member of the Soviet Academy) and two professors.

Unfortunately, the new Department of Physics and Technology ("Fiztekh," in the Russian abbreviation) had many powerful enemies, and not only among adherers of the traditional Highest Education system. More dangerously, Fiztekh was constantly criticized on "ideological" grounds for "elitism," for so-called "cosmopolitanism," and for other "-isms." Finally, in 1951, several leading figures, such as Nobel Laureates: P. Kapitza, Lev Landau, several Academicians and other professors — the founders — were fired. Out of the department of Moscow University, "Fiztekh" was transformed into the independent Moscow Institute for Physics and Technology (MPTI).

Nonetheless, after years of the devastating war (1941–1945), the country was in desperate need of a new generation of scientists and engineers so that the basic principles of the "Fiztekh" educational system which were preserved at MPTI. Among them was the tradition of oral entrance exams. For the most part, the remarkable success of MPTI was due to selecting promising young fellows via the interview system.

Historically, to the extent one can know it now; the Fiztech's initiative belonged to Petr (Pyotr) Kapitza who put it forward even before the Second World War. The initiative was enthusiastically supported by a group of the Academy members and several leading figures from the industry as progress in the atomic project and in the modern aviation industry did hinge on specialists with education equal to world standards.

Years after the War was a somber period in the Soviet Union. The ideological pressure, growing anti-Semitism, a sinister atmosphere in general — especially in years before Stalin's death in 1953 — the notorious crackdown on literature in 1947, all kept everyone among the so-called intelligentsia uneasy. As for science, while biology and cybernetics were in ruins, physics and mathematics were spared. For physicists, mathematicians and engineers, it was participation in the atomic projects that had boosted the prestige and secured them a comparative independence in Soviet society.

I cannot say that the students' community had keen sense of outside events. Life of a student at Fiztekh was interesting, but it was busy. As for me, I soon realized that I had no qualifications (and no wish) to become an experimentalist or an engineer, but "Theory" as a specialty was not on the official list.

The imperative was to find the adviser. My first choice was Academician N. N. Bogolyubov and his theory group at the Steklov Mathematical Institute of the Soviet Academy of Sciences. I succeeded in talking with Bogolyubov personally and he included me in his group as a student. All went well until one day a friend of mine, a bright mathematician himself working at the Steklov Institute, was arrested on a political accusation. He previously chanced to commend me once or twice; him doing so prompted the Bogolyubov group to show that they had no interest in me anymore.

After the affront, I started studying the so-called "Theoretical Minimum" of Landau. Since Landau himself was fired from Fiztekh, there was no question about asking him to be my official adviser. Besides, as I mentioned, in 1951 Fiztekh was reorganized into the Moscow Institute for Physics and Technology and my group was transferred from MPTI to the Moscow Mechanical Institute. Luckily for me, the very transfer was such a bureaucratic mess that without saying a word I managed to finish as a theorist, under the auspice of one of Landau's disciples, Professor Ya. A. Smorodinskii. My official master diploma qualifies me as an "engineer–physicist proficient in build-up of experimental equipment and its exploitation."

I should have mentioned first that although the education was nominally free, there was still such thing as postgraduates' work assignments, i.e. the obligation to work for two years. The rule usually was not strictly enforced, however, making matters worse, the Mechanical Institute belonged to the atomic project, so it seemed almost inevitable to be assigned to one of the classified centers. A nervous time it was, but in 1953, having the support of one or two influential members of the Academy Landau succeeded in getting permission to enroll me as a postgraduate at the Institute of Physical Problems of the Academy of Sciences of the USSR.

1.2 Science in Russia after the World War II; two schools in theory

The leading role in most of the fundamental research belonged to the Academy of Sciences of the USSR. At least in non-classified fields, the organization and the very research were confined to facilities of the Academy Institutes, rather different from the university system in the West.

Though exact sciences in the USSR turned out isolated from the West after the war, physics and mathematics remained at the level of the highest world standards; in that time the pre-war generation of physicists, such figures as Petr (Pyotr) Kapiza, Lev Landau, Igor Tamm, to name just a few, were active and at the height of their powers. More new institutes of the Highest Education Ministry had been organized, in addition to MPTI to help in bringing up a whole new generation of young scientists.

As direct contacts between Russian and Western scientists were cut to a minimum on both sides, the situation as a whole was, of course, far from being wholesome. In ridiculous attempts to save on foreign currency, Soviet bureaucrats even strictly cut subscriptions of the Institutes of the Russian Academy to Western journals. Buying any modern experimental equipment was next to impossible due to the restrictions imposed by the USA on trade with the USSR. No doubt these restrictions had their negative impact, although, in a sense, they also stimulated the traditional originality of Russian science. As an example, in fundamental research, as it concerned superfluidity and superconductivity, the Soviet physicists at that time were at the forefront. Palpable lagging behind in the area of experiments and in computations started to be evident only in the early 60s.

Important roles in many areas of physics, theoretical physics and mathematics belonged to scientific Schools. As it is not my intention to draw any complete picture, in what follows I stay briefly only on two Schools in the theoretical physics: the Landau School and the group headed by N. N. Bogolyubov.

Gravitating to Landau School were many internationally acclaimed scientists; many were in charge of the theoretical groups at the Institutes of the Academy of the USSR. Common to the Landau School group was a broadness of interests, an active approach to physics, and a high professionalism based, in particular, on sound mathematical background.

Affiliation with the School was open to everyone after passing exams of the so-called Landau "Theoretical Minimum." The famous *Course of Theoretical Physics* by Landau and Lifshitz originated from the efforts by Landau to educate his disciples that ran back to his stay in Kharkov in the 1930s. Landau himself was one of the last universally educated scientists and was strong in practically every field of theoretical physics. Naturally, he did not expect all his people to be geniuses, but he wanted them to at least understand what he was saying or doing. But as a matter of principle for him, Landau would never agree to discuss physics with newcomers before they had passed all exams of the "Minimum." The exams comprised a considerable part of all ten current volumes of the Landau and Lifshitz course of theoretical physics; to pass the "Theoretical Minimum" it would usually take between one to two years.

By the time in 1953 when I started as a postgraduate, the only official position Landau had held was that of the head of a small theoretical department at the Institute for Physical Problems in Moscow. Actually, for a long time his "group" consisted of only two people: E. M. Lifshitz and I. M. Khalatnikov (A. A. Abrikosov was admitted around 1950). Then there were three postgraduates: L. P. Gor'kov, I. E. Dzyaloshinskii and L. P. Pitaevskii. Nominally, each of us had either E. M. Lifshitz or I. M. Khalatnikov as the official thesis advisor; Landau did not want any bureaucratic burdens, but his role in our making as scientists was truly critical. We all joined staff of the Landau group in the mid-1950s.

The circle of scientists who worked in wake of the Bogolyubov's ideas seemed to be smaller. With my unlucky personal experience, I can hardly say how Bogolyubov selected his pupils. He was of course one of the most outstanding figures in theoretical physics in those days, but his personal style and interests in physics were somewhat of a more mathematical character. N. N. Bogolyubov had numerous administrative positions. In the mid-1950s, he was the head of the Theoretical Department at the Steklov Mathematical Institute of the Soviet Academy of Sciences in Moscow, while still preserving posts in Kiev, Ukraine. Bogolyubov became the first Director of the Joint Institute for Nuclear Research (JINR) that the Soviet Government organized in 1956, in Dubna city, not far from Moscow. JINR was an international institution open to scientists from countries belonging to the so-called Eastern Soviet Block. The participation of European countries with somewhat more liberal traditions, such as Poland, Hungary and Czechoslovakia, made it possible for scientists at JINR to enjoy a relative amount of freedom in their contacts with the West. Unlike Landau himself or members of his group, scientists from JINR used to travel abroad.

1.3 Back to me again

Years at the Kapiza Institute for Physical Problems were most significant for each of us: Dzyaloshinskii, Pitaevskii and I. Everyday interactions with Lev Landau, his personality, his advice and, most importantly, his critique were crucial as we grew as scientists. Besides, he was the ultimate moral authority of everyone at his School and not only in Science, but in the Ethics, too.

A seemingly trifling, but characteristic example: Landau strictly enforced the alphabetic order of the authors' names in all joint publications, never exempting himself from the rule. He read every paper from the group before approving it for publication.

The ill-fated car accident in 1962 was a terrible disaster for the School. Although after much suffering Landau had partially recovered physically, in a year or two it became obvious that the Great Lev Landau would never return to Physics. The group of four (A. A. Abrikosov, L. P. Gor'kov, I. E. Dzyaloshinskii, and I. M. Khalatnikov) decided to organize an independent Theory institute in a hope to at least preserve the best traditions of the School. This institute, later the L. D. Landau Institute for Theoretical Physics of the Academy of Sciences (ITP), was established in 1966 at the Noginsk Scientific Center of the USSR Academy in Chernogolovka, not too far from Moscow. Over many years the Institute educated generations of students, mainly from MPTI, in the spirit of the Landau School. Undeniably, ITP was a success. Very soon the Institute earned broad international acclaim and became an attraction to students, and to many professionals for its unordinary scientific atmosphere. Its relative independence owing to a farness from the countless Moscow bureaucrats paved the way for truly creative and fruitful work. For that matter, the Institute was unique in the USSR. However, already in the 1970s and especially by the end of the 1980s, with the beginning of the so-called "Perestroika," centrifugal tendencies started to show up; the outside "pressure" weakened and even our students began moving away from the country.

In 1991 I was invited to the USA for a one year visit as the Miller Professor at the University at Urbana-Champaign of Illinois. The breakdown of the Soviet Union took place just few months later after my arrival. For everyone who paid attention to the political undercurrents in the country and sensed a tense atmosphere of the preceding years, the future for science in Russia looked gloomy, to say the least, so I accepted position of a Professor at the National High Magnetic Field Laboratory at the Florida State University.

1.4 Selecting papers for the volume

My PhD thesis was on "*Quantum Electrodynamics of Particles with Zero Spin*." I published few papers on its results (some with my official advisor I. M. Khalatnikov). After doing my PhD I worked on several topics in hydrodynamics, liquid helium, and the second order phase transitions.

My scientific career started in earnest at the end of 1957 with my first results in superconductivity. Therefore the book begins with the early publications on Gor'kov equations. The equations are based on the concept of the pair wave function as the symmetry order parameter, which I dare say, have played the fundamental role in all further developments in the microscopic theory of superconductivity. Together with Quantum Field Theory methods in modern statistical physics that were created jointly with A. Abrikocov and I. Dzyaloshinskii, I count these results as my basic accomplishments.

As it was rather common in the Landau School, I worked in many diverse directions, including other areas of condensed matter physics, although superconductivity has ever remained among my interests. Another longtime interest I also had was in the physics of low-dimensional systems, including organic conductors and superconductors.

While preparing this volume I tried to shorten and to clear the lists of my publications. I must confess, it was not always simple. Often, even if I knew that one or another paper was obsolete (or may even be erroneous), either the physical problem by itself or the very idea seemed worthy of attention and thc paper would remain on the main list. There are also other papers in relation to which I have lively reminiscences, often with recollections about the people who are no more.

Howbeit, I loosely sorted papers according to the main topics and grouped them together. Each group is forestalled by a short summary of the results that seemed to me new and important, at least at that time.

I could not find a proper spot for several papers of a more speculative character. As the exact results are such rarity in the physics of strongly correlated systems, nowadays in many cases it is necessary to restrict oneself by qualitative considerations guided by experiment.

Chapter 2

Fundamentals of Theory of Superconductivity

2.1 Advent of the microscopic theory

2.1.1 *Landau criterion of superfluidity.* Superfluidity and superconductivity were among the priorities at the Kapiza Institute for Physical Problems (IPP). Superfluidity of *liquid II He*, i.e. the capability of helium to flow without viscosity at low temperatures (below $T_\lambda = 2.17K$) along the narrow capillary was discovered by Kapiza in 1938. Superconductivity, when defined similarly as the disappearance of resistivity below some temperature of the transition, seems to have much in common with superfluidity.

As is known, the theory of *liquid II He* developed by L. D. Landau in 1940–1941 relates to the phenomenon of superfluidity to specific features of the quasiparticles' energy spectrum of helium. The idea of quasiparticles is one of the most fundamental concepts that Landau brought into statistical physics. That is, Landau suggested that properties of any macroscopic system, whatever the interactions between the particles of which it consists; at low enough temperatures can be interpreted in terms of the properties of a gas of excitations called *quasiparticles* and can be characterized by an energy spectrum, $\varepsilon(\vec{p})$. That spectrum has little or nothing in common with the spectrum of the non-interacting particles.

In the late 40s to early 50s the importance of *quasiparticles* as a key concept in condensed matter physics was already well-realized, especially by the Landau circle in Moscow and few other low temperature groups, in particular, in Khar'kov.

I briefly repeat here the Landau arguments explaining the mechanisms of the superfluidity of *II He*. Consider the system at rest and at $T = 0$; there are no excitations. Then consider helium moving with respect to the capillary walls with a constant velocity, V. Choose an excitation with the energy $\varepsilon(\vec{p})$ and the momentum $\vec{p}$; in the initial reference system, its wave function is proportional to: $\psi(\vec{r}, t) \propto \exp[-i\varepsilon(\vec{p})t + i(\vec{p}\vec{r})]$.

In the *moving* reference system the wave function of the very same excitation after the substitution of $\vec{r} \Rightarrow \vec{r} - \vec{V}t$ becomes $\psi(\vec{r}, t) \propto \exp[-i(\varepsilon(\vec{p}) + (\vec{p}\vec{V}))t + i(\vec{p}\vec{r})])$. Thus, the Galilean transformation of the energy of a quasiparticle transforms as $\varepsilon(\vec{p}) \Rightarrow \varepsilon(\vec{p}) + (\vec{p}\vec{V})$. Thereby, once the helium starts moving, the energy cost of an excitation can be *lower* if $(\vec{p}\vec{V}) = -pV < 0$. Above the threshold $\varepsilon(\vec{p}) - pV_{cr} = 0$ excitations start emerge spontaneously; the threshold V_{cr} signifies the onset of dissipation i.e. the non-zero viscosity.

The famous Landau expression for the critical velocity is $V_{cr} = \min[\varepsilon(p)/p]$. As the energy spectrum of helium starts with the acoustic branch $\varepsilon(p) \approx cp$ (c — the sound velocity), one could suppose $V_{cr} = c$. Actually, V_{cr} for *II He* is smaller, being determined by the so-called "*roton minimum*" in the experimental quasiparticles' spectrum.

If superconductivity was indeed the "superfluidity of superconducting currents," there should be an analogous mechanism. The microscopic BCS theory gave such mechanism directly. In fact, while in the normal phase electrons as the Fermi excitations possess the continuous spectrum $\varepsilon(\vec{p}) = v_F(p - p_F) \equiv \xi$

with p_F and v_F for the Fermi momentum and velocity, for the energy spectrum of excitations in the superconducting state the theory finds $E(p) = \sqrt{\xi^2 + |\Delta|^2}$, i.e. the spectrum has the gap $|\Delta|$. That is, non-dissipative currents can indeed flow in the system as the Landau criterion provides for the critical "velocity" a non-zero value$V_{cr} \approx \Delta/p_F$. (One should, of course, distract oneself from the fact that electrical currents generate magnetic fields by considering thin wires).

Evidences of a finite gap in the energy spectrum of superconductors grew, accumulating in the pre-BCS experimental literature in 1955–1957.

2.1.2 *BCS theory and further developments.* Leaving aside many important features, such as the retarded character of interactions inherent in the phonon mechanism, from the above standpoint the main principle result of the BCS theory is namely the non-zero gap in the energy spectrum; superconductivity itself, as phenomenon of the non-dissipative currents then follows from the Landau arguments.

As a theoretical paper in itself, BCS with its variational approach was rather cumbersome. Soon after the BCS publication, N. N. Bogolyubov proposed his method of the canonical transformation that was considerably more transparent and instantly acquired popularity.

Shortcomings both of the BCS and the Bogolyubov methods were that they were applicable only to clean superconductors. Neither formulation has provided a definition of the true order parameter, i.e. the symmetry of superconducting phase; none would allow an immediate generalization to account for the behavior of a superconductor in the presence of strong magnetic fields, neither could be extended to the case of superconducting alloys. Besides, the accuracy of the so-called weak coupling approximation and how to go beyond it if necessary was not clear.

Quantum Field Theory (QFT) methods that I describe below allowed to overcome these difficulties.

2.1.3 *Bogolyubov gives seminar.* Years that immediately followed publication of the BCS paper in 1957 were most exciting for the whole experimental and theoretical condensed matter community around the globe. As for me, I can date the beginning of my interest in superconductivity precisely: it was the day of the Bogolyubov's talk at the Landau seminar in the Kapitza Institute and I cannot help but recollect some details.

It was the fall of 1957 when the rumor spread in Moscow that Academician N. N. Bogolyubov had completed a paper on the microscopic theory of superconductivity; and some day in October, Landau invited him to give a talk. Unfortunately, without any introduction, Bogolyubov at once started with the so-called 'truncated' Hamiltonian and the method of canonical transformation, entering then into details of his rather weird "principle of compensation of 'most dangerous' diagrams". That was the worst way of delivering scientific results at the Landau Seminar, and for many in the audience the Seminar immediately became a spectacle. All the time Landau kept asking questions, sometimes impatient questions; first of all, of course, about what was the physics behind the mathematics of the Bogolyubov's canonical transformation and reasons for him choosing his ground state.

I never could understand how Bogolyubov, a man of such great caliber, could have chosen such an unfortunate way of presenting his results. Petty bickering took the whole seminar hour. Howbeit, the seminar time was two hours and after the break, N. N. Bogolyubov finally mentioned Cooper's paper and reproduced briefly the calculations on the blackboard. That was a bit embarrassing, as no one from our group had noticed the Cooper paper published in 1956, or the short BCS letter in *Physical Review* (1957).

At the time I was just a junior scientist in the Landau group. As I was listening, it came to my mind that the Cooper's instability of the Fermi sea at any *arbitrary* weak attractive electron–electron interaction implies spontaneous formation of pairs and, hence, the emergence of a boson degree of freedom. On the same day after the seminar I set myself to the task of thinking over the idea.

2.2 Gor'kov equations: publications on basics of superconductivity

In the papers of this Section the microscopic theory of superconductivity was built anew on basis of the Quantum Field Theory (QFT) methods. The equations known as *Gor'kov* equations were first published in the English translation in 1958.

2.2.1 *Quantum theory approach at* $T = 0$. The main conceptual result of the paper [2a] was the identification of the wave function of the Cooper pair as the true symmetry order parameter of the superconducting phase. The new system of equations had also demonstrated the gradient invariance of the microscopic theory.

As point of the departure was taken the BCS Hamiltonian, but rewritten in the real space as:

$$H = \int \left\{ -\left(\hat{\psi}^+ \frac{\Delta}{2m} \hat{\psi} \right) + \frac{g}{2} (\hat{\psi}^+ (\hat{\psi}^+ \hat{\psi}) \hat{\psi}) \right\} d^3 r. \tag{1}$$

(Here g stands for the strength of the point-like interaction). The equations for the field operators $\hat{\psi}$, $\hat{\psi}^+$ in the Heisenberg representation:

$$\{i\partial/\partial t + \Delta/2m\}\hat{\psi}(x) - g(\hat{\psi}^+(x)\hat{\psi}(x))\hat{\psi}(x) = 0; \{i\partial/\partial t - \Delta/2m\}\hat{\psi}^+(x) + g\hat{\psi}^+(x)(\hat{\psi}^+(x)\hat{\psi}(x)) = 0$$

were then applied to the Green function $G_{\alpha\beta}(x - x') = -i\langle T(\hat{\psi}_\alpha(x)\hat{\psi}^+_\beta(x'))\rangle$. In doing so in the *normal* metal phase, the emerging block of four operators $\langle T(\hat{\psi}_\alpha(x_1)\hat{\psi}_\beta(x_2)\hat{\psi}^+_\gamma(x_3)\hat{\psi}^+_\delta(x_4)\rangle$ can be rearranged in terms of the well-known self-energy part in the diagrammatic expansion of the Green function.

The innovative idea in [Ia] was to single out the *bosonic* degree of freedom related to the Cooper pairs condensed at $T = 0$ by presenting the block of the four field operators in the following form:

$$\begin{aligned} &\langle T(\hat{\psi}_\alpha(x_1)\hat{\psi}^+_\gamma(x_3))\rangle\langle T(\hat{\psi}_\beta(x_2)\hat{\psi}^+_\delta(x_4))\rangle + \langle T(\hat{\psi}_\alpha(x_1)\hat{\psi}^+_\delta(x_4))\rangle\langle T(\hat{\psi}_\beta(x_2)\hat{\psi}^+_\gamma(x_4))\rangle \\ &+ \langle N|T(\hat{\psi}_\alpha(x_1)\hat{\psi}_\beta(x_2))|N+2\rangle\langle N+2|T(\hat{\psi}^+_\gamma(x_3)\hat{\psi}^+_\delta(x_4))|N\rangle \end{aligned} \tag{2}$$

The first two terms are the aforementioned self-energy contributions that would bring nothing new, at least in the weak coupling limit. The last averages define the two *anomalous* functions: $F_{\alpha\beta}(x)$ and $F^+_{\alpha\beta}(x)$ functions inherent *only* to the superconducting state:

$$\begin{aligned} \langle N|T(\hat{\psi}_\alpha(x)\hat{\psi}_\beta(x'))|N+2\rangle &= \exp(-2i\mu t)F_{\alpha\beta}(x - x') \\ \langle N+2|T(\hat{\psi}^+_\alpha(x)\hat{\psi}^+_\beta(x'))|N\rangle &= \exp(2i\mu t)F^+_{\alpha\beta}(x - x') \end{aligned} \tag{3}$$

(Product of the two last terms in (2) is proportional to the density of the Cooper pairs in the condensate).

The total number of particles is not as convenient as the thermodynamic variable. Changing to the chemical potential $H \Rightarrow H - \mu N$ results in the system of equations:

$$\begin{aligned} &\{i\partial/\partial t + \Delta/2m + \mu\}\hat{G}(x - x') - ig\hat{F}(0+)\hat{F}^+(x - x') = \delta(x - x') \\ &\{i\partial/\partial t - \Delta/2m - \mu\}\hat{F}^+(x - x') + ig\hat{F}^+(0+)\hat{G}(x - x') = 0, \end{aligned} \tag{4}$$

where in (4) $\hat{F}_{\alpha\beta}(0+) \equiv \langle \hat{\psi}_\alpha(x)\hat{\psi}_\beta(x)\rangle)$. The operators with the non-equal spin indexes anti-commute and define the spin structure of $\hat{F}_{\alpha\beta}(0+) = i(\hat{\sigma}_y)_{\alpha\beta}F$. ($\hat{F}^+_{\alpha\beta}(0+) = -i(\hat{\sigma}_y)_{\alpha\beta}F^+$). Here the antisymmetric matrix $(\hat{\sigma}_y)_{\alpha\beta}$ appears as the wave function for the spin variables of two electrons the *singlet* pairing (the only one considered in the BCS days).

For a homogeneous superconductor the energy spectrum $E(p) = \pm\sqrt{[v_F^2(p-p_F)^2 + |\Delta|^2]}$ follows from (4) written in the momentum representation; the gap is $|\Delta| = |gF(0)|$. Note that at the derivation neither the variation procedure of BCS, nor the Bogolyubov's "principle of compensation of dangerous diagrams" is necessary.

Superconductors in external magnetic fields are not homogeneous. The true order parameters are the two anomalous functions $F_{\alpha\beta}(x,x')$ and $F^+_{\alpha\beta}(x,x')$, but more common is to take for that the *complex* functions of one spatial coordinate from (4) either $\Delta(\vec{r}) = gF(\vec{r},\vec{r};0)$, or $\Delta^+(\vec{r}) = gF^+(\vec{r},\vec{r};0)$. The symmetry broken at T_c is $U(1)$ is the gauge invariance symmetry. Obviously, the system of equations (4) is invariant at the substitution $-i\vec{\partial} \Rightarrow (-i\vec{\partial} - (e/c)\vec{\nabla}\varphi(r))$. Strange as it may appear now, but the gradient invariance of the microscopic theory was subject to prolonged debates in the West.

2.2.2 *Superconducting alloys.* To superconductivity in metallic alloys, the above ideas were extended in two papers [2b1,2].

In a normal metal or a semiconductor, defects play a most significant role in kinetic phenomena such as charge transport, especially at low temperatures when phonons are frozen. Relevant parameters are the matrix elements for the scattering of quasiclassical electrons on the potential of a defect.

The kinetics approach is not a way to treat thermodynamics and other equilibrium properties of superconductors in the presence of the magnetic fields. In regards to the electromagnetic properties of a superconductor in general, note that the new quasiparticles cannot be characterized by a charge equal to the charge of electron.

Randomness in spatial distribution of defects, first of all, impedes the use of momentum representation. To overcome that difficulty, in the first paper [2b1] we elaborate the diagrammatic technique that implicates scattering on defects directly into the equations for the Green functions. Matrix elements for scattering on defects get marked in a diagram by "crosses", hence the name: the "cross-technique". After averaging over positions of impurities, new equations contain only characteristics of the normal metal, such as the inverse mean free time.

It was assumed at the derivation that averaging over positions of defects in the diagrams for the Green functions and averaging the superconducting gap could be independently performed. Validity of that assertion is obvious for defects with sizes on the atomic scale. Most physical properties of a superconducting alloy depend on the so called "transport" scattering time that, in the normal phase, characterizes transfer of the momentum to the lattice.

Effectiveness of the new method was demonstrated by calculating the penetration depth and the surface impedance of a superconducting alloy as a function of the defects' concentration. Only the case of zero temperatures was considered in the first paper.

(As an alternative to the Boltzmann equation in normal metals, the diagrammatic cross-technique was formulated independently in Ref. 1)

In the second paper, with use of the newly elaborated thermodynamic technique (see below) the theory of superconducting alloys was extended to finite temperatures. The significant *qualitative* result was that *thermodynamics* of a superconductor is not sensitive to the presence of defects. The corollary concerns, first of all, were to do with the temperature of the superconducting transition. Experimentally, changes of T_c at alloying were known to be minor, but why it was so remained unclear for a long time.

Independence of T_c on the concentration of defects was obtained first in our paper; in the West it is known under the name of the Anderson's theorem. As the matter of fact, the result is rigorous only for the so-called s-wave (BCS) pairing and for the isotropic energy spectrum.

2.2.3 *BCS and GL Equations.* Interrelation between the microscopic theory and the Ginsburg-Landau theory (GL) was established in [2c].

The GL theory was the most elaborated phenomenology of superconductivity. Put forward in 1950, it remained unknown for years in the West, being published in Russian "ZhETF" (Journal of Experimental and Theoretical Physics). The GL theory allows addressing practically every question of experimental interest from thermodynamics of superconducting transition in finite size superconductors to the properties of superconductors in the presence of strong magnetic fields. The theory gave definitions to the critical currents and the critical magnetic fields, introduced the notion of the surface energy tensor at the boundary between normal and superconducting (NS) phases and calculated its value, and much more. Its predictions were broadly used by the Russian experimentalists for interpretation and presenting data.

In essence, it was a generalization of the theory of the second order symmetry transitions developed by Landau previously, but with a very significant new feature. That is, as the order parameter was chosen a *complex* function $\Psi(\vec{r})$. The phenomenological free energy functional has the form:

$$F_S = F_N + \alpha|\Psi|^2 + \frac{\beta}{2}|\Psi|^4 + \frac{1}{2m^*}\left|\left(-i\hbar\vec{\partial} - \frac{e^*}{c}\vec{A}\right)\Psi\right|^2 + \frac{(rot\vec{A})^2}{8\pi}.$$

Its minimization leads to the equations that have the formal resemblance to the non-linear Schrödinger equation for some "giant atom".

Here $\vec{A}$ is the vector potential and e^* is a charge. As the old story goes, Landau insisted that, not to violate the gauge invariance, the "effective charge" e^* can be chosen only as an integer of the electronic charge. As there were no reasons to think otherwise in that time, e^* was postulated to be equal to the charge of the electron.

Equations of the *microscopic theory* are too complicated. Analytical solutions are limited by a few problems. The attractiveness of the phenomenological GL theory lies in its transparency and in relative simplicity of the mathematics. Unlike the original Landau theory of the second order phase transitions, the GL theory is applicable only in a vicinity of temperature of the superconducting transition T_c.

In the first paper [2c] the GL equations were derived only for clean superconductors. The GL wave function $\Psi(\vec{r})$ turns out be proportional to the order parameter $\Delta(\vec{r}) = gF(\vec{r}, \vec{r}; 0)$. Besides providing the microscopic expressions for all coefficients in the GL phenomenological expansion, the derivation has lead the double charge $e^* = 2e$ in the free energy functional and in the GL equations themselves, thus once again confirming role of the pair wave function as the symmetry order parameter.

Experience of the following years has shown that *qualitatively* the ideas and the results of the GL theory describe most phenomena in superconductors away from T_c, even down to the lowest temperatures. In the first place, it concerns the behavior of superconductors in high magnetic fields.

The dimension analysis made in the original GL paper has shown that there exists only one dimensionless constant in the theory, κ. The latter is expressed via combination $\kappa = 2\sqrt{2}(e/hc)\delta_L^2(T)H_c(T)$ of the following experimentally measurable parameters: $\delta_L(T)$, the (London) penetration depth in weak fields, and $H_c(T)$, the thermodynamic critical field. As the temperature dependence fully drops out from this combination, the value of the parameter κ predetermines most of the qualitative properties of a superconductor in magnetic fields at all temperatures.

In the microscopic theory $\kappa \approx 0.96\delta_L(T = 0)/\xi_0$, where $\xi_0 = 0.18\hbar v_F/k_bT_c$ is the BCS coherence length that defines the "size" of the Cooper pair [1c].

2.2.4 *Type II superconductors.* In the original GL paper it was found that the surface tension at NS boundary changes sign at $\kappa = 1/\sqrt{2}$ and becomes negative at larger κ. The immediate consequence of the negative surface energy should be instability of the boundary leading to a spontaneous increase of the interface area. GL has not discussed this case being not aware of any relevant data (the phenomenon has actually been seen in 1937 by the Schubnikov group in Kharkov during experiments on alloys).

Properties of such alloys in magnetic fields were studied theoretically by A. Abrikosov. He dubbed superconductors with $\kappa > 1/\sqrt{2}$ Type II superconductors. Characteristic of the Type II group is the existence of a "low critical field" H_{c1} above which the Meissner screening becomes incomplete and at $H > H_{c1}$ the magnetic field gradually penetrates into the bulk forming a rare vortex structure.

Microscopic GL equations were derived for the superconducting alloys in my paper [2e] in which it was found also dependence of all parameters on the concentration of defects. For applications, especially important is the limiting case of the so called "dirty alloys", $l \ll \xi_0$ where l is the mean free path. At $l \ll \xi_0$ the penetration depth is $\delta \propto l^{-1/2}$ and, correspondingly $\kappa \propto 1/l$. Remarkably, for "dirty alloys" the GL parameter κ can be expressed exclusively in terms of experimental parameters of the normal phase $\kappa = 0.065ec\gamma^{1/2}/k_B\sigma$. Here γ is the coefficient in front of the linear in temperature electronic specific heat (the Sommerfeld coefficient), σ — the residual conductivity of the normal phase.

Before the BCS era, most studied were Type I superconductors ($\kappa < 1/\sqrt{2}$). To this group, with few exceptions, belong all elemental metals. Characteristic of the Type I class is the *non-local* connection between the current and the magnetic field. Unlike the Type II in which case the current and the vector potential are merely proportional to each other in every point of the space: $\vec{j}(\vec{r}) = -\delta_L^{-2}\vec{A}(\vec{r})$, for the Type I the current and the vector potential are coupled by an integral relation that involves integration over the spatial region with sizes larger than the penetration depth.

"Non-local" Type I superconductors with $\kappa \ll 1$ small are also known as the *Pippard* superconductors. Those of Type II are then usually named as the "*London*" superconductors. Alloying increases κ and, as shown in [2e], it may transform a clean Pippard superconductor into that of the London type. (There are other distinctions between properties of the two groups; see in [2c, e]).

The vast majority of more recent materials, such as intermetallic compounds, low dimensional superconductors, the so-called "heavy fermions" and many others belong to the Type II already in their pristine form. Those of the Type I nowadays are the rarity.

Together, the GL phenomenological theory, the Abrikosov's GL solution for Type II superconductors and the above microscopic results are known in the literature as the GLAG theory (after Ginsburg, Landau, Abrikosov and Gor'kov).

2.2.5 *Upper critical field.* The self-consistency equation for the superconducting parameter $\Delta(\vec{r})$ as in the case of a spatially inhomogeneous problem has the form of a nonlinear integral equation; namely this last circumstance makes solving equations of the microscopic theory difficult.

At that time I was tempted to find a solution for at least one simple case as I wanted to understand better a couple of minor technical details, e.g. such as relation between the upper cut-off in the momentum space (as in the equation for T_c) and cut off in the integral equations at the short distances in real space. I have chosen the problem known in old literature as "supercooling" in magnetic fields [2d].

The physics is this: in a strong enough magnetic field, superconductors are in the normal phase. As the field is reduced, sooner or later, the normal phase will inevitably become unstable to nucleation of the superconducting phase. As I mentioned above, at the time (1959) all known clean superconductors were of Type I. For the Type I superconductors the thermodynamic phase transition in the magnetic field is of the 1st order. With notation H_{c1} for the "supercooling" critical field, the notation H_{c2} was reserved for the "superheating" field, another critical field above which would be unstable for the superconducting phase. The two fields, H_{c1} and H_{c2}, hence, mean the low and the upper hysteresis boundaries, respectively, for the 1st phase order transition.

Type II superconductors behave in the magnetic field differently. As discussed in **2.2.4**, at low enough fields even a Type II superconductor at first expels the field, but above some threshold H_{c1} the magnetic fields begin entering into the bulk forming a rarefied lattice of the magnetic vortices. Density of the vortex lines increases with the further increase of the field, until at some upper field denoted as H_{c2} fully suppressed is the superconducting parameter. Thereby, notations H_{c1} and H_{c2} for Type II in current literature are reciprocal to the notations H_{c1} and H_{c2} for the above hysteresis problem.

Retrospectively, as far as the field of "supercooling" defines the boundary *for stability of the normal phase* in the magnetic field, I have for the first time calculated H_{c2} in clean superconductors [2d].

I solved the linear equation for $\Delta(\vec{r})$ by the variation method and found $H_{c2}(0)\xi_0^2 \simeq 1.3(\varphi_0/2\pi)$ (at $T = 0$). Near T_c, $H_{c2}(T)\xi_0^2 \simeq 1.8(\varphi_0/2\pi)[1-(T/T_c)]$; $\varphi_0 = c/2eh \approx 2\times 10^{-7} Gauss$ is the flux quantum (notice the double charge).

The critical field $H_{c2}(T)$ in the whole temperature interval (as for clean superconductors and as for alloys) was later calculated in Ref. 2. (My expression for $H_{c2}(0)$ was the exact result).

Notice in the paper the useful transformation: $\tilde{G}_\omega(\vec{r},\vec{r}') = \exp\{-i(eH/2c\hbar)(y+y')(x-x')\}\tilde{G}'_\omega(\vec{r}-\vec{r}')$ for the Green function of a normal metal in the presence of the homogenous magnetic field; $\tilde{G}'_\omega(\vec{r}-\vec{r}')$ is translation invariant. (Such transformation is, of course, not limited by choice of the Landau gauge in [2d]).

2.2.6 *Paramagnetic impurities.* Paramagnetic centers are foreign atoms or ions that, in the metallic matrix, posses non-zero spin $\vec{S}$. Potential for scattering of electrons on a paramagnetic center has the form: $V(\vec{p}-\vec{p}') \Rightarrow V(\vec{p}-\vec{p}') + \tilde{V}(\vec{p}-\vec{p}')(\vec{\sigma}\cdot\vec{S})$. One immediately sees that the presence of such centers must be detrimental to superconductivity because foreign spins $\vec{S}$ on the impurity sites violate the time reversal symmetry for electrons [2f]. As each of the two electrons comprising the Cooper pair may *independently* scatter on the local spin, the $(\vec{\sigma}\cdot\vec{S})$-term tends to misalign their anti-parallel orientation. Equation $\ln(T_{c0}/T_c) = \psi[1/2 + (T_c\tau_S)^{-1}] - \psi(1/2)$ describes how the transition temperature $T_c(x)$ decreases with the increase of the concentration of paramagnetic impurities x. Here $\tau_S^{-1} \propto x$ is the inverse mean time for flipping the electronic spin. The critical concentration at which $T_c(x)$ vanishes is given by $1/\tau_{S,crit} = \pi T_{c0}/2\gamma \equiv \Delta_0/2$. (In the above $\psi(x)$ is derivative of the logarithm of the Γ-function, $\psi(x) = \Gamma'(x)/\Gamma(x)$).

The energy gap in the density of state in the superconducting phase decreases as well. Most remarkably though, that the gap gets closed *before* superconductivity is completely destroyed at $1/\tau_{S,crit}$. In a narrow interval of concentrations $1/\tau_S < 1/\tau_{S,crit}$ superconductivity persists in spite of the energy gap in the density of states being zero. The result is of the principal importance, for it shows that superconductivity, i.e. the capability of electrons to carry persistent currents is not always controlled by a finite gap in the spectrum, but by the non-zero *order parameter*, Δ. Thereby, the Landau criterion discussed in the introductory part has a lesser generality.

2.2.7 *Superfluidity of imperfect Fermi gas.* The paper has recently acquired fresh interest in connection with the physics of cold atoms. The problem under discussion is crossover between the Cooper-type pairing and the Bose-Einstein condensation (BEC) of pairs of the bound fermions. In the first case the Cooper pairs in condensate strongly overlap; in the BEC limit the pairs exist already in the normal phase and condensate at low temperatures in the same fashion as a gas of the Bose particles.

The 1961 paper [2g] dealt with an interesting but a mainly theoretical question at the time. Namely, in the BCS model the phonon mediated electron-electron attraction involves electrons only from the "thin energy layer" $\omega_D \ll \varepsilon_F$ near the Fermi surface, where ω_D is a typical phonon energy scale. The latter serves as the natural *upper energy cutoff* in the integral determining temperature of the transition.

In case of a rare Fermi *gas* with a short range attractive interaction, there is *no* other energy scale except the Fermi energy ε_F itself. To obtain the closed expression for T_c from diverging integrals in the Cooper channel it was necessary to properly renormalize the scattering amplitude. When this is done the integrals become convergent. The temperature for the onset of superfluidity in a rare gas of neutral fermions was found in [2g] rigorously. (Of course, at that time everyone kept in mind possible superfluidity of ^3He, but *helium-three* is the liquid, not a gas).

In cold atoms physics, as is known, the strength of the interaction between two atoms can be varied by the magnetically tuned Feshbach resonance. Thereby, for a rarified gas of the Fermi atoms, it is possible for the gradual crossover from weakly coupled Cooper pairs to the strongly bound pairs to which the Bose-Einstein statistics should apply. Results [2g] on the side of the Cooper pairs were confirmed by the numerical analysis.

2.2.8 *GL equations for anisotropic superconductors.* That the energy spectrum of electrons is not isotropic in most superconductors, even in the elemental metals, is the well–known fact. Surprisingly, the *isotropic* BCS model, as a rule, agreed with the experimental data within the accuracy often even better than $\sim$10%.

Anisotropic GL equations derived in [2h] reveal the important intertwining between non-local features of the microscopic theory and the anisotropy of the Fermi surface. That is, there are no straightforward relations between parameters of the normal and superconducting phases. For instance, the inverse tensor of effective masses m_{ik}^{-1} is found proportional to $\langle v_i v_k |\Psi(\vec{p})|^2\rangle_{FS}$ where the average is over the Fermi surface and $\Psi(\vec{p})$ is a solution for the *anisotropic* superconducting order parameter of the non-local linear self-consistency equation defining the temperature of transition T_c. Thereby, in particular, the masses extracted from the de Haas-van Alphen style experiments are not the same masses that figure in the expression for the superconducting currents.

While the anisotropic GL theory does not provide us with much new experimental information, the analysis [2h] divulge the reasons for the surprisingly high accuracy of the BSC predictions. In [2h] all other physical parameters appear also weighted and averaged over the Fermi surface with $|\Psi(\vec{p})|$ in some power thus smoothing the anisotropy of the Fermi surfaces of a normal metal.

References

1. N. N. Bogolyubov, *J. Phys., U.S.S.R.* **11**, 32 (1947)
2. S. F. Edwards, *Phil. Mag.* **3**, 1020 (1958)
3. E. Helfand and N. R. Werthamer, *Phys. Rev. Lett.* **13**, 626 (1964); *Phys. Rev.* **147**, 288 (1966)

List of original papers

[2a] L. P. Gor'kov, *On the energy spectrum of superconductors*, Sov. Phys. JETP, 34(7), 505 (1958)

[2b1, 2] A. A. Abrikosov and L. P. Gor'kov, *On the theory of superconducting alloys: I. The electrodynamics of alloys at absolute zero*, Sov. Phys. JETP, (35)8, 1090 (1958); *II. Superconducting alloys at finite temperatures*, Sov. Phys. JETP, (36)9, 220 (1959)

[2c] L. P. Gor'kov, *Microscopic derivation of the Ginsburg-Landau equations in the theory of superconductivity*, Sov. Phys. JETP, (36)9, 1364 (1959)

[2d] L. P. Gor'kov, *The critical supercooling field in superconductivity theory*, Sov. Phys. JETP, (37)10, 593 (1960)

[2e] L. P. Gor'kov, *Theory of superconducting alloys in strong magnetic field near critical temperature*, Sov. Phys. JETP, (37)10, 998 (1960)

[2f] A. A. Abrikosov and L. P. Gor'kov, *Contribution to the theory of superconducting alloys with paramagnetic impurities*, Sov. Phys. JETP, 12, 1243 (1961)

[2g] L. P. Gor'kov and T. K. Melik-Barkhudarov, *Contribution to the theory of superfluidity of an imperfect Fermi gas*, Sov. Phys. JETP, 13, 1018 (1961)

[2h] L. P. Gor'kov and T. K. Melik-Barkhudarov, *Microscopic derivation of the Ginsburg-Landau equations in anisotropic superconductors*, Sov. Phys. JETP, 18, 1964 (1964)

SOVIET PHYSICS JETP VOLUME 34(7), NUMBER 3 SEPTEMBER, 1958

ON THE ENERGY SPECTRUM OF SUPERCONDUCTORS

L. P. GOR' KOV

Institute for Physical Problems, Academy of Sciences, U.S.S.R.

Submitted to JETP editor November 18, 1957

J. Exptl. Theoret. Phys. (U.S.S.R.) **34**, 735-739 (March, 1958)

A method is proposed, based on the mathematical apparatus of quantum field theory, for the calculation of the properties of a system of Fermi particles with attractive interaction.

It was shown in the work of Cooper[1] that if the interaction of electrons in a metal leads to an effective mutual attraction for two electrons close to the Fermi surface, then the pair of particles which possess mutually opposite momenta and spins can have bound states with negative coupling energies. In the works of Bardeen, Cooper and Schreiffer[2,3] and of Bogoliubov[4] a systematic theory of superconductivity has been erected on this principle. It was shown that the ground state of a system of interacting Fermi particles is located below the normal state with a filled Fermi sphere and, in consequence, is separated from the excited states by a gap in order of magnitude equal to the energy of coupling of the individual pair.

In the present work, a method is proposed, based on the physical idea of Cooper, which permits us, with the help of the apparatus of quantum field theory, to obtain all the results by a short and simple method.

We shall start out from a Hamiltonian in the form[2] which is written in the case of second quantization:

$$\hat{H} = \int \left\{ -\left(\psi^+ \frac{\Delta}{2m} \psi\right) + \frac{g}{2} (\psi^+ (\psi^+ \psi) \psi) \right\} d^3x, \tag{1}$$

where

$$\psi_\alpha(\mathbf{x}) = V^{-1/2} \sum_{k\sigma} a_{k\sigma} s_{\sigma\alpha} e^{i\mathbf{kx}}; \quad \psi_\beta^+(\mathbf{x}') = V^{-1/2} \sum_{k\sigma} a_{k\sigma}^+ s_{\beta\sigma}^* e^{-i\mathbf{kx}'}$$

satisfy the usual commutation relations:

$$\{\psi_\alpha(\mathbf{x}), \psi_\beta^+(\mathbf{x}')\} = \delta_{\alpha\beta}\,\delta(\mathbf{x}-\mathbf{x}'),$$
$$\{\psi_\alpha(\mathbf{x}), \psi_\beta(\mathbf{x}')\} = \{\psi_\alpha^+(\mathbf{x}), \psi_\beta^+(\mathbf{x}')\} = 0. \tag{2}$$

We shall consider the interaction to be equal to zero everywhere except in a region of energy of the particles 2κ around the Fermi surface, from $\epsilon_F - \kappa$ to $\epsilon_F + \kappa$.

We transform to the Heisenberg representation, in which the operators ψ and ψ^+ depend on the time and satisfy the following equations:

$$\{i\partial/\partial t + \Delta/2m\}\psi(x) - g(\psi^+(x)\psi(x))\psi(x) = 0,$$
$$\{i\partial/\partial t - \Delta/2m\}\psi^+(x) + g\psi^+(x)(\psi^+(x)\psi(x)) = 0. \tag{3}$$

We determine the Green's function $G_{\alpha\beta}(x - x')$ as an average over the ground state of the system:

$$G_{\alpha\beta}(x-x') = -i\langle T(\psi_\alpha(x),\ \psi_\beta^+(x'))\rangle, \tag{4}$$

where T is time-ordering operator.

For the derivation of the equation for the function $G(x-x')$ we take it into consideration that the ground state of the system differs from the usual state with a filled Fermi sphere by the presence of bound pairs of electrons. In the ground state, all the pairs are at rest as a whole. (This means that the interaction between particles is considered only insofar as it enters into the formation of the bound pairs. We neglect scattering effects.) A sort of "Bose condensation" of pairs takes place in the case in which the momentum of their motion as a whole is equal to zero, just as in a Bose gas such a condensation takes place by virtue of the statistics for the particles themselves. This circumstance permits us to write down in a definite way the mean form $\langle T(\psi(x_1)\psi(x_2) \times \psi^+(x_3)\psi^+(x_4))\rangle$, which appears in the equations for $G(x-x')$ by virtue of (3).

For example, we have

$$\begin{aligned}\langle T(\psi_\alpha(x_1)\psi_\beta(x_2)\psi_\gamma^+(x_3)\psi_\delta^+(x_4))\rangle &= \\ -\langle T(\psi_\alpha(x_1)\psi_\gamma^+(x_3))\rangle\langle T(\psi_\beta(x_2)\psi_\delta^+(x_4))\rangle & \\ +\langle T(\psi_\alpha(x_1))\psi_\delta^+(x_4))\rangle\langle T(\psi_\beta(x_2)\psi_\gamma^+(x_3))\rangle & \\ +\langle N|T(\psi_\alpha(x_1)\psi_\beta(x_2))|N & \\ +2\rangle\langle N+2|T(\psi_\gamma^+(x_3)\psi_\delta^+(x_4))|N\rangle, &\end{aligned} \tag{5}$$

where $|N\rangle$ and $|N+2\rangle$ are the ground states of the system with numbers of particles N and N + 2. The quantity

$$\langle N|T(\psi\psi)|N+2\rangle\langle N+2|T(\psi^+\psi^+)|N\rangle, \tag{5a}$$

evidently has the order of the density of the number of pairs, while $\langle T(\psi\psi^+)\rangle$ is the particle number density.

It is easy to show that the quantities thus introduced can be written in the form

$$\langle N|T(\psi_\alpha(x)\psi_\beta(x'))|N+2\rangle = e^{-2i\mu t}F_{\alpha\beta}(x-x'),$$
$$\langle N+2|T(\psi_\alpha^+(x)\psi_\beta^+(x'))|N\rangle = e^{2i\mu t}F_{\alpha\beta}^+(x-x'). \quad (6)$$

The function $G(x-x')$ depends only on the difference $x-x'$, because of the homogeneity of the problem. So far as the additional dependence on t in Eq. (6) is concerned, its origin is seen from the general quantum mechanical formula for the time derivative of an arbitrary operator $\hat{A}(t)$:

$$\frac{\partial}{\partial t}\langle N|\hat{A}(t)|N+2\rangle = i(E_N - E_{N+2})\langle N|\hat{A}(t)|N+2\rangle.$$

The value of the energy difference $E_{N+2} - E_N$ is obviously equal to $2\mu\ (\partial E/\partial N = \mu)$.

Making use of Eq. (3), we obtain equations for the functions $\hat{G}(x-x')$ and $\hat{F}(x-x')$:

$$\{i\partial/\partial t + \Delta/2m\}\hat{G}(x-x')$$
$$-ig\hat{F}(0+)\hat{F}^+(x-x') = \delta(x-x'),$$
$$\{i\partial/\partial t - \Delta/2m - 2\mu\}\hat{F}^+(x-x') \quad (7)$$
$$+ig\hat{F}^+(0+)\hat{G}(x-x') = 0.$$

Here terms are omitted which correspond to the first two terms in Eq. (5), inasmuch as they only change μ, which one can neglect, and the notations

$$F_{\alpha\beta}^+(0+) = e^{-2i\mu t}\langle\psi_\alpha^+(x)\psi_\beta^+(x)\rangle \equiv \lim_{x\to x'(t>t')} F_{\alpha\beta}^+(x-x')$$

and introduced, and correspondingly,

$$F_{\alpha\beta}(0+) = e^{2i\mu t}\langle\psi_\alpha(x)\psi_\beta(x)\rangle.$$

The complex conjugate yields

$$(F_{\alpha\beta}^+(0+))^* = -F_{\alpha\beta}(0+). \quad (8)$$

We transform in Eqs. (7) to the Fourier components of all functions, for example,

$$G_{\alpha\beta}(x-x') = (2\pi)^{-4}\int G_{\alpha\beta}(p\omega)\exp\{ip(x-x') - i\omega(t-t')\}\,d\omega d^3p.$$

Denoting $\omega - \mu = \omega'$, we find

$$(\omega' - \xi_p)\hat{G}(p\omega) - ig\hat{F}(0+)\hat{F}^+(p\omega) = 1,$$
$$(\omega' + \xi_p)\hat{F}^+(p\omega) + ig\hat{F}^+(0+)\hat{G}(p\omega) = 0, \quad (9)$$

where

$$\xi_p = p^2/2m - \mu \approx v_F(p - p_F),$$

and p_F is the Fermi momentum. In what follows, we shall only have ω' in the formulas, hence we shall omit the prime.

It follows from Eq. (8) that $\hat{F}(0+)$ and $\hat{F}^+(0+)$ have the following matrix form:

$$\hat{F}^+(0+) = J\begin{pmatrix}0 & 1\\ -1 & 0\end{pmatrix} \equiv J\hat{I};\ \hat{F}(0+) = -J\hat{I},$$
$$\hat{I}^2 = -\hat{E}; \quad (10)$$

equal both to $\hat{F}^+(p\omega)$ and $\hat{F}(p\omega)$, and the Green's function, in accord with (9), is proportional to the unit matrix. Substituting (10) in (9), we obtain

$$(\omega^2 - \xi_p^2 - g^2J^2)F^+(p\omega) = -igJ,$$
$$(\omega - \xi_p)G(p\omega) = 1 + igJF^+(p\omega). \quad (11)$$

Hence

$$F^+(p\omega) = -ig\frac{J}{\omega^2 - \xi_p^2 - \Delta^2};$$

$$G(p\omega) = \frac{\omega + \xi_p}{\omega^2 - \xi_p^2 - \Delta^2}, \quad (12)$$

where

$$\Delta^2 = g^2J^2.$$

As is evident from the first equation of (11), $F^+(p\omega)$ is determined with accuracy up to a solution of the homogeneous equation of the form $A(p)\delta(\omega^2 - \xi_p^2 - \Delta^2)$. It is not difficult to verify the fact that this term reduces in the expression for $G(p\omega)$ to an arbitrary imaginary part. In other words, Eqs. (11) determine only the real part of the Green's function.

We determine the rule for circling the poles in (12), making use of a theorem given by Landau,[5] according to which the imaginary part of the Green's function of a Fermi system is positive for $\omega < 0$ and changes sign in the transition from negative to positive frequencies.

As a result, we obtain

$$F^+(p\omega) = -igJ/(\omega - \varepsilon_p + i\delta)(\omega + \varepsilon_p - i\delta), \quad (13)$$

$$G(p\omega) = u_p^2(\omega - \varepsilon_p + i\delta)^{-1} + v_p^2(\omega + \varepsilon_p - i\delta)^{-1}, \quad (14)$$

where $\epsilon_p = \sqrt{\xi_p^2 + \Delta^2}$ and the functions μ_p^2 and v_p^2 are equal to

$$u_p^2 = \frac{1}{2}\left(1 + \frac{\xi_p}{\varepsilon_p}\right);\quad v_p^2 = \frac{1}{2}\left(1 - \frac{\xi_p}{\varepsilon_p}\right) \quad (14')$$

For the determination of the quantity Δ, we make use of the fact that

$$J = (2\pi)^{-4}\int F^+(p\omega)\,d\omega\,d^3k. \quad (15)$$

Substituting (13), we obtain the equation

$$1 = -\frac{g}{2(2\pi)^3}\int\frac{d^3k}{\sqrt{\xi_k^2 + \Delta^2}}\ (|\xi| < \varkappa). \quad (16)$$

For small $g < 0$ (attraction), this equation has a solution of the form

$$\Delta = 2\varkappa e^{-1/2\rho},$$

where

$$\rho = p_F |g| m / 2\pi^2.$$

The positive pole in (14) determines the excitation spectrum which, as is shown, has a gap of magnitude Δ. These results coincide with those obtained in Refs. 2 — 4.

The chemical potential μ is connected with the particle number density by the relation

$$N/V = \langle \psi^+(x)\psi(x)\rangle = -i(2\pi)^{-4}\int G_{\alpha\alpha}(p\omega)e^{i\omega\delta}d\omega d^3p, \quad (17)$$

with accuracy up to small exponential terms: $\mu = \epsilon_F$.

The method laid out also permits us to make use of it for temperatures differing from absolute zero. In this case we consider the mean Green's function (thermodynamically averaged)

$$G_{\alpha\beta}(x-x')$$

$$= -i\sum_n \exp\left\{\frac{\Omega + \mu N - E_n}{T}\right\}\langle n|T(\psi_\alpha(x)\psi_\beta^+(x'))|n\rangle,$$

where Ω is the thermodynamic potential in the variables T, V, μ. As is well known, the result of averaging does not depend on whether it is carried out with the use of a Gibbs distribution or over the stationary state with a given energy. This corresponds to the choice of the quantity $\overline{E}$ as a thermodynamic variable, in place of the temperature T. Taking the averaging in such a fashion we get the earlier equations (11) for the quantities G, F^+ and F, with this difference, that the corresponding averaging of T products are taken not over the ground state of the system but over a state with total energy E equal to the energy of the system at a given temperature. Equations (11), as we have already noted above, determine uniquely only the real part of the Green's function $G(p\omega)$, which is evidently equal to the real part of Eq. (14). We write down the general solution for the function $F^+(p\omega)$:

$$F^+(p\omega) = -igJ/(\omega - \varepsilon_p + i\delta)(\omega + \varepsilon_p - i\delta)$$
$$+ A_1(p, T)\delta(\omega - \varepsilon_p) + A_2(p, T)\delta(\omega + \varepsilon_p). \quad (18)$$

We have made use of the fact that $1/(x \mp i\delta) = 1/x \pm \pi i\delta(x)$. The value of the quantities $A_1(p,T)$ and $A_2(p,T)$ can be obtained from the relations between the real and imaginary parts of the Green's function,[5] which has the following form at temperatures different from zero:

$$\operatorname{Re} G(\omega) = -\frac{1}{\pi}\int_{-\infty}^{+\infty}\coth\frac{x}{2T}\frac{\operatorname{Im} G(x)}{\omega - x}dx.$$

We obtain

$$A_1(p, T) = A_2(p, T) = -(\pi\Delta/\varepsilon_p)n(\varepsilon_p)$$

and for the Green's function,

$$G(p\omega) = u_p^2(\omega - \varepsilon_p + i\delta)^{-1} + v_p^2(\omega + \varepsilon_p - i\delta)^{-1}$$
$$+ 2\pi i n(\varepsilon_p)[u_p^2\delta(\omega - \varepsilon_p) - v_p^2\delta(\omega + \varepsilon_p)], \quad (19)$$

where $n(\epsilon_p)$ has the form of the Fermi distribution of excitations at the given temperature:

$$n(\varepsilon_p) = [\exp(\varepsilon_p/T) + 1]^{-1}.$$

The excitation spectrum is then

$$\varepsilon_p = \sqrt{\xi_p^2 + \Delta^2}, \quad (20)$$

where Δ is a function of the temperature. Condition (15), upon substitution of $F^+(p\omega)$ in the form (18) in it, gives a relation which determines the magnitude of the gap in its temperature dependence:

$$1 = \frac{|g|}{2(2\pi)^3}\int\frac{d^3k(1 - 2n(\varepsilon_k))}{\sqrt{\xi_k^2 + \Delta^2(T)}}, \quad (|\xi| < \varkappa). \quad (21)$$

Equation (21) was obtained by Bardeen, Cooper and Schrieffer[3]; it was found that the magnitude of the gap $\Delta(T)$ vanishes at $T = T_c \sim \Delta(0)$. We shall show briefly how calculation of thermodynamical quantities is carried out by our method.

The heat capacity per unit volume is equal to

$$Vc_V = (\partial\overline{E}/\partial T)_V.$$

The quantity

$$\overline{E}/V = \left\langle\left\{-\left(\psi^+\frac{\Delta}{2m}\psi\right) + \frac{g}{2}(\psi^+(\psi^+\psi)\psi)\right\}\right\rangle$$

is expressed by the function of G, F^+ and F with accuracy up to inconsequential (constant in temperature) terms in the following fashion:

$$\overline{E} = 2V(2\pi)^{-3}\int\varepsilon_p[v_p^2(1 - n(\varepsilon_p)) + u_p^2 n(\varepsilon_p)]d^3p + gVJ^2.$$

The total contribution to the heat capacity at such temperatures gives the interval $|\xi_p| \ll \kappa$. Substituting here for the functions u_p^2 and v_p^2 their expressions (14′), and making use of Eq. (21), we get for the heat capacity

$$c_V = 2(2\pi)^{-3}\int\varepsilon_p\frac{\partial n(\varepsilon_p)}{\partial T}d^3p,$$

i.e., the ordinary formula for heat capacity of a gas of Fermi excitations with a spectrum (20). Computation of the heat capacity and the value of the gap in its temperature dependence was given in Ref. 3.

The author expresses his gratitude to Academician L. D. Landau for his valued advice and interest in the work. The author also acknowledges A. A. Abrikosov, I. M. Khalatnikov, I. E. Dzialoshinskii and L. P. Pitaev for discussion of the results of this work.

[1] L. Cooper, Phys. Rev. **104**, 1189 (1956).

[2] Bardeen, Cooper and Schrieffer, Phys. Rev. **106**, 162 (1957).

[3] Bardeen, Cooper and Schrieffer, Phys. Rev. **108**, 5 (1957).

[4] N. N. Bogoliubov, J. Exptl. Theoret. Phys. (U.S.S.R.) **34**, 58 (1958); Soviet Phys. JETP **7**, 41 (1958).

[5] L. D. Landau, J. Exptl. Theoret. Phys. (U.S.S.R.) **34**, 262 (1958); Soviet Phys. JETP **7**, 182 (1958).

Translated by R. T. Beyer
138

SOVIET PHYSICS JETP VOLUME 35(8), NUMBER 6 JUNE, 1959

ON THE THEORY OF SUPERCONDUCTING ALLOYS
I. THE ELECTRODYNAMICS OF ALLOYS AT ABSOLUTE ZERO

A. A. ABRIKOSOV and L. P. GOR'KOV

Institute for Physics Problems, Academy of Sciences, U.S.S.R.

Submitted to JETP editor July 16, 1958

J. Exptl. Theoret. Phys. (U.S.S.R.) **35**, 1558-1571 (December, 1958)

In this paper we give the theory of superconductors containing impurities at the absolute zero. The dependence of the penetration depth on the impurity concentration is considered for small atomic concentrations. We have obtained the electrodynamic equations in an alternating field for superconductors with a mean free path which is smaller than the correlation length.

It is well known that Bardeen, Cooper, and Schrieffer[1] succeeded in formulating the electrodynamics of superconductors with the aid of their theory of superconductivity. They obtained equations connecting the current with the vector potential which replaced the old equations of the phenomenological theory of H. and F. London. Of most interest was the non-local character of the connection between the current and the field for the majority of pure superconductors.

The non-local character is put into the very basis of the theory which is based on the physical idea of Cooper's[2] about the formation of bound singlet pairs of electrons near the Fermi surface under the action of the phonon interaction. The dimensions of such a pair causes the existence of a correlation between electrons at distances of the order $\xi_0 \sim 10^{-4}$, which becomes apparent in the form of a non-local connection between the field and the current, if the penetration depth of the field is less than ξ_0. The majority of pure superconductors belong to that class.

The creation of a theory of superconductivity made it possible to consider a whole series of phenomena. With its aid they could construct the thermodynamics and electrodynamics of superconductors[1] and, in particular, determine the behavior of superconductors in a high-frequency field.[3] A very interesting problem is that of "alloys," i.e., of superconductors with impurity atoms of other elements or other lattice defects. In the normal state these defects cause the so-called residual resistance.

At very small concentrations the role of impurities is small. Increasing the concentration, however, leads to a decrease of the spatial correlation of the electrons in the superconductor. For a sufficiently concentrated alloy the role of the correlation parameter goes over from ξ_0 to the electron mean free path. Thanks to this, it is natural to expect the appearance of a local connection between the current and the vector potential for such concentrations where the mean free path becomes less than the penetration depth. The difference with the Londons' theory will lie in a change in the constant of proportionality of **j** to **A**. Below we shall consider this problem, assuming, however, that the atomic impurity concentration is small.

The problem of the penetration of a weak electromagnetic field into a medium is solved if we know the Green function of the field in the medium

$$D_{ik}(x-x') = \frac{4\pi i}{c}\langle T(A_i(x)A_k(x'))\rangle. \tag{1}$$

The function D_{ik} is directly connected with experimentally observed quantities. Let $D_{ik}(\mathbf{k}, \omega_0)$ be the Fourier transform of $D_{ik}(x-x')$:

$$D_{ik}(x-x') = \frac{4\pi}{(2\pi)^4}\int \exp\{i\mathbf{k}\cdot(\mathbf{x}-\mathbf{x}') - i\omega_0(t-t')\}\, D_{ik}(\mathbf{k}, \omega_0)d^3\mathbf{k}\,d\omega_0.$$

In view of the symmetry of the tensor D_{ik} we have

$$D_{ik}(\mathbf{k}, \omega_0) = D_1(\mathbf{k}^2, \omega_0)(\delta_{ik} - k_ik_k/k^2) + D_2(\mathbf{k}^2, \omega_0)k_ik_k/k^2.$$

The function D_2 which determines the longitudinal part of D_{ik} is connected with the gauge of the vector potential and is arbitrary. The physical quantities depend only on the transverse part, i.e., on D_1.

With the aid of the general theory of the anomalous skin effect we can arrive at the following formula for the surface impedance in the case of diffuse scattering at the surface*

$$Z = \frac{4\pi^2 i\omega_0}{c^2} \Bigg/ \int\limits_0^\infty \ln\left[\mathbf{k}^2 D_1(\mathbf{k}^2, \omega_0)\right] d\,|\,\mathbf{k}\,|. \quad (2)$$

We recall that the penetration depth of the field is

$$\delta = \mathrm{Re}\left[\frac{c^2}{4\pi\omega_0}\, iZ\right]. \quad (3)$$

Usually, instead of finding the function D_{ik} one finds the connection between the current and the vector potential

$$j_i = -Q_{ik}(\mathbf{k}, \omega_0)\, A_k(\mathbf{k}, \omega_0). \quad (4)$$

To find the connection between the tensors D_{ik} and Q_{ik}. It is sufficient to substitute (4) into the Maxwell equation and to write down with its aid an equation for D_{ik}. This gives

$$(D^{-1})_{ik} = (\mathbf{k}^2\delta_{ik} - k_i k_k) + \frac{4\pi}{c} Q_{ik}. \quad (5)$$

The first term corresponds to the "zero" value of the tensor D_{ik}, obtained by neglecting the interaction of the field with the electrons. In Eqs. (2) to (5) we have implied $\omega_0 > 0$.

From Eq. (5) it is clear that the tensor Q_{ik} must be symmetric. Also, since the current is conserved ($k_i j_i = 0$) it follows from Eq. (4) that it must be transverse.† Equation (5) determines thus only the transverse part of D_{ik} and its longitudinal part remains arbitrary, as it should be.

We evaluate below the function $D_{ik}(\mathbf{k}, \omega_0)$ for a superconductor with impurities, using the techniques of the quantum field theory‡ (see reference 4, which in contradistinction to other methods[1,5] makes it possible to state the problem very clearly). We assume that we have already taken into account the interaction of the electrons with the impurities and consider their interaction with the electromagnetic field. The Hamiltonian of this interaction is

$$H_1' = \frac{e^2}{mc^2}\psi^+(x)\psi(x)\mathbf{A}^2(x) + \frac{ie\mathbf{A}(x)}{2mc}(\nabla_{\mathbf{x}} - \nabla_{\mathbf{x}'})_{x\to x'}\psi^+(x')\psi(x). \quad (6)$$

Dyson's equation for the function D will be of the form

$D^{-1} = D^{-1}_{(0)} +$  , (7)

where a solid line corresponds to an electron and a wavy one to a photon. More complicated diagrams with internal photon lines can be neglected in the non-relativistic approximation.

Comparing Eqs. (7) and (5) we see that, apart from a factor $4\pi/c$ the term added to $D_{(0)}$ corresponds to the tensor Q_{ik}. The first term of this addition gives simply $(Ne^2/mc)\,\delta_{ik}$ in Q_{ik}. The second diagram is more complicated. For the following it is convenient to separate, as is usually done, one of the photon vertices, for instance p_i, from the remaining part which, apart from a coefficient, can be considered as the result of including the photon vertex p_k in the electron line. To find the coefficient is not difficult (the factor 2 comes from taking the trace over the spin). In this way we get

$$Q_{ik}(\mathbf{k}, \omega_0) = \frac{Ne^2}{mc}\delta_{ik} - \frac{2ie^2}{m^2c}\int p_i \Pi_k(p_+, p_-)\frac{d^4p}{(2\pi)^4}, \quad (8)$$

where $p_\pm = (\mathbf{p} \pm \mathbf{k}/2,\ \omega \pm \omega_0/2)$.

As far as the interaction of the electrons with the impurities is concerned, the corresponding term in the Hamiltonian is of the form

$$H_2' = \sum_a H_a, \qquad H_a = u(\mathbf{x} - \mathbf{x}_a)\psi^+(x)\psi(x). \quad (9)$$

where $u(\mathbf{x} - \mathbf{x}_a)$ is the interaction potential with an impurity atom at the point $\mathbf{x}_a$. We shall assume that the impurity atoms are randomly distributed in the metal and that their distance apart is large compared to the lattice period. We assume, of course, that the impurity concentration is so small that the interaction between the electrons is not changed. We shall denote the impurity vertex in the Feynman diagrams by a cross. It corresponds to a factor

$$u(\mathbf{q})\, e^{i\mathbf{q}\cdot\mathbf{x}_a}\,\delta(\omega_1 - \omega_2), \quad (10)$$

where $u(\mathbf{q})$ is the Fourier component of the potential $u(\mathbf{x})$ and $\mathbf{q}$ is the momentum transferred.

For the sake of clarity we first give the calculations for the application to a normal metal with a residual resistance (Sec. 1). We then consider superconducting alloys (Sec. 2). All calculations refer only to the absolute zero.

*The correspondence of this equation with the usual expression is easily established using Eqs. (4) and (5).

† The method of finding the tensor Q_{ik} usually employed in various problems of the theory of superconductivity (see references 1 and 3) leads to an expression which is not purely transverse. One can, however, show by means of a more rigorous calculation based on the equations of reference 4 that to get the correct result for the given problem it is sufficient to retain only the transverse part of the tensor Q_{ik} obtained by the usual method, or to use a vector potential satisfying the condition div $\mathbf{A} = 0$.

‡ We must emphasize that the method used here is suitable only at $T = 0$. In the case where $T \neq 0$ there will occur, instead of simple time-ordered products, more complicated combinations which make this method practically innapplicable to the problem under consideration.

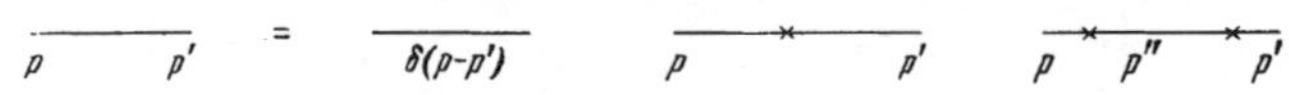

FIG. 1

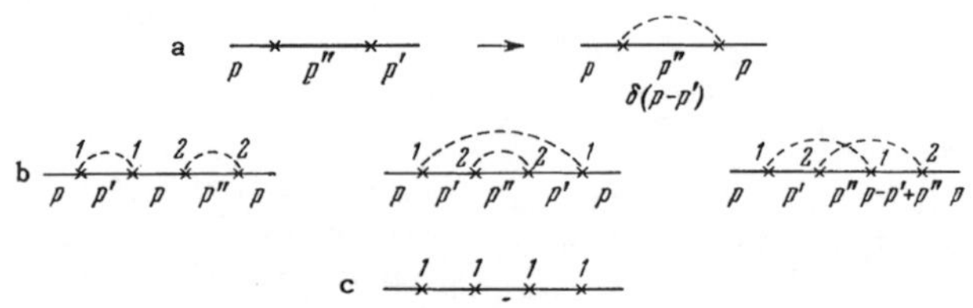

FIG. 2

1. RESIDUAL RESISTANCE OF A NORMAL METAL

We introduce the Green function of a system of fermions

$$G_{\alpha\beta}(x, x') = -i\langle T(\psi_\alpha(x)\,\psi_\beta^+(x'))\rangle. \tag{11}$$

For a pure metal we have:*

$$\begin{gathered} G_{0\alpha\beta}(x-x') = \delta_{\alpha\beta}(2\pi)^{-4}\int G_0(p)\,e^{ip(x-x')}d^4p, \\ G_0(p) = (\omega - \xi + i\delta\,\mathrm{sign}\,\omega)^{-1}, \end{gathered} \tag{12}$$

where $\xi = v(|\mathbf{p}| - p_0)$, $v = p_0/m$, $\delta \to +0$. When impurities are present the Green function (11) is not the same as expression (12). We shall write it in the form

$$\begin{gathered} G_{\alpha\beta}(x, x') = \delta_{\alpha\beta}(2\pi)^{-7}\int G(\mathbf{p}, \mathbf{p}', \omega) \\ \times \exp\{i\mathbf{p}\cdot\mathbf{x} - i\mathbf{p}'\cdot\mathbf{x}' - i\omega(t - t')\}\,d^3\mathbf{p}d^3\mathbf{p}'\,d\omega. \end{gathered} \tag{13}$$

The function $G(\mathbf{p}, \mathbf{p}', \omega)$ is expressed as a sum of the diagrams given in Fig. 1 according to the well-known rules of field theory. Each line corresponds to $G_0(\mathbf{p})$.

Summing over the diagrams gives an integral equation for $G(\mathbf{p}, \mathbf{p}', \omega)$:

$$\begin{gathered} G(\mathbf{p}, \mathbf{p}', \omega) = \delta(\mathbf{p} - \mathbf{p}')\,G_0(p) \\ + (2\pi)^{-3}\int G_0(p)\sum_a u(\mathbf{p} - \mathbf{p}'')\,e^{i(\mathbf{p}-\mathbf{p}'')\mathbf{x}_a}\,G(\mathbf{p}'', \mathbf{p}', \omega)\,d^3\mathbf{p}'' \end{gathered} \tag{14}$$

We are not interested in the exact solution of (14), since the impurity atoms are randomly distributed over the metal. We must average all expressions over the positions of all impurity atoms. It is then essential that by virtue of our assumption about the smallness of the atomic concentration the average distance apart of the impurity atoms is much larger than the atomic distances in the metallic lattice, so that the averaging can take place over volumes with dimensions large compared to the interatomic distances. After such an averaging, the Green function $G(\mathbf{p}, \mathbf{p}', \omega)$ is clearly of the form

$$\overline{G(\mathbf{p}, \mathbf{p}', \omega)} = G(\mathbf{p}, \omega)\,\delta(\mathbf{p} - \mathbf{p}'). \tag{15}$$

The absolute magnitude of the momenta $\mathbf{p}$, $\mathbf{p}'$ which are of interest to us are of the order of magnitude of the momentum $\mathbf{p}_0$ at the Fermi surface which in turn is of the order of the inverse of the interatomic distances. This fact makes it at once easy to carry out the averaging.

We shall perform the calculation in the Born approximation, i.e., we shall assume that $p_0^3\int u(\mathbf{x})\,d^3\mathbf{x} \ll E_{ph}$. Once can show that the final results expressed in terms of the collision time will also be correct in the general case.

The simplest diagram for G contains only one cross. The value obtained by averaging over the position of the impurity atom is a constant $\overline{u(\mathbf{q})\,e^{i(\mathbf{q}\cdot\mathbf{x}_\alpha)}} = u(0)$, which can be included in the ground state energy and which in the following we assume to be equal to zero. The next diagram, as far as complexity is concerned, contains two crosses referring to one atom (Fig. 2a). The average value of these diagrams (without external G_0) is equal to

$$\frac{1}{V}\int |u(\mathbf{p} - \mathbf{p}')|^2\,G_0(p')\,\frac{d^3\mathbf{p}'}{(2\pi)^3}, \tag{16}$$

where V is the volume of the system. In the following we shall be interested in values of $|\mathbf{p}|$ which in absolute magnitude are close to p_0. The integral in Eq. (16) can be split into two parts: an integral over values of $|\mathbf{p}'|$ far from the Fermi surface and one over values near the Fermi surface (the limits of the second integral over $|\mathbf{p}'|$ can be taken symmetric with respect to $|\mathbf{p}'| = p_0$). The integral over the far region gives a real constant which together with $u(0)$ is a renormalization of the chemical potential and need not be considered. One may assume that in the second integral $u(\mathbf{p} - \mathbf{p}')$ is a slowly-varying function. Substituting Eq. (12) for $G_0(\mathbf{p})$ and summing over the impurity atoms (which simply means multiplying by the number of atoms) we get the essen-

*Here and henceforth we use units with $\hbar = 1$.

tial contribution to the G-function:

$$i \operatorname{sign} \omega / 2\tau,$$

where

$$\frac{1}{\tau} = n \frac{p_0 m}{(2\pi)^2} \int |u(\theta)|^2 d\Omega \qquad (17)$$

is the time between collisions, n the number of impurity atoms per unit volume. From this it is clear that the region near the Fermi surface plays the leading role in the integrals.

From this point of view not all diagrams are equivalent. Let us compare, for instance, the three diagrams of Fig. 2b. (Dotted lines connect crosses referring to one atom.) We see easily that in the first two expressions the integration over **p'** and **p"** can be performed over arbitrary angles between the momenta near the Fermi surface. On the other hand, in the third integral the requirement that all arguments of the G-functions are near the Fermi surface leads to a limitation of the angles. As a result the contribution from such a graph turns out to be less by a factor $(vp_0G)^{-1}$ as compared to the others. Since in the following we need values of ω and $\xi \sim 1/\tau$ where τ is the time between collisions, the smallness of the "intersecting" graphs can be estimated to be $(p_0 l)^{-1}$ where $l = v\tau$ is the mean free path.

One can easily show that a small contribution is also given by diagrams containing more than two crosses from one impurity atom. Let us, for instance, compare the total contribution (from all impurity atoms) of diagrams of the kind of the first one of Fig. 2b with diagrams of the type 2c. The first ones give a quantity of the order of magnitude

$$\frac{1}{\tau^2} G_0 \sim \frac{1}{\tau}.$$

The second ones give, as far as order of magnitude is concerned

$$\frac{1}{\tau} \frac{u^2(q)\, p_0^4}{v^2} \sim \frac{1}{\tau}\left[\frac{\int u dV p_0^3}{E_{\text{ph}}}\right]^2 \ll \frac{1}{\tau}.$$

It is thus clear that we need consider only diagrams containing up to two crosses on one impurity atom.

Summing all essential diagrams (i.e., only "pair" diagrams and those which do not contain "intersections" like the third diagram of Fig. 2b) we get the following expression for the G-function:

$$G(p) = G_0(p) + nG_0(p) \int |u(\mathbf{p}-\mathbf{p}')|^2 G(p') \frac{d^3p'}{(2\pi)^3} G(p). \qquad (18)$$

If we do not wish to use the Born approximation, it is necessary to take into account diagrams with many crosses at one impurity atom. One can show

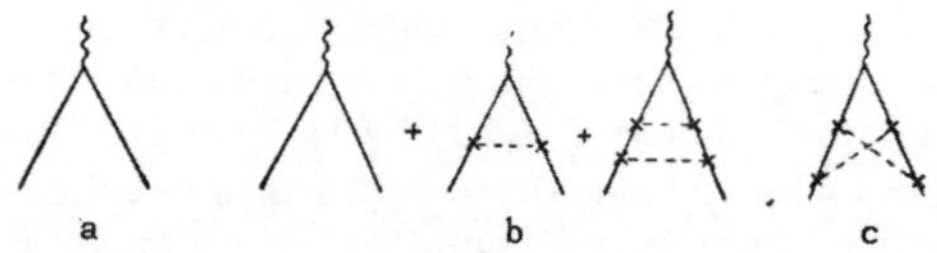

FIG. 3

that all changes which this entails are taken into account by replacing the Born amplitude $u(\theta)$ by the total scattering amplitude. This remains true also for all calculations in the following. We can therefore imply in all formulae by $u(\theta)$ the total scattering amplitude.

The solution of Eq. (18) is

$$G(p) = \frac{1}{\omega - \xi - \overline{G}(\omega)},$$

where $\overline{G}(\omega)$ satisfies the equation

$$\overline{G}(\omega) = n \int |u(\mathbf{p}-\mathbf{p}')|^2 \frac{1}{\omega - \xi - \overline{G}(\omega)} \frac{d^3\mathbf{p}'}{(2\pi)^3}$$

Assuming $\overline{G}(\omega)$ to be purely imaginary ($\overline{G}(\omega) = -i\beta$) and applying to the integral on the right hand side the same approach as was applied to (16) we find

$$\beta = \operatorname{sign} \beta / 2\tau,$$

where τ is determined by Eq. (17). Comparing G which is obtained in this manner with the case of small amounts of impurities ($G \rightarrow G_0$) we find that $\beta = \operatorname{sign} \omega / 2\tau$ or

$$G = (\omega - \xi + i \operatorname{sign} \omega / 2\tau)^{-1}. \qquad (19)$$

Going over to the x-representation we see easily that the whole change in G as compared to G_0 is obtained by multiplying the latter by an exponential screening factor

$$G = G_0 e^{-|\mathbf{x}-\mathbf{x}'|/2l}, \qquad (20)$$

where $l = v\tau$.

We now proceed to evaluate Q_{ik}. Putting the photon vertex into the electron Green function $G(\mathbf{p}, \mathbf{p}', \omega)$ we get:

$$\Pi(p_+, p_-) \qquad (21)$$
$$= (2\pi)^{-3} \int G\left(\mathbf{p}_+, \mathbf{p}_+', \omega + \frac{\omega_0}{2}\right) \mathbf{p}' G\left(\mathbf{p}_-', \mathbf{p}_-, \omega - \frac{\omega_0}{2}\right) d^3\mathbf{p}'.$$

The function $G(\mathbf{p}, \mathbf{p}', \omega)$ entering here corresponds to the sum of the diagrams in Fig. 1 and is described by Eq. (14). To average over the positions of the impurity atoms we must take into account that the average of two Green functions is not equal to the product of the two average values.

For a pure metal expression (21) corresponds to the diagram of Fig. 3a. After averaging over the positions of the impurity atoms, it turns out that apart from the simple diagrams corresponding to the transition from the zeroth Green functions $G_0(p)$ to the function $G(p)$ of (19) the essential diagrams for the quantity (21) are those of Fig. 3b. The large contribution from these corrections is connected with the fact that the photon momentum occurring in the vertex $|\mathbf{k}| \ll p_0$. Diagrams of another kind, for instance the graph of Fig. 3c, are far smaller since one of the integrations in them is over a region of momenta far from the Fermi surface. The averaging of the quantity (21) is thus performed by the summation of the "ladder" diagrams of Fig. 3b.

The integral equation for $\Pi(p_+, p_-)$ is

$$\Pi(p_+, p_-) = G(p_+)G(p_-)\left[\mathbf{p} + \frac{n}{(2\pi)^3}\int |u(\mathbf{p}-\mathbf{p}')|^2\,\Pi(p'_+, p'_-)\,d^3\mathbf{p}'\right]. \tag{22}$$

There are two limiting cases possible:

(a) The skin effect is anomalous, $(kv) \gg \tau$; it is easy to verify that the integral on the right hand side of (22) will be negligibly small in that case $(\sim 1/kv\tau)$.

(b) The skin effect is normal, $(kv) \ll 1/\tau$; this case is just of the greatest interest to us. In Eq. (22) we can then assume $\mathbf{p}_+ = \mathbf{p}_-$.

The vector obtained from the integral on the right hand side of (22) will clearly be parallel to $\mathbf{p}$. We introduce the notation

$$\mathbf{p}\Lambda(\omega, \omega_0) = \frac{n}{(2\pi)^3}\int |u(\mathbf{p}-\mathbf{p}')|^2\,\Pi(p'_+, p'_-)\,d^3\mathbf{p}'. \tag{23}$$

Since $|\mathbf{p}| \approx p_0$, we can assume $\Lambda(\omega, \omega_0)$ to be independent of $|\mathbf{p}|$. We multiply (22) by $n(2\pi)^{-3} \times |u(\mathbf{1}-\mathbf{p})|^2$ and integrate over $d^3\mathbf{p}$:

$$\mathbf{l}\Lambda(\omega, \omega_0) = \frac{n}{(2\pi)^3}\int |u(\mathbf{l}-\mathbf{p})|^2\,\mathbf{p}\,G(p_+)\,G(p_-)\,[1 + \Lambda(\omega, \omega_0)]\,d^3\mathbf{p}. \tag{24}$$

Replacing $G(p)$ by Eq. (19) we find easily that $\Lambda(\omega, \omega_0)$ is only different from zero if $|\omega| < \omega_0/2$, since in the opposite case both poles in (24) for the integration over ξ according to (20) lie in the same half-plane. In that integral Λ does not depend on ω. Integrating over ξ and using the relation

$$\cos\theta = \cos\theta'\cos\theta'' + \sin\theta'\sin\theta''\cos(\varphi' - \varphi''), \tag{25}$$

we get

$$\Lambda(\omega, \omega_0) = \begin{cases} \dfrac{i}{\tau_1}\,\dfrac{1}{\omega_0 + i/\tau_{tr}} & \omega^2 < \omega_0^2/4, \\ 0, & \omega^2 > \omega_0^2/4, \end{cases} \tag{26}$$

where

$$\frac{1}{\tau_1} = \frac{1}{\tau} - \frac{1}{\tau_{tr}}; \quad \frac{1}{\tau_{tr}} = \frac{nmp_0}{(2\pi)^2}\int |u(\theta)|^2(1 - \cos\theta)\,d\Omega. \tag{27}$$

After substituting (26) into (22) and (7) and integrating, assuming that $\omega_0\tau \ll 1$, we find

$$Q_{ik} = -\frac{4\pi i\omega_0\sigma}{c^2}\,\delta_{ik}. \tag{28}$$

The "transport" collision time enters into the conductivity $\sigma = Ne^2\tau_{tr}/m$, as should be the case.

2. SUPERCONDUCTING ALLOYS

The method given above will now be applied to a study of a superconductor with impurities. The main difference as compared to what was given above lies in the fact that a superconductor is described by three Green functions:

$$\begin{aligned} \delta_{\alpha\beta}G(x, x') &= -i\langle T(\psi_\alpha(x)\psi^+_\beta(x'))\rangle; \\ \hat{I}_{\alpha\beta}F^+(x, x') &= \langle T(\psi^+_\alpha(x)\psi^+_\beta(x'))\rangle; \\ -\hat{I}_{\alpha\beta}F(x, x') &= \langle T(\psi_\alpha(x)\psi_\beta(x'))\rangle, \end{aligned} \tag{29}$$

where $\hat{I}_{\alpha\beta} = -\hat{I}_{\beta\alpha}$, $\hat{I}^2 = -E$. These functions were introduced in reference 4 for a pure superconductor and are equal to

$$\begin{aligned} G(x - x') &= (2\pi)^{-4}\int G(p)\,e^{ip(x-x')}\,d^4p, \\ F^+(x - x') &= (2\pi)^{-4}\int F^+(p)\,e^{ip(x-x')}\,d^4p, \\ F(x - x') &= F^+(x - x'), \end{aligned}$$

where

$$G_0(p) = \frac{\omega + \xi}{\omega^2 - \xi^2 - \Delta^2 + i\delta}; \quad F_0^+(p) = \frac{i\Delta}{\omega^2 - \xi^2 - \Delta^2 + i\delta} \quad (\delta \to +0). \tag{30}$$

The occurrence of three Green functions makes it necessary to change the diagram technique.* The functions G, F^+, and F can be represented by the lines with arrows given in Fig. 4a. If vertices are included in an electron line it is necessary to put an arrow next to the vertex. If the interaction Hamiltonian includes products $\psi^+_\alpha(y')\psi_\alpha(y)$ (this always occurs in our problem) the arrows around the vertex must be aimed in the same direction. There are then two possibilities: both arrows can be directed towards the left or towards the right. From Fig. 4c it is clear what will be the result of including vertices $\psi^+\psi$ in the functions G, F^+, and F. The result can be written in the form

$$\begin{aligned} G(x, x') &\to G(x, y')\,G(y, x') + F(x, y)\,F^+(y', x'); \\ F^+(x, x') &\to F^+(x, y')\,G(y, x') + G(y, x)\,F^+(y', x'); \\ F(x, x') &\to G(x, y')\,F(y, x') + F(x, y)\,G(x', y'). \end{aligned} \tag{31}$$

---

*The technique with three Green functions was worked out by S. T. Beliaev[6] in an application to the Bose gas.

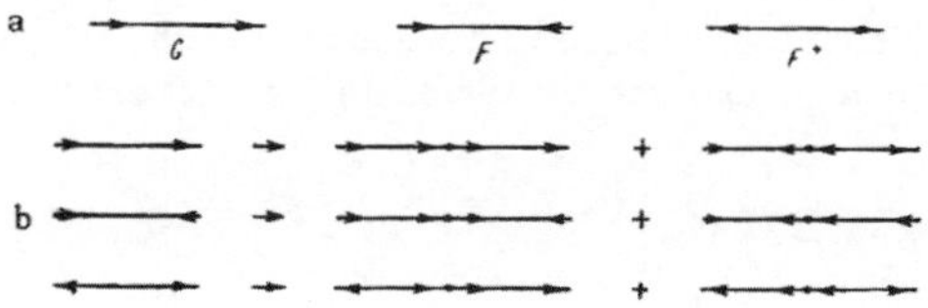

FIG. 4

The inclusion of vertices leads in this way to the appearance of F and F^+ lines in the diagrams for G and similarly to G lines in the diagrams for F and F^+. The equations for G, F^+, and F are therefore more complicated than Eq. (14), obtained by summing the diagrams of Fig. 1 with one G-function.

The derivation of these equations and their solution are given in the Appendix. It turns out that the averaging over the positions of the impurity atoms of the functions G(p) and $F^+(p) = F(p)$ can be obtained from the expressions (30) for a pure superconductor by the substitution

$$\omega;\ \Delta \rightarrow \omega\eta_i;\quad \Delta\eta_i, \tag{32}$$

where

$$\eta_i = \begin{cases} 1 + i/2\tau\sqrt{\omega^2 - \Delta^2}; & \omega^2 > \Delta^2, \\ 1 + 1/2\tau\sqrt{\Delta^2 - \omega^2}; & \omega^2 < \Delta^2; \end{cases}$$

τ is the collision time for the metal in its normal state. One can easily verify that, as in Eq. (18) of Sec. 1, these equations signify in coordinate space a multiplication of the function $G_0(x)$ and $F_0^+(x)$ by $e^{-|\mathbf{x}|/2l}$

We now turn to the determination of the function Q_{ik}. This time we have two diagrams corresponding to the simple loop for a normal metal. They are given in Fig. 5. This means that instead of a product of two G's we have now the combination $GG - FF^+$ (the sign is due to the fact that the operator $\nabla_{\mathbf{y}} - \nabla_{\mathbf{y}'}$ occurs in the electromagnetic interaction, which changes sign if the direction of the arrows at the vertex is changed). For the following it will again be convenient for us, taking the two loops of Fig. 5 together, and taking one of the photon vertices p_i separately, to consider the remaining factor as the result of including the photon vertex p_k in the function G. After this Q_{ik} will be of the form (8) and the problem lies in evaluating the vector $\mathbf{\Pi}$.

To average $\mathbf{\Pi}$ over the impurities we note at once that since superconductivity changes the Green functions only near the Fermi surface we need again only sum over ladder graphs of the type given in Fig. 3b. Owing to the modified diagram technique, we must consider, as can easily be seen from Fig. 6, not one, but four different vectors $\mathbf{\Pi}^{(i)}$ $(i = 1, \ldots, 4)$ which occur as a result of including

FIG. 5

photon vertices in G, F^+, $G^{(-)}$, and F (where $G^{(-)}$ denotes G with the arguments interchanged). The first of these quantities enters directly into the integral (8) for Q_{ik}.

We obtain easily for the $\mathbf{\Pi}^{(i)}$ a set of four integral equations. The first one of them is obtained with the aid of Fig. 6 and has the form:

$$\begin{aligned} \mathbf{\Pi}^{(1)}(p_+, p_-) &= \mathbf{p}\,[G(p_+)\,G(p_-) - F(p_+)\,F(p_-)] \\ &+ G(p_+)\,G(p_-)\,\Lambda^{(1)} + F(p_+)\,G(p_-)\,\Lambda^{(2)} \\ &+ F(p_+)\,F(p_-)\,\Lambda^{(3)} + G(p_+)\,F(p_-)\,\Lambda^{(4)}. \end{aligned} \tag{33}$$

The quantities $\Lambda^{(i)}$ $(\omega, \omega_0, \mathbf{k}, \mathbf{p})$ are the integrals (23) of the $\mathbf{\Pi}^{(i)}(p_+, p_-)$. The remaining equations are obtained in a similar way. We shall not write them down since, as was pointed out above, we only need $\mathbf{\Pi}^{(1)}(p_+, p_-)$.

As was already noted, for high impurity concentrations the role of the correlation length is assumed here not by $\xi_0 = v/\Delta$ but by the mean free path l. As the impurity concentration increases this length decreases. We shall consider here the case where l is much less than the penetration depth. It is then natural to expect a local connection between the current and the vector potential. The case considered is characterized by the condition $kl \ll 1$, which leads to a considerably simplified equation.

Indeed, neglecting $\mathbf{k}$, we obtain $p_+ = p_-$. It is easy to check that in that case $\Lambda^{(i)}(\omega, \omega_0, \mathbf{k}, \mathbf{p}) \approx \mathbf{p}\Lambda^{(i)}(\omega, \omega_0)$ and

$$\Lambda^{(3)}(\omega, \omega_0) = -\Lambda^{(1)}(\omega, \omega_0); \quad \Lambda^{(2)}(\omega, \omega_0) = \Lambda^{(4)}(\omega, \omega_0). \tag{34}$$

The set of equations for $\Lambda^{(1,2)}$ takes the form

$$\begin{aligned} \mathbf{p}\Lambda^{(1)}(\omega, \omega_0) &= \frac{n}{(2\pi)^3}\int |u(\mathbf{p} - \mathbf{p}')|^2\,\mathbf{p}'\,\{[G(p'_+)\,G(p'_-) \\ &\quad - F(p'_+)\,F(p'_-)] \times [1 + \Lambda^{(1)}(\omega, \omega_0)] \\ &+ [F(p'_+)\,G(p'_-) + G(p'_+)\,F(p'_-)]\,\Lambda^{(2)}(\omega, \omega_0)\}\,d^3\mathbf{p}; \\ \mathbf{p}\Lambda^{(2)}(\omega, \omega_0) &= \frac{n}{(2\pi)^3}\int |u(\mathbf{p} - \mathbf{p}')|^2\,\mathbf{p}'\,\{[F(p'_+)\,G(p'_-) \\ &\quad - G(-p'_+)\,F(p'_-)]\,[1 + \Lambda^{(1)}(\omega, \omega_0)] \\ &+ [G(-p'_+)\,G(p'_-) + F(p'_+)\,F(p'_-)]\,\Lambda^{(2)}(\omega, \omega_0)\}\,d^3\mathbf{p}. \end{aligned} \tag{35}$$

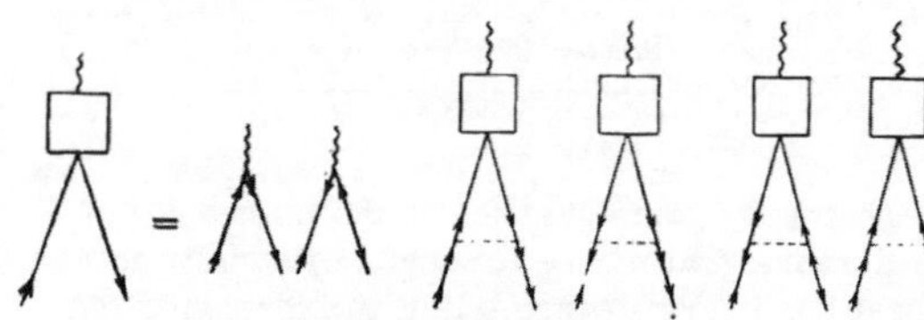

FIG. 6

Substituting into (35) the expressions (30) and (32) for the functions G and F, integrating over ξ, and using the relation (25) we arrive at the following expressions for the quantities of interest to us:

$$\Lambda^{(1)}(\omega, \omega_0) = \frac{i}{2\tau_1} \frac{1-(\omega_+\omega_- + \Delta^2)/\sqrt{\omega_+^2-\Delta^2}\sqrt{\omega_-^2-\Delta^2}}{\sqrt{\omega_+^2-\Delta^2}+\sqrt{\omega_-^2-\Delta^2}+i/\tau_{tr}};$$

$$\Lambda^{(2)}(\omega, \omega_0) = \frac{\Delta}{2\tau_1} \frac{2\omega}{\sqrt{\omega_+^2-\Delta^2}\sqrt{\omega_-^2-\Delta^2}\left[\sqrt{\omega_+^2-\Delta^2}+\sqrt{\omega_-^2-\Delta^2}+i/\tau_{tr}\right]}. \tag{36}$$

Substituting (34) and (36) into (33) we find $\Pi^{(1)}(p_+, p_-)$ and substituting this into Eq. (8) we get an expression for Q:

$$Q_{ik}(\omega_0) = \delta_{ik}\left\{\frac{Ne^2}{mc} - \frac{ie^2N}{2\pi mc}\int d\xi \int_C d\omega [((\omega_+\omega_- + \Delta^2)\eta_+\eta_- + \xi^2)\right.$$

$$\times (1 + \Lambda^{(1)}(\omega, \omega_0)) + 2i\Delta\omega\Lambda^{(2)}(\omega, \omega_0)] \tag{37}$$

$$\times \frac{1}{[\xi^2-(\omega_+^2-\Delta^2)\eta_+^2][\xi^2-(\omega_-^2-\Delta^2)\eta_-^2]}.$$

The contour around the singular points in the integral over ω is given in Fig. 7a for the case $\omega_0 < 2\Delta$ and in Fig. 7c for the case $\omega_0 > 2\Delta$. Before performing the integration (37) we note that the integral of the first term in (33) diverges. One must therefore observe the well-known caution in the problem of the order of integration over ω and ξ. A closer inspection shows that one must first integrate by parts.

We first find the penetration depth of a static magnetic field into a superconductor. The latter is connected with Q_{ik} by the relation

$$\delta = \sqrt{\frac{c}{4\pi Q(0)}}; \quad \text{if } Q_{ik} = \delta_{ik}Q). \tag{38}$$

The contour C in Fig. 7a can in the case $\omega_0 = 0$ conveniently be deformed so that the integration over ω is performed along the imaginary axis (this is possible because of the decrease of the expression under the integral sign as $\omega \to \infty$):

$$Q_{ik}(0) = \delta_{ik}Q(0) = \frac{Ne^2}{mc}\delta_{ik}\left\{1\right.$$

$$+ \frac{1}{2\pi}\iint_{-\infty}^{\infty} d\xi\, d\omega \frac{1 + \dfrac{\Delta^2}{2\tau_1(\omega^2+\Delta^2)\left(\sqrt{\omega^2+\Delta^2}+1/2\tau_{tr}\right)}}{\xi^2 + (\sqrt{\omega^2+\Delta^2}+1/2\tau_{tr})^2}$$

$$\left. - \frac{1}{\pi}\iint_{-\infty}^{\infty} \frac{d\xi\, d\omega\, \omega^2\left(1+\dfrac{1}{2\tau\sqrt{\omega^2+\Delta^2}}\right)^2}{[\xi^2+(\sqrt{\omega^2+\Delta^2}+1/2\tau_{tr})^2]^2}\right\}.$$

Integrating by parts over ω in the first term of the first integral makes it possible formally to cancel the diverging terms in that expression. In the remaining double integral one can change the order of integration. The integral over ξ is at once performed:

$$Q(0) = \frac{Ne^2}{mc}\left\{1 - \frac{\Delta^2}{4\tau_{tr}}\int_{-\infty}^{\infty} \frac{d\omega}{(\omega^2+\Delta^2)^{3/2}(\sqrt{\omega^2+\Delta^2}+1/2\tau_{tr})}\right\}.$$

After that we integrate over ω and find

$$Q(0) = \frac{Ne^2}{mc}\tau_{tr}\Delta\begin{cases}\pi - \dfrac{4\tau_{tr}\Delta}{\sqrt{1-(2\tau_{tr}\Delta)^2}}\cosh^{-1}\dfrac{1}{2\tau_{tr}\Delta}; & 2\tau_{tr}\Delta < 1,\\[2ex] \pi - \dfrac{4\tau_{tr}\Delta}{\sqrt{(2\tau_{tr}\Delta)^2-1}}\arccos\dfrac{1}{2\tau_{tr}\Delta}; & 2\tau_{tr}\Delta > 1.\end{cases} \tag{39}$$

Substituting this into Eq. (38) we can from this obtain the penetration depth. If $\tau_{tr}\Delta \ll 1$ it is (in the usual units) of the form

$$\delta = \frac{c}{2\pi}\sqrt{\hbar/\sigma\Delta}, \tag{40}$$

where $\sigma = Ne^2\tau_{tr}/m$ is the conductivity of the metal in the normal state. In the case $\tau_{tr}\Delta \gg 1$ the expression for the penetration depth goes over into the usual London formula:

$$\delta = \sqrt{mc^2/4\pi Ne^2}.$$

It is well known[1] that pure superconductors fall into two classes: in one we have $\delta > v/\Delta$ for all temperatures; they can be called London superconductors. In that case there is no limit whatever for the quantity $\tau_{tr}\Delta$. For the other superconductors, which can be called Pippard superconductors, we have $\delta < v/\Delta$ at $T = 0$. In that case from the condition $l \ll \delta$ follows always that $\tau_{tr}\Delta \ll 1$. The Pippard superconductors are thus described by Eq. (40) for $l \ll \delta$. In the case when the mean free path $l \gg \delta$ the theory developed in reference 1 is clearly correct for Pippard superconductors,

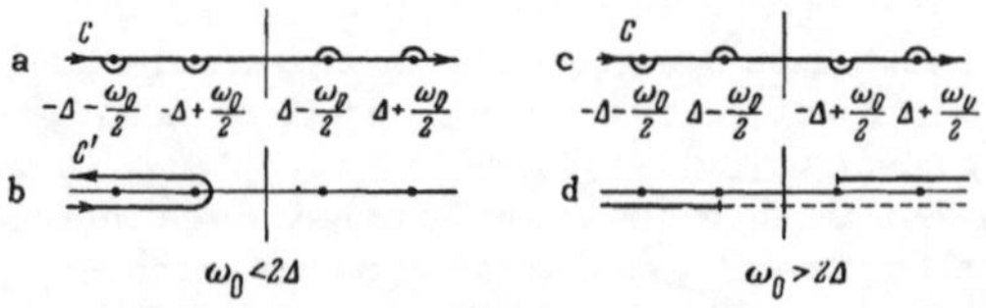

FIG. 7

independent of the relation between l and v/Δ. From this theory we obtain

$$Q(k, 0) = \frac{3\pi^2}{4mc}\frac{Ne^2\Delta}{vk}. \tag{41}$$

Comparison of (40) and (41) shows that Eq. (40) must be used, if

$$\delta / l_{tr} \gg 3\pi/4. \tag{42}$$

We now proceed to the case of an alternating field. It is of interest to follow the transition from a superconductor to a normal metal as the frequency ω_0 increases from zero to $\omega_0 \gg \Delta$. We consider only the case $\tau_{tr}\Delta \ll 1$. We subtract from the expression under the integral sign its value at $\omega_0 = 0$. The remainder can then be integrated first over ξ. In the case when $\omega_0 < 2\Delta$ we go over to the contour C′ (Fig. 7b) for the integration over ω. We get

$$Q(\omega_0) = \frac{Ne^2}{mc}\left\{\int_{\Delta+\omega_0/2}^{\infty} d\omega\left[1 - \frac{\omega_+\omega_- + \Delta^2}{\sqrt{\omega_+^2 - \Delta^2}\sqrt{\omega_-^2 - \Delta^2}}\right]\right.$$
$$\times \frac{\sqrt{\omega_+^2 - \Delta^2} + \sqrt{\omega_-^2 - \Delta^2}}{\left(\sqrt{\omega_+^2 - \Delta^2} + \sqrt{\omega_-^2 - \Delta^2}\right)^2 + 1/\tau_{tr}^2}$$
$$\left. + \tau_{tr}\int_{\Delta-\omega_0/2}^{\Delta+\omega_0/2} \frac{\omega_+\omega_- + \Delta^2}{\sqrt{\omega_+^2 - \Delta^2}\sqrt{\omega_-^2 - \Delta^2}}\right\}.$$

The contribution from the first integral is of the order $(\tau\Delta)^2$ and we can neglect it. The second integral leads to a complete elliptic integral.[7] The final expression has the following form for the case $\omega_0 < 2\Delta$

$$Q(\omega_0) = \frac{2\sigma\Delta}{c} E\left(\frac{\omega_0}{2\Delta}\right). \tag{43}$$

In particular, we get for $\omega_0 \ll 2\Delta$

$$Q(\omega_0) = \frac{\pi\sigma\Delta}{c}\left[1 - \left(\frac{\omega_0}{4\Delta}\right)^2\right]. \tag{44}$$

In the case $\omega_0 > 2\Delta$ it is convenient to complete the integral to go from $-\infty$ to $+\infty$ in the lower half-plane (Fig. 7c). This part can then be discarded, since there are no poles in the expression under the integral sign. This makes it possible to simplify the expressions considerably. As a result, Q can be expressed as the sum of elliptic integrals, if $\tau\Delta \ll 1$:

$$Q(\omega_0) = \frac{\sigma}{c}\left\{\omega_0 E(\varkappa) - \frac{\omega_0^2 - (2\Delta)^2}{\omega_0} K(\varkappa) - i\left[\omega_0 E(\varkappa_1) - \frac{(2\Delta)^2}{\omega_0} K(\varkappa_1)\right]\right\}, \tag{45}$$

where $\kappa = 2\Delta/\omega_0$; $\kappa_1 = \sqrt{1-\kappa^2}$.

In particular if $\omega_0 \gg 2\Delta$

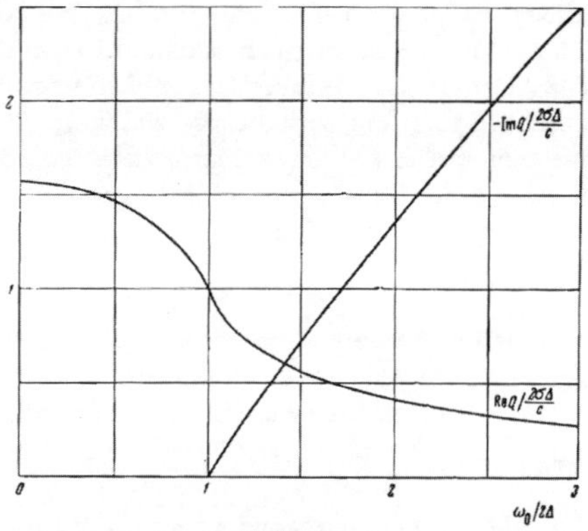

FIG. 8

$$Q(\omega_0) = -\frac{\sigma\omega_0}{c}\left\{i\left[1 - \frac{2\Delta^2}{\omega_0^2}\left(\ln\frac{2\omega_0}{\Delta} + \frac{1}{2}\right)\right] - \frac{\pi\Delta^2}{\omega_0^2}\right\}. \tag{46}$$

From Eq. (46) it is clear that if $\omega_0 \gg \Delta$ the superconducting properties disappear; the usual formula for the normal skin effect turns up. It is of interest that the imaginary part of the current corresponding to absorption occurs only for $\omega_0 > 2\Delta$; the connection of this threshold energy and the binding energy of the pairs is evident. Near the threshold $|\omega_0 - 2\Delta| \ll \Delta$ we have:

$$Q = \frac{\sigma}{c}\left[2\Delta + \left(\Delta - \frac{\omega_0}{2}\right)\left(\ln\frac{16\Delta}{2\Delta - \omega_0} - 1\right)\right]; \quad \omega_0 < 2\Delta, \tag{47}$$
$$Q = \frac{\sigma}{c}\left[2\Delta - \left(\frac{\omega_0}{2} - \Delta\right)\left(\ln\frac{16\Delta}{\omega_0 - 2\Delta} - 1 + i\pi\right)\right]; \quad \omega_0 > 2\Delta.$$

In Fig. 8 we have given the complete curves of the dependence of the real and imaginary parts of Q on ω_0. Along the ordinate axis are put $\mathrm{Re}\left[Q\Big/\frac{2\Delta\sigma}{c}\right]$ and $-\mathrm{Im}\left[Q\Big/\frac{2\sigma\Delta}{c}\right]$ in their dependence on $\omega_0/2\Delta$. The impedance of a bulk specimen with a plane surface is obtained with the aid of Eq. (4) and is of the form

$$Z = -\frac{i\omega_0}{c}\sqrt{\frac{4\pi}{cQ(\omega_0)}}; \tag{48}$$

If Q is complex the radical means the analytical continuation of that quantity which is positive for real Q.

The Eqs. (39), (40), and (43) to (48) describe completely the behavior of alloys in a constant or alternating weak electromagnetic field at T = 0. Apart from alloys, these expressions can also be applied to films of a pure substance with a thickness less than the penetration depth. Such films are as a rule polycrystalline and the dimensions of the crystallites are not more than the film thickness. The presence of boundaries between the crystallites will lead to a break-up of the correlation. The local character of the connection in the case $\delta \gg l$ makes it possible to apply the results obtained above to films without any changes. It may then turn out to be convenient to use the complex dielectric susceptibility which is connected with Q through the relation

$$\varepsilon = -4\pi c Q/\omega^2 \tag{49}$$

In conclusion the authors use this opportunity to express their gratitude to Academician L. D. Landau for his constant attention and valuable remarks.

APPENDIX

In the case of the superconducting alloy the equations replacing (14) are easily found by drawing the corresponding diagrams. They are of the form

$$G(\mathbf{p}, \mathbf{p}', \omega) = G_0(p)\,\delta(\mathbf{p}-\mathbf{p}') + \frac{1}{(2\pi)^3}\Big\{G_0(p)\int u(\mathbf{p}-\mathbf{p}'')\sum_a e^{i(\mathbf{p}-\mathbf{p}'')\mathbf{x}_a}\,G(\mathbf{p}'', \mathbf{p}', \omega)\,d^3\mathbf{p}'' + F_0(p)\int u(\mathbf{p}-\mathbf{p}'')\sum_a e^{i(\mathbf{p}-\mathbf{p}'')\mathbf{x}_a}\,F^+(\mathbf{p}'', \mathbf{p}', \omega)\,d^3\mathbf{p}''\Big\};$$

$$F^+(\mathbf{p}, \mathbf{p}', \omega) = F_0^+(p)\,\delta(\mathbf{p}-\mathbf{p}') + \frac{1}{(2\pi)^3}\Big\{F_0(p)\int u(\mathbf{p}-\mathbf{p}'')\sum_a e^{i(\mathbf{p}-\mathbf{p}'')\mathbf{x}_a}\,G(\mathbf{p}'', \mathbf{p}', \omega)\,d^3\mathbf{p}'' + G_0(-p)\int u(\mathbf{p}-\mathbf{p}'')\sum_a e^{i(\mathbf{p}-\mathbf{p}'')\mathbf{x}_a}\,F^+(\mathbf{p}'', \mathbf{p}', \omega)\,d^3\mathbf{p}''\Big\} \tag{A.1}$$

In principle it is necessary to consider also the equations for the function $-I_{\alpha\beta}F(x, x') = \langle T(\psi_\alpha(x)\psi_\beta(x'))\rangle$.

For a pure superconductor $F(x, x') = F^+(x, x')$. This property will be true also for alloys after Eq. (A.1) is averaged over the positions of the impurity atoms.

The averaging is performed exactly as in Sec. 1. Using the equations[4] that are satisfied by the functions $G_0(p)$ and $F_0(p)$, this set can be put into a very simple form:

$$(\omega - \xi - \overline{G_\omega})\,G(p) + (i\Delta - \overline{F_\omega^+})\,F^+(p) = 1,$$
$$(\omega + \xi - \overline{G_\omega})\,F^+(p) - (i\Delta - \overline{F_\omega^+})\,G(p) = 0.$$

The solution of this set is

$$G(p) = \frac{\omega - \overline{G_\omega} + \xi}{(\omega - \overline{G_\omega})^2 + (i\Delta - \overline{F_\omega^+})^2 - \xi^2};$$
$$F^+(p) = \frac{i\Delta - \overline{F_\omega^+}}{(\omega - \overline{G_\omega})^2 + (i\Delta - \overline{F_\omega^+})^2 - \xi^2}. \tag{A.2}$$

Remembering the definition of $\overline{G_\omega}$ we see that as before $\overline{G_\omega}$ contains a constant term which means a renormalization of the chemical potential and which arises from the integration far from the Fermi surface:

$$\delta\mu = \frac{n}{(2\pi)^3}\int \frac{|u(\mathbf{p}-\mathbf{p}')|^2\,\xi\,d\xi}{(\omega - \overline{G_\omega})^2 + (i\Delta - \overline{F_\omega^+})^2 - \xi^2}.$$

After subtracting this term, the remaining part of $\overline{G_\omega}$ is apart from a coefficient determined by the same integral as $\overline{F_\omega^+}$, as is clear from (A.2). In this way

$$\overline{G_\omega}/\omega = \overline{F_\omega^+}/\Delta.$$

After this it is easy to find that the solution of the set (A.2) gives Eqs. (30) and (32).

[1] Bardeen, Cooper, and Schrieffer, Phys. Rev. **108**, 1175 (1957).

[2] L. N. Cooper, Phys. Rev. **104**, 1189 (1956).

[3] Abrikosov, Gor'kov, and Khalatnikov, J. Exptl. Theoret. Phys. (U.S.S.R.) **35**, 265 (1958), Soviet Phys. JETP **8**, 182 (1959); D. C. Mattis and J. Bardeen, Phys. Rev. **111**, 412 (1958).

[4] L. P. Gor'kov, J. Exptl. Theoret. Phys. (U.S.S.R.) **34**, 735 (1958), Soviet Phys. JETP **7**, 505 (1958).

[5] N. N. Bogoliubov, J. Exptl. Theoret. Phys. (U.S.S.R.) **34**, 58, 73 (1958), Soviet Phys. JETP **7**, 41, 51 (1958); V. V. Tolmachev and S. V. Tiablikov, J. Exptl. Theoret. Phys. (U.S.S.R.) **34**, 66 (1958), Soviet Phys. JETP **7**, 46 (1958).

[6] S. T. Beliaev, J. Exptl. Theoret. Phys. (U.S.S.R.) **34**, 417 (1958), Soviet Phys. JETP **7**, 289 (1958).

[7] E. Jahnke and F. Emde, Tables of Functions, Dover, New York, 1945.

Translated by D. ter Haar
322

SUPERCONDUCTING ALLOYS AT FINITE TEMPERATURES

A. A. ABRIKOSOV and L. P. GOR'KOV

Institute for Physical Problems, Academy of Sciences, U.S.S.R.

Submitted to JETP editor July 16, 1958

J. Exptl. Theoret. Phys. (U.S.S.R.) **36**, 319-320 (January, 1959)

WE have recently[1] formulated the electrodynamics of superconductors that contain small atomic concentrations of impurities at $T = 0$. The method of reference 1 is not applicable at finite temperatures.

The authors of the present paper obtained, together with I. E. Dzyaloshinskii, a generalization of the method applied at $T = 0$. It was basically founded upon the formulation of the thermodynamic theory proposed by Matsubara.[2] An exposition of this method will be given in a separate paper. We note here only that the evaluation of the basic functions at $T \neq 0$ is formally very similar to the calculations at $T = 0$. In the case of equilibrium practically the only change involved is the replacement of the integrals over the frequency by sums over the discrete variable

$$\frac{1}{2\pi}\int_{-\infty}^{\infty} d\omega f(\omega) \rightarrow \frac{iT}{\hbar}\sum_{n=-\infty}^{\infty} f(i\omega_n), \qquad (1)$$

where $\omega_n = (\pi T/\hbar)(2n + 1)$, and T is the temperature in energy units.

This method was applied by us to study the equilibrium properties of superconducting alloys at finite temperatures. All calculations are exactly the same as the evaluations performed[1] for $T = 0$ apart from the transformation (1).

The evaluation of the functions $G(x, x')$ and $F(x, x')$ leads to the conclusion that in alloys, as at $T = 0$, these functions are simply multiplied by an exponential factor

$$\begin{aligned} G(x, x') &= G_0(x, x')\exp\{-|\mathbf{x}-\mathbf{x}'|/2l\}, \\ F(x, x') &= F_0(x, x')\exp\{-|\mathbf{x}-\mathbf{x}'|/2l\}, \end{aligned} \qquad (2)$$

where l is the mean free path in the normal state.

To find the thermodynamic functions it is sufficient to determine the particle density $N(\mu, T)$ as a function of the chemical potential and temperature:

$$N = \langle\psi^+(x)\psi(x)\rangle = -i\,[G(x, x')]_{x=x';\,t=t'-0}. \qquad (3)$$

One sees easily that the function $N(\mu, T)$ is the same as in the case of a pure superconductor. In the model under consideration, where one assumes that the impurities do not influence the interaction between the electrons, but only scatter them, the occurrence of impurities will therefore not change the thermodynamic functions. In particular, the critical temperature T_c is not shifted. We emphasize once again that this result is, of course, valid only if the impurity concentration is supposed to be small.

We have considered the behavior of alloys in a constant magnetic field. The connection between the current and the vector potential is local in the London case. At finite temperatures it is of the form

$$\mathbf{j} = -Q\mathbf{A}, \qquad (4)$$

$$Q = \frac{Ne^2}{mc}2\pi T\Delta^2\sum_0^{\infty} 1/[(\hbar\omega_n)^2 + \Delta^2](\sqrt{(\hbar\omega_n)^2 + \Delta^2} + \hbar/2\tau_{tr}) \qquad (5)$$

where $\omega_n = (\pi T/\hbar)(2n+ 1)$; $\Delta(T)$ the temperature dependent "gap" occurring in the paper by Bardeen, Cooper, and Schrieffer;[3] $\tau_{tr} = [\int w(\theta)(1 - \cos\theta)\,d\Omega]^{-1}$ is the transport collision time for the metal in its normal state, and N is the electron density.

From Eq. (4) together with the Maxwell equation one obtains the connection between Q and the penetration depth:

$$\delta = \sqrt{c/4\pi Q}. \qquad (6)$$

In the limiting case of large free paths, $\Delta\tau_{tr}/\hbar \gg 1$, this formula goes over into the usual result for a pure London superconductor, i.e., a superconductor with $\delta \gg \hbar v/\Delta$:

$$\delta = \sqrt{mc^2/4\pi N_s e^2}, \qquad (7)$$

where N_s is the number of superconducting electrons (see Ref. 3),

$$N_s/N_0 = 2\pi T\Delta^2\sum_n[(\hbar\omega_n)^2 + \Delta^2]^{-3/2} = \frac{1}{4T}\int_\Delta^\infty \frac{\varepsilon\, d\varepsilon}{\cosh^2\frac{\varepsilon}{2T}\sqrt{\varepsilon^2 - \Delta^2}}. \qquad (8)$$

In the opposite limiting case, $\Delta\tau_{tr}/\hbar \ll 1$,

$$\delta = (c/2\pi)\,[\hbar/\Delta\sigma\tanh(\Delta/2T)]^{1/2}, \qquad (9)$$

where $\sigma = Ne^2\tau_{tr}/m$ is the conductivity of the alloy in the normal state.

It is well known that superconductors fall into two classes. For London superconductors $\delta > \hbar v/\Delta$. In that case the connection between Δ and $\hbar/\tau_{tr}$ can be arbitrary. In the case of the Pippard superconductors, for which $\delta < \hbar v/\Delta$, the condition $\delta \gg l$ leads to $\Delta\tau_{tr}/\hbar \ll 1$. Hence, only Eq. (9) refers to Pippard superconductors.

As was noted in reference 1, all formulae obtained for δ can by analogy with this be applied for the characteristics of films of thickness $d \ll \delta$.

In that case the quantity δ no longer plays the role of the penetration depth, but this quantity goes over into the effective dielectric constant for low frequencies and determines the magnetic susceptibility. Since the connection between $\mathbf{j}$ and $\mathbf{A}$ is local, ϵ and χ, expressed in terms of δ, have the usual, London form:

$$\varepsilon = -c^2/\omega^2\delta^2, \tag{10}$$

$$\chi = \frac{1}{4\pi}\left(1 - \frac{2\delta}{d}\tanh\frac{d}{2\delta}\right) \approx -\frac{1}{12\pi}\left(\frac{d}{2\delta}\right)^2. \tag{11}$$

In conclusion the authors express their gratitude to Academician L. D. Landau for discussions of this paper.

[1] A. A. Abrikosov and L. P. Gor'kov, J. Exptl. Theoret. Phys. (U.S.S.R.) **35**, 1558 (1958), Soviet Phys. JETP **8**, 1090 (1959).

[2] T. Matsubara, Progr. Theoret. Phys. **14**, 351 (1955).

[3] Bardeen, Cooper, and Schrieffer, Phys. Rev. **108**, 1175 (1957).

Translated by D. ter Haar
48

SOVIET PHYSICS JETP VOLUME 36(9), NUMBER 6 DECEMBER, 1959

MICROSCOPIC DERIVATION OF THE GINZBURG-LANDAU EQUATIONS IN THE THEORY OF SUPERCONDUCTIVITY

L. P. GOR'KOV

Institute for Physical Problems, Academy of Sciences, U.S.S.R.

Submitted to JETP editor February 3, 1959

J. Exptl. Theoret. Phys. (U.S.S.R.) 36, 1918-1923 (June, 1959)

It is shown that the phenomenological Ginzburg-Landau equations follow from the theory of superconductivity in the London temperature region in the neighborhood of T_c. In these equations there occurs, however, twice the electronic charge; this is related to the physical meaning of $\Psi(x)$ as the wave function for Cooper pairs. The constant κ turns out to be small. The problem of the surface energy for the boundary between the normal and superconducting phases in the neighborhood of T_c is discussed.

It is well known that the behavior of superconductors in a magnetic field in the neighborhood of the critical temperature is qualitatively well described by Ginzburg and Landau's phenomenological theory.[1] We shall show in the present paper that equations of the same type as those in Ginzburg and Landau's theory follow indeed from the theory of superconductivity.

We shall start from the equations for the thermodynamic Green functions[3,4] which we obtained earlier.[2] These equations are in a magnetic field of the form

$$\left\{-\frac{\partial}{\partial\tau}+\frac{1}{2m}\left(\frac{\partial}{\partial\mathbf{r}}-ie\mathbf{A}(\mathbf{r})\right)^2+\mu\right\}G(x,x')+\Delta(\mathbf{r})F^+(x,x')=\delta(x-x'),$$

$$\left\{\frac{\partial}{\partial\tau}+\frac{1}{2m}\left(\frac{\partial}{\partial\mathbf{r}}+ie\mathbf{A}(\mathbf{r})\right)^2+\mu\right\}F^+(x,x')-\Delta^*(\mathbf{r})G(x,x')=0; \qquad (1)$$

where

$$\Delta^*(\mathbf{r})=gF^+(\tau,\mathbf{r};\tau,\mathbf{r}). \qquad (1')$$

The last relation expresses $\Delta(\mathbf{r})$ in terms of the coupling constant and the value of the function $F^+(x, x')$ for equal arguments. If there is no field, Δ = const. The quantity 2Δ is in that case the energy gap in the spectrum. $F^+(x, x')$ tends, in fact, logarithmically to infinity as $x = x'$; this corresponds to the well known logarithmic divergence in the equation determining the gap when one integrates in momentum space. In the papers by Bardeen, Cooper, and Schrieffer[5] and Bogolyubov[6] this divergence was removed by cutting-off the energy of the interacting electrons at distances of the order of the Debye frequency $\tilde{\omega}$ from the Fermi surface; we shall do the same in the following. One verifies easily that such a cut-off corresponds in the coordinate representation to a "spread out" in Eq. (1') over a distance of the order $\hbar v/\tilde{\omega}$.

Let us go over to the Fourier components of the Green functions $G(x, x')$ and $F^+(x, x')$ with respect to the difference of the variables[4] $u = \tau-\tau'$, for instance,

$$G(\mathbf{r},\mathbf{r}';u)=T\sum_n e^{-i\omega_n u}\mathfrak{G}_\omega(\mathbf{r},\mathbf{r}'),$$

$$\mathfrak{G}_\omega(\mathbf{r},\mathbf{r}')=\frac{1}{2}\int_{-1/T}^{+1/T}e^{i\omega_n u}G(\mathbf{r},\mathbf{r}';u)\,du,$$

where $\omega_n = \pi(2n + 1)T$ ($n = \ldots -1, 0, 1, \ldots$). Equations (1) go over into the following equations for the Fourier components $\mathfrak{G}_\omega(\mathbf{r},\mathbf{r}')$ and $\mathfrak{F}^+_\omega(\mathbf{r},\mathbf{r}')$:

$$\left\{i\omega_n+\frac{1}{2m}\left(\frac{\partial}{\partial\mathbf{r}}-ie\mathbf{A}(\mathbf{r})\right)^2+\mu\right\}\mathfrak{G}_\omega(\mathbf{r},\mathbf{r}')+\Delta(\mathbf{r})\mathfrak{F}^+_\omega(\mathbf{r},\mathbf{r}')=\delta(\mathbf{r}-\mathbf{r}'),$$

$$\left\{-i\omega_n+\frac{1}{2m}\left(\frac{\partial}{\partial\mathbf{r}}+ie\mathbf{A}(\mathbf{r})\right)^2+\mu\right\}\mathfrak{F}^+_\omega(\mathbf{r},\mathbf{r}')-\Delta^*(\mathbf{r})\mathfrak{G}_\omega(\mathbf{r},\mathbf{r}')=0, \qquad (2)$$

and condition (1') can be written in the form

$$\Delta^*(\mathbf{r})=T\sum_n\mathfrak{F}^+_\omega(\mathbf{r},\mathbf{r}). \qquad (2')$$

We introduce, finally, $\tilde{\mathfrak{G}}_\omega(\mathbf{r},\mathbf{r}')$, the Fourier component of the Green function for an electron in the normal metal in a magnetic field. The equation satisfied by $\tilde{\mathfrak{G}}_\omega(\mathbf{r},\mathbf{r}')$ can be written in two ways:

$$\left\{i\omega_n+\frac{1}{2m}\left(\frac{\partial}{\partial\mathbf{r}}-ie\mathbf{A}(\mathbf{r})\right)^2+\mu\right\}\tilde{\mathfrak{G}}_\omega(\mathbf{r},\mathbf{r}')=\delta(\mathbf{r}-\mathbf{r}') \qquad (3)$$

or

$$\left\{i\omega_n + \frac{1}{2m}\left(\frac{\partial}{\partial r'} + ie\mathbf{A}(\mathbf{r}')\right)^2 + \mu\right\}\widetilde{\mathfrak{G}}_\omega(\mathbf{r}, \mathbf{r}') = \delta(\mathbf{r} - \mathbf{r}').$$

It is convenient to use the function $\widetilde{\mathfrak{G}}_\omega(\mathbf{r}, \mathbf{r}')$ to write Eqs. (2) in integral form, using the second of Eqs. (3):

$$\mathfrak{G}_\omega(\mathbf{r}, \mathbf{r}') = \widetilde{\mathfrak{G}}_\omega(\mathbf{r}, \mathbf{r}') - \int \widetilde{\mathfrak{G}}_\omega(\mathbf{r}, \mathbf{s})\Delta(\mathbf{s})\mathfrak{F}^+_\omega(\mathbf{s}, \mathbf{r})\, d^3s,$$

$$\mathfrak{F}^+_\omega(\mathbf{r}, \mathbf{r}') = \int \mathfrak{G}_\omega(\mathbf{s}, \mathbf{r}')\Delta^*(\mathbf{s})\widetilde{\mathfrak{G}}_{-\omega}(\mathbf{s}, \mathbf{r})\, d^3s. \quad (4)$$

Before we proceed, we find out what the function $\widetilde{\mathfrak{G}}_\omega(\mathbf{r}, \mathbf{r}')$ is equal to. If there is no magnetic field, one sees from (3) that

$$\widetilde{\mathfrak{G}}^0_\omega(\mathbf{r} - \mathbf{r}') = -\frac{m}{2\pi R}\exp\left\{\left[i\,\mathrm{sign}\,\omega p_0 - \frac{|\omega_n|}{v}\right]R\right\}. \quad (5)$$

Here

$$\mathrm{sign}\,\omega = \omega/|\omega|, \quad R = |\mathbf{r} - \mathbf{r}'|, \quad \mu = p_0^2/2m \gg |\omega|.$$

$\widetilde{\mathfrak{G}}_\omega(\mathbf{r}, \mathbf{r}')$ in a magnetic field differs from (5) by a phase factor. Indeed,* because p_0 is a large quantity we can use the quasi-classical approximation to determine $\widetilde{\mathfrak{G}}_\omega(\mathbf{r}, \mathbf{r}')$, i.e., write it in the following form:

$$\widetilde{\mathfrak{G}}_\omega(\mathbf{r}, \mathbf{r}') = e^{i\varphi(\mathbf{r}, \mathbf{r}')}\widetilde{\mathfrak{G}}^0_\omega(\mathbf{r} - \mathbf{r}'), \quad \varphi(\mathbf{r}, \mathbf{r}')|_{\mathbf{r}=\mathbf{r}'} = 0. \quad (6)$$

Substituting (6) into (3) we find for the addition to the action of $\varphi(\mathbf{r}, \mathbf{r}')$

$$(\mathbf{n}, \nabla_r\varphi(\mathbf{r}, \mathbf{r}')) = \frac{e}{c}(\mathbf{n}, \mathbf{A}(\mathbf{r})), \quad \mathbf{n} = (\mathbf{r} - \mathbf{r}')/|\mathbf{r} - \mathbf{r}'|. \quad (7)$$

Near the transition temperature $\Delta(\mathbf{r})$ (and at the same time $\mathfrak{F}^+_\omega(\mathbf{r}, \mathbf{r}')$) is small. We shall therefore expand Eq. (4) in powers of Δ and Δ^* and express $\mathfrak{F}^+_\omega(\mathbf{r}, \mathbf{r}')$ up to terms of the fourth order, and $\mathfrak{G}_\omega(\mathbf{r}, \mathbf{r}')$ up to terms of second order in Δ, inclusive. The substitution of the expression obtained for $\mathfrak{F}^+_\omega(\mathbf{r}, \mathbf{r}')$ into (2′) gives one equation connecting $\Delta(\mathbf{r})$ and $\mathbf{A}(\mathbf{r})$:

$$\Delta^*(\mathbf{r}) = gT\sum_n\int\widetilde{\mathfrak{G}}_\omega(\mathbf{r}, \mathbf{r}')\widetilde{\mathfrak{G}}_{-\omega}(\mathbf{r}, \mathbf{r}')\Delta^*(\mathbf{r}')\, d^3r'$$

$$- gT\sum_n\iiint\widetilde{\mathfrak{G}}_\omega(\mathbf{s}, \mathbf{r})\widetilde{\mathfrak{G}}_{-\omega}(\mathbf{s}, \mathbf{l})\widetilde{\mathfrak{G}}_\omega(\mathbf{m}, \mathbf{l})\widetilde{\mathfrak{G}}_{-\omega}(\mathbf{m}, \mathbf{r})$$

$$\times \Delta(\mathbf{s})\Delta^*(\mathbf{l})\Delta^*(\mathbf{m})\, d^3\mathbf{s}\, d^3\mathbf{l}\, d^3\mathbf{m}. \quad (8)$$

The important distances in Eq. (8) are of the order $\xi = \hbar v/T_c$, since it can be seen from (5) and (6) that the Green function $\widetilde{\mathfrak{G}}_\omega(\mathbf{r}, \mathbf{r}')$ decreases exponentially for $R \gg \xi_0$. At the same time, the changes in the gap $\Delta(\mathbf{r})$ and the field $\mathbf{A}(\mathbf{r})$ take place at distances of the order of the penetration depth which is much larger than ξ_0 near the critical temperature. For the same reason the phase $\varphi(\mathbf{r}, \mathbf{r}')$ in (6) is small near T_c and equal to

$$\varphi(\mathbf{r}, \mathbf{r}') \approx \frac{e}{c}(\mathbf{A}(\mathbf{r}'), \mathbf{r} - \mathbf{r}'). \quad (9)$$

*This was pointed out by L. D. Landau.

Using these facts we can perform the integration in Eq. (8).

The first term on the right hand side is of the form

$$g\int K(\mathbf{r}, \mathbf{r}')\Delta^*(\mathbf{r}')\, d^3r',$$

where the kernel

$$K(\mathbf{r}, \mathbf{r}') = K_0(\mathbf{r} - \mathbf{r}')\exp\left\{\frac{2ie}{c}(\mathbf{A}(\mathbf{r}'), \mathbf{r} - \mathbf{r}')\right\}$$

has a singularity for $R = |\mathbf{r} - \mathbf{r}'| = 0$:

$$K_0(R) = m^2T/(2\pi R)^2 \sinh(2\pi TR/v).$$

This singularity is connected, as was pointed out already in the foregoing, with the cut-off of the interaction at the Debye frequencies $\widetilde{\omega}$.

Introducing this cut-off we get

$$\int K_0(R)\, d^3R = \frac{mp_0}{2\pi^2}\int_0^{\widetilde{\omega}}\frac{\tanh(\xi/2T)}{\xi}d\xi$$

$$= \frac{mp_0}{2\pi^2}\ln\left(\frac{2\gamma}{\pi}\frac{\widetilde{\omega}}{T}\right) = \frac{mp_0}{2\pi^2}\left(\frac{1}{g} + \ln\frac{T_c}{T}\right), \quad (10)$$

where the transition temperature T_c is connected to the gap in the energy spectrum at absolute zero in the well-known way[5]:

$$\frac{\Delta_0\gamma}{\pi} = T_c \qquad \left(\frac{1}{g} = \frac{mp_0}{2\pi^2}\ln\frac{2\widetilde{\omega}}{\Delta_0}\right).$$

Expanding the exponents in the Green functions in terms of $\varphi(\mathbf{r}, \mathbf{r}')$ using the fact that $\Delta(\mathbf{r})$ and $\mathbf{A}(\mathbf{r})$ vary slowly near T_c, and eliminating $\widetilde{\omega}$ and g by using (10) we get the following equation

$$\left\{\frac{1}{2m}\left(\frac{\partial}{\partial \mathbf{r}} + 2ie\mathbf{A}(\mathbf{r})\right)^2 + \frac{1}{\lambda}\left[\frac{T_c - T}{T_c} - \frac{7\zeta(3)}{8(\pi T_c)^2}|\Delta(\mathbf{r})|^2\right]\right\}\Delta^*(\mathbf{r}) = 0, \quad (11)$$

where $\lambda = 7\zeta(3)\epsilon_F/12(\pi T_c)^2$ ($\zeta(x)$ is here Riemann's zeta function).

To get a second equation connecting Δ and $\mathbf{A}$ we shall now calculate what the current density $\mathbf{j}(\mathbf{r})$ is equal to. To do this we use the relation expressing $\mathbf{j}(\mathbf{r})$ in terms of a Green function:

$$\mathbf{j}(\mathbf{r}) = \left\{\frac{ie}{m}(\nabla_{r'} - \nabla_r)G(x, x') - \frac{2e^2}{mc}\mathbf{A}(\mathbf{r})G(x, x')\right\}_{t'=t+0;\ r=r'}$$

(the additional factor of 2 is due to a summation over spin). We note that if we substitute everywhere the Green function for the normal metal in the form (6) and then sum over frequencies we get no contribution to the expression for the current, as should be the case. The total current is thus equal to

$$\mathbf{j}(\mathbf{r}) = \frac{ie}{m}(\nabla_{r'} - \nabla_r)\delta G(x, x')|_{x=x'}, \quad (12)$$

where

$$\delta G(x, x')|_{x=x'} = -T\sum_n \iint \mathfrak{G}_\omega(\mathbf{r}, \mathbf{s})\mathfrak{G}_\omega(\mathbf{l}, \mathbf{r})\mathfrak{G}_{-\omega}(\mathbf{l}, \mathbf{s})\Delta^*(\mathbf{l})\Delta(\mathbf{s})\,d^3s\,d^3l. \quad (13)$$

Restricting ourselves in the expression for the current to terms linear in $\mathbf{A}$ and assuming $\mathbf{A}(\mathbf{r})$ to vary slowly with $\mathbf{r}$ in the London temperature range near T_c we get

$$\mathbf{j}(\mathbf{r}) = \left\{\frac{ie}{m}\left(\Delta\frac{\partial\Delta^*}{\partial\mathbf{r}} - \Delta^*\frac{\partial\Delta}{\partial\mathbf{r}}\right) - \frac{4e^2|\Delta|^2}{mc}\mathbf{A}\right\}\frac{7\zeta(3)N}{16(\pi T)^2} \quad (14)$$

We shall introduce a "wave function" $\Psi(\mathbf{r})$, proportional to $\Delta(\mathbf{r})$:

$$\Psi(\mathbf{r}) = \Delta(\mathbf{r})\sqrt{7\zeta(3)N}/4\pi T_c. \quad (15)$$

The set of Eqs. (11) and (14) takes then a form completely analogous to the equations of the phenomenlogical Ginzburg-Landau theory:

$$\left\{\frac{1}{2m}\left(\frac{\partial}{\partial\mathbf{r}} - ie^*A(\mathbf{r})\right)^2 + \frac{1}{\lambda}\left[\frac{T_c - T}{T_c} - \frac{2}{N}|\Psi(\mathbf{r})|^2\right]\right\}\Psi(\mathbf{r}) = 0,$$

$$\mathbf{j}(\mathbf{r}) = -\frac{ie^*}{2m}\left(\Psi^*\frac{\partial\Psi}{\partial\mathbf{r}} - \Psi\frac{\partial\Psi^*}{\partial\mathbf{r}}\right) - \frac{e^{*2}}{mc}\mathbf{A}|\Psi|^2, \quad (16)$$

where $e^* = 2e$, N is the number density of the electrons in the normal metal and λ was defined after Eq. (11). There is a difference, however, in that (16) contains the electronic charge e twice, clearly owing to the physical meaning of $\Psi(\mathbf{r})$ as the wave function of a Cooper pair.

These equations arise thus in a natural way in the theory of superconductivity in the London region* near T_c. The phenomenological constant defined in the same way as in the old theory, but with $e^* = 2e$, can now be evaluated and turns out to be equal to

$$\varkappa = 3T_c(\pi/v)^{1/2}(c/ep_0)\sqrt{2/7\zeta(3)}. \quad (17)$$

The region of applicability of Eqs. (16) is small for sharply expressed Pippard metals and occurs very near T_c; at large distances from T_c the local relation (14) ceases to be valid.

For metals of the London type the region of applicability of (16) is determined solely by the smallness of the ratio $(T - T_c)/T_c$.

The possibility to apply (16) to describe the properties of thin films in a magnetic field with the microscopic value (17) of the constant κ is limited by the fact that usually films are polycrystalline specimens with grain sizes of the order of the film thickness. The presence of boundaries between different crystals will act like the presence of foreign impurities. One must therefore possibly describe real films in the microscopic theory from the same point of view as superconducting alloys.

*In reference 5 the assumption was stated that some quantity related to the gap should play the role of the expansion parameter Ψ in the theory of reference 1.

We note now that to evaluate the difference in the free energy of the metal in the normal state and in the superconducting state in a magnetic field we can use the following formula:

$$F_{sH} - F_n = -(1/\lambda N)\int|\Psi(\mathbf{r})|^4 d^3\mathbf{r}. \quad (18)$$

Indeed, we shall use the thermodynamic relation

$$\delta\Omega/\delta g(\mathbf{r}) = -\langle H_{int}(\mathbf{r})\rangle/g(\mathbf{r}),$$

where according to (5)

$$H_{int}(\mathbf{r}) = {}^1/_2\, g(\mathbf{r})(\hat{\Psi}^+(\mathbf{r})(\hat{\Psi}^+(\mathbf{r})\hat{\Psi}(\mathbf{r}))\hat{\Psi}(\mathbf{r})),$$

Ω is the thermodynamic potential [we introduce here for the sake of convenience a variable "interaction constant" $g(\mathbf{r})$]. Taking averages we get

$$\Omega_s - \Omega_n = -\int d^3\mathbf{r}\int_0^g \frac{\delta g(\mathbf{r})}{g^2(\mathbf{r})}|\Delta(\mathbf{r})|^2. \quad (19)$$

Equation (8) determines $1/g(\mathbf{r})$ as a functional of $\Delta(\mathbf{r})$ [for the sake of simplicity we shall consider the real solution for $\Delta(\mathbf{r})$ as is usually done in one-dimensional problems]. Using (8) to integrate (19), and using the fact that small additions to all thermodynamic potentials in the corresponding variables are equal, we get Eq. (18).

The constant κ of (17) is connected with the magnitude of the critical field H_{cm} of a bulk specimen and the penetration depth δ_0 by the relation (which contains the charge e^*):

$$\varkappa = \frac{\sqrt{2}e^*}{\hbar c}H_{cm}\delta_0^2. \quad (20)$$

For metals of the London type or such intermediate metals as tin the region of applicability of Eqs. (16) is fairly large and in that case we should consistently use the experimental data* for H_{cm} and δ_0^2 to compare the theory with experiments. For Pippard metals δ_0 is the penetration depth only in a very narrow range of temperatures near T_c. It is thus convenient in this case to turn to the theoretical formulae for a definition of κ. It can easily be verified that (17) can be written in the form

$$\varkappa \approx 0.96\,\delta_L/\xi_0, \quad (21)$$

where $\delta_L^{-1} = (4\pi Ne^2/mc^2)^{1/2}$ is the London penetration depth and $\xi_0 = 0.18\,\hbar v/kT_c$ is the non-localization parameter introduced in Bardeen, Cooper, and Schrieffer's paper. Using the data given in reference 5 we get from (21) for tin $\kappa \approx 0.14$, and for aluminum $\kappa \approx 0.01$.

We note an interesting consequence of (21). The

*A detailed study of experimental data is given in a paper by Ginzburg.[7]

condition that a superconductor be a London one requires $\kappa \approx \delta_L/\xi_0 \gg 1$. At the same time if $\kappa \gg 1/\sqrt{2}$ a metal displays in a strong field the same behavior as for alloys. A metal which in weak fields at T = 0 must be assigned to the London class will thus behave in strong fields near T_c as an "alloy." At the same time the case of small κ corresponds to a Pippard metal.

Let us now stop a moment at the problem of the surface energy at the boundary between the normal and the superconducting phase. For strongly expressed Pippard superconductors we can, because κ is small, use the limiting solution[1]

$$\sigma_{ns}(8\pi / H_{cm}^2) = 1.89\,\delta_0 / \varkappa, \tag{22}$$

where

$$\delta_0 = \delta_L(T = 0) / \sqrt{2(1 - T/T_c)}.$$

We note that for these metals the applicability of Eq. (22) in an appreciable temperature interval near T_c is just connected with the fact that κ is small, since, as was shown in reference 1, in that case the quantity σ_{ns} does not depend on the distribution of the field in the transition region.

For tin the accuracy of this equation is insufficient and one must use the results of numerical calculations. Estimating σ_{ns} from Ginzburg's calculations[8] we get for $\kappa \approx 0.14$

$$\sigma_{ns}(8\pi / H_n^2) \approx (2 \div 3)\cdot 10^{-5} / \sqrt{1 - T/T_c}.$$

According to Sharvin's data[9]

$$\sigma_{ns}(8\pi / H_c^2) = 2.5\cdot 10^{-5} / \sqrt{1 - T/T_c},$$

and according to Faber's data[10]

$$\sigma_{ns}(8\pi / H_c^2) = 1.88\cdot 10^{-5} / \sqrt{1 - T/T_c}.$$

These results do, apparently, not contradict the theory (See reference 7).

In conclusion the author expresses his gratitude to Academician L. D. Landau for valuable advice and to V. L. Ginzburg for discussing the results of this paper and for helpful comments.

[1] V. L. Ginzburg and L. D. Landau, J. Exptl. Theoret. Phys. (U.S.S.R.) **20**, 1064 (1950).

[2] L. P. Gor'kov, J. Exptl. Theoret. Phys. (U.S.S.R.) **34**, 735 (1958), Soviet Phys. JETP **7**, 505 (1958).

[3] T. Matsubara, Progr. Theoret. Phys. **14**, 352 (1954).

[4] Abrikosov, Gor'kov, and Dzyaloshinskiĭ, J. Exptl. Theoret. Phys. (U.S.S.R.) **36**, 900 (1959), Soviet Phys. JETP **9**, 636 (1959).

[5] Bardeen, Cooper, and Schrieffer, Phys. Rev. **108**, 1175 (1957).

[6] N. N. Bogolyubov, J. Exptl. Theoret. Phys. (U.S.S.R.) **34**, 58 (1958), Soviet Phys. JETP **7**, 41 (1958).

[7] V. L. Ginzburg, J. Exptl. Theoret. Phys. (U.S.S.R.) **36**, 1930 (1959), Soviet Phys. JETP this issue, p 1372.

[8] V. L. Ginzburg, J. Exptl. Theoret. Phys. (U.S.S.R.) **30**, 593 (1956), Soviet Phys. JETP **3**, 621 (1956).

[9] Yu. V. Sharvin, J. Exptl. Theoret. Phys. (U.S.S.R.) **33**, 1341 (1957), Soviet Phys. JETP **6**, 1031 (1958).

[10] T. E. Faber, Proc. Roy. Soc. (London) **A248**, 460 (1958).

Translated by D. ter Haar
380

SOVIET PHYSICS JETP VOLUME 37(10), NUMBER 3 MARCH, 1960

THE CRITICAL SUPERCOOLING FIELD IN SUPERCONDUCTIVITY THEORY

L. P. GOR'KOV

Institute for Physical Problems, Academy of Sciences, U.S.S.R.

Submitted to JETP editor April 23, 1959

J. Exptl. Theoret. Phys. (U.S.S.R.) **37**, 833-842 (September, 1959)

The magnitude of the critical supercooling field H_{c1} in superconductivity theory[1] is determined. It is found that the field H_{c1} is larger than or smaller than the critical field H_c, depending on whether the superconductor in a weak field is of the London or of the Pippard type. The superconductor in the first case must in a strong field display a behavior similar to that of alloys. The magnitude of the ratio H_{c1}/H_c depends weakly on the temperature in the whole temperature range.

It is well known that superconducting metals in a magnetic field undergo at some value of the field a phase transition from the normal to the superconducting phase. For a bulk specimen this transition is a first-order transition; the value of the critical magnetic field H_c can thus be obtained from thermodynamic considerations and was evaluated by Bardeen, Cooper, and Schrieffer (BCS) in their theory of superconductivity.[1] Along with the thermodynamic field there exist, apparently, for a given temperature, still two other critical field values corresponding to the so-called "superheating" field and the "supercooling" field H_{c1}. These fields determine a region of possible hysteresis: for fields above H_c, but below the superheating field the superconducting phase is metastable and, on the other hand, for field values below H_c, but above H_{c1} the normal phase is metastable. (We understand by metastability, as always, an instability with respect to a finite perturbation.) To determine the magnitude of these fields thermodynamic considerations are insufficient and one must turn to the microscopic theory of superconductivity. Using an earlier developed method[2] the existence in the BCS theory of a "supercooling" field is proved in the present paper and its magnitude is found.

In the method mentioned above, the superconductor is described by two Green functions $G(x, x')$ and $F^+(x, x')$, and equations for these in a magnetic field can in the usual manner be obtained from the field-free equations. We shall first consider the absolute zero. These equations are then of the form

$$\left\{i\frac{\partial}{\partial t} + \frac{1}{2m}\left(\frac{\partial}{\partial \mathbf{r}} - ie\mathbf{A}(\mathbf{r})\right)^2 + \mu\right\} G(x, x') + i\Delta(\mathbf{r}) F^+(x, x') = \delta(x - x'),$$

$$\left\{i\frac{\partial}{\partial t} - \frac{1}{2m}\left(\frac{\partial}{\partial \mathbf{r}} + ie\mathbf{A}(\mathbf{r})\right)^2 - \mu\right\} F^+(x, x') - i\Delta^*(\mathbf{r}) G(x, x') = 0, \tag{1}$$

where

$$\Delta^*(\mathbf{r}) = |g| F^+(x, x), \tag{2}$$

In a constant magnetic field $G(x, x')$ and $F^+(x, x')$ are functions of the difference in the time variables, $t - t'$; expanding all quantities in Fourier integrals in this difference we get for the Fourier components $G_\omega(\mathbf{r}, \mathbf{r}')$ and $F^+_\omega(\mathbf{r}, \mathbf{r}')$ the following set of equations

$$\left\{\omega + \frac{1}{2m}\left(\frac{\partial}{\partial \mathbf{r}} - ie\mathbf{A}(\mathbf{r})\right)^2 + \mu\right\} G_\omega(\mathbf{r}, \mathbf{r}') + i\Delta(\mathbf{r}) F^+_\omega(\mathbf{r}, \mathbf{r}') = \delta(\mathbf{r} - \mathbf{r}'),$$

$$\left\{-\omega + \frac{1}{2m}\left(\frac{\partial}{\partial \mathbf{r}} + ie\mathbf{A}(\mathbf{r})\right)^2 + \mu\right\} F^+_\omega(\mathbf{r}, \mathbf{r}') + i\Delta^*(\mathbf{r}) G_\omega(\mathbf{r}, \mathbf{r}') = 0; \tag{1'}$$

$$\Delta^*(\mathbf{r}) = |g|(2\pi)^{-1} \int_{-\infty}^{+\infty} d\omega F^+_\omega(\mathbf{r}, \mathbf{r}). \tag{2'}$$

The value H_{c1} of the supercooling field is the boundary itself of the metastability of the normal phase; at smaller fields the normal phase is absolutely unstable with respect to the appearance of a superconducting phase. We are, of course, not talking of the appearance of small layers of the superconducting phase but about the possibility that there appears that specific correlation between electrons which is caused by the interaction between them and which leads to the superconductivity phenomenon, and which finds an expression in the existence of a non-vanishing function $F^+(x, x')$. The metastability of the normal phase for fields $H > H_{c1}$ means instability with respect to the occurrence of finite values of $F^+(x, x')$ and $\Delta(\mathbf{r})$. In the point H_{c1} there occurs for the first time a solution with infinitesimal $\Delta(\mathbf{r})$ and $F^+(x, x')$. This makes it possible to simplify the set (1′) essentially, retaining in the equations only the first nonvanishing terms for small $\Delta(\mathbf{r})$. Furthermore, the magnetic field is uniform in the normal phase; it is convenient to take the vector potential $\mathbf{A}(\mathbf{r})$ in the form $A_x = -Hy$, $A_y = A_z = 0$. The second of Eqs. (1′) takes the following form

$$\left\{-\omega + \frac{1}{2m}\left[\left(\frac{\partial}{\partial x} - \frac{ieHy}{c}\right)^2 + \frac{\partial^2}{\partial y^2} + \frac{\partial^2}{\partial z^2}\right] + \mu\right\} F^+_\omega(\mathbf{r}, \mathbf{r}') - i\Delta^*(\mathbf{r})\,\tilde{G}_\omega(\mathbf{r}, \mathbf{r}') = 0. \tag{3}$$

In this equation $\tilde{G}_\omega(\mathbf{r}, \mathbf{r}')$ is the Green function of the electrons in the normal metal:

$$\left\{\omega + \frac{1}{2m}\left[\left(\frac{\partial}{\partial x} + \frac{ieHy}{c}\right)^2 + \frac{\partial^2}{\partial y^2} + \frac{\partial^2}{\partial z^2}\right] + \mu\right\} \tilde{G}_\omega(\mathbf{r}, \mathbf{r}') = \delta(\mathbf{r} - \mathbf{r}'). \tag{4}$$

If we rewrite (4) in such a way that the differentiation is with respect to the variables $\mathbf{r}'$:

$$\left\{\omega + \frac{1}{2m}\left[\left(\frac{\partial}{\partial x'} - \frac{ieHy'}{c}\right)^2 + \frac{\partial^2}{\partial y'^2} + \frac{\partial^2}{\partial z'^2}\right] + \mu\right\} \tilde{G}_\omega(\mathbf{r}, \mathbf{r}') = \delta(\mathbf{r} - \mathbf{r}'), \tag{4'}$$

we can use (4′) and (3) to write the function $F^+_\omega(\mathbf{r}, \mathbf{r}')$ in the following form

$$F^+_\omega(\mathbf{r}, \mathbf{r}') = -i\int \tilde{G}_\omega(\mathbf{s}, \mathbf{r}')\,\tilde{G}_{-\omega}(\mathbf{s}, \mathbf{r})\,\Delta^*(\mathbf{s})\,d^3s. \tag{5}$$

Substitution of (5) into (2) leads to an integral equation for $\Delta^*(\mathbf{s})$:

$$\Delta^*(\mathbf{r}) = -i|g|\frac{1}{2\pi}\int_{-\infty}^{+\infty} d\omega \int \tilde{G}_\omega(\mathbf{s}, \mathbf{r})\,\tilde{G}_{-\omega}(\mathbf{s}, \mathbf{r})\,\Delta^*(\mathbf{s})\,d^3s. \tag{6}$$

The value of the magnetic field for which there occurs for the first time (coming from large fields) a nonvanishing solution of this equation is then just the value of the supercooling field H_{c1} which interests us.

For what follows, it is necessary to determine the form of the function $\tilde{G}_\omega(\mathbf{r}, \mathbf{r}')$. We perform the transformation

$$\tilde{G}_\omega(\mathbf{r}, \mathbf{r}') = \exp\left\{-\frac{ieH}{2c}(y + y')(x - x')\right\}\tilde{G}'_\omega(\mathbf{r} - \mathbf{r}'). \tag{7}$$

After substituting this expression into (4) the coefficients in the equation for $\tilde{G}_\omega(\mathbf{r} - \mathbf{r}')$ are functions of the difference, namely,

$$\left\{\omega + \frac{1}{2m}\left(\frac{\partial}{\partial \mathbf{r}} - \frac{ie}{2c}\mathbf{H}\times[\mathbf{r} - \mathbf{r}']\right)^2 + \mu\right\}\tilde{G}'_\omega(\mathbf{r} - \mathbf{r}') = \delta(\mathbf{r} - \mathbf{r}'). \tag{8}$$

Thanks to this $\tilde{G}'_\omega(\mathbf{r} - \mathbf{r}')$ clearly also depends only on the difference $\mathbf{R} = \mathbf{r} - \mathbf{r}'$. If there is no field the Green function $G_{0\omega}(\mathbf{R})$ is, as can easily be checked, of the form

$$G_{0\omega}(\mathbf{R}) = \begin{cases} -(m/2\pi R)\exp[ip_0R + i\omega R/v], & \omega > 0 \\ -(m/2\pi R)\exp[-ip_0R + i\omega R/v], & \omega < 0 \end{cases} \quad (|\omega| \ll \mu). \tag{9}$$

Such a choice of the solutions of (8) without a field was made in accordance with the requirement that the imaginary part of the Green function changes sign when the frequency changes sign.[3] We shall show that the function $\tilde{G}'_\omega(\mathbf{R})$ in (7) is the same as its value (9) without a field. Looking ahead we note that in the following we require distances R of the order of ξ_0 where ξ_0 is the parameter in the BCS theory which is connected with the value Δ_0 of the gap in the spectrum at $T = 0$ and which is equal to $\xi_0 \sim \hbar v/\Delta_0$. It turns out that one can neglect at those distances the term quadratic in the field in (8), or in other words, that the curving of the electrons in the fields of interest to us is very small. Under those conditions it is convenient, because of the large value of p_0, to look for solutions of (8) in the form $\exp\{i\varphi(\mathbf{R})\}G_{0\omega}(\mathbf{R})$, where $G_{0\omega}(\mathbf{R})$ is the Green function (9) when there is no field present. Substituting this expression into (8) leads to the following equation for $\varphi(\mathbf{R})$, which contains only quadratic terms in the field:

$$\mp p_0\partial\varphi/\partial R - (e^2/4c^2)([\mathbf{H}\times\mathbf{R}])^2 = 0,$$

[the ± sign in this equation depends on the sign of the frequency in accordance with (9)]. As will become clear below, the fields of interest to us, are such that $eH\xi_0^2/c \sim 1$. Therefore, $\varphi(\mathbf{R})$ which is determined by this equation will give a correction in the phase factor in (7) of the order of $1/p_0\xi_0$, i.e., an insignificant quantity of the order 10^{-4}.

Substituting (7) and (9) into (6) and integrating over the frequency we get the following equation

$$\Delta^*(\mathbf{r}) = \frac{mp_0}{2\pi^2}\frac{|g|}{2\pi}\int \frac{\exp\{(ieH/c)(y + y')(x - x')\}}{|\mathbf{r} - \mathbf{r}'|^3}\Delta^*(\mathbf{r}')\,d^3r'. \tag{10}$$

The kernel of this equation leads to a logarithmic divergence when we integrate the right hand side for $\mathbf{r} = \mathbf{r}'$, which is connected with the fact that the relation (2) determines the gap $\Delta(\mathbf{r})$ as the value of the function F^+ for equal arguments. This definition is somewhat inaccurate. The fact is that in the BCS model the interaction Hamiltonian which leads to superconductivity was chosen in the form

$$H_{int} = \frac{g}{2} \sum_{p_1+p_2=p_3+p_4} a^+_{p_1\sigma_1} a^+_{p_2\sigma_2} a_{p_3\sigma_2} a_{p_4\sigma_1} \quad (11)$$

with a constant coupling constant for all electrons with energies in a narrow layer $\hbar\tilde{\omega}$ in the neighborhood of the Fermi surface. This cutting off is realized in (11) by the functions θ_p:

$$\theta_p = \begin{cases} 1, & |\varepsilon_p - \varepsilon_F| < \tilde{\omega} \\ 0, & |\varepsilon_p - \varepsilon_F| > \tilde{\omega} \end{cases}.$$

The quantity $\tilde{\omega}$ enters logarithmically in the determination of the gap Δ_0, in the Fermi spectrum:

$$1 = |g| (mp_0 / 2\pi^2) \ln (2\tilde{\omega} / \Delta_0).$$

To the cut-off in momentum representation there corresponds, in coordinate space, some spreading out over distances of the order $\hbar v/\tilde{\omega}$. Using (11) for the interaction Hamiltonian and repeating the derivation of the equations in reference 2, we find easily that the quantity $\Delta^*(\mathbf{r})$ is connected with the function $F^+(x, x')$ in the following manner:

$$\Delta^*(\mathbf{r}) = \iint \theta(\mathbf{r}-\mathbf{s})\,\theta(\mathbf{r}-\mathbf{m})\,F^+(t, \mathbf{s}; t, \mathbf{m})\,d^3s\,d^3m, \quad (2'')$$

where $\theta(R)$ is a smearing out function corresponding to the cut-off function θ_p in the momentum representation:

$$\theta(\mathbf{R}) = (2\pi)^{-3} \int e^{i\mathbf{p}\mathbf{R}} \theta_p d^3p = \frac{p_0}{\pi^2} \frac{\sin p_0 R}{R} \frac{\sin (\tilde{\omega} R / v)}{R}. \quad (12)$$

On this basis we must, strictly speaking, write Eq. (6) in the form

$$\Delta^*(\mathbf{r}) = i|g| \int_{-\infty}^{+\infty} \frac{d\omega}{2\pi} \int \overline{G_\omega(\mathbf{s},\mathbf{r})}\,\overline{G_{-\omega}(\mathbf{s},\mathbf{r})}\,\Delta^*(\mathbf{s})\,d^3s, \quad (6')$$

denoting by $\overline{G}_\omega(\mathbf{r}, \mathbf{r}')$ the integral over the Green function (7):

$$\overline{G_\omega(\mathbf{r},\mathbf{r}')} = \int \theta(\mathbf{r}-\mathbf{s})\,G_\omega(\mathbf{s},\mathbf{r}')\,d^3s.$$

The field dependent phase factor in (7) can be taken out from under the integral sign in the last expression since $\theta(\mathbf{r}-\mathbf{s})$ has the character of a δ-function. It amounts thus practically to replacing $G_{0\omega}(R)$ in the previous calculations [see Eq. (9)] by

$$\overline{G_{0\omega}(R)} = (2\pi)^{-3} \int e^{i\mathbf{p}\mathbf{R}} G_{0\omega}(\mathbf{p})\,\theta_p d^3p.$$

Following this, all expressions stay finite. We have succeeded in eliminating the cut-off quantity using the relation (2″), which determines the gap in the energy spectrum at $T = 0$ when there is no field. One verifies easily, by evaluating the corresponding function $F_0^+(t, \mathbf{r}; t, \mathbf{r}')$ through the Fourier components found in reference 2, that

$$F_0^+(t, \mathbf{r}; t, \mathbf{r}') = \frac{m\Delta_0}{2\pi^2} \frac{\sin p_0 R}{R} K_0\left(\frac{R\Delta_0}{v}\right), \quad (13)$$

where K_0 is a Bessel function of imaginary argument which tends logarithmically to infinity for small R.

Substituting (13) into (2″) and noting that the function $\theta(\mathbf{r})$ possesses the following δ-function property

$$\int \theta(\mathbf{r}-\mathbf{s})\,\theta(\mathbf{s})\,d^3s = \theta(\mathbf{r})$$

(this equation can be most easily checked by the Fourier components) we get, using the form (12) for the function $\theta(\mathbf{R})$ and averaging the fast oscillating factors:

$$1 = |g| \frac{mp_0}{4\pi^4} \int K_0\left(\frac{R\Delta_0}{v}\right) \frac{\sin (\tilde{\omega} R / v)}{R^3} d^3R$$

$$= \frac{m^2|g|}{4\pi^4} \int \left(\int_0^{\tilde{\omega}} \frac{\cos (R\xi / v)}{R^2} d\xi \right) K_0\left(\frac{R\Delta_0}{v}\right) d^3R.$$

Integrating over $d|\mathbf{R}|$ in the last expression by parts we find

$$\frac{1}{|g|} = -\frac{m^2\Delta_0}{4\pi^4} \int_0^{\tilde{\omega}} \frac{d\xi}{\xi} \int \frac{\sin (R\xi / v)}{R^2} K_1 (R\Delta_0 / v)\,d^3R.$$

This expression for $1/|g|$ is convenient because the integral over d^3R posssesses the same singularity as Eq. (10) when $\tilde{\omega} \to \infty$.

After substituting this expression into (6) we get an equation in which the logarithmically diverging terms are cancelled after which $\tilde{\omega}$ can tend to infinity. Omitting the intermediate calculations we shall give the final form of (10), taking into account that the function $\Delta^*(\mathbf{r})$ depends on the coordinate y only:

$$\Delta^*(y) \ln \left(\frac{e\gamma\Delta_0}{v} \varepsilon\right)$$

$$= -\frac{1}{2} \int_{-\infty}^{\infty} \frac{\exp\{-|(eH/c)(y^2 - y'^2)|\}}{|y-y'|} \Delta^*(y')\,d_\varepsilon y'.$$

[We shall show below that the general form of the function $\Delta^*(\mathbf{r})$ corresponding to a given value of H_{c1} can be found by a simple method in terms of the solution of this equation.] Here γ is Euler's constant, ϵ an infinitesimal constant introduced to cause the integration on the right hand side to be performed only over a distance $|y - y'| \geq \epsilon$. Thanks to the logarithmic character of the singularity on the right hand side, ϵ disappears from the final results. It is convenient to introduce dimensionless variables $\xi = y\sqrt{eH/c}$ in which the equation becomes

$$\Delta^*(\xi)\ln\left(\frac{e\gamma\Delta_0\delta}{v\sqrt{eH/c}}\right) = -\frac{1}{2}\int\limits_{-\infty}^{\infty}\frac{\Delta(\xi')\exp\{-|\xi^2-\xi'^2|\}}{|\xi-\xi'|}\,d\xi'. \tag{14}$$

The largest value of the magnetic field, for which a solution that decreases at infinity exists for (14), is just the value of the critical field H_{c1}, the "supercooling" field. Equation (14) can only be solved numerically. We shall show that (14) has a maximum eigenvalue for the magnetic field and we shall estimate its magnitude. To do this we add to and subtract from $\Delta(\xi')$ under the integral on the right hand side of (14) the value $\Delta(\xi)$. After this (14) is transformed to the following form, which no longer contains the infinitesimal constant δ:

$$\Delta^*(\xi)\ln\left(\frac{\Delta_0 e}{v}\sqrt{\frac{\gamma c}{eH}}\right) = \Phi(\xi)\Delta^*(\xi) - \frac{1}{2}\int\limits_{-\infty}^{\infty}\frac{\exp\{-|\xi^2-\xi'^2|\}}{|\xi-\xi'|}[\Delta^*(\xi')-\Delta^*(\xi)]\,d\xi'. \tag{15}$$

The function $\Phi(\xi)$ is here equal to

$$\Phi(\xi) = e^{-\xi^2}\int\limits_0^{|\xi|} e^{\xi'^2}\ln\frac{|\xi|+\xi'}{|\xi|-\xi'}\,d\xi'.$$

Multiplying (15) by $\Delta(\xi)$ and integrating over ξ we get

$$|\Delta|^2\ln\left(\frac{e\Delta_0}{v}\sqrt{\frac{\gamma c}{eH}}\right) = \int\limits_{-\infty}^{\infty}\Phi(\xi)|\Delta(\xi)|^2\,d\xi + \frac{1}{4}\int\int\limits_{-\infty}^{\infty}\frac{\exp\{-|\xi^2-\xi'^2|\}}{|\xi-\xi'|}|\Delta(\xi)-\Delta(\xi')|^2\,d\xi'd\xi. \tag{15'}$$

Since $\Phi(\xi) > 0$ it follows from the fact that the right-hand side is positive that

$$eH/c \leqslant \Delta_0^2 e^2\gamma/v. \tag{16}$$

We can obtain the value of H_{c1} approximately (and conceivably with great accuracy) by a variational method if we choose $\Delta(\xi)$ in the form of $\exp(-\alpha\xi^2)$. With such a form for $\Delta(\xi)$ all integrals in the variational principle for (14) can be evaluated and the maximum value of H arises when $\alpha = 1$ and is equal to

$$H_{c1} \approx (e^2\gamma/2)(c\Delta_0^2/ev).$$

One can show that a lower limit is then obtained for the value of H_{c1}. The theory gives for the thermodynamic critical field H_c at T = 0

$$H_c = \Delta_0\sqrt{2mp_0/\pi}.$$

The ratio of the two fields is

$$H_{c1}/H_c = 1.77(3\pi T_c mc/e)(2\pi m/7\zeta(3)p_0^5)^{1/2} \tag{17}$$

[$\zeta(x)$ is Riemann's zeta-function; $\zeta(3) = 1.202$]. It is convenient to write this quantity as follows. We have shown earlier[4] that in a sufficiently narrow region of temperatures near T_c the equations of the phenomenological Ginzburg-Landau theory[5] with a double charge follow from the theory of superconductivity. The phenomenological constant κ of this theory was defined microscopically and turned out to be equal to

$$\varkappa = (3\pi T_c mc/e)(2\pi m/7\zeta(3)p_0^5)^{1/2}. \tag{18}$$

If the ratio (17) is expressed in terms of this constant it can be written as follows

$$H_{c1}/H_c = 1.77\varkappa. \tag{19}$$

Near T_c the result of the Ginzburg-Landau theory[5] must hold, i.e.,

$$H_{c1}/H_c = \sqrt{2}\varkappa. \tag{20}$$

We see that the change in the ratio H_{c1}/H_c is small in the whole of the temperature range and does not exceed 25%.

It is interesting that according to (18) — (20) the magnitude of the field H_{c1} does by no means have to be less than H_c, not even for pure superconductors. Moreover, it is well known that in the new theory we can distinguish among the pure superconductors two classes depending on their properties in a weak magnetic field. For superconductors of the first class the penetration depth δ is at all temperatures appreciably larger than the parameter ξ_0 of the BCS theory. Such superconductors satisfy the electrodynamics of the Londons and may be called London superconductors (see the survey by Abrikosov and Khalatnikov[6]). The criterion for this case $\delta \gg \xi_0$ can be written as the condition $\kappa \gg 1$ (more accurately $\kappa \gg 0.4$). The opposite case may be called the Pippard case. Here, the penetration depth is in the whole of the temperature range, except in the immediate vicinity of T_c, much less than ξ_0, and the London equation is replaced by a more complicated nonlocal relation. The corresponding criterion will be $\kappa \ll 1$. The majority of the known pure superconductors belongs to the last or to an intermediate type. It is clear from (18) and (20) that if a superconductor belongs to the Pippard class the condition $H_{c1} \ll H_c$ is sure to be satisfied. In the London case, however, H_{c1} must be larger than H_c. In other words, a metal the behavior of which in a weak field shows the London character must display in a strong field the characteristic properties of alloys: a negative surface energy, a smeared out transition, and so on. Abrikosov[7] studied alloys in the phenomenological Ginzburg-Landau theory from this point of view and showed that superconductors with $\kappa > 1/\sqrt{2}$ possess such properties.

Apart from T_c, the combination of quantities that make up the constant κ according to (18) contains also the effective electron mass m at the Fermi surface and the Fermi momentum p_0; expressing κ in terms of the density of the number of "free electrons" $n = p_0^3/3\pi^2$ we get

$$\varkappa = 0.485\, kT_c c m^{3/2} / e\hbar^2 n^{2/3}.$$

To determine m and n from experimental data we choose as one of the quantities the value of the critical magnetic field H_{c0} at $T = 0$, which is expressed as follows in terms of m and n:

$$H_{c0} = 2.48\, \hbar^{-1} m^{1/2} n^{1/6} kT_c.$$

We can take as the second quantity, for instance, data on the anomalous skin-effect in the normal state. The magnitude of the active part of the resistance R yields n directly:[8]

$$n = (\pi^{2/3} 3^{1/6} \omega^3 \hbar^{3/2} / c^6 e^3)\, R^{-3/2}.$$

When we evaluate the processed data on the specific heat in the normal state and the magnitude of R for the anomalous skin effect in reference 1 (see reference 9) we find $\kappa = 0.011$ for aluminum and $\kappa = 0.135$ for tin. At the same time we must note that this method is rather unreliable, since the result is very sensitive to a change in the magnitude of R (R^6 enters into κ). It is therefore advisable to use other data. The evaluation must proceed differently for Pippard and for London superconductors.

It is convenient to use for London superconductors the fact that the equations of the Ginzburg-Landau theory hold for them over a wide range near the critical temperature. We can thus use the well-known relation[5] with the doubled charge[4]

$$\varkappa = (\sqrt{2}\, 2e / \hbar c)\, (H_{cT} \delta_T^2)_{T=T_c}, \qquad (21)$$

where H_{cT} is the critical field at the given temperature and δ_T the penetration depth. We can also use this relation for metals intermediate between London and Pippard types, since in that case the region of applicability of (21) is sufficiently wide. The majority of the most studied superconductors belongs apparently to the intermediate type. Tin[10] ($\kappa = 0.158$), lead[11] ($\kappa = 0.234$) and indium[11] ($\kappa = 0.22$) are, for instance, such superconductors.

The Ginzburg-Landau equations are, finally, applicable also for Pippard metals in the immediate vicinity of T_c, but this neighborhood is very small ($\Delta T/T_c \sim \kappa^2$). For aluminum the London temperature region begins at $\Delta T/T_c \sim 10^{-4}$. Equation (21) can therefore not be used to determine κ in the range of temperatures which is easily accessible. The value of κ can in that case be determined from data on the magnitude of the magnetic field H_{c0} and the penetration depth δ_0 at $T = 0$. The theoretical expression for δ_0 in the Pippard case is according to the BCS theory of the following form:

$$\delta_0 = (\sqrt{3}/2)\, (\pi^2 p_0^2 kT_c / \gamma \hbar^4 c^3)^{-1/3}.$$

Expressing the constant κ in terms of δ_0 and H_{c0} we get

$$\varkappa = 213\, (eH_{c0} / c\hbar)^3 \delta_0^6. \qquad (22)$$

For aluminum we find in this way $\kappa \approx 0.012$.

The available experimental data of Faber's on the magnitude of the supercooling field refer to a temperature range near T_c.[12] For tin, the value of H_{c1}/H_c as given by (20), with $\kappa = 0.16 - 0.226$, agrees very well with the experimental[12] value 0.232. For indium with $\kappa = 0.22$ the theoretical value $H_{c1}/H_c = 0.32$ is nearly twice as large as the experimental[12] one, 0.16. It is difficult to understand the causes of such a discrepancy.

At first sight Eq. (20) for aluminum pertains only to that temperature range where the Ginzburg-Landau equations are applicable. This range is very narrow, but we shall see in the following that the region of applicability of (20) is for a Pippard metal appreciably wider than the London temperature region* near T_c. It is, in particular, helpful to note that at $T = 0$ the character of the solution of (14) does not depend on whether the metal is a Pippard or a London metal in a weak field. With the value of κ found above for aluminum near T_c we have $H_{c1}/H_c \approx 0.017$ while the experimental[12] value is 0.036, i.e., larger than the theoretical one by a factor two. Such a discrepancy should not cause any surprise for a Pippard metal if we take into account how rough the model is and the high powers of the experimental quantities which enter into (22). At the same time Eq. (20) can itself serve to determine κ for Pippard metals. The smallness of the ratio H_{c1}/H_c (or, what is the same, the large magnitude of the surface energy) is a criterion for a metal being of the Pippard type.

In conclusion we wish to discuss some additional points. First we touch upon the general form of the solution of (6). Up to now we assumed that the magnitude of the gap $\Delta(\mathbf{r})$ depended only on y. Such an assumption is permissible, since an appropriate choice for the vector potential **A** had been made in the equations of the basic set (1'). It is also

*This fact was already noted by Ginzburg.[13]

rather obvious from uniformity considerations that the value of the field equal to H_{c1} will correspond to a constant solution $\Delta(x, y)$ in the direction of the field. At the same time the problem is degenerate in the x, y plane, i.e., many solutions $\Delta(x,y)$ correspond to one value of H_{c1}. One can easily verify with the aid of (3) and (4) that if $\Delta(y)$ is a solution of (6) the function

$$\Delta(x, y) = \exp(ieHax/c)\,\Delta(y-a)$$

is also a solution, and the general form of the function which corresponds to a given value of H_{c1} will thus be

$$\Delta(x, y) = \sum_i \exp(ieHxa_i/c)\,A_i\Delta(y-a_i)$$

with arbitrary a_j and A_j.

We shall finally dwell on the problem of the temperature dependence of the field H_{c1}. If the temperature is different from absolute zero, one must use instead of (1') the appropriate thermodynamic technique.[14] We applied this technique earlier[4] to a superconductor; we shall here only indicate to what differences from the foregoing results it leads. The mathematical change in the formalism used above consists of the following: it is sufficient to replace ω by $i\omega_n$ [where $\omega_n = \pi T(2n+1)$, n is an integer] in all equations, beginning with (4), and to replace the integration over $d\omega$ by summation over n, as follows:

$$-\frac{i}{2\pi}\int_{-\infty}^{\infty} d\omega\,(\ldots)_\omega \to T\sum_{-\infty}^{+\infty}(\ldots)_{i\omega_n}.$$

The field dependence of the Green functions remains as before, and the further calculations lead instead of to Eq. (14) to

$$\Delta_T(\xi)\ln\left(\frac{c\gamma\Delta_0}{v}\sqrt{\frac{c}{eH}}\,\delta\right) = -\frac{1}{2}\int_{-\infty}^{\infty} K_T(\xi,\xi')\,\Delta_T(\xi')\,d\xi', \quad (23)$$

where

$$K_T(\xi,\xi') = \frac{2\pi T}{v}\sqrt{\frac{c}{eH}}\int_1^{\infty}\frac{J_0[(\xi^2-\xi'^2)\sqrt{u^2-1}]}{u\sinh(2\pi T v^{-1}\sqrt{c/eH}\,u\,|\xi-\xi'|)}\,du. \quad (24)$$

Because of the complexity of this kernel one can in the general case only obtain the dependence of H_{c1} on T by a numerical integration. We shall consider limiting expressions.

The kernel (24) can for low temperatures be written in the following form:

$$K_T(\xi,\xi') = K_0(\xi,\xi') - \frac{2\pi^2T^2c^2}{3eH_{c1}v^2}\,\frac{\exp(-|\xi^2-\xi'^2|)}{|\xi+\xi'|}, \quad (25)$$

where $K_0(\xi, \xi')$ is the kernel of (14). This expansion loses its validity in the neighborhood of $\xi = -\xi'$, where the temperature dependent correction term in (25) has a singularity. In that point one cannot expand the hyperbolic sine in (24) in powers of T since it just guarantees the convergence of (24) for large u. One sees easily that the width of the region $\Delta\xi = |\xi+\xi'|$ where (25) is inapplicable is of the order $\Delta\xi \sim (2\pi T/v)\sqrt{c/eH_{c1}}$. We rewrite (23) in the following form:

$$\lambda(\xi)\ln\left(\frac{c\gamma\Delta_0}{v}\sqrt{\frac{c}{eH_{c1}}}\,\delta\right) + \frac{1}{2}\int_{-\infty}^{+\infty} K_0(\xi,\xi')\,\lambda(\xi')\,d\xi'$$
$$= \Delta_0(\xi)\frac{\delta H_T}{H_{c1}} - \frac{1}{2}\int_{-\infty}^{+\infty}\delta K_T(\xi,\xi')\,\Delta_0(\xi')\,d\xi',$$

where

$$\lambda(\xi) = \Delta_T(\xi) - \Delta_0(\xi).$$

One must determine $\delta H_T/H_{c1}$ from the condition that the right hand side be orthogonal to the solution of the homogeneous Eq. (14). If we use an approximate solution for $\Delta_0(\xi) \sim \exp(-\xi^2)$ we can find the main term in the additional term in H_{c1}/H_{cT} at low temperatures, using (25), which is of the form $(T^2/T_c^2)\ln(T/T_c)$. Integrating with logarithmic accuracy, we get

$$H_{c1}/H_{cT} = 1.77\times[1 + 0.65\,(T/T_c)^2\ln(\mathrm{const}\cdot T/T_c)] \quad (26)$$

with an unknown constant [if $T \ll T_c$ we have:[6] $H_{cT} = H_{c0}(1-\gamma^2T^2/3T_c^2)$].

Near the transition temperature the field H_{c1} is small, and the equation becomes considerably simpler. It is clear from (23) and (24) that in that case in the integration the essential $|\xi-\xi'|$ are of the order of $\xi_0\sqrt{eH/c}$ because of the exponential character of the kernel (24), while it is natural to assume that $\Delta(\xi)$ changes little over those distances. Using this, it turns out that near T_c the integral equation (23) is changed into a differential equation which to a first approximation is the same as the corresponding equation for the same problem in the Ginzburg-Landau theory with a doubled charge,[4,5] the known solution of which leads to (20).

Taking the next terms into account, we can find after some calculations the temperature dependence of the ratio H_{c1}/H_c near T_c:

$$H_{c1}/H_c = \sqrt{2}\,\varkappa\,(1+0.41\,t); \quad t = 1 - T/T_c. \quad (27)$$

We should like to note here that the condition for the applicability of (20) and (27) follows from the derivation: $\Delta T \ll T_c$, whereas the applicability of the general equations of the Ginzburg-Landau theory obtained in reference 4 near T_c was confined to a much narrower temperature range near T_c. In discussing the penetration depth problem, for instance, for these equations to be applicable it is necessary that not only $\Delta T/T_c$ be small, but also that the superconductor be in the London region at the temperatures considered, i.e., that the penetration depth be large compared to ξ_0. For this it is necessary that the condition $\sqrt{\Delta T/T_c} \ll \kappa$ is satisfied which for superconductors with a small κ (Pippard superconductors) may be an appreciably stronger restriction. In the problem of the supercooling field, therefore, the question whether a superconductor belongs to the Pippard or to the London class is not a decisive factor over the whole temperature range.

We saw already that the change in the ratio H_{c1}/H_c is small in the whole of the temperature range. We can thus believe that an interpolation formula, combining (26) and (27) can represent adequately the variation of H_{c1}/H_c over the whole temperature range. Since the logarithm in (26) is a slowly-varying function, we can take simply for such a formula the polynomial

$$H_{c1}/H_c = \kappa\,[1.77 - 0.43\,(T/T_c)^2 + 0.07\,(T/T_c)^4].$$

Returning, in conclusion, to the problem of the possibility whether there exist pure superconductors with alloy properties, i.e., with $\kappa > 0.56$, we note that among the best studied superconductors, lead and indium, have the largest κ, about 0.23. To increase κ it is necessary to have a larger value of H_c and a larger penetration depth. From this point of view, La, V, U, and Nb are worthy of attention. Unfortunately there are, apparently, no data about the penetration depth for these superconductors. There are in reference 11 some indications about the anomalous properties of these metals which are similar to the properties of alloys, although it is not clear whether these anomalies are caused by the presence of impurities. There is at any rate considerable interest in clarifying the problem of the existence of pure superconductors with $\kappa > 0.56$, since they should possess all anomalous properties of alloys.

In conclusion the author expresses his gratitude to Academician L. D. Landau for his interest in this work and for valuable advice and to Professor V. L. Ginzburg for remarks made.

[1] Bardeen, Cooper, and Schrieffer, Phys. Rev. **108**, 1175 (1957).

[2] L. P. Gor'kov, JETP **34**, 735 (1958), Soviet Phys. JETP **7**, 505 (1958).

[3] L. D. Landau, JETP **34**, 262 (1958), Soviet Phys. JETP **7**, 182 (1958).

[4] L. P. Gor'kov, JETP **36**, 1918 (1959), Soviet Phys. JETP **9**, 1364 (1959).

[5] V. L. Ginzburg and L. D. Landau, JETP **20**, 1064 (1950); V. L. Ginzburg, JETP **31**, 541 (1956), Soviet Phys. JETP **4**, 594 (1957).

[6] A. A. Abrikosov and I. M. Khalatnikov, Usp. Fiz. Nauk **65**, 551 (1958), Adv. Phys. **8**, 45 (1959).

[7] A. A. Abrikosov, JETP **32**, 1442 (1957), Soviet Phys. JETP **5**, 1174 (1957).

[8] G. E. H. Reuter and E. H. Sondheimer, Proc. Roy. Soc. (London) **A195**, 336 (1948).

[9] T. E. Faber and A. B. Pippard, Proc. Roy. Soc. (London) **A231**, 336 (1955).

[10] V. L. Ginzburg, JETP **36**, 1930 (1959), Soviet Phys. JETP **9**, 1372 (1959).

[11] D. Shoenberg, Superconductivity, (Cambridge Univ. Press, 1952).

[12] T. E. Faber, Proc. Roy. Soc. (London) **A241**, 531 (1957).

[13] V. L. Ginzburg, Physica, **24**, S42 (1958).

[14] Abrikosov, Gor'kov, and Dzyaloshinskiĭ, JETP **36**, 900 (1959), Soviet Phys. JETP **9**, 636 (1959); E. S. Fradkin, JETP **36**, 1286 (1959), Soviet Phys. JETP **9**, 912 (1959).

Translated by D. ter Haar
154

SOVIET PHYSICS JETP VOLUME 37 (10), NUMBER 5 MAY, 1960

THEORY OF SUPERCONDUCTING ALLOYS IN A STRONG MAGNETIC FIELD NEAR THE CRITICAL TEMPERATURE

L. P. GOR'KOV

Institute for Physical Problems, Academy of Sciences, U.S.S.R.

Submitted to JETP editor June 17, 1959

J. Exptl. Theoret. Phys. (U.S.S.R.) **37**, 1407-1416 (November, 1959)

The equations of the phenomenological Ginzburg-Landau theory near T_c have been deduced from the Bardeen-Cooper-Schrieffer (BCS) theory. As in the case of pure superconductors, a double charge is encountered in the equations. The relation between the constant κ of the alloy with κ_0 of the pure superconductor has been found under the assumption that the shift in T_c is small. For a sufficiently "impure" alloy, κ depends only on the coefficient in the linear law for the electronic specific heat of the metal. Agreement between the theory and experiments has been found to be satisfactory.

IT has been shown by us[1] that it follows from the superconductivity theory of Bardeen, Cooper and Schrieffer (BCS)[2] that, close to the critical temperature, the behavior of pure superconductors in a magnetic field is described by the equations of the phenomenological theory of Ginzburg-Landau.[3] Below we shall carry out a similar consideration for superconducting alloys. As usual, we assume the concentration of the impurity to be so small that we can neglect any change in the critical temperature. This can be done, for while the shift of T_c is a comparatively slow effect, the magnetic properties of the superconducting alloys depend very strongly on the concentration of the impurity.

The role of the impurities in a superconductor, as is well known, lies in the fact that the scattering of the electrons by the atoms of the impurity leads to the disappearance of the correlation between electrons which is established in the transition to the superconducting state. For appreciable concentration of impurity, the role of the correlation length will be played by the mean free path. For sufficiently "impure" superconductors, the penetration depth is greater than the mean free path; consequently, its behavior in the magnetic field will be described by local "London" equations. As is known, one can distinguish two groups among pure conductors which differ, according to their properties, in a weak magnetic field.[2,4] The so-called "London" superconductors are characterized by the fact that for them the penetration depth over the whole temperature range is larger than the characteristic correlation parameter of superconducting interaction. Such a parameter in the BCS theory is the size of the pair coupling $\xi_0 \sim \hbar v/kT_c$. To this case there correspond the electrodynamics of London. A "Pippard" metal represents the opposite limiting case. For such superconductors, at almost all temperatures, the penetration depth is much smaller than ξ_0. Most of the known superconductors belong to the Pippard type and to the intermediate type. The ordinary Pippard metal has a London interval of temperatures only in the immediate vicinity of the critical temperature. This region is very small. A superconducting alloy like a London or intermediate metal has a much larger region of temperatures close to T_c where their properties in a magnetic field (as is shown) can be described by equations corresponding completely to the equations of the Ginzburg-Landau theory.

As in reference 1, we shall in the derivation begin with the equations[5] for the Fourier components of the thermodynamic Green's function:

$$\left\{i\omega_n + \frac{1}{2m}\left(\frac{\partial}{\partial \mathbf{r}} - ie\mathbf{A}(\mathbf{r})\right)^2 + V(\mathbf{r}) + \mu\right\} \times \mathfrak{G}_\omega(\mathbf{r}, \mathbf{r}') + \Delta(\mathbf{r})\mathfrak{F}^+_\omega(\mathbf{r}, \mathbf{r}') = \delta(\mathbf{r} - \mathbf{r}'). \quad (1)$$

$$\left\{-i\omega_n + \frac{1}{2m}\left(\frac{\partial}{\partial \mathbf{r}} + ie\mathbf{A}(\mathbf{r})\right)^2 + V(\mathbf{r}) + \mu\right\} \times \mathfrak{F}^+_\omega(\mathbf{r}, \mathbf{r}') - \Delta^*(\mathbf{r})\mathfrak{G}_\omega(\mathbf{r}, \mathbf{r}') = 0,$$

$$\Delta^*(\mathbf{r}) = gT\sum_n \mathfrak{F}^+_\omega(\mathbf{r}, \mathbf{r}), \qquad \omega = \pi T(2n + 1),$$

$\mathbf{A}(\mathbf{r})$ is the vector potential (below we shall use everywhere the standard div $\mathbf{A} = 0$), $V(\mathbf{r})$ is the potential energy of interaction with all atoms of the impurity:

$$V(\mathbf{r}) = \sum_a u(\mathbf{r} - \mathbf{r}_a), \quad (2)$$

where the sum runs over all atoms of the impurity arbitrarily distributed over a lattice. The smallness of the concentration means that the mean distance between atoms of the impurity is large in comparison with the interatomic distances in the lattice.

Repeating the derivation of reference 1, expanding $\mathfrak{G}$ and $\mathfrak{F}^+$ near T_c in powers of $|\Delta|$, we obtain the following equation:

$$\Delta^*(\mathbf{r}) = gT\sum_n \int \widetilde{\mathfrak{G}}_{V\omega}(\mathbf{r},\mathbf{r}')\widetilde{\mathfrak{G}}_{V-\omega}(\mathbf{r},\mathbf{r}')\Delta^*(\mathbf{r}')d^3r' - gT$$

$$\times \sum_n \iiint \widetilde{\mathfrak{G}}_{V\omega}(\mathbf{s},\mathbf{r})\,\widetilde{\mathfrak{G}}_{V-\omega}(\mathbf{s},\mathbf{l})\,\widetilde{\mathfrak{G}}_{V\omega}(\mathbf{m},\mathbf{l})\,\widetilde{\mathfrak{G}}_{V-\omega}(\mathbf{m},\mathbf{r})$$

$$\times \Delta(\mathbf{s})\Delta^*(\mathbf{l})\Delta^*(\mathbf{m})\,d^3s\,d^3m\,d^3l. \quad (3)$$

This equation differs from Eq. (8) of reference 1 by the fact that the Green's function $\widetilde{\mathfrak{G}}_{V\omega}$ appearing in it represents the Green's function of electrons in a normal metal in the absence of impurity atoms. The corresponding quantity in the pure metal $\widetilde{\mathfrak{G}}_{\omega}$, as was found in reference 1, depends on the magnetic field in simple fashion:

$$\widetilde{\mathfrak{G}}_{\omega}(\mathbf{r},\mathbf{r}') = \exp\{ie(\mathbf{A}(\mathbf{r}),\ \mathbf{r}-\mathbf{r}')/c\}\,\mathfrak{G}_{\omega}(\mathbf{r}-\mathbf{r}'), \quad (4)$$

where $\mathfrak{G}_{\omega}(\mathbf{r}-\mathbf{r}')$ is the Green's function of normal electrons in the absence of a field. The dependence of $\widetilde{\mathfrak{G}}_{V\omega}$ on the field has precisely the same form

$$\widetilde{\mathfrak{G}}_{V\omega}(\mathbf{r},\mathbf{r}') = \exp\{ie(\mathbf{A}(\mathbf{r}),\ \mathbf{r}-\mathbf{r}')/c\}\,\mathfrak{G}_{V\omega}(\mathbf{r},\mathbf{r}'), \quad (4')$$

where $\mathfrak{G}_{V\omega}(\mathbf{r},\mathbf{r}')$ is the Green's function in the normal metal in the absence of impurities, but without a magnetic field. Actually, the twisting of electrons in a magnetic field is very small, the radius of twisting being much larger than the free path length. Therefore, scattering on impurities takes place independent of the motion in the magnetic field.

Equation (3) should be averaged over the position of each atom of the impurities at distances large in comparison with interatomic distances. The function $\Delta(\mathbf{r})$ is slowly changing since the functions $\mathfrak{G}_{V\omega}$ oscillate rapidly at atomic distances ($\sim 1/p_0$); therefore, averaging in (3) leads to an average of the corresponding combinations of functions $\widetilde{\mathfrak{G}}_{V\omega}(\mathbf{r},\mathbf{r}')$.

In accordance with what has been said about the dependence on a magnetic field, one can take the phase multipliers in (4′) out from under the averaging sign. After averaging, Eq. (3) has a much simpler form. Getting ahead of ourselves, we can point out that the essential distances in the integration will be distances of the order of the free path length l if $l \ll \xi_0 = \hbar v/2\pi k T_c$, and of the order ξ_0 if $l \gg \xi_0$. Further, close to T_c the change of Δ and of the field takes place at distances much larger than when the same quantity Δ is small. Repeating the discussions of reference 1, we can, for example, take all Δ and Δ^* in terms of third order in (3) out from under the integral at the point $\mathbf{r}$ and neglect the dependence on the field (4′), after which it is shown to be necessary to compute the constant

$$B = T_c\sum_n \iiint \overline{\mathfrak{G}_{V\omega}(\mathbf{s},\mathbf{r})\,\mathfrak{G}_{V-\omega}(\mathbf{s},\mathbf{l})\,\mathfrak{G}_{V\omega}(\mathbf{m},\mathbf{l})\,\mathfrak{G}_{V-\omega}(\mathbf{m},\mathbf{r})}$$

$$\times\, d^3s\,d^3m\,d^3l, \quad (5)$$

in this term, where the bar indicates averaging over the position of the impurity atoms.

If we introduce the notation

$$Q(\mathbf{r}-\mathbf{r}') = T\sum_n \overline{\mathfrak{G}_{V\omega}(\mathbf{r},\mathbf{r}')\,\mathfrak{G}_{V-\omega}(\mathbf{r},\mathbf{r}')}, \quad (6)$$

then, by consideration of (5) and (6), we can write Eq. (3) in the form

$$g^{-1}\Delta^*(\mathbf{r}) = \int Q(\mathbf{r}-\mathbf{r}')\exp\{2ie(\mathbf{A}(\mathbf{r}),\ \mathbf{r}-\mathbf{r}')/c\}$$

$$\times\, \Delta^*(\mathbf{r}')\,d^3r' - B\,|\Delta(\mathbf{r})|^2\Delta^*(\mathbf{r})$$

Expanding the expression under the integral in the second term on the right side in terms of second order in $\mathbf{r}-\mathbf{r}'$, we get

$$g^{-1}\Delta^*(\mathbf{r}) = \Delta^*(\mathbf{r})\int Q(R)\,d^3\mathbf{R} + \frac{1}{6}\left(\frac{\partial}{\partial \mathbf{r}} + 2ie\mathbf{A}(\mathbf{r})\right)^2\Delta^*(\mathbf{r})$$

$$\times \int Q(R)\,R^2 d^3\mathbf{R} - B\,|\Delta(\mathbf{r})|^2\,\Delta^*(\mathbf{r}). \quad (7)$$

To obtain the constants in (7) it is consequently necessary to calculate the average in (5) and (6).

A diagram technique has been developed in the work of Abrikosov and the author, and also in that of Edwards,[6] by means of which it is convenient to perform the averaging over the location of the impurity atoms. Each scattering of an electron by an impurity atom leads to a factor $u(\mathbf{q}) \times \exp\{i\mathbf{q}\cdot\mathbf{r}_a\}$ in the expression for the Green's function in the momentum representation, where $\mathbf{q}$ is the forward momentum, and $u(\mathbf{q})$ is the Fourier component of the potential of the impurity atom. In averaging over $\mathbf{r}_a$, inasmuch as $|\mathbf{q}| \sim p_0$ essentially, this factor vanishes only if there is no term in the averaged expression corresponding to scattering by this very atom, but with $\mathbf{q}' = -\mathbf{q}$; in this case there appears $|u(\mathbf{q})|^2$ — a quantity proportional in the Born approximation to the probability of scattering with exchange of momentum $\mathbf{q}$. Thus, in the averaging, it is necessary to throw away pairwise all scatterings by identical atoms. If we represent scattering by each atom by a cross on the diagram, then in the averaging it is necessary to connect two crosses, corresponding to scattering by identical atoms, by a line. In the matrix element $n\,|u(\mathbf{q})|^2$ corresponds to such a line, where n is the number density of the impurity atoms. According to reference 6, if the dotted line on the diagram includes a vertex with a large transfer r of momentum, then the corresponding contribution from this diagram is small ($\sim 1/p_0 l$) and it can be discarded.

Averaging of the Green's function itself leads to the factor $\exp\{-|\mathbf{r}-\mathbf{r}'|/2l\}$, which must be multiplied by the Green's function of the pure metal. Here the mean free path $l = v\tau$ in the Born approximation is directly connected with the potential of the impurity atom:

$$\frac{1}{\tau} = \frac{nmv^2}{(2\pi)^2}\int |u(\mathbf{q})|^2\, do.$$

Averaging of the product of several functions, which takes place in (5) and (6), is somewhat more complicated. Let us first consider the averages:

$$K_\omega(\mathbf{r}-\mathbf{s}, \mathbf{r}'-\mathbf{s}) = \overline{\mathfrak{G}_{V\omega}(\mathbf{r}, \mathbf{s})\,\mathfrak{G}_{V-m}(\mathbf{r}', \mathbf{s})}. \qquad (8)$$

We introduce the Fourier components

$$K_\omega(\mathbf{x}, \mathbf{y}) = (2\pi)^{-6}\iint e^{i\mathbf{p}_1\mathbf{x}+i\mathbf{p}_2\mathbf{y}} K_\omega(\mathbf{p}_1, \mathbf{p}_2)\, d^3\mathbf{p}_1 d^3\mathbf{p}_2.$$

The product of the two Green's functions in (8)

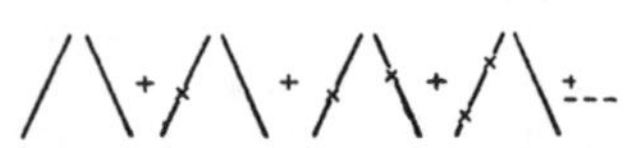

FIG. 1

is shown graphically in Fig. 1, where the crosses, as indicated above, express scattering by individual impurities. In the averaging, in correspondence with the program that we have outlined, we shall connect the crosses with one another pairwise in all processes by dotted lines, which corresponds to scattering by one and the same impurity atom.

FIG. 2

In Fig. 2 diagrams are shown of lower orders which arise as a result of such averaging. As has been pointed out, diagrams with an intersection of dotted lines are not important in this group. All dotted lines joining points together that are located on a single electron line evidently give the following Green's function for the corresponding line:

$$\mathfrak{G}^{\tau}_{\omega}(\mathbf{r}-\mathbf{r}') = \exp\{-|\mathbf{r}-\mathbf{r}'|/2l\}\,\mathfrak{G}_\omega(\mathbf{r}-\mathbf{r}'). \qquad (9)$$

Taking this into account, it is easy to write down the integral equation which $K_\omega(\mathbf{p}_1, \mathbf{p}_2)$ satisfies:

$$K_\omega(\mathbf{p}_1, \mathbf{p}_2) = \mathfrak{G}^{\tau}_{\omega}(\mathbf{p}_1)\,\mathfrak{G}^{\tau}_{\omega}(\mathbf{p}_2)$$
$$\times\left[1 + \frac{n}{(2\pi)^3}\int |u(\mathbf{p}_1 - \mathbf{p}_1')|^2 K_\omega(\mathbf{p}_1', \mathbf{p}_2')\, d^3\mathbf{p}_1'\right], \qquad (10)$$

where $\mathbf{p}_1 + \mathbf{p}_2 = \mathbf{p}_1' + \mathbf{p}_2' = \mathbf{q}$. This equation sums the set of diagrams shown in Fig. 3, where the electron line must now be compared with the average Green's function $\mathfrak{G}^{\tau}_{\omega}(\mathbf{p})$.

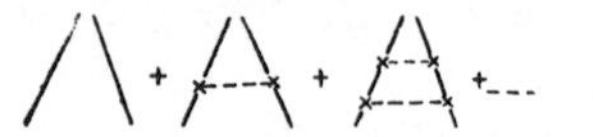

FIG. 3

The kernel $Q(\mathbf{r}-\mathbf{r}')$, in terms of which the first two coefficients in Eq. (7) are expressed, is evidently connected with (8). We write down the expression for the Fourier component $Q_\mathbf{q}$ in terms of $K_\omega(\mathbf{p}_1, \mathbf{p}_2)$:

$$Q_\mathbf{q} = T(2\pi)^{-3}\sum_\omega \int K_\omega(\mathbf{p}, \mathbf{p}-\mathbf{q})\, d^3\mathbf{p};$$

Then, as is seen from (7), the constant which we must determine is simply

$$Q_0 = \int Q(R)\, d^3\mathbf{R}, \qquad -\partial^2 Q_q/\partial q_i^2|_{q=0} = \int Q(R) R^2 d^3\mathbf{R}.$$

Therefore, in the calculation of $K_\omega(\mathbf{p}, \mathbf{p}-\mathbf{q})$ we must consider $\mathbf{q}$ to be small and limit ourselves in the expression for $K_\omega(\mathbf{p}, \mathbf{p}-\mathbf{q})$ to terms of second order in $\mathbf{q}$.

For the solution of Eq. (10), it is useful to introduce the notation

$$L_{\omega,\mathbf{q},\mathbf{p}} = n(2\pi)^{-3}\int |u(\mathbf{p}-\mathbf{p}')|^2 K_\omega(\mathbf{p}', \mathbf{p}'-\mathbf{q})\, d^3\mathbf{p}'.$$

The equation for this quantity is obtained in obvious fashion after multiplying (10) by $n(2\pi)^{-3} \times |u(\mathbf{1}-\mathbf{p})|$ and integration over $\mathbf{p}$:

$$L_{\omega,\mathbf{q},\mathbf{p}} = n(2\pi)^{-3}\int |u(\mathbf{p}-\mathbf{p}')|^2 \mathfrak{G}^{\tau}_{\omega}(\mathbf{p}')\,\mathfrak{G}^{\tau}_{-\omega}(\mathbf{q}-\mathbf{p}')$$
$$\times\,[1 + L_{\omega,\mathbf{p}',\mathbf{q}-\mathbf{p}'}]\, d^3\mathbf{p}'. \qquad (11)$$

In accordance with reference 6, we have the following simple expression for the components of the Green's function $\mathfrak{G}^{\tau}_{\omega}(\mathbf{p})$:

$$\mathfrak{G}^{\tau}_{\omega}(\mathbf{p}) = [i\eta\omega - \xi]^{-1}, \qquad \xi = v(p - p_0), \quad \eta = 1 + 1/2\tau|\omega|$$

The product of the two Green's functions under the integral in (11) quickly falls off upon increasing distance from the Fermi surface. Integration over ξ can easily be completed. It is evident that because of the slow dependence of $u(\mathbf{p}-\mathbf{p}')$ on the quantity $\mathbf{p}$, only the dependence on the direc-

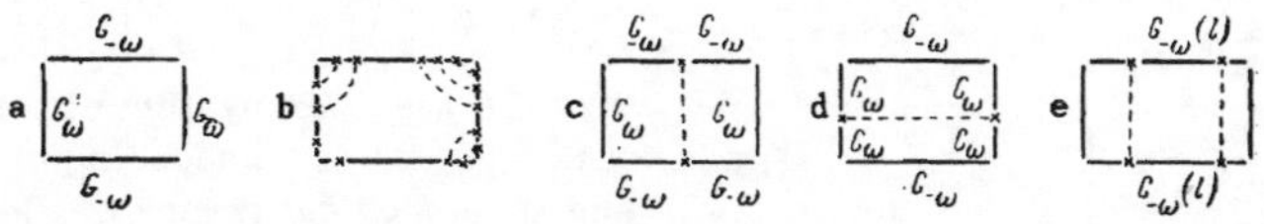

FIG. 4

tion of the vector $\mathbf{p}$ ($|\mathbf{p}| \approx p_0$) enters into $L_{\omega,\mathbf{q},\mathbf{p}}$. The equation takes the form

$$L_{\omega,\mathbf{q},\mathbf{p}} = nv\int (d\sigma/do_{\mathbf{p}'})\,\nu_{\omega,\mathbf{q},\mathbf{p}'}\,[1 + L_{\omega,\mathbf{q},\mathbf{p}'}]\,do_{\mathbf{p}'} \qquad (12)$$

where

$$\nu_{\omega,\mathbf{q},\mathbf{p}} = (2|\omega\eta|)^{-1}\,[1 - \mathbf{q}\mathbf{p}/2im\omega\eta]^{-1} \qquad (12')$$

is the result of integration of the product of two $\mathfrak{G}$ functions in (11) over ξ.

We limit ourselves only to calculation of Q_0. In (11), $\mathbf{q} = 0$, $\nu_{\omega,0,\mathbf{p}'} = \frac{1}{2}|\omega\eta|$ and $L_{\omega,0,\mathbf{p}} = \frac{1}{2}\tau|\omega|$. Then, in accord with (10),

$$K_\omega(\mathbf{p},\mathbf{p}) = \eta\,\mathfrak{G}^\tau_\omega(\mathbf{p})\,\mathfrak{G}^\tau_{-\omega}(\mathbf{p}). \qquad (13)$$

In the calculation of Q_0, before summing over all frequencies in the expression

$$Q_0 = T\sum(2\pi)^{-3}\int d^3p\,\mathfrak{G}^\tau_\omega(\mathbf{p})\,\mathfrak{G}^\tau_{-\omega}(\mathbf{p})\,\eta$$

we must satisfy ourselves, by integrating the difference $\mathfrak{G}^T_\omega(\mathbf{p})\mathfrak{G}^T_{-\omega}(\mathbf{p})\eta - \mathfrak{G}_\omega(\mathbf{p})\mathfrak{G}_{-\omega}(\mathbf{p})$, that Q_0 for alloys coincides with the corresponding integral for the pure metal:

$$Q_0 = T\sum_\omega(2\pi)^{-3}\int d^3p\,\mathfrak{G}_\omega(\mathbf{p})\,\mathfrak{G}_{-\omega}(\mathbf{p}) = \frac{mTp_0}{2\pi^2}\sum_\omega\int\frac{d\xi}{\omega^2+\xi^2}$$

$$(\omega = \pi T(2n+1)).$$

Completing the summation, we find

$$Q_0 = \frac{mp_0}{4\pi^2}\int d\xi\,\frac{\tanh(\xi/2T)}{\xi}$$

The divergence in this integral for large ξ must be eliminated by a cut corresponding to the fact that only electrons found in a layer within a distance $\tilde{\omega}$ from the Fermi surface interact in the model under consideration.[2] In accordance with reference 2.

$$Q_0 = \frac{mp_0}{2\pi^2}\ln\frac{2\tilde{\omega}\gamma}{\pi T} = \frac{mp_0}{2\pi^2}\left(\frac{1}{g} + \ln\frac{T_c}{T}\right).$$

No difficulty arises in finding $-\partial^2 Q_\mathbf{q}/\partial q_i^2|_{\mathbf{q}=0}$; differentiating Eqs. (10) and (12) with respect to $\mathbf{q}$, we obtain, after some calculation,

$$-\partial^2 Q_\mathbf{q}/\partial q_i^2|_{\mathbf{q}=0} = (mp_0/2\pi^2)(\pi v^2 T/2)\sum_\omega \omega^{-3}\eta_{tr}^{-1},$$

where

$$\eta_{tr} = 1 + 1/2\tau_{tr}|\omega|, \qquad \tau_{tr}^{-1} = nv\int(d\sigma/do)(1-\cos\theta)\,do.$$

Finally, the average (5) is computed in the following fashion. The product of the four Green's functions in the integral (5) is drawn in Fig. 4. Their average over the position of the impurity atoms, in accord with the method employed, leads first to the diagram of Fig. 4a, in which now the average Green's functions appear. There will be other diagrams such as those in Fig. 4b, for example. These diagrams represent a modification of the corresponding vertices and are summed by means of Eqs. (10) and (12). In addition, there are two types of diagrams shown in Figs. 4c and 4d. It is not necessary to consider other diagrams, inasmuch as the diagrams with the intersection of dotted lines make a small contribution, according to the results of reference 6, while diagrams such as those shown in Fig. 4e contribute nothing in the integration over the internal momentum l, since the product of two $\mathfrak{G}$ functions appearing under the integral has a pole for ξ only in a single halfplane.

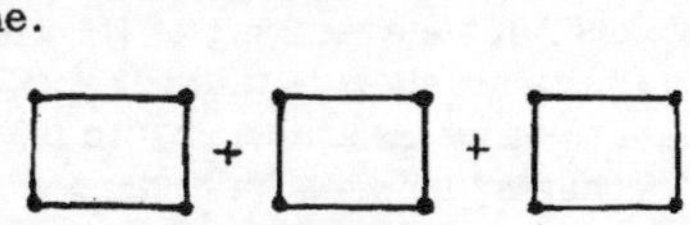

FIG. 5

Figure 5 shows the final form of the diagrams which are to be computed; the heavy dots at the vertices correspond to summation of all corrections to the vertex according to Fig. 3. Inasmuch as the momentum $\mathbf{q} = \mathbf{p}_1 + \mathbf{p}_2 = 0$ in Eq. (5) at all vertices, this means that the heavy dot corresponds to the factor η, as was found in (13). The integral and sums over the frequencies are easily computed:

$$B = (mp_0/2\pi^2)\,(7\zeta(3)/8(\pi T_c)^2),$$

where $\zeta(x)$ is the Riemann function, $\zeta(3) = 1.202$.

Substituting the results obtained above in Eq. (7), we find the following equation:

$$\left\{\frac{1}{2m}\left(\frac{\partial}{\partial \mathbf{r}} - 2ie\mathbf{A}(\mathbf{r})\right)^2 + \frac{1}{\lambda_\tau}\left[\frac{T_c - T}{T_c} - \frac{7\zeta(3)}{8(\pi T_c)^2}|\Delta(\mathbf{r})|^2\right]\right\}\Delta(\mathbf{r}) = 0,$$

which differs from the corresponding equation of reference 1 only in the expression for λ_T:

$$\lambda_\tau = \lambda_0\chi(\rho), \qquad (14)$$

where λ_0 is the corresponding coefficient for a pure superconductor, $\lambda_0 = 7\zeta(3)\epsilon_F/12\pi^2T_c^2$, and

$$\chi(\rho) = \frac{8}{7\zeta(3)} \sum_{0}^{\infty}{}' \frac{1}{(2n+1)^2[2n+1+\rho]}$$
$$= \frac{8}{7\zeta(3)}\left(\frac{1}{\rho}\right)\left[\frac{\pi^2}{8} + \frac{1}{2\rho}\left\{\psi\left(\frac{1}{2}\right) - \psi\left(\frac{1}{2}+\rho\right)\right\}\right] \qquad (15)$$

(for $\rho \to 0$, $\chi(\rho) \to 1$, while for $\rho \to \infty$, $\chi(\rho) \approx \pi^2/7\zeta(3)\rho$). Here, $\rho = \frac{1}{2}\pi T_c \tau_{tr}$, $\psi(x)$ is the logarithmic derivative of the Γ function.

For the determination of the value of the electromagnetic current, it is necessary to know the contribution of second order in $|\Delta|$ to the Green's function. Repeating the derivation given in reference 1, we obtain

$$\delta\mathfrak{G}_\omega(\mathbf{r}, \mathbf{r}') = -\int \tilde{\mathfrak{G}}_{V\omega}(\mathbf{r}, \mathbf{s})\, \Delta(\mathbf{s})\, \tilde{\mathfrak{G}}_{V\omega}(\mathbf{l}, \mathbf{r}')$$
$$\times \Delta^*(\mathbf{l})\, \tilde{\mathfrak{G}}_{V-\omega}(\mathbf{l}, \mathbf{s})\, d^3 l\, d^3 s. \qquad (16)$$

As above, the averaging over the sites of the impurity atoms of a combination of three Green's functions of the normal metal does not show any dependence of the integrand on the magnetic field.

In the determination of the current,

$$\mathbf{j}(\mathbf{r}) = + \left[\frac{ie}{m}(\nabla_r - \nabla_{r'})\, G(x, x') + \frac{2e^2}{mc}\mathbf{A}(\mathbf{r})\, G(xx')\right]_{r=r',\, t'=t+0} \qquad (17)$$

a Green's function appears with coinciding arguments; therefore, in the averaging of (16), it is sufficient to limit ourselves to this case. Considering the field to be weak, and $\Delta(\mathbf{r})$ to be of the order of l, changing slowly with distance, we carry out the corresponding simplification in (16) and (17). We shall not linger over the calculations, limiting ourselves only to the scheme of averaging over the impurity in Fig. 6, which represents the corresponding diagram which must be summed.

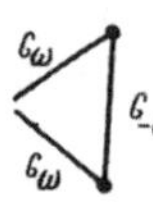

FIG. 6

We note that the corrections to the vertex into which enter two Green's functions with the same frequency are absent, inasmuch as such corrections lead to the integration of products of two $\mathfrak{G}$ functions with poles located in one halfplane. After computation we obtain the following expression for the current:

$$\mathbf{j}(\mathbf{r}) = \left\{\frac{ie}{m}\left(\Delta(\mathbf{r})\frac{\partial\Delta^*(\mathbf{r})}{\partial\mathbf{r}} - \Delta^*(\mathbf{r})\frac{\partial\Delta(\mathbf{r})}{\partial\mathbf{r}}\right)\right.$$
$$\left. - \frac{4e^2}{mc}\mathbf{A}(\mathbf{r})\,|\Delta(\mathbf{r})|^2\right\}\chi(\rho)\frac{7\zeta(3)N}{16\pi^2 T_c^2}.$$

If we introduce the "wave function" $\psi(\mathbf{r})$ connected with $\Delta(\mathbf{r})$ by the relation

$$\psi(\mathbf{r}) = [\chi(\rho)\, 7\zeta(3)\, N/16\pi^2 T_c^2]^{1/2}\, \Delta(\mathbf{r}),$$

then the resultant system of equations can be written in a form corresponding to the equations of the phenomenological theory of Ginzburg-Landau:[3]

$$\left\{\frac{1}{2m}\left(\frac{\partial}{\partial\mathbf{r}} - 2ie\mathbf{A}(\mathbf{r})\right)^2 + \frac{1}{\lambda_\tau}\left[\frac{T_c - T}{T_c} - \frac{2}{N\chi(\rho)}|\psi(\mathbf{r})|^2\right]\right\}\psi(\mathbf{r}) = 0,$$
$$\mathbf{j}(\mathbf{r}) = \frac{ie}{m}\left(\psi\frac{\partial\psi^*}{\partial\mathbf{r}} - \psi^*\frac{\partial\psi}{\partial\mathbf{r}}\right) - \frac{4e^2}{mc}\mathbf{A}(\mathbf{r})\,|\psi|^2. \qquad (18)$$

The most interesting part of this result is the two-fold charge which figures in the equations. The existence of bound pairs of electrons in the BCS theory makes this result entirely understandable. In this light, the meaning of the wave function $\psi(\mathbf{r})$ of the Ginzburg-Landau theory as a quantity proportional to the wave function of a bound pair (more precisely, the coordinates of its center of mass) becomes clarified. An identical result was obtained by us for a pure superconductor.[1]

The difference of alloys from the case of a pure metal lies, as is well known, in the constant κ which appears in the phenomenological theory of Ginzburg-Landau, and which is determined (with consideration of the double charge) by the relation

$$\varkappa = \sqrt{2}\, 2eH_{cT}\delta_T^2/\hbar c, \qquad (19)$$

where H_{cT} is the critical field, and δ_T is the penetration depth close to T_c which is temperature dependent.

The equations (18) permit us to find the connection of the constant κ for alloys with κ_0 of a pure superconductor, and to express it in terms of the parameters of the BCS theory. The constant κ for alloys is equal to

$$\varkappa = \varkappa_0/\chi(\rho),$$

where the function $\chi(\rho)$ has been introduced above in (15), while κ_0 for a pure metal is expressed (according to reference 1) in the following form:

$$\varkappa_0 = (3\pi T_c mc/e)\sqrt{2\pi m/7\zeta(3)\, p_0^5} \qquad (\zeta(3) = 1.202). \qquad (20)$$

In the relations that have been obtained and in the determination of $\rho = \frac{1}{2}\pi T_c \tau_{tr}$, four parameters generally appear: T_c, m, p_0 for the pure metal, and the "transport" time of the free flight τ_{tr}. For pure superconductors of the London type and for intermediates, i.e., such for which throughout the whole temperature range the penetration depth is larger or comparable to the correlation length $\xi_0 = hv/2\pi T_c$, an appreciable temperature region close to T_c exists where one can use the relations (19) with the experimental values of H_c and δ_0 for the determination of κ_0. The theoretical expression[2] for δ_0 close to T_c has the form

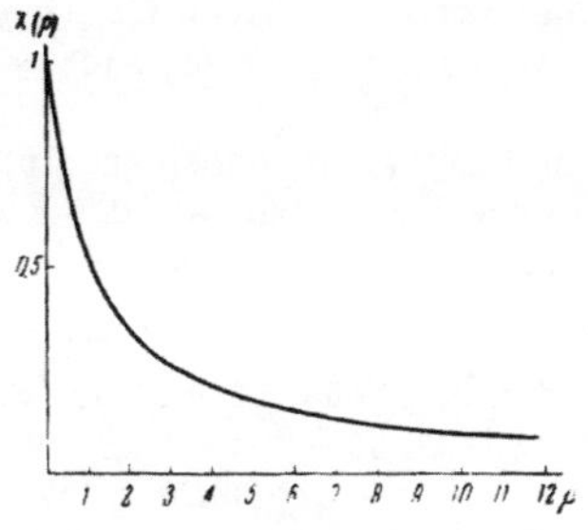

FIG. 7

$$\delta_0 = \delta_{0L}/\sqrt{2(1-T/T_c)} \qquad (\delta_{0L} = \sqrt{mc^2/4\pi Ne^2}) \quad (21)$$

In the isotropic model under consideration, the time between collisions τ_{tr} can be expressed in terms of δ_{0L} and the conductivity σ:

$$\tau_{tr} = 4\pi\sigma\delta_{0L}^2/c^2. \quad (22)$$

In terms of these quantities

$$\rho = \hbar c^2/8\pi^2 kT_c\sigma\delta_{0L}^2. \quad (23)$$

The function $\chi(\rho)$ is shown graphically in Fig. 7. We also give the formula which connects the penetration depth of a weak magnetic field for an alloy with the penetration depth for the pure superconductor

$$\delta = \delta_0/\sqrt{\chi(\rho)}. \quad (24)$$

If the quantity of impurity is sufficiently large, so that $l \ll \xi_0$, the equations are simplified. In this case, κ for alloys can be expressed only in terms of the parameters of the normal metal: the coefficient in the linear law for the heat capacity γ and the conductivity σ:

$$\varkappa = (ec\gamma^{1/2}/\sigma k\pi^3)\sqrt{21\zeta(3)/2\pi} = 0.065ec\gamma^{1/2}/\sigma k \quad (25)$$

For metals of the Pippard type, i.e., those for which at low temperatures the penetration depth is much smaller than the parameter ξ_0, chief interest lies in the latter formula. This is connected with the fact that for a pure superconductor of the Pippard type, the London region of temperatures close to T_c, where the Ginzburg-Landau equations hold, is very narrow (for aluminum, $\Delta T/T_c \sim 10^{-4}$). The presence of impurities enlarges this region. In order that this region be experimentally accessible, it is necessary that $l \ll \xi_0$.

Equations (20) and (25) permit us to determine the critical concentration of impurities for which the superconductor in a strong field begins to reveal the characteristic peculiarities of alloys. As is well known, the surface energy on the boundary of separation between the normal and superconducting phases vanishes for $\kappa_{cr} = 1/\sqrt{2}$. At the same time, the critical field of supercooling is made equal to the critical thermodynamic field, while for large κ it exceeds it. Therefore, for $\kappa > 1/\sqrt{2}$, the transition in the magnetic field from the normal to the superconducting state begins at fields larger than the critical field H_{cT} of the pure superconductor, namely at the field $H_{c1} = \sqrt{2}\,\kappa H_{cT}$. This peculiarity of alloys was pointed out by Abrikosov.[7] For Pippard superconductors, $\kappa = 1/\sqrt{2}$ is achieved at such concentrations of impurities that $l \ll \xi_0$; therefore, it follows from (25) that for the critical amount of impurities, the residual conductivity is connected only with the coefficient γ in the linear law for electronic specific heat $\sigma_{cr} = 0.092\,ec\gamma^{1/2}/k$.

The linear dependence of H_{cT}/H_{c1} on the path length l for contaminated alloys on the basis of phenomenological considerations was first pointed out by Pippard.[8] For $l \ll \xi_0$, the formula for the penetration depth was obtained in the work of Abrikosov and the author.[6]

We note that, as experiment has shown,[8,9] the actual region $\kappa \sim 0.7$ corresponds, for example, in the case of an alloy of tin and indium, to $l \sim 5 \times 10^{-6}$ cm (or a concentration of indium ~3 percent). For such concentrations of impurities, there is a certain insignificant shift in the temperature[9] ($\Delta T_c/T_c \sim 5$ percent). We shall neglect this effect, assuming the number of impurities to be so small that we do not have to take into consideration their effect on the fundamental lattice. At high concentrations, it appears that such an approximation is improper. However, it is possible to assume, that the relation (25) will still hold in such alloys, inasmuch as there appear in it only quantities which refer to the normal metal. The coefficient in the linear law for the specific heat γ must be introduced in this case for the alloy itself.

The available experimental data refer to tin.[8,9] In the work of Dodge,[9] the transition in a magnetic field from a normal phase to the superconducting phase was studied for an alloy of tin and indium. For concentrations of indium of about 3 percent, this transition has the usual character; for higher concentrations, the appearance of the superconducting phase occurs at fields larger than the critical thermodynamic field. The ratio H_{cT}/H_{c1}, i.e., the value $\chi(\rho)/\sqrt{2}\,\kappa_0$ is plotted in Fig. 8 as a function of the mean free path length l. (We have made use of the result of the research of Chambers[10] for tin: $\sigma/l = 8.5 \times 10^{22}$, by means of which one can express ρ in terms of the mean free path length: $\rho = 2 \times 10^{-5}/l$.) For pure tin, $\kappa_0 = 0.158$,[11] the value of κ of the alloy (0.707) corresponds in the calculations to $l \sim 5.5 \times 10^{-6}$ cm, while the experimental value is somewhat less than 8×10^{-6} cm.

Taking into account the known crudeness in the determination of δ_{0L} (we have used $\delta_{0L} = \delta_{00}/\sqrt{2}$, where δ_{00} is the coefficient in the law for the skin depth

$$\delta = \delta_{00}/\sqrt{1-(T/T_c)^4},$$

is equal to $\delta_{00} = 5.1 \times 10^{-6}$ cm), one can consider the agreement to be excellent.

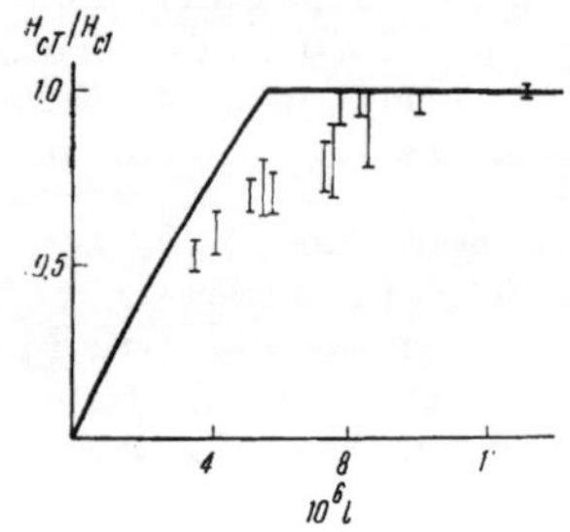

FIG. 8

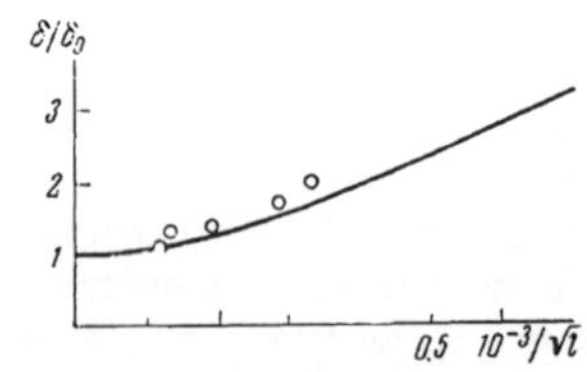

FIG. 9

Results are shown in Fig. 9 of a comparison of the formula (24) for the skin depth with the results of Pippard,[8] which apply to an alloy of indium and tin with a concentration up to 3 percent of indium. We note that the proportionality of the penetration depth $l^{-1/2}$ is still not attained for such concentrations.

In conclusion I express my deep gratitude to Acad. L. D. Landau for his interest in the work and for the comments that he has made.

[1] L. P. Gor'kov, JETP **36**, 1918 (1959), Soviet Phys. JETP **9**, 1364 (1959).

[2] Bardeen, Cooper, and Schrieffer, Phys. Rev. **108**, 1175 (1957).

[3] V. L. Ginzburg and L. D. Landau, JETP **20**, 1064 (1950).

[4] A. A. Abrikosov and I. M. Khalatnikov, Usp. Fiz. Nauk **65**, 551 (1958).

[5] L. P. Gor'kov, JETP **34**, 735 (1958), Soviet Phys. JETP **8**, 505 (1958).

[6] A. A. Abrikosov and L. P. Gor'kov, JETP **35**, 1558 (1958); **36**, 319 (1959); Soviet Phys. JETP **8**, 1090 (1959); **9**, 220 (1959).

[7] A. A. Abrikosov, JETP **32**, 1442 (1957), Soviet Phys. JETP **5**, 1174 (1957);

[8] A. B. Pippard, Proc. Roy. Soc (London) **A216**, 547 (1953).

[9] P. R. Dodge, Phil. Trans. Roy. Soc. (London) **A248**, 553 (1956).

[10] R. G. Chambers, Proc. Roy. Soc. (London) **A215**, 481 (1952).

[11] V. L. Ginzburg, JETP **36**, 1930 (1959); Soviet Phys. JETP **9**, 1372 (1959).

Translated by R. T. Beyer
278

SOVIET PHYSICS JETP VOLUME 12, NUMBER 6 JUNE, 1961

CONTRIBUTION TO THE THEORY OF SUPERCONDUCTING ALLOYS WITH PARAMAGNETIC IMPURITIES

A. A. ABRIKOSOV and L. P. GOR'KOV

Institute for Physics Problems, Academy of Sciences, U.S.S.R.

Submitted to JETP editor July 25, 1960

J. Exptl. Theoret. Phys. (U.S.S.R.) **39**, 1781-1796 (December, 1960)

We show in the framework of superconductivity theory that the transition temperature T_c is lowered in comparison with that of a pure metal when paramagnetic impurities are introduced into a superconductor. We have determined the dependence of T_c on the impurity concentration. Superconductivity vanishes at a certain critical concentration. We have made a detailed study of the thermodynamic and magnetic properties of superconducting alloys in the concentration range close to critical. We have found that in that range there is no exponential term in the expression for the heat capacity as $T \to 0$. The concentration dependence of the gap in the energy spectrum of such a system is determined. It is shown that the gap disappears at a concentration $n' \approx 0.9\, n_{cr}$, while at higher concentrations the energy spectrum of the system is continuous.

A large number of recent papers are devoted to a study of the influence of paramagnetic impurities on the transition of a metal into a superconducting state. It was observed in experiments by Matthias et al.[1] that rare earth impurities in lanthanum in the range of concentrations of about 1% appreciably lower the superconducting transition temperature. Herring[2] and also Suhl and Matthias[3] have in this connection expressed the hypothesis that the occurrence of this effect might be explained on the basis of an exchange interaction between the conduction electrons and the impurity atoms. A little later Matthias et al.[4] found also that a small amount of ferromagnetic impurities in titanium can lead to an increase of its superconducting transition temperature.

The present paper is devoted to a study of this problem on the basis of a microscopic superconductivity theory. It is well known that the mechanism responsible for superconductivity is the formation of bound electron pairs in a singlet state. The exchange interaction between electrons and spinning impurity atoms leads to nonconservation of the electron spin, which may affect the formation of Cooper pairs. It is most probable that the spin of the impurity should inhibit the appearance of the superconducting correlation, i.e., cause a decrease in T_c. The rigorous calculation given in the following confirms this supposition.

It is well known that the transition temperature of superconductors with nonmagnetic impurities remains practically unchanged in the region of low concentrations, and this was also shown in a paper by the present authors.[5] The significant change in T_c is in such cases caused by the influence of the foreign atoms on the basic lattice. One can neglect this effect in the low-concentration region in comparison with the change in the electromagnetic properties of the alloy, which are completely due to the scattering of the electrons by impurity atoms. After the metal has changed into the superconducting state, there arises a specific interelectron correlation which acts at distances of the order of $\xi_0 = \hbar v/2\pi T_c$; these distances have the physical meaning of the average dimensions of a Cooper pair. The scattering of the electrons by the impurities decreases this correlation, restricting it at sufficiently high concentrations to distances of the order of the mean free path.

In the present paper we shall assume that there is a term due to exchange in the interaction between an electron and an impurity atom

$$\hat{v}(\mathbf{r}) = u_1(\mathbf{r}) + u_2(\mathbf{r})(\mathbf{S}\hat{\sigma}), \tag{1}$$

where S is the moment of the impurity atom and $\hat{\sigma}$ are the electron spin matrices. The Hamiltonian of the interaction between the electrons and the impurity atoms is in the second quantization representation of the form

$$H_{int} = (\psi^+ \hat{V} \psi) = (\psi^+ \sum_a \hat{v}(\mathbf{r} - \mathbf{r}_a)\psi).$$

1. THE DEPENDENCE OF THE TRANSITION TEMPERATURE ON THE IMPURITY CONCENTRATION

We consider first the problem of the dependence of the transition temperature on the impurity concentration. To do this we note that one can describe a superconductor by means of two Green's functions $\mathfrak{G}(x, x')$ and $\mathfrak{F}^+(x, x')$ as was shown by one of the authors.[6] In the thermodynamic approach the equations for the Fourier components of these functions are of the form

$$(i\omega_n + \nabla^2/2m + \mu)\hat{\mathfrak{G}}_\omega(\mathbf{r}, \mathbf{r}') - \hat{V}\hat{\mathfrak{G}}_\omega(\mathbf{r}, \mathbf{r}') + \hat{\Delta}(\mathbf{r})\hat{\mathfrak{F}}^+_\omega(\mathbf{r}, \mathbf{r}') = \delta(\mathbf{r}-\mathbf{r}'),$$
$$(i\omega_n - \nabla^2/2m - \mu)\hat{\mathfrak{F}}^+_\omega(\mathbf{r}, \mathbf{r}') + \hat{V}^t\hat{\mathfrak{F}}^+_\omega(\mathbf{r}, \mathbf{r}') + \hat{\Delta}^*(\mathbf{r})\hat{\mathfrak{G}}_\omega(\mathbf{r}, \mathbf{r}') = 0,$$
$$V^t_{\alpha\beta} = V_{\beta\alpha}, \quad \omega_n = (2n+1)\pi T. \tag{2}$$

We used the fact that in the constant field of the impurities the functions $\mathfrak{G}$ and $\mathfrak{F}^+$ will be functions of the difference between the "time" variables $\tau = \tau_1 - \tau_2$ only, and we have performed a Fourier transformation with respect to this difference. In particular,

$$\hat{\Delta}(\mathbf{r}) = |g| T \sum_\omega \hat{\mathfrak{F}}_\omega(\mathbf{r}, \mathbf{r}). \tag{3}$$

Unlike reference 7, we have written Eqs. (2) in matrix form as far as the spin variables are concerned, and the product $\hat{V}\hat{\mathfrak{G}}$, for instance, indicates a matrix product. Since $\hat{\Delta}(\mathbf{r})$ and $\hat{\Delta}^*(\mathbf{r})$ are expressed in terms of averages over $\psi\psi$ or $\psi^+\psi^+$ in one point,

$$\hat{\Delta}_{\alpha\beta}(\mathbf{r}) = \langle\psi_\alpha(x)\psi_\beta(x)\rangle, \quad \Delta^*_{\alpha\beta}(\mathbf{r}) = \langle\psi^+_\alpha(x)\psi^+_\beta(x)\rangle, \tag{4}$$

it is clear that the relation

$$\Delta_{\alpha\beta}(\mathbf{r}) = -\Delta_{\beta\alpha}(\mathbf{r}), \tag{5}$$

and also

$$\Delta^*_{\alpha\beta}(\mathbf{r}) = -\Delta^*_{\beta\alpha}(\mathbf{r}) \tag{6}$$

are always valid. When the metal is in its normal state with $T > T_c$, Eqs. (2) do not possess solutions with non-vanishing $\mathfrak{F}^+$ and Δ. The temperature T_c, at which the transition into the superconducting phase occurs, is determined by the fact that at $T = T_c$ there exist for the first time solutions of Eqs. (2) with non-vanishing, though arbitrary small, $\mathfrak{F}^+$ (and Δ). Retaining in (2) only the terms which are linear in $\mathfrak{F}^+$ (and Δ) we get

$$(i\omega_n + \nabla^2/2m + \mu)\hat{\mathfrak{G}}'(\mathbf{r}, \mathbf{r}') - \hat{V}(\mathbf{r})\hat{\mathfrak{G}}'_\omega(\mathbf{r}, \mathbf{r}') = \delta(\mathbf{r}-\mathbf{r}'),$$
$$(i\omega_n - \nabla^2/2m - \mu)\hat{\mathfrak{F}}^+_\omega(\mathbf{r}, \mathbf{r}') + \hat{V}^t(\mathbf{r})\hat{\mathfrak{F}}^+_\omega(\mathbf{r}, \mathbf{r}') + \hat{\Delta}^*(\mathbf{r})\hat{\mathfrak{G}}'_\omega(\mathbf{r}, \mathbf{r}') = 0. \tag{7}$$

The first of these equations determines the Green's function $\mathfrak{G}'_\omega(\mathbf{r}, \mathbf{r}')$ of the normal metal when there are impurity atoms present. One checks easily that the Green's function $\mathfrak{G}'_\omega(\mathbf{r}, \mathbf{r}')$, on the other hand, also satisfies the equation

$$(i\omega_n + \nabla'^2/2m + \mu)\hat{\mathfrak{G}}'_\omega(\mathbf{r}, \mathbf{r}') - \hat{\mathfrak{G}}'_\omega(\mathbf{r}, \mathbf{r}')\hat{V}(\mathbf{r}') = \delta(\mathbf{r}-\mathbf{r}'),$$

where the prime on the ∇^2 indicates that the differentiation is with respect to the coordinates $\mathbf{r}'$ in $\mathfrak{G}'_\omega(\mathbf{r}, \mathbf{r}')$. Comparing this last equation with the second of Eqs. (7) we find that $\mathfrak{F}^+_\omega(\mathbf{r}, \mathbf{r}')$ can be expressed in terms of Δ^* as follows

$$(\hat{\mathfrak{F}}^+_\omega(\mathbf{r}, \mathbf{r}'))_{\alpha\beta} = \int \mathfrak{G}'_{-\omega\gamma\alpha}(\mathbf{s}, \mathbf{r})\Delta^*_{\gamma\lambda}(\mathbf{s})\mathfrak{G}'_{\omega\lambda\beta}(\mathbf{s}, \mathbf{r}')\,d^3s. \tag{8}$$

Substituting this expression into (3) we get an equation for the function $\hat{\Delta}^*(\mathbf{r})$. The temperature at which this equation has for the first time a non-vanishing solution is just the transition temperature T_c which interests us:

$$\Delta^*_{\alpha\beta}(\mathbf{r}) = |g| T \sum_\omega \int \mathfrak{G}'_{-\omega\gamma\alpha}(\mathbf{s}, \mathbf{r})\Delta^*_{\gamma\lambda}(\mathbf{s})\mathfrak{G}'_{\omega\lambda\beta}(\mathbf{s}, \mathbf{r})\,d^3s. \tag{9}$$

We must average this equation over the positions of the impurity atoms. We developed in reference 5 a method which enabled us to perform such an averaging (see also reference 7). If on the diagrams for the Green's functions or for the other quantities we designate the scattering by an impurity atom by a cross, we see easily that to each such cross there corresponds a factor $v(\mathbf{q})e^{i\mathbf{q}\cdot\mathbf{r}_a}$ where $\mathbf{q}$ is the momentum transferred and $\mathbf{r}_a$ the position of the impurity atom.* The average over $\mathbf{r}_a$ vanishes when $\mathbf{q} \neq 0$ unless this diagram contains at least one scattering at $\mathbf{r}_a$ with $\mathbf{q}' = -\mathbf{q}$. In the latter case the result of the averaging will be

$$n|v(\mathbf{q})|^2, \tag{10}$$

where n is the number of impurity atoms per unit volume. It thus turns out that the averaging over the impurity-atom positions can be put in the form of a special diagram technique, by joining two identical atoms on the diagram with a dotted line, and this corresponds to a factor (10) in the matrix elements.

When taking the average of Eq. (9) one can independently take the average of $\Delta^*(\mathbf{s})$ and of the product of the two Green's functions on the right-hand side. One can directly check that this is correct. It can, however, also be seen from the fact that $\Delta^*(\mathbf{s})$, which is a constant when there

*We use here the Born approximation for the scattering of the electrons by an impurity atom. The physical results are independent of this approximation and we can therefore replace everywhere the Fourier component of the impurity potential by the corresponding scattering amplitude.

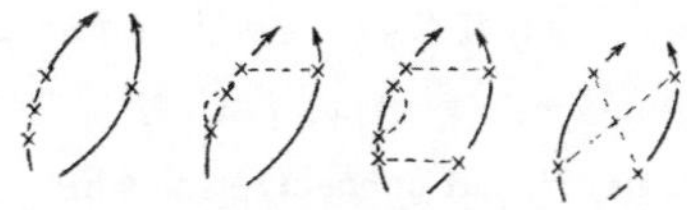

FIG. 1

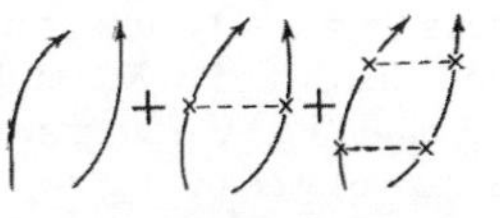

FIG. 2

are no impurities, is only changed near the impurity atoms while $\mathfrak{G}_\omega(\mathbf{r}, \mathbf{s})$ is a fast oscillating function. The result is

$$\overline{\Delta^*_{\alpha\beta}} = |g| T \sum_\omega \int \overline{\Delta^*_{\gamma\lambda} \mathfrak{G}'_{-\omega\gamma\alpha}(\mathbf{s}, \mathbf{r}) \mathfrak{G}'_{\omega\lambda\beta}(\mathbf{s},\mathbf{r})}\, d^3s. \quad (11)$$

The product of the two Green's functions $\mathfrak{G}'$ in (11) is depicted in Fig. 1a. The averaging over the impurity positions leads to the diagrams of Figs. 1b, c, and d. In this case, however, the dotted line no longer corresponds to a factor (10), but to a factor

$$n v_{\alpha\beta}(\mathbf{q}) v_{\gamma\lambda}(-\mathbf{q}),$$

since the electron — impurity interaction contains, according to (1), an exchange part.

Apart from averaging over the impurity-atom coordinates, we must still average over the orientation of their spins. We assume that the spins are oriented arbitrarily relative to the lattice and that there is no correlation whatever between them. Therefore

$$\overline{S} = 0; \qquad \overline{S_i S_k} = \tfrac{1}{3} S(S+1)\delta_{ik}.$$

We shall discuss later on the basis of such an assumption. Assuming that the averaging sign in (11) includes also an average over the orientation of the spins of all impurity atoms, we shall let each dotted line in Fig. 1 correspond to a factor

$$n\left(|u_1(\mathbf{q})|^2 + \tfrac{1}{3} S(S+1)|u_2(\mathbf{q})|^2 \sigma_{\alpha\beta}\sigma_{\gamma\lambda}\right). \quad (12)$$

It was shown in reference 5 that diagrams such as the ones of Fig. 1d, which contain intersections of dotted lines, and also any diagrams in which a dotted line includes a vertex with a momentum transfer of an electron line of the order of p_0, give a small contribution $\sim 1/p_0 l$, where l is the mean free path, and can therefore be discarded. Dotted lines such as the ones of Fig. 1b which join two crosses on the same electron line give a contribution to the averaged Green's function $\mathfrak{G}'_\omega(\mathbf{r}, \mathbf{r}')$ [we bear in mind that here $\mathfrak{G}'_\omega(\mathbf{r}, \mathbf{r}')$ refers to the normal metal]. One can easily check that in complete analogy with the result of reference 5 the sum of all such diagrams leads to the Green's function

$$\overline{\mathfrak{G}'_\omega(\mathbf{r}, \mathbf{r}')} = \widetilde{\mathfrak{G}}_\omega(\mathbf{r} - \mathbf{r}') = \mathfrak{G}_\omega(\mathbf{r} - \mathbf{r}') \exp(-|\mathbf{r} - \mathbf{r}'|/2l_1), \quad (13)$$

where $\mathfrak{G}_\omega(\mathbf{r} - \mathbf{r}')$ is the Green's function of the normal metal when there are no impurity atoms. The Fourier components of the function

$$\widetilde{\mathfrak{G}}_\omega(\mathbf{r} - \mathbf{r}') = (2\pi)^{-3} \int \widetilde{\mathfrak{G}}_\omega(\mathbf{p}) e^{i\mathbf{p}(\mathbf{r}-\mathbf{r}')} d^3p$$

are equal to

$$\widetilde{\mathfrak{G}}_\omega(\mathbf{p}) = [i\omega\eta_1 - \xi]^{-1}, \quad (14)$$

where

$$\eta_1 = 1 + 1/2\tau_1|\omega|, \qquad \xi = v(p - p_0), \qquad v = p_0/m,$$

$$\frac{1}{\tau_1} = \frac{nmp_0}{(2\pi)^2} \int \left\{|u_1(\mathbf{q})|^2 + |u_2(\mathbf{q})|^2 \frac{1}{3} S(S+1)\right\} do, \quad l_1 = v\tau_1.$$

The evaluation of the average of the product of the two Green's functions in (11) can thus be reduced to a summation of the "ladder" diagrams depicted in Fig. 2. In those diagrams the averaged electron Green's functions (13) and (14) occur as electron lines.

It follows from (5) and (6) that $\Delta^*_{\alpha\beta}$ is of the form

$$\Delta^*_{\alpha\beta} = \Delta^* g_{\alpha\beta}, \qquad g_{\alpha\beta} = \begin{pmatrix} 0 & 1 \\ -1 & 0 \end{pmatrix}_{\alpha\beta}, \quad (15)$$

where $g_{\alpha\beta}$ is the covariant metric tensor in spinor algebra.[8] We can now cancel the constant Δ^* in (11). We introduce the symbol $K_{\omega\alpha\beta}(\mathbf{p}_1, \mathbf{p}_2)$ for the Fourier component of the expression

$$\overline{\mathfrak{G}_{-\omega\gamma\alpha}(\mathbf{s}, \mathbf{r}) g_{\gamma\lambda} \mathfrak{G}_{\omega\lambda\beta}(\mathbf{s},\mathbf{r}')} = (2\pi)^{-6} \int K_{\omega\alpha\beta}(\mathbf{p}_1, \mathbf{p}_2) \exp\{i\mathbf{p}_1(\mathbf{s} - \mathbf{r}) + i\mathbf{p}_2(\mathbf{s} - \mathbf{r}')\}\, d^3p_1 d^3p_2. \quad (16)$$

The sum of all diagrams of Fig. 2 can be obtained by solving the following obvious integral equation

$$K_{\omega\alpha\beta}(\mathbf{p}_1, \mathbf{p}_2) = \widetilde{\mathfrak{G}}_\omega(\mathbf{p}_1) \widetilde{\mathfrak{G}}_{-\omega}(\mathbf{p}_2) \Big[g_{\alpha\beta} + \frac{n}{(2\pi)^3} \int \Big[|u_1|^2 + \frac{1}{3} S(S+1)|u_2|^2 (\sigma_{\lambda\alpha}\sigma_{\rho\beta})\Big] K_{\lambda\rho\omega}(\mathbf{p}'_1, \mathbf{p}'_2)\, d^3p'\Big],$$

$$\mathbf{p}_1 - \mathbf{p}'_1 = \mathbf{p}'_2 - \mathbf{p}_2. \quad (17)$$

It is clear from this that $K_{\omega\alpha\beta}$ is proportional to $g_{\alpha\beta}$. One verifies easily by using spinor algebra that

$$\hat{g}^{-1}\hat{\sigma}^t\hat{g} = -\hat{\sigma}. \quad (18)$$

If $K_{\omega\alpha\beta}(\mathbf{p}_1, \mathbf{p}_2) = K_\omega(\mathbf{p}_1, \mathbf{p}_2) g_{\alpha\beta}$, we have thus

$$K_\omega(\mathbf{p}_1, \mathbf{p}_2) = \widetilde{\mathfrak{G}}_\omega(\mathbf{p}_1) \widetilde{\mathfrak{G}}_{-\omega}(\mathbf{p}_2) \Big[1 + \frac{n}{(2\pi)^3} \int \Big[|u_1|^2 - \frac{S(S+1)}{4}|u_2|^2\Big] K_\omega(\mathbf{p}'_1, \mathbf{p}'_2) d^3p'\Big]. \quad (19)$$

We are, according to (11), interested in the values for $p_1 = -p_2$. Substituting Eq. (14) for $\mathfrak{G}_\omega(p)$ into (19) and denoting the integral on the right-hand side of this expression by L_ω we find

$$K_\omega(p, q-p) = \mathfrak{G}_\omega(p)\tilde{\mathfrak{G}}_{-\omega}(q-p)[1+L_\omega]. \qquad (20)$$

For L_ω we get the following expression

$$L_\omega = \frac{1}{2\tau_2\omega\eta_1}\left[1-\frac{1}{2\tau_2\omega\eta_1}\right]^{-1} = \frac{1}{2\tau_2\omega\eta_s}, \qquad (21)$$

where

$$\frac{1}{\tau_2} = \frac{nmp_0}{(2\pi)^2}\int\left[|u_1|^2 - \frac{1}{4}S(S+1)|u_2|^2\right]do,$$

$$\eta_s = 1 + 1/\tau_s|\omega|; \qquad 1/\tau_s = 1/2\tau_1 - 1/2\tau_2 > 0.$$

Substituting the expression for L_ω into (20) and then substituting $K_\omega(p_1, p_2)$ into the kernel of Eq. (11) we get, if we use (21), the following relation

$$1 = \frac{|g|T_c}{(2\pi)^3}\sum_\omega\int\frac{\eta_1/\eta_s}{\omega^2\eta_1^2+\xi^2}d^3p; \qquad \omega_n = (2n+1)\pi T_c.$$

The sum and the integral over d^3p in this expression diverge for large ω and ξ. In the papers by Bardeen, Cooper, and Schrieffer[9] and Bogolyubov[10] this divergence was removed by cutting off the electron energies at $|\xi| = \omega_D$. It is convenient to subtract from the right-hand side of the relation which we have obtained the term that does not contain τ. The remaining difference can be integrated over ξ. Putting $1/|g| = (mp_0/2\pi^2)\ln(2\omega_D\gamma/\pi T_{c0})$, where T_{c0} is the transition temperature of the pure superconductor, we get

$$\frac{1}{2}\ln\frac{T_{c0}}{T_c} = \frac{1}{\pi T_c\tau_s}\sum_{n>0}\frac{1}{(2n+1)(2n+1+1/\tau_s\pi T_c)}.$$

We can express the expression on the right-hand side in terms of $\psi(x)$, the derivative of the logarithm of the Γ-function

$$\ln\frac{T_{c0}}{T_c} = \psi\left(\frac{1}{2}+\frac{\rho}{2}\right) - \psi\left(\frac{1}{2}\right), \qquad (22)$$

where $\rho = 1/\pi\tau_s T_c$. The shift in T_c is thus connected only with the exchange part of the interaction and depends on the parameter $1/\tau_s > 0$. Introducing f_+ and f_-, the scattering amplitudes of an electron with total angular momentum $S + 1/2$ and $S - 1/2$, we can express $1/\tau_s$ in the form

$$\frac{1}{\tau_s} = \frac{nmp_0}{2\pi^2}\frac{S(S+1)}{(2S+1)^2}\int|f_+ - f_-|^2 do.$$

If $\rho \ll 1$, i.e., if the impurity concentration is low, we have

$$T_c = T_{c0} - \pi/4\tau_s. \qquad (23)$$

At low impurity concentrations the transition temperature decreases proportionally to the concentration.

If $\rho \gg 1$, i.e., if $T_c/T_{c0} \ll 1$, we have

$$T_c^2 = (6/\pi^2\tau_s^2)\ln(\pi T_{c0}\tau_s/2\gamma). \qquad (24)$$

At a certain critical concentration which is determined by the condition

$$\tau_{s\,cr} = 2\gamma/\pi T_{c0}, \qquad (25)$$

superconductivity vanishes at all temperatures in a system with paramagnetic impurities. The "critical mean free path" $l_{s\,cr} = v\tau_{s\,cr} \sim 10^{-4}$ cm. One should, however, bear in mind that the magnitude of the exchange interaction u_2 in (1) is always less than the non-exchange part so that the value of l_s is roughly two orders of magnitude longer than the mean free path which is determined, say, by the conductivity of the normal metal. Thus, $l_s \sim 10^{-4}$ cm corresponds to $l \sim 10^{-6}$ cm, or, an impurity concentration of about 1%.

The experimental data of Matthias, Suhl, and Corenzwit[1] indicate apparently the linear dependence of the change in the critical temperature of a solution of gadolinium in lanthanum. If we use the slope of their straight line to construct the theoretical dependence according to Eqs. (23) and (22), the disagreement with the experimental data is nowhere more than 10%. One must regard such an agreement as good. At the same time there exists another paper by the same authors[4] in which they report that for titanium with several ferromagnetic elements as impurities the transition temperature increases with increasing number of impurity atoms at comparatively low concentrations (1 – 5%). This result is in direct contradiction with the results obtained in the foregoing. The reason for this is not clear at the present time. Of course, one must note the well-known difficulties of experiments with alloys particularly in connection with the poor solubility of elements of the iron type in titanium.* Moreover, the concentrations used in reference 4 were of the order of several percent.

Generally speaking, the theory given in the foregoing contained an expansion in terms of the order a/l, where a is the interatomic distance and l the mean free path, and it should therefore be valid at sufficiently low concentrations when the presence of impurity atoms should lead solely to the scattering of the conduction electrons. It is possible that the increase in T_c observed in reference 4 is a secondary effect caused by different interactions between the impurity atoms which are important at the concentrations studied. In any

*The authors are grateful to A. I. Shal'nikov for drawing their attention to this fact.

case, the actual nature of the impurity atoms was inessential for our consideration.

Let us now consider in more detail the possibility of averaging over the orientation of the spin of a separate atom. There have appeared recently a number of papers[1,11] in which the existence of "impurity" ferromagnetism was detected in a number of alloys and, in particular, it was asserted in reference 11 that in some cases it was possible that there existed simultaneously superconductivity and ferromagnetism caused by paramagnetic impurities. If, indeed, some ferromagnetic ordering of spins were possible, the averaging over the spin orientation which was used by us should be completely unfounded. In that case it would, in general, not be possible to consider the spins on separate impurity atoms to be independent. In the framework of the model considered here, however, we see no positive basis whatever for such an ordering at low impurity concentrations.

Ruderman and Kittel[12] have considered the interaction between two paramagnetic atoms caused in a metal by the exchange of conduction electrons. The fact that the long-range part of this interaction oscillates over distances of the order of atomic distances, is typical of this interaction. At low concentrations, where the distances between the impurity atoms are sufficiently large, this interaction will thus vanish on the average for all atoms and, even if it leads to a ferromagnetic transition, this will occur with an exponentially small Curie temperature.* As far as the direct dipole interaction between impurities is concerned, this is also too small.

On these grounds, our averaging over the spins is fully valid for low concentrations. Apparently, this is precisely the situation with the experiments by Matthias et al.[1] from which one can see that for low concentrations the dependence of the ferromagnetic transition temperature on the concentration is not at all linear. The crossing of the ferromagnetic and the superconducting transition temperatures of an alloy, which was observed in reference 11, occurred when the concentration of one of the components was about 6%. Such a concentration is no longer low and all that has been said in the foregoing is inapplicable to it. As far as the nature of impurity ferromagnetism is concerned for such a system, it is probable that this is caused by the above-mentioned exchange interaction between impurities, which is due to the conduction electrons.

We must draw attention to the fact that order-of-magnitude estimates of all the quantities lead, as was shown in the foregoing, to the conclusion that the whole phenomenon breaks up in the concentration range on the order of 1%. The magnitude of the interval in which T_c changes with the impurity concentration as determined from our theory depends thus strongly on the nature of the effect studied or, to be more precise, on the magnitude of the exchange interaction u_2 in (1). The larger u_2, the faster will the T_c corresponding to the superconducting transition decrease and the lower the concentration that corresponds to the critical value $\tau_{s\,cr}$ of (25). It follows in any case from (23) that for low concentrations T_c must decrease linearly with increasing concentration. As we have said already, the fact that this was not observed in reference 11 shows that it is necessary to study those alloys at even lower concentrations.

2. THE THERMODYNAMIC AND ELECTROMAGNETIC PROPERTIES OF ALLOYS NEAR THE CRITICAL CONCENTRATION

We now turn to a study of the properties of alloys with paramagnetic impurities. Of most interest will be the results obtained in the concentration range where the temperature of the superconducting transition is small compared with T_{c0} of a pure superconductor, i.e., when $\tau_s \approx \tau_{s\,cr}$. In that range the quantities Δ and $\mathfrak{F}^+$ are small ($\Delta\tau_{s\,cr} \ll 1$). We write Eqs. (2) in yet another form, including in them an additional magnetic field

$$\left\{i\omega_n + \frac{1}{2m}\left(\frac{\partial}{\partial \mathbf{r}} - ie\mathbf{A}\right)^2 + \mu\right\}\hat{\mathfrak{G}}_\omega(\mathbf{r},\mathbf{r}') - \hat{V}\hat{\mathfrak{G}}_\omega(\mathbf{r},\mathbf{r}') + \hat{\Delta}(\mathbf{r})\hat{\mathfrak{F}}^+_\omega(\mathbf{r},\mathbf{r}') = \delta(\mathbf{r}-\mathbf{r}'),$$

$$\left\{i\omega_n - \frac{1}{2m}\left(\frac{\partial}{\partial \mathbf{r}} + ie\mathbf{A}\right)^2 - \mu\right\}\hat{\mathfrak{F}}^+_\omega(\mathbf{r},\mathbf{r}') + \hat{V}'\hat{\mathfrak{F}}^+_\omega(\mathbf{r},\mathbf{r}') + \hat{\Delta}^*(\mathbf{r})\hat{\mathfrak{G}}_\omega(\mathbf{r},\mathbf{r}') = 0. \quad (26)$$

In the region under consideration we can expand Eqs. (26) in terms of the quantity Δ. The corresponding calculations are similar to the derivation, by one of the authors,[7] of the equations of the Ginzburg-Landau theory[14] near T_c for superconducting alloys.

Introducing the Green's function $\mathfrak{G}'_\omega(\mathbf{r},\mathbf{r}')$ of the normal metal, which satisfies the equation

$$\left\{i\omega_n + \frac{1}{2m}\left(\frac{\partial}{\partial \mathbf{r}} - ie\mathbf{A}\right)^2 + \mu\right\}\hat{\mathfrak{G}}'_\omega(\mathbf{r},\mathbf{r}') - \hat{V}\hat{\mathfrak{G}}'_\omega(\mathbf{r},\mathbf{r}') = \delta(\mathbf{r}-\mathbf{r}'),$$

*We feel in this connection that the attempt by Brout[13] to explain impurity ferromagnetism by a model where the Hamiltonian contains a constant term proportional to $\Sigma S_i S_k$, where S_i are the spins of the impurity atoms, is incorrect.

we can use it to write for $\mathfrak{F}^+_\omega$ and $\mathfrak{G}_\omega$ up to terms of the third and second order in $|\Delta|$ respectively

$$\mathfrak{F}^+_{\omega\alpha\beta}(\mathbf{r},\mathbf{r}') = \int \mathfrak{G}'_{-\omega\lambda\alpha}(\mathbf{s},\mathbf{r})\,\Delta^*_{\lambda\nu}(\mathbf{s})\,\mathfrak{G}'_{\omega\nu\beta}(\mathbf{s},\mathbf{r}')\,d^3s$$
$$-\iiint \mathfrak{G}'_{-\omega\lambda\alpha}(\mathbf{s},\mathbf{r})\,\Delta^*_{\lambda\nu}(\mathbf{s})\,\mathfrak{G}'_{\omega\nu k}(\mathbf{s},\mathbf{l})\,\Delta_{ki}(\mathbf{l})\,\mathfrak{G}'_{-\omega\varkappa i}(\mathbf{m},\mathbf{l})$$
$$\times\Delta^*_{\varkappa\nu}(\mathbf{m})\,\mathfrak{G}'_{\omega\nu\beta}(\mathbf{m},\mathbf{r}')\,d^3m\,d^3l\,d^3s,$$

$$\mathfrak{G}_{\omega\alpha\beta}(\mathbf{r},\mathbf{r}') = \mathfrak{G}'_{\omega\alpha\beta}(\mathbf{r},\mathbf{r}') - \iint \mathfrak{G}'_{\alpha k}(\mathbf{r},\mathbf{l})\,\Delta_{ki}(\mathbf{l})\,\mathfrak{G}'_{-\omega\varkappa i}$$
$$\times(\mathbf{m},\mathbf{l})\,\Delta^*_{\varkappa\rho}(\mathbf{m})\,\mathfrak{G}'_{\omega\rho\beta}(\mathbf{m},\mathbf{r}')\,d^3m\,d^3l. \qquad (27)$$

Substituting Eq. (27) for $\mathfrak{F}^+$ into (3) we get an equation to determine $\hat{\Delta}^*(\mathbf{r})$

$$\Delta^*_{\alpha\beta}(\mathbf{r}) = |g|\,T\sum_\omega\Big\{\int \mathfrak{G}'_{-\omega\lambda\alpha}(\mathbf{s},\mathbf{r})\,\Delta^*_{\lambda\nu}(\mathbf{s})\,\mathfrak{G}'_{\omega\nu\beta}(\mathbf{s},\mathbf{r})\,d^3s$$
$$-\iiint \mathfrak{G}'_{-\omega\lambda\alpha}(\mathbf{s},\mathbf{r})\,\Delta^*_{\lambda\nu}(\mathbf{s})\,\mathfrak{G}'_{\omega\nu k}(\mathbf{s},\mathbf{l})\,\Delta_{ki}(\mathbf{l})\,\mathfrak{G}'_{-\omega\varkappa i}(\mathbf{m},\mathbf{l})$$
$$\times\,\Delta^*_{\varkappa\nu}(\mathbf{m})\,\mathfrak{G}'_{\omega\nu\beta}(\mathbf{m},\mathbf{r})\,d^3m\,d^3l\,d^3s\Big\}. \qquad (28)$$

As far as the dependence of the Green's function $\mathfrak{G}'_{\omega\alpha\beta}(\mathbf{s},\mathbf{r})$ on the magnetic field is concerned, it was established in reference 7 that since the screening radius of the electron in a magnetic field is large compared with the mean free path, one can write it in the form

$$\mathfrak{G}'_{\omega\alpha\beta}(\mathbf{s},\mathbf{r}) = \exp\{iec^{-1}(\mathbf{A},\mathbf{s}-\mathbf{r})\}\,\mathfrak{G}^0_{\omega\alpha\beta}(\mathbf{s},\mathbf{r}), \qquad (29)$$

where $\mathfrak{G}^0_{\omega\alpha\beta}(\mathbf{s},\mathbf{r})$ is the Green's function of the normal electrons when there is no field. The concentration range $n \sim n_{cr}$ is characterized, as we have seen, by the quantities $1/\tau_s \sim T_{c0}$. As has been stated above, the magnitude of the exchange interaction is appreciably less than the non-exchange part in (1). The normal mean free path l is thus in that case appreciably less than the correlation parameter $\xi \sim \hbar v/2\pi T_{c0}$.

It is natural to assume that under those conditions a localized situation occurs, i.e., that the function $\Delta(\mathbf{r})$ and the field $A(\mathbf{r})$ change over distances which are larger than the characteristic distances in the integrals of Eq. (28). For this reason we can in the third-order term neglect the dependence of $\Delta(\mathbf{r})$ on the coordinates and that of the functions $\mathfrak{G}'$ on the field, as is confirmed by the final result, and in the second term we can as far as this dependence is concerned expand in a Taylor series about the point $\mathbf{r}$, limiting ourselves to terms of the second order in $|\mathbf{r}-\mathbf{s}|$. Using (29) we get as a result

$$\frac{1}{|g|}\Delta^*(\mathbf{r})\,g_{\alpha\beta} = \Delta^*(\mathbf{r})\,Q^{(2)}_{\alpha\beta}$$
$$+\frac{1}{6}\Big(\frac{\partial}{\partial\mathbf{r}} + 2ie\mathbf{A}(\mathbf{r})\Big)\Delta^*(\mathbf{r})\,Q^{(2)}_{\alpha\beta} + B_{\alpha\beta}|\Delta(\mathbf{r})|^2\Delta^*(\mathbf{r});$$
$$Q^{(1)}_{\alpha\beta} = T\sum_\omega\int \overline{\mathfrak{G}^0_{-\omega\lambda\alpha}(\mathbf{s},\mathbf{r})\,\hat{g}_{\lambda\nu}\,\mathfrak{G}^0_{\omega\nu\beta}(\mathbf{s},\mathbf{r})}\,d^3s, \qquad (30)$$

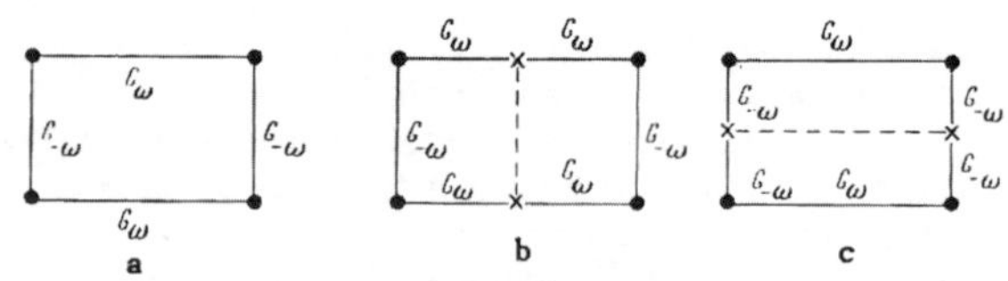

FIG. 3

$$Q^{(2)}_{\alpha\beta} = T\sum_\omega\int |\mathbf{r}-\mathbf{s}|^2\,\overline{\mathfrak{G}^0_{-\omega\lambda\alpha}(\mathbf{s},\mathbf{r})\,g_{\lambda\nu}\,\mathfrak{G}^0_{\omega\nu\beta}(\mathbf{s},\mathbf{r})}\,d^3s, \quad B_{\alpha\beta} = T$$
$$\times\sum_\omega\iiint \overline{\mathfrak{G}^0_{-\omega\lambda\alpha}(\mathbf{s},\mathbf{r})\,g_{\lambda\nu}\,\mathfrak{G}^0_{\omega\nu k}(\mathbf{s},\mathbf{l})\,g_{ki}\,\mathfrak{G}^0_{-\omega\varkappa i}(\mathbf{m},\mathbf{l})\,g_{\varkappa\nu}\,\mathfrak{G}^0_{\omega\nu\beta}(\mathbf{m},\mathbf{r})}$$
$$\times\,d^3m\,d^3l\,d^3s. \qquad (31)$$

As before, the bar indicates averaging over the impurity atom positions and over the orientation of their spins. The first two coefficients $Q^{(1)}_{\alpha\beta}$ and $Q^{(2)}_{\alpha\beta}$ can easily be connected with the kernel $K_{\omega\alpha\beta}(\mathbf{p}_1,\mathbf{p}_2)$ of (16). Indeed, we have

$$Q^{(1)}_{\alpha\beta} = \frac{T}{(2\pi)^3}\sum_\omega\int K_{\omega\alpha\beta}(\mathbf{p},-\mathbf{p})\,d^3p,$$

$$Q^{(2)}_{\alpha\beta} = -\frac{T}{(2\pi)^3}\sum_\omega\int\Big[\frac{\partial^2}{\partial q_i^2}K_{\omega\alpha\beta}(\mathbf{p},\mathbf{q}-\mathbf{p})\Big]_{\mathbf{q}=0}d^3p.$$

To find them we must thus solve Eqs. (19) and (20) up to terms of second order in q.

The product of four $\mathfrak{G}$-functions in B is schematically depicted in Fig. 3a. It follows from (20) and (21) that we average over the impurity atom positions first by replacing the lines in Fig. 3a by averaged Green's functions $\tilde{\mathfrak{G}}_\omega(\mathbf{p})$, and by replacing unity in the vertex by the ratio η_1/η_s. Moreover, the averaging gives rise to the two new diagrams of Figs. 3b and c, in which a dotted line joins two Green's functions with the same frequency. One verifies easily that all other diagrams are either equal to zero, (because all the poles lie in one half-plane) when integrating over $\xi = v(p - p_0)$ or contain intersecting dotted lines and their contribution can be neglected for the reasons given at the start of this paper. The actual determination of the coefficients $Q^{(1)}$, $Q^{(2)}$, and B requires elementary, but somewhat cumbersome algebraic calculations which we shall omit here.

As in (18), it is extremely essential to maintain the correct order of the spinor indices in the calculations. If we assume that in (1) the exchange part of the interaction is small compared with the non-exchange part, we get the final form of Eq. (30) in the temperature range $T \ll T_{c0}$ and range of concentrations $|n - n_{cr}| \ll n_{cr}$

$$\left\{\pi^2 (T^2 - T_c^2) + \frac{|\Delta|^2}{2}\right\} \Delta^* + 2\left(\frac{\partial}{\partial r} + 2ie\mathbf{A}\right)^2 \Delta^* v^2 \frac{\tau_{tr}}{\tau_s} = 0, \quad (32)$$

where τ_{tr} is the "transport" collision time, which enters in the conductivity of the normal metal

$$\sigma = Ne^2\tau_{tr}/m.$$

In the approximation considered we have

$$\frac{1}{\tau_{tr}} = \frac{nmp_0}{(2\pi)^2}\int |u_1|^2 (1 - \cos\theta)\, do.$$

When there is no field

$$\Delta^2 = 2\pi^2 (T_c^2 - T^2). \quad (33)$$

To determine the thermodynamic characteristics of the system we use the well-known formula for the variation of the thermodynamic potential

$$\delta\Omega / \delta g = g^{-1} \langle H_{int} \rangle.$$

We get from this[15]

$$\Omega_s - \Omega_n = \int_0^{|g|} \delta(1/|g|)\, \Delta^2. \quad (34)$$

Using the relation $\delta(1/|g|) = -(mp_0/2\pi^2)\delta T_{c0}/T_{c0}$ and Eq. (32) we get

$$F_s - F_n = -\frac{mp_0}{2\pi^2}\frac{\tau_s^2 \Delta^4}{24} = -\frac{mp_0}{2\pi^2}\frac{\tau_s^2 \pi^4}{6}(T_c^2 - T^2)^2. \quad (35)$$

The critical field, defined in the usual way, is equal to

$$H_c = \frac{2\gamma}{T_{c0}}\sqrt{\frac{2mp_0\pi}{3}}\left(1 - \frac{T^2}{T_c^2}\right) = H_{c0}\left(1 - \frac{T^2}{T_c^2}\right).$$

We find from (35) the ratio of the heat capacity in the superconducting phase to the heat capacity in the normal phase at $T = T_c$

$$\begin{aligned} C_s(T)/C_n(T_c) &= T/T_c - (\pi^2\tau_s^2/T_c)(T_c^2 T - 3T^3) \\ &= T/T_c - 12.4\,(T_c/T_{c0})^2 (T/T_c - 3T^3/T_c^3). \end{aligned} \quad (36)$$

When there is a field one must add to Eq. (34) the equation of the electrodynamics of superconductors. To do this one must substitute in the definition of the electromagnetic current

$$\mathbf{j}(\mathbf{r}) = \left\{\frac{ie}{m}(\nabla_r - \nabla_{r'})\mathfrak{G}(x, x') + \frac{2e^2\mathbf{A}(\mathbf{r})}{mc}\mathfrak{G}(x, x')\right\}_{r=r',\, t'=t+0}$$

the expression (27) for the Green's function. We shall not dwell on the calculations since they are, apart from the existence of spin at the impurities, similar to the calculations in reference 7. Performing the averaging and restricting ourselves to the first non-vanishing terms in $\mathbf{A}$ and Δ, we get

$$\mathbf{j} = \left\{\frac{ie}{m}\left(\Delta\frac{\partial\Delta^*}{\partial \mathbf{r}} - \Delta^*\frac{\partial\Delta}{\partial \mathbf{r}}\right) - \frac{4e^2}{mc}\mathbf{A}|\Delta|^2\right\}\frac{N\tau_s\tau_{tr}}{2\pi^2}. \quad (37)$$

Equations (32) and (37) together have the same form as the set of equations in the phenomenological Ginzburg-Landau theory.[14] The origin of the doubled elementary charge in these equations has a distinct physical meaning. It is related to the existence of pair correlations in present-day superconductivity theory.

In contrast to the case of pure superconductors, and also to the case of non-magnetic impurities,[7,15] these equations are applicable here in the whole temperature range of the superconducting phase. The temperature dependence enters into the different characteristics of the system through the factor $1 - T^2/T_c^2$ in Eq. (32). In particular, we get for the penetration depth of a weak magnetic field in superconductors

$$\begin{aligned} \delta &= \left[\frac{16\pi^3 Ne^2\tau_s\tau_{tr}T_c^2}{mc^2}\left(1 - \frac{T^2}{T_c^2}\right)\right]^{-1/2} \\ &= \frac{c}{4\pi T_c}\left[\frac{2\sigma\gamma}{T_{c0}}\left(1 - \frac{T^2}{T_c^2}\right)\right]^{-1/2}. \end{aligned} \quad (38)$$

For Eqs. (32) and (37) we can construct one dimensionless constant

$$\varkappa = 2\sqrt{2}eH\delta^2/\hbar c = ec(\pi\gamma)^{1/2}/4\sigma,$$

where γ is the coefficient in the linear law for the electronic part of the heat capacity of the normal metal. It is worth noting that as in reference 7 κ depends only on the parameters of the normal metal, while the dependence on τ_s disappears.

The most interesting result is that there is no exponential term in the heat capacity (36) as $T \to 0$. This indicates that there is no gap in the spectrum of such superconductors. This conclusion is confirmed also by the absence of a threshold in the absorption of electromagnetic radiation at $T = 0$. We can, namely, obtain by a method which we have used before[5] the following expression for the dielectric constant of paramagnetic alloys

$$\varepsilon(\omega) = -c^2/\omega^2\delta^2 + 4\pi i\sigma/\omega. \quad (39)$$

We study this problem now in more detail.

3. THE DEPENDENCE OF THE GAP IN THE SPECTRUM AT T = 0 ON THE IMPURITY CONCENTRATION AND THE HEAT CAPACITY AT LOW TEMPERATURES

Migdal and Galitskiĭ[16] have shown that when $\omega > 0$ the imaginary part of the Fourier component of the electron Green's function

$$G(x, x') = -i\langle T(\psi(x), \psi^+(x'))\rangle$$

can at absolute zero be written in the form

$$\operatorname{Im} G_\omega(\mathbf{p}) = \sum_n \rho_n(\mathbf{p})\,\delta(\omega - \omega_{n0}), \quad (\omega > 0),$$

where $\rho_n(\mathbf{p})$ is some real spectral function, and

$\omega_{n0} = E_n - E_0$ is the difference in energy between the n-th excited state and the ground state of the system. When there are impurities present the system has no single-particle spectrum, which is determined from the poles of the function $G_\omega(\mathbf{p})$. All the same, however, if the first excited state is separated from the ground state by a gap, the magnitude of the gap is determined by the value of the frequency ω at which the Green's function has a non-vanishing imaginary part for the first time.

One can show in the case of a superconductor that the same holds for the function $iF_\omega(\mathbf{p})$, where $F_\omega(\mathbf{p})$ is the Fourier component of $F^+(x, x') = \langle T(\psi^+(x), \psi^+(x'))\rangle$, or for the function $iF^+_{R\omega}(\mathbf{p})$, which is the Fourier component of the so-called retarded anticommutator of the two ψ^+ (the two functions are the same at $T = 0$ for $\omega > 0$). Bearing in mind the further calculation of the heat capacity, we consider the function $F^+_{R\omega}(\mathbf{p})$. Its characteristic property is its direct connection with the temperature-dependent function $\mathfrak{F}^+_{\omega_n}(\mathbf{p})$. Using methods similar to the ones used in reference 17, one can show that in the upper half-plane $iF^+_{R\omega}(\mathbf{p})$ is an analytic function that coincides with $\mathfrak{F}^+_{\omega_n}(\mathbf{p})$ in the points $\omega_n = i\pi T(2n+1)$ (when $n > 0$). We have solved the equations for the temperature-dependent Green's functions in the limiting case where Δ and $\mathfrak{F}^+$ were small. These equations can be obtained in the general case by the same method as was used by us for the case of normal superconductors.[5] A new aspect is here the occurrence of two relaxation times τ_1 and τ_2 and the dependence of the quantity $\Delta^* = |g|\,\mathfrak{F}^+(x, x)$ on the concentration of the paramagnetic impurities. The functions $\mathfrak{G}_{\omega_n}$ and $\mathfrak{F}^+_{\omega_n}$ obtained in this way are of the form

$$\mathfrak{G}_{\omega_n}(\mathbf{p}) = -\frac{\tilde{\omega}_n + \xi}{\tilde{\omega}_n^2 + \xi^2 + \tilde{\Delta}_n^2}, \qquad \mathfrak{F}^+_{\omega_n}(\mathbf{p}) = \frac{\tilde{\Delta}_n}{\tilde{\omega}_n^2 + \xi^2 + \tilde{\Delta}_n^2}, \tag{40}$$

where

$$\tilde{\omega}_n = \omega_n + \frac{1}{2\tau_1}\frac{u_n}{\sqrt{u_n^2+1}}, \qquad \tilde{\Delta}_n = \Delta + \frac{1}{2\tau_2}\frac{1}{\sqrt{u_n^2+1}}, \tag{41}$$

$$u_n = \frac{\tilde{\omega}_n}{\tilde{\Delta}_n}, \qquad \frac{\omega_n}{\Delta} = u_n\left(1 - \frac{1}{\tau_s\Delta}\frac{1}{\sqrt{u_n^2+1}}\right). \tag{42}$$

One can obtain similar equations using the technique applicable at absolute zero. We can determine the function $F^+_{R\omega}(\mathbf{p})$ which interests us as follows. If we formally replace ω_n by $-i\omega$ and at the same time $\tilde{\omega}_n$ by $-i\tilde{\omega}$, u_n by $-iu$, and $\tilde{\Delta}_n$ by $\tilde{\Delta}$ we get for u the following equation

$$\frac{\omega}{\Delta} = u\left(1 - \frac{1}{\tau_s\Delta}\frac{1}{\sqrt{1-u^2}}\right). \tag{42'}$$

The radical must be defined here as the analytical continuation of the positive square root $+\sqrt{1-u^2}$ for real u and $|u| < 1$ in the upper half-plane of the variable u. (One can verify that this definition is correct by comparing these formulae with the results obtained by the usual technique at the absolute zero.)

In actual fact we need for the following not the whole of the function $F^+_{R\omega}(\mathbf{p})$ but only some of its properties. It follows from Eq. (42′) that when $\tau_s\Delta > 1$ and when ω is sufficiently small there exists a real solution for u, with $|u| < 1$. The connection with the limiting case $1/\tau_s \to 0$ shows that in our case we must also choose just that solution. Studying the right-hand side of (42′) one sees easily that it has a maximum for

$$u_0 = [1 - (1/\tau_s\Delta)^{2/3}]^{1/2},$$

$$\omega_0 = \Delta[1 - (1/\tau_s\Delta)^{2/3}]^{3/2}, \tag{43}$$

and that therefore all solutions of Eq. (42′) are complex when $\omega > \omega_0$.

Since we must take the real solution of Eq. (42′) when ω is small, it follows that u becomes complex and $F_{R\omega}(\mathbf{p})$ acquires thereby an imaginary part when $\omega > \omega_0$. The quantity ω_0 thus determines the size of the gap. When the concentration is increased, the value of ω_0 diminishes and tends to zero when $1/\tau_s\Delta = 1$. The magnitude of Δ is thus not at all equal to the gap in the energy spectrum for paramagnetic alloys.

We find now the dependence of Δ on the impurity concentration. The definition of $\Delta^* = |g|\,\mathfrak{F}^+(x, x)$ yields the equation

$$\Delta = \frac{|g|T}{(2\pi)^3}\sum_n \int \frac{d^3p\,\tilde{\Delta}_n}{\tilde{\omega}_n^2 + \xi^2 + \tilde{\Delta}_n^2}. \tag{44}$$

Since we are interested in the case $T = 0$, we can change from summation over n to integration, using the formula

$$T\sum_n \to (2\pi)^{-1}\int d\Omega.$$

After this it is convenient to add and subtract under the integral sign the expression $\Delta_0/(\Omega^2 + \xi^2 + \Delta_0^2)$ (Δ_0 is the value of Δ for $T = 0$). The result is

$$\Delta_0 \ln\frac{\Delta_0}{\Delta_{00}} = \int_0^\infty d\Omega\left\{\frac{1}{\sqrt{u^2+1}} - \frac{1}{\sqrt{\Omega^2/\Delta_0^2+1}}\right\}$$

(Δ_{00} is the value of Δ_0 for a pure superconductor). It is convenient to use Eq. (42) and change in this integral the integration over $d\Omega$ to one over du. The result is

$$\ln \frac{\Delta_0}{\Delta_{00}} = -\frac{\pi}{4\tau_s\Delta_{00}}, \qquad \frac{1}{\tau_s\Delta_0} < 1;$$

$$\ln \frac{\Delta_0}{\Delta_{00}} = -\ln\left[\frac{1+\sqrt{1-(\tau_s\Delta_0)^2}}{\tau_s\Delta_0}\right] + \frac{1}{2}\sqrt{1-(\tau_s\Delta_0)^2} - \frac{1}{2\tau_s\Delta_0}\arcsin\tau_s\Delta_0,$$

$$\frac{1}{\tau_s\Delta_0} > 1. \qquad (45)$$

We have $1/\tau_{s\,cr} = \pi T_{c0}/2\gamma = \Delta_{00}/2$. When $\tau_s\Delta_0 = 1$ we have $\Delta_0 = e^{-\pi/4}\Delta_{00}$. From this it is clear that there is some value of the concentration which is lower than the critical one for which the gap in the energy spectrum (43) vanishes

$$n' = 2e^{-\pi/4}n_{cr} \approx 0.91\, n_{cr}. \qquad (46)$$

The fact that there exists a superconductor without a gap needs some explanations. We started from Eqs. (2) in our derivation. The possibility of describing a superconductor by such a set of equations for the functions $\mathfrak{G}$ and $\mathfrak{F}^+$ is based upon the hypothesis that several pair formations of electrons show a Bose condensation.[6] In a pure superconductor such formations are Cooper pairs, which are condensed in a state with total momentum equal to zero [$\mathfrak{F}^+(x, x')$ depends only on the difference between the coordinates]. Since the number of such pairs in an infinitely large system is arbitrarily large, the creation or annihilation of such a pair would not change the ground state of the system. This fact is indeed reflected in the existence of non-vanishing averages

$$\mathfrak{F}^+(x-x') = \langle T(\psi^+(x), \psi^+(x'))\rangle.$$

We have shown that when n is larger than the n′ given by (46) the functions $\mathfrak{F}^+$ are non-vanishing and finite, which confirms the validity of the assumption of a Bose condensation; the change in the gap in the spectrum (46) shows, however, that we are dealing here with a more complicated situation than in the case of a pure superconductor. There remains something obscure in the overall picture.

It is of interest to trace how the vanishing of the gap with increasing impurity concentration affects the temperature dependence of the electronic heat capacity in the range $T \ll T_c$. We first transform somewhat Eq. (34) for the thermodynamic potential. We subtract from (34) the analogous equation for a metal in which a different value for the coupling constant has been taken, $|g_1| > |g|$. This leads to

$$\Omega_s - \Omega_{s1} = -\int_{|g|}^{|g_1|} \delta\left(\frac{1}{|g|}\right)\Delta^2.$$

We use the equation for the gap, which we write in the form

$$\Delta = \frac{|g|\, mp_0}{2\pi^2} f_T(\Delta), \qquad (47)$$

and we change from an integral over $|g|$ to one over Δ. Clearly

$$\Omega_s - \Omega_{s1} = -\frac{mp_0}{2\pi^2}\int_{\Delta}^{\Delta_1} [f_T'(y)\,y - f_T(y)]\,dy.$$

We now subtract from this relation its value for T = 0.

We assume that $\Delta_1(0) \gg \Delta(0)$. In that case $\Omega_{s1}(T) - \Omega_{s1}(0) \to 0$. We get thus

$$\Omega_s(T) - \Omega_s(0) = -\frac{mp_0}{2\pi^2}\left\{\int_{\Delta}^{\Delta_1}[f_T'(y)\,y - f_T(y)]\,dy - \int_{\Delta_0}^{\Delta_1}[f_0'(y)\,y - f_0(y)]\,dy\right\}_{\Delta_1\to\infty},$$

where $\Delta_0 \equiv \Delta(0)$ and the $f_0(y)$ are the values of the corresponding functions at T = 0. At low temperatures we can write

$$f_T(\Delta) = f_0(\Delta) + \delta f_T(\Delta), \qquad \Delta = \Delta_0 + \delta\Delta(T);$$

the additional temperature-dependent terms are small compared with the values at absolute zero. Using Eq. (47) we can express $\delta\Delta(T)$ in terms of $\delta f_T(\Delta)$:

$$\delta\Delta(T) = \Delta_0\delta f_T(\Delta)/(f_0 - f_0'\Delta_0).$$

Substituting this into the equation for the thermodynamic potential we find

$$\Omega_s(T) - \Omega_s(0) = \frac{mp_0}{\pi^2}\int_{\Delta_0}^{\infty}\delta f_T(y)\,dy. \qquad (48)$$

It is thus sufficient for us to find merely the temperature-dependent addition to the right-hand side of Eq. (44) for Δ.

We start with the simplest case $\tau_s\Delta_0 < 1$. We note that the expression under the summation sign in (44) is symmetric with respect to a change in the sign of n. As before we go over from a summation to an integration but we now take into account a correction term using the well-known formula

$$T\sum_{n>0}\varphi(\omega_n) \to \frac{1}{2\pi}\left\{\int_0^{\infty}\varphi(\Omega)\,d\Omega - \frac{\pi^2T^2}{6}[\varphi'(\infty) - \varphi'(0)]\right\}.$$

The first term here is temperature-independent and the correction is proportional to T^2. After the necessary calculations we get

$$\delta f_T(\Delta) = -\pi^2T^2\tau_s/6\sqrt{(1/\tau_s\Delta)^2 - 1}.$$

When substituting this into (48) we must take into account (vide infra) that the region where $\delta f \sim T^2$

extends only to $\Delta = 1/\tau_s$ and furthermore that this function depends exponentially on T, i.e., it tends to zero, even when we take terms in T^2 into account. The integral over Δ must thus only be taken up to that limit. As a result we get for the heat capacity

$$C_s = \frac{1}{3}\, mp_0 \sqrt{1-(\tau_s\Delta_0)^2}\; T. \tag{49}$$

When $\Delta_0\tau_s \ll 1$ this formula is the same as (36) in the limit $T \ll T_c$.

In the case $\Delta_0\tau_s > 1$ the method given here can not be applied, since it yields only an extra term which is quadratic in T in the thermodynamic potential and which is not present here. If we were to apply this method to the case we are now considering, we would only get the result that there is no such extra term. This is not surprising, since the presence of a gap in the energy spectrum compels us to assume that when $\Delta_0\tau_s > 1$ the extra term will be exponential. Because of this we shall apply another method.

If in Eq. (44) we change to a summation over positive n only, we can replace this sum by a contour integral. The right-hand side of Eq. (44) will be of the form

$$-\frac{1}{2\pi i}\int_C d\omega \tanh\frac{\omega}{2T}\int_{-\omega_D}^{\omega_D} d\xi\, \frac{mp_0}{2\pi^2}\,\mathfrak{F}(-i\omega, \xi).$$

The contour of integration is given in Fig. 4. We now distort the contour and take it along the real axis. This is possible since it was shown that the integral over ω can be considered to be the integral of the function $iF_{R\omega}(\mathbf{p})$ which is analytical in the upper half-plane. For real ω which are such that $|\omega| < \omega_0$ this function is real and symmetric with respect to replacing ω by $-\omega$. From this it follows that the real part of $iF_{R\omega}(\mathbf{p})$ on the real axis is symmetric with respect to a change in sign in ω while its imaginary part is antisymmetric. All that remains of this integral is therefore

$$\frac{1}{\pi}\,\mathrm{Im}\left\{\int_0^\infty d\omega \tanh\frac{\omega}{2T}\int_{-\omega_D}^{\omega_D}\frac{mp_0}{2\pi^2}\,d\xi\,\frac{\tilde{\Delta}}{-\tilde{\omega}^2+\xi^2+\tilde{\Delta}^2}\right\}.$$

From the foregoing it follows that the imaginary part of the function $iF_{R\omega}(\mathbf{p})$ appears only starting with $\omega = \omega_0$. If $T \ll \omega_0$ we can thus replace $\tanh \omega/2T$ by $1 - e^{-\omega/T}$. The term 1 gives a contribution to $f_T(\Delta)$ which is independent of the temperature. $\delta f_T(\Delta)$ is thus determined by substituting $-2e^{-\omega/T}$ for $\tanh \omega/2T$. The integral over ξ and ω is then formally convergent and we can first integrate over ξ. In this way we find

$$\delta f_T(\Delta) = -2\int_{\omega_0}^{\infty} e^{-\omega/T}\,\mathrm{Im}\left[\frac{1}{\sqrt{1-u^2}}\right] d\omega.$$

The $\sqrt{1-u^2}$ under the integral sign must be defined, as a function of ω, as the analytical continuation of its real positive value with a cut $\omega < \omega_0$ through the upper half-plane. Because of the presence of an exponential factor the most essential part of the integral will come from the neighborhood of the point ω_0. Expanding Eq. (42′) in a power series in $u - u_0$ and taking the definition of $\sqrt{1-u^2}$ for $\omega > \omega_0$ into account, we get

$$\mathrm{Im}\left[\frac{1}{\sqrt{1-u^2}}\right] = \sqrt{\omega-\omega_0}\left(\frac{3\Delta_0}{2}\right)^{-1/2}\left[1-\left(\frac{1}{\tau_s\Delta_0}\right)^{2/3}\right]^{1/4}(\Delta_0\tau_s)^{2/3}.$$

Substituting this into the integral for $\delta f_T(\Delta)$ we find

$$\delta f_T(\Delta) = -\sqrt{\frac{2\pi}{3}}\, T^{3/2}\tau_s^{1/2}\left[(\Delta_0\tau_s)^{2/3}-1\right]^{1/4} e^{-\omega_0/T}.$$

Substituting this expression into (48) and integrating over Δ we can find the temperature dependent extra term in the thermodynamic potential. We get for the heat capacity

$$C_s(T) = \left(\frac{mp_0}{\pi^2}\right)\sqrt{\frac{2\pi}{3}}\,\tau_s^{2/3}\Delta_0^{1/3}\left[\Delta_0^{2/3} - \tau_s^{-2/3}\right]^{11/4}\exp\left\{-\frac{1}{T}\left[\Delta_0^{2/3}-\tau_s^{-2/3}\right]^{3/2}\right\}. \tag{50}$$

One sees easily that this formula does not go over into the formula for the pure superconductor in the limit as $\tau_s \to \infty$. This is connected with the fact that the possibility of expanding Eq. (42′) in powers of $u - u_0$ imposes a restriction on the region of applicability of Eq. (50). The latter is obtained when one estimates the next term of such an expansion. This gives

$$T \ll \Delta_0^{1/3}\tau_s^{-2/3}.$$

At low concentrations this condition is stronger than the condition $T \ll \Delta_0$ which is necessary to separate off the exponential region of the heat capacity. The connection between Eq. (50) and the equation for a pure superconductor could be obtained by considering the case $\tau_s\Delta_0 \gg 1$, $T \ll \Delta_0$, but we shall not perform here the appropriate calculations as we are interested in the basic influence of the vanishing of the gap in the spectrum on the heat capacity.

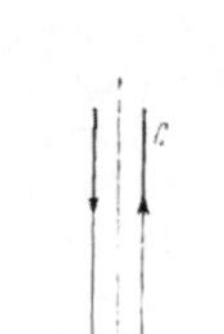

FIG. 4

It follows from Eq. (50) that when the impurity concentration increases the absolute magnitude of the exponent decreases while it does not vanish at $\Delta_0\tau_s = 1$. In the point $\Delta_0 = 1/\tau_s$ itself, where the exponential and the linear dependence come together one finds from calculating it that $C_s \sim T^{4/3}$. In view of the fact that this single point is not of great interest we shall not give the corresponding calculations or write out the coefficients.

In conclusion we express our gratitude to Academician L. D. Landau for many useful discussions of this paper.

[1] Matthias, Suhl, and Corenzwit, Phys. Rev. Letters **1**, 92 (1958).

[2] C. Herring, Physica **24**, S 184 (1958).

[3] H. Suhl and B. T. Matthias, Phys. Rev. **114**, 977 (1959).

[4] Matthias, Compton, Suhl, and Corenzwit, Phys. Rev. **115**, 1597 (1959).

[5] A. A. Abrikosov and L. P. Gor'kov, JETP **35**, 1558 (1958), Soviet Phys. JETP **8**, 1090 (1959); JETP **36**, 319 (1959), Soviet Phys. JETP **9**, 220 (1959).

[6] L. P. Gor'kov, JETP **34**, 735 (1958), Soviet Phys. JETP **7**, 505 (1958).

[7] L. P. Gor'kov, JETP **37**, 1407 (1959), Soviet Phys. JETP **10**, 998 (1960).

[8] L. D. Landau and E. M. Lifshitz, Quantum Mechanics, Pergamon Press, London, 1958, Sec. 54.

[9] Bardeen, Cooper, and Schrieffer, Phys. Rev. **108**, 1175 (1957).

[10] N. N. Bogolyubov, JETP **34**, 58 (1958), Soviet Phys. JETP **7**, 41 (1958).

[11] Matthias, Suhl, and Corenzwit, Phys. Rev. Letters **1**, 449 (1958).

[12] M. A. Ruderman and C. Kittel, Phys. Rev. **96**, 99 (1954).

[13] R. Brout, Phys. Rev. **115**, 824 (1959).

[14] V. L. Ginzburg and L. D. Landau, JETP **20**, 1064 (1950).

[15] L. P. Gor'kov, JETP **36**, 1918 (1959), Soviet Phys. JETP **9**, 1364 (1959).

[16] A. B. Migdal and V. M. Galitskiĭ, JETP **34**, 139 (1958), Soviet Phys. JETP **7**, 96 (1958).

[17] Abrikosov, Gor'kov, and Dzyaloshinskiĭ, JETP **36**, 900 (1959), Soviet Phys. JETP **9**, 636 (1959).

Translated by D. ter Haar
325

SOVIET PHYSICS JETP VOLUME 13, NUMBER 5 NOVEMBER, 1961

CONTRIBUTION TO THE THEORY OF SUPERFLUIDITY IN AN IMPERFECT FERMI GAS

L. P. GOR'KOV and T. K. MELIK-BARKHUDAROV

Institute of Physics Problems, Academy of Sciences, U.S.S.R.

Submitted to JETP editor December 21, 1960

J. Exptl. Theoret. Phys. (U.S.S.R.) **40**, 1452-1458 (May, 1961)

The Cooper effect in a low-density Fermi gas is studied. The magnitude of the transition temperature into the superfluid state is determined.

It is well known that the basis of superconductivity is the so-called Cooper effect[1] which consists in that the existence of an effective attraction between the electrons in a metal leads to the formation of electron pairs in a singlet state. The exchange of phonons between the electrons is the mechanism that causes the presence of such an attraction. We consider in the present paper the phenomenon of superfluidity in a low-density Fermi gas. The main methodological difference in this case consists in that in the theory of superconductivity the transition temperature is expressed in terms of a quantity of the order of the Debye temperature $\tilde{\omega} \ll \mu$, which is used as the natural parameter for a cutoff, while in a low-density Fermi-gas the only small parameter is its density, or more accurately, the quantity $f_0 p_0^2/v$, where f_0 is the s-wave scattering amplitude for interparticle scattering, and p_0 and v are respectively the limiting momentum and the velocity on the Fermi surface.

For our investigation it is convenient to use quantum field-theoretical methods. We shall assume, starting from Cooper's idea, that the attraction between the particles leads to the formation of bound pairs. Such a pair is a Bose formation. The temperature-dependent Green's function of a free boson $\mathfrak{G}(\mathbf{p}\omega)$ is of the form[2,3]

$$(i\omega_n - \varepsilon(\mathbf{p}) + \mu)^{-1}, \qquad \omega_n = 2\pi n T$$

and for $\omega_n = 0$ becomes infinite as soon as the so-called Bose condensation takes place $[\mu(T_c) = 0]$.

The two-particle Green's function plays the same role for a bound pair as the Green's function of a boson. If one studies the dependence of the two-particle Green's function on the total energy and momentum of the two particles (i.e., on the energy and momentum of the bound pair as a whole) one can find the temperature of the transition of the system into the superfluid state from the condition that this quantity should become infinite when the transition takes place. At lower temperatures bound pairs begin to form and to "condense." The normal ground state of attractive Fermi-particles is thus unstable. The tendency to form bound pairs will at absolute zero cause a change in the ground state of non-interacting fermions under the influence of their mutual attraction. We shall determine the temperature at which the transition to the superfluid state takes place and we shall also study the characteristic singularities of the vertex part of a Fermi system of weakly mutually attracting particles at T = 0 which appear because of the above-mentioned instability of the usual ground state. For the sake of convenience we shall start with the study of the vertex part at absolute zero.

The vertex function $\Gamma_{\alpha\beta\gamma\delta}(p_1, p_2, p_3, p_4)$ is defined through the Fourier component of the two-particle Green's function

$$G_{\alpha\beta\gamma\delta}(x_1, x_2, x_3, x_4) = \langle T(\psi_{1\alpha}\psi_{2\beta}\psi^+_{3\gamma}\psi^+_{4\delta})\rangle \tag{1}$$

by the following relation

$$\begin{aligned} G_{\alpha\beta\gamma\delta}(p_1, p_2, p_3, p_4) &= (2\pi)^8 G(p_1) G(p_2) [\delta(p_1 - p_3)\delta(p_2 - p_4)\delta_{\alpha\gamma}\delta_{\beta\delta} \\ &- \delta(p_1 - p_4)\delta(p_2 - p_3)\delta_{\alpha\delta}\delta_{\beta\gamma}] \\ &+ i(2\pi)^4 G(p_1)G(p_2)G(p_3)G(p_4)\Gamma_{\alpha\beta\gamma\delta}(p_1, p_2, p_3, p_4)\delta(p_1 \\ &+ p_2 - p_3 - p_4). \end{aligned} \tag{2}$$

The Hamiltonian of the interparticle interaction is of the form

$$H_{int} = \frac{1}{2}\int V(x-y)\psi_1^+(x)\psi_2^+(y)\psi_2(y)\psi_1(x)\, d^4x\, d^3y,$$
$$V(x-y) = V(\mathbf{x}-\mathbf{y})\delta(x_0 - y_0). \tag{2a}$$

It was shown by Landau[4] that the singularities of Γ for small values of the momentum transfer ($q = p_3 - p_1$ or $p_4 - p_1$) are connected with the existence of so-called "zero sound." We shall study the singularities of the vertex function* as far as the vari-

*A study of these singularities in general form in the theory of a Fermi liquid was made in an unpublished paper by A. A. Abrikosov, L. P. Gor'kov, L. D. Landau, and I. M. Khalatnikov. In the model considered the same approach allowed us to obtain more detailed results.

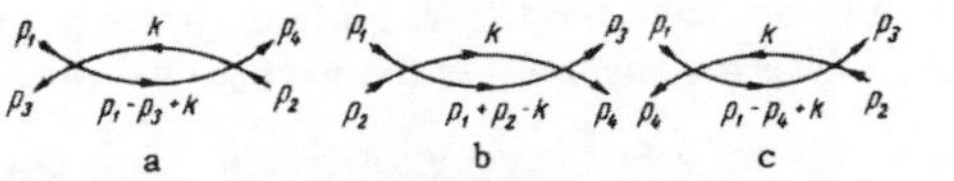

a b c

FIG. 1

able $q = p_1 + p_2$ is concerned. To do this we consider the diagrams of the first orders in perturbation theory (Fig. 1).

Singularities of the "zero-sound" type are connected with the diagrams of Fig. 1a, c. As far as the diagram of Fig. 1b is concerned, it contains an integral of two Green's functions of the form $f_0^2 \int G(k) G(q-k) d^4k$. Substituting here for $G(k)$ the Green's function of a perfect Fermi gas

$$G(k) = [\omega + \mu - \varepsilon(\mathbf{k}) + i\delta \operatorname{sign}(\varepsilon(\mathbf{k}) - \mu)]^{-1},$$

we get after integrating over the frequency

$$-2\pi i f_0^2 \int \frac{d^3k}{\varepsilon + 2\mu - \varepsilon(\mathbf{k}) - \varepsilon(\mathbf{q}-\mathbf{k}) + i\delta}, \quad \varepsilon(\mathbf{k}) > \mu, \quad \varepsilon(\mathbf{q}-\mathbf{k}) > \mu,$$

$$2\pi i f_0^2 \int \frac{d^3k}{\varepsilon + 2\mu - \varepsilon(\mathbf{k}) - \varepsilon(\mathbf{q}-\mathbf{k}) - i\delta}, \quad \varepsilon(\mathbf{k}) < \mu, \quad \varepsilon(\mathbf{q}-\mathbf{k}) < \mu. \tag{3}$$

This integral diverges at large k; this is connected with the fact that in that range it is the same as the Born correction to the scattering amplitude of two particles in vacuo. One can thus remove the divergence for large k by renormalizing the scattering amplitude. However, at small $q \ll p_0$ and $\epsilon \ll \mu$ the result obtained diverges logarithmically also near the Fermi surface for $k \sim p_0$. The integration in (3) leads thus when q and ϵ are small to a logarithmic term of the order $(p_0^2 f_0^2/v) \times \ln\{qv/\mu, \epsilon/\mu\}$. The large magnitude of the logarithm can be compensated by the small parameter $p_0^2 f_0/v \ll 1$.

To evaluate the vertex part in that range of $\{qv, \epsilon\}$ values perturbation theory turns out to be insufficient and we must sum a whole "ladder" of diagrams shown in Fig. 1b. We write down the equation for the vertex part in such a way that we explicitly separate off special integrations of the type (3):

$$\Gamma_{\alpha\beta\gamma\delta}(p_1, p_2, p_3, p_4) = \tilde{\Gamma}_{\alpha\beta\gamma\delta}(p_1, p_2, p_3, p_4) + \frac{i}{2(2\pi)^4} \int \tilde{\Gamma}_{\alpha\beta\xi\eta}(p_1, p_2, k, q-k) G(k) G(q-k) \Gamma_{\xi\eta\gamma\delta}(k, q-k, p_3, p_4) d^4k, \tag{4}$$

where $\tilde{\Gamma}_{\alpha\beta\gamma\delta}(p_1, p_2, p_3, p_4)$ is the totality of those diagrams which are irreducible in the sense that they cannot be divided by a vertical line into two parts which are joint by two fermion lines directed to one side.

The above-mentioned divergence in the large momentum region can be removed as follows. We define a vertex function for particles, interacting in vacuo, using the equation (Belyaev[5] and Galitskii[6] introduced this quantity)

$$\Gamma^{(0)}_{\alpha\beta\gamma\delta}(p_1, p_2, p_3, p_4) = \Gamma^{(1)}_{\alpha\beta\gamma\delta}(p_1, p_2, p_3, p_4) + \frac{i}{2(2\pi)^4} \int \Gamma^{(1)}_{\alpha\beta\xi\eta}(p_1, p_2, k, q-k) G^{(0)}(k) G^{(0)}(q-k) \Gamma^{(0)}_{\xi\eta\gamma\delta} \times (k, q-k, p_3, p_4) d^4k, \tag{5}$$

where $\Gamma^{(1)}_{\alpha\beta\gamma\delta}(p_1, p_2, p_3, p_4)$ is the term of first order in the interaction, which is equal to $V(p_3 - p_1)\delta_{\alpha\gamma}\delta_{\beta\delta} - V(p_3 - p_2)\delta_{\alpha\delta}\delta_{\beta\gamma}$ while $G^{(0)}(k)$ is the vacuum particle Green function:

$$G^{(0)}(k) = (\omega - \varepsilon(\mathbf{k}) + i\delta)^{-1}.$$

If we write (5) in the form

$$\Gamma^{(1)}_{\alpha\beta\gamma\delta}(p_1, p_2, p_3, p_4) = L\Gamma_{\alpha\beta\gamma\delta}(p_1, p_2, p_3, p_4) \equiv \Gamma^{(0)}_{\alpha\beta\gamma\delta}(p_1, p_2, p_3, p_4) - \frac{i}{2(2\pi)^4} \int \Gamma^{(1)}_{\alpha\beta\xi\eta}(p_1, p_2, k, q-k) \times G^{(0)}(k) G^{(0)}(q-k) \Gamma^{(0)}_{\xi\eta\gamma\delta}(k, q-k, p_3, p_4) d^4k$$

and then subtract from both sides of Eq. (4) the integral

$$\frac{i}{2(2\pi)^4} \int \Gamma^{(1)}_{\alpha\beta\xi\eta}(p_1, p_2, k, q-k) G^{(0)}(k) G^{(0)}(q-k) \Gamma_{\xi\eta\gamma\delta}(k, q-k, p_3, p_4) d^4k,$$

we get

$$L\Gamma_{\alpha\beta\gamma\delta}(p_1, p_2, p_3, p_4) = L\Gamma^{(0)}_{\alpha\beta\gamma\delta}(p_1, p_2, p_3, p_4) + \tilde{\Gamma}^*_{\alpha\beta\gamma\delta}(p_1, p_2, p_3, p_4) + \frac{i}{2(2\pi)^4} \int L\Gamma^{(0)}_{\alpha\beta\xi\eta}(p_1, p_2, k, q-k) [G(k) G(q-k) - G^{(0)}(k) G^{(0)}(q-k)] \Gamma_{\xi\eta\gamma\delta}(k, q-k, p_3, p_4) d^4k + \frac{i}{2(2\pi)^4} \int \tilde{\Gamma}^*_{\alpha\beta\xi\eta}(p_1, p_2, p_3, p_4) G(k) G(q-k) \times \Gamma_{\xi\eta\gamma\delta}(k, q-k, p_3, p_4) d^4k, \tag{6}$$

where $\tilde{\Gamma}^*_{\alpha\beta\gamma\delta}(p_1, p_2, p_3, p_4)$ stands for all irreducible diagrams from the second order in the interaction onwards.

We get the vertex part $\Gamma_{\alpha\beta\gamma\delta}(p_1, p_2, p_3, p_4)$ up to terms of order $p_0^2 f_0^2/v$ inclusive. It is clear that as the diagrams for $\tilde{\Gamma}$ do not contain logarithmic integrations we can restrict ourselves in this order in the expression for $\tilde{\Gamma}^*$ to terms of second order in f_0 (Fig. 2)

Applying to both sides of Eq. (6) the operator L^{-1} we are led to the following equation

$$\Gamma_{\alpha\beta\gamma\delta}(p_1, p_2, p_3, p_4) = \Gamma^{(0)}_{\alpha\beta\gamma\delta}(p_1, p_2, p_3, p_4) + \frac{i}{2(2\pi)^4} \int \Gamma^{(0)}_{\alpha\beta\xi\eta} \times (p_1, p_2, k, q-k) [G(k) G(q-k) - G^{(0)}(k) G^{(0)}(q-k)] \times \Gamma_{\xi\eta\gamma\delta}(k, q-k, p_3, p_4) d^4k + \frac{i}{2(2\pi)^4} \int \tilde{\Gamma}^*_{\alpha\beta\xi\eta}(p_1, p_2, k, q-k) G(k) G(q-k) \Gamma_{\xi\eta\gamma\delta}(k, q-k, p_3, p_4) d^4k. \tag{7}$$

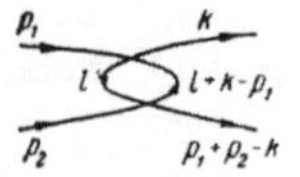

FIG. 2

Within the accuracy which we have adopted $L\tilde{\Gamma}^* = \tilde{\Gamma}^*$ and we have neglected in the terms which do not contain logarithmic integrations quantities of second order in f_0. As far as $\Gamma^{(0)}$ is concerned in the integral on the right hand side of (7), we must evaluate it up to terms in second order in f_0:

$$\Gamma^{(0)}_{\alpha\beta\xi\eta}(p_1, p_2, k, q-k) = V(\mathbf{k}-\mathbf{p}_1)\delta_{\alpha\xi}\delta_{\beta\eta} - V(\mathbf{k}-\mathbf{p}_2)\delta_{\alpha\eta}\delta_{\beta\xi}$$
$$-\frac{i}{2(2\pi)^4}\int [V(\mathbf{l}-\mathbf{p}_1)V(\mathbf{k}-\mathbf{l})$$
$$+V(\mathbf{l}-\mathbf{p}_2)V(\mathbf{l}-\mathbf{q}+\mathbf{k})]G^{(0)}(l)G^{(0)}(q-l)\delta_{\alpha\xi}\delta_{\beta\eta}d^4l$$
$$-\frac{i}{2(2\pi)^4}\int [V(\mathbf{l}-\mathbf{p}_1)V(\mathbf{k}-\mathbf{q}+\mathbf{l})$$
$$+V(\mathbf{l}-\mathbf{p}_2)V(\mathbf{k}-\mathbf{l})]G^{(0)}(l)G^{(0)}(q-l)\delta_{\alpha\eta}\delta_{\beta\xi}d^4l. \quad (8)$$

If we use the relation between the scattering amplitude for two particles in vacuo and the interaction potential

$$V(\mathbf{k}_1-\mathbf{k}_2) = f(\mathbf{k}_1, \mathbf{k}_2) + \frac{m}{(2\pi)^3}\int \frac{f(\mathbf{k}_1, \mathbf{l})f^*(\mathbf{k}_2, \mathbf{l})\,d^3\mathbf{l}}{l^2-k_2^2-i\delta} \quad (9)$$

and the well-known relation between the imaginary part of the scattering amplitude and the effective scattering cross section

$$\operatorname{Im} f(\mathbf{k}_1, \mathbf{k}_2) = -\frac{mk_1}{16\pi^2}\int d\mathbf{n}\, f(\mathbf{k}_1, k_1\mathbf{n})f^*(\mathbf{k}_2, k_1\mathbf{n}), \quad (10)$$

we get after substituting (9) and (10) into Eq. (8) an expression for $\Gamma^{(0)}_{\alpha\beta\xi\eta}(p_1, p_2, k, q-k)$:

$$\Gamma^{(0)}_{\alpha\beta\xi\eta}(p_1, p_2, k, q-k) = f_0(1 - if_0p_0^2/4\pi v)(\delta_{\alpha\xi}\delta_{\beta\eta} - \delta_{\alpha\eta}\delta_{\beta\xi}). \quad (11)$$

When one uses a Green's function in $G(p)$ one should, strictly speaking, evaluate its value up to terms of second order in f_0. Galitskii[6] has, however, shown that the correction of second order in f_0 to the perfect Fermi-gas Green's function leads merely to a renormalization of the Fermi energy. We shall thus in the following use the perfect Fermi-gas Green's functions assuming μ to be the renormalized Fermi energy.

The contribution to $\Gamma^*_{\alpha\beta\xi\eta}(p_1, p_2, k, q-k)$ in the intervals $\epsilon \ll \mu$ and $q \ll p_0$ which are of interest to us is equal to

$$\frac{if_0^2}{(2\pi)^4}\int G(l)G(p_1-k+l)\,d^4l\,(\delta_{\alpha\xi}\delta_{\beta\eta} - \delta_{\alpha\eta}\delta_{\beta\xi}).$$

Substituting $\Gamma^{(0)}_{\alpha\beta\xi\eta}(p_1, p_2, k, q-k)$ and $\tilde{\Gamma}^*_{\alpha\beta\xi\eta}(p_1, p_2, k, q-k)$ into Eq. (7) and taking into account that of the terms of second order in f_0 we need only the logarithmic ones, we get the following expression for the vertex function:

$$\Gamma_{\alpha\beta\gamma\delta}(p_1, p_2, p_3, p_4) = \Gamma^{(0)}_{\alpha\beta\gamma\delta}(p_1, p_2, p_3, p_4)$$
$$\times\left\{1 - \frac{if_0}{(2\pi)^4}\left(1 - \frac{if_0p_0^2}{4\pi v}\right)\int [G(k)G(q-k)\right.$$
$$-G^{(0)}(k)G^{(0)}(q-k)]\,d^4k$$
$$\left.+\frac{f_0^2}{(2\pi)^8}\int G(l)G(p_1-k+l)G(k)G(q-k)\,d^4l\,d^4k\right\}^{-1}. \quad (12)$$

The first of these integrals gives in the denominator of (12) a term

$$-\frac{p_0^2f_0}{(2\pi)^2v}\left(1 - \frac{if_0p_0^2}{4\pi v}\right)\left(2 + \ln\frac{\varepsilon^2 - q^2v^2 + i\delta}{64\mu^2} + \frac{\varepsilon}{qv}\ln\frac{\varepsilon + vq - i\delta}{\varepsilon - vq + i\delta}\right).$$

As we need only take into account the logarithmic term we must, when evaluating the second integral, put $p_1 = k = p_0$, $\epsilon_1 = \omega = \mu$ after integrating over l and we must average over the direction of the vector $\mathbf{k}$. We get

$$\frac{2f_0^2p_0^4}{3(2\pi)^4v^2}(1 + 2\ln 2)\ln\frac{\varepsilon^2 - q^2v^2}{64\mu^2}.$$

Substituting the results of these calculations into (12) we get

$$\Gamma_{\alpha\beta\gamma\delta}(p_1, p_2, p_3, p_4) = -(2\pi)^2vp_0^{-2}(\delta_{\alpha\gamma}\delta_{\beta\delta}$$
$$-\delta_{\alpha\delta}\delta_{\beta\gamma})\{(2\pi)^2v/p_0^2|f_0| - i\pi + \frac{8}{3} - \frac{14}{3}\ln 2$$
$$+\ln\frac{\varepsilon^2 - q^2v^2 + i\delta}{\mu^2} + \frac{\varepsilon}{qv}\ln\frac{\varepsilon + qv - i\delta}{\varepsilon - qv + i\delta}\}^{-1}. \quad (13)$$

We shall consider (13) to be the analytical continuation of $\Gamma_{\alpha\beta\gamma\delta}(p_1, p_2, p_3, p_4)$ in the upper half-plane of the variable ϵ. Let $q = 0$ to begin with. We see that $\Gamma_{\alpha\beta\gamma\delta}(p_1, p_2, p_3, p_4)$ has a pole at $\epsilon_0 = i\Delta$, where $\Delta = \mu(2/e)^{7/3}\exp\{-2\pi^2v/|f_0|p_0^2\}$. If we express the vertex part $\Gamma_{\alpha\beta\gamma\delta}(p_1, p_2, p_3, p_4)$ in terms of ϵ_0 it has the form

$$\Gamma_{\alpha\beta\gamma\delta}(p_1, p_2, p_3, p_4) = -(2\pi)^2vp_0^{-2}(\delta_{\alpha\gamma}\delta_{\beta\delta} - \delta_{\alpha\delta}\delta_{\beta\gamma})$$
$$\times\left\{-2 + \ln\frac{\varepsilon^2 - q^2v^2 + i\delta}{\varepsilon_0^2} + \frac{\varepsilon}{qv}\ln\frac{\varepsilon + vq - i\delta}{\varepsilon - vq + i\delta}\right\}^{-1}.$$

For small $qv \ll \Delta$ the position of the pole $\epsilon(\mathbf{q})$ approaches the real axis: $\epsilon(\mathbf{q}) = i\Delta(1 - q^2v^2/6\Delta^2)$. When q increases further $|\epsilon(\mathbf{q})|$ decreases and it tends to zero when $vq_{max} = e\Delta$ as

$$\varepsilon = (2e\Delta i/\pi)\ln(e\Delta/vq) \approx 2e\Delta i\,(q_{max} - q)/\pi q_{max}.$$

The Cooper phenomenon, i.e., the instability of a fermion system when there are attractions between the particles, leads thus to the occurrence in the vertex part of poles in the upper half-plane of the complex variable $\epsilon = \omega_1 + \omega_2$. Only those particles for which the total momentum is relatively small ($q \ll p_0$) will here display a tendency to form bound pairs. The quantity Δ has clearly the meaning of the reciprocal of the relaxation time of the system.

We turn now to the problem of determining the temperature at which the system makes the transition to the superfluid state. Because we cannot apply the usual technique at finite temperatures we must use the Matsubara quantities.[2,3] In the Matsubara method we have instead of (2) the following connection between the two-particle Green's function and the vertex part

$$\mathfrak{G}_{\alpha\beta\gamma\delta}(\omega_1 p_1, \omega_2 p_2, \omega_3 p_3, \omega_4 p_4)$$
$$= (2\pi)^3 T^{-1} \{(2\pi)^3 T^{-1} [\mathfrak{G}(\omega_1 p_1)\mathfrak{G}(\omega_2 p_2)$$
$$\times \delta_{\omega_1\omega_4}\delta(p_1 - p_4)\delta_{\alpha\delta}\delta_{\beta\gamma} - \mathfrak{G}(\omega_1 p_1)\mathfrak{G}(\omega_2 p_2)\delta_{\omega_1\omega_3}\delta(p_1$$
$$- p_3)\delta_{\alpha\gamma}\delta_{\beta\delta}] - \tfrac{1}{2}\mathfrak{G}(\omega_1 p_1)\mathfrak{G}(\omega_2 p_2)\mathfrak{G}(\omega_3 p_3)\mathfrak{G}(\omega_4 p_4)$$
$$\times \mathfrak{T}_{\alpha\beta\gamma\delta}(\omega_1 p_1, \omega_2 p_2, \omega_3 p_3, \omega_4 p_4)\}$$
$$\times \delta_{\omega_1+\omega_2-\omega_3-\omega_4}\delta(p_1 + p_2 - p_3 - p_4). \qquad (14)$$

At the beginning of this paper we have already remarked that one can define the critical temperature as the temperature at which the Bose "condensation" of bound pairs sets in. At that point the two-particle Green's function. or what amounts to the same, according to (14), the vertex part, tends to infinity for the first time. We note that the dependence of the temperature dependent quantity $\mathfrak{T}_{\alpha\beta\gamma\delta}(\omega_1 \mathbf{p}_1, \omega_2 \mathbf{p}_2, \omega_3 \mathbf{p}_3, \omega_4 \mathbf{p}_4)$ on the variable $\epsilon = \omega_1 + \omega_2$ is defined only in the discrete points $\epsilon_n = 2\pi n T i$, $n = 0, \pm 1, \pm 2, \ldots$ At the critical temperature the position of the pole is $\epsilon = 0$. Moreover, it is clear from similarity considerations that the Fourier component of the Matsubara quantity $\mathfrak{T}_{\alpha\beta\gamma\delta}(\omega_1 + \omega_2 = 0,\ \mathbf{p}_1 + \mathbf{p}_2 = 0)$ becomes infinite at the transition.

The remainder of our discussion will refer to the case $\epsilon = \omega_1 + \omega_2 = 0$, $\mathbf{q} = \mathbf{p}_1 + \mathbf{p}_2 = 0$.

The diagram of Fig. 1b contains in second order in f_0 an integral over the product of two Matsubara Green's functions $\mathfrak{G}(\omega\mathbf{k})$ for the perfect Fermi gas of the form

$$2\pi i f_0^2 T \sum_{\omega_n} \int \mathfrak{G}(\omega_n \mathbf{k})\mathfrak{G}(-\omega_n, -\mathbf{k})\, d^3k,$$

which after summation over ω_n becomes

$$\pi i f_0^2 \int \frac{d^3k}{\mu - \varepsilon(k)} \operatorname{th}\frac{\mu - \varepsilon(k)}{2T}. \qquad (15)^*$$

The integral (15) diverges for large k. The divergence of this integral can be removed in the same way as at absolute zero by renormalizing the scattering amplitude. The integral diverges, however, also near the Fermi surface where its magnitude is $\sim (f_0 p_0^2/v)\ln(T/\mu)$ which is compensated by the smallness of the parameter $f_0 p_0^2/v \ll 1$. To evaluate $\mathfrak{T}_{\alpha\beta\gamma\delta}(\omega_1 \mathbf{p}_1, \omega_2 \mathbf{p}_2, \omega_3 \mathbf{p}_3, \omega_4 \mathbf{p}_4)$ we must therefore again sum a "ladder" of diagrams such as Fig. 1b. We write down an equation for $\mathfrak{T}_{\alpha\beta\gamma\delta}(\omega_1 \mathbf{p}_1, \omega_2 \mathbf{p}_2, \omega_3 \mathbf{p}_3, \omega_4 \mathbf{p}_4)$ which is similar to Eq. (5):

*th = tanh.

$$\mathfrak{T}_{\alpha\beta\gamma\delta}(\omega_1 p_1, \omega_2 p_2, \omega_3 p_3, \omega_4 p_4)$$
$$= \widetilde{\mathfrak{T}}_{\alpha\beta\gamma\delta}(\omega_1 p_1, \omega_2 p_2, \omega_3 p_3, \omega_4 p_4)$$
$$- \frac{T}{2(2\pi)^3} \sum_{\omega_n} \int \widetilde{\mathfrak{T}}_{\alpha\beta\xi\eta}(\omega_1 p_1, \omega_2 p_2, \omega_n k, \varepsilon - \omega_n, q - k)$$
$$\times \mathfrak{G}(\omega_n k)\mathfrak{G}(\varepsilon - \omega_n, q - k)$$
$$\times \mathfrak{T}_{\xi\eta\gamma\delta}(\omega_n k, \varepsilon - \omega_n, q - k, \omega_3 p_3, \omega_4 p_4)\, d^3k. \qquad (16)$$

We obtain an equation to determine the temperature at which the transition into the superfluid state takes place from (16) if we take into account that $\widetilde{\mathfrak{T}}$ has no singularities of the kind (15) and that we need only to know of the terms in $\widetilde{\mathfrak{T}}^{(2)}$ that one which gives a contribution $\sim (f_0^2 p_0^4/v^2) \times \ln(T/\mu)$. Near the point where $\mathfrak{T}_{\alpha\beta\gamma\delta}(\epsilon, \mathbf{q})$ becomes infinite Eq. (16) becomes of the form

$$1 = -\frac{T_c}{(2\pi)^3} \sum_{\omega_n} \int V(p_1 - k)\mathfrak{G}(\omega_n k)\mathfrak{G}(-\omega_n, -k)\, d^3k$$
$$- \frac{T_c}{(2\pi)^3} \sum_{\omega_n} \int \widetilde{\mathfrak{T}}^{(2)}(\omega_1 p_1, \omega_2 p_2, \omega_n k, -\omega_n, -k)\mathfrak{G}(\omega_n k)$$
$$\times \mathfrak{G}(-\omega_n, -k)\, d^3k. \qquad (17)$$

We have here taken into account that $\mathfrak{T}^{(1)}_{\alpha\beta\gamma\delta}(\omega_1 \mathbf{p}_1, \omega_2 \mathbf{p}_2, \omega_n \mathbf{k}, -\omega_n, -\mathbf{k})$ is independent of the fourth components of the momenta and is thus the same as the quantity $\Gamma^{(1)}_{\alpha\beta\gamma\delta}(\mathbf{p}_1, \mathbf{p}_2, \mathbf{k}, -\mathbf{k})$.

As at $T = 0$, the divergence in the first integral in (17) far from the Fermi surface can be removed by renormalizing the scattering amplitude; after that Eq. (17) is reduced to the form

$$1 = -\frac{2f_0}{(2\pi)^3}\left(1 - \frac{i f_0 p_0^2}{4\pi v}\right)\int\left[\frac{\operatorname{th}(\mu - \varepsilon(k))/2T_c}{\mu - \varepsilon(k)}\right.$$
$$\left. + \frac{1}{\mu - \varepsilon(k) + i\delta}\right] d^3k - \frac{T_c}{(2\pi)^3}\sum_{\omega_n}\int \widetilde{\mathfrak{T}}^{(2)}(\omega_1 p_1, \omega_2 p_2, \omega_n k,$$
$$-\omega_n, -k)\mathfrak{G}(\omega_n k)\mathfrak{G}(-\omega_n, -k)\, d^3k. \qquad (18)$$

The contribution to $\mathfrak{T}^{(2)}$ which contains all irreducible diagrams of second order in the interaction is given by the expression

$$-\frac{f_0^2 T_c}{(2\pi)^3}\sum_{\varepsilon_n}\int \mathfrak{G}(\varepsilon_n, l)\mathfrak{G}(\varepsilon_n + \omega_1 - \omega_n, l + p_1 - k)\, d^3l.$$

The first integral in (18) gives the contribution

$$-\frac{p_0^2 f_0}{2\pi^2 v}\left(1 - \frac{i f_0 p_0^2}{4\pi v}\right)\left(\ln\frac{8\gamma\mu}{T_c\pi} - 2 - \frac{i\pi}{2}\right), \qquad (19)$$

where $\ln\gamma$ is Euler's constant.

The evaluation of the second integral in (18) proceeds taking the same considerations into account as at the absolute zero. The result of these calculations is

$$-\frac{p_0^4 f_0^2}{12\pi^4 v^2}(1+2\ln 2)\ln\frac{8\gamma\mu}{T_c\pi}. \qquad (20)$$

Substituting (19) and (20) into (18) we get

$$-\frac{2\pi^2 v}{|f_0|\, p_0^2} = \ln\left[\frac{T_c\pi}{\gamma\mu}\left(\frac{e}{2}\right)^{1/3}\right].$$

The temperature of the transition into the superfluid state is thus determined by the relation

$$T_c = (\gamma\mu/\pi)(2/e)^{1/3}\exp\{-2\pi^2 v/p_0^2|f_0|\}$$

or, $T_c = \gamma\Delta/\pi$, where Δ is the previously determined quantity which has the meaning of the reciprocal of the relaxation time of the system.

It is of interest to note that as in the model considered the integration practically is performed near the Fermi surface the relation between the "gap" width and the transition temperature must be the same as in the theory of superconductivity, and it follows thus that the gap width at $T = 0$ is equal to the reciprocal of the relaxation time of the system.

In conclusion we express our gratitude to Academician L. D. Landau for valuable discussions of the results of this paper.

[1] L. Cooper, Phys. Rev. **104**, 1189 (1956).

[2] Abrikosov, Gor'kov, and Dzyaloshinskii, JETP **36**, 900 (1959), Soviet Phys. JETP **9**, 636 (1959).

[3] E. S. Fradkin, JETP **36**, 1286 (1959), Soviet Phys. JETP **9**, 912 (1959).

[4] L. D. Landau, JETP **35**, 97 (1958), Soviet Phys. JETP **8**, 70 (1959).

[5] S. T. Belyaev, JETP **34**, 417 (1958), Soviet Phys. JETP **7**, 328 (1958).

[6] V. M. Galitskii, JETP **34**, 151 (1958), Soviet Phys. JETP **7**, 104 (1958).

Translated by D. ter Haar
248

SOVIET PHYSICS JETP VOLUME 18, NUMBER 4 APRIL, 1964

MICROSCOPIC DERIVATION OF THE GINZBURG-LANDAU EQUATIONS FOR AN ANISOTROPIC SUPERCONDUCTOR

L. P. GOR'KOV and T. K. MELIK-BARKHUDAROV

Institute of Physics Problems, Academy of Sciences, U.S.S.R.

Submitted to JETP editor April 28, 1963

J. Exptl. Theoret. Phys. (U.S.S.R.) **45**, 1493-1498 (November, 1963)

A microscopic derivation of the Ginzburg-Landau equations for an anisotropic superconductor is presented. The meaning of the constants encountered in the phenomenological theory and defined in terms of the gap values at absolute zero is clarified.

THE equations of superconductivity theory can be reduced to the Ginzburg-Landau equations[1] near the critical temperature T_c. This is possible because the theory is local in this region, i.e., all the quantities change noticeably over distances that are much larger than the correlation parameter ($\delta \gg \xi_0$, where $\xi_0 \sim v/T$ and δ is the depth of penetration of the field). It is interesting to ascertain what this can lead to in the anisotropic case. It turns out that this results in the equations proposed by Ginzburg[2]. The microscopic theory allows us to clarify the meaning of the quantities in these equations.

In the derivation we assume that the potential of the interaction, spectrum, etc are anisotropic. Such a procedure is a simple generalization of the usual one and was proposed in many places (in particular by Bogolyubov, Tolmachev, and Shirkov[3] and in the papers of Pokrovskiĭ[4,5]). It is convenient to rewrite it in a form where the superconductor is described by three Green's functions[6]:

$$\delta_{\alpha\beta}\,\mathfrak{G}(\mathbf{p}_1, \mathbf{p}_2; \tau_1 - \tau_2) = -\langle T\{a_{\mathbf{p}_1,\alpha}(\tau_1)\, a^+_{\mathbf{p}_2,\beta}(\tau_2)\}\rangle,$$

$$I_{\alpha\beta}\,\mathfrak{F}^+(\mathbf{p}_1, \mathbf{p}_2; \tau_1 - \tau_2) = \langle T\{a^+_{\mathbf{p}_1,\alpha}(\tau_1)\, a^+_{\mathbf{p}_2,\beta}(\tau_2)\}\rangle,$$

$$I_{\alpha\beta}\,\mathfrak{F}(\mathbf{p}_1, \mathbf{p}_2; \tau_1 - \tau_2) = -\langle T\{a_{\mathbf{p}_1,\alpha}(\tau_1)\, a_{\mathbf{p}_2,\beta}(\tau_2)\}\rangle,$$

where

$$a_{\mathbf{p}\alpha}(\tau) = e^{-(\mu N - H\tau)}\, a_{\mathbf{p}\alpha} e^{(\mu N - H\tau)},$$

$$a^+_{\mathbf{p}\alpha}(\tau) = e^{-(\mu N - H)\tau}\, a^+_{\mathbf{p}\alpha} e^{(\mu N - H)\tau}.$$

The specific form of the Hamiltonian is of no interest to us.

Introducing the Fourier components for the functions $\mathfrak{G}$, $\mathfrak{F}$, $\mathfrak{F}^+$

$$\mathfrak{G}(\mathbf{p}_1, \mathbf{p}_2; \tau) = T\sum_\omega \mathfrak{G}_\omega(\mathbf{p}_1, \mathbf{p}_2)\, e^{-i\omega\tau},$$

$$\mathfrak{F}(\mathbf{p}_1, \mathbf{p}_2; \tau) = T\sum_\omega \mathfrak{F}_\omega(\mathbf{p}_1, \mathbf{p}_2)\, e^{-i\omega\tau},$$

$$\mathfrak{F}^+(\mathbf{p}_1, \mathbf{p}_2; \tau) = T\sum_\omega \mathfrak{F}^+_\omega(\mathbf{p}_1, \mathbf{p}_2)\, e^{-i\omega\tau}, \qquad \omega = 2\pi(n + 1/2),$$

we can write down for them equations[6] that account for the anisotropy in the presence of the magnetic field. The gauge is chosen such that the scalar potential is equal to zero. The corresponding equations are of the form

$$\{i\omega - \xi(\mathbf{p})\}\mathfrak{G}_\omega(\mathbf{p}, \mathbf{p}') + e\sum_{\mathbf{q}} \mathbf{A}_{\mathbf{q}}\, \mathbf{v}_{\mathbf{p}} \mathfrak{G}_\omega(\mathbf{p} - \mathbf{q}, \mathbf{p}') + \sum_{\mathbf{k}} \Delta(\mathbf{p}, \mathbf{k})\, \mathfrak{F}^+_\omega(\mathbf{k}, \mathbf{p}') = \delta_{\mathbf{p},\mathbf{p}'},$$

$$\{i\omega + \xi(\mathbf{p})\}\mathfrak{F}^+_\omega(\mathbf{p}, \mathbf{p}') - e\sum_{\mathbf{q}} \mathbf{A}_{\mathbf{q}} \mathbf{v}_{\mathbf{p}}\, \mathfrak{F}^+_\omega(\mathbf{p} + \mathbf{q}, \mathbf{p}') + \sum_{\mathbf{k}} \Delta^*(\mathbf{p}, \mathbf{k})\, \mathfrak{G}_\omega(\mathbf{k}, \mathbf{p}') = 0, \qquad (1)$$

$$\Delta^*(\mathbf{p}, \mathbf{k}) = gT \sum_{\mathbf{k}_1 \mathbf{k}_2, \omega} V(\mathbf{k}_1, \mathbf{k}_2; \mathbf{p}, \mathbf{k})\, \mathfrak{F}^+_\omega(\mathbf{k}_1, \mathbf{k}_2), \qquad \mathbf{k}_1 + \mathbf{k}_2 = \mathbf{p} + \mathbf{k}. \qquad (1')$$

We have used here the following notation: $\xi(\mathbf{p}) = \epsilon(\mathbf{p}) - \epsilon_F$, where $\epsilon(\mathbf{p})$ — arbitrary electron dispersion law and ϵ_F — Fermi energy, $\mathbf{v}_\mathbf{p} = \nabla_\mathbf{p}\epsilon(\mathbf{p})$, $\mathbf{A}_\mathbf{q}$ — Fourier component of the vector potential, $V(\mathbf{k}_1, \mathbf{k}_2, \mathbf{p}, \mathbf{k})$ — dimensionless matrix elements of the interaction energy[1]

The summation in (1′) is over those $\mathbf{k}_1$ and $\mathbf{k}_2$, for which $|\xi(\mathbf{k}_1)| \le \tilde{\omega}(\mathbf{k}_1/k_1)$ and $|\xi(\mathbf{k}_2)| \le \tilde{\omega}(\mathbf{k}_2/k_2)$, where $\tilde{\omega}(\mathbf{k}/k)$ is a quantity on the order of the Debye energy. The magnetic field is introduced, as usual, by making the substitution $\mathbf{p} \to \mathbf{p} - e\mathbf{A}$ in $\epsilon(\mathbf{p})$ and by expanding $\epsilon(\mathbf{p} - e\mathbf{A})$ in powers of $\mathbf{A}$. Only the term linear in $\mathbf{A}$ was left in the series, since in the field of interest to us the electron torsion radius ep_0/H is large

[1]The fact that we make use of Fourier components of all quantities, disregarding the fact only a quasi-momentum and not a momentum exists in the lattice, will not affect the final results, since the field A and Δ change at distances that are large compared with the period of the lattice.

compared with the depth of penetration ($p_0 \gg eA \sim eH\delta$). In addition, account is taken in (1) of the fact that the change in the field occurs over distances considerably larger than the inter-atomic distances. Equations (1) for an arbitrary potential are gauge-invariant, as are the corresponding equations of the theory of Bardeen, Cooper, and Schrieffer, accurate to terms $T/\widetilde{\omega}$, which are always small in real superconductors. Therefore, by calculating all the quantities accurate to terms $T/\widetilde{\omega} \ll 1$, we obtain the final expression in gauge-invariant form.

We introduce the Green's function of the electrons in the normal metal in the presence of the field $\widetilde{\mathfrak{G}}_\omega(p_1, p_2)$. The equation satisfied by this function can be written in two ways:

$$\{i\omega - \xi(\mathbf{p})\}\widetilde{\mathfrak{G}}_\omega(\mathbf{p}, \mathbf{p}') + e\sum_{\mathbf{q}} \mathbf{A}_\mathbf{q}\mathbf{v}_\mathbf{p}\mathfrak{G}_\omega(\mathbf{p}-\mathbf{q}, \mathbf{p}') = \delta_{\mathbf{p},\mathbf{p}'}, \tag{2}$$

$$\{i\omega - \xi(\mathbf{p}')\}\widetilde{\mathfrak{G}}_\omega(\mathbf{p}, \mathbf{p}') + e\sum_{\mathbf{q}} \mathbf{A}_\mathbf{q}\mathbf{v}_{\mathbf{p}}\mathfrak{G}_\omega(\mathbf{p}, \mathbf{p}'+\mathbf{q}) = \delta_{\mathbf{p},\mathbf{p}'}. \tag{2'}$$

With the aid of the Green's function $\widetilde{\mathfrak{G}}_\omega(p_1, p_2)$ and Eqs. (2) and (2'), we write the system (1) for $\mathfrak{G}$, $\mathfrak{F}$, and $\mathfrak{F}^+$ in integral form

$$\mathfrak{G}_\omega(\mathbf{p}, \mathbf{p}') = \widetilde{\mathfrak{G}}_\omega(\mathbf{p}, \mathbf{p}') - \sum_{\mathbf{l},\mathbf{k}} \widetilde{\mathfrak{G}}_\omega(\mathbf{p}, \mathbf{l})\,\Delta(\mathbf{l}, \mathbf{k})\,\mathfrak{F}^+_\omega(\mathbf{k}, \mathbf{p}'), \tag{3}$$

$$\mathfrak{F}^+_\omega(\mathbf{p}, \mathbf{p}') = \sum_{\mathbf{l},\mathbf{k}} \widetilde{\mathfrak{G}}_{-\omega}(\mathbf{l}, \mathbf{p})\,\Delta^*(\mathbf{l}, \mathbf{k})\,\mathfrak{G}_\omega(\mathbf{k}, \mathbf{p}'). \tag{3'}$$

From (2) we obtain $\widetilde{\mathfrak{G}}_\omega(p, p')$ accurate to terms that are quadratic in $\mathbf{A}\cdot\mathbf{v}$ inclusive:

$$\widetilde{\mathfrak{G}}_\omega(\mathbf{p}, \mathbf{p}') = \mathfrak{G}_\omega(\mathbf{p})\,\delta_{\mathbf{p},\mathbf{p}'} - e\mathfrak{G}_\omega(\mathbf{p})\,\mathbf{A}_{\mathbf{p}-\mathbf{p}'}\,\mathbf{v}_\mathbf{p}\mathfrak{G}_\omega(\mathbf{p}')$$
$$+ e^2\mathfrak{G}_\omega(\mathbf{p})\,\mathfrak{G}_\omega(\mathbf{p}')\sum_{\mathbf{q}} \mathbf{A}_\mathbf{q}\mathbf{v}_\mathbf{p}(\mathbf{A}_{\mathbf{p}-\mathbf{p}'-\mathbf{q}}\,\mathbf{v}_\mathbf{p})\,\mathfrak{G}_\omega(\mathbf{p}-\mathbf{q}), \tag{4}$$

where $\mathfrak{G}_\omega(p) = \{i\omega - \xi(p)\}^{-1}$ is the Green's function of the normal metal in the absence of the field.

Inasmuch as it is our aim to derive an equation relative to Δ in the vicinity of T_c, where $\Delta/T_c \sim \sqrt{1 - T/T_c} \ll 1$, expressions (3) and (3') can be expanded in powers of Δ. It is seen from (3') and (1') that the expansion of the function $\mathfrak{F}^+$ can be carried out accurate to terms of third order in Δ, exclusive. The function $\mathfrak{G}$ must be known accurate to second-order terms in Δ:

$$\mathfrak{G}_\omega(\mathbf{p}, \mathbf{p}') = \widetilde{\mathfrak{G}}_\omega(\mathbf{p}, \mathbf{p}')$$
$$- \sum \widetilde{\mathfrak{G}}_\omega(\mathbf{p}, \mathbf{l})\,\Delta(\mathbf{l}, \mathbf{k}_1)\,\widetilde{\mathfrak{G}}_{-\omega}(\mathbf{l}', \mathbf{k}_1)\,\Delta^*(\mathbf{l}', \mathbf{k}_1')\,\widetilde{\mathfrak{G}}_\omega(\mathbf{k}_1', \mathbf{p}') \tag{5}$$

(the summation is over $\mathbf{l}$, $\mathbf{l}'$, $\mathbf{k}_1$, and $\mathbf{k}_1'$). Substituting (5) in (3') we obtain the required expansion for $\mathfrak{F}^+$, with the aid of which we get from (1') an equation for $\Delta^*(p_1, p_2)$:

$$\Delta^*(\mathbf{p}_1, \mathbf{p}_2) = gT\sum V(\mathbf{k}_1, \mathbf{k}_2; \mathbf{p}_1, \mathbf{p}_2)\,\widetilde{\mathfrak{G}}_{-\omega}(\mathbf{l}, \mathbf{k}_1)\,\Delta^*(\mathbf{l}, \mathbf{k}_3)$$
$$\times \widetilde{\mathfrak{G}}_\omega(\mathbf{k}_3, \mathbf{k}_2) - gT\sum V(\mathbf{k}_1, \mathbf{k}_2; \mathbf{p}_1, \mathbf{p}_2)$$
$$\times \widetilde{\mathfrak{G}}_{-\omega}(\mathbf{l}, \mathbf{k}_1)\,\Delta^*(\mathbf{l}, \mathbf{k}_3)\,\widetilde{\mathfrak{G}}_\omega(\mathbf{k}_3, \mathbf{l}')\,\Delta(\mathbf{l}', \mathbf{k}')\,\widetilde{\mathfrak{G}}_{-\omega}(\mathbf{l}', \mathbf{k}')$$
$$\times \Delta^*(\mathbf{l}'', \mathbf{k}'')\,\widetilde{\mathfrak{G}}_\omega(\mathbf{k}'', \mathbf{k}_2) \tag{6}$$

(the summation is over all the repeated indices, and $\mathbf{k}_1 + \mathbf{k}_2 = \mathbf{p}_1 + \mathbf{p}_2$). Using expression (4) for $\widetilde{\mathfrak{G}}_\omega(p_1, p_2)$, we obtain for the first term of the right half of (6)

$$T\sum_{\mathbf{k}_1\mathbf{k}_2} V(\mathbf{k}_1, \mathbf{k}_2; \mathbf{p}_1, \mathbf{p}_2)\,\{\mathfrak{G}_{-\omega}(\mathbf{k}_1)\,\Delta^*(\mathbf{k}_1, \mathbf{k}_2)\,\mathfrak{G}_\omega(\mathbf{k}_2)$$
$$- e\sum_{\mathbf{k}_3} \mathfrak{G}_{-\omega}(\mathbf{k}_1)\,\Delta^*(\mathbf{k}_1, \mathbf{k}_3)\,\mathfrak{G}_\omega(\mathbf{k}_3)(\mathbf{A}_{\mathbf{k}_3-\mathbf{k}_2}\mathbf{v}_{\mathbf{k}_3})\,\mathfrak{G}_\omega(\mathbf{k}_2)$$
$$- e\sum_{\mathbf{l}} \mathfrak{G}_{-\omega}(\mathbf{l})\,(\mathbf{A}_{\mathbf{l}-\mathbf{k}_1}\mathbf{v}_\mathbf{l})\,\mathfrak{G}_{-\omega}(\mathbf{k}_1)\,\Delta^*(\mathbf{l}, \mathbf{k}_2)\,\mathfrak{G}_\omega(\mathbf{k}_2)$$
$$+ e^2\sum_{\mathbf{k}_3\mathbf{q}} \mathfrak{G}_{-\omega}(\mathbf{k}_1)\,\Delta^*(\mathbf{k}_1, \mathbf{k}_3)\,\mathfrak{G}_\omega(\mathbf{k}_3)\,(\mathbf{A}_\mathbf{q}\mathbf{v}_{\mathbf{k}_3})$$
$$\times (\mathbf{A}_{\mathbf{k}_3-\mathbf{k}_2-\mathbf{q}}\mathbf{v}_{\mathbf{k}_3})\,\mathfrak{G}_\omega(\mathbf{k}_2) + e^2\sum_{\mathbf{l},\mathbf{q}} \mathfrak{G}_{-\omega}(\mathbf{l})\,(\mathbf{A}_\mathbf{q}\mathbf{v}_\mathbf{l})\,(\mathbf{A}_{\mathbf{l}-\mathbf{q}-\mathbf{k}_1}\mathbf{v}_\mathbf{l})$$
$$\times \mathfrak{G}_\omega(\mathbf{l}-\mathbf{q})\,\mathfrak{G}_\omega(\mathbf{k}_1)\,\mathfrak{G}_\omega(\mathbf{k}_2)\,\Delta^*(\mathbf{l}, \mathbf{k}_2) + e^2\sum_{\mathbf{k}_3,\mathbf{l}} \mathfrak{G}_{-\omega}(\mathbf{l})$$
$$\times (\mathbf{A}_{\mathbf{l}-\mathbf{k}_1}\mathbf{v}_\mathbf{l})\,(\mathbf{A}_{\mathbf{k}_3-\mathbf{k}_2}\mathbf{v}_{\mathbf{k}_3})\,\mathfrak{G}_{-\omega}(\mathbf{k}_1)\,\mathfrak{G}_\omega(\mathbf{k}_3)\,\Delta^*(\mathbf{l}, \mathbf{k}_8)\,\mathfrak{G}_\omega(\mathbf{k}_2). \tag{7}$$

In the absence of a field $\Delta(p_1, p_2) = \Delta(p_1)\,\delta_{p_1,-p_2}$, so that $k = p_1 + p_2$ characterizes the inhomogeneity of the gap. Starting from this, we shall henceforth write $\Delta(p_1, p_2) = \Delta_k(p_1)$. Since the important quantities in Δ_k and A_q are $k, q \sim 1/\delta$, and $\xi \sim T_c$, we can use the fact that $T/v\delta \ll 1$ in the case of a London superconductor in which we are interested and expand (7) up to second powers of $\mathbf{v}\cdot\mathbf{k}$ and $\mathbf{v}\cdot\mathbf{q}$ inclusive. Recognizing also that the changes in all the quantities occur at distances much larger than interatomic, so that $k \ll p$ and $q \ll p$, we obtain as a result

$$\Delta^*_\mathbf{k}(\mathbf{p}) = \frac{gT}{(2\pi)^3}\sum_\omega \int U(\mathbf{k}_1, \mathbf{p})\,\{[\mathfrak{G}_\omega(\mathbf{k}_1)\,\mathfrak{G}_\omega(\mathbf{k}_1)$$
$$- \mathfrak{G}_\omega(\mathbf{k}_1)\,\mathfrak{G}_{-\omega}(\mathbf{k}_1)\,(\mathbf{v}_{\mathbf{k}_1}\mathbf{k})^2]\,\Delta^*_\mathbf{k}(\mathbf{k}_1) + e\sum_\mathbf{q} (\mathbf{A}_\mathbf{q}\mathbf{v}_{\mathbf{k}_1})$$
$$\times (\mathbf{q}\mathbf{v}_{\mathbf{k}_1} + 2\mathbf{k}\mathbf{v}_\mathbf{k})\,\mathfrak{G}^2_\omega(\mathbf{k}_1)\,\mathfrak{G}^2_{-\omega}(\mathbf{k}_1)\,\Delta^*_{\mathbf{k}+\mathbf{q}}(\mathbf{k}_1)$$
$$+ e^2\sum_{\mathbf{q},\mathbf{q}'} (\mathbf{A}_\mathbf{q}\mathbf{v}_{\mathbf{k}_1})\,(\mathbf{A}_{\mathbf{q}'}\mathbf{v}_{\mathbf{k}_1})\,\Delta^*_{\mathbf{k}+\mathbf{q}+\mathbf{q}'}(\mathbf{k}_1)\,[2\mathfrak{G}^3_{-\omega}(\mathbf{k}_1)\,\mathfrak{G}_\omega(\mathbf{k}_1)$$
$$- \mathfrak{G}^2_{-\omega}(\mathbf{k}_1)\,\mathfrak{G}^2_\omega(\mathbf{k}_1)] - \mathfrak{G}^2_\omega(\mathbf{k}_1)\,\mathfrak{G}^2_{-\omega}(\mathbf{k}_1)$$
$$\times \sum_{\mathbf{p}_1,\mathbf{p}_2} \Delta^*_{\mathbf{p}_1}(\mathbf{k}_1)\,\Delta^*_{\mathbf{p}_2}(\mathbf{k}_2)\,\Delta_{\mathbf{p}_1+\mathbf{p}_2-\mathbf{k}}(\mathbf{k}_1)\}\,\frac{d\xi_1\,d\sigma_{1F}}{v_{1F}}. \tag{8}$$

Here $U(\mathbf{k}_1, \mathbf{p}) \equiv V(\mathbf{k}_1, -\mathbf{k}_2; \mathbf{p}, -\mathbf{p})$ and the integration with respect to σ_F is carried over the Fermi surface.

Summing over ω and integrating over ξ_1 we get

$$\Delta^{\bullet}_{\mathbf{k}}(\mathbf{p}) = \frac{g}{(2\pi)^3} \ln \frac{2\widetilde{\omega}\gamma}{T\pi} \int U(\mathbf{k}_1, \mathbf{p}) \frac{d\sigma_{1F}}{v_{1F}} \Delta^{\bullet}_{\mathbf{k}}(\mathbf{k}_1)$$
$$+ \frac{k_i k_j g\lambda}{2(2\pi)^3} \int U(\mathbf{k}, \mathbf{p}) v_i v_j \Delta^{\bullet}_{\mathbf{k}}(\mathbf{k}_1) \frac{d\sigma_{1F}}{v_{1F}} + \frac{e\lambda g}{(2\pi)^3}$$
$$\times \sum_{\mathbf{q}} (\mathbf{A}_{\mathbf{q}})_i (\mathbf{q} + 2\mathbf{k})_j \int U(\mathbf{k}_1, \mathbf{p}) v_i v_j \Delta^{\bullet}_{\mathbf{k}+\mathbf{q}}(\mathbf{k}_1) \frac{d\sigma_{1F}}{v_{1F}}$$
$$- \frac{2e^2\lambda}{(2\pi)^3} \sum_{\mathbf{q},\mathbf{q}'} (\mathbf{A}_{\mathbf{q}})_i (\mathbf{A}_{\mathbf{q}'})_j \int U(\mathbf{k}_1, \mathbf{p}) v_i v_j \Delta^{\bullet}_{\mathbf{k}+\mathbf{q}+\mathbf{q}'}(\mathbf{k}_1) \frac{d\sigma_{1F}}{v_{1F}}$$
$$- \frac{g\lambda}{(2\pi)^3} \sum_{\mathbf{p}_1\mathbf{p}_2} \int U(\mathbf{k}_1, \mathbf{p}) \Delta^{\bullet}_{\mathbf{p}_1}(\mathbf{k}_1) \Delta^{\bullet}_{\mathbf{p}_2}(\mathbf{k}_1) \Delta_{\mathbf{p}_1+\mathbf{p}_2-\mathbf{k}}(\mathbf{k}_1) \frac{d\sigma_{1F}}{v_{1F}},$$
$$\lambda = \frac{7\zeta(3)}{8(\pi T)^2}. \qquad (9)$$

Here $\ln \widetilde{\omega}$ is connected with $\widetilde{\omega}(\mathbf{p}/p)$ by relation (9) of [4]

$$\ln \widetilde{\omega} = \int \psi^2(\mathbf{p}) \ln \widetilde{\omega}\left(\frac{\mathbf{p}}{p}\right) \frac{d\sigma_F}{v_F} \Big/ \int \frac{d\sigma_F}{v_F}$$

(the meaning of the quantity $\psi(\mathbf{p})$ will be clarified later).

Pokrovskiĭ [4,5] has shown for the equilibrium case when $\Delta(\mathbf{p})$ does not depend on k, that in the weak-coupling model we have for arbitrary temperatures $\Delta^*(\mathbf{p}) = Q^*\psi(\mathbf{p})$, where $Q^* \equiv Q^*(T)$ is a function of the temperature only. Here $T_c = (2\widetilde{\omega}\gamma/\pi)e^{-\Lambda/g}$, and $\psi(\mathbf{p})$ and Λ are determined from the integral equation [4]

$$\psi(\mathbf{p}) = \Lambda \int U(\mathbf{k}_1, \mathbf{p}) \psi(\mathbf{k}_1) \frac{d\sigma_{1F}}{v_{1F}}.$$

The function $\psi(\mathbf{p})$ characterizes the anisotropy gap at absolute zero $\Delta(\mathbf{p}, T = 0) = 2\widetilde{\omega}\psi(\mathbf{p})e^{-\Lambda/g}$, where $\psi(\mathbf{p})$ is normalized by the condition

$$\int \psi^2(\mathbf{p}) \frac{d\sigma_F}{v_F} = \int \frac{d\sigma_F}{v_F}.$$

We analogously seek $\Delta\mathbf{k}(\mathbf{p})$ in the form $\Delta\mathbf{k}(\mathbf{p}) = Q^*(\mathbf{k})\psi(\mathbf{p})$. To obtain an equation for $Q^*(\mathbf{k})$, it is necessary to seek for $\Delta\mathbf{k}(\mathbf{p})$ the following approximation

$$\Delta^{\bullet}_{\mathbf{k}}(\mathbf{p}) = Q^{\bullet}(\mathbf{k}) \psi(\mathbf{p}) + \psi_1(\mathbf{p}).$$

The equation obtained in this manner for $\psi_1(\mathbf{p})$ can be solved if

$$\frac{T_c - T}{T_c} Q^{\bullet}_{\mathbf{k}} \int \psi^2(\mathbf{k}_1) \frac{d\sigma_{1F}}{v_{1F}} + \lambda \int v_i v_j \psi^2(\mathbf{k}_1) \frac{d\sigma_{1F}}{v_{1F}} \Big[\frac{k_i k_j}{2} Q^{\bullet}_{\mathbf{k}}$$
$$+ e \sum_{q} (\mathbf{A}_{\mathbf{q}})_i (\mathbf{q} + 2\mathbf{k})_j Q^{\bullet}_{\mathbf{k}+\mathbf{q}} - 2e^2 \sum_{\mathbf{q},\mathbf{q}'} (\mathbf{A}_{\mathbf{q}})_i (\mathbf{A}_{\mathbf{q}'})_j Q^{\bullet}_{\mathbf{k}+\mathbf{q}+\mathbf{q}'} \Big]$$
$$- \lambda \sum_{\mathbf{p}_1\mathbf{p}_2} \int Q^{\bullet}_{\mathbf{p}_1} Q^{\bullet}_{\mathbf{p}_2} Q_{\mathbf{p}_1+\mathbf{p}_2-\mathbf{k}} \psi^4(\mathbf{k}_1) \frac{d\sigma_{1F}}{v_{1F}} = 0. \qquad (10)$$

Introducing the notation

$$\frac{1}{m_{ij}} = \frac{2}{2\varepsilon_F} \int v_i v_j \psi^2(\mathbf{k}_1) \frac{d\sigma_{1F}}{v_{1F}} \Big/ \int \frac{d\sigma_{1F}}{v_{1F}},$$
$$f = \frac{7\zeta(3)}{8(\pi T)^2} \int \psi^4(\mathbf{k}_1) \frac{d\sigma_{1F}}{v_{1F}}, \qquad \eta = \frac{7\zeta(3)\varepsilon_F}{12(\pi T)^2} \qquad (11)$$

and taking Fourier transforms with respect to **k**, we obtain ultimately an equation for $Q(\mathbf{x})$:

$$\sum_{j,k} \frac{1}{2m_{jk}} (\nabla_j + 2ieA_j(\mathbf{x}))(\nabla_k + 2ieA_k(\mathbf{x})) Q^{\bullet}(\mathbf{x})$$
$$+ \frac{1}{\eta} \left\{ \frac{T_c - T}{T_c} - f \,|Q(\mathbf{x})|^2 \right\} Q^{\bullet}(\mathbf{x}) = 0.$$

Let us proceed to calculate the current density. In anisotropic conductors the current density is given by

$$j_i(\mathbf{q}) = 2eT \sum_{\omega,\mathbf{p}} \mathfrak{G}_{\omega}(\mathbf{p} + \mathbf{q}, \mathbf{p}) v_i(\mathbf{p})$$
$$- 2e^2T \sum_{\mathbf{p},\mathbf{k},\omega} w_{ij}(\mathbf{p}) A_j(\mathbf{k}) \mathfrak{G}_{\omega}(\mathbf{p} + \mathbf{q} - \mathbf{k}, \mathbf{p}), \qquad (12)$$

where $w_{ij}(\mathbf{p}) = \partial^2\epsilon(\mathbf{p})/\partial p_i \partial p_j$. Since in a normal metal the current in a constant magnetic field is zero, we have

$$2eT \sum_{\omega,\mathbf{p}} \widetilde{\mathfrak{G}}_{\omega}(\mathbf{p} + \mathbf{q}, \mathbf{p}) v_i(\mathbf{p})$$
$$- 2e^2T \sum_{\mathbf{p},\mathbf{k},\omega} w_{ij}(\mathbf{p}) A_j(\mathbf{k}) \widetilde{\mathfrak{G}}_{\omega}(\mathbf{p} + \mathbf{q} - \mathbf{k}, \mathbf{p}) = 0. \qquad (13)$$

Subtracting (13) from (12) we obtain for current density

$$j_i(\mathbf{q}) = 2eT \sum_{\omega,\mathbf{p}} \mathfrak{G}^{(2)}_{\omega}(\mathbf{p} + \mathbf{q}, \mathbf{p}) v_i(\mathbf{p})$$
$$- 2e^2T \sum_{\omega,\mathbf{p},\mathbf{k}} w_{ij}(\mathbf{p}) A_j(\mathbf{k}) \mathfrak{G}^{(2)}_{\omega}(\mathbf{p} + \mathbf{q} - \mathbf{k}, \mathbf{p}), \qquad (14)$$

where $\mathfrak{G}^{(2)}_{\omega}$ is a correction quadratic in Δ to $\mathfrak{G}_{\omega}$.

We seek the current density accurate to terms linear in $\mathbf{q}\cdot\mathbf{v}$, $\mathbf{k}\cdot\mathbf{v}$, and $\mathbf{A}\cdot\mathbf{v}$, inclusive. With this accuracy, the second term in (14) vanishes and

$$j_i(\mathbf{q}) = 2eT \sum_{\omega,\mathbf{p}} \mathfrak{G}^{(2)}(\mathbf{p} + \mathbf{q}, \mathbf{p}) v_i(\mathbf{p}). \qquad (15)$$

Substituting $\mathfrak{G}^{(2)}$ from (5) in (15) and carrying out calculations similar to those before, we obtain for $\mathbf{j}(\mathbf{x}) = \sum \mathbf{j}_{\mathbf{q}} e^{i\mathbf{q}\cdot\mathbf{x}}$ the expression

$$j_k(\mathbf{x}) = \left\{ \frac{ie}{m_{kj}} [Q(\mathbf{x}) \nabla_j Q^{\bullet}(\mathbf{x}) - Q^{\bullet}(\mathbf{x}) \nabla_j Q(\mathbf{x})] - \frac{4e^2 |Q(\mathbf{x})|^2 A_j(\mathbf{x})}{m_{kj}} \right\} C,$$
$$C = \frac{7\zeta(3)\varepsilon_F}{12(2\pi)^3 T^2} \int \frac{d\sigma_F}{v_F}.$$

The system of equations obtained corresponds to the phenomenological scheme obtained by Ginzburg [2]. The connection between the function $\psi(\mathbf{x})$ introduced in [2] and our $Q(\mathbf{x})$ is given by the relation $\psi(\mathbf{x}) = \sqrt{C}\,Q(\mathbf{x})$. An important factor in the equations obtained, however, is the doubled charge 2e, which corresponds to the fact that $\psi(\mathbf{x})$ has the meaning of a wave function of a Cooper pair. Taking this circumstance into account, we can use directly Ginzburg's formulas [2] for the physical quantities, in which we make the substitution $e \to 2e$.

The best method of determining the mass tensor seems to be measurement of the anisotropy of the depth of penetration near T_c. There are apparently no such data at present. The data on the anisotropy of the surface tension Δ are less definite in this respect. According to Sharvin and Gantmakher[7] the quantity $(\Delta_{[110]} - \Delta_{[001]})/\Delta$ for tetragonal tin is 20–25%. Inasmuch as $\Delta \backsim m^{-1/2}$, we get $\Delta m/m \sim 50\%$.

Several recent results (for example, data on the absorption of ultrasound[8]) show that the anisotropy of the energy gap itself is quite small. From Hohenberg's estimates[9] of the impurity dependence of T_c it follows that for aluminum $[\overline{\Delta^2(n)} - (\overline{\Delta(n)})^2]/(\overline{\Delta(n)})^2$ is of the order of 10^{-2}. In this case the anisotropy of the mass tensor characterizes the anisotropy of the energy spectrum of the normal metal. In light of this, the large anisotropy m_{ik} obtained by Sharvin and Gantmakher[8] becomes quite understandable.

The authors are grateful to Yu. V. Sharvin for a useful discussion.

[1] V. L. Ginzburg and L. D. Landau, JETP **20**, 1064 (1950).

[2] V. L. Ginzburg, JETP **23**, 236 (1952).

[3] Bogolyubov, Tolmachev, and Shirkov, Novyĭ metod v teorii sverkhprovodimosti (New Method in Superconductivity Theory) AN SSSR, 1958.

[4] V. L. Pokrovskiĭ, JETP **40**, 641 (1961), Soviet Phys. JETP **13**, 447 (1961).

[5] V. L. Pokrovskiĭ and N. S. Ryvkin, JETP **43**, 92 (1962), Soviet Phys. JETP **16**, 67 (1963).

[6] L. P. Gor'kov, JETP **34**, 735 (1958), Soviet Phys. JETP **7**, 505 (1958).

[7] Yu. V. Shavrin and V. F. Gantmakher, JETP **38**, 1456 (1960), Soviet Phys. JETP **11**, 1052 (1960).

[8] Morse, Olsen, and Gavenda, Phys. Rev. Lett. **3**, 15 (1959).

[9] P. Hohenberg, JETP **45**, 1208 (1963), Soviet Phys. JETP 18, 834,(1964).

Translated by J. G. Adashko
245

Chapter 3

Statistical Physics and Thermodynamics

3.1 Quantum field theory methods in statistical physics

Except for a few theoreticians of the older generations, probably nobody now remembers the awkwardness of the quantum-mechanical perturbation approach to the problems of Statistical Physics at finite temperatures. Therefore any further progress in the microscopic theory of superconductivity and the physics of condensed matter seemed unthinkable without an effective diagrammatic technique. Quantum Field Theory methods [3a] constitute the basics of modern Statistical Physics.

The first step was made in 1955 by Matsubara (Ref. 1) who pointed out on the formal analogy between the so-called $\hat{S}$-matrix in Quantum Field Theory and the statistical $\hat{S}$-matrix in the Gibbs distribution: $\exp[(\mu\hat{N} - \hat{H})/T] = \exp[(\mu\hat{N} - \hat{H}_0)/T] \times \hat{S}(1/T)$, (integration in $\widehat{S}(1/T) = \widehat{T}_\tau \exp(-\int_0^{1/T} \widehat{H}_{\rm int}(\tau)d\tau)$ is over the "imaginary time" τ:$0 < \tau < 1/T$).

Continuing the analogy, Matsubara introduced the "Heisenberg" operators: $\hat{\hat{\psi}}(x) = e^{\tau(\hat{H}-\mu\hat{N})}\hat{\psi}(\vec{r})$ $e^{-\tau(\hat{H}-\mu\hat{N})}$ and $\hat{\hat{\psi}}^+(x) = e^{-\tau(\hat{H}-\mu\hat{N})}\hat{\psi}^+(\vec{r})e^{\tau(\hat{H}-\mu\hat{N})}$ $(x \equiv (\vec{r}, \tau))$, then the operators $\hat{\psi}(x)$, $\hat{\psi}^+(x)$ in the "interaction representation" (now $\hat{H} \Rightarrow \hat{H}_0$ in the exponents) and finally he defined the new Green function $\bar{G}(1,2) = -\langle\langle \hat{T}(\hat{\psi}(1)\hat{\psi}^+(2)\hat{S})\rangle\rangle / \langle\langle \hat{S}\rangle\rangle$.

Double brackets $\langle\langle \ldots \rangle\rangle$ in $\bar{G}(1,2)$ mean the average over the *grand canonical ensemble*. All field operators now depend on the imaginary "time" coordinate, the expression for $\bar{G}(1,2)$ admit the diagrammatic expansion because the Wick Theorem applies to the thermodynamic averaging of the pairs of the operators $\hat{\psi}(x_1)$, $\hat{\psi}^+(x_2)$. In the spatial and *time* τ coordinates the diagrammatic expansion is similar to that of $T = 0$ and, in principle; one can now proceed to calculations of the thermodynamic parameters.

While each separate term in the expansion for $\bar{G}(1,2)$ over $\hat{H}_{\rm int}$ can be explicitly written down, the degree of complexity in the expressions rapidly grows with the perturbation order; as τ varies inside the *finite* interval $(0 < \tau < 1/T)$ transition to the Fourier "time"-components is not possible and automatism of the zero temperature diagrammatic technique is lost.

Expanding the Green functions $\bar{G}(\vec{R}, \tau) = T\int [d\vec{p}/(2\pi)^3] \sum_n \bar{G}(\vec{p}; \omega_n) \exp\{i(\vec{p}\vec{R} - i\omega_n\tau)\}$ in the *Fourier series* [3a] has lead to the new diagrammatic technique with the rules analogous to ones in the zero temperature case.

The Green function $\bar{G}(\vec{p}; \omega_n)$ is defined as the discrete "frequencies" $\omega_n = \pi T n$ (n is *even* for the Bose system and *odd* in the Fermi case). As to the Green functions on the *real* (physical) frequency axis, they can be found with the help of the analytical continuation from the infinite manifold of the discrete points in the imaginary axis $z_n = i\omega_n$ [3a]. (Most results [3a] were also obtained somewhat later in Ref. 2).

In literature, the new diagrammatic method is called the *Matsubara* or *thermodynamic* technique. The technique [3a] accomplishes formulation of Statistical Physics in terms of Quantum Field Theory. The new methods and the applications were summarized in the book: A. A. Abrikosov, L. P. Gor'kov, and I. E. Dzyaloshinskii, *Methods of Quantum Field Theory in Statistical Physics*; Prentice-Hall, New York, 1963.

3.2 Statistics of levels in a small metallic particle

In a macroscopic system, the momentum components of an electron are quantized by imposing periodic boundary conditions on the walls of a large cube of the linear size $L: pL = 2\pi n$. Electrons in a metal occupy all states below the Fermi energy $E_F = p_F^2/2m$, where p_F is the Fermi momentum. Correspondingly, the distance between the two levels nearest in the energy is $\Delta\varepsilon = \nu(E_F)[2\pi/L]$, where the density of states (per unit volume) in the bulk is equal to $\nu(E_F) = (mp_F/2\pi^2)$. For a metallic particle of a very small size, even crystalline and of a regular shape, such systematic of the levels becomes not applicable due to an atomic roughness at the surface, a. With sizes of the surface irregularities $d \sim a \sim p_F^{-1}$ the estimate for strength of the random surface potentials gives $\Delta\varepsilon \sim (v_F/a) \sim E_F$. Therefore, the only meaningful number for such particle with a volume V is the *total* density of states $\bar{\nu} = V\nu(E_F)$. This value defines the average energy interval between two neighboring energy levels $\Delta = \bar{\nu}^{-1}$.

Estimates (see [3b]) show that the value of Δ can be quite reasonable; with $N \sim 10^4 - 10^5$ particles in the volume $V\Delta = \bar{\nu}^{-1}$ is of the order of $\Delta \sim E_F/N \sim 1 - 0.1K$. That is, discreteness of the energy levels can be observed, at least in principle in the *nano*-size particles. The important distinction in the averaging of the bulk properties is that the *grand canonical ensemble* cannot be used at low temperatures, for in *isolated* particles the number of electrons is fixed. Indeed, the energy cost of charging by removing or adding one electron is the Coulomb energy $\sim e^2V^{-1/3} \gg \Delta$. Thanks to that, one may expect observing experimentally the differences between systems with odd and even numbers of electrons with such properties as the low temperature specific heat, magnetization, superconductivity etc.

The problem itself was not new and was first discussed as early as in 1937 by Fröhlich (Ref. 3) and in 1962 by Kubo (Ref. 4). While Fröhlich merely assumed the equidistant levels, Kubo has chosen a more realistic looking Poisson distribution $P(\Delta)d\Delta = \bar{\nu}\exp[-\Delta\bar{\nu}\}d\Delta$, where $P(\Delta)d\Delta$ is the probability of finding the nearest energy level inside the interval $\{\Delta, \Delta + d\Delta\}$. However, the distinctive feature of the Poisson distribution is the non-zero value of $P(\Delta = 0) = \bar{\nu}$; from the physical point of view that means attraction between the two levels. As there is no reason for degeneracy in the spectrum of electrons moving in a *random* field, we were looking for a more satisfactory form of distribution.

In nuclear physics, the notion of a probability distribution for the energy distance between the nearest levels was first put forward by E. Wigner in 1955 in relation to systematic spectra of complex nuclei in the range of high quantum numbers. As a mathematical problem, it was put forward and solved in 1962 by F. Dyson and M. Mehta (see references in [3b]). The basic idea in [3b] was that randomness of the electronic potential in small metallic particles warrants the realization of one of the Dyson ensembles: 1) for the orthogonal ensemble spins of electrons are preserved; the system is invariant with respect to the time and space inversion; 2) if the spin-orbit interaction on a defect is strong, the symplectic ensemble is realized; the spin of an electron is not preserved and the remaining invariance is only with respect to the time reversal; 3) for the unitary ensemble the time reversal invariance is broken as, say, in the presence of a strong magnetic field.

Our hypothesis was confirmed in 1982 when K. Efetov found the exact solution for the single particle spectra in disordered metallic particles (Ref. 5).

Implications, e.g. for the specific heat or magnetization could be indeed non-trivial, but the effects are difficult to observe experimentally, in the first place, because of smallness of the total particles' volume in a solution. We thought of studying the electromagnetic properties of metallic colloids, but the prediction of the large effect in [3b] turned out wrong because the Fermi-Thomas screening of the electrical field at the surface drastically decreases the value of the electric field inside the particle. More technical problems lie in the difficulty of producing particles homogeneous in size. (Worth noticing in [3b] is the quasiclassical method of calculations of the dipole matrix elements).

3.3 Fluctuations in branching processes

Common sense suggests that fluctuations in a cascade process could be large, but I have never seen any quantitative discussion on the subject in condensed matter literature. In our paper on fragmentation [3c] we have a implicitly modeled cascade created by a high energy phonon that may decay into two phonons of smaller energy via the third order anharmonic terms in the Hamiltonian. I remember, that I heard it first from Landau how to address the problem in principle, but to the best of my knowledge, he had never published anything. Therefore, I think it was worth explaining briefly the key ideas in our paper.

Let $N(E,t;E_0,0)$ be the number of phonons at time t created by a phonon with the energy E_0 at $t=0$. Denote as $W(E;E_1,E_2)dE_1$ the probability distribution for the phonon with the energy E to decay into two phonons with the energies E_1 and $E_2(E_1+E_2=E)$. In the standard approach the master equation describes the multiplication process in the cascade initiated by the phonon with energy E at $t=0$.

Now introducing, more generally, the *total distribution* $W_{tot}(N,E,t;1,E_0,0)$ as the probability of finding N phonons in the cascade initiated by the first phonon. In the new method $W_{tot}(N,E,t;1,E_0,0)$ is expanded over momenta $\langle N^K(E,t;1,E_0,0)\rangle \equiv \sum_N N^K W_{tot}(N,E,t;1,E_0,0)$ of the arbitrary order in N. The mathematics is nothing but calculating the Mellin transform.

Compare now the two cascades that start at $t=0$ and $t+\delta t$, correspondingly; with the probability $dt\int_0^{E_0} W(E_0;E_1,E-E_1)dE_1$ no decay occurs in the time interval dt: the phonon numbers are same. Should the first decay have occurred during the time interval dt, the event creates the *two cascades* with the new initial energies E_1 and $E_2(E_1+E_2=E)$ which proceed *independently*.

As an illustration, we can calculate the second momentum, i.e. the average *square* of the number of phonons keeping in mind that after the *first* decay the total number of phonons becomes the sum from *two* cascades: $N(E,t;E_0,0)=N(E,t;E_1,0)+N(E,t;E_0-E_1,0)$.

The master equation for the second moment $< N^2(E,t;E_0,0) >$ has the form:

$$
\begin{aligned}
d\langle N^2(E,t;E_0,0)\rangle/dt &= -\langle N^2(E,t;E_0,0)\rangle \int_0^{E_0} W(E_0;E_1,E-E_1)dE_1 \\
&\quad +2\int_E^{E_0} W(E_o;E_1,E-E_1)\langle N^2(E,t;E_1,0)\rangle dE_1 \\
&\quad +2\int_E^{E_0-E} W(E_0;E_1,E-E_1)\langle N(E,t;E_1,0)\rangle\langle N(E,t;E_0-E_1,0)\rangle dE_1
\end{aligned}
$$

That is, the second moment $\langle N^2(E,t;E_0,0)\rangle$ can be expressed through the first order moments. The difference: $\sqrt{\langle N^2(E,t;E_0,0)\rangle} - \langle N(E,t;E_0,0)\rangle$ shows the scale of fluctuations.

The method applies to different systems that evolve along a tree in the phase space. Formally, the "averaging" above is the averaging over many realizations of a cascade. The important physical question, therefore, is whether the cascade at its late stages becomes self-averaging. We considered a steady state (with a "source" of phonons) as well. I shall not dwell upon the details (see in [3c]).

References

1. T. Matsubara, *Prog. Theor. Phys.* **14**, 351 (1955)
2. E. S. Fradkin, *Sov. Phys. JETP* **9**, 912 (1959)
3. H. Fröhlich, *Physica (Utr)* **4**, 406 (1937)]
4. R. Kubo, *J. Phys. Soc. Japan* **17**, 975 (1962)
5. K. B. Efetov, *Sov. Phys. JETP* **56**, 467 (1982)

List of original papers

[3a] A. A. Abrikosov, L. P. Gor'kov, and I. E. Dzyaloshinski, *On the application of quantum field theory methods to problems of quantum statistics at finite temperatures*, Sov. Phys. JETP, (36)9, 636 (1959)

[3b] L. P. Gor'kov and G. M. Eliashberg, *Minute metallic particles in an electromagnetic field*, Sov. Phys. JETP, 21, 940 (1965)

[3c] S. E. Esipov, L. P. Gor'kov, and T. J. Newman, *Fluctuation in Fragmentation Process*, Journal of Phys. A26, 787 (1993)

SOVIET PHYSICS JETP VOLUME 36 (9), NUMBER 3 SEPTEMBER, 1959

ON THE APPLICATION OF QUANTUM-FIELD-THEORY METHODS TO PROBLEMS OF QUANTUM STATISTICS AT FINITE TEMPERATURES

A. A. ABRIKOSOV, L. P. GOR'KOV, and I. E. DZYALOSHINSKIĬ

Institute of Physical Problems, Academy of Sciences, U.S.S.R.

Submitted to JETP editor December 4, 1958

J. Exptl. Theoret. Phys. (U.S.S.R.) **36**, 900-908 (March, 1959)

A formulation of thermodynamic perturbation theory is proposed which makes it possible to make full use of quantum-field-theory methods in quantum statistics at finite temperatures. The method is a generalization of the Matsubara technique and is based on the expansion of the Green's functions in Fourier series in the "imaginary time" variable. The technique so obtained differs from the usual diagram technique for $T = 0$ by the replacement of integration over the frequencies by summation over discrete values of the imaginary frequency. The analytic properties of the Fourier components of the Green's functions are examined. It is shown that owing to the possibility of analytic continuation a knowledge of the corresponding equilibrium Green's functions is sufficient for the solution of various kinetic and nonstationary processes.

1. INTRODUCTION

THE methods of quantum field theory have recently been successfully applied to problems of statistics.[1-4] The application is based on the fact that the apparatus of quantum field theory is developed through the wide use of the "diagram technique," which provides a very intuitive representation of the structure and character of any approximation. Earlier methods of quantum statistics, which started from ideal-gas approximations, could scarcely take the interaction between the particles into account beyond one or two approximations because of the complexity of the older quantum perturbation theory. This is utterly inadequate for the actual many-body problem; to obtain any approximation of physical significance it is necessary to sum over an infinite set of different terms of the perturbation-theory series. The diagram technique is extraordinarily useful in such problems, since it formulates simple rules by which any term of the perturbation theory can be written down.

The basis of the quantum-field-theory methods is the calculation of the so-called Green's function of a particle, which for the case of temperatures different from zero is defined as

$$G(1,2) = -i\mathrm{Sp}\{\exp[(\Omega + \mu N - \hat{H})/T]\cdot T(\tilde{\psi}(x_1)\tilde{\psi}^+(x_2))\}, \quad (1)$$

where $\tilde{\psi}$, $\tilde{\psi}^+$ are second-quantization operators in the Heisenberg representation. Knowledge of the Green's function makes it possible easily to calculate all the thermodynamic quantities of the system. For example, the particle number density as a function of the chemical potential μ and the temperature is connected with G by the relation

$$N(\mu, T) = \mp iG(x, x)|_{t'=t+0}$$

(the minus sign is for Fermi statistics, the plus for Bose statistics).

At the absolute zero of temperature the Green's function is calculated by going over to the so-called "interaction representation." In this representation $G(1, 2)$ has the form

$$G(1,2) = -i\langle T(\psi_1\psi_2^+ S)\rangle / \langle S\rangle, \quad (1a)$$

where $\langle\ldots\rangle$ means the average over the ground state of the system, and S is the well known S matrix of the quantum field theory. Expanding S in powers of the interaction constant, we obtain the usual Feynman diagram technique.

For temperatures other than zero it is already impossible to represent G in a form like Eq. (1a), and consequently it is impossible to calculate G by means of the diagram technique.

Several years ago Matsubara[4] proposed a new formulation of the thermodynamic perturbation theory which is almost completely analogous to the diagram technique of field theory. The present paper is devoted to the description of a technique which is a further development of the Matsubara method. This technique makes it possible to calculate the thermodynamic quantities at finite temperatures with almost the same simplicity with which this is done for $T = 0$. It will further be shown that by establishing the connection of the quantities that figure in this technique with the corresponding quantities in the usual technique one can determine all the kinetic characteristics of the system.*

2. THE MATSUBARA METHOD

The basic idea of the Matsubara method is the passage from the time t to a certain imaginary "time" τ. We note that the ordinary Green's function $G(1, 2)$ can be represented in the form

$$G(1,2)=\begin{cases} -i\mathrm{Sp}\left\{\exp\left[\frac{\Omega+\mu N-\hat{H}}{T}\right] \times\exp\left[\frac{i(\hat{H}-\mu N)}{\hbar}(t_1-t_2)\right]\psi(\mathbf{r}_1) \times\exp\left[-\frac{i(\hat{H}-\mu N)}{\hbar}(t_1-t_2)\right]\psi^+(\mathbf{r}_2)\right\}, & t_1>t_2, \\ \pm i\mathrm{Sp}\left\{\exp\left[\frac{\Omega+\mu N-\hat{H}}{T}\right] \times\exp\left[-\frac{i(\hat{H}-\mu N)}{\hbar}(t_1-t_2)\right]\psi^+(\mathbf{r}_2) \times\exp\left[\frac{i(\hat{H}-\mu N)}{\hbar}(t_1-t_2)\right]\psi(\mathbf{r}_1)\right\}, & t_1<t_2, \end{cases} \tag{2}$$

where ψ, ψ^+ are the second-quantization operators in the Schrödinger representation.

In the Matsubara method one introduces a temperature-dependent Green's function $\mathfrak{G}$, which is obtained from G by the replacement $t \rightarrow -i\tau\hbar$:

$$\mathfrak{G}(1,2)=\begin{cases} -\mathrm{Sp}\left\{\exp\left[\frac{\Omega+\mu N-\hat{H}}{T}\right] \times\exp\left[(\hat{H}-\mu N)(\tau_1-\tau_2)\right]\psi(\mathbf{r}_1) \times\exp\left[-(\hat{H}-\mu N)(\tau_1-\tau_2)\right]\psi^+(\mathbf{r}_2)\right\}, & \tau_1>\tau_2, \\ \pm\mathrm{Sp}\left\{\exp\left[\frac{\Omega+\mu N-\hat{H}}{T}\right] \times\exp\left[-(\hat{H}-\mu N)(\tau_1-\tau_2)\right]\psi^+(\mathbf{r}_2) \times\exp\left[(\hat{H}-\mu N)(\tau_1-\tau_2)\right]\psi(\mathbf{r}_1)\right\}, & \tau_1<\tau_2. \end{cases} \tag{3}$$

Matsubara showed that in the formula (3) one can go over to a sort of "interaction representation". Namely, let us represent the statistical matrix $\hat{\rho} = \exp\{(\mu N - \hat{H})/T\}$ in the form

$$\exp\left[\frac{\mu N-\hat{H}}{T}\right]=\exp\left[\frac{\mu N-\hat{H}_0}{T}\right]\cdot\hat{S}(1/T) \tag{4}$$

and introduce the new operators $\overline{\psi}(\mathbf{r}, \tau)$, $\psi(\mathbf{r}, \tau)$ in the "interaction representation" by the formulas

$$\psi(\mathbf{r},\tau)=\exp\left[\tau(\hat{H}_0-\mu N)\right]\psi(\mathbf{r})\exp\left[-\tau(\hat{H}_0-\mu N)\right],$$

$$\overline{\psi}(\mathbf{r},\tau)=\exp\left[\tau(\hat{H}_0-\mu N)\right]\psi^+(\mathbf{r})\exp\left[-\tau(\hat{H}_0-\mu N)\right]. \tag{5}$$

Then, using the formal resemblance of the equation satisfied by the matrix $\hat{S}(1/T)$, $-\partial\hat{S}(\tau)/\partial\tau = H(\tau)\hat{S}(\tau)$, with the Schrödinger equation, we write $\hat{S}(1/T)$, in complete analogy with the usual S matrix, in the form

$$\hat{S}(1/T)=T_\tau\exp\left\{-\int_0^{1/T}\hat{H}_{\mathrm{int}}(\tau)\,d\tau\right\},$$

where T_τ is the time-ordering operator with respect to the imaginary "time" τ. The interaction Hamiltonian $H_{\mathrm{int}}(\tau)$ now depends on the operators $\psi(\mathbf{r}, \tau)$ and $\overline{\psi}(\mathbf{r}, \tau)$ in the same way that it formerly depended on $\psi(\mathbf{r})$ and $\psi^+(\mathbf{r})$.

Now, using the possibility of cyclic permutation of the operators under the sign Sp in Eq. (2), we can show that $\mathfrak{G}$ can be written in a form analogous to that used in the field theory:

$$\mathfrak{G}(1,2)=-\frac{\mathrm{Sp}\{\exp[(\mu N-\hat{H}_0)/T]T_\tau[\psi_1\overline{\psi}_2\hat{S}(1/T)]\}}{\mathrm{Sp}\{\exp[(\mu N-\hat{H}_0)/T]\hat{S}(1/T)\}}, \tag{6}$$

It is not hard to see that for the potential Ω we can write the following formula:

$$\Omega=\Omega_0-T\ln\left[\frac{\mathrm{Sp}\left\{\exp\left[\frac{\mu N-\hat{H}_0}{T}\right]\hat{S}(1/T)\right\}}{\mathrm{Sp}\left\{\exp\left[\frac{\mu N-\hat{H}_0}{T}\right]\right\}}\right]. \tag{6a}$$

The expressions (6) and (6a) have the forms of the corresponding formulas of field theory:

$$\mathfrak{G}(1,2)=-\langle T(\psi_1\overline{\psi}_2S)\rangle/\langle S\rangle,\quad \Omega=\Omega_0-T\ln\langle S\rangle, \tag{7}$$

if we understand the averaging symbol $\langle\ldots\rangle$ to mean the operation

$$\mathrm{Sp}\left\{\exp\left[\frac{\mu N-\hat{H}_0}{T}\right]\ldots\right\}\Big/\mathrm{Sp}\left\{\exp\left[\frac{\mu N-\hat{H}_0}{T}\right]\right\}.$$

Matsubara showed that for such averages of T-products of several ψ-operators there exists a "Wick's theorem," according to which the average in question breaks up into a sum of products of averages by pairs of operators:

*We have learned that similar results have been obtained by E. S. Fradkin (J. Exptl. Theoret. Phys. (U.S.S.R.), in press).

$$\mathfrak{G}_0(1,2) = -\langle T_\tau(\psi_1\bar{\psi}_2)\rangle.$$

Of course this does not apply to Bose systems below the point of the Bose condensation. Similarly, the Wick's theorem is not applicable to Fermi systems in which superconductivity exists. For the construction of a thermodynamic perturbation theory in this range of temperatures one must use in the first case an obvious generalization of some work of Belyaev,[5] and in the second case, work of Gor'kov.[6]

Expanding the expression for $S(1/T)$ appearing in Eq. (7) in a series of powers of the interaction Hamiltonian and using Wick's theorem, we get the usual diagram technique, in which each line corresponds to a zeroth-order Green's function $\mathfrak{G}_0(1, 2)$. For example, in the case of the four-fermion interaction with

$$\hat{H}_{\text{int}} = \frac{g}{2}\Lambda_{\alpha\beta;\gamma\delta}\bar{\psi}_\alpha(x)\bar{\psi}_\beta(x)\psi_\gamma(x)\psi_\delta(x),$$

where $x = \{\mathbf{r}, \tau\}$ and $\Lambda_{\alpha\beta;\gamma\delta}$ is some function of the spin indices antisymmetric in $\alpha\beta$ and $\gamma\delta$,

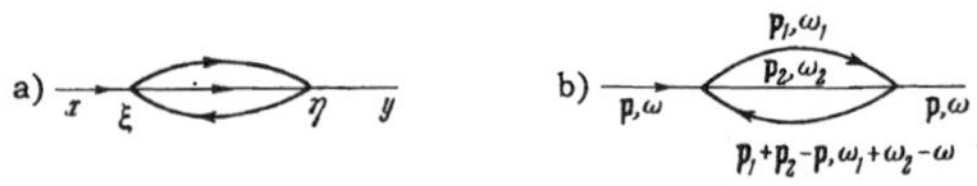

FIG. 1

we find for the second-approximation correction to the Green's function, corresponding to the diagram of Fig. 1a

$$2g^2\Lambda_{\lambda\mu\nu\tau}\Lambda_{\rho\pi\kappa\omega}\int_0^{1/T}d\tau_\xi\int_0^{1/T}d\tau_\eta\int d\xi\int d\eta\,\mathfrak{G}^0_{\alpha\lambda}(x-\xi)$$
$$\times\,\mathfrak{G}^0_{\nu\rho}(\xi-\eta)\,\mathfrak{G}^0_{\tau\pi}(\xi-\eta)\mathfrak{G}^0_{\kappa\mu}(\eta-\xi)\,\mathfrak{G}^0_{\omega\beta}(\eta-y). \quad (8)$$

In the case of the electron-phonon interaction[1,4]

$$\hat{H}_{\text{int}} = g\bar{\psi}\psi\varphi,$$

the first correction to the electron Green's function, corresponding to the diagram of Fig. 2a, has

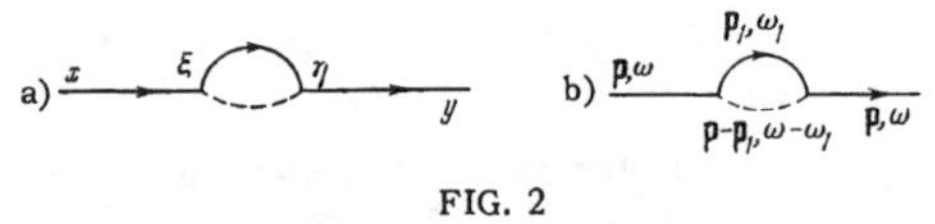

FIG. 2

the form (the dotted line indicates the phonon Green's function $\mathfrak{D}_0 = -\langle T_\tau(\varphi_1\varphi_2)\rangle$, which is the analogue of the ordinary D function of the phonon):

$$-g^2\int_0^{1/T}d\tau_\xi\int_0^{1/T}d\tau_\eta\int d\xi\int d\eta\,\mathfrak{G}^0$$
$$\times(x-\xi)\,\mathfrak{G}^0(\xi-\eta)\,\mathfrak{D}^0(\xi-\eta)\,\mathfrak{G}^0(\eta-y). \quad (9)$$

The actual calculation of expressions of types (8), (9) and of the more complicated expressions in the higher approximations is, however, very laborious in the Matsubara technique. The point is that the success of field-theory methods in statistics for $T = 0$ is due to the highly automatic nature of the calculations, which is achieved by the use of Fourier expansions of all the quantities with respect to all four coordinates. It is obvious that in the Matsubara method the automatic feature is lacking because of the finiteness of the range of τ, from zero to $1/T$. Indeed in the coordinate representations (as regards τ) $\mathfrak{G}^0$ (and $\mathfrak{G}$) is a discontinuous function of the variable τ; therefore in fact all the integrals over τ break up into integrals over a very large number of regions. The number of such regions obviously increases rapidly with increase of the order of the approximation.

3. THE EXPANSION IN FOURIER SERIES

The Matsubara technique can be decidedly improved by the exploitation of certain general properties of the thermodynamic Green's functions.

As we have shown, by Eq. (3) the Green's function $\mathfrak{G}(1, 2)$ is a function of the difference $\tau_1 - \tau_2$, specified in the interval from $-1/T$ to $1/T$. It is therefore expedient to continue it periodically and expand it in Fourier series in the variable τ:*

$$\mathfrak{G}(\tau) = T\sum_n e^{-i\omega_n\tau}\mathfrak{G}(\omega_n),\ \omega_n = \pi Tn,$$

$$\mathfrak{G}(\omega_n) = \frac{1}{2}\int_{-1/T}^{1/T} e^{i\omega_n\tau}\mathfrak{G}(\tau)\,d\tau. \quad (10)$$

One major property of the function $\mathfrak{G}$ is essential for the transformation of the perturbation-theory series. It follows from Eq. (3) that the function $\mathfrak{G}$ for negative values of τ is simply related to the $\mathfrak{G}$ for $\tau > 0$, namely we have the relation

$$\mathfrak{G}(\tau) = \mp\,\mathfrak{G}(\tau + 1/T),\ \tau < 0, \quad (11)$$

where the minus sign is for Fermi statistics and the plus for Bose statistics. The formula (11) is easily derived if we use the fact that one may cyclically permute the operators under the sign of the trace. The relation (11) is of course also valid for the free Green's functions. For the functions as periodically continued this relation is valid for any value of τ.

*Recently I. M. Khalatnikov called our attention to the fact that the idea of expanding the temperature-dependent Green's functions of Matsubara in Fourier series is also contained in a paper by Ezawa, Tomozawa, and Umezawa.[9] These writers, however, being interested only in the application of the Matsurbara method to the problem of multiple production, confined themselves to the case of zero chemical potention. Meanwhile the construction of the technique is nontrivial just in the case $\mu \neq 0$.

If we note further that at each vertex of a Feynman diagram an even number of fermion lines come together, it is easy to see that all the integrals $\int_0^{1/T} \ldots d\tau$ in formulas of the type of Eqs. (8) and (9) can be replaced by $\frac{1}{2}\int_{-1/T}^{1/T} \ldots d\tau$. When this is done the Fourier transformation (10) is easily carried out in all terms of the perturbation-theory series.

The relation (11) also has the consequence that in the Fourier expansion of a fermion Green's function only the components with $\omega_n = \pi T(2n+1)$ are different from zero, and the expansion of a boson Green's function has only components with $\omega_n = \pi T \cdot 2n$. We shall present the expressions for the zeroth-order Green's functions $\mathfrak{G}_0(\omega_n, \mathbf{p})$ (here the Fourier integral transformation for the space coordinates has also been carried out).

Fermions:

$$\mathfrak{G}_0(\omega_n, \mathbf{p}) = [i\omega_n + \mu - \varepsilon(\mathbf{p})]^{-1}, \quad \omega_n = (2n+1)\pi T$$

(here $\epsilon(\mathbf{p}) = p^2/2m$, and μ is the chemical potential); bosons above the Bose condensation point:

$$\mathfrak{G}_0(\omega_n, \mathbf{p}) = [i\omega_n + \mu - \varepsilon(\mathbf{p})]^{-1}, \quad \omega_n = 2n\pi T;$$

phonons:

$$\mathfrak{D}_0(\omega_n, \mathbf{k}) = -\omega^2(\mathbf{k})/[\omega^2(\mathbf{k}) + \omega_n^2], \quad \omega_n = 2n\pi T$$

($\omega(\mathbf{k})$ is the energy of a phonon).

Carrying out the Fourier transformation in all the terms of the expansion of the Green's function in the perturbation-theory series, we can verify that the technique so obtained is entirely equivalent to the diagram technique in the momentum space at absolute zero. To each line of a diagram there corresponds a zeroth-order Green's function $\mathfrak{G}_0(\omega_n, \mathbf{p})$, and to each vertex a δ function expressing the conservation laws $\Sigma \mathbf{p} = 0$, $\Sigma \omega_n = 0$. The only difference is that in our case we have instead of an integration over frequency a summation over discrete imaginary "frequencies" $i\omega_n$. For example, the formula (8) for the correction to the electron Green's function from the four-fermion interaction is transformed into the form

$$\Lambda_{\alpha\mu\nu\tau}\Lambda_{\nu\tau\mu\beta}\frac{2g^2T^2}{(2\pi)^6}\sum_{\omega_1\omega_2}\int d^3\mathbf{p}_1 d^3\mathbf{p}_2\,[i\omega_1 + \mu - \varepsilon(\mathbf{p}_1)]^{-1}$$
$$\times [i\omega_2 + \mu - \varepsilon(\mathbf{p}_2)]^{-1}\,[i(\omega_1 + \omega_2 - \omega) + \mu - \varepsilon(\mathbf{p}_1 + \mathbf{p}_2 - \mathbf{p})]^{-1}$$
$$\times [i\omega + \mu - \varepsilon(\mathbf{p})]^{-2},$$

corresponding to the diagram of Fig. 1b.

The expression (9) takes the form

$$\delta\mathfrak{G}(\omega, \mathbf{p}) = \frac{g^2T}{(2\pi)^3}\Big\{\sum_{\omega'}\int d^3\mathbf{p}'\,[i\omega' + \mu - \varepsilon(\mathbf{p}')]^{-1}\,\omega^2(\mathbf{p}-\mathbf{p}')$$
$$\times [\omega^2(\mathbf{p}-\mathbf{p}') + (\omega-\omega')^2]^{-1}\Big\}\,[i\omega + \mu - \varepsilon(\mathbf{p})]^{-2},$$

which corresponds to the diagram of Fig. 2b.

On passage to the absolute zero of temperature, large values of n play the main part in the sums. Therefore in the perturbation-theory formulas we must make the replacement

$$T\sum_{\omega_n} \to \frac{1}{2\pi}\int_{-\infty}^{\infty} \ldots d\omega.$$

Furthermore, it is obvious that when one replaces all frequencies by imaginary values, $\omega \to -i\omega'$, integrals with respect to $d\omega$ from $-\infty$ to $+\infty$ of products of several factors of the form

$$(i\omega + \mu - \varepsilon(\mathbf{p}))^{-1}$$

go over into integrals of products of factors

$$(\omega' + \mu - \varepsilon(\mathbf{p}))^{-1}, \qquad (12)$$

which corresponds to a rotation of the path of integration to real ω. In the case of Fermi Green's functions $\epsilon(\mathbf{p}) - \mu$ can be either positive or negative. Therefore after the indicated transformation of the contour of integration the path around the poles is chosen as shown in Fig. 3a. For Bose particles the chemical potential μ of an ideal gas

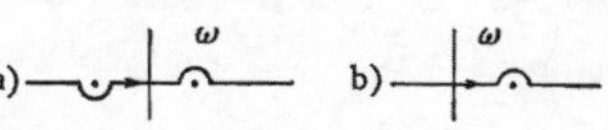

FIG. 3

is always negative; therefore the poles in Eq. (11) lie in the lower half-plane of ω' (Fig. 3b). Conversely, it can be seen from this how from the well known rules of the diagram technique for $T = 0$ one can write down the corresponding term of the perturbation theory for $T \neq 0$. To do this we first make the replacement $\omega' \to i\omega_n$ ($n = (2m+1)$ for Fermi particles and $n = 2m$ for Bose systems) in the corresponding expressions for $T = 0$, and must then go from integration over the frequencies to summation:

$$\int_{-\infty}^{+\infty} d\omega \to -2\pi T i \sum_{\omega_n}.$$

Everything said up to now has referred to calculations by perturbation theory. In quite a number of problems, however, calculations by perturbation theory have been found to be inadequate. For these purposes it is essential to know certain exact equations connecting the quantities with each other. One such relation is the Dyson equation. As is well known, in field theory one can write an equation connecting the Green's function with the so-called vertex part (the Dyson equation). The analogue of this equation can of course be obtained in the technique proposed here. For the case of the four-fermion interaction it is shown schematically, in the language of the diagrams, in Fig. 4. Here heavy

FIG. 4

lines mean the complete Green's function $\mathfrak{G}$, and the rhomb means the complete vertex part, i.e., the sum of all connected diagrams having four external fermion lines. In the momentum representation the Dyson equation corresponding to this diagram has the form:

$$\mathfrak{G}^{-1}_{\alpha\beta}(\omega, \mathbf{p}) = \mathfrak{G}^{0-1}_{\alpha\beta}(\omega, \mathbf{p}) - \frac{2gT}{(2\pi)^3} \sum_{\omega'} \int d^3\mathbf{p}\,\mathfrak{G}_{\nu\mu}(\omega', \mathbf{p}')\,\Lambda_{\alpha\mu;\nu\beta}$$

$$- \frac{2gT^2}{(2\pi)^6} \Lambda_{\alpha\lambda,\mu\nu} \sum_{\omega_1\omega_2} \int d^3\mathbf{p}_1 d^3\mathbf{p}_2 \mathfrak{G}_{\mu\tau}(\omega_1, \mathbf{p}_1)\,\mathfrak{G}_{\nu\rho}(\omega_2, \mathbf{p}_2)$$

$$\times \mathfrak{G}_{\varkappa\lambda}(\omega_1 + \omega_2 - \omega; \mathbf{p}_1 + \mathbf{p}_2 - \mathbf{p})$$

$$\times \mathfrak{T}_{\tau\rho,\varkappa\beta}(\omega_1, \mathbf{p}; \omega_2, \mathbf{p}_2; \omega_1 + \omega_2 - \omega, \mathbf{p}_1 + \mathbf{p}_2 - \mathbf{p}; \omega, \mathbf{p}).$$

Here $\mathfrak{T}$ is the complete vertex part already mentioned. It is connected with the Fourier components of the two-particle Green's function

$$\mathfrak{G}_{\alpha\beta,\gamma\delta}(x_1, x_2; x_3, x_4) = \langle T(\psi_\alpha(x_1)\psi_\beta(x_2)\bar{\psi}_\gamma(x_3)\bar{\psi}_\delta(x_4)\hat{S})\rangle / \langle \hat{S}\rangle$$

by the relation

$$\mathfrak{G}_{\alpha\beta,\gamma\delta}(\omega_1, \mathbf{p}_1, \omega_2, \mathbf{p}_2; \omega_3, \mathbf{p}_3, \omega_4, \mathbf{p}_4)$$

$$= \frac{(2\pi)^3}{T} \Big\{ \frac{(2\pi)^3}{T} \Big[\mathfrak{G}_{\alpha\delta}(\omega_1, \mathbf{p}_1) \mathfrak{G}_{\beta\gamma}(\omega_2, \mathbf{p}_2)$$

$$\times \delta_{\omega_1\omega_4} \delta(\mathbf{p}_1 - \mathbf{p}_4) - \mathfrak{G}_{\alpha\gamma}(\omega_1, \mathbf{p}_1)\,\mathfrak{G}_{\beta\delta}(\omega_2, \mathbf{p}_2)\,\delta_{\omega_1\omega_3}\delta(\mathbf{p}_1 - \mathbf{p}_3) \Big]$$

$$- \frac{1}{2} \mathfrak{G}_{\alpha\lambda}(\omega_1, \mathbf{p}_1)\,\mathfrak{G}_{\beta\mu}(\omega_2, \mathbf{p}_2)\,\mathfrak{T}_{\lambda\mu,\nu\tau}(\omega_1, \mathbf{p}_1; \omega_2, \mathbf{p}_2; \omega_3, \mathbf{p}_3; \omega_4, \mathbf{p}_4)$$

$$\times \mathfrak{G}_{\nu\gamma}(\omega_3, \mathbf{p}_3)\mathfrak{G}_{\tau\delta}(\mathbf{p}_4, \omega_4) \Big\} \delta_{\omega_1 + \omega_2 - \omega_3 - \omega_4} \delta(\mathbf{p}_1 + \mathbf{p}_2 - \mathbf{p}_3 - \mathbf{p}_4).$$

We shall not write out the analogous Dyson equation for the case of the electron-phonon interaction.

Knowing the thermodynamic Green's function $\mathfrak{G}$ one can find thermodynamic functions of the system. For this purpose one can use, for example, the following simple relation

$$\frac{\partial\Omega}{\partial g} = T\frac{n}{g} \sum_{\omega} \int \frac{d^3\mathbf{p}}{(2\pi)^3} [\mathfrak{G}^{0-1}_{\alpha\beta}(\mathfrak{G}_{\beta\alpha} - \mathfrak{G}_{\beta\alpha}{}^0)].$$

Here n is a coefficient depending on the form of the interaction. For the interaction with phonons n = 1; for the four-fermion interaction $n = \frac{1}{2}$.

4. THE ANALYTICAL PROPERTIES OF THE GREEN'S FUNCTIONS AND THE TRANSITION TO TIME-DEPENDENT QUANTITIES

As has already been mentioned, a knowledge of the Green's function is sufficient for the determination of the thermodynamic characteristics of a system. But for the study of various kinetic phenomena, such as scattering or absorption of light or sound, diffusion, and so on, one must know the ordinary time-dependent functions. We shall show how, knowing the thermodynamic one-particle function $\mathfrak{G}(\omega_n, \mathbf{p})$, one can find the time-dependent $G(\omega, \mathbf{p})$.

Beginning with the coordinate space, and comparing Eqs. (2) and (3), we see that in the range of τ from $-1/T$ to $1/T$ the transition from $\mathfrak{G}$ to G is accomplished by replacing τ by it. It is much more interesting, however, to find the connection between the Fourier components $G(\omega, \mathbf{p})$ and $\mathfrak{G}(\omega_n, \mathbf{p})$.

As has been shown by Landau,[7] when the function $G(\omega, \mathbf{p})$ is expanded as a Fourier integral, its Fourier transform is a nonanalytic function of ω. Its real and imaginary parts are connected by the relation

$$G'(\omega, \mathbf{p}) = \frac{1}{\pi} \int_{-\infty}^{+\infty} \coth\frac{x}{2T} \frac{G''(x, \mathbf{p})}{x - \omega} dx$$

for the Fermi statistics, and the relation

$$G'(\omega, \mathbf{p}) = \frac{1}{\pi} \int_{-\infty}^{+\infty} \tanh\frac{x}{2T} \frac{G''(x, \mathbf{p})}{x - \omega} dx$$

for Bose statistics.

It can be shown that the Fourier transform of the so-called retarded Green's function $G^R(\mathbf{r}_1 - \mathbf{r}_2, t_1 - t_2)$,

$$G^R(1,2) = \begin{cases} -i\,\mathrm{Sp}\left\{\exp\left[\frac{\Omega + \mu N - \hat{H}}{T}\right] \times [\tilde{\psi}(x_1)\tilde{\psi}^+(x_2) \pm \tilde{\psi}^+(x_2)\tilde{\psi}(x_1)]\right\}, & t_1 > t_2, \\ 0, & t_1 < t_2. \end{cases}$$

is analytic in the upper half-plane. Here the plus sign is for Fermi statistics, the minus for Bose statistics. It turns out that the Fourier transform $G^R(\omega, \mathbf{p})$ of this function is simply related to $G(\omega, \mathbf{p})$. Namely, their real parts are identical, and

$$G''^R(\omega, \mathbf{p}) = \coth(\omega/2T)\,G''(\omega, \mathbf{p})$$

for the Fermi statistics, and

$$G''^R(\omega, \mathbf{p}) = \tanh(\omega/2T)\,G''(\omega, \mathbf{p})$$

for the Bose statistics. For $G^R(\omega, \mathbf{p})$ we can write an expansion of the type used by Lehmann[8]

$$G^R(\omega, \mathbf{p}) = \int_{-\infty}^{+\infty} \frac{\rho(x, \mathbf{p})}{x - \omega - i\delta} dx, \tag{13}$$

where

$$\rho(\omega, \mathbf{p}) = -(2\pi)^3 \sum_{n,m} \exp\left[\frac{\Omega + \mu N_n - E_n}{T}\right] \times |\varphi_{mn}(0)|^2 \left(1 \pm \exp\left[-\frac{\omega_{mn}}{T}\right]\right) \times \delta(\mathbf{p} - \mathbf{p}_{mn}) \delta(\omega - \omega_{mn}). \tag{14}$$

Here $\omega_{mn} = E_m - E_n - \mu$, $\mathbf{p}_{mn} = \mathbf{p}_m - \mathbf{p}_n$; E_n, $\mathbf{p}_n$ are the energy and momentum in the n-th state of the system. It can be seen from Eq. (14) that $\rho(\omega, \mathbf{p})$ is a real function, for which the integral $\int_{-\infty}^{+\infty} \rho(\omega, \mathbf{p}) d\omega$ is finite.

Let us write the analogous Lehmann expansion for $G(\omega_n, \mathbf{p})$. By the use of Eq. (6) it is easy to show, by a method analogous to that of reference 7, that

$$\mathfrak{G}(\omega_n, \mathbf{p}) = \int_{-\infty}^{\infty} \frac{\rho(x)\, dx}{x - i\omega_n}. \tag{15}$$

Comparing Eqs. (13) and (15), we see that

$$\mathfrak{G}(\omega_n) = G^R(i\omega_n), \quad \omega_n > 0. \tag{16}$$

On the other hand, let us use the integral (15), regarded formally as a function of the complex variable $i\omega_n$, to determine a function $\mathfrak{G}(\omega)$ analytic in the upper half-plane. In virtue of Eq. (16) and the well known theorem of the theory of functions of a complex variable regarding the analytic continuation of a function given at an infinite set of discrete points possessing a point of concentration, we find from the above that G^R is, apart from a constant factor, the analytic continuation of the function $\mathfrak{G}(-i\omega_n)$ into the upper half-plane. Namely, we have

$$G^R(\omega) = \mathfrak{G}(-i\omega). \tag{17}$$

The converse problem of finding $\mathfrak{G}(\omega_n)$ from a given $G^R(\omega)$ is solved by Eq. (16) and the relation

$$\mathfrak{G}(\omega_n < 0) = \mathfrak{G}^*(|\omega_n|), \tag{16a}$$

which follows from Eq. (15).

Relations of the types (16) and (17) can in principle be obtained also for the vertex parts Γ and $\mathfrak{T}$. The general relations will, however, be more complicated. For physical applications one ordinarily needs not Γ and $\mathfrak{T}$ themselves, but the correlation functions, i.e., certain integrals of these quantities, and the practical procedure is to establish the connections between them separately for each concrete case.

In conclusion we express our gratitude to Academician L. D. Landau and L. P. Pitaevskiĭ for helpful discussions of the results of this work.

[1] A. Salam, Prog. Theor. Phys. **9**, 550, 1953.
[2] A. B. Migdal, J. Exptl. Theoret. Phys. (U.S.S.R.) **34**, 1438 (1958), Soviet Phys. JETP **7**, 996 (1958).
[3] V. M. Galitskiĭ and A. B. Migdal, J. Exptl. Theoret. Phys. (U.S.S.R.) **34**, 139 (1958), Soviet Phys. JETP **7**, 96 (1958).
[4] T. Matsubara, Prog. Theor. Phys. **14**, 351 (1955).
[5] C. T. Belyaev, J. Exptl. Theoret. Phys. (U.S.S.R.) **34**, 517 (1958), Soviet Phys. JETP **7**, 289 (1958).
[6] L. P. Gor'kov, J. Exptl. Theoret. Phys. (U.S.S.R.) **34**, 735 (1958), Soviet Phys. JETP **7**, 505 (1958).
[7] L. D. Landau, J. Exptl. Theoret. Phys. (U.S.S.R.) **34**, 262 (1958), Soviet Phys. JETP **7**, 182 (1958).
[8] H. Lehmann, Nuovo cimento **11**, 342 (1954).
[9] Ezawa, Tomozawa, and Umezawa, Nuovo cimento **5**, 810 (1957).

Translated by W. H. Furry
162

SOVIET PHYSICS JETP VOLUME 21, NUMBER 5 NOVEMBER, 1965

MINUTE METALLIC PARTICLES IN AN ELECTROMAGNETIC FIELD

L. P. GOR'KOV and G. M. ÉLIASHBERG

Institute of Chemical Physics, Academy of Sciences, U.S.S.R.

Submitted to JETP editor December 2, 1964

J. Exptl. Theoret. Phys. (U.S.S.R.) **48**, 1407-1418 (May, 1965)

We consider some properties of metallic particles with dimensions so small that the spectrum of the electronic excitations becomes discrete. The existence of microscopic roughnesses on the particle surfaces makes it impossible to calculate the detailed level distribution in the spectrum of each individual particle. At the same time, the mean level density is still determined by the macroscopic characteristics of the metal. This circumstance allows us to describe the level distribution statistically, in a manner similar to that employed in nuclear physics to find the distribution of the highly excited levels of the atomic nucleus. The formulas obtained for the electric polarizability in a high-frequency field, especially the part of the polarizability responsible for the absorption, contain explicitly the binary correlation function introduced in Dyson's well-known papers. It is shown that all three types of level statistics proposed by Dyson are realized in the objects under study under different conditions. It therefore becomes possible to observe in the level scheme of a random system a long-range order that leads to strong oscillations of the absorption when the field frequency is varied. Formulas are also obtained for the specific heat and for the paramagnetic-resonance intensity in minute metallic particles. The possibility of experimentally observing the phenomena in question is discussed.

1. STATEMENT OF PROBLEM

A system of many particles in a finite volume has discrete levels, the distances between which, generally speaking, decreases exponentially with increasing number of particles. The systematics of the low-excitation levels, to which it is sufficient to confine oneself at low temperatures, has a unique character. As is presently well known from statistics, it is possible to use at low temperatures the notion of excitations or quasiparticles obeying a definite dispersion law (which as a rule is phenomenological). The energy of the entire system is represented by the sum of the energies of the individual quasiparticles.

Of special interest for what follows are excitations of the Fermi type or, more accurately speaking, electronic excitations in metals. The density of the number of states per unit energy interval does not vanish in this case at zero excitation energy, but is equal to a constant, whose values for an isotropic dispersion law is

$$\nu = Vm^* p_0 / 2\pi^2, \qquad (1)$$

where V is the volume of the system, m^* the effective mass, and p_0 the limiting momentum of the Fermi surface ($\hbar = 1$). The reciprocal $\Delta = \nu^{-1}$ is thus the average distance between the levels near the Fermi surface. Usually the level systematics of particles in a large volume is obtained by considering the system in a large cube with periodic boundary conditions, with the final levels corresponding to the values of the momentum quantized in intervals $\Delta p \sim 1/a$, where a is the dimension of the system. This is perfectly adequate for a description of a macroscopic system, but becomes incorrect if we consider seriously the levels of sufficiently minute metallic particles. Imperfections in the shapes of such particles, which cannot be controlled at all, and which arise when the metal is sputtered, cause the distances between levels to be mainly of the order of Δ, and certainly not of the order p_0/m^*a.

This circumstance was already pointed out by Kubo[1], who introduced the phenomenological level-distribution probability density, on the basis of which he investigated the thermodynamic properties of such systems, namely the specific heat and the magnetic susceptibility. We shall show below that Kubo's phenomenological function leads to incorrect relations for these quantities, but the most interesting from our point of view is a study of the properties of minute particles in an alternating field, where the concepts developed below lead to non-trivial consequences.

It is important to note that the distribution of the levels should remain random even if the particles have the same volume and a good shape, say spherical particles of equal size. The point is that electrons in the metal have a wavelength on the order of atomic dimensions. Therefore surface irregularities of atomic size are sufficient to make the level distribution perfectly random. This distribution ceases to be random if the electron reflection is not difuse but specular.

Thus, if the quasiparticle concept is valid in this range of dimensions and temperatures, then the quasiparticle levels are determined by the perfectly random interactions with the walls, and the average distance between levels is $\Delta = \nu^{-1}$ with ν given by (1).

The situation is therefore completely analogous with that which takes place for the distribution of high-excitation levels of the nucleus in nuclear physics. The complexity of the real interactions in the nucleus leads to the possibility of considering the distribution of the levels statistically. In the nucleus, the averaging is carried out over different nuclei with respect to the average density of the levels of the nucleus, or over different sections of the spectrum of the nucleus. In our problem, the averaging reduces to a study of a large assemblage of metallic particles of equal size.

The concepts of nuclear-level statistics based on complete randomness of the interactions were apparently first developed by Wigner[2]. Recent papers by Dyson[3] and Dyson and Mehta[4] presented a mathematical formulation of these concepts and formulas for the correlation functions between these levels, which we shall henceforth simply borrow from these papers.

According to [3], the level distribution coincides with the distribution on the unit circle of the eigenvalues of matrices from different ensembles. There are three such ensembles: 1) orthogonal, which describes systems that are invariant against time and space inversion, or with integer spin; 2) symplectic—for random interactions, invariant against time reversal; the total spin of the system is half-integer; 3) unitary—for systems that are not invariant against time reversal.

In nuclear physics, apparently, only the first ensemble is realized. If we employ this approach to describe the spectrum of an individual Fermi excitation in the field of a random potential, then the orthogonal ensemble pertains to the case of small spin-orbit coupling, when the spin is a quantum number; the symplectic ensemble is realized when the spin-orbit coupling is sufficient to mix levels with different spins. Finally, the unitary ensemble is obtained from the symplectic one when the magnetic field is turned on and $\mu H \gg \Delta$.

The value of Δ is inversely proportional to the volume of the particle and for a linear law of electronic specific heat it is connected with the coefficient by

$$\Delta = 1.1 \cdot 10^{-20} \frac{A}{a^3 \rho \gamma} [°K] \sim \frac{\varepsilon_F}{N} \qquad (2)$$

(A is the atomic weight, ρ is the density, and γ is expressed in mJ/mole-deg^2). For aluminum with $a = 4.2 \times 10^{-7}$ cm we have $\Delta \approx 1°$. In general, it is seen from (2) that when $N \sim 10^4 - 10^5$ the average distance is $\Delta \sim 1 - 0.1°$, that is, it is quite large when the system dimensions are still sufficiently large to be able to use the macroscopic characteristics of the metal.

To conclude this section, let us discuss the main question: to what extent is the excitation concept applicable to the case of such small dimensions or such low temperatures (as $\Delta \to 0$). We note immediately that acoustic oscillations, both volume and surface, can be excluded from consideration, since long-wave sound has a strictly discrete set of frequencies, starting with low values of $\omega \sim u/a$ (u—speed of sound). The ratio

$$\omega / \Delta \sim u (p_0 a)^2 / v_F$$

can always be made sufficiently large. For example, when $a \sim 10^{-6}$ we have $\omega/\Delta \sim 10$. For this reason, at sufficiently low temperatures $T \sim \Delta$, the sound oscillations are not excited at all, and at higher temperatures the interaction between the electrons and the phonons is less effective because of the discreteness of the spectrum.

The reason why the excitation concept may turn out to be incorrect is the attenuation connected with the interaction between the electrons themselves. In bulk metal at frequency ω, the attenuation is on the order of ω^2/ϵ_F or T^2/ϵ_F. It is therefore clear that at temperatures

$$T \ll \sqrt{\Delta \varepsilon_F} \sim 10\text{—}100° \text{ K} \qquad (3)$$

the attenuation becomes smaller than the level spacings Δ. Finally, heat exchange with the medium is also too weak to lead to a noticeable broadening of the levels. By virtue of these considerations, we assume that it is meaningful to speak of Fermi excitations in the objects of interest to us.

2. PARTICLES IN AN ELECTROMAGNETIC FIELD

So long as the frequencies ω of the field are small compared with the reciprocal of the time the electron travels from wall to wall ($\sim v/a$), and the field intensities are small compared with the internal fields

$$eEa \ll \Delta,$$

the interaction of a "metallic" small particle with the field can be considered by using perturbation theory for a quantum system. In this range of fields the particle behaves like an atom with a certain polarizability and does not manifest its metallic properties at all. Inasmuch as the field changes little over the dimensions of the particle, the interaction reduces to the term

$$\hat{U} = -\mathbf{E}\hat{\mathbf{d}},$$

where $\hat{\mathbf{d}}$ is the dipole moment of the system.

The average dipole moment

$$\langle d_\alpha \rangle = i\int_{-\infty} \langle [d_\alpha(t),\ d_\beta(t')] \rangle E_\beta(t')\,dt',$$

([,]—commutator) can be easily expressed in terms of matrix elements of the coordinates of the excitations and the occupation numbers of the individual levels n_k:

$$d_{\alpha\omega} = -e^2 E_{\beta\omega} \sum_{k,l} \frac{n_k - n_l}{\varepsilon_k - \varepsilon_l + \omega + i\delta} \mathbf{r}_{kl}{}^{\alpha} \mathbf{r}_{lk}{}^{\beta}. \quad (4)$$

In the last formula we have changed over to the Fourier components of all the quantities in accordance with

$$d(t) = \frac{1}{2\pi} \int_{-\infty}^{\infty} d_\omega e^{-i\omega t}\, d\omega.$$

Separating in (4) the static part of the polarizability α, we obtain

$$\alpha_\omega = \alpha_0 - \frac{e^2\omega}{3} \sum_{k,l} \frac{n_k - n_l}{\varepsilon_k - \varepsilon_l} \frac{\mathbf{r}_{kl}\mathbf{r}_{lk}}{(\varepsilon_k - \varepsilon_l)^2 - (\omega + i\delta)^2}, \quad (5)$$

$$\alpha_0 = -\frac{e^2}{3} \sum_{k,l} \mathbf{r}_{kl}\mathbf{r}_{lk} \frac{n_k - n_l}{\varepsilon_k - \varepsilon_l} = \frac{2}{3} \nu e^2 \sum_{l} \mathbf{r}_{k_0 l}\mathbf{r}_{l k_0}. \quad (6)$$

As will be shown below, the sum over l in (6) converges when $\epsilon_k - \epsilon_l \sim v/a$, that is, near the Fermi surface. As a result,

$$\alpha_0 = {}^2/{}_3 e^2 \nu \langle r^2 \rangle = 4e^2 a^5 m p_0 / 15\pi = 4a^5 e^2 \gamma / 5\pi,$$

where γ is the coefficient in the linear law of specific heat: $C_e = V\gamma T$. The order of this quantity is

$$\alpha_0 \sim a^3 (p_0 a)^2 \sim \alpha_{met} (p_0 a)^2,$$

where $\alpha_{met} = a^3$ denotes the polarizability of the equivalent metallic particle in the field.

For the objects considered, a polarizability $\alpha \sim \alpha_{met}$ is attained in fields $eEa \gg \Delta$, sufficiently strong to mix the levels of the system. In these fields the system will have the usual metallic properties. As far as we know, no one has called attention to the anomalously large values of the polarizability of minute metallic particles in weak fields. Yet, apparently, this circumstance uncovers a possibility of producing artificial dielectrics with large controllable dielectric constants.

In formulas (4) and (5) n_k are the equilibrium occupation numbers for which, however, we cannot in general use the known formulas with a chemical potential for the Fermi distributions. When $T \sim \Delta$ the number of excitations is very small and it is therefore necessary to take into account the well known premise of the Fermi-liquid theory concerning the conservation of the total number of quasiparticles. An analysis for $T = 0$ and for large values of the temperatures (3), when we can write

$$n_k = \{\exp[(\varepsilon_k - \mu)/T] + 1\}^{-1}$$

yields the same result. That the results which follow from (5) are completely independent of the temperature can be demonstrated by using the more general method of Fowler and Darwin[5], in which the conservation of the number of excitations is explicitly taken into account. Without dwelling on this proof, we present a derivation for the case of high temperatures (3), since it is of greatest experimental interest.

Let us average (5) over the level distribution. To this end we multiply each term in the sum (5) by the probability

$$R(\varepsilon_1 - \varepsilon_2)\,\nu d\varepsilon_1 \nu d\varepsilon_2 \quad (7)$$

that two levels will be in the interval of values $d\epsilon_1$ and $d\epsilon_2$, and integrated with respect to ϵ_1 and ϵ_2:

$$\alpha_\omega - \alpha_0 = -\frac{e^2\omega^2}{3} \nu^2 \int_{-\infty}^{\infty}\int d\varepsilon_1 d\varepsilon_2 \frac{n_1 - n_2}{\varepsilon_1 - \varepsilon_2} \frac{(\mathbf{r}_{12}\mathbf{r}_{21}) R(|\varepsilon_1 - \varepsilon_2|)}{(\varepsilon_1 - \varepsilon_2)^2 - (\omega + i\delta)^2}. \quad (8)$$

The integral over the energy difference converges when $\epsilon_1 - \epsilon_2 \sim \omega$. If $\omega \ll v/a$, as assumed, then $\mathbf{r}_{21} \cdot \mathbf{r}_{12} \approx (\mathbf{r}_{21} \cdot \mathbf{r}_{12})_0$ is constant. Replacing $(n_1 - n_2)/(\epsilon_1 - \epsilon_2)$ by $\partial n/\partial\epsilon$ for $\epsilon \sim \omega \ll T$, we obtain

$$\alpha_\omega = \alpha_0 + \frac{1}{3} e^2 (\mathbf{r}_{21}\mathbf{r}_{12})_0 \nu A(\omega), \quad (9)$$

$$A(\omega) = \omega^2 \nu \int_{-\infty}^{\infty} \frac{R(|\varepsilon|)\,d\varepsilon}{\varepsilon^2 - (\omega + i\delta)^2}. \quad (10)$$

We now calculate the quasiclassical matrix element $(\mathbf{r}_{12} \cdot \mathbf{r}_{21})_0$. For the case of diffuse reflection of the electrons from the surface, Shapoval[6] developed a general method of calculating such quantities, which we shall use with slight modification. The average of two quantities $\hat{f}(t)$ and $\hat{g}(t)$ at different instants of time, taken over the state $|ni\rangle$, is

$$\langle ni\,|\hat{f}(t)\,\hat{g}(t+\tau)|\,ni\rangle = \sum_{jk}\langle ni\,|\hat{f}(t)|\,kj\rangle\,\langle kj\,|\hat{g}(t)|\,ni\rangle\,e^{-i\omega_{kn}\tau},$$

where n and k denote the energies, while i and j are the other quantum numbers. It follows therefore that the matrix element averaged over i and j, which we shall denote by $f_{nk}g_{kn}$, is

$$f_{nk}g_{kn} = \frac{1}{2\pi\nu}\int_{-\infty}^{\infty} e^{i\omega_{kn}\tau}\,\langle f(t)\,g(t+\tau)\rangle d\tau.$$

Going over to the quasiclassical approach, we obtain

$$f_{nk}g_{kn} = \frac{1}{2\pi\nu}\,\frac{1}{2T_1}\overline{\int_{-T_1}^{T_1} dt\int_{-\infty}^{\infty} d\tau\, e^{i\omega_{kn}\tau} f(t)\,g(t+\tau)}.$$

Here the integration with respect to t corresponds to averaging along the given trajectory, and the bar over the integral denotes averaging over all the quasiclassical trajectories of the particle. For the quantity of interest to us, using the fact that the correlations with respect to τ are even, we obtain

$$\mathbf{r}_{kl}\mathbf{r}_{lk} = \frac{1}{\pi\nu}\,\frac{1}{2T_1}\overline{\int_{-T_1}^{T_1} dt\int_{0}^{\infty} d\tau\cos(\omega_{kl}\tau)\,(\mathbf{r}(t)\,\mathbf{r}(t+\tau))}. \quad (11)$$

As can be seen, expression (11) is actually cut off at $\omega_{kl} \sim v/a$.

Let $0, t_1, t_2, \ldots, t_n \ldots$ be the times of collisions of the particles with the wall. Let the time t lie in the interval $0 < t < t_1$, and then $t + \tau$ can lie either in the same interval or in the next interval, corresponding to one, two, etc. reflections. It is therefore sufficient to calculate the integrals

$$\int_0^{t_1}\mathbf{r}(t)\,dt\left\{\int_0^{t_1-t}\mathbf{r}(t+\tau)\,d\tau + \int_{t_1-t}^{t_2-t}\mathbf{r}(t+\tau)\,d\tau + \ldots\right\} \quad (12)$$

and to average them over all the reflection angles and the initial point of the trajectory $\mathbf{r}_0$. Total integration with respect to t reduces to multiplication of (12) by a number of segments of the trajectory N equal to the number of collisions of the particle with the surface during the time $2T_1$:

$$N = 2T_1 vS/4V.$$

The probability of a given trajectory is

$$d\Sigma_{\text{traject}} = \frac{dS}{S}\prod_i \frac{\cos\vartheta_i}{\pi}\,dO_i,$$

where ϑ_i is the angle of reflection of the particle after the i-th collision. Finally,

$$(\mathbf{r}_{kl}\mathbf{r}_{lk})_0 = \frac{v}{4\pi\nu V}\int dS\int\int\ldots\int\prod_i\frac{\cos\vartheta_i}{\pi}\,dO_i\int_0^{t_1}\mathbf{r}(t)\,dt$$

$$\times\left\{\int_0^{t_1-t}\mathbf{r}(t+\tau)\,d\tau + \int_{t_1-t}^{t_2-t}\mathbf{r}(t+\tau)\,d\tau + \ldots\right\}. \quad (13)$$

For a sphere we obtain after elementary calculations

$$(\mathbf{r}_{kl}\mathbf{r}_{lk})_0 = \frac{139}{150}\,\frac{1}{2\pi\nu}\,a^2\frac{a}{v}. \quad (14)$$

A somewhat different approach is necessary if the range of the electrons in the sample is smaller than the sample dimensions. In the integration with respect to τ, the characteristic time in formula (11) is determined by the diffusion from the point $\mathbf{r}(t)$ to the point $\mathbf{r}(t+\tau)$:

$$\tau_{\text{eff}} \sim a^2/D \sim a^2/lv.$$

In this case we always have $\Delta\cdot\tau_{\text{eff}} \sim ap_0 lp_0 \gg 1$. The matrix element of interest to us can be written in the form

$$(\mathbf{r}_{kl}\mathbf{r}_{lk})_0 = \frac{1}{\pi\nu V}\int d^3\mathbf{r}_0\int d^3\mathbf{r}\,(\mathbf{r}_0\mathbf{r})\int_0^{\infty} W_{\mathbf{r}_0}(\mathbf{r}, t)\,dt, \quad (15)$$

where $W_{\mathbf{r}_0}(\mathbf{r}, t)$ is the probability of finding the particle in the point $\mathbf{r}$ at the instant t, if its position at $t = 0$ is $\mathbf{r}_0$:

$$\frac{\partial W_{\mathbf{r}_0}(\mathbf{r}, t)}{\partial t} = D\Delta W_{\mathbf{r}_0}(\mathbf{r}, t), \quad W_{\mathbf{r}_0}(\mathbf{r}, t=0) = \delta(\mathbf{r}-\mathbf{r}_0).$$

Integrating the diffusion equation with respect to t, expanding $W_{\mathbf{r}_0}(\mathbf{r}, t)$ in Legendre polynomials and using the fact that $W_{\mathbf{r}_0}(\mathbf{r}, t=\infty) = 1/V$, we obtain

$$\frac{1}{V} - \delta(\mathbf{r}-\mathbf{r}_0) = D\sum_l\left\{\frac{1}{r^2}\frac{d}{dr}r^2\frac{d}{dr} - \frac{l(l+1)}{r^2}\right\}$$

$$\times\int_0 W_{r_0}{}^l(r, t)\,P_l(\cos\vartheta)\,dt.$$

Putting

$$\frac{f(r)}{r} = \int\frac{d^3\mathbf{r}_0}{V}\frac{(\mathbf{r}_0\mathbf{r})}{r}\int_0^{\infty} W_{\mathbf{r}_0}(\mathbf{r}, t)\,dt,$$

we obtain for $f(r)$ the elementary equation

$$-\frac{r}{V} = D\left(\frac{f''}{r} - \frac{2}{r^3}f\right), \qquad \frac{d}{dr}\frac{f(r)}{r}\Big|_{r=a} = 0,$$

from which, after calculations, we finally obtain

$$(\mathbf{r}_{kl}\,\mathbf{r}_{lk})_0 = \frac{48}{175}\,\frac{1}{2\pi\nu}\,\frac{a^4}{D}. \quad (16)$$

We note here that whereas for a pure sample

the frequency-dependent part of the polarizability (including Im α, which determines the absorption) is of the order of $\alpha - \alpha_0 \sim \alpha_{met}$, in the case of a small mean free path we have

$$\alpha - \alpha_0 \approx \alpha_{met} a / l \gg \alpha_{met}.$$

We now proceed to expression (10) for $A(\omega)$. As already mentioned, the orthogonal ensemble is realized in the case when the spin-orbit interaction can be neglected, that is, for light metals. According to [3], we have in this case

$$R^{ort}(\pi\nu\varepsilon) \equiv R^{ort}(x) = 1 - \frac{\sin^2 x}{x^2} + \int_0^1 \frac{\sin xt}{t} dt \times \frac{d}{dx}\left(\frac{\sin x}{x}\right) - \frac{\pi}{2}\frac{d}{dx}\left(\frac{\sin x}{x}\right). \quad (17)$$

Breaking up the part of (17) that is symmetrical in x into two branches $f_+ + f_-$, which are analytic in the upper and lower half planes respectively, we can calculate the integral in (10) in explicit form

$$A(\omega\pi\nu) \equiv A(z) = 2 - \frac{\sin 2z}{z} - \frac{2\,\mathrm{Ci}(z)}{z}(\sin z - z\cos z) + i\left[2z - \frac{1}{z} + \frac{\cos 2z}{z} - \frac{\mathrm{Si}(z)}{z}(\sin z - z\cos z)\right], \quad (18)$$

where Ci(z), and Si(z) are the integral sine and cosine, respectively.[7]

In Fig. 1 is shown a plot of the real and imaginary parts of the ratio $A(z)/2z = [A_1(z) + iA_2(z)]/2z$. This dependence is far from trivial. In particular, we see that when $\omega \lesssim \Delta$ the absorption acquires essentially a quantum character. As $\omega \to 0$ we get

$$A(z) \approx {}^2/_3 z^2 (2 - \ln(\gamma z)) + {}^1/_3 i\pi z^2.$$

In the opposite limiting case, as $z \to \infty$, we have

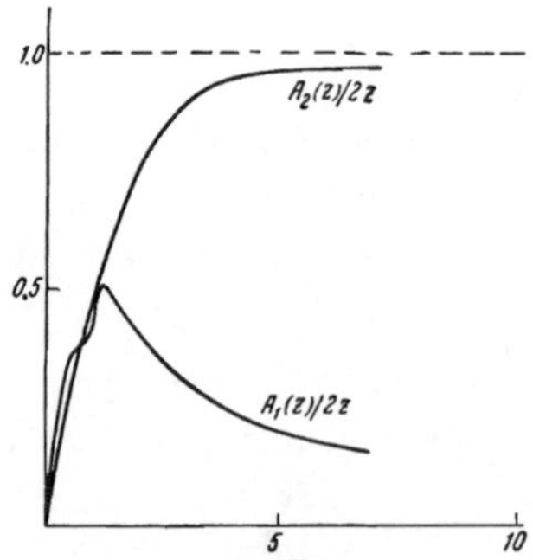

FIG. 1.

$$\frac{A(z)}{2z} \approx i\left(1 - \frac{1}{z^2} + \frac{3}{2z^4} + \frac{\cos 2z}{2z^4}\right) + \frac{1}{z} - \frac{1}{z^3} - \frac{\sin 2z}{2z^4}.$$

The oscillating term denotes the presence of a long-range correlation in the arrangements of the levels—the effect of "repulsion" between levels, that is, the existence of an appreciable probability of a periodic arrangement of the levels[2,3]. Unfortunately, these oscillations are too small to be observed experimentally. We therefore proceed now to discuss the results for heavier metals or for larger particles, where the spin-orbit interaction cannot be neglected, and which we therefore describe by the so called "symplectic ensemble"[4].

The function

$$R^{simpl}(2\pi\nu\varepsilon) \equiv R^{simpl}(x)$$

is given by

$$R^{simpl}(x) = 1 - \frac{\sin^2 x}{x^2} - \int_0^1 \frac{\sin xt}{t} dt \frac{d}{dx}\left(\frac{\sin x}{x}\right).$$

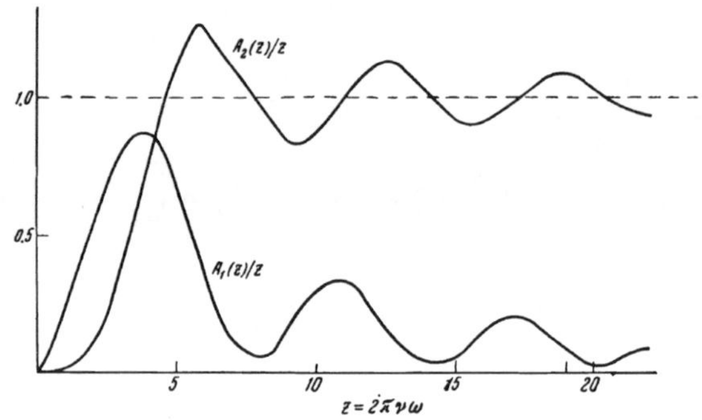

FIG. 2.

Figure 2 shows the behavior of the ratio $A(z)/z$:

$$A(2\pi\nu\omega) \equiv A(z) = 2 - \frac{\sin 2z}{2z} - \left(\frac{\pi}{2} + \mathrm{Si}(z)\right)\frac{\cos z + z\sin z}{z} + i\left[z - \frac{\sin^2 z}{z} + \left(\frac{\pi}{2} + \mathrm{Si}(z)\right)\frac{z\cos z - \sin z}{z}\right]. \quad (19)$$

We see here the strongly pronounced oscillations of both the real and the imaginary parts of $A(z)/z$. Indeed, as $z \to \infty$ we have

$$\frac{A(z)}{z} \approx \frac{2}{z} - \frac{\pi}{2z}\sin z + i\left(1 + \frac{\pi}{2z}\cos z\right), \quad (20)$$

that is, the oscillations have a much larger amplitude than in lighter substances. The physical reason for this effect, as is well known, is that the spin-orbit interaction lifts all but the Kramers degeneracy. Therefore a random appearance of two closely-lying levels is even less probable than in an orthogonal ensemble, and accordingly a stronger "repulsion" is produced between levels.

We present also an asymptotic expression for low frequencies:

$$A(z) \approx {}^2/_9 z^2 + {}^1/_{45} z^5,$$

from which we see that absorption is very small in this case and has a very steep variation with increasing frequency.

To conclude this section we present formulas describing the absorption of electromagnetic radiation, when, on the one hand, the spin-orbit interaction is sufficiently strong, and on the other there is an external constant magnetic field sufficient to mix the levels ($\mu H \gg \Delta$) and to produce a "unitary ensemble"[3,4]:

$$A^{unit}(2\pi\nu\omega) \equiv A^{unit}(z) = 2 - \frac{\sin 2z}{2z} + i\left[z - \frac{\sin 2z}{2z}\right]. \quad (21)$$

Although the oscillations here are weaker than in the symplectic ensemble, they are also quite noticeable.

3. SPECIFIC HEAT AND SINGULARITIES OF THE BEHAVIOR IN A MAGNETIC FIELD

At high temperatures $T \gg \Delta$, all the thermodynamic quantities, naturally, have the same form as for bulk metal. We confine ourselves therefore here only to the case of low temperatures $T \ll \Delta$. The situation is the simplest with specific heat, for at such low temperatures the appearance of more than one excitation per system has little likelihood. For the orthogonal and symplectic ensembles, which both have doubly degenerate levels, the level filling scheme is shown in Fig. 3. The dashed line denotes the free levels, the solid line with circle the occupied levels.

Figure 3a corresponds to "even" filling. The total partition function is equal to

$$Z_{\text{even}} = 1 + 4\exp\left[-\frac{\varepsilon_1 - \varepsilon_0}{T}\right] + \exp\left[-\frac{2(\varepsilon_1 - \varepsilon_0)}{T}\right].$$

The state on Fig. 3b is doubly degenerate. Leaving again in the partition function only those excitations which affect only one level, we obtain

$$Z_{\text{odd}} = 2 + 2\exp\left[-\frac{\varepsilon_0 - \varepsilon_{-1}}{T}\right] + 2\exp\left[-\frac{\varepsilon_{+1} - \varepsilon_0}{T}\right].$$

Averaging $F = -T \ln Z$ over both possibilities, and also integrating over the probability $P(\Delta)$ of

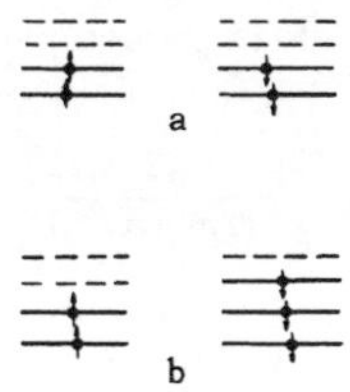

FIG. 3.

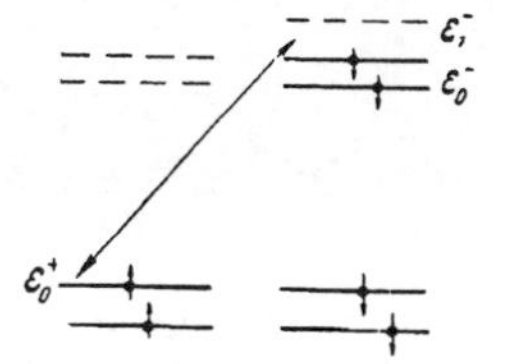

FIG. 4.

a given spacing Δ between neighboring levels, we obtain

$$F = -\frac{T}{2}\ln 2 - 3T\int_0^\infty e^{-\Delta/T}P(\Delta)\,d\Delta - \frac{T}{2}\int_0^\infty e^{-2\Delta/T}P(\Delta)\,d\Delta.$$

The last formula leads to the results of Kubo[1], if we use for $P(\Delta)$ his expression which, however, does not take into account the repulsion of the levels. The correct functions lead to a specific heat which is proportional at low temperatures to T^2, T^5, and T^4 for the orthogonal, symplectic, and unitary ensembles, respectively.

We proceed to the magnetic susceptibility. For the symplectic ensemble, as expected, the susceptibility always coincides with the value for the bulk metal. Curious results can be obtained, however, for the orthogonal ensemble when the spin-orbit coupling is small and the spin remains a good quantum number. In this case we can speak separately of electronic excitations in which the spin is directed with and against the field. Let the field H have an arbitrary value (μH comparable with Δ). Figure 4 shows the level scheme of an individual particle in the field. Let ϵ_0^+ and ϵ_0^- be respectively the last filled levels having spin parallel and antiparallel to the field, respectively.

Let us increase the field to a value $H + dH$. It may turn out that following this it is energetically easier for the excitation to go from the level ϵ_0^+ to the level ϵ_1^-. The condition for this is

$$\varepsilon_1^- - \varepsilon_0^+ - 2\mu H - 2\mu dH < 0.$$

The moment of the entire system changes by an amount -2μ. The probability of such an event is equal to the probability of two levels being separated by a distance $2\mu H$ in the energy interval $2\mu dH$. Therefore

$$\chi_\parallel = dM/dH = 4\mu^2\nu R^{ort}(2\pi\nu\mu H). \quad (22)$$

In weak fields

$$dM/dH \approx \chi_\infty {}^1/_3 \pi^2 \nu\mu H$$

vanishes. Of course, when $\mu H \gg \Delta$, Eq. (22) goes over to the value of the paramagnetic susceptibility of the bulk metal. In this case $\chi_\parallel$ experiences small oscillations of the same nature as described in Sec. 2:

$$\frac{dM}{dH} \approx \chi_\infty \left\{ 1 - \left(\frac{1}{2\pi\nu\mu H} \right)^2 + \frac{1 + \cos^2(2\pi\nu\mu H)}{(2\pi\nu\mu H)^4} \right\}.$$

The smaller the role of the spin-orbit interaction, the more accurately these relations are satisfied for the orthogonal ensemble. In particular, it can be thought that paramagnetic resonance would become observable, with good accuracy, in minute particles made of light metals. It is easy to obtain in the usual manner an expression for the perpendicular component of the paramagnetic susceptibility

$$\chi_\perp = -\mu^2 \sum_{k,\,p} \frac{n_k - n_p}{\varepsilon_k - \varepsilon_p + \omega + i\delta} \sigma_{kp}{}^\perp \sigma_{pk}{}^\perp.$$

In the absence of spin-orbit coupling, the spin precesses about the direction of the magnetic field with frequency $2\mu H$. From formulas of the type (11) it follows immediately that the only nonvanishing transitions are those with $\omega_{kp} = 2\mu H \equiv \omega_0$:

$$\chi_\perp = -2\mu^2 \frac{\omega_0}{\omega_0^2 - (\omega + i\delta)^2} \sum_k [n(\varepsilon_k + \mu H) - n(\varepsilon_k - \mu H)].$$

The meaning of this formula becomes perfectly clear if we look at Fig. 4: $\chi_\perp$ is proportional to the number of places which are free on the left in this figure. Finally,

$$\chi_\perp = \frac{\omega_0^2}{\omega_0^2 - \omega^2} \frac{1}{H} \int_0^H \chi_\parallel(H)\, dH. \tag{23}$$

Thus, experiments will disclose that the intensity of the paramagnetic resonance lines experiences a quantum variation with varying magnetic field, changing in accordance with (22) and (23) from zero at small fields to a value characteristic of bulk metal when $\mu H \gg \Delta$.

We emphasize once more that all the formulas of this section, including (22) and (23), are valid only at low temperatures $T \ll \Delta$.

4. CONCLUSION

We now discuss briefly the possibility of experimentally observing these effects. Measurements of the specific heat seem to us of little likelihood, primarily because of the smallness of the specific heat itself. It can be readily understood that when $T \sim \Delta$ the specific heat is of the order of k per particle, that is, is comparable with the contribution made to the specific heat by the vibrations in the host matrix. For analogous reasons, it is difficult to measure the magnetic susceptibility directly. The point is that, when we considered the bulk properties we disregarded completely the existence of different surface centers with localized magnetic moments. Yet the contribution made to the susceptibility by such a center is of the order of μ^2/T, and is much larger than (22) when $T \ll \Delta$. An analogous role is played by various paramagnetic impurities. Nevertheless, it seems to us that a direct measurement of the paramagnetic susceptibility is possible by determining the field dependence of the Knight shift of the frequency of the nuclear magnetic resonance. (We recall that only the conduction electrons are responsible for the Knight shift.) This dependence $d\Delta\omega/dH$ is given directly by formula (22). When $\Delta \sim 1°$ the required temperature is 0.1° and the required field is $H \sim 10^4$ Oe.

Of course, local centers can make a contribution to the paramagnetic resonance line, but in this case there is no field dependence of the line intensity. Therefore such measurements are also desirable.

The foregoing experiments necessitate, however, the use of low temperatures. Therefore, the most convenient for the confirmation of the proposed laws seem to be, in our opinion, the singularities of the behavior of small metallic particles in an electric field. We note, first, that it would be generally of interest to demonstrate the quantum character of such systems, and that even the observation of large polarizability of the systems in a weak static field is sufficient for this purpose. When $a \sim 10^{-6}$ we have $\Delta \sim 0.1°$ and the permissible fields ($eAe \ll \Delta$) are $E \ll 10$ V/cm.

At low concentrations, the dielectric constant is

$$\varepsilon = 1 + 4\pi N\alpha, \tag{24}$$

where N is the number of particles per unit volume. Thus, a unique nonlinear effect exists even in relatively weak electric fields, in that the polarizability α, the order of magnitude of which is $(p_0 a)^2 a^3 \gg a^3$ when $eEa \ll \Delta$ begins to decrease strongly when $eEa \sim \Delta$, and reaches finally the usual value a^3 at $eEa \gg \Delta$. The field at which the nonlinear variation of α begins decreases rapidly with increasing particle dimensions ($\sim a^{-4}$). It is necessary to bear in mind here that the larger the particles the lower the temperature at which this effect can exist (see the end of Sec. 1).

Of fundamental interest, of course, is the quantitative verification of the formulas of Sec. 2, which follow from the general fundamental laws of the statistics of complex interactions. A principal factor here is the condition that the dispersed particles be sufficient homogeneous in their dimensions and shapes. As can be seen from the diagram in Fig. 2, for a symplectic ensemble the absorption oscillations come into play when the volume or the frequency change by a factor 2 or 3. This imposes the condition that the dimension fluctuations not exceed several percent. In addition, the polarizability in (24) can, generally speaking, depend not only on the dimensionless quantity $z = \omega/\Delta$, but may contain, in accordance with (16), an extra power of a if the mean free path is shorter than the particle dimensions. It is difficult to control the mean free path and it would therefore be preferable to measure the frequency dependence of the dielectric constant (24). It can be assumed, however, that for dispersed particles of a sufficiently pure metal the mean free path will be of the order of the sample dimensions, and by virtue of this α is a function of only the variable $z = \omega/\Delta$, which enters in $A(\omega)$ (10).

We are indebted to V. I. Petinov and I. F. Shchegolev for calling our attention to this question.

[1] R. Kubo, J. Phys. Soc. Japan **17**, 975 (1962).

[2] E. P. Wigner, Ann. Math. **53**, 36 (1951) and **62**, 548 (1955).

[3] F. J. Dyson, J. Math. Phys. **3**, 140, 157, 166 (1962).

[4] M. L. Mehta and F. J. Dyson, ibid **4**, 713 (1963).

[5] R. H. Fowler and E. A. Guggenheim, Statistical Thermodynamics, Cambridge, 1939.

[6] E. A. Shapoval, JETP **47**, 1007 (1964), Soviet Phys. JETP **20**, 675 (1965).

[7] E. Jahnke and F. Emde, Tables of Functions, Dover, 1943.

Translated by J. G. Adashko
206

J. Phys. A: Math. Gen. 26 (1993) 787–806. Printed in the UK

Fluctuations in fragmentation processes

Sergei E Esipov†‡, L P Gor'kov†§|| and T J Newman†

† Department of Physics, University of Illinois at Urbana-Champaign, 1110 West Green Street, Urbana, IL 61801, USA

‡ Institute of Solid State Physics, Chernogolovka, Moscow Province, 142432, Russia

§ Landau Institute for Theoretical Physics, Chernogolovka, Moscow Province, 142432, Russia

|| National High Magnetic Field Laboratory and Department of Physics, Florida State University, 1800 East Paul Dirac Drive, Tallahassee, FL 32306, USA

Received 21 July 1992, in final form 11 November 1992

Abstract. We investigate the role of fluctuations in fragmentation processes using a simple analytic method to verify the stability of the mean-field similarity solution. For the fragmentation process which slows down with decreasing fragment size, the cascade becomes self-averaging in the late stages of branching. For the fragmentation which speeds up with decreasing fragment size, fluctuations dominate. In both cases the deviation from the mean-field solution is neither Gaussian nor symmetric. The steady-state regime in the presence of an external source is investigated and compared with the above results. Some exact solutions are obtained for the case of homogeneous fragmentation and for the discrete Yule–Ferry cascade process of decay into two equal fragments. The connection of the kinetic method with the combinatorial-type analysis of branching is indicated. Applications are made to the process of phonon decay.

1. Introduction

In what follows we address some problems related to the role of fluctuations in cascade-like phenomena [1, 2]. We investigate the stability of the mean-field solutions in fragmentation [2, 3] with respect to fluctuations, develop the picture of the steady-state regime of fragmentation, and solve a sample problem to indicate a possible mapping between the combinatorial description [4–6] of the tree process and the kinetic approach. We also discuss the physical realization of the cascade process by studying phonon decay [7, 8].

In cascade-like phenomena the appearance of self-averaging (and the time required for this) or, in contrast, the appearance of non-self-averaging (and the amplitude of fluctuations) have not been addressed so far in detail. We think there is a need for a quantitative description. This is the motivation of the present work.

For brevity we denote by the phenomena of fragmentation an inter-disciplinary approach for describing many different physical systems. Generally speaking, it includes any kinetic process with scattering, decay, breaking, or splitting of an initial state into two (or several) new states (fragments), etc. A characteristic feature of any cascade process is that the fragments then proceed independently and in the same fashion, i.e. decay to smaller pieces. This is therefore a linear system with the generic feature known as ultrametricity [9]. We may cite phonon down-conversion

0305-4470/93/040787+20$07.50

in solid state physics [7], particle decay in nuclear physics [10], radiation disrupture of condensed matter [6], cosmic showers [11], atomic collision cascades [12], neutron noise in a multiplying medium [11], electron transport and even cell division [11] as examples.

It is rather obvious that fluctuations are inherent in any fragmentation process: for any given realization the probability of the distribution over fragment sizes, say, at some late stage of the process is the result of the repetition of numerous breakups, each of them being weighted with the cross-section of the single act of decay. If this cross-section is not somehow peaked in a deterministic way—if each decay corresponds to the breakup into two fragments of comparable size (energy)—the probability distribution with a significant width would be predetermined just by the spread of the very first decay act.

Thus, the important issue is related to the information that one can extract from the study of the decay products of single species. We address this problem in section 2. The averaging over many realizations is formulated in section 3 as the steady-state regime. The possibility of having a steady-state solution is not obvious from the very beginning, since the breakup probability may depend on the current fragment size (energy) in two qualitatively different manners: the decay may either slow down or speed up with diminishing fragment size. Here the discrete model of the Yule–Ferry cascade [11] reconsidered by some of us [13] proves to be useful for understanding the convergence of the probability distribution to the steady state. In section 4 we discuss the correspondence between the combinatorial approach along a tree and the kinetic process developing along the same tree. For the simple discrete model [5, 6] the one-to-one equivalence can be found, unlike the case of more complicated situations. In section 5 we apply the formalism of section 2 to the process of phonon frequency-down conversion in the simplest homogeneous case. Section 6 briefly summarizes the results.

2. Fragmentation

In this section we consider the typical example of the decay process—the fragmentation model by Kolmogorov–Filippov studied in [1–3]. Let us introduce the kernel $K(x,y)$ describing the rate density of splitting of the initial state y into two fragments of sizes x and $y-x$; the total mass (energy) is thus conserved. We define the conditional probability (with continuous argument x) $W(N,x,t|1,x_0,t_0)$ of having N fragments of size x at time t given there was one initial 'fragment' of size x_0 at time t_0. For the case of binary fragmentation (see [14] and appendix B) the kernel $K(x,y)$ in [2,3] was taken to have the form

$$K(x,y) = y^{\beta} \tag{2.1}$$

and the so-called mean-field solution was obtained, i.e. the Boltzmann equation for the first moment

$$\langle N(x,x_0,t)\rangle = \sum_N NW(N,x,t|1,x_0,0)$$

i.e. the equation

$$\partial_t\langle N(x,x_0,t)\rangle = -\langle N(x,x_0,t)\rangle \int_0^{x_0} \mathrm{d}y\, K(y,x_0) + 2\int_x^{x_0} \mathrm{d}y\, K(y,x_0)\langle N(x,y,t)\rangle \tag{2.2}$$

was solved analytically in terms of the Kummer function

$$\langle N(x, x_0, t)\rangle = e^{-tx_0^{\beta+1}}\left[\delta(x - x_0) + 2tx_0^{\beta}\ {}_1F_1\left(\frac{\beta+3}{\beta+1};2;(x_0^{\beta+1} - x^{\beta+1})t\right)\right]. \quad (2.3)$$

The dependence of kernel (2.1) upon β gives the relaxation rate of mass (energy) y to be

$$\frac{dy}{dt} = -\frac{y}{\tau(y)} = -y\int_0^y dx\, K(x,y) = -y^{\beta+2}$$

and describes two situations: at $\beta > -1$ the relaxation is infinitely long, while at $\beta < -1$ the fragmentation process speeds up and completes in finite time of the order of the very first breakup time. The latter results in a 'shattering' kinetic transition—part of the initial mass (energy) starts (formally) to accumulate at the low limit, $x \to 0$.

The solution (2.3) has a simple asymptotic form [2]. Above shattering, at $\beta > -1$ and $x < x_0$, $tx_0^{\beta+1} \gg 1$, it reduces to

$$\langle N(x, x_0, t)\rangle = \frac{2x_0}{\Gamma[(\beta+3)/(\beta+1)]} t^{2/(\beta+1)} e^{-tx^{\beta+1}} \quad (2.4)$$

(the so-called similarity solution). Below shattering, at $\beta < -1$ and $x < x_0$, $tx^{\beta+1} \gg 1$ we have

$$\langle N(x, x_0, t)\rangle = \frac{2x_0^{\beta} x^{-\beta-3}}{\Gamma[(\beta-1)/(\beta+1)]} t^{-2/(\beta+1)} e^{-tx_0^{\beta+1}}. \quad (2.5)$$

The immediate question arises whether this solution is stable with respect to fluctuations, as has been said above. In other words, the issue is whether the solutions (2.3)–(2.5) can be seen in a single experiment. This question has been addressed mathematically by Filippov [2]. We shall use a different method which allows us to derive equations for the entire distribution and its moments and obtain some explicit solutions together with their physical meaning. To the best of our knowledge the method employed goes back to unpublished works by Landau on cosmic showers (at the end of the 1940s) but has, since that time, been re-discovered by other researchers [12] (there exist other methods making use of more standard approaches [11]). In the method which we use one writes down the explicit equation for $W(x, x_0, t|1, x_0, 0)$ in the form

$$\begin{aligned}\partial_t W(N, x, t|1, x_0, 0) &= -W(N, x, t|1, x_0, 0)\int_0^{x_0} dy\, K(y, x_0 - y) \\ &+ \sum_{N'}\int_x^{x_0-x} dy\, K(y, x_0) W(N', x, t|1, y, 0) W(N-N', x, t|1, x_0-y, 0) \\ &+ \int_0^x dy\, K(y, x_0) W(N, x, t|1, x_0 - y, 0) \\ &+ \int_{x_0-x}^{x_0} dy\, K(y, x_0) W(N, x, t|1, y, 0)\,. \end{aligned} \quad (2.6)$$

which makes use of the fact that a shower (cascade) can be described in terms of the very first act of decay and can be derived from the backward Chapman–Kolmogorov equation (see appendix A). (The linear terms take into account the situation when only the larger fragment can contribute to $W(N, x, t|1, x_0, 0)$ at given $x, x < x_0$.)

The corresponding equation for the generating function

$$G(x, x_0, t; z) = \sum_N W(N, x, t|1, x_0, 0)\exp(zN) \tag{2.7}$$

reads

$$\begin{aligned}\partial_t G(x, x_0, t; z) = &-G(x, x_0, t; z)\int_0^{x_0} \mathrm{d}y\, K(y, x_0) \\ &+ \int_x^{x_0 - x} \mathrm{d}y\, K(y, x_0) G(x, y, t; z) G(x, x_0 - y, t; z) \\ &+ \int_0^{x} \mathrm{d}y K(y, x_0) G(x, x_0 - y, t; z) \\ x, x \qquad &+ \int_{x_0 - x}^{x_0} \mathrm{d}y\, K(y, x_0) G(x, y, t; z)\,.\end{aligned} \tag{2.8}$$

The moments can be defined via

$$\langle N^k(x, x_0, t)\rangle = \partial^k G(x, x_0, t; z)/\partial z^k{}_{|z=0} \tag{2.9}$$

and the cumulants can be determined by the connection formulae, the first three of which are

$$c_1 = \langle N\rangle \qquad c_2 = \langle N^2\rangle - \langle N\rangle^2 \qquad c_3 = \langle N^3\rangle - 3\langle N^2\rangle\langle N\rangle + 2\langle N\rangle^3\,. \tag{2.10}$$

From (2.8) the equation for the second moment is

$$\begin{aligned}\partial_t \langle N^2(x, x_0, t)\rangle = &-\langle N^2(x, x_0, t)\rangle \int_0^{x_0} \mathrm{d}y\, K(y, x_0) \\ &+ 2\int_x^{x_0} \mathrm{d}y\, K(y, x_0)\langle N^2(x, y, t)\rangle \\ &+ 2\int_x^{x_0 - x} \mathrm{d}y\, K(y, x_0)\langle N(x, y, t)\rangle\langle N(x, x_0 - y, t)\rangle\,.\end{aligned} \tag{2.11}$$

Higher equations can easily be written.

Comments should be made about the normalization in the continuum limit. Since the probability W is dimensionless together with all its moments, the continuum limit can be conveniently defined through the discrete version if we express all the masses in terms of the unit mass step of the discrete formulation. Then the increase of the number of discrete states affects the value of x_0 and defines the continuum limit. This means that all powers of δ-functions should be considered as Kronecker symbols.

To investigate fluctuations, we solve (2.11) with the kernel (2.1) using the solution (2.3) for $\langle N(x, x_0, t)\rangle$ as the Green function:

$$\langle N^2(x, x_0, t)\rangle = \langle N(x, x_0, t)\rangle + 2\int_0^t \mathrm{d}t' \int_{2x}^{x_0} \mathrm{d}y\, y^\beta \langle N(y, x_0, t')\rangle$$
$$\times \int_x^{y-x} \mathrm{d}z \langle N(x, z, t - t')\rangle \langle N(x, y - z, t - t')\rangle\,. \qquad (2.12)$$

At arbitrary β the corresponding integrals for the Kummer function are not known analytically, and the only answer can be written in terms of multiple sums, which is not useful for further investigation. However, in the case $\beta = 0$ the exact answer is given in appendix C. Asymptotical behaviour is studied below. Namely, some progress can be made if we confine ourselves to the region of validity of (2.4), (2.5). Fluctuations are different above and below the shattering transition $\beta = -1$.

2.1. Above shattering, $\beta > -1$

We return to (2.11) and use (2.4) to evaluate the inhomogeneity under conditions leading to the similarity solution (2.4) at $x \ll x_0$. The last term in (2.11) gives

$$\frac{4}{3\Gamma^2[(\beta+3)/(\beta+1)]^2} x_0^{\beta+1} t^{4/(\beta+1)} \mathrm{e}^{-2tx^{\beta+1}}\,. \qquad (2.13)$$

The solution of (2.11) can be found to be in the form $\langle N^2(x, x_0, t)\rangle = x_0^2 f_0(x, t)$. Plugging this ansatz into (2.11) one can see that it corresponds to the quasi-steady regime ($\partial_t f_0 \approx 0$) and finally in the main order

$$\langle N^2(x, x_0, t)\rangle = \langle N(x, x_0, t)\rangle^2\,. \qquad (2.14)$$

This result can be directly found from (2.12): one notices that $t' \ll t$, then integration over time can be performed with the first Kummer function, while the other two are to be replaced by the similarity solutions. Equation (2.14) suggests that the similarity solution corresponds to effective averaging of fluctuations. One can extend this statement by showing that to the same order the generation function is simply

$$G(x, x_0, t; z) = \exp\{z\langle N(x, x_0, t)\rangle\}\,. \qquad (2.15)$$

This can be done term by term after expanding the exponential (2.15) into series and using (2.8). However, to this order, there is no information whatsoever about the fluctuation amplitude, since all the cumulants above the first one are equal to zero.

One can therefore compare then two different sources of corrections to (2.14). The first correction comes from the next order in the expansion of the Kummer function and exact limits of integration in (2.11). Denoting the similarity solution (2.4) as $\langle N_0(x, x_0, t)\rangle$ we obtain

$$\langle N(x, x_0, t)\rangle - \langle N_0(x, x_0, t)\rangle = \langle N_0(x, x_0, t)\rangle \left[\frac{\beta-1}{\beta+1}\left(\frac{x}{x_0}\right)^{\beta+1} + \frac{2(1-\beta)}{(\beta+1)^2}\frac{1}{tx_0^{\beta+1}}\right]\,. \qquad (2.16)$$

Comparison of two terms on the right-hand side of (2.16) gives two different cases $\beta > 1$ and $-1 < \beta < 1$. These cases can be traced further in (2.11) with explicit account taken for the integration limits. In the former case one can use the ansatz

$$\langle N^2(x, x_0, t)\rangle = \langle N_0(x, x_0, t)\rangle^2 \left[1 + \left(\frac{x}{x_0}\right)^2 f_>(x, t)\right] \tag{2.17}$$

which results in

$$f_>(x, t) = \frac{2}{\beta + 1}\left[\beta - 1 + \frac{4}{(\beta + 1)tx^{\beta+1}}\right]. \tag{2.18}$$

The latter case $-1 < \beta < 1$ requires the ansatz

$$\langle N^2(x, x_0, t)\rangle = \langle N_0(x, x_0, t)\rangle^2 \left[1 + \left(\frac{x}{x_0}\right)^{\beta+1} f_<(x, t)\right] \tag{2.19}$$

and gives

$$f_<(x, t) = \frac{2(\beta - 1)}{(\beta + 1)^2}\left[1 - \frac{2}{(\beta + 1)tx^{\beta+1}}\right]. \tag{2.20}$$

The second source of corrections is due to the fluctuations caused by the initial condition, which we have seen in the right-hand side of (2.12). It is simply equal to $\langle N(x, x_0, t)\rangle$. Comparing both types of corrections one can easily check that the second type dominates. This can be traced for higher moments, and we get

$$\langle N^n(x, x_0, t)\rangle = \langle N_0(x, x_0, t)\rangle^n + \langle N(x, x_0, t)\rangle. \tag{2.21}$$

Fluctuations can be conveniently characterized by the signal-to-noise ratio, $c_1/c_2^{1/2}$, which to our approximation is just the square root of the mean-field solution. Therefore, near the maximum of the mean-field solution at $t_m = 2x^{-\beta-1}/(\beta + 1)$, we have $c_1/c_2^{1/2} \sim x_0 x^{-2}$. This ratio is large with respect to unity and indicates that eventually the cascade enters into a self-averaging regime. Computer simulation of the system considered here shows that numerically this averaging takes place at a rather late stage (or small enough x), where there appears a 'window' with $\langle N(x, x_0, t)\rangle \gg 1$. Outside of the window fluctuations dominate. From figure 1 we can see how small are the corresponding fragment sizes x. The third cumulant which describes the asymmetry of the distribution near $\langle N(x, x_0, t)\rangle$ (sometimes we shall refer to this asymmetry as a non-Gaussian distribution) follows from (2.21) and to the main order is equal to $c_3 = \langle N_0(x, x_0, t)\rangle - 3\langle N_0(x, x_0, t)\rangle^2$. At small and large times this difference is positive and the ratio $c_3^{1/3}/c_2^{1/2}$ is as large as $\langle N_0(x, x_0, t)\rangle^{-1/6}$. The third cumulant passes through zero near the edges of the window and is negative (and large again) inside the window, so that the mentioned ratio equals $-3^{1/3}\langle N_0(x, x_0, t)\rangle^{1/6}$. This behaviour is illustrated below on the example of the discrete model (see figure 3).

Equation (2.21) explicitly shows the self-averaging at the late stage of branching. We did not study this self-averaging for multiple (triple, etc) fragmentation processes—its existence for slowing fragmentation was proven mathematically by Filippov [2].

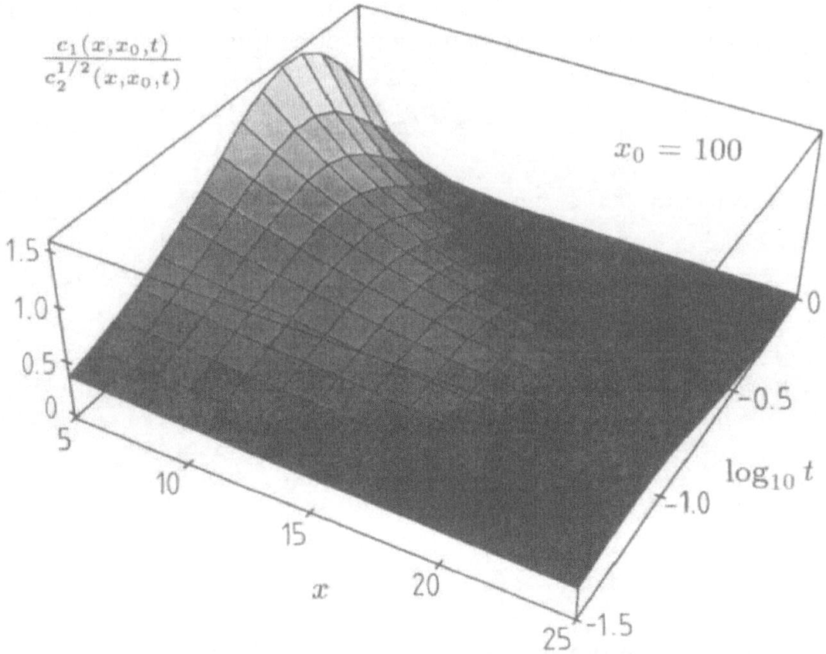

Figure 1. Solution of (2.12) for the second moment with kernel (2.1) (at $\beta = 0$) plotted as the signal-to-noise ratio. For the analytic expression of the second moment, see appendix C.

2.2. *Below shattering, $\beta < -1$*

In this case fluctuations dominate. To show this we may evaluate the triple integral (2.12) using the asymptotical solution (2.5). Some study shows that the integral gains its value at $z \sim y$, $t' \sim t$, $y \sim x_0$ if we assume $tx_0^{\beta+1} \sim 1$, about the maximum of occupation for a given x. The result is

$$\langle N^2(x, x_0, t \sim x_0^{-\beta-1})\rangle \sim x^{-2\beta-6} x_0^{2\beta+4} \sim \langle N(x, x_0, t)\rangle^2 \qquad (2.22)$$

and the second cumulant is of the same order. One concludes that the signal-to-noise ratio is of order of unity here. The situation changes at $\beta < -2$, where $\langle N(x, x_0, t)\rangle \ll 1$. Again the first term on the right-hand side of (2.12) (and the very first splitting) becomes important and fluctuations dominate strongly. The signal-to-noise ratio is as small as $(x_0/x)^{(\beta+2)/2}$. Finally, the mass of dust (which is defined as the initial mass minus the mass of all fragments with essentially positive mass, $1 - \int_0^{x_0} \mathrm{d}x\, N(x, x_0, t)$) is subject to the fluctuations mentioned.

3. The steady-state regime

As has been shown the fluctuations may be large and any observation makes sense only if the results are expressed in terms of the values averaged over a number of independent repetitions, larger than c_2/c_1^2. In many physical systems (cosmic showers, electron, phonon transport) one deals instead with a source of incoming 'particles' having the initial mass (energy) x_0 and collects the outgoing fragments after many breakups. The question now concerns establishing the steady state in such an experiment and the minimum source intensity required to suppress the fluctuations which may still be present in the steady state.

We shall introduce the simple procedure (analogous to [15]) which should provide such an averaging (steady state). At the very first state (x_0) there is assumed to be a constant discrete source of new 'particles', which produces new cascades with a time delay τ between each two; this time is a random variable distributed with the Poisson probability $P(\tau) = e^{-\tau/\bar{\tau}}$. The parameter $\bar{\tau}$ is inversely proportional to the source intensity. In our system the evolution of each particle is independent, and the logarithmic generating function $Z = \ln(G)$ is an additive variable. If there exists a steady-state regime, these functions should be equal to

$$Z(x, x_0; z) = \sum_{k=0}^{\infty} Z\left(x, x_0, t - \sum_{m=0}^{k} \tau_m; z\right). \tag{3.1}$$

Here τ_m denotes the mth realization of the random value τ. We can average the above-written sum over the distribution $P(\tau)$ to find

$$Z(x, x_0; z) = \frac{1}{\bar{\tau}} \int_0^{\infty} dt\, Z(x, x_0, t; z). \tag{3.2}$$

Clearly, the same result is valid for the average properties of the same system with a periodic source where the period is $\bar{\tau}$. In the steady-state regime it is natural to work with cumulants (2.10), which are connected with the time-dependent cumulants due to (3.2)

$$c_n(x, x_0) = \frac{1}{\bar{\tau}} \int_0^{\infty} dt\, c_n(x, x_0, t). \tag{3.3}$$

The corresponding equations for moments can be derived from here. Note that the equations for the time-dependent cumulants (obtainable from (2.8)) allow one to reduce the integration in (3.3) to an integral equation, whose usage may sometimes be easier. For example, by integrating the backward kinetic equation over time from 0 to ∞ we find the integral equation for c_1 ($= \langle N \rangle$)

$$c_1(x, x_0) \int_0^{x_0} dy\, K(y, x_0) = 2 \int_x^{x_0} dy\, K(y, x_0) c_1(x, y). \tag{3.4}$$

The equation for $c_2(x, x_0)$ can be derived from (2.12), etc. The criterion on the minimum source intensity to have the signal-to-noise ratio large in the steady-state regime is given by $1/\bar{\tau} \gg c_2/c_1^2$ and depends upon x, x_0. Below we set the time-delay constant $\bar{\tau}$ to unity—it can be easily retrieved by rescaling cumulants.

The existence of the steady state is now connected with the convergence of the integrals (3.3). In principle it could happen that some of these integrals converge (that for c_1, for example) but others do not. It does not take place for the linear systems considered here, since the maximum occupation occurs at finite time. Nevertheless, we shall find that the *time* required to establish the steady state depends upon the order of the cumulants of distribution—the higher the order, the longer it may take for the corresponding integral to converge. The distribution under study is far from Gaussian, so that high cumulants may be of interest. Usually it is instructive enough to study the convergence for low-order cumulants c_n, $n = 1,2,3$.

To get some experience concerning the properties of different quantities it is useful to investigate in some detail the simple Yule–Ferry discrete model [11] solved in the formulation following [13]. Namely, consider a set of generations $n \geqslant 0$. One initial particle ($n = 0$) decays creating two particles in the first generation (breakup to two halves); each particle of this generation ($n = 1$) decays independently creating two particles in the second generation and so on. Let the probability $W(N, n, t|1, 0, 0)$ describe the presence of N particles in generation n at time t given there was one particle at the very beginning (cf section 2). The generating function for W is

$$G_n(t; z) = \sum_N W(N, n, t|1, 0, 0) \exp(zN) \tag{3.5}$$

where we omit the indexes describing our initial condition. It obeys the equation (cf (2.8))

$$\partial_t G_n(t; z) = -G_n(t; z) + G_{n-1}^2(t; z) . \tag{3.6}$$

It follows that the equations for the first three moments of W are

$$\begin{aligned}
\partial_t \langle N_n \rangle &= -\langle N_n \rangle + 2\langle N_{n-1} \rangle \\
\partial_t \langle N_n^2 \rangle &= -\langle N_n^2 \rangle + 2\langle N_{n-1}^2 \rangle + 2\langle N_{n-1} \rangle^2 \\
\partial_t \langle N_n^3 \rangle &= -\langle N_n^3 \rangle + 2\langle N_{n-1}^3 \rangle + 6\langle N_{n-1}^2 \rangle \langle N_{n-1} \rangle
\end{aligned} \tag{3.7}$$

where $\langle N_n^k(t) \rangle = \partial^k G_n(t; z)/\partial z^k{}_{|z=0}$. Equations (3.7) can be recurrently solved by using Laplace transform in time. Solutions for $N_{1,n}$ and $N_{2,n}$ have been obtained in [13] by another technique. The Laplace transform with the initial condition $\langle N_n(0) \rangle = \delta_{n,0}$ gives us the Poisson-type solution for the occupation numbers

$$\langle N_n(t) \rangle = \frac{2^n}{n!} t^n \mathrm{e}^{-t} . \tag{3.8}$$

The maximum mean occupation at a given site n occurs after a time $t = n$. We would like to study how fluctuations affect this simple picture. The same transform gives the (rather cumbersome) solution for the mean square occupation number at a given site:

$$\begin{aligned}
\langle N_n^2(t) \rangle = \frac{2^n}{n!} t^n \mathrm{e}^{-t} &+ \mathrm{e}^{-t} \sum_{m=1}^{n} \sum_{l=0}^{n-m} \frac{(-1)^l 2^{n+m-1} (2m + l - 2)!}{[(m-1)!]^2 (n - m - l)! (l)!} t^{n-m-l} \\
&+ \mathrm{e}^{-2t} \sum_{m=1}^{n} \sum_{l=0}^{2m-2} \frac{(-1)^{n-m+1} 2^{n+m-1} [2(m-1)]! (n - m + l)!}{(n-m)! [(m-1)!]^2 (2m - l - 2)! (l)!} t^{2m-l-2} .
\end{aligned} \tag{3.9}$$

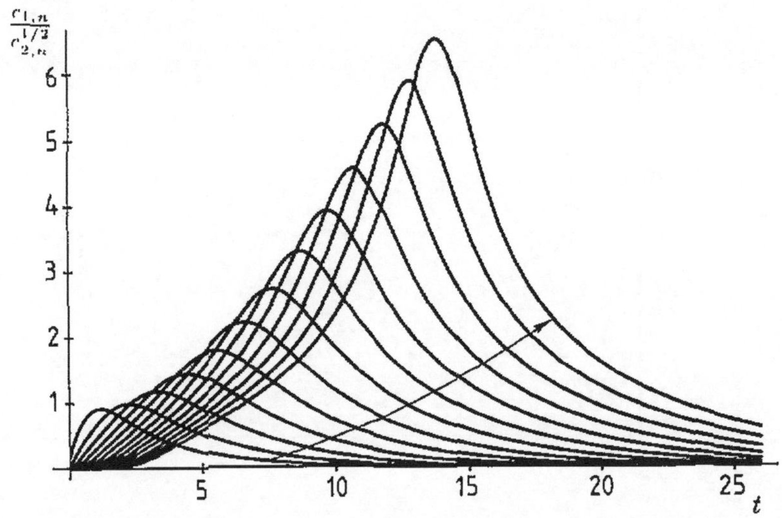

Figure 2. Signal-to-noise ratios at different sites $n \in [1, 13]$ versus time in the simple cascade Yule–Ferry model. The arrow shows the increase of the generation number.

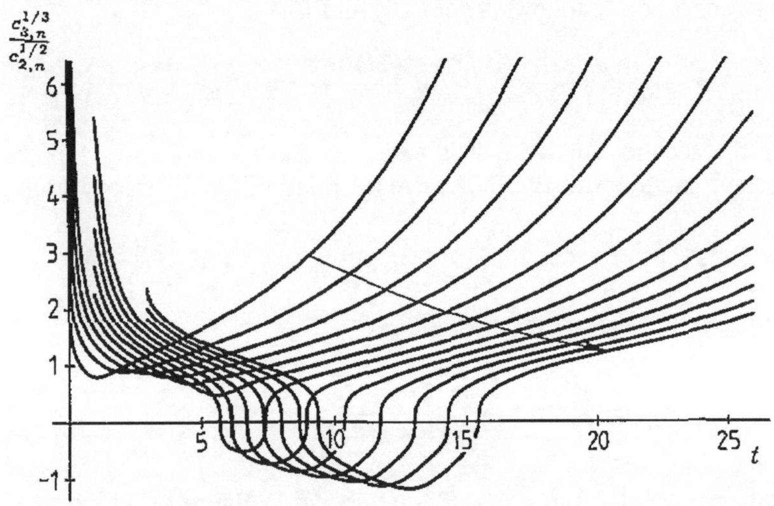

Figure 3. Cubic root of third cumulant divided by square root of second cumulant—the non-Gaussian property of occupation numbers. Sites and times are as in figure 2.

The signal-to-noise ratio $c_1/c_2^{1/2}$ is shown in figure 2. The arrow indicates the increase of n. In the discrete model, fluctuations are also small near the maximum of occupation for $n \gg 1$. Figure 3 shows the ratios $c_3^{1/3}/c_2^{1/2}$ showing that fluctuations in our simple cascade are entirely non-Gaussian and non-symmetric. The behaviour of the third cumulant is quite rich and analogous to that found in section 2. The asymmetry is almost always large and is initially right-sided, then left-sided in the window, and finally right-sided again.

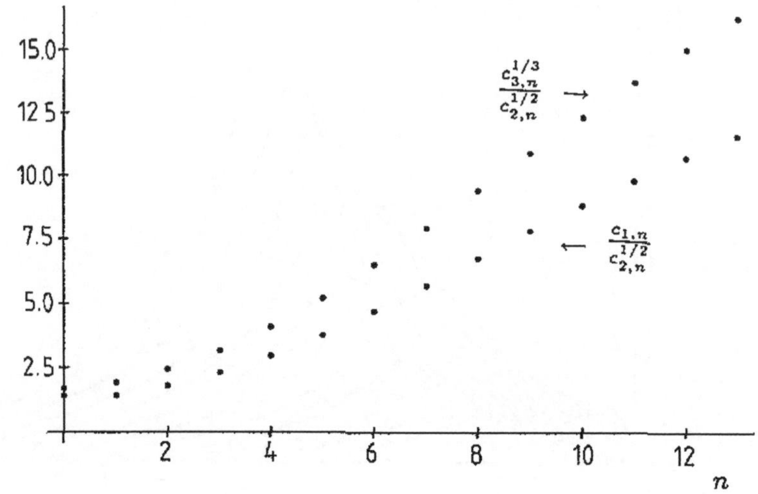

Figure 4. Steady-state occupancies in the simple cascade model at different generations. The source intensity is unity.

We return to the steady-state regime now. The steady-state occupation number for the generation n follows from (3.8) and (3.3):

$$c_{1,n} = \langle N_n \rangle = \int_0^\infty \mathrm{d}t \, \frac{(2t)^n \mathrm{e}^{-t}}{n!} = 2^n \, . \tag{3.10}$$

As for the second cumulant, it is easier to make use of (3.7) and the equation for the second cumulant to find a recurrence relation for the steady-state cumulants

$$c_{2,n} = 2c_{2,n-1} + \int_0^\infty \mathrm{d}t \, (2c_{1,n-1} - c_{1,n})^2 = 2c_{2,n-1} + \frac{(2n-2)!}{n!(n-1)!} \, . \tag{3.11}$$

The solution is given by the expression

$$c_{2,n} = 2^{n-1} + \sum_{k=0}^{n-1} \frac{2^k (2k)!}{k!(k+1)!} \, . \tag{3.12}$$

The ratios $c_{1,n}/c_{2,n}^{1/2}$ and $c_{3,n}^{1/3}/c_{2,n}^{1/2}$ are shown in figure 4. The convergence of time integrals (3.3) for c_1, c_2, c_3 is shown in figure 5. We conclude that the steady-state regime is well defined in this model and fluctuations become less and less important with the generation number. However, the distribution is again non-symmetric around its maximum. It is also interesting that the time required for the third cumulant to converge is larger than that for c_1, c_2.

Keeping the above results in mind let us return to continuum models with the random source acting at the initial point x_0 (section 2). The stationary solution of the kinetic equation can be found by either direct integration of its time-dependent solution (2.3) in (3.3), or by solving the integral equation (3.4):

$$x_0 c_1(x, x_0) = 2 \int_x^{x_0} \mathrm{d}y \, c_1(x, y) \tag{3.13}$$

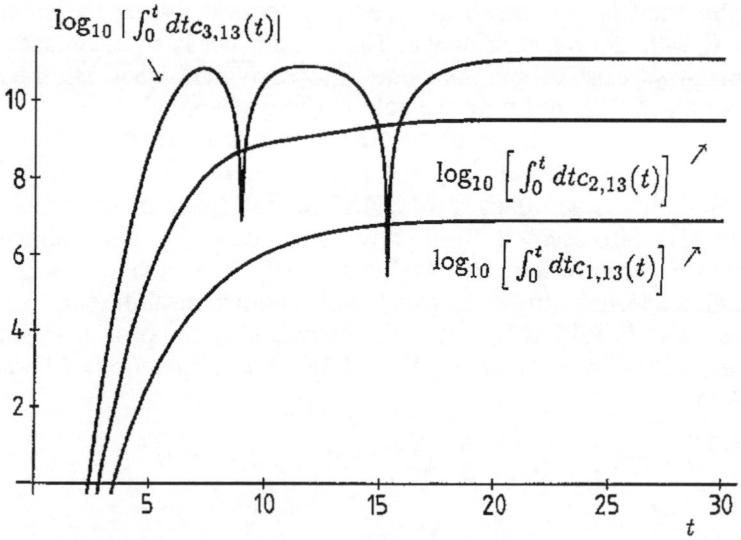

Figure 5. Convergence of the first, second and third cumulants to their steady-state values from figure 4 at the site (generation) $n = 13$.

which can be converted into a differential equation. The solution is

$$c_1(x, x_0) = x_0^{-\beta-1}\delta(x - x_0) + 2x_0 x^{-\beta-3}. \tag{3.14}$$

Note that the total mass of all finite fragments is infinite if $\beta \geqslant -2$, and it would take an increasingly long time, $t \sim \max(x^{-\beta-1}, x_0^{-\beta-1})$, for establishing such a distribution for small fragment sizes.

Integrating the expression for the second cumulant following from (2.21) one finds at $\beta > -1$, $x \ll x_0$ that

$$c_2(x, x_0) = c_1(x, x_0) \tag{3.15}$$

and, recalling the source intensity $1/\bar{\tau}$, the steady-state signal-to-noise ratio is large at

$$1/\bar{\tau} \gg x^{\beta+3}/x_0\,. \tag{3.16}$$

In the region $-2 < \beta < -1$, $x \ll x_0$ the estimate (2.21) gives the x-independent ratio $c_1(x, x_0)/c_2^{1/2}(x, x_0) \sim x_0^{-1/2}$, and $1/\bar{\tau} \gg x_0$. At $\beta < -2$ (3.15) and therefore (3.16) are again valid if $x \ll x_0$.

4. Connection with the combinatorial approach

We discuss here the connection between two different methods to study fragmentation processes which exist in the literature [3, 16, 17, 6]. The alternative to the kinetic method is the approach (we call it combinatorial) to declare that a given definite amount of fragments $\overline{N}$ is found in an experiment [6]. The problem would be to describe what is the mass distribution of these observed fragments and then to compare it with the measurements. The solution leads to combinatorial analysis of how many ways exist to get the same number $\overline{N}$ and what are their probabilities. The time passed does not play any role in the problem.

Generally speaking, the combinatorial approach is artificial (a second experiment would give another number $\overline{N}'$) and does not reflect the proper behaviour of the system. While both approaches sample the phase space of the problem, the kinetic approach takes into account all the possible states, while the combinatorial one only takes care of those which have the same number $\overline{N}$ and it is clear that only the mean-field properties can be captured. Fluctuations are different.

Indeed, Grady [6] implemented the combinatorial approach to study fragmentation in the simple case of uniform breakup ($\beta = 0$ in terms of kernel (2.1)). The distribution

$$\int_x^{x_0} \mathrm{d}x' \langle N(x', x_0, \overline{N}) \rangle = \overline{N} \left(\frac{x_0 - x}{x_0} \right)^{\overline{N}-1} \tag{4.1}$$

was obtained in the case of a continuous finite body. For the same problem the solution (2.3) for $\beta = 0$ implies

$$\int_x^{x_0} \mathrm{d}x' \langle N(x', x_0, t) \rangle = (1 - tx + tx_0) \mathrm{e}^{-tx} \,. \tag{4.2}$$

We expect the combinatorial approach to become meaningful only at large times or, equivalently, at small sizes ($x \ll x_0$). Indeed, from (4.1) at $\overline{N} \gg 1$, $x \ll x_0$ the asymptotic form follows $\overline{N} \exp(-x\overline{N}/x_0)$, which agrees with the asymptotic form $tx_0 \exp(-tx)$ for (4.2) if we establish the physically obvious connection $t = \overline{N}/x_0$. Above shattering there is an equivalence between the regions of validity of combinatorial and similarity solutions.

Below we shall address the abovementioned connection for the final or steady-state regimes. As time proceeds so does the fragmentation and finally a collection of objects of the smallest possible size would result. At this stage the system arrives at the final deterministic state. However, if one adds an additional rule that some of the intermediate fragments cannot proceed further, the final state may be distributed. Such an example is given by the random breaking of an interval studied by Derrida and Flyvbjerg [5]. In each binary breakup one of the two fragments is not allowed to proceed further and contributes to the final distribution. The other fragment breaks up and the scheme repeats. The model has very rich properties.

Although time does not appear in this problem we shall show below how to introduce it and derive the Derrida and Flyvbjerg equation for the final distribution. Correlators of the distribution can also be obtained by the same method. Let $P(N, x, t|x_0, 0)$ denote the probability density to have N fragments of size (mass, energy) x at time t given that the process starts with one 'particle' at $x = x_0$, $t = 0$. By analogy with (2.6) the function P obeys the following (backward) equation

$$\begin{aligned} \partial_t P(N, x, t|x_0, 0) &= -P(N, x, t|x_0, 0) \int_0^{x_0} \mathrm{d}y\, K(y, x_0) \\ &+ \tfrac{1}{2} \int_0^{x_0 - x} \mathrm{d}y\, K(y, x_0) P(N, x, t|x_0 - y, 0) \end{aligned}$$

$$+\tfrac{1}{2}\int_x^{x_0} dy\, K(x_0-y,x_0)P(N,x,t|y,0)$$
$$+\tfrac{1}{2}\Theta\left(\frac{x_0}{2}-x\right)(1-\delta_{N,0})[K(x,x_0)+K(x_0-x,x_0)]$$
$$\times[P(N-1,x,t|x_0-x,0)-P(N,x,t|x_0-x,0)]$$
$$+\tfrac{1}{2}\Theta\left(x-\frac{x_0}{2}\right)\delta_{N,1}[K(x,x_0)+K(x_0-x,x_0)][1-P(N,x,t|x_0-x,0)]\,. \tag{4.3}$$

Here the first integral on the right-hand side deals with the case when no breakup occurs at the very first state. The second and third integrals describe breakup when the first fragments of size y and x_0-y contribute to the final distribution, while the complementary one proceeds further and may contribute to P later. There are special cases when the fragment y (or x_0-y) arrives exactly at the point x and contributes explicitly to P. If $x > x_0/2$, this is the only way for the point x to become occupied (multiplier $\delta_{N,1}$), else the N argument of the P-function is changed to $N-1$. The non-integral P-terms with minus signs take care of the situation when the exact arrival is contained in the integral terms, and should be excluded.

The equation for the generation function defined by analogy with (2.7) reads

$$\partial_t G(x,t|x_0,0;z) = -G(x,t|x_0,0;z)\int_0^{x_0} dy\,\bar{K}(y,x_0)$$
$$+\int_0^{x_0} dy\,\bar{K}(y,x_0)G(x,t|y,0;z)$$
$$+\bar{K}(x,x_0)(e^z-1)\left[\Theta\left(\frac{x_0}{2}-x\right)G(x,t|x_0,0;z)+\Theta\left(x-\frac{x_0}{2}\right)\right] \tag{4.4}$$

where $\bar{K}(x,y) = [K(x,y)+K(y-x,y)]/2$. For the first moment (occupation numbers) we have

$$\partial_t\langle N(x,t|x_0,0)\rangle = -\langle N(x,t|x_0,0)\rangle\int_0^{x_0} dy\,\bar{K}(y,x_0)$$
$$+\int_0^{x_0} dy\,\bar{K}(y,x_0)\langle N(x,t|y,0)\rangle+\bar{K}(x,x_0)\,. \tag{4.5}$$

Omitting the time derivative we get the integral equation for the final distribution. If we consider the self-similar kernels $\bar{K}(x,x_0) = K(x/x_0)$, then the final distribution also becomes self-similar, $\langle N(x,x_0)\rangle = \langle N(x/x_0)\rangle$. Introducing the new variable of integration $y = x_0 - x$, we get the equation equivalent to that obtained in [5]:

$$\langle N(x)\rangle = \frac{1}{x_0}K(x)+\int_0^{1-x} dy\,K(y)\left\langle N\left(\frac{x}{x_0-y}\right)\right\rangle\,. \tag{4.6}$$

Note, that the kernel $K(x)$ has to be necessarily symmetric as is shown above. This means that the solutions obtained in [5] for a non-symmetric kernel do not apply to the considered fragmentation process. Other characteristics of randomly broken objects [5] can be studied by this approach.

5. Fluctuations in phonon decay

In this section we deal with the fragmentation process for the example of phonons. In order to be specific we consider the process of the phonon frequency down-conversion studied by Kazakovtsev and Levinson [7]. The physical situation is the following. A population of hot phonons is created in a semiconducting or insulating sample. Somewhere inside there is a device which measures (at least in principle) the energy distribution of the arriving low-frequency phonons. The energy evolution of each phonon ϵ_0 is the familiar process of anharmonic splitting into two daughter phonons, ϵ, $\epsilon_0 - \epsilon$; the conventional approximation for the kernel reads [7]

$$K(\epsilon, \epsilon_0) = A\epsilon^2(\epsilon_0 - \epsilon)^2 . \tag{5.1}$$

If the occupation numbers (here in the quantum sense) are smaller than unity, the phonon cascade is linear, each phonon is independent. As for the space propagation, it can easily be included since the energy evolution is independent of that in space, and the problem is to 'spread' the distribution in space in accordance to its energy content.

We may therefore immediately make use of (2.11) which now reads

$$\begin{aligned} \partial_t \langle N^2(\epsilon, t|\epsilon_0, 0)\rangle &= -\langle N^2(\epsilon, t|\epsilon_0, 0)\rangle \int_0^{\epsilon_0} \mathrm{d}\epsilon'\, K(\epsilon', \epsilon_0) \\ &+ 2c \int_{\epsilon}^{\epsilon_0} \mathrm{d}\epsilon'\, K(\epsilon', \epsilon_0)\langle N^2(\epsilon, t|\epsilon', 0)\rangle \\ &+ 2c \int_{\epsilon}^{\epsilon_0 - \epsilon} \mathrm{d}\epsilon'\, K(\epsilon', \epsilon_0)\langle N(\epsilon, t|\epsilon', 0)\rangle\langle N(\epsilon, t|\epsilon_0 - \epsilon', 0)\rangle \end{aligned} \tag{5.2}$$

where c is a number resulting from the presence of density of states, $\rho(\epsilon) \propto \epsilon^2$. We make use of the similarity solution of the kinetic equation, which has the form [7]

$$\langle N(\epsilon, r, t|\epsilon_0, 0, 0)\rangle = C\epsilon^{-4} f\left(\frac{t}{\tau(\epsilon)}\right) . \tag{5.3}$$

Here the normalization to one phonon is assumed, $C = \epsilon_0^3/\rho(\epsilon_0)V$ with V being the normalization volume, and $\tau^{-1}(\epsilon)$ is the inverse phonon lifetime, which is defined as $\int_0^{\epsilon} \mathrm{d}\epsilon'\, K(\epsilon', \epsilon)$. We deal here with the fragmentation above shattering, and may immediately apply the result (2.21) of section 2

$$\langle N^2(\epsilon, t|\epsilon_0, 0)\rangle = \langle N(\epsilon, t|\epsilon_0, 0)\rangle^2 + \langle N(\epsilon, t|\epsilon_0, 0)\rangle \tag{5.4}$$

and corresponding formulae for higher cumulants. Here $\langle N\rangle$ has the meaning of the number of phonons in a given 'box' v in space and $\rho(\epsilon)\Delta\epsilon$ in energy space. As has been said in section 2, the result (5.4) shows that fluctuations are not large around the maximum of the similarity solution (5.3), where $v\rho(\epsilon)\Delta\epsilon\langle N\rangle \gg 1$ and are given simply by the square root of this average number of phonons in the box. These fluctuations are entirely non-Gaussian in the same sense as we have explained above.

6. Conclusion

We have presented the application of the kinetic approach of backward equations to the study of different systems evolving along tree-like paths in the phase space (Bethe lattice, Cayley tree and ultrametric manifold are the other names used). For the systems below the shattering threshold our results show that fluctuations dominate and the mean-field solution can be 'measured' only by using the given large number of repetitions. For larger systems, the fluctuations are exceedingly strong *if* they dominate. We have shown that the similarity solution is the result of self-averaging along the branching cascade if it slows down. The steady-state regime with an external source has been considered, which may suppress fluctuations even below shattering due to the presence of many particles proceeding simultaneously. We have also shown that the combinatorial and kinetic approaches provide equivalent results only for very simple models in the mean-field approximation. Exact solutions have been given for some models. The results obtained for the slowing-down fragmentation have been used to study the phonon frequency-down conversion.

Acknowledgments

LPG acknowledges communications with D Pines, G Toulouse and B Derrida. SEE and TJN are thankful to N Goldenfeld, P Goldbart, and Y Oono for their interest with respect to this work. SEE is especially grateful to D L Maslov for stimulating discussions and several key references. SEE and LPG express their gratitude to AT&T for the quality of telephone connections.

The work of SEE and TJN was supported by NSF Grants NSF-DMR-90-15791 and NSF-DMR-89-20538. LPG was supported by NSF Grant NSF-INT-89-18665.

Appendix A

We present here the formal derivation of our basic equation of (2.6) type for the discrete Yule–Ferry cascade model. The continuous case of (2.6) can be treated in an analogous way. Let us introduce the vector of state $\mathbf{N} = \{N_0, N_1, N_2, \ldots, N_n, \ldots\}$ which describes the occupation numbers at the sites $0, 1, 2, \ldots, n, \ldots$. The probability of the system having a given state $\mathbf{N}$ at time t provided it has had the state $\mathbf{N}'$ at time t' is denoted as $P(\mathbf{N}, t|\mathbf{N}', t')$. It is connected with the 'short' description $W(N, n, t|1, 0, 0)$ due to the reduction relation

$$W(N, n, t|1,0,0) = \sum_{\text{all } N_k,\ k \neq n} P(\mathbf{N}, t|\mathbf{I}, 0) \tag{A1}$$

where $\mathbf{I} = \{1, 0, 0, \ldots\}$. The probability P obeys the forward and backward Chapman–Kolmogorov equations [11]. The forward equation is equivalent here to the master equation and can be written as

$$\partial_t P(\mathbf{N}, t|\mathbf{N}', t') = \sum_{\text{all } \mathbf{N}''} [w(\mathbf{N}'' \to \mathbf{N}) P(\mathbf{N}'', t|\mathbf{N}', t') - w(\mathbf{N} \to \mathbf{N}'') P(\mathbf{N}, t|\mathbf{N}', t)] \tag{A2}$$

where $w(\mathbf{N} \to \mathbf{N}')$ is the transition rate from the state $\mathbf{N}$ to the state $\mathbf{N}'$. In our case it is equal to N_n for the transition when the occupation number N_n diminishes by one and the next generation occupation number N_{n+1} increases by two, and is zero otherwise. We apply the reduction formula (A1) attempting to derive a closure for the 'short' description and immediately find that the sum

$$\sum_{N_{n-1}} (N_{n-1}+1) P(\{\ldots, N_{n-1}+1, N_n-2, \ldots\}, t|\ldots) \tag{A3}$$

which appears in the first term on right-hand side of (A2) where summation over all N_k have been performed except for $k = n-1$, does not reduce to the short description, because we cannot perform the last summation in terms of W functions. One can see that the forward time equation always generates the higher-order correlators.

The situation is quite different for the backward Chapman–Kolmogorov equation,

$$\partial_{t'} P(\mathbf{N}, t|\mathbf{N}', t') = \sum_{\text{all } \mathbf{N}''} w(\mathbf{N}' \to \mathbf{N}'')\left[P(\mathbf{N}, t|\mathbf{N}', t') - P(\mathbf{N}, t|\mathbf{N}'', t)\right]. \tag{A4}$$

Applying the summation (A1) and the above-mentioned transition rate w we derive the equation analogous to that of Williams [12],

$$\partial_{t'} W(N, n, t|1,0,0) = W(N, n, t|1,0,0) - W(N, n, t|2,1,0) \tag{A5}$$

which with the help of the identity

$$W(N, n, t|2,1,0) = \sum_{N'} W(N', n, t|1,1,0) W(N-N', n, t|1,1,0) \tag{A6}$$

the change of $\partial_{t'} = -\partial_t$, shifting the argument $(N, n, t|1,1,0) \to (N, n-1, t|1,0,0)$, and using the generation function representation (3.5) gives (3.6).

Appendix B

In the case of more than binary fragmentation the generalization of (2.8) is straightforward. Let us first write down the equation for the triple fragmentation; the triple kernel $K_3(x, y, x_0)$ describes the breakup rate of the initial particle x_0 into three fragments $x, y, x_0 - x - y$, the corresponding symmetry is assumed:

$$\begin{aligned}
\partial_t G(x, x_0, t; z) = & -G(x, x_0, t; z) \int \mathcal{D} \\
& + \int \mathcal{D} G(x, x', t; z) G(x, y', t; z) G(x, x_0 - x' - y', t; z) \\
& + \int \mathcal{D} G(x, y', t; z) G(x, x_0 - x' - y', t; z) \\
& + \int \mathcal{D} G(x, x', t; z) G(x, x_0 - x' - y', t; z) \\
& + \int \mathcal{D} G(x, x', t; z) G(x, y', t; z) \\
& + \int \mathcal{D} G(x, x', t; z) + \int \mathcal{D} G(x, y', t; z) + \int \mathcal{D} G(x, x_0 - x' - y', t; z)
\end{aligned} \tag{B1}$$

where $\mathcal{D} = \mathrm{d}x'\,\mathrm{d}y'K_3(x', y', x_0)$. The integration regions should be organized in such a way that all the second arguments (of G-functions being integrated) have to lie between x and $x_0 - x$ altogether. For example, the integral containing the triple product of G-functions is to be performed over the region $\{x' \geqslant x, y' \geqslant x, x' + y' \leqslant x_0 - x\}$. All the integration regions are supposed to be positive.

It is simple to modify (B1) for k-pieces breakup. There is the main k-product of all G-functions and subsequent terms describing that only some of the k-pieces (those which have sizes from x to x_0 in all the possible combinations) can contribute to the distribution at the point x.

The next step is to take the sum of all these equations for arbitrary k and finally compute the time derivative of the generation function,

$$\partial_t G(x, x_0, t; z) = \partial_t \sum_{k=2}^{\infty} G_k(x, x_0, t; z)\,. \tag{B2}$$

Although this equation is rather cumbersome, the corresponding equations for the moments simplify considerably. For example, studying the first moment, we have to take the first derivative of (B2) with respect to z. This gives the backward kinetic equation

$$\begin{aligned}\partial_t\langle N(x, x_0, t)\rangle = &-\langle N(x, x_0, t)\rangle \int_0^{x_0} \mathrm{d}y\, \mathcal{K}_{\langle N\rangle}(y, x_0)\\ &+ \int_x^{x_0} \mathrm{d}y\, \mathcal{K}_{\langle N\rangle}(y, x_0)\langle N(x, y, t)\rangle\end{aligned} \tag{B3}$$

with the *first* effective kernel

$$\begin{aligned}\mathcal{K}_{\langle N\rangle}(x, x_0) = &\,2K_2(x, x_0) + 3\int_0^{x_0} \mathrm{d}y_1\, K_3(x, y, x_0) + \cdots\\ &+ (n+2)\int\cdots\int \mathrm{d}y_1\cdots\mathrm{d}y_n\, K_{n+2}(x, y_1, \ldots, y_n, x_0) + \cdots\end{aligned} \tag{B4}$$

where the integration is performed over the region of non-negative $\{y_1, \ldots, y_n\}$ satisfying the condition $x + y_1 + \cdots + y_n \leqslant x_0$.

The second moment $\langle N^2\rangle$ obeys the following equation:

$$\begin{aligned}\partial_t\langle N^2(x, x_0, t)\rangle = &-\langle N^2(x, x_0, t)\rangle \int_0^{x_0} \mathrm{d}y\, \mathcal{K}_{\langle N\rangle}(y, x_0)\\ &+ \int_x^{x_0} \mathrm{d}y\, \mathcal{K}_{\langle N\rangle}(y, x_0)\langle N^2(x, y, t)\rangle\\ &+ \int_x^{x_0 - x} \mathrm{d}y\, \mathcal{K}_{\langle N^2\rangle}(y, x_0)\langle N(x, y, t)\rangle\langle N(x, x_0 - y, t)\rangle\end{aligned} \tag{B5}$$

with the *second* effective kernel

$$\begin{aligned}\mathcal{K}_{\langle N^2\rangle}(x, x_0) = &\,2K_2(x, x_0) + 6\int_0^{x_0} \mathrm{d}y_1\, K_3(x, y, x_0) + \cdots\\ &+ (n+1)(n+2)\int\cdots\int \mathrm{d}y_1\cdots\mathrm{d}y_n\, K_{n+2}(x, y_1, \ldots, y_n, x_0) + \cdots\end{aligned} \tag{B6}$$

having the same rules for the integration region as (B4). Therefore, the power-law assumption [3] for the first argument x of the kernel $\mathcal{K}_{\langle N\rangle}(x, x_0)$ does not specify higher-order effective kernels, except for the case of binary fragmentation. An independent power-law assumption can be made for $\mathcal{K}_{\langle N^2\rangle}(x, x_0)$ following Filippov.

Appendix C

The solution for the second moment $\langle N^2(x, x_0, t)\rangle$ in the case $\beta = 0$ can be obtained from (2.12) by inserting into it the first moment (2.3) which now reads

$$\langle N(x, x_0, t)\rangle = \mathrm{e}^{-tx_0}\delta(x - x_0) + t(2 + t(x_0 - x))\mathrm{e}^{-tx} . \tag{C1}$$

The following expression results by applying a symbolic manipulation program:

$$\langle N^2(x, x_0, t)\rangle = \langle N(x, x_0, t)\rangle + t^2(6 - 12tx + 4t^2x^2 + 6tx_0 - 4t^2xx_0 + t^2x_0^2)\mathrm{e}^{-2tx} \tag{C2}$$

where $x \leqslant x_0/2$. In the region $x > x_0/2$ the second moment is not affected by the inhomogeneity in (2.11) and is just equal to the first moment. Note that the continuum limit corresponds to $x_0 \gg 1$. However, at any finite x_0 there are discontinuities in $\langle N^2(x, x_0, t)\rangle$, i.e. at $x = x_0, x_0/2$ the latter one becomes small at large x_0.

If we subtract $\langle N(x, x_0, t)\rangle^2$ from $\langle N^2(x, x_0, t)\rangle$ given by (C2) (or simply by $\langle N(x, x_0, t)\rangle$ at $x > x_0/2$) and integrate over time, we obtain the steady-state value of the second cumulant

$$c_2(x, x_0) = \begin{cases} 2x_0/x^3 - 1/4x^3 - 3x_0^2/4x^5 & x > x_0/2 \\ 2x_0/x^3 - 1/4x^3 - 3x_0/4x^4 & x < x_0/2 \end{cases} \tag{C3}$$

where the first term in both of the expressions (C3) results from the integration of $\langle N(x, x_0, t)\rangle$ over time.

References

[1] Kolmogorov A N 1941 *DAN USSR (Doklady)* **31** 99
[2] Filippov A F 1961 *Theor. Probab. Appl. (USSR)* **6** 275
[3] McGrady E D and Ziff R M 1987 *Phys. Rev. Lett.* **58** 892
[4] Derrida B 1985 *J. Physique Lett.* **46** L401
Derrida B and Gardner E 1986 *J. Phys. C: Solid State Phys.* **19** 2253
[5] Derrida B and Flyvbjerg H 1987 *J. Phys. A: Math. Gen.* **20** 5273
[6] Grady D E 1990 *J. Appl. Phys.* **68** 6099
[7] Kazakovtsev D V and Levinson Y B 1979 *Phys. Status Solidi* b **96** 117
Levinson Y B 1986 *Nonequilibrium Phonons in Nonmetallic Crystals* ed W Eisenmenger and A A Kaplyanskii (Amsterdam: North-Holland)
[8] Esipov S E and Gor'kov L P, in preparation
[9] Rammal R, Toulouse G and Virasoro M A 1986 *Rev. Mod. Phys.* **58** 765
[10] Jaqaman H R and Gross D H E 1991 *Nucl. Phys.* A **524** 321
Ploszajczak M and Tucholski A 1990 *Phys. Rev. Lett.* **24** 1539
[11] See the following books and references therein:

Bharucha-Reid A T 1960 *Elements of the Theory of Markov Processes and Their Applications* (New York: McGraw-Hill)
Harris T 1963 *The Theory of Branching Processes* (Berlin: Springer)
Cox D R and Miller H D 1972 *The Theory of Stochastic Processes* (London: Chapman and Hall)
Shrinivasan S K 1969 *Stochastic Theory and Cascade Processes* (New York: Elsevier)
See also the following recent article:
Boziev S N 1990 *Sov. J. Nucl. Phys.* **52** 320
[12] Williams M M R 1977 *Proc. R. Soc.* A **358** 105;
Pazsit I and Williams M M R 1980 *J. Phys. D: Appl. Phys.* **13** 351
[13] Esipov S E and Newman T J, in preparation
[14] In the case of multiple breakup, the kernel entering the integral equations for high-order moments does not coincide with that in the kinetic equation (see appendix B). Therefore, the kernel (2.1) *implies* the binary fragmentation if we apply it to fluctuations.
[15] For a more general discussion see:
Pazsit I 1987 *J. Phys. D: Appl. Phys.* **20** 151
[16] Baumann G, Freyberger M, Glöckle W G and Nonnenmacher T F 1991 *J. Phys. A: Math. Gen.* **24** 5085
[17] Cheng Z and Redner S 1990 *J. Phys. A: Math. Gen.* **23** 1233

Chapter 4

Quantum Mechanics

4.1 Asymptotic of exchange interaction in hydrogen molecule

Split in the energy between the symmetrical (spin-*singlet*) and the antisymmetrical (spin-*triplet*) terms of the hydrogen molecule define the exchange interaction between the two hydrogen atoms $\hat{H}_{exc}(R) = -2J(R)(\vec{\hat{s}}_1 \cdot \vec{\hat{s}}_2)$. (Below R is the distance between atoms in units of the Bohr radius a_B). In textbooks the value of exchange integral J is calculated in the Heitler-London scheme that amounts to mere calculating of the Coulomb integrals on the symmetrized and antisymmetrized wave functions of the two electrons on the neutral atoms. We have proven [4a] that this approximation is not correct even at large distances between the atoms.

Inconsistency of the perturbation Heitler-London scheme is hidden in the fact that inside of the potential barrier on distances of order $R \gg a_B$, the interaction energy of one electron with the other and each electron with an extraneous nucleus is of the same order as the Coulomb interaction energy between two nuclei. Therefore, the Schrödinger equation for the wave functions of electrons in that area had to be solved anew. When the solution is obtained, the problem can be reduced to calculating an integral over the surface of a hyper-plane in the 6-dimensional coordinate space, analogous to how it was done by Landau in 1961 for the H_2^+-ion (see Ref. 1). The final result is: $J(R) = -0.821R^{5/2}e^{-2R}$ (By negligence, the wave functions of Eq. (6) in [4a] were not properly symmetrized; the numerical factor in [4a] was corrected in Ref. 2).

Method [4a] was modified in [4b] and applied to the calculation of the asymmetric exchange interactions between the two hydrogen-like donors in the presence of a weak spin-orbit interaction in semiconductors lacking the inversion symmetry. In Ref. 3 the general method [4b] is extended to include the excited states of shallow centers on doped semiconductor in the dilute limit.

4.2 Mott exciton in strong field

In classic semiconductors the name "Mott exciton" stands for the weakly bound hydrogen-like state of a conduction electron and a hole in the valence band. Paper [4c] addresses properties of excitons in strong magnetic fields. A single exciton can freely move along the crystal. In particular, in the presence of a homogenous magnetic field the exciton, being electrically neutral, regarding its motion in perpendicular-to-field-direction possesses the Galilean invariance and, hence, must be characterized by some analog of the transverse momentum. Dependence of the vector potential on coordinates makes construction of the momentum operator quite non-trivial [4c].

The free electron and the hole in the magnetic field rotate in the opposite directions. Correspondingly, there is the question of separating the exciton's center-of-mass motion before calculating the internal spectrum of exciton in high magnetic fields.

References

1. L.D. Landau and E. M. Lifshitz, *Quantum Mechanics*, Butterworth-Heinemann, Oxford (1998); *problem*, p. 314
2. C. Herring and M. Flicker, *Phys. Rev.* **134**, A362 (1964)
3. L. P. Gor'kov and P. L. Krotkov, *Phys. Rev.* B 68, 155206 (2003)

List of original papers

[4a] L. P. Gor'kov and L. P. Pitaevskii, *Energy terms splitting of the hydrogen molecule*, Sov. Phys. Doklady, 8, 788 (1964)

[4b] L. P. Gor'kov and P. L. Krotkov, *Spin relaxation and antisymmetric exchange in n-doped III-V semiconductors*, Phys. Rev. B67, 033203 (2003)

[4c] L. P. Gor'kov and I. E. Dzyaloshinski, *Contribution to the Theory of the Mott Exciton in a Strong Magnetic Field*, Sov. Phys. JETP, 26, 449 (1968)

SOVIET PHYSICS—DOKLADY VOL. 8, NO. 8 FEBRUARY, 1964

PHYSICS

TERM SPLITTING ENERGY OF THE HYDROGEN MOLECULE

L. P. Gor'kov and L. P. Pitaevskii

Institute of Physics Problems, Academy of Sciences, USSR
(Presented by Academician P. L. Kapitsa, March 28, 1963)
Translated from Doklady Akademii Nauk SSSR, Vol. 151, No. 4, pp. 822-825, August, 1963
Original article submitted March 20, 1963

The electron term energies of the hydrogen molecule are usually calculated on the Heitler-London scheme, where the inital wave functions are taken as symmetrized and antisymmetrized combinations of the electron wave functions in the neutral atoms. We wish to show in the present paper that this approximation is untrue even at large distances between the atoms. Below we shall calculate the asymptotic value of the splitting energy of the singlet and triplet terms of the hydrogen molecule. The singlet term has a symmetric $\psi_S(\mathbf{r}_1, \mathbf{r}_2)$ wave function while the triplet has an antisymmetric wave function $\psi_A = \{\mathbf{r}_1, \mathbf{r}_2\}$. The Schrödinger equation in atomic units is of the form

$$\Big\{-\frac{\Delta_1}{2}-\frac{\Delta_2}{2}-\sum_i\frac{1}{R_{1i}} -\sum_i\frac{1}{R_{2i}}+\frac{1}{r_{12}}+\frac{1}{2a}\Big\}\Psi_{S,A}=E_{S,A}\Psi_{S,A}, \quad (1)$$

where R_{1i}, and R_{2i} are the distances between the ith electrons in the first or second nucleus, which we place at the points $\pm a$ on the x axis.

In the Heitler-London theory, the splitting energy is expressed in terms of the integrals over the overlap region of the product of the wave functions $\Psi_a(\mathbf{r}_1) \times \Psi_{-a}(\mathbf{r}_1)\Psi_a(\mathbf{r}_2)\Psi_{-a}(\mathbf{r}_2)$, where $\Psi_{\pm a}(\mathbf{r})$ is the Bohr hydrogen atom function $\exp\{-|\mathbf{r}\pm\mathbf{a}|\}$. It is easily seen where the important range is located in these integrals. Actually, if we expand $|\mathbf{r}\pm\mathbf{a}|\simeq |x\pm a|+\rho^2/2|x\pm a|$ in products of the functions

$$\Psi_a(\mathbf{r})\Psi_{-a}(\mathbf{r})\simeq\exp\Big\{-2a+\frac{\rho^2}{2(x+a)}+\frac{\rho^2}{2(x-a)}\Big\},$$

where ρ is the distance to the x axis, it immediately follows that what is important is the cylindrical region where x is of the order a while $\rho\sim\sqrt{a}\ll x$. In this range of distances, the interaction energy of an electron with another one and of an electron with an extraneous nucleus is of the same order as its Coulomb interaction energy with the nucleus at which it is assumed to be localized. It is accordingly unjustifiable to treat these terms by perturbation theory, as is done in the Heitler-London theory. Instead of this, we have to solve the Schrödinger equation anew, in order to get a sufficiently accurate determination of the electron wave functions in the vicinity of the potential barrier. This fact, as applied to the problem of the molecular hydrogen ion, was first pointed out in 1961 by L. D. Landau [1].*

The molecular term splitting, E_S-E_A, is a quantity that is exponentially small at large distances between atoms: $E_S-E_A\sim e^{-4a}$. Nevertheless, it is known that two neutral atoms interact with one another by van der Waals forces, i.e., by a power law of the form $\sim a^{-6}$. It is accordingly very important to obtain a formula for the term splitting which has the exponential accuracy that we require. To this end, we construct in the six dimensional space $\{\mathbf{r}_1, \mathbf{r}_2\}$ the hyperplane $S(x_1 = x_2)$, which divides it into two parts. Consider now the integrals $\iint_\Omega \Psi_A E_S \Psi_S\, d\mathbf{r}_1\, d\mathbf{r}_2$ and $\iint_\Omega \Psi_S E_A \Psi_A\, d\mathbf{r}_1\, d\mathbf{r}_2$, in which the integration is extended over the regions to

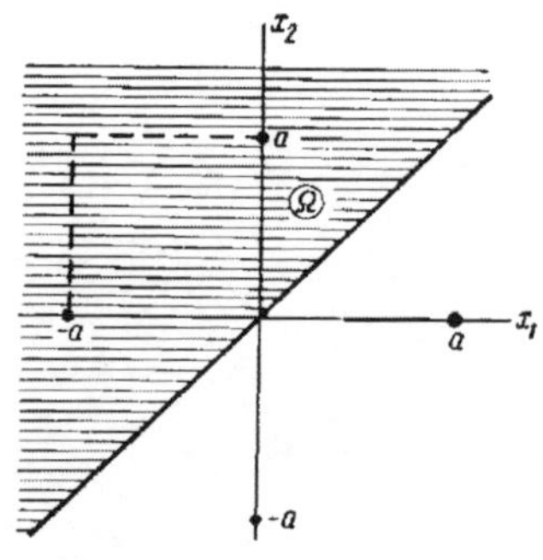

the "left" of the hyperplane $S(x_1 = x_2)$ (this region is shown schematically in the figure). By means of the Schrödinger equation (1) we obtain

*A critique of the Heitler-London model and a calculation for the molecular hydrogen ion are also given in the recently published paper by Herring [2].

$$(E_S - E_A)\iint_\Omega \Psi_A\Psi_S dr_1\, dr_2$$

$$= \frac{1}{2}\iint_\Omega \{\Psi_S(\Delta_1+\Delta_2)\Psi_A - \Psi_A(\Delta_1+\Delta_2)\Psi_S\}\, dr_1 dr_2$$

$$= \iint_\Omega \nabla_1(\Psi_S\nabla_1\Psi_A - \Psi_A\nabla_1\Psi_S)\, dr_1\, dr_2$$

$$= \oiint_S (\Psi_S\nabla_1\Psi_A - \Psi_A\nabla_1\Psi_S)\, dS.$$

Introduce combinations of the functions $\Psi_1 = {}^1/_2(\Psi_S + \Psi_A)$, $\Psi_2 = {}^1/_2(\Psi_S - \Psi_A)$, which correspond to states where, for large $a \gg 1$, the electron is principally localized at its own atom. We take ψ_1 and ψ_2 such that

$$\Psi_1(\mathbf{r}_1, \mathbf{r}_2) \simeq \frac{1}{\pi}\exp[-|\mathbf{r}_1 + \mathbf{a}| - |\mathbf{r}_2 - \mathbf{a}|]$$
$$(\mathbf{r}_1 \to -\mathbf{a};\ \mathbf{r}_2 \to \mathbf{a}), \tag{2}$$
$$\Psi_2(\mathbf{r}_1, \mathbf{r}_2) \simeq \frac{1}{\pi}\exp[-|\mathbf{r}_1 - \mathbf{a}| - |\mathbf{r}_2 + \mathbf{a}|]$$
$$(\mathbf{r}_1 \to \mathbf{a};\ \mathbf{r}_2 \to -\mathbf{a}),$$

i.e., under the above conditions they will reduce to a product of normalized Bohr hydrogen atom functions.

In the integral $\iint_\Omega \Psi_A\Psi_S dr_1 dr_2 = \iint_\Omega (\Psi_1^2 - \Psi_2^2) dr_1 dr_2$ the integration with respect to $\mathbf{r}_1$ is done mainly in the vicinity of $-\mathbf{a}$, while that with respect to $\mathbf{r}_2$ is around $\mathbf{a}$. With an accuracy up to small van der Waals corrections, in this range of the functions ψ_1 and ψ_2 are given by (2). Because of this, to the above accuracy we have $\iint_\Omega \Psi_A\Psi_S dr_1\, dr_2 = 1$. As a result we obtain

$$E_S - E_A = 2\oiint_S (\Psi_2\nabla_1\Psi_1 - \Psi_1\nabla\Psi_2)\, dS. \tag{3}$$

Thus, $E_S - E_A$ is expressed in terms of the integral of the product $\psi_2\psi_1$ over the hypersurface $S(x_1 = x_2)$. It is immediately seem from (3) that $E_S - E_A \sim e^{-4a}$. The important region of integration here, as is easily seen from (2), is $x_1 = x_2 \sim a$ and $\rho_1, \rho_2 \sim \sqrt{a}$. We now pass to the determination of ψ_1 and ψ_2 in this range.

Note first that for $a \gg 1$, we can neglect the van der Waals interaction in the energies E_S and E_A, i.e., $E_S \simeq E_A \simeq -1$. We shall look for ψ_1 and ψ_2 in the form

$$\Psi_1(\mathbf{r}_1, \mathbf{r}_2) = \frac{\chi_1(\mathbf{r}_1, \mathbf{r}_2)}{\pi}\exp[-|\mathbf{r}_1 + \mathbf{a}| - |\mathbf{r}_2 - \mathbf{a}|],$$
$$\Psi_2(\mathbf{r}_1, \mathbf{r}_2) = \frac{\chi_2(\mathbf{r}_1, \mathbf{r}_2)}{\pi}\exp[-|\mathbf{r}_1 - \mathbf{a}| - |\mathbf{r}_2 + \mathbf{a}|]. \tag{4}$$

In the range that we are interested in, $x_1, x_2 \sim a$, $\rho_1, \rho_2 \sim \sqrt{a}$ the functions $\chi_{1,2}(\mathbf{r}_1, \mathbf{r}_2)$ change slowly in comparison with the exponentials:

$$\exp\{-|\mathbf{r}_1 + \mathbf{a}| - |\mathbf{r}_2 - \mathbf{a}|\} \simeq \exp\left\{-2a - x_1 + x_2 + \frac{\rho_1^2}{2(x_1+a)} + \frac{\rho_2^2}{2(x_2-a)}\right\}.$$

Substituting (4) in (1), and limiting ourselves in differentiation to the principal terms, we obtain the equation for χ_1:

$$\left\{\frac{\partial}{\partial x_1} - \frac{\partial}{\partial x_2} + \frac{1}{2a} - \frac{1}{a - x_1} - \frac{1}{a + x_2} + \frac{1}{|\mathbf{r}_1 - \mathbf{r}_2|}\right\}\chi_1 = 0. \tag{5}$$

Substituting $x_1 = \xi + \eta/2$ and $x_2 = \xi - \eta/2$, we rewrite the equation in the more convenient form

$$\left\{\frac{\partial}{\partial\eta} - \frac{1}{2(a-\xi)-\eta} - \frac{1}{2(a+\xi)-\eta} + \frac{1}{2\sqrt{\eta^2 + \rho_{12}^2}} + \frac{1}{4a}\right\}\chi_1 = 0.$$

The solution of the last equation is

$$\chi_1 = \frac{C(\xi, \rho_{12})\left[\sqrt{\eta^2 + \rho_{12}^2} - \eta\right]^{1/2}}{[2(a-\xi)-\eta][2(a+\xi)-\eta]}e^{-\eta/4a},$$

where $C(\xi, \rho_{12})$ is an arbitrary function of the variables ξ and ρ_{12}.

To find the function, we make use of the obvious condition that χ_1 approaches unity when either $x_1 \to -a$ or $x_2 \to a$. As a matter of fact, let, for example, $x_1 \to -a$ (x_2 arbitrary). Here, the electron charge is neutralized by the charge of the ion at the point $-a$, and the motion of the second electron occurs only in the field of the ion at the point $+a$. This fact is also immediately obvious from (5). After simple calculations we obtain

$$\Psi_1(\mathbf{r}_1, \mathbf{r}_2) = \frac{2a\exp\left[-2a + x_2 - x_1 + \dfrac{\rho_1^2}{2(x_1+a)} + \dfrac{\rho_2^2}{2(x_2-a)}\right]}{\pi(a - x_1)(a + x_2)}$$
$$\times\begin{cases}(2a + x_1 + x_2)\exp\left[-\dfrac{a + x_1}{2a}\right]\left[\dfrac{\sqrt{(x_1 - x_2)^2 + \rho_{12}^2} + x_2 - x_1}{\sqrt{(2a + x_1 + x_2)^2 + \rho_{12}^2} + x_1 + x_2 + 2a}\right]^{1/2} \\ (x_1 + x_2 < 0); \\ (2a - x_1 - x_2)\exp\left[-\dfrac{a - x_2}{2a}\right]\left[\dfrac{\sqrt{(x_1 - x_2)^2 + \rho_{12}^2} + x_2 - x_1}{\sqrt{(x_1 + x_2 - 2a)^2 + \rho_{12}^2} + 2a - x_1 - x_2}\right]^{1/2} \\ (x_1 + x_2 > 0).\end{cases} \tag{6}$$

The function $\Psi_2(\mathbf{r}_1, \mathbf{r}_2)$ is obtained from (6) by making the substitution $1 \leftrightarrow 2$.

We return now to Eq. (3). Substituting (4) in it, and differentiating only the exponential, we transform it to the form

$$E_S - E_A = -8\int_0^a dx\, d^2\rho_1 d^2\rho_2\, [\Psi_2\Psi_1]_{\substack{x_1+x_2>0\\ x_1=x_2}}$$

(here $[\Psi_2\Psi_1]_{\substack{x_1+x_2>0\\ x_1=x_2}}$ means that the function (6) is taken for $x_1 \to x_2$, but $x_1 + x_2 > 0$).

As a result of the calculations we obtain

$$E_S - E_A = -\frac{4}{e}\sqrt{\pi}\,(2a)^{5/2}e^{-4a}\int_0^1 \frac{e^x dx}{x+1}. \quad (7)$$

If $R = 2a$ is the distance between the nuclei in atomic units, (7) is conveniently rewritten in the form

$$E_S - E_A = -2.94\, R^{5/2} e^{-2R}. \quad (8)$$

The result obtained from the Heitler-London scheme [3] is

$$E_S - E_A = -e^{-2R}\,\frac{4}{15}\,R^3\,(\ln R - 4.34). \quad (9)$$

The two expressions differ in the factor multiplying the exponential. Note however that in contrast with (9), Eq. (8) is asymptotically correct for $R \gg 1$. The relative accuracy with which (8) is calculated is determined by the one-dimensional approximation used in Eq. (5) and is of the order R^{-1}.

We emphasize once more in conclusion that for $R \to \infty$, the term splitting energy is exponentially small in comparison with the power law decrease in the van der Waals interaction energy: $E_{V\text{-}i} \simeq -6.5R^{-6}$. Comparing this quantity with (8) shows, however, that both energies become of the same order of magnitude for fairly large values of $R \sim 5$. For small values of R, the exchange part of the interaction energy of the atoms predominates.

LITERATURE CITED

1. L. D. Landau and E. M. Lifshits, Quantum Mechanics [in Russian], Moscow (1963).
2. C. Herring, Rev. Mod. Phys., 34, 631 (1962).
3. H. Bethe, Quantum Mechanics of the Simplest Systems [Russian translation], Moscow–Leningrad (1935).

PHYSICAL REVIEW B **67**, 033203 (2003)

Spin relaxation and antisymmetric exchange in *n*-doped III-V semiconductors

L. P. Gor'kov and P. L. Krotkov
National High Magnetic Field Laboratory, Florida State University, Tallahassee, Florida 32310
and L. D. Landau Institute for Theoretical Physics, Russian Academy of Sciences, 2 Kosygina Street, Moscow 117334, Russia
(Received 20 September 2002; published 17 January 2003)

Recently, Kavokin [Phys. Rev. B **64**, 075305 (2001)] suggested that the Dzyaloshinskii-Moriya interaction between localized electrons governs slow spin relaxation in *n*-doped GaAs in the regime close to the metal-insulator transition. We derive the correct spin Hamiltonian and apply it to the determination of spin dephasing time using the method of moments expansion. Our estimates would give longer than the observed values of the spin-relaxation time.

DOI: 10.1103/PhysRevB.67.033203 PACS number(s): 71.70.Gm, 71.70.Ej, 76.30.Da

Observation of exceedingly long spin-relaxation times for electrons optically pumped into the conduction band of *n*-doped GaAs (Ref. 1) prompted suggestions for the use of polarized spins in quantum computing.[2] No consensus has been reached, however, on the mechanisms governing the spin-relaxation processes. Low-temperature spin dephasing time τ_s in Ref. 1 behaves nonmonotonically as a function of donor concentration n_D, reaching values above 100 ns at $n_D = 10^{16}$ cm^{-3}. This concentration still belongs to the non-metallic regime.

In an interesting paper[3] the author introduced a new dephasing mechanism acting between the spins of electrons localized on randomly distributed shallow donors. According to Ref. 3, dephasing would occur each time the spins of the two electrons on neighboring centers rotate against each other in an effective spin-orbital field in the process of tunneling between the two centers during the spin flip-flop act caused by the exchange interaction. This notion and the way it was treated in Ref. 3 in terms of the exchange narrowing does not seem to be fully self-consistent. The point is that "random effective fields" originating in Ref. 3 from the antisymmetric Dzyaloshinskii-Moriya (DM) interaction[4,5] and the dynamical averaging due to the isotropic exchange, both, in fact, arise from the interaction of the spins of the very same pair and hence are not independent.

As a more consistent approach to the problem, we tried its formulation in terms of the electron paramagnetic resonance (EPR) line shape for the spin Hamiltonian that for semiconductors without an inversion center is a sum of the conventional isotropic and the antisymmetric DM exchange terms. The latter is responsible for nonconservation of the total spin of the system.

The antisymmetric interaction

$$\hat{\mathcal{H}}_{\rm DM} = \mathbf{d}(\mathbf{R})(\hat{\mathbf{s}}_1 \times \hat{\mathbf{s}}_2) \tag{1}$$

in semiconductors lacking the inversion symmetry (III-V and II-VI compounds) is of interest in itself, but has not been properly calculated yet. In Ref. 3 the asymptotic behavior of $\mathbf{d}(\mathbf{R})$ at large distances between centers $R \gg a_{\rm B}$, where $a_{\rm B}$ is the effective Bohr radius, for shallow (hydrogenlike) donor centers was calculated in the Heitler-London approximation. But it has been shown[6,7] that the Heitler-London theory does not account for the Coulomb correlations inside the "tunneling corridor" connecting the two centers, which dramatically affects the prefactor of the isotropic exchange.

Below we derive the asymptotic form of $\mathbf{d}(\mathbf{R})$ at $R \gg a_{\rm B}$. We have slightly modified the approach of Ref. 6,7 to include the initial spin degeneracy of the problem and its lifting by interactions in the presence of a weak spin-orbital (so) interaction

$$\hat{\mathcal{H}}_{\rm so} = \mathbf{h}(\hat{\mathbf{p}})\hat{\mathbf{s}}. \tag{2}$$

Our theoretical result for the R dependence of the antisymmetric exchange $\mathbf{d}(\mathbf{R})$ differs significantly from Ref. 3,8,9. In Ref. 8 the authors concluded that their experimental results for the spin-relaxation time τ_s together with the data[1,10] agree well with the mechanism[3] in the concentration range $n_D = 2\times10^{15} - 2\times10^{16}$ cm^{-3}. Upon substituting into the same formulas of Refs. 3,8, our corrected DM term increases the estimate for τ_s by a factor of 4. Although this does not rule out completely the relaxation mechanism[3] because of the phenomenological character of the analysis of the experimental data in Ref. 8, the possibility that the antisymmetric exchange alone accounts for the values of τ_s observed in Refs. 1,8,10 in this doping interval seems rather unlikely.

The essence of the method is as follows. The envelope of the wave function of an electron localized on a donor impurity in the effective-mass theory satisfies Schrödinger equation with the Hamiltonian

$$\hat{\mathcal{H}} = \mathbf{p}^2/2m_e + \hat{\mathcal{H}}_{\rm so} + e^2/\kappa r, \tag{3}$$

where m_e is the effective electron mass and κ is the permittivity. (For GaAs, $m_e \approx 0.072 m_0$ and $\kappa \approx 12.5$.)

The spin-orbital Hamiltonian (2) in zinc-blende semiconductors has the form[11,12]

$$h_x(\mathbf{p}) = \frac{\alpha_{\rm so}}{m_e\sqrt{2m_eE_g}} p_x(p_y^2 - p_z^2). \tag{4}$$

$h_y(\mathbf{p})$ and $h_z(\mathbf{p})$ are obtained by permutation in Eq. (4). Here E_g is the band gap and $\alpha_{\rm so}$ is a phenomenological parameter. For GaAs, $E_g \approx 1.43$ eV and $\alpha_{\rm so} \approx 0.07$.

Without the spin-orbital interaction the ground state had the energy $\frac{1}{2}E_0 = -{\rm Ry} = -m_e e^4/2\kappa^2\hbar^2$. ($E_0 \approx -12.6$ meV for GaAs.) The corresponding eigenfunction of Eq. (3) were $\varphi_\mu^\alpha(\mathbf{r}) = \varphi_0(r)\delta_\mu^\alpha$, where $\delta_\mu^\alpha \equiv \delta_{\uparrow,\downarrow}^\alpha$ denotes the spinors for up- and down-spin projection,

$$\varphi_0(r)=\frac{1}{\sqrt{\pi a_B^3}}e^{-r/a_B}, \tag{5}$$

and $a_B=\kappa\hbar^2/m_e e^2$ is the Bohr radius ($a_B\approx 92$ Å for GaAs). In the presence of a spin-orbital interaction, the spin projection no longer commutes with Eq. (3), but the twofold (Kramers) degeneracy of the ground state remains. For small spin-orbital interaction (4), eigenfunctions of Eq. (3) are

$$\varphi_\mu^\alpha(\mathbf{r})\approx\varphi_0(r)(e^{im_e\hbar\mathbf{h}(\hat{\mathbf{r}})r\sigma/2a_B^2})_{\alpha\mu}, \tag{6}$$

where $\hat{\mathbf{r}}=\mathbf{r}/r$. The phase factor turns into unity when $r\to 0$. Thus the Kramers index μ takes on the meaning of a spin projection of an electron on the center.

The two-electron Hamiltonian is obtained by combining two single-electron Hamiltonians (3) and the Coulomb interaction of electrons between themselves and with the alien ions:

$$\hat{\mathcal{H}}_{tot}=(\mathbf{p}_1^2+\mathbf{p}_2^2)/2m_e+\hat{\mathcal{H}}_{so1}+\hat{\mathcal{H}}_{so2}$$
$$+\frac{e^2}{\kappa}\left(\frac{1}{r_{12}}+\frac{1}{R}-\frac{1}{r_{1A}}-\frac{1}{r_{1B}}-\frac{1}{r_{2A}}-\frac{1}{r_{2B}}\right). \tag{7}$$

Here $r_{1A,B}$ and $r_{2A,B}$ denote the distances between the electrons and the two donor centers A and B.

For isolated donors, i.e., on large distances $R\gg a_B$ between two centers a two-electron wave function $\psi_{\mu\nu}^{\alpha\beta}$ is merely a product $\varphi_\mu^\alpha(\mathbf{r}_{1A})\varphi_\nu^\beta(\mathbf{r}_{2B})$ of two one-electron wave functions (6). For large but finite R, these functions have to be corrected to account for potentials of the Coulomb interactions of electrons between themselves and with the alien ions (7). Following Refs. 6,7, one starts with the Schrödinger equation for the corrected functions $\psi_{\mu\nu}^{\alpha\beta}$

$$\hat{\mathcal{H}}_{tot}^{\alpha\alpha',\beta\beta'}\psi_{\mu\nu}^{\alpha'\beta'}=E_0\psi_{\mu\nu}^{\alpha\beta}. \tag{8}$$

Let $\hat{\boldsymbol{\zeta}}$ denote the direction from the donor A to the donor B, positioned at $\zeta=\mp R/2$, respectively. It turns out[6] that for the calculation of exchange interaction, it is enough to know $\psi_{\mu\nu}$ on the median hyperplane $\zeta_1=\zeta_2$ far from the centers. There one may seek for the functions $\psi_{\mu\nu}$ approximately in the form

$$\psi_{\mu\nu}^{\alpha\beta}(\mathbf{r}_1,\mathbf{r}_2)=\chi^{\alpha\alpha',\beta\beta'}(\mathbf{r}_1,\mathbf{r}_2)\varphi_\mu^{\alpha'}(\mathbf{r}_{1A})\varphi_\nu^{\beta'}(\mathbf{r}_{2B}), \tag{9}$$

where $\chi^{\alpha\alpha',\beta\beta'}$ varies on the scale of the order of R.

Substituting Eqs. (9) into (8) yields a differential equation on $\chi^{\alpha\alpha',\beta\beta'}$. The boundary condition is determined from the condition that when either $\mathbf{r}_{1A}\to 0$ or $\mathbf{r}_{2B}\to 0$, the function $\psi_{\mu\nu}^{\alpha\beta}$ should convert to $\varphi_\mu^\alpha(\mathbf{r}_{1A})\varphi_\nu^\beta(\mathbf{r}_{2B})$, i.e., $\chi^{\alpha\alpha',\beta\beta'}\to\delta^{\alpha\alpha'}\delta^{\beta\beta'}$. The principal terms in spin-orbital interaction are included into definition (6), and the equation for $\chi^{\alpha\alpha',\beta\beta'}$ turns out to coincide with the one in the absence of the spin orbital interaction,[6]

$$\left(\partial_{\zeta_1}-\partial_{\zeta_2}-\frac{2}{R-2\zeta_1}-\frac{2}{R+2\zeta_2}+\frac{1}{R}+\frac{1}{r_{12}}\right)\chi=0. \tag{10}$$

Here we used that $\partial\varphi\sim\varphi/a_B$ and hence one can neglect all but the first derivatives of the function χ which varies for distances of order $R\gg a_B$.

Therefore $\chi^{\alpha\alpha',\beta\beta'}=\chi_0\delta^{\alpha\alpha'}\delta^{\beta\beta'}$, where on the median five-dimensional hyperplane $\zeta_1=\zeta_2\equiv\zeta$ (Ref. 9)

$$\chi_0=\frac{2R}{R+2|\zeta|}\sqrt{\frac{2\rho_{12}}{R-2|\zeta|}}e^{(2|\zeta|-R)/2R}. \tag{11}$$

Here $\rho_{12}=|\boldsymbol{\rho}_1-\boldsymbol{\rho}_2|$, and $\boldsymbol{\rho}_{1,2}$ are the transverse radii vectors of the electrons.

The two-electron eigenfunction $\psi^{\alpha\beta}(\mathbf{r}_1,\mathbf{r}_2)$ of $\hat{\mathcal{H}}_{tot}$ adjusted for the exchange is then sought in the form

$$\psi^{\alpha\beta}=\sum_{\mu,\nu}c_{\mu\nu}\psi_{\mu\nu}^{\alpha\beta}+\sum_{\mu,\nu}c_{\nu\mu}^{(P)}\psi_{\nu\mu}^{\alpha\beta}, \tag{12}$$

where $c_{\mu\nu}$ and $c_{\mu\nu}^{(P)}$ are unknown coefficients. $\psi_{\nu\mu}^{\alpha\beta}$ stands for the interchanged states of the two electrons, when the first electron is now almost localized on atom B in the state ν and the second one on atom A in the state μ,

$$\psi_{\nu\mu}^{\alpha\beta}(\mathbf{r}_1,\mathbf{r}_2)\equiv\psi_{\mu\nu}^{\beta\alpha}(\mathbf{r}_2,\mathbf{r}_1). \tag{13}$$

Because the total fermionic wave function should be antisymmetric in particle permutation $\psi^{\alpha\beta}(\mathbf{r}_1,\mathbf{r}_2)=-\psi^{\beta\alpha}(\mathbf{r}_2,\mathbf{r}_1)$, the coefficients $c_{\mu\nu}^{(P)}=-c_{\mu\nu}$.

We can now proceed to the derivation of the total exchange Hamiltonian. Multiplying the Schrödinger equation $E\psi=\hat{\mathcal{H}}\psi$ by $\psi_{\mu\nu}^*$ (for brevity, we omit spinor structure) and the Schrödinger equation (8) $\hat{\mathcal{H}}^*\psi_{\mu\nu}^*=E_0\psi_{\mu\nu}^*$ by ψ, and subtracting term by term and integrating over the region $\zeta_1<\zeta_2$, we find

$$(E-E_0)c_{\mu\nu}=\int_{\zeta_1<\zeta_2}(\psi_{\mu\nu}^*\hat{\mathcal{H}}\psi-\psi\hat{\mathcal{H}}^*\psi_{\mu\nu}^*)d^6\mathbf{r}_{1,2}. \tag{14}$$

The kinetic part of $\hat{\mathcal{H}}$ can be readily reduced to a surface integral,[6,7]

$$(E-E_0)c_{\mu\nu}=\frac{1}{2}\int_{\zeta_1=\zeta_2}(\psi\boldsymbol{\partial}\psi_{\mu\nu}^*-\psi_{\mu\nu}^*\boldsymbol{\partial}\psi)d\mathbf{S} \tag{15}$$

over the median plane. Here $d\mathbf{S}=(\hat{\boldsymbol{\zeta}}_1,-\hat{\boldsymbol{\zeta}}_2)d\zeta d^4\boldsymbol{\rho}_{1,2}$ is its surface element and $\boldsymbol{\partial}=(\boldsymbol{\partial}_1,\boldsymbol{\partial}_2)$ is the six-dimensional gradient.

We then substitute Eq. (12) into Eq. (15) neglecting derivatives of χ_0 compared to those of φ_μ:

$$\partial_{\zeta_{1,2}}\psi_{\mu\nu}\approx\mp\psi_{\mu\nu},\quad \partial_{\zeta_{1,2}}\psi_{\nu\mu}\approx\pm\psi_{\nu\mu}. \tag{16}$$

After simple calculations, Eq. (15) results in

$$(E-E_0)c_{\mu\nu}=-2c_{\mu'\nu'}^{(P)}\int_{\zeta_1=\zeta_2}\psi_{\mu\nu}^*\psi_{\nu'\mu'}d\zeta d^4\boldsymbol{\rho}_{1,2}$$
$$=J(R)(e^{-i\gamma\sigma/2})_{\mu\nu'}(e^{i\gamma\sigma/2})_{\nu\mu'}c_{\mu'\nu'}^{(P)} \tag{17}$$

with the angle

$$\boldsymbol{\gamma}=m_e\hbar\mathbf{h}(\hat{\boldsymbol{\zeta}})R/a_B^2. \tag{18}$$

The exchange integral equals[6,7]

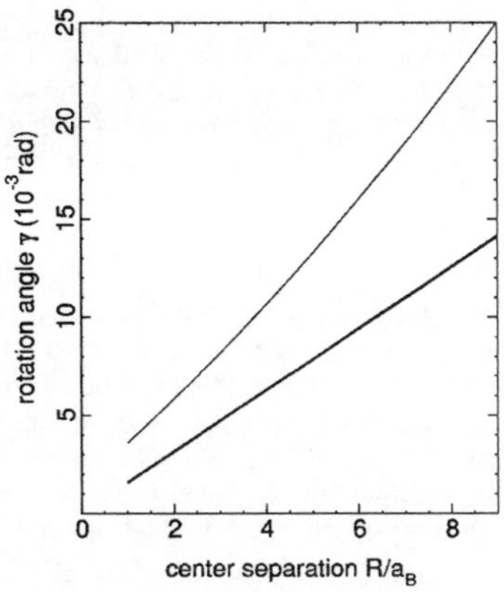

FIG. 1. Mean-square rotation angles $\overline{\gamma}$ of the asymmetric interaction in GaAs as functions of the intercenter separations as calculated in the present paper (thick line) and according to the interpolation Eq. (7) in Ref. 8 (thin curve).

$$J(R)=-2\int_{\zeta_1=\zeta_2}\chi_0^2\varphi_0^*(r_{1A})\varphi_0^*(r_{2B})\varphi_0(r_{2A})\varphi_0(r_{1B})\times d\zeta d^4\boldsymbol{\rho}_{1,2}\approx-0.818\times 2\text{Ry}(R/a_B)^{5/2}e^{-2R/a_B}. \tag{19}$$

Using

$$c_{\mu\nu}^{(P)}=-c_{\mu\nu}\equiv-\tfrac{1}{2}(\boldsymbol{\sigma}_{\mu\mu'}\boldsymbol{\sigma}_{\nu\nu'}+\delta_{\mu\mu'}\delta_{\nu\nu'})c_{\nu'\mu'}, \tag{20}$$

we finally arrive at the equation for the coefficients $c_{\mu\nu}$, $(E-E_0)c=\hat{\mathcal{H}}_{ex}c$, with the Hamiltonian

$$\hat{\mathcal{H}}_{ex}=-\tfrac{1}{2}Je^{-i\gamma(\sigma_1-\sigma_2)/2}(\hat{\boldsymbol{\sigma}}_1\hat{\boldsymbol{\sigma}}_2+1)\equiv-\tfrac{1}{2}J[\hat{\boldsymbol{\sigma}}_1\overleftrightarrow{R}(\boldsymbol{\gamma})\hat{\boldsymbol{\sigma}}_2+1]\approx-\tfrac{1}{2}J[\hat{\boldsymbol{\sigma}}_1\hat{\boldsymbol{\sigma}}_2+1-\boldsymbol{\gamma}(\hat{\boldsymbol{\sigma}}_1\times\hat{\boldsymbol{\sigma}}_2)], \tag{21}$$

where the Pauli matrices $\hat{\boldsymbol{\sigma}}_1$ and $\hat{\boldsymbol{\sigma}}_2$ are supposed to act in the space of the Kramers indices, and $\overleftrightarrow{R}(\boldsymbol{\gamma})$ is the three-dimensional rotation matrix on the angle γ around the axis $\hat{\boldsymbol{\gamma}}$.

The rotation angle γ (18) gives the strength of the DM term relative to the isotropic exchange. It is perpendicular to axis $\hat{\boldsymbol{\zeta}}$ since $\mathbf{h}(\hat{\boldsymbol{\zeta}})\hat{\boldsymbol{\zeta}}=0$. Taking the square mean of $\mathbf{h}(\hat{\boldsymbol{\zeta}})$ over the directions of $\hat{\boldsymbol{\zeta}}$, we find that for GaAs,

$$\overline{\gamma}=\langle\gamma^2\rangle_{\hat{\zeta}}^{1/2}=\frac{2\alpha_{so}}{\sqrt{35E_g/\text{Ry}}}\frac{R}{a_B}\approx 0.001\,57\frac{R}{a_B}. \tag{22}$$

This $\overline{\gamma}$ is shown in Fig. 1 against the R dependence of $\overline{\gamma}$ from Ref. 8 [Eq. (7)] for comparison. In the region of interest $[R\sim(3-7)a_B]$, our result for $\overline{\gamma}$ is about one-half that of Ref. 8 [Eq. (7)].

We now turn to the question that how the antisymmetric exchange would manifest itself as a spin-relaxation mechanism. It was suggested in Ref. 3,8 to write down the corresponding relaxation time τ_{sa} in terms of a precession mechanism:[12]

$$\tau_{sa}=\tfrac{3}{2}\tau_c\overline{\gamma}^{-2}, \tag{23}$$

where now, unlike Ref. 12, $1/\tau_c$ is a rate of the flip-flop transition caused by isotropic exchange. Inconsistency of the physical picture of Ref. 3 in the use of the dynamical averaging in Eq. (23) was mentioned above. Another objectionable point is to write τ_c in the form[8]

$$\tau_c\approx\hbar/\xi J(R_c), \tag{24}$$

where $R_c\approx\beta n_D^{-1/3}$, ξ and β being the fitting parameters of the order of unity. The choice for the latter one, $\beta\approx 0.65$, interpolates the maxima in the Poisson distribution for the nearest and next nearest-neighbors. Such a form is assumed in Ref. 8 to phenomenologically account for large average distances between donors. But since the maximum in the Poisson distribution is rather broad, it appears erroneous to merely substitute R_c for R in the exponential dependence $J(R)$.

At $n_D\approx 10^{16}$ cm^{-3}, the distance $R_c\approx 3.3a_B$, and from Fig. 1 for the angles $\overline{\gamma}$ (Ref. 8, [Eq. (7)]), one obtains $\overline{\gamma}(R_c)\approx 0.009$, while the corrected value in Eq. (22) is $\overline{\gamma}\approx 0.005$. The additional factor ≈ 4 in Eq. (23) immediately erodes the claimed quantitative agreement with experimental data in Refs. 1,8.

With the applicability of Eq. (23) being questioned and the experimental value of τ_c not being determined in Ref. 8 unambiguously, we try to explore spin relaxation due to DM terms as a question of the EPR line shape for a dilute system of paramagnetic spins interacting via Hamiltonian (21). This appears a proper approach if all other sources of inhomogeneous line broadening are neglected and the limit is taken of the Larmor frequency $\omega_L=g\mu_B B\to 0$.

In Ref. 13, such a problem dealing with DM interaction has already been addressed for Mn-based II-VI compound diluted magnetic semiconductors in the framework of *high-temperature moment expansion* of the linear response function

$$\chi''(\omega)=\frac{1}{2}\int_{-\infty}^{\infty}e^{-i\omega t}\langle[M_x(t),M_x(0)]\rangle dt. \tag{25}$$

Here square brackets denote quantum commutator of the magnetization operator $\mathbf{M}=g\mu_B\Sigma_n\mathbf{s}_n$ (summation runs over all donors). Angle brackets denote both thermodynamical average and averaging over disorder.

The spin-relaxation time is then given by[13]

$$\tau_{sa}=\sqrt{2I_4/\pi I_2^3}, \tag{26}$$

where the frequency moments of the response function $I_n=\int_{-\infty}^{\infty}\omega^{n-1}\chi''(\omega)d\omega/\pi\chi(0)$ obtained in Ref. 13 as a high-temperature expansion, are

$$I_2=-\text{Tr}([\hat{\mathcal{H}}_{DM},M_x]^2)/\text{Tr}(M_x^2)+O(T^{-1}), \tag{27}$$

$$I_4 = \mathrm{Tr}([\hat{\mathcal{H}}_{\mathrm{ex}},[\hat{\mathcal{H}}_{\mathrm{DM}},M_x]]^2)/\mathrm{Tr}(M_x^2) + O(T^{-1}). \quad (28)$$

Calculating the commutators, we find

$$I_2 = 4n_D \int J^2(r)[\gamma^2(\hat{\mathbf{r}}) - \gamma_x^2(\hat{\mathbf{r}})]d^3\mathbf{r} + O(T^{-1}), \quad (29)$$

$$I_4 \approx 32n_D \int J^4(r)[3\gamma^2(\hat{\mathbf{r}}) - \gamma_x^2(\hat{\mathbf{r}})]d^3\mathbf{r} + O(T^{-1}). \quad (30)$$

[For an order-of-magnitude estimate below, we neglected in Eq. (30) the terms with double integration under the pretext that they are of the "second order" in n_D. Averaging over $\hat{\mathbf{r}}$ we use $\langle \gamma_i(\hat{\mathbf{r}})\gamma_j(\hat{\mathbf{r}})\rangle_{\hat{r}} = \frac{1}{3}\overline{\gamma^2}\delta_{ij}$.]

The high-temperature expansions (29) and (30), where one would integrate over all randomly distributed centers, applies to weakly interacting spins only. Equations (29) and (30) need to be corrected to take into account that if two spins are so close that their exchange interaction $J(R) \gg T$, they would form a singlet state and drop out of the system's thermodynamics. A proper calculation of the temperature effects in the EPR line due to clusterization should proceed in the framework of a scaling theory.[14] The problem is difficult and remains unsolved. We take the effects into account by qualitatively introducing a cutoff in the integrals (29) and (30) at short distances a_T found from the condition $J(a_T) = T$. (At $T=2$ K, this radius $a_T \approx 3.7a_{\mathrm{B}}$; and at $T=4.2$ K, $a_T \approx 3.1a_{\mathrm{B}}$.)

Calculation of Eqs. (29) and (30) yields the answer in terms of incomplete gamma functions $\Gamma(n,x)$:

$$\tau_{sa} = \frac{6}{\pi n_D a_{\mathrm{B}}^3 0.818\mathrm{Ry}\overline{\gamma^2}(a_T)} \sqrt{\frac{2\Gamma(13,8a_T/a_{\mathrm{B}})}{\pi\Gamma^3(8,4a_T/a_{\mathrm{B}})}}. \quad (31)$$

For GaAs at $n_D = 10^{16}$ cm^{-3}, at $T=2$ K we find $\tau_{sa} \approx 290$ ns, and at $T=4.2$ K we find $\tau_{sa} \approx 170$ ns, as compared to $\tau_s \approx 80$ ns found experimentally. [Using expression (23) with corrected value for γ from Fig. 1 would give the temperature-independent value of $\tau_{sa} \approx 330$ ns at $n_D = 10^{16}$ cm^{-3}.]

The method of taking a spin-orbital interaction into account, developed in the present paper to shallow donor centers in *bulk* zinc-blende semiconductors, may be easily generalized. In a low-dimensional environment, e.g., in quantum dots, spin-orbital interaction is described by a linear dependence on **p**: $h_i(\mathbf{p}) = h_{ik}p_i$, where h_{ik} is a tensor. The spin structure of an exchange Hamiltonian (21) acting on the coefficients of expansion (12) again decouples from the coordinate dependence $J(R)$. Although explicit expression for $J(R)$, of course, now depends on the potential of the quantum dots, the form of the "rotated" exchange Hamiltonian (21) with the rotation angle $\gamma_k = m_e \hbar h_{ik} \hat{\zeta}_k R/a_{\mathrm{B}}^2$ instead of Eq. (18) remains. For a specific choice of the linear in **p** spin-orbital interaction, this result may be obtained by an exact unitary transformation, as has been recently shown in Ref. 15.

Finally, we comment on the role of the spin-orbital interaction of electrons with the electric field of alien ions. For a homogeneous field, it has the form[16] $\alpha_E \mu_{\mathrm{B}} g \hat{\mathbf{s}}(\hat{\mathbf{p}} \times \mathbf{E})2c/E_g$, the coefficient $\alpha_E \sim 1$ depends on the band structure. This term is proportional to $\sim \alpha_E(\mathrm{Ry}^2/E_g)(m_e/m_0)$, and is much smaller than the term (4) considered above, which is proportional to $\sim \alpha_{\mathrm{so}}\mathrm{Ry}(\mathrm{Ry}/E_g)^{1/2}$.

To summarize, we derived the asymptotically correct form of the spin Hamiltonian for two hydrogenlike donors in the *n*-doped GaAs. We applied the EPR line-shape formalism to analyze spin dephasing times τ_{sa} due to the antisymmetric DM exchange. The rough estimate for τ_{sa} (neglecting the detailed low-temperature spin clusterization that would further increase τ_{sa}) yields values exceeding those observed experimentally. Although the concentration range $n_D^{-1/3} \sim a_{\mathrm{B}}$ is very difficult for theory, our estimates may suggest that the anisotropic exchange between the localization centers is not a prevailing mechanism of the spin relaxation in this concentration range.

The authors are thankful to V. Korenev and K. Kavokin for discussions and sharing the details of interpretation of their data in Ref. 8. L.P.G. thanks V. Dobrosavljevic, who attracted his attention to Ref. 14, and gratefully acknowledges fruitful conversations on the subject with V. Kresin and E. Rashba. The work was supported (L.P.G.) by the NHMFL through the NSF cooperative agreement (Grant No. DMR-9527035) and the State of Florida, and (P.L.K.) by the DARPA through the Naval Research Laboratory Grant No. N00173-00-1-6005.

[1] J.M. Kikkawa and D.D. Awschalom, Phys. Rev. Lett. **80**, 4313 (1998).

[2] See, e. g., G.A. Prinz, Science **282**, 1660 (1998) on the problem of quantum computing.

[3] K.V. Kavokin, Phys. Rev. B **64**, 075305 (2001).

[4] I.E. Dzyaloshinskii, Phys. Chem. Solids **4**, 241 (1958).

[5] T. Moriya, Phys. Rev. **120**, 91 (1960).

[6] L.P. Gor'kov and L.P. Pitaevskii, Dokl. Acad. Nauk SSSR **151**, 822 (1963) [Sov. Phys. Dokl. **8**, 788 (1964)].

[7] C. Herring and M. Flicker, Phys. Rev. **134**, A362 (1964).

[8] R.I. Dzhioev *et al.*, cond-mat/0208083 (unpublished).

[9] Our paper has already been completed when we found the preprint (Ref. 7). We added now a few comments regarding the interpretation of Ref. 7 of the experimental data in terms of the DM mechanism.

[10] R.I. Dzhioev *et al.*, Phys. Rev. Lett. **88**, 256801 (2002).

[11] G. Dresselhaus, Phys. Rev. **100**, 580 (1955).

[12] M.I. Dyakonov and V.I. Perel, Fiz. Tverd. Tela **13**, 3581 (1971) [Sov. Phys. Solid State **13**, 3023 (1972)]; Zh. Exp. Teor. Fiz. **65**, 362 (1973) [Sov. Phys. JETP **38**, 177 (1974)].

[13] B.E. Larson and H. Ehrenreich, Phys. Rev. B **39**, 1747 (1989).

[14] R.N. Bhatt and P.A. Lee, J. Appl. Phys. **52**, 1703 (1981); Phys. Rev. Lett. **48**, 344 (1982).

[15] K. Kavokin, cond-mat/0212347 (unpublished).

[16] P. Nozières and C. Lewiner, J. Phys. (Paris) **34**, 901 (1973).

SOVIET PHYSICS JETP VOLUME 26, NUMBER 2 FEBRUARY, 1968

CONTRIBUTION TO THE THEORY OF THE MOTT EXCITON IN A STRONG MAGNETIC FIELD

L. P. GOR'KOV and I. E. DZYALOSHINSKIĬ

Institute of Theoretical Physics, Academy of Sciences, U.S.S.R.

Submitted March 16, 1967

Zh. Eksp. Teor. Fiz. 53, 717–722 (August, 1967)

The dependence of the energy of a Mott exciton on its transverse momentum is calculated in the case when the exciton is in a strong magnetic field (the distance between the Landau levels $e\hbar H/\mu c$ exceeds the Coulomb energy $\mu e^4/\varepsilon^2\hbar^2$, where μ is the reduced mass and ε the dielectric constant). The dependence of the probability for exciton production on the transverse component of the photon momentum and on the electric field is also determined.

THE theory of a Mott exciton in a strong magnetic field was developed by Elliott and Loudon[1] and by Hasegawa and Howard,[2] who confined themselves, however, to the case of a zero exciton momentum. We shall investigate the influence of the exciton motion on its spectrum and on the production probability.

In a strong magnetic field, when the distances between the Landau levels $e\hbar H/\mu c$ (μ = reduced mass) exceed the Coulomb energy $\mu e^4/\varepsilon^2\hbar^2$ (ε = dielectric constant), the question arises of separating the motion of the exciton mass center. Although such a problem was solved in principle by Lamb,[3] its exposition is very intricate. We therefore present here a more lucid derivation of Lamb's result, and use the derivation to obtain formulas for the exciton production probability.

We confine ourselves to the case of an isotropic dispersion law for electrons and holes. Then the Hamiltonian $\mathcal{H}$ of an exciton situated in homogeneous electric and magnetic fields $\mathscr{E}$ and H is

$$\frac{1}{2m_1}\left(-i\hbar\nabla_1+\frac{e}{c}\mathbf{A}_1\right)^2+\frac{1}{2m_2}\left(-i\hbar\nabla_2-\frac{e}{c}\mathbf{A}_2\right)^2+e\mathscr{E}(\mathbf{r}_1-\mathbf{r}_2)-\frac{e^2}{\varepsilon|\mathbf{r}_1-\mathbf{r}_2|}=\mathcal{H}. \tag{1}$$

We introduce the vector operator

$$\hat{\mathbf{P}}=-i\hbar\nabla_1-i\hbar\nabla_2+\frac{e}{c}(\mathbf{A}_1-\mathbf{A}_2)-\frac{e}{c}[\mathbf{H},\mathbf{r}_1-\mathbf{r}_2], \tag{2*}$$

which plays the role of the momentum of the exciton in a magnetic field. It is easy to verify that the operator $\hat{\mathbf{P}}$ commutes with the Hamiltonian (1)[1)] and, moreover, all three of its components commute with one another. This circumstance makes it possible to obtain the dependence of the eigenfunctions of $\mathcal{H}$ on the coordinate of the center of gravity of the exciton:

$$\mathbf{R}=(m_1\mathbf{r}_1+m_2\mathbf{r}_2)/(m_1+m_2).$$

Without loss of generality we can choose a definite gauge for **A**:

$$\mathbf{A}={}^1/_2[\mathbf{Hr}].$$

Then (2) takes the form

$$\hat{\mathbf{P}}=-i\hbar\nabla_{\mathbf{P}}-\frac{e}{2c}[\mathbf{Hr}],\qquad \mathbf{r}=\mathbf{r}_1-\mathbf{r}_2. \tag{3}$$

The eigenfunctions of (3) are

$$\Psi_{\mathbf{P}}(\mathbf{r}_1,\mathbf{r}_2)=\exp\left\{i\left(\mathbf{P}+\frac{e}{2c}[\mathbf{Hr}]\frac{\mathbf{R}}{\hbar}\right)\right\}\Psi_{\mathbf{P}}(\mathbf{r}), \tag{4}$$

where $\Psi_{\mathbf{P}}(\mathbf{r})$ is an arbitrary function of **r**, and **P** is the exciton momentum. Finally, substituting (4) in (1) we obtain the equation of relative motion of the electron and hole:

$$\left\{-\frac{\hbar^2}{2\mu}\Delta+\frac{ie\hbar}{2\mu c}\gamma\mathbf{H}[\mathbf{r}\nabla]+\frac{e^2}{8\mu c^2}[\mathbf{Hr}]^2+\frac{e}{Mc}[\mathbf{PH}]\mathbf{r}+e\mathscr{E}\mathbf{r}-\frac{e^2}{\varepsilon r}+\frac{\mathbf{P}^2}{2M}\right\}\Psi_{\mathbf{P}}(\mathbf{r})=E\Psi_{\mathbf{P}}(\mathbf{r}). \tag{5}$$

Here

$$\mu=\frac{m_1m_2}{m_1+m_2},\quad M=m_1+m_2,\quad \gamma=\frac{m_2-m_1}{m_2+m_1}. \tag{6}$$

The quantity **P** which enters in this equation has all the properties of a momentum. In particular, the average exciton velocity **V** is determined from the usual relation

$$\mathbf{V}=\partial E/\partial\mathbf{P}. \tag{7}$$

For the proof we note that, according to footnote [1)]

$$\hat{\mathbf{P}}=M\hat{\mathbf{V}}-\frac{e}{c}[\mathbf{Hr}],$$

where $\hat{\mathbf{V}}=d\mathbf{R}/dt$ is the operator of the c.m.s. velocity. On the other hand, $\hat{\mathbf{V}}$ coincides with the derivative of the Hamiltonian of (5), $\mathcal{H}_{\mathbf{P}}$, with respect to the momentum

$$\frac{\partial\mathcal{H}_{\mathbf{P}}}{\partial\mathbf{P}}=\frac{1}{M}\left(\hat{\mathbf{P}}+\frac{e}{c}[\mathbf{Hr}]\right)=\hat{\mathbf{V}}.$$

We shall henceforth confine ourselves to the case of an electric field and momentum perpendicular to the magnetic field; the latter will be assumed directed along the z axis. The transverse components will be designated by the index ρ.

*$[\mathbf{H},\mathbf{r}_1-\mathbf{r}_2]\equiv\mathbf{H}\times(\mathbf{r}_1-\mathbf{r}_2)$.

1)To this end it is sufficient to note that $\hat{\mathbf{P}}=m_1\hat{\mathbf{v}}_1+m_2\hat{\mathbf{v}}_2-ec^{-1}\mathbf{H}(\mathbf{r}_1-\mathbf{r}_2)$, where $\hat{\mathbf{v}}_{1,2}$ are the electron and hole velocity operators, and to write the operator equations of motion

$$m_1\frac{d\hat{\mathbf{v}}_1}{dt}=-e\mathscr{E}+\frac{e}{c}[\mathbf{H}\hat{\mathbf{v}}_1],\qquad m_2\frac{d\hat{\mathbf{v}}_2}{dt}=e\mathscr{E}-\frac{e}{c}[\mathbf{H}\hat{\mathbf{v}}_2],$$

from which it follows directly that $d\hat{\mathbf{P}}/dt=0$.

By using the transformation

$$\Psi(\mathbf{r}) = \Phi(\mathbf{r} - \rho_0)\exp(i\gamma\mathbf{r}\mathbf{P}'/2\hbar), \tag{8}$$

where

$$\mathbf{P}' = \mathbf{P} + \frac{Mc}{H^2}[\mathbf{H}\mathscr{E}], \qquad \rho_0 = \frac{c}{eH^2}[\mathbf{H}\mathbf{P}'], \tag{9}$$

we can reduce (5) to the form

$$\left\{-\frac{\hbar^2}{2\mu}\Delta_\rho + \frac{ie\hbar}{2\mu c}\gamma\mathbf{H}[\rho\nabla] - \frac{e^2}{8\mu c^2}H^2\rho^2 - \frac{\hbar^2}{2\mu}\frac{d^2}{dz^2} - \frac{e^2}{\varepsilon[(\rho+\rho_0)^2+z^2]^{1/2}} + \frac{\mathbf{P}^2}{2M} - \frac{\mathbf{P}'^2}{2M}\right\}\Phi = E\Phi. \tag{10}$$

In the case of a strong field we can neglect in the zeroth approximation the Coulomb interaction. Then the dependence of the energy on $\mathbf{P}$ and $\mathscr{E}$ is determined by the expression

$$\frac{\mathbf{P}^2 - \mathbf{P}'^2}{2M} = -\frac{c}{H^2}\mathbf{P}[\mathbf{H}\mathscr{E}] - \frac{Mc^2}{2}\frac{\mathscr{E}^2}{H^2}.$$

From this follow the physically obvious relations for the drift velocity

$$\mathbf{v} = \frac{c}{H^2}[\mathscr{E}\mathbf{H}] \tag{11a}$$

and for the dipole moment of the exciton

$$\mathbf{d} = \frac{Mc^2}{H^2}\mathscr{E} + \frac{c}{H^2}[\mathbf{P}\mathbf{H}]. \tag{11b}$$

As expected, the total effective mass of the free electron and hole, corresponding to motion across the field, is infinite. On the other hand, the nonzero values of the momentum (in the absence of $\mathscr{E}$) describe the average distance between particles

$$\rho_0 = -d/e.$$

A finite transverse mass occurs only when the Coulomb interaction is taken into account.

As shown by Elliott and Loudon,[1] the wave function of the exciton Φ from (10) can be represented in first approximation in the small parameter of the theory $(\mu e^4/\varepsilon^2\hbar^2)/(eH\hbar/\mu c)$ in the form

$$\Phi(\mathbf{r}) = \varphi(\rho)\psi(z),$$

where $\varphi(\rho)$ describes free transverse motion, and $\psi(z)$ satisfies the equation that results from the averaging of (10) with the aid of $\varphi(\rho)$.

We confine ourselves to an exciton in the zeroth Landau band. Then

$$\varphi_0(\rho) = \frac{1}{\sqrt{2\pi}\,r_0}\exp\left\{-\left(\frac{\rho^2}{4r_0^2}\right)\right\}, \tag{12}$$

where $r_0 = \sqrt{c\hbar/eH}$. The function $\psi(z)$ satisfies the equation

$$\left\{-\frac{\hbar^2}{2\mu}\frac{d^2}{dz^2} + U(z)\right\}\psi = W\psi, \tag{13}$$

where

$$U(z) = -\frac{e^2}{2\pi\varepsilon r_0^2}\int\frac{d\rho}{[(\rho+\rho_0)^2+z^2]^{1/2}}\exp\left\{-\frac{\rho^2}{2r_0^2}\right\}. \tag{14}$$

The energy of the exciton (with allowance for the spin) is written here in the form

$$E = \Delta + \frac{\mathbf{P}^2 - \mathbf{P}'^2}{2M} + \frac{e\hbar H}{2\mu c} + W, \tag{15}$$

where Δ is the distance between the electron and hole bands.

Equation (13) can be easily solved by the method used in [2]. For the exciton ground state we obtain

$$W_0 = -\frac{\hbar^2}{2\mu r_B^2}\lambda^2, \quad r_B = \frac{\varepsilon\hbar^2}{\mu e^2}, \tag{16}$$

where λ is the solution of the equation

$$\lambda = 2\ln\left(\frac{r_B}{\sqrt{2}\lambda r_0}\right) - C - \Lambda\left(\frac{r_0^2P'^2}{2\hbar^2}\right); \tag{17}$$

$$\Lambda(x) = \int_0^x dy e^{-y}\ln\frac{x}{y}, \tag{18}$$

with $\Lambda(x) \approx x$ as $x \to 0$ and $\Lambda(x) = \ln x - C$ as $x \to \infty$; C is Euler's constant.

The excited states of the discrete spectrum are given by the formula $(\nu = 1, 2, \ldots)$

$$W_\nu = -\frac{\hbar^2}{2\mu r_B^2}\frac{1}{\nu^2} + \frac{\hbar^2}{2\mu r_B^2}\frac{1}{\nu^3}\upsilon_\nu, \tag{19}$$

$$\frac{1}{\upsilon_\nu} = \ln\left(\frac{r_B\nu}{\sqrt{2}\,r_0}\right) - \psi(\nu) - \frac{1}{2\nu} - \frac{3}{2}C - \frac{1}{2}\Lambda\left(\frac{r_0^2P'^2}{2\hbar^2}\right) \tag{20}$$

with Λ from (18). Here $\psi(\nu)$ is the ψ-function.

In addition, the exciton has a hydrogen series of levels, corresponding to wave functions that are antisymmetrical in z.[2] All the states are not degenerate with respect to the momentum.

At small momenta P' $(r_0P'/\hbar \ll 1)$, the P'-dependent addition to W_0 is equal to

$$\delta W_0 = \frac{1}{2\mu}\frac{r_0^2}{r_B^2}\lambda_0P'^2 \equiv \frac{P'^2}{2M_0}, \qquad M_0 = \mu\frac{r_B^2\lambda_0}{r_0^2}, \tag{21}$$

where λ_0 is the solution of (17) with $P' = 0$. The corresponding addition to W_ν is

$$\delta W_\nu = \frac{1}{4\mu}\frac{r_0^2}{r_B^2}\upsilon_{\nu_0}^2\frac{P'^2}{\nu^3} \equiv \frac{P'^2}{2M_\nu}, \qquad M_\nu = 2\mu\frac{r_B^2}{r_0^2}\frac{\nu^3}{\upsilon_{\nu_0}^2}; \tag{22}$$

$v_{\nu 0}$ is given by formula (20) with $P' = 0$.

At large momenta $(r_0P'/\hbar \gg 1)$, Eq. (17) takes the form

$$\lambda = 2\ln\left(\frac{r_B\hbar}{\lambda r_0^2P'}\right) - 2C, \tag{23}$$

and formula (20) for v_ν goes over into

$$\frac{1}{\upsilon_\nu} = \ln\frac{\nu\hbar r_B}{r_0^2P'} - \psi(\nu) - \frac{1}{2\nu} - 2C. \tag{24}$$

All the foregoing formulas are valid for momenta that are not too large: $r_0^2P'/\hbar \ll r_B$.

At small values of P', the part of the energy that depends on the momentum $\mathbf{P}$ and on the electric field $\mathscr{E}$ is $(\nu = 0, 1, 2, \ldots)$

$$\delta E_\nu = \frac{\mathbf{P}^2 - \mathbf{P}'^2}{2M} + \frac{\mathbf{P}'^2}{2M_\nu} = \frac{\mathbf{P}^2}{2M_\nu} - \left(1 - \frac{M}{M_\nu}\right)\frac{c}{H^2}[\mathbf{H}\mathscr{E}]\mathbf{P} - \frac{1}{2}\left(1 - \frac{M}{M_\nu}\right)\frac{Mc^2}{H^2}\mathscr{E}^2, \tag{25}$$

where M_ν is given by (21) and (22). For the velocity and the dipole moment we have

$$\mathbf{V} = \frac{\mathbf{P}}{M_\nu} + \left(1 - \frac{M}{M_\nu}\right)\frac{c}{H^2}[\mathscr{E}\mathbf{H}],$$
$$\mathbf{d} = \left(1 - \frac{M}{M_\nu}\right)\frac{Mc^2}{H}\mathscr{E} + \left(1 - \frac{M}{M_\nu}\right)\frac{c}{H^2}[\mathbf{PH}]. \quad (26)$$

From (26) we get the polarizability at a specified exciton momentum **P**:

$$\alpha_P = \left(1 - \frac{M}{M_\nu}\right)\frac{Mc^2}{H^2}. \quad (27)$$

Its order of magnitude is $\varepsilon r_0^4/r_B$, i.e., it is very small. The polarizability at a specified velocity α_ν, which coincides with the polarizability of the exciton at rest, is

$$\alpha_\nu = \frac{(M_\nu - M)c^2}{H^2} \approx \frac{M_\nu c^2}{H^2}. \quad (28)$$

Its order of magnitude is $\varepsilon r_0^2 r_B$, as expected from physical considerations.

We present one more formula for the addition to the energy, expressed in terms of **V** and $\mathscr{E}$:

$$\delta E_\nu = \frac{1}{2}M_\nu \mathbf{V}^2 - \frac{1}{2}\alpha_\nu \mathscr{E}^2 \approx \frac{1}{2}M_\nu\left(\mathbf{V}^2 - \frac{c^2}{H^2}\mathscr{E}^2\right).$$

At large values of $\mathbf{P}'$, the dependence of the exciton energy on the momentum and on $\mathscr{E}$ has a more complicated form. We present the corresponding expression for the ground state only:

$$\delta E_0 = -\frac{2\hbar^2}{\mu r_B^2}\ln^2\left(\frac{r_B\hbar}{r_0^2 P'}\right) + \frac{\mathbf{P}^2 - \mathbf{P}''}{2M} \quad (29)$$

(the formula is given with logarithmic accuracy). When $P' \ll \hbar/r_B$, the principal term is the logarithmic one. To the contrary, when $P' \gg \hbar/r_B$ the logarithmic term becomes a small addition.

In conclusion, let us discuss the dependence of the probability of exciton production on the electric field and on the momentum of the absorbed photon. As is well known (see, e.g., [4]), this dependence is determined by a factor $|\Psi(0)|^2$, where $\Psi(\mathbf{r})$ is the wave function of the relative motion of the exciton (8). According to (8), (9), and (12) we have in the zeroth Landau band

$$|\Psi_\nu(0)|^2 = \varphi_0(\rho_0)^2\psi_\nu^2(0) = \frac{1}{2\pi r_0^2}\exp\left\{-\frac{r_0^2 P''}{2\hbar^2}\right\}\psi_\nu^2(0). \quad (30)$$

The expressions for $\psi_\nu^2(0)$ can be readily obtained by the same method.[2] For the ground state we have

$$|\Psi_0(0)|^2 = \frac{\lambda}{2\pi r_0^2 r_B}\exp\left\{-\frac{r_0^2 P''}{2\hbar^2}\right\}, \quad (31)$$

and for the excited states (ν = 1, 2, ...)

$$|\Psi_\nu(0)|^2 = \frac{1}{2\pi r_0^2 r_B}\frac{v_\nu^2}{\nu^3}\exp\left\{-\frac{r_0^2 P''}{2\hbar^2}\right\} \quad (32)$$

with λ and v_λ from (17) and (20).

$\mathbf{P}'$ is given by formula (9), in which we must substitute for **P** the transverse component of the photon wave vector $\hbar\mathbf{k}_\perp$:

$$\mathbf{P}' = \hbar\mathbf{k}_\perp + \frac{Mc}{H^2}[\mathbf{H}\mathscr{E}]. \quad (33)$$

The dependence of $|\Psi(0)|^2$ on the momentum and on $\mathscr{E}$ is determined essentially by the exponential factor. In the absence of an electric field, it takes the form (λ is the photon wavelength)

$$\exp\left\{-\frac{2\pi^2 r_0^2}{\lambda^2}\sin^2\theta\right\} \quad (34)$$

where θ is the angle between the magnetic field and the photon direction. In the optical band, the numerical value of the argument of the exponential is $(1 - 0.5)H^{-1}$ (H is in kOe). Thus, in fields on the order of several kOe and weaker, the exponential factor plays an important role.

The influence of the electric field comes into play for fields

$$\mathscr{E} \sim 2\pi\hbar H/\lambda Mc$$

or (when M is on the order of the mass of the free electron)

$$\mathscr{E}\ [\text{в/см}] \sim H\ [\text{кэ}].$$

In still stronger field, the exponential takes the form

$$\exp\left\{-\frac{M^2c^3}{2eH^3\hbar}\mathscr{E}^2\right\}, \quad (35)$$

and the intensity starts to decrease exponentially. The numerical value of the argument of the exponential is $\sim 0.5\mathscr{E}^2/H^2$ ($\mathscr{E}$ in V/cm, H in kOe, M is of the order of the electron mass).

We are grateful to E. I. Rashba for valuable discussions.

[1] R. J. Elliott and R. Loudon, J. Phys. Chem. Solids 15, 196 (1960).

[2] H. Hasegawa and R. E. Howard, J. Phys. Chem. Solids 21, 179 (1961).

[3] W. Lamb, Phys. Rev. 85, 259 (1952).

[4] J. Elliott and R. Loudon, Phys. Chem. Solids 8, 382 (1959).

Translated by J. G. Adashko
81

Chapter 5

Theory of Metals and Superconductors

5A. Metals

5A.1 Quantum oscillation in fermi liquid

By the end of the 50's, the consensus was that the basic principles of the Landau Fermi liquid (FL) theory of helium-three (^3He) equally apply to the thermodynamics of electrons in metals, yet, it was less clear for the features that depend on charge of the carriers. As is known, the electronic spectrum in metals is usually studied in the experiments of the de Haas-van Alphen (dHvA)-type that measure the so-called quantum oscillations (QO) of different quantities in the magnetic field. The quasiclassical theory by Lifshitz and Kosevich (LK) commonly accepted at the time served as the main theoretical instrument at reconstructing the shapes of Fermi surfaces on basis of the experimental data on QO. As the LK theory was formulated for non-interacting quasiparticles, the outstanding issue was whether the Fermi liquid interactions may significantly change its predictions.

Role of the FL interactions in the dHvA effect was investigated by J. Luttinger (Ref. 1) and, independently in our paper [5Aa]. Our approach was based on several identical diagrammatic relations first derived by L. Pitaevskii and used by Landau and him for the microscopic derivation of the FL theory. With these identities we could analyze the self energy diagrams to all orders in interactions. In most instances our conclusions coincided with those by Luttinger. In particular, it was confirmed the one-to-one LK correspondence between frequencies of the QO oscillations and areas of the Fermi surface cross-section in metals.

As Luttinger was analyzing the self energy part only to the first order in the interaction potential, he left out the change of the oscillations' phase in the LK expression arising from the FL renormalization of the paramagnetic susceptibility: $\cos(\pi l \frac{m^*}{m_o}) \Rightarrow \cos(\pi l \frac{m^*}{m_o} \zeta)$ (l is the harmonic's number).

5A.2 Zero-sound in metals

The ordinary acoustic sound in a neutral FL attenuates at lowering the temperature. At zero temperature, its place is taken over by the so-called *zero sound*, an entirely quantum effect owed to the FL interactions between quasiparticles. Oscillations in the zero-sound wave involve oscillations in the quasipartices' occupation numbers at the Fermi surface. In helium-three the zero sound waves are coupled to the density oscillations. The latter are prohibited for the charged carriers in metals.

It was speculated in [5Ab] that in a crystal the zero sound-type oscillations could exist for the wave vectors $\vec{k}$ along the high symmetry directions, say, along the four- or three-fold crystalline axis. To be the case, the zero sound oscillations should be expressed in terms of a non-identical representation of the small group of the vector $\vec{k}$, not to cause modulations in the electronic density. I think that there were no attempts of verifying the suggestion experimentally, but results [5Ab] have their merits from a theoretical point of view.

5A.3 Ferromagnetism of paramagnetic alloys

Paper [5Ac] addresses the possibility of the ferromagnetic transition in a dilute system of paramagnetic centers embedded into the metallic matrix. Interactions between the local spins in a metal are usually described in terms of the well-known RKKY interaction. The latter rapidly oscillates and decreases as $1/R^3$ on large distances $Rk_F \gg 1$ (k_F is the Fermi vector). Therefore any ferromagnetic ordering between the local spins at the average distances $R \approx n^{-1/3}$ is deemed impossible. Meanwhile, such transition was observed in few experiments (see references in [5Ac]).

We pointed out that the RKKY interaction has nothing to do with the transition, for mechanism of the ferromagnetic ordering bears a collective character. That is, suppose that metallic electrons were magnetized and their average spin is $\vec{\sigma}_{av} \neq 0$. It would cost energy because of the misbalance between the Fermi momentums for spins up and down. But spins of the metallic electrons interact with spins on the impurity centers and, being ordered, build on the latter an average mean spin $\vec{S}_{av}$ via the $J(\vec{\sigma} \cdot \vec{S})$–exchange. Below some temperatures, spins of the two sub-systems can be mutually oriented so as to give gains in the total energy at *any* sign of J.

Reference

1. J. M. Luttinger, *Phys. Rev.* **121**, 1251 (1961)

List of original papers

[5Aa] Yu. A. Bychkov and L. P. Gor'kov, *Quantum oscillations of the thermodynamic values of metals in magnetic fields in the Fermi liquid model*, Sov. Phys. JETP, **14**, 1132 (1962)

[5Ab] L. P. Gor'kov and I. E. Dzyaloshinski, *Feasibility of zero-sound oscillations in metals*, Sov. Phys. JETP, **17**, 1111 (1963)

[5Ac] A. A. Abrikosov and L. P. Gor'kov, *On the nature of impurity ferromagnetism*, Sov. Phys. JETP, **16**, 1575 (1963)

SOVIET PHYSICS JETP VOLUME 14, NUMBER 5 MAY, 1962

QUANTUM OSCILLATIONS OF THE THERMODYNAMIC QUANTITIES OF A METAL IN A MAGNETIC FIELD ACCORDING TO THE FERMI-LIQUID MODEL

Yu. A. BYCHKOV and L. P. GOR'KOV

Institute of Physics Problems, Academy of Sciences, U.S.S.R.

Submitted to JETP editor May 31, 1961

J. Exptl. Theoret. Phys. (U.S.S.R.) **41**, 1592-1605 (November, 1961)

Quantum field theory methods are used to study the influence of Fermi-liquid effects on the oscillations of the thermodynamic quantities of a metal in a magnetic field. It is shown that the periods of the oscillations can be calculated by the usual scheme of I. Lifshitz and Kosevich. Deviations from the usual results appear in the amplitudes of the oscillations and are due to the change of the effective magneton of an excitation owing to the interaction between the electrons.

1. INTRODUCTION

THE theory of the oscillations of the magnetic susceptibility in metals, which was first constructed in a paper by I. Lifshitz and Kosevich,[1] is based on definite ideas about the structure of the energy levels of a metal in a magnetic field. It is assumed that these levels can be constructed on the basis of a quantization of the levels of individual excitations treated as an ideal Fermi gas. However, the electrons in a metal are a sort of Fermi liquid. The theory of the Fermi liquid, constructed by Landau[2,3] in its application to liquid He^3, shows that in such a system effects of the interaction between the excitations play a large role. The correlations that thus arise are of the order of interatomic distances. The question arises as to how the presence of these interactions between the excitations affects the various quantum phenomena in the theory of metals in magnetic fields.

We shall here treat the de Haas-van Alphen effect for the electrons in a metal on the model of the isotropic Fermi liquid. Generally speaking, the situation in the case of electrons is complicated by specific features of the Coulomb interaction. We shall assume that the long-range part of the Coulomb interaction is already screened off. The final formulas obtained below contain only the characteristics of the free-electron spectra (without a field), and therefore, in our opinion, cannot depend on this assumption. The same arguments also apply to the question of the effects of anisotropy.

The results obtained in this paper show that when the spin susceptibility is not taken into account the expression for the oscillating part of the magnetic moment can be obtained on the basis of the usual concept of a system of electrons as a gas of quasiparticles. This same result has been obtained in a paper by Luttinger[4] which has recently appeared. In that paper, however, the analysis of the Green's functions of the electrons in the magnetic field was made only in perturbation theory to terms of first order in the interaction, i.e., terms that give only a trivial renormalization of the chemical potential. Also Luttinger did not investigate the question of the effect of the paramagnetic susceptibility. The whole difference from previously known formulas arises in including the paramagnetic susceptibility, which, as is well known,[2] depends strongly on the Fermi-liquid properties of the system.

Our further study will be made by the methods of quantum theory. We shall be interested only in the quantum oscillations of all quantities. The value of the susceptibility in a weak magnetic field cannot be expressed in terms of the characteristics of the spectrum, since its diamagnetic part depends also on electrons that are located "deep" below the Fermi surface. For simplicity we shall confine ourselves to the temperature absolute zero.

2. THE ENERGY SPECTRUM

We shall begin with a study of the properties of the Green's functions of electrons in a magnetic field. As usual,[3] the Green's function $G(x, x')$ is defined as an average over the ground state of the system:

$$G(\mathbf{r}, \mathbf{r}'; t-t')\,\delta_{\alpha\beta} = -i\langle T(\psi_\alpha(\mathbf{r}, t)\,\psi_\beta^+(\mathbf{r}', t'))\rangle. \quad (1)$$

The particle field operators $\psi_\alpha(x)$, $\psi_\beta^+(x')$ include the dependence on the magnetic field H; we choose the vector potential $\mathbf{A}(\mathbf{r})$ of the field in the form

$$\mathbf{A}(\mathbf{r}) = \{-Hy, 0, 0\}. \quad (2)$$

The dependence of the Green's function (1) on the coordinates can be represented in the following way:

$$G(x, x') = \exp\{-i(eH/2c)(y+y')(x-x')\} \times G(\mathbf{r}-\mathbf{r}'; t-t'). \quad (3)$$

This follows from gauge invariance, since when we make a displacement of the origin of coordinates, $y \to y + b$, the operators $\psi(x)$ and $\psi^+(x)$ transform according to the law

$$\psi \to \psi e^{-ieHbx/c}, \quad \psi^+ \to \psi^+ e^{ieHbx/c}. \quad (4)$$

We shall be dealing with $G(\mathbf{r}, \mathbf{r}'; \epsilon)$, the Fourier component of (1) with respect to the time difference $t - t'$. In the absence of a magnetic field

$$G^0(\mathbf{r}-\mathbf{r}'; \varepsilon) = \frac{1}{(2\pi)^3}\int G^0(\mathbf{p}, \varepsilon)\, e^{i\mathbf{p}(\mathbf{r}-\mathbf{r}')}\, d\mathbf{p}.$$

For small ϵ the function $G^0(\mathbf{p}, \epsilon)$ has a pole near the Fermi surface of the form[3]

$$G^0(\mathbf{p}, \varepsilon) = a/(\varepsilon - v(p-p_0) + i\delta(\varepsilon)). \quad (5)$$

The value of $\epsilon = v(p - p_0)$ determines the spectrum of the Fermi liquid.

In a magnetic field the form of the Green's function near the Fermi surface is decidedly altered owing to the quantization of the levels. We shall show, however, that the energy spectrum of the electrons in the magnetic field can be obtained from the expression (5) by the usual rules of quasi-classical quantization, as was also suggested in the paper by I. Lifshitz and Kosevich.[1]

To prove this assertion we write the Dyson equation satisfied by the Green's function $G(\mathbf{r}, \mathbf{r}'; \epsilon)$ in coordinate space:

$$\left[\varepsilon + \mu - \frac{1}{2m}\left(\hat{\mathbf{p}} - \frac{e}{c}\mathbf{A}\right)^2\right] G(\mathbf{r}, \mathbf{r}'; \varepsilon) - \int \Sigma(\mathbf{r}, \mathbf{r}''; \varepsilon)\, G(\mathbf{r}'', \mathbf{r}'; \varepsilon)\, d^3r'' = \delta(\mathbf{r}-\mathbf{r}'). \quad (6)$$

Here $\hat{\mathbf{p}} = -i\partial/\partial\mathbf{r}$, μ is the chemical potential of the electrons in the magnetic field, and $\Sigma(\mathbf{r}, \mathbf{r}''; \epsilon)$ is the so-called self-energy part arising from the interactions between the particles in the Fermi liquid. We shall not specify concrete forms for these interactions. The spectrum of the system is determined by the eigenvalues of the operator which appears in square brackets in Eq. (6).

In the notation of second quantization the Hamiltonian for the interaction of the electrons with the magnetic field (2) takes the form

$$H_{int} = \int \psi^+(\mathbf{r}')\left[\frac{e}{2mc}(\hat{\mathbf{p}} - \hat{\mathbf{p}}')_x + \frac{e^2Hy}{2mc^2}\right] Hy\psi(\mathbf{r})\, d^3\mathbf{r}, \quad (7)$$

where $\hat{p}_x$ and $\hat{p}_{x'}$ denote differentiation ($\hat{p}_x = -i\partial/\partial x$) with respect to the corresponding arguments in the limit $\mathbf{r} \to \mathbf{r}'$. We shall now investigate the dependence of the self-energy part $\Sigma(\mathbf{r}, \mathbf{r}'; \epsilon)$ on the magnitude of the magnetic field. According to Eq. (7) an increment of the magnetic field, $H \to H + \delta H$, is equivalent to an additional interaction Hamiltonian:

$$\delta H_{int} = \delta H \int \psi^+(\mathbf{r}')\left[\frac{e}{2mc}(\hat{\mathbf{p}} - \hat{\mathbf{p}}')_x + \frac{e^2Hy}{mc^2}\right] y\psi(\mathbf{r})\, d^3\mathbf{r}. \quad (7')$$

Using the usual diagram technique to calculate from Eq. (7') the change $\delta\Sigma(\mathbf{r}, \mathbf{r}'; \epsilon) = (\partial\Sigma/\partial H)\delta H$, we can assign to this quantity the diagram shown in Fig. 1. The cross denotes the operator in square brackets in the integrand in Eq. (7'); the circle denotes the complete vertex part in the magnetic field, $\Gamma_{\alpha\beta,\gamma\delta}(\xi_1, \xi_2; \xi_3, \xi_4)$.

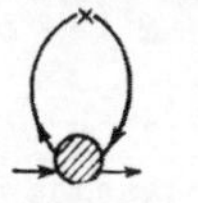

FIG. 1

The vertex part is defined as usual by the relation

$$\langle T(\psi_\alpha(\xi_1)\psi_\beta(\xi_2)\psi_\gamma^+(\xi_3)\psi_\delta^+(\xi_4))\rangle = G_{\alpha\gamma}(\xi_1, \xi_3)\, G_{\beta\delta}(\xi_2, \xi_4) - G_{\alpha\delta}(\xi_1, \xi_4)\, G_{\beta\gamma}(\xi_2, \xi_3) + i\int G_{\alpha\alpha'}(\xi_1, \xi_1')\, G_{\beta\beta'}(\xi_2, \xi_2') \times \Gamma_{\alpha'\beta',\gamma'\delta'}(\xi_1', \xi_2'; \xi_3', \xi_4')\, G_{\gamma'\gamma}(\xi_3', \xi_3)\, G_{\delta'\delta}(\xi_4', \xi_4) \times d^4\xi_1' d^4\xi_2' d^4\xi_3' d^4\xi_4'.$$

In the absence of the magnetic field $\Gamma_{\alpha\beta,\gamma\delta}(\xi_1, \xi_2; \xi_3, \xi_4)$ has Fourier components of the form $(2\pi)^4\Gamma_{\alpha\beta,\gamma\delta}(p_1, p_2; p_3, p_4) \times \delta(p_1 + p_2 - p_3 - p_4)$. As is well known,[3] the vertex part plays an important role in the theory of the Fermi liquid.

The diagram of Fig. 1 gives the following result for the derivative $\partial\Sigma/\partial H$:

$$\delta_{\alpha\beta}\frac{\partial\Sigma(\mathbf{r}, \mathbf{r}'; \varepsilon)}{\partial H} = \frac{i}{2\pi}\int d\omega' \int d^3r_1 d^3r_2 d^3l \left[\frac{e}{2mc}(\hat{\mathbf{p}} - \hat{\mathbf{p}}')_{l_x} + \frac{e^2H}{mc^2} l_y\right] l_y\, \Gamma_{\alpha\gamma,\gamma\beta}(\mathbf{r}, \varepsilon; \mathbf{r}_1, \omega'; \mathbf{r}_2, \omega'; \mathbf{r}', \varepsilon)\, G(\mathbf{l}, \mathbf{r}_1; \omega') \times G(\mathbf{r}_2, \mathbf{l}'; \omega'); \quad \hat{p}_x = -i\frac{\partial}{\partial l_x}. \quad (8)$$

It is convenient to make a transformation of this expression. To do so we note that under the infinitesimal transformation (4) $\Sigma(\mathbf{r}, \mathbf{r}'; \epsilon)$ changes by the quantity

$$-i\,(eH/c)\,\delta b\,(x-x')\,\Sigma(\mathbf{r}, \mathbf{r}'; \varepsilon).$$

At the same time, under the displacement of the origin, $y \to y + \delta b$, the Hamiltonian (7) receives the increment

$$\delta H_{int} = \delta b \int \psi^{+}(\mathbf{r}') \left[\frac{e}{2mc} (\hat{\mathbf{p}} - \hat{\mathbf{p}}')_x + \frac{e^2 H y}{mc^2} \right]_{\mathbf{r}' \to \mathbf{r}} \psi(\mathbf{r})\, d^3\mathbf{r}.$$

Then, in analogy with Eq. (8), we get

$$-\frac{ie(x-x')}{c} \delta_{\alpha\beta} \Sigma(\mathbf{r}, \mathbf{r}'; \varepsilon) = \frac{i}{2\pi} \int d\omega' \int d^3\mathbf{r}_1 d^3\mathbf{r}_2 d^3\mathbf{l} \left[\frac{e}{2mc} (\hat{\mathbf{p}} - \hat{\mathbf{p}}')_{l_x} + \frac{e^2 H}{mc^2} l_y \right] \times \Gamma_{\alpha\gamma,\gamma\beta}(\mathbf{r}, \varepsilon; \mathbf{r}_1, \omega'; \mathbf{r}_2, \omega'; \mathbf{r}', \varepsilon) \times G(\mathbf{l}, \mathbf{r}_1; \omega')\, G(\mathbf{r}_2, \mathbf{l}'; \omega').$$

Combining this last relation with Eq. (8), we find

$$\frac{\partial \Sigma_{\alpha\beta}(\mathbf{r}, \mathbf{r}'; \varepsilon)}{\partial H} = -i \frac{e(x-x')(y+y')}{2c} \Sigma_{\alpha\beta}(\mathbf{r}, \mathbf{r}'; \varepsilon) + M_{\alpha\beta}(\mathbf{r}, \mathbf{r}', \varepsilon), \tag{9}$$

$$M_{\alpha\beta}(\mathbf{r}, \mathbf{r}'; \varepsilon) = \frac{i}{2\pi} \int d\omega' \int d^3\mathbf{r}_1 d^3\mathbf{r}_2 d^3\mathbf{l} \left[\frac{e}{2mc} (\hat{\mathbf{p}} - \hat{\mathbf{p}}')_{l_x} + \frac{e^2 H l_y}{mc^2} \right] [l_y - (y+y')/2]\, G(\mathbf{l}, \mathbf{r}_1; \omega')\, G(\mathbf{r}_2, \mathbf{l}'; \omega') \times \Gamma_{\alpha\gamma,\gamma\beta}(\mathbf{r}, \varepsilon; \mathbf{r}_1, \omega'; \mathbf{r}_2, \omega'; \mathbf{r}', \varepsilon). \tag{10}$$

We shall show below that the terms in $\Sigma_{\alpha\beta}(\mathbf{r}, \mathbf{r}'; \epsilon)$ that come from $M_{\alpha\beta}(\mathbf{r}, \mathbf{r}'; \epsilon)$ are of the order $H^{3/2}$ in the magnetic field, whereas for our purposes it is enough to know the spectrum of the system correct to terms of the order H. Neglecting the last term in Eq. (9), we get

$$\Sigma_{\alpha\beta}(\mathbf{r}, \mathbf{r}'; \varepsilon) = \exp\left\{-\tfrac{1}{2} ieHc^{-1}(x-x')(y+y')\right\} \delta_{\alpha\beta} \Sigma^0(\mathbf{r} - \mathbf{r}'; \varepsilon), \tag{11}$$

where $\Sigma^0(\mathbf{r} - \mathbf{r}'; \epsilon)$ is the self-energy part in the absence of the magnetic field.

The expression (11) for $\Sigma(\mathbf{r}, \mathbf{r}')$ can be written symbolically in the form

$$\Sigma(\mathbf{r}, \mathbf{r}') = \hat{\Sigma}^0(\hat{\mathbf{p}} - e\mathbf{A}/c).$$

In fact, let us apply the "Hamiltonian"

$$\hat{h} = \frac{1}{2m}\left(\hat{\mathbf{p}} - \frac{e}{c}\mathbf{A}\right)^2 + \hat{\Sigma}^0\left(\hat{\mathbf{p}} - \frac{e}{c}\mathbf{A}\right)$$

to an arbitrary function $\psi(\mathbf{r})$ and go over to the p representation. As Zil'berman[5] has shown, in the momentum representation the operator $\hat{h}$ can be written in the following form:

$$\hat{h}\psi_p = \left\{ \frac{1}{2m}\left(p_x + i\frac{eH}{c}\frac{d}{dp_y}\right)^2 + \frac{1}{2m}(p_y^2 + p_z^2) + \int \Sigma^0(\mathbf{s}; \varepsilon) \exp\left[-i\left(p_x + \frac{ieH}{c}\frac{d}{dp_y}\right)s_x - ip_y s_y - ip_z s_z\right] d^3s \right\} \psi_p. \tag{12}$$

The expression (12) contains definite prescriptions as to the order of the operators p_y and d/dp_y (the requirement of complete symmetry of the Hamiltonian[5]). In the isotropic model of the Fermi liquid, which we are using in this paper, the self-energy part $\Sigma^0(\mathbf{p}, \epsilon)$ is a function of $|\mathbf{p}|^2$ only. At the same time it is obvious that under the condition that $\Sigma^0(\hat{\mathbf{p}} - e\mathbf{A}/c)$ is to remain Hermitian different requirements as to the order of the operators p_y and d/dp_y in $\Sigma^0(\hat{\mathbf{p}} - e\mathbf{A}/c)$ lead to Hamiltonians (12) which differ by terms of order in the energy eigenvalues not lower than H^2. For this reason it is more convenient to choose $\Sigma^0(\hat{\mathbf{p}} - e\mathbf{A}/c)$ in the form $\Sigma^0(\hat{\mathbf{p}} - e\mathbf{A}/c) \equiv \Sigma^0(|\hat{\mathbf{p}} - e\mathbf{A}/c|^2)$.

The exact eigenvalues and eigenfunctions of the operator

$$\left(\hat{\mathbf{p}} - \frac{e\mathbf{A}}{c}\right)^2 = \left(\hat{p}_x + \frac{ieH}{c}\frac{d}{dp_y}\right)^2 + \hat{p}_y^2 + \hat{p}_z^2$$

are well known:[6]

$$\left(\hat{\mathbf{p}} - \frac{e\mathbf{A}}{c}\right)^2 \psi_{p,n} = [p_z^2 + (2n+1)\,eH/c]\,\psi_{p,n},$$

$$\psi_{p,n} = \exp\{-icp_x p_y/eH - cp_y^2/2eH\} H_n(p_y\sqrt{c/eH}),$$

where $H_n(x)$ are the Hermite polynomials. Therefore the Green's function of a particle in a magnetic field can be written near the Fermi surface (in the p representation) in the following way:

$$G(\mathbf{p}, \mathbf{p}'; \varepsilon) = \sum_n \frac{\psi_n(\mathbf{p})\, \psi_n^*(\mathbf{p}')\, \delta(p_x - p_x')\, \delta(p_z - p_z')}{\varepsilon + \mu - (n + 1/2)\, eH/mc - p_z^2/2m - \Sigma^0(p_z^2 + 2(n + 1/2)\, eH/c) + i\delta(\varepsilon)}. \tag{13}$$

According to Eq. (5), when the magnetic field H = 0,

$$G^0(\mathbf{p}, \mathbf{p}'; \varepsilon) = \frac{\delta(\mathbf{p} - \mathbf{p}')}{\varepsilon + \mu - p^2/2m - \Sigma^0(p^2, \varepsilon) + i\delta(\varepsilon)} = \frac{a\,\delta(\mathbf{p} - \mathbf{p}')}{\varepsilon - v(p - p_0) + i\delta(\varepsilon)}. \tag{14}$$

Here a is a renormalization constant, and p_0 is the Fermi limiting momentum, which is determined from the equation

$$p_0^2/2m + \Sigma^0(p_0, 0) = \mu.$$

As is well known,[1] the electrons important for the de Haas-van Alphen effect are those near the Fermi surface i.e., the electrons in the region in which $(2n+1)\,eH/c + p_z^2 \approx p_0^2$. Therefore if we introduce the notation m^* for the "effective" mass ($v = p_0/m^*$), there follows directly from Eqs. (13) and (14) the following expression for the Green's function of electrons in a magnetic field near the Fermi surface:

$$G(\mathbf{p}, \mathbf{p}'; \varepsilon) = \sum_n{}' \psi_n(\mathbf{p}) \psi_n^*(\mathbf{p}') \delta(p_x - p_x') \ \delta(p_z - p_z') G_n(p_z, \varepsilon),$$

$$G_n(p_z, \varepsilon) = a / (\varepsilon + p_0^2 / 2m^* - (n + 1/2)\omega^* - p_z^2 / 2m^* + i\delta(\varepsilon)), \quad (15)$$

where $\omega^* = eH/m^*c$. As is shown below, the constants a, m^*, and p_0 depend on the field H only through terms of order $H^{3/2}$.

The Greens' function far from the Fermi surface $[|(2n+1)eH/c + p_z^2 - p_0^2| \sim p_0^2]$ contains no sharp poles; this is due to the strong damping of the excitations in this region (for small ϵ the damping $\delta(\epsilon)$ of the excitations in Eq. (15) increases like $|\epsilon|\epsilon/\mu$). Therefore the contributions to the various quantities from these distant regions give no oscillating singularities and can be expanded in powers of the field strength.

Let us discuss the behavior of the Green's function $G(\mathbf{r}, \mathbf{r}'; \epsilon)$ in the coordinate representation. In the absence of a magnetic field

$$G^0(\mathbf{r} - \mathbf{r}'; \varepsilon) = \frac{1}{(2\pi)^3} \int G^0(\mathbf{p}) e^{i\mathbf{p}(\mathbf{r}-\mathbf{r}')} d^3\mathbf{p}.$$

The Green's function $G(\mathbf{R}, \epsilon)$ expresses the correlation of the electrons at different points of space. At distances large in comparison with atomic distances the only electrons that contribute are those moving like "free" electrons, i.e., those from the region near the Fermi surface. Electronic excitations far from the Fermi surface are strongly damped because of collisions with each other and contribute to the Green's function $G(\mathbf{r} - \mathbf{r}'; \epsilon)$ only at atomic distances. If in the function $G(\mathbf{r} - \mathbf{r}'; \epsilon)$ we separate out the term (5) that has a singularity near the Fermi surface,

$$G^0(\mathbf{p}, \varepsilon) = \frac{a}{\varepsilon - v(p - p_0) + i\delta(\varepsilon)} + g(\mathbf{p}, \varepsilon)$$

[so that $g(\mathbf{p}, \epsilon)$ has no singularities], it is easy to see that for small ϵ the behavior of the function $G^0(\mathbf{R}, \epsilon)$ at distances $Rp_0 \gg 1$ is given by

$$G^0(\mathbf{R}, \varepsilon) = \frac{am^*}{2\pi R} \begin{cases} \exp(ip_0 R + i\varepsilon R/v), & \varepsilon > 0 \\ \exp(-ip_0 R - i\varepsilon R/v), & \varepsilon < 0. \end{cases} \quad (16)$$

In the case of large ϵ the function $G^0(\mathbf{R}, \epsilon)$ is rapidly damped at distances $Rp_0 \sim \epsilon^2/\mu^2$.

Returning to the Green's function in a magnetic field, we conclude that its behavior at large distances is also determined by the singular part of the expression (15). It is easy to verify that in the coordinate representation $G(\mathbf{r}, \mathbf{r}'; \epsilon)$ (for $|\mathbf{r} - \mathbf{r}'| \gg 1/p_0$) can be written in the following form:

$$G(\mathbf{r}, \mathbf{r}'; \varepsilon) = \exp[-i(eH/2c)(x - x')(y + y')]$$
$$\times \frac{eH}{c(2\pi)^2} \sum_n e^{-eH\rho^2/4c} L_n\left(\frac{eH}{2c}\rho^2\right)$$
$$\times \int \frac{a e^{ip_z(z-z')} dp_z}{\varepsilon + p_0^2/2m^* - (n + 1/2)\omega^* - p_z^2/2m^* + i\delta(\varepsilon)}$$
$$= \exp\{-i(eH/2c)(x - x')(y + y')\} \overline{G}(\mathbf{R}, \varepsilon), \quad (17)$$

$L_n(x)$ are the Laguerre polynomials. [Here and in what follows we use the notation $\rho^2 = (x - x')^2 + (y - y')^2$]. As can be seen from Eq. (17), generally speaking the magnetic field causes a decided change in the character of the dependence of $G(\mathbf{r}, \mathbf{r}'; \epsilon)$ on $\mathbf{R}$. In the limit of weak magnetic field, $\omega^* \ll p_0^2/2m^*$, for large n and $\rho \ll cp_0/eH$,

$$e^{-eH\rho^2/4c} L_n(eH\rho^2/2c) \approx J_0(\sqrt{2eHn/c}\,\rho).$$

By replacing the summation over n by an integration, we would again arrive at the result (16).

A special role is played in Eq. (17) by terms with values of $(n + \frac{1}{2})\omega^*$ close to $p_0^2/2m^*$ (small p_z). Let $\Delta < \omega^*$ and $p_0^2/2m^* = \Delta + \omega^*(N_0 + \frac{1}{2})$; then each such term in Eq. (17) contributes to $G(\mathbf{R}, \epsilon)$ a small quantity of the order

$$\sim \frac{eH}{c} e^{-eH\rho^2/4c} L_{N_0}\left(\frac{eH}{2c}\rho^2\right) e^{i\Delta|z-z'|} \Delta^{-1/2}.$$

This expression is exponentially damped for $\rho \gg cp_0/eH$ and is a plane wave in its dependence on $|z - z'|$. This part of the G-function determines the oscillations of all the quantities in which we are interested.

Coming now to an estimate of the term $M(\mathbf{r}, \mathbf{r}'; \epsilon)$ dropped from Eq. (9), let us rewrite this term in a somewhat different form, by substituting Green's functions in the formula (3):

$$M_{\alpha\beta}(\mathbf{r}, \mathbf{r}'; \varepsilon) = \frac{i}{2\pi} \int d\omega \int d^3\mathbf{r}_1 d^3\mathbf{r}_2 d^3\mathbf{l} \left(l_y - \frac{y + y'}{2}\right)$$
$$\times \exp\{-i(eH/2c)[(r_{1y} + l_y)(r_{1x} - l_x)$$
$$+ (r_{2y} + l_y)(l_x - r_{2x})]\} \Gamma_{\alpha\gamma,\gamma\beta}(\mathbf{r}, \varepsilon; \mathbf{r}_1, \omega; \mathbf{r}_2, \omega; \mathbf{r}', \varepsilon)$$
$$\times \Big\{ (e^2H/2mc^2) G(\mathbf{l} - \mathbf{r}_1, \omega) G(\mathbf{r}_2 - \mathbf{l}, \omega)$$
$$\times \left(l_y - \frac{r_{1y} + r_{2y}}{2}\right) + \frac{ie}{2mc} \left[\frac{\partial G(\mathbf{l} - \mathbf{r}_1, \omega)}{\partial r_{1x}} G(\mathbf{r}_2 - \mathbf{l}, \omega)\right.$$
$$\left.\left. - G(\mathbf{l} - \mathbf{r}_1, \omega) \frac{\partial G(\mathbf{r}_2 - \mathbf{l}, \omega)}{\partial r_{2x}}\right]\right\}. \quad (18)$$

This expression involves integrals containing G-functions of the differences of coordinates between the various points of the diagram of Fig. 1. The region of integration in which these differences are of the order of interatomic distances are of no interest, since their contribution to Eq. (18) is of the order H (and of order H^2 in Σ).

Let the distances $|\mathbf{l} - \mathbf{r}_i|$ be large in comparison with atomic distances. In this region we can use for the Green's functions the asymptotic expression (17). In estimating the quantity (18) we shall assume that the magnetic field is weak, and set it equal to zero wherever possible. In particular, for the vertex part we use the expression for zero field. Let us rewrite this expression in the following way:

$$\Gamma_{\alpha\gamma,\gamma\beta}(\mathbf{r}, \varepsilon; \mathbf{r}_1\omega; \mathbf{r}_2, \omega; \mathbf{r}', \varepsilon) = \frac{1}{(2\pi)^9}\int \Gamma^0_{\alpha\gamma,\gamma\beta}(\mathbf{p}_1, \varepsilon; \mathbf{p}_2, \omega; \mathbf{p}_2 + \mathbf{k}, \omega; \mathbf{p}_1 - \mathbf{k}, \varepsilon) \exp[i\mathbf{p}_1(\mathbf{r}-\mathbf{r}') + i\mathbf{p}_2(\mathbf{r}_1 - \mathbf{r}_2) + i\mathbf{k}(\mathbf{r}' - \mathbf{r}_2)]\, d^3\mathbf{p}_1 d^3\mathbf{p}_2 d^3\mathbf{k}. \quad (19)$$

The Fourier component $\Gamma^0_{\alpha\gamma,\gamma\beta}(\mathbf{p}_1, \mathbf{p}_2; \mathbf{p}_3, \mathbf{p}_4)$ in Eq. (19) involves momenta p of the order of the Fermi momentum and has important variations in this region only for changes $\Delta p \sim p_0$. Besides this, in the region of integration over l in Eq. (18) where $|\mathbf{l}-\mathbf{r}|$ is large the important values of k in Eq. (19) are small, and therefore we can neglect the dependence of the vertex part on k.

Thus as a function of $|\mathbf{r}_2 - \mathbf{r}_1|$ and $|\mathbf{r} - \mathbf{r}'|$ the expression (19) is rapidly oscillating at atomic distances. [Close to the Fermi surface $|\mathbf{p}_1|$, $|\mathbf{p}_2| \sim p_0$ the Fourier components $\Gamma_{\alpha\gamma,\gamma\beta}(\mathbf{p}_1, \mathbf{p}_2; \mathbf{p}_2, \mathbf{p}_1)$ in Eq. (19) depend only on the angle between the vectors $\mathbf{p}_1$ and $\mathbf{p}_2$]. Consequently, for purposes of the integration over $\mathbf{r}_1$ and $\mathbf{r}_2$ in Eq. (18) for $|\mathbf{l}-\mathbf{r}_1| \gg 1/p_0$, for which values only the Fourier components with momenta close to the Fermi surface are important in the functions $G(\mathbf{l}-\mathbf{r}_1, \omega)$, the vertex part (19) is essentially a δ function of $\mathbf{r}_1 - \mathbf{r}_2$ and $\mathbf{r} - \mathbf{r}'$. This fact is the expression of a fundamental physical assumption of the theory of the Fermi liquid,[2] according to which the interaction between the particles is a short-range one, and all of the correlations that arise between them fall off rapidly at atomic distances. (For the electrons in a metal the Coulomb interaction is also screened off at distances of the order of atomic distances.) Nevertheless this assertion is not completely rigorous.

As has been shown by Landau,[3] in a number of cases the vertex part can have a "long-range" part. Such singularities are due to the diagrams in the vertex part which are shown in Fig. 2, a. In this diagram the squares denote irreducible vertex parts which have no singularities in the direction 1 — 2. In fact, substituting in these diagrams the expressions (16) for the Green's functions and integrating over the frequencies of the internal lines, we get ($R_{12} \gg 1/p_0$)

$$\int G^0(\mathbf{R}_{12}, \omega)\, G^0(\mathbf{R}_{12}, \omega - \varepsilon)\, d\omega \sim \frac{a^2 m^{*2}\varepsilon}{(2\pi R_{12})^2} e^{i\varepsilon R_{12}/v},$$

where ϵ is the small frequency transfer. Figure 2, b shows this same diagram in the momentum representation, and from it we can see that the slowly decreasing dependence in the vertex part is due to small frequency and momentum transfers in the Fourier component $\Gamma^0(\mathbf{p}_1, \mathbf{p}_2; \mathbf{p}_3, \mathbf{p}_4)$. In Eq. (19) such a small transfer can correspond

a b

FIG. 2

either to a small value of k or to a small transfer $|\mathbf{p}_3 - \mathbf{p}_1| \approx |\mathbf{p}_2 - \mathbf{p}_1| \ll p_0$. Obviously in the latter case the contribution from these singularities in Eq. (19) is small, because of the smallness of the region of integration over $\mathbf{p}_1$ and $\mathbf{p}_2$. As for the small values of k, the question of these singularities arises owing to the fact that in Eq. (18) there is a factor $l_y - r_y = l_y - r_{1y} + r_{1y} - r_y$, and the quantity $|r_{1y} - r_y|$ may not be small. It is easy to see, however, that in zeroth order in the field we have by considerations of symmetry

$$\int (s_y - r_y)\, G^0(\mathbf{s} - \mathbf{r})\, G^0(\mathbf{s} - \mathbf{r})\, d^3\mathbf{s} \equiv 0$$

and therefore in Eq. (18) the factor $l_y - (y + y')/2$ can be replaced by $l_y - r_{1y}$. We shall not present the detailed proof that all of these assumptions correspond to dropping in $\mathbf{M}(\mathbf{r}, \mathbf{r}')$ terms that are of higher order in the field strength H as compared with the estimate that we get on the assumption that the vertex part is equal to the expression for this part in the absence of a magnetic field and is of short-range character.

Accordingly, in Eq. (18) we now take $|\mathbf{r}_2 - \mathbf{r}_1| \sim |\mathbf{r}_1 - \mathbf{r}| \sim 1/p_0$. We can drop the exponential factors in Eq. (18). By considerations of symmetry the term in the square brackets that contains derivatives of the Green's functions is identically zero. Therefore instead of Eq. (18) we get

$$M_{\alpha\beta}(\mathbf{r}, \mathbf{r}') \approx \frac{ie^2H}{4\pi mc^2}\int d\omega \int d^3\mathbf{r}_1 d^3\mathbf{r}_2 d^3\mathbf{l} \left(l_y - \frac{r_{1y} + r_{2y}}{2}\right)^2 \times G(\mathbf{l} - \mathbf{r}_1, \omega)\, G(\mathbf{r}_2 - \mathbf{l}, \omega)\, \Gamma^0_{\alpha\gamma,\gamma\delta} \times (\mathbf{r}, \;; \mathbf{r}_1, \omega; \mathbf{r}_2, \omega; \mathbf{r}', \varepsilon). \quad (20)$$

According to the foregoing the quantity $\mathbf{M}(\mathbf{r}, \mathbf{r}')$ of Eq. (20) is determined by the region of integration $|\mathbf{l} - \mathbf{r}_1| \gg 1/p_0$ and $\omega \ll \mu$. For small ϵ and ω we can neglect the frequency dependence of Γ^0 in Eq. (19). We shall denote the corresponding Fourier component by $\Gamma^0_{\alpha\gamma,\gamma\beta}(\mathbf{p}_1, \mathbf{p}_2)$. Thus it follows from Eq. (20) that

$$M_{\alpha\beta}(\mathbf{r}, \mathbf{r}') \approx \frac{ie^2H}{4\pi mc^2}\left[\int \frac{d^3\mathbf{p}}{(2\pi)^6} e^{i\mathbf{p}_1(\mathbf{r}-\mathbf{r}')} \int \frac{d\Omega}{4\pi} \Gamma_{\alpha\gamma,\gamma\beta}(\mathbf{p}_1, \mathbf{p}_2)\right]\frac{1}{2} \times \int G^2(\mathbf{R}, \omega)\, \rho^2 d^3\mathbf{R}. \quad (21)$$

Proceeding to the calculation of this last integral, we substitute in it the Green's function (17):

$$\int G^2(\mathbf{R}, \omega)\, \rho^2 d^3\mathbf{R}$$
$$= \frac{a^2}{(2\pi)^2}\left(\frac{eH}{c}\right)^2 \iint \sum_{n,m} \frac{dp_z d\omega}{(\omega - E_n + i\delta(\omega))(\omega - E_m + i\delta(\omega))}$$
$$\times \int_0^\infty \rho^2 d\rho\, e^{-eH\rho^2/4c} L_n\left(\frac{eH}{2c}\rho^2\right) L_m\left(\frac{eH}{2c}\rho^2\right),$$
$$E_n = (n + 1/2)\,\omega^* + p_z^2/2m^* - p_0^2/2m^*,$$
$$E_m = (m + 1/2)\,\omega^* + (p_z - q)^2/2m^* - p_0^2/2m^*. \qquad (21')$$

In the expression for E_m we have introduced a small quantity q. The point is that the integral in question is not uniquely defined: the expression

$$\int d\omega \int G(\mathbf{R}, \omega)\, G(\mathbf{R}, \omega - \omega_0)\, \rho^2 e^{iqz} d^3\mathbf{R}$$

has different limits for ω_0, $q \to 0$, depending on whether $\omega_0/q \to 0$ or $q/\omega_0 \to 0$. For the case of no magnetic field the singularities of such integrals have been studied by Landau.[3] To give a meaning to the integral (21′) we must take the limit $\omega_0/q \to 0$, which corresponds to the fact that $\mathbf{M}(\mathbf{r}, \mathbf{r}'; \epsilon)$ is being calculated in a field that is constant in time but weakly nonuniform along the z axis. Therefore, strictly speaking, the quantity given by Eq. (21) is a definite one of the two vertex parts Γ^k and Γ^ω introduced by Landau,[3] namely Γ^k. For our estimates this is of course unimportant.

In calculating the integral of Laguerre polynomials in Eq. (21′) it is helpful to use the exact relation

$$xL_n(x) = (2n+1)\,L_n(x) - (n+1)\,L_{n+1}(x) - nL_{n-1}(x). \qquad (22)$$

Therefore m = n ± 1. Integrating Eq. (21′) over the frequencies ω, we get

$$Y = \frac{i}{2\pi}\int d\omega G^2(\mathbf{R}, \omega)\, \rho^2 d^3\mathbf{R}$$
$$= \frac{a^2}{2\pi^2}\sum_{n,m}\int \frac{dp_z}{E_n - E_m}\int_0^\infty xL_n(x)\, L_m(x)\, e^{-x}dx$$

($E_n > 0$, $E_m < 0$). Using Eq. (22) and performing the integration over p_z, we find after simple manipulations

$$Y = (2m^*)^{1/2}\frac{a^2}{\pi^2}\sum_n\left\{\frac{1}{2}\frac{n + 1/2}{[p_0^2/2m^* - \omega^*(n + 1/2)]^{1/2}}\right.$$
$$\left. - \frac{1}{\omega^*}\sqrt{\frac{p_0^2}{2m^*} - \omega^*(n + 1/2)}\right\}.$$

The summation is taken over all values of n for which the radicand is positive.

It is not hard to verify that this quantity can be represented in the following way:

$$\frac{eH}{c}Y = -\frac{\sqrt{2m^*}}{\omega^*}\frac{a^2}{\pi^2}\frac{\partial}{\partial H}\left\{\frac{eH}{c}\sum_n\sqrt{\frac{p_0^2}{2m^*} - \omega^*(n + 1/2)}\right\}. \qquad (23)$$

The sum in the curly brackets has been repeatedly

FIG. 3

investigated in the literature.[7,8] Using the expression obtained by Sondheimer and Wilson,[8] which is accurate to terms of order H^2,

$$\frac{1}{2\pi^2}\frac{eH}{c}\sum_n\sqrt{\mu - \omega^*(n + 1/2)}$$
$$= \frac{p_0^3}{3\pi^2}\left\{1 + \frac{3}{2\pi}\left(\frac{\omega^*}{\mu}\right)^{3/2}\sum_{r=1}^\infty \frac{(-1)^r}{r^{3/2}}\sin\left(2\pi r\frac{\mu}{\omega^*} - \frac{\pi}{4}\right)\right\}$$

and substituting it in Eq. (23), we get finally

$$\frac{eH}{c}Y = -\frac{\sqrt{2m^*}}{\omega^*\pi^3}a^2\frac{\partial}{\partial H}\left\{\left(\frac{\omega^*}{\mu}\right)^{3/2}\varphi\left(\frac{\mu}{\omega^*}\right)\right\},$$

where $\varphi(x)$ is a rapidly oscillating function. Thus in actual fact the terms that correspond to $\mathbf{M}(\mathbf{r}, \mathbf{r}'; \epsilon)$ give in Eq. (14) a correction to the energy levels of the excitations of the order $H^{3/2}$.

3. THE OSCILLATING SINGULARITIES OF THE THERMODYNAMIC FUNCTIONS

Let us now proceed to the derivation of the formula for the thermodynamic potential. In order to separate out the small terms that have an oscillating character, it is more convenient to start from the expression for the derivative $\partial N/\partial\mu$ of the particle-number density with respect to the chemical potential of the system. The particle-number density is connected in a simple way with the Green's function (1) of the system:

$$N = -iG_{\alpha\alpha}(x, x')_{x' \to x,\, t' \to t+0}.$$

The expression for the derivative $\partial N/\partial\mu$ can be obtained by the following arguments (cf. [9]). Let us place the system in a weak and slowly varying potential field $\delta U(\mathbf{r})$. Then the quantity $\mu + \delta U = \mu_0$ is conserved throughout the system. On the other hand, the interaction of the system with the field $\delta U(\mathbf{r})$ is described by the Hamiltonian

$$H_{int} = \int \psi^+(x)\, \delta U \psi(x)\, d^3\mathbf{r}.$$

According to the usual rules of the diagram technique the change of the Green's function in first order in δU can be represented by the first diagram of Fig. 3. In the limit of a δU independent of the coordinates, $\delta U = -\delta\mu$, we get

$$\frac{\partial N}{\partial\mu} = i\int G_{\alpha\gamma}(x, l)\, G_{\gamma\alpha}(l, x)\, d^4l - \int d^4\xi_1 d^4\xi_2 d^4\xi_3 d^4\xi_4 d^4l\, G_{\alpha\alpha_1}$$
$$\times (x, \xi_1)\, G_{\beta\alpha_2}(l, \xi_2)\, G_{\alpha_3\beta}(\xi_3, l)\, G_{\alpha_4\alpha}(\xi_4, x')$$
$$\times \Gamma_{\alpha_1\alpha_2,\, \alpha_3\alpha_4}(\xi_1, \xi_2; \xi_3, \xi_4)$$

or, when we go over to Fourier components with respect to the difference of the time coordinates,

$$\frac{\partial N}{\partial \mu} = \frac{i}{2\pi}\int d\omega \int G_{\alpha\gamma}(\mathbf{r}, \mathbf{l}; \omega)\, G_{\gamma\alpha}(\mathbf{l}, \mathbf{r}; \omega)\, d^3\mathbf{l}$$

$$- \frac{1}{(2\pi)^2}\iint d\omega d\omega' \int d^3\mathbf{s}_1 d^3\mathbf{s}_2 d^3\mathbf{s}_3 d^3\mathbf{s}_4 d^3\mathbf{l}\, G_{\alpha\alpha_1}(\mathbf{r}, \mathbf{s}_1; \omega)\, G_{\alpha_4\alpha}$$

$$\times (\mathbf{s}_4, \mathbf{r}; \omega)\, \Gamma_{\alpha_1\alpha_2, \alpha_3\alpha_4}(\mathbf{s}_1, \omega; \mathbf{s}_2, \omega'; \mathbf{s}_3, \omega'; \mathbf{s}_4, \omega)\, G_{\beta\alpha_2}$$

$$\times (\mathbf{l}, \mathbf{s}_2; \omega')\, G_{\alpha_3\beta}(\mathbf{s}_3, \mathbf{l}; \omega').$$

For simplicity let us first investigate the singularities of the first term in the right member of Eq. (24). These singularities are due to electrons near the Fermi surface, and therefore in this case it is sufficient to use the asymptotic expression (17) for the Green's functions. Substituting this expression in the integral and carrying out the integration, we get

$$\frac{i}{2\pi}\int d\omega \int G_{\alpha\gamma}(\mathbf{r}, \mathbf{l}; \omega)\, G_{\gamma\alpha}(\mathbf{l}, \mathbf{r}; \omega)\, d^3\mathbf{l}$$

$$= \frac{2i}{(2\pi)^3}\left(\frac{eH}{c}\right)^2 a^2 \sum_n \int d\omega \int \frac{dp_z}{(\omega - E_n + i\delta(\omega))(\omega - E_n' + i\delta(\omega))}$$

(here in $E_n' = \omega^*(n + 1/2) + (p_z - q)^2/2m^* - p_0^2/2m^*$ we have introduced a small momentum; this corresponds to the fact that in deriving Eq. (24) we started from the condition for equilibrium of the system in a field $\delta U(\mathbf{r})$ constant in time but slowly varying in space). Integrating the resulting expression first over ω, and then over p_z, we arrive at the expression

$$\frac{\sqrt{2m^*}}{2\pi^2}\left(\frac{eH}{c}\right)a^2 \sum_n \left(\frac{p_0^2}{2m^*} - \omega^*(n + 1/2)\right)^{-1/2}. \tag{25}$$

In the sum (25) a special role is played by the values of n for which the radicand is small. These terms give the nonanalytic part of the integral under consideration. At the same time the main contribution as regards magnitude comes from values of n for which $n\omega^* \sim p_0^2/2m^*$. For a weak field the summation over the main region can be replaced by an integration, and from this we find

$$a^2 m^* p_0/\pi^2. \tag{26}$$

(A contribution to $\partial N/\partial \mu$ of this same order of magnitude is given by the distant regions.) Let us again introduce the number N_0, the integer part of $p_0^2/2m^*$ in units ω^*: $p_0^2/2m^* = (N_0 + 1/2)\omega^* + \Delta$ ($\Delta > 0$). The terms in the sum (25) for which $(N_0 - n) \ll N_0$ are of the order of

$$(\sqrt{2m^*}/2\pi^2)(eH/c)(a^2/\sqrt{\Delta})$$

and remain small in comparison with (26) provided that

$$\Delta \gg \omega^*\left(\omega^* \Big/ \frac{p_0^2}{2m^*}\right). \tag{27}$$

The condition (27) imposes a restriction on the study of the detailed structure of the oscillations in the immediate neighborhood of the Fermi surface. It must be remembered, however, that the actual region of interest for the de Haas-van Alphen effect just coincides with the condition (27), since in the magnetic fields that are obtainable up to the present time one begins to get a violation of the condition (27) only at the very lowest temperatures. Therefore in our later discussion of the quantum case we shall assume that the condition (27) is always satisfied.

In order to separate out the irregular part of the expression (25) we substitute $n \to N_0 - k$. Subtracting from the new sum the terms that diverge at the upper limit, we get

$$\frac{\sqrt{2m^*}}{2\pi^2}\left(\frac{eH}{c}\right)a^2\left[\sum_0^{N_0}(\sqrt{\Delta + \omega^* k})^{-1/2} - 2\sqrt{N_0}/\sqrt{\omega^*}\right]_{N_0\to\infty}$$

$$= \frac{\sqrt{2m^*}}{2\pi^2} m^* a^2 \sqrt{\omega^*}\,\zeta\left(\frac{1}{2}, \frac{\Delta}{\omega^*}\right), \tag{28}$$

where $\zeta(s, x)$ is the generalized Riemann zeta function[10]:

$$\zeta(s, x) = -\frac{\Gamma(1-s)}{2\pi i}\int_\infty^{(0+)} \frac{(-z)^{s-1} e^{-xz}}{1 - e^{-z}}\, dz.$$

Thus apart from regular terms of higher orders in eH/cp_0^2 the integration of the two G-functions in the loop of Fig. 4 a leads to the appearance of small terms which are rapidly oscillating functions of the ratio cp_0^2/eH.

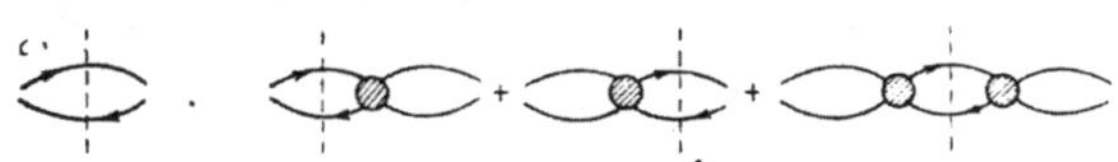

a b

FIG. 4

Returning to Eq. (24), we see that singularities (28) in $\partial N/\partial \mu$ arise in every loop (cf. Fig. 3) in which there are two horizontal lines (with their arrows in opposite directions). Since these terms are small, we have to separate each such loop only once. In particular, the loop can belong to the diagram for a vertex part. As the result of the separation of the singular terms the situation arises which is shown graphically in Fig. 4 b. The loop from which the singularity is separated out is marked with a vertical dashed line. Since these terms are small, we can set the magnetic field strength equal to zero in all the other parts of the diagram.

In this connection we recall once again that the singularity (28) is due to the behavior at large distances $\mathbf{r} - \mathbf{l}$ of the Green's function in the loop of Fig. 4 a. Therefore, using the same procedure with the vertex part as in the derivation of Eq. (21), we get instead of Eq. (24) the following result:

$$\frac{\partial N}{\partial\mu}=\frac{\sqrt{2m^*}\,m^*}{2\pi^2}\sqrt{\omega^*}\,\zeta\left(\frac{1}{2},\frac{\Delta}{\omega^*}\right)\Phi^2,$$

$$\Phi=a\left\{1+\frac{i}{2(2\pi)^4}\int\Gamma^{0k}_{\alpha\gamma,\,\gamma\alpha}(p_1,p_2)\,G^0(p_1,\omega)\,G^0(p_2,\omega)\,d^4p\right\} \tag{29}$$

(the vector p_1 is taken on the Fermi surface). In accordance with what was said earlier, our choice between the two limits Γ^{0k} and $\Gamma^{0\omega}$ introduced by Landau[3] must be to take the limit Γ^{0k}, which means that the potential field used in the derivation of Eq. (24) is strictly independent of the time.

The connection of the renormalization factor Φ with physical quantities has been established by Pitaevskii.[9] It turns out that in the absence of a magnetic field

$$\Phi=a\left(\frac{\partial G^{-1}}{\partial\mu}\right)_{p=p_0,\ \omega=0}=\frac{p_0}{m^*}\frac{dp_0}{d\mu}. \tag{29'}$$

Substituting this result in the expression obtained earlier, we get

$$\frac{\partial N}{\partial\mu}=\frac{1}{\sqrt{2m^*}\,4\pi^2}\sqrt{\omega^*}\left(\frac{dp_0^2}{d\mu}\right)^2\zeta\left(\frac{1}{2},\frac{\Delta}{\omega^*}\right). \tag{30}$$

We must calculate the thermodynamic quantities N and Ω (the potential) in terms of the variables μ and V. According to Eqs. (14) and (23), apart from terms of the order $H^{3/2}$ the quantity $p_0(\mu)$ which appears in Eq. (29) and also in Eq. (25) is a function of the chemical potential as defined on the Fermi surface in the absence of a magnetic field. Taking for the oscillating part of Ω only the double integral of the rapidly oscillating terms $\zeta(1/2, \Delta/\omega^*)=\varphi(cp_0^2/eH)$, we get

$$\delta\Omega_{\rm osc}=-\frac{4m^{*2}\omega^{*5/2}}{3\sqrt{2m^*}\,\pi^2}\zeta\left(-\frac{3}{2},\frac{\Delta}{\omega^*}\right)=\frac{m^{*3/2}\omega^{*5/2}}{4\pi^4}\sum_1^\infty r^{-5/2}\cos\left(2\pi r\frac{\Delta}{\omega^*}-\frac{\pi}{4}\right),$$

that is, an expression which agrees with the results of I. Lifshitz and Kosevich[1] (in the isotropic model).

4. THE EFFECT OF THE SPIN ON THE OSCILLATIONS

In the scheme that has been expounded it is easy to include also the interaction of the magnetic field with the spin magnetic moment of the electron. The paramagnetic susceptibility of a Fermi liquid has been calculated by Landau.[2] In a metal, however, one cannot separate the paramagnetic part of the susceptibility from the diamagnetic part. At the same time the latter, as we have already said, is due to all of the electrons, and not just to those near the Fermi surface, and therefore cannot be expressed in terms of the characteristics of the spectrum. Owing to this we shall concern ourselves here only with the effect of the spin susceptibility on the quantum oscillations of the thermodynamic quantities.

Let us find how the Green's function (15) changes its form when we include in the Hamiltonian (7) the additional term

$$-\beta\int\psi^+(\mathbf{r})(\sigma\mathbf{H})\psi(\mathbf{r})\,d^3r.$$

To do this we again consider the derivative $\partial\Sigma_{\alpha\beta}(\mathbf{r},\mathbf{r}';\epsilon)/\partial H$. Using the results obtained earlier, we get instead of Eq. (9)

$$\partial\Sigma_{\alpha\beta}(\mathbf{r},\mathbf{r}';\varepsilon)/\partial H=-(ie/2c)(x-x')(y+y')\,\Sigma_{\alpha\beta}(\mathbf{r},\mathbf{r}';\varepsilon)+\frac{i}{2\pi}\beta\int d\omega\int d^3\mathbf{r}_1d^3\mathbf{r}_2d^3\mathbf{l}\,\Gamma_{\alpha\rho,\,\tau\beta}(\mathbf{r},\varepsilon;\mathbf{r}_1,\omega;\mathbf{r}_2,\omega;\mathbf{r}',\varepsilon)\times G_{\lambda\rho}(\mathbf{l},\mathbf{r}_1;\omega)\,G_{\tau\varkappa}(\mathbf{r}_2,\mathbf{l};\omega)\,(\sigma\mathbf{n})_{\varkappa\lambda} \tag{31}$$

($\mathbf{n}$ is a vector in the direction of the field). It is not hard to verify that in the last term we can set H = 0 from the very beginning. Substituting here the expression (19), we get

$$\frac{1}{(2\pi)^6}\int e^{i\mathbf{p}_1(\mathbf{r}-\mathbf{r}')}d^3\mathbf{p}_1\,\frac{i}{2\pi}\int d\omega d^3\mathbf{p}_2\Gamma^k_{\alpha\rho,\tau\beta}(\mathbf{p}_1,\varepsilon;\mathbf{p}_2,\omega;\mathbf{p}_2,\omega;\mathbf{p}_1,\varepsilon)\times G^2(\mathbf{p}_2,\omega)(\sigma\mathbf{n})_{\rho\tau}.$$

Because the free functions are isotropic and because for momenta $\mathbf{p}_1$ close to the Fermi surface the integral in brackets (sic) is a slowly varying function of $|\mathbf{p}_1|$, we can write the second term in the right member of Eq. (31) in the form

$$A\,(\sigma\mathbf{n})_{\alpha\beta}\delta(\mathbf{r}-\mathbf{r}'), \tag{32}$$

where A is a constant. It follows from Eqs. (31) and (32) that as before the Green's function is of the form (15), but the energy of the excitations is compounded of the orbital part $\omega^*(n+1/2)+p_z^2/2m^*$ and the spin part $-\xi\sigma\cdot\mathbf{H}$, where ξ is the "effective" magnetic moment of the electron spin: $\xi=a(\beta+A)$. The connection between ξ and the magnitude of the paramagnetic susceptibility is given by relations obtained by Landau.[2]

We can now proceed to the calculation of $\partial N/\partial\mu$. Here the calculations are of a nature quite analogous to that of those done in the preceding section. We first concern ourselves with the calculation of the first term in the formula (24):

$$\frac{i}{2\pi}\int d\omega d^3\mathbf{l}\,\mathrm{Sp}\,\hat{G}(\mathbf{r},\mathbf{l};\omega)\hat{G}(\mathbf{l},\mathbf{r};\omega)=\frac{i}{(2\pi)^3}\left(\frac{eH}{c}\right)^2a^2\sum_n\iint d\omega dp_z\left[(\omega-E_n)^2+\frac{1}{4}\xi^2H^2\right]\times[(\omega-E_n+\xi H/2+i\delta)(\omega-E_n-\xi H/2+i\delta)\times(\omega-E_n'-\xi H/2+i\delta)(\omega-E_n'+\xi H/2+i\delta)]^{-1}.$$

As before, we first integrate this expression over

$d\omega$. Here we note that the integral vanishes both in the case in which all of the poles of the integrand are on one side of the axis of ω, and also in the case in which there are two poles above the axis and two below. The integral is different from zero only in the region of $E_n = \pm\xi H/2$. Defining the integral in a suitable way as a k-limit, we finally get for the first term

$$\frac{1}{4\pi^2}\sqrt{2m^*}\left(\frac{eH}{c}\right)a^2\sum_n\left\{[p_0^2/2m^* - \omega^*(n + 1/2) - \xi H/2]^{-1/2} + \left[\frac{p_0^2}{2m^*} - \omega^*\left(n+\frac{1}{2}\right) + \xi H/2\right]^{-1/2}\right\}, \quad (33)$$

that is, the only difference from the formulas written before is that the single sum over n is replaced by the average of two terms with $p_0^2/2m^* \to p_0^2/2m^* \pm \xi H/2$.

Let us now recall that in the second term of Eq. (24) the singularity (33) can be separated out in three ways, as shown in Fig. 4 b. Since the terms involving the magnetic field are small, we can set the field equal to zero in all quantities except the two Green's functions marked with the dashed line in the diagram. In the isotropic model the integral is

$$\int\Gamma_{\alpha\rho,\rho\beta}(p_1,p_2;p_2,p_1)G(p_2)G(p_2)d^4p_2 = \frac{1}{2}\delta_{\alpha\beta}\int\Gamma_{\gamma\rho,\rho\gamma}(p_1,p_2;p_2,p_1)G^2(p_2)d^4p_2.$$

When this fact is used it is not hard to show that the sum of all of the loops of Fig. 4 b gives a singularity of the form (33) multiplied by the renormalization factors (29). Thus instead of Eq. (30) we get as the final result for $\partial N/\partial\mu$

$$\frac{\partial N}{\partial\mu} = \frac{\sqrt{\omega^*}}{8\pi^2\sqrt{2m^*}}\left(\frac{dp_0^2}{d\mu}\right)^2\left\{\zeta\left(\frac{1}{2},\frac{\Delta}{\omega^*}+\frac{\xi H}{2\omega^*}\right)+\zeta\left(\frac{1}{2},\frac{\Delta}{\omega^*}-\frac{\xi H}{2\omega^*}\right)\right\},$$

and for the oscillating part of the thermodynamic potential

$$\sigma\Omega_{osc} = -\frac{2m^{*2}\omega^{*5/2}}{3\pi^2\sqrt{2m^*}}\left\{\zeta\left(-\frac{3}{2},\frac{\Delta+\xi H/2}{\omega^*}\right) + \zeta\left(-\frac{3}{2},\frac{\Delta-\xi H/2}{\omega^*}\right)\right\}. \quad (34)$$

The oscillating parts of all the thermodynamic quantities are usually stated in the form of series of harmonics.[1] For example, by going from Eq. (34) to the oscillating part of the free energy, we get for the magnetic moment the expression

$$M_{osc} = -\frac{m^{*3/2}(\beta^*H)^{1/2}\mu}{2\pi^3\hbar^3 H}\sum_1^\infty\frac{(-1)^r}{r^{3/2}}\cos\left(\frac{\xi}{\beta^*}\pi r\right)\times\sin\left(\pi r\frac{cp_0^2}{e\hbar H}-\frac{\pi}{4}\right),$$

where $\beta^* = e\hbar/m^*c$, and ξ can be connected with the paramagnetic susceptibility χ and with the coefficient γ in the linear term in the heat capacity per unit volume $c = \gamma T$, if we use the results obtained in Landau's paper[2]:

$$\xi/\beta^* = 4\pi^2\chi/3\beta\beta^*\gamma. \quad (36)$$

In this last formula $\beta = e\hbar/mc$ is the magnetic moment of the free electron. Although the factor with the cosine has indeed been written previously, it was then assumed that the spin energy of the electron was equal to $-\beta\boldsymbol{\sigma}\cdot\mathbf{H}$; that is, the Fermi-liquid properties were not taken into account, and the result of this was that the argument of the cosine was written $\pi r m^*/m$.

We also mention that according to Landau's results[2]

$$\frac{1}{\chi} = \beta^{-2}\left(\bar{\zeta} + \frac{4\pi^2}{3\gamma}\right), \quad \bar{\zeta} = \frac{1}{4\pi}\int\zeta(\Omega)\,d\Omega.$$

$\bar{\zeta}$ is the integral of the spin part of the function f[2] of the Fermi liquid:

$$f_{\sigma\sigma'}(\mathbf{p},\mathbf{p}') = f(\mathbf{p},\mathbf{p}') + \zeta(\mathbf{p},\mathbf{p}')(\sigma\sigma')$$

(all of the notations are taken from the paper of Abrikosov and Khalatnikov[11]). Therefore from this one could get an estimate of the size of the Fermi-liquid effects in a metal. In particular, the sign of $\bar{\zeta}$ is very interesting, since for $\zeta > 0$ there could exist in a metal spin oscillations of the type of zeroth sound.[2,11]

In conclusion the writers express their gratitude to Academician L. D. Landau for a discussion of the results of this work and for his comments.

[1] I. M. Lifshitz and A. M. Kosevich, JETP **29**, 743 (1955), Soviet Phys. JETP **2**, 646 (1956).

[2] L. D. Landau, JETP **30**, 1058 (1956), Soviet Phys. JETP **3**, 920 (1956).

[3] L. D. Landau, JETP **35**, 97 (1958), Soviet Phys. JETP **8**, 70 (1959).

[4] J. M. Luttinger, Phys. Rev. **121**, 1251 (1961).

[5] G. E. Zil'berman, JETP **32**, 296 (1957), Soviet Phys. JETP **5**, 208 (1957).

[6] L. D. Landau and E. M. Lifshitz, (Quantum Mechanics), Pergamon, 1948.

[7] Yu. B. Rumer, JETP **18**, 1081 (1948).

[8] E. Sondheimer and A. Wilson, Proc. Roy. Soc. **A210**, 173 (1951).

[9] L. P. Pitaevskii, JETP **37**, 1794 (1959), Soviet Phys. JETP **10**, 1267 (1960).

[10] Higher Transcendental Functions, Bateman Ms Project, New York, 1953, Vol. 1, p. 24.

[11] A. A. Abrikosov and I. M. Khalatnikov, UFN **66**, 177 (1958), Soviet Physics Uspekhi 1, 68 (1959).

Translated by W. H. Furry
267

SOVIET PHYSICS JETP VOLUME 17, NUMBER 5 NOVEMBER, 1963

FEASIBILITY OF ZERO-SOUND OSCILLATIONS IN METALS

L. P. GOR'KOV and I. E. DZYALOSHINSKIĬ

Institute of Physics Problems, Academy of Sciences, U.S.S.R.

Submitted to JETP editor December 20, 1962

J. Exptl. Theoret. Phys. (U.S.S.R.) **44**, 1650-1660 (May, 1963)

Zero-sound electron oscillations in an anisotropic metal are studied on the basis of the theory of Fermi liquids. Spin as well as non-spin oscillations are possible. The latter apparently exist in any type of metal and possess a linear dispersion law throughout the whole frequency range. Spinless waves can exist if some restrictions are imposed on the magnitude of the Fermi-liquid interaction; these restrictions can be appreciably relaxed for symmetric directions in the crystal. The non-spin oscillations have two linear dispersion regions in the radio-frequency and infrared ranges. The possibility of observing zero sound in metals is discussed.

ONE of the most important results of the Landau theory of the Fermi liquid [1] was the existence of the so-called "zero" sound oscillations in He^3. It has turned out that ordinary acoustic oscillations of the Fermi liquid attenuate at sufficiently low temperatures. Instead, waves with a linear dispersion law and with a speed higher than that of ordinary sound can propagate in He^3 at T = 0. The existence of zero-sound oscillations is connected with the self-consistent interaction of the excitations.

According to [1], the energy change $\delta\epsilon(\mathbf{p})$ of an excitation with momentum **p**, resulting from variation of the distribution function $\delta n(\mathbf{p}')$, is equal to

$$\delta\varepsilon(\mathbf{p}) = \int f(\mathbf{p}, \mathbf{p}')\, \delta n(\mathbf{p}')\, d\tau'. \qquad (1)$$

The function f, which is connected with the amplitude of the mutual zero-angle scattering of the excitations [2], determines the propagation velocity of the zero-sound oscillations.

Owing to experimental difficulties, zero sound has not yet been observed in He^3. In the present note we discuss the possibility of producing zero-sound oscillations in metals whose conduction electrons also form a sort of Fermi liquid. In addition to general physical interest that attaches to this phenomenon, the study of zero sound in metals is attractive also because one can attempt to excite it by radio-frequency and optical methods.

A specific feature of the electron Fermi liquid in metals is the presence of Coulomb interactions between the electrons and between the electrons and the ions. This creates a situation wherein there are no low-excitation-energy oscillations, accompanied by electron-density oscillations in the electron liquid. The expected zero-sound oscillations in He^3 (see also [3]) are accompanied by changes of the density and therefore cannot exist in a metal. Zero sound in a degenerate isotropic electron plasma was investigated by Silin [4]. Unfortunately, Silin's result cannot be ascribed to a real metal, since the presence of anisotropy, as will be shown below, leads to a whole series of essential singularities. We shall also discuss the possibility of observing zero sound in metals.

Let us consider the kinetic equation for electronic excitations in a metal:

$$\frac{\partial n}{\partial t} + \frac{\partial \varepsilon}{\partial \mathbf{p}}\frac{\partial n}{\partial \mathbf{r}} - \frac{\partial \varepsilon}{\partial \mathbf{r}}\frac{\partial n}{\partial \mathbf{p}} = J_{st},$$

which we write for small deviations δn from the equilibrium state:

$$\frac{\partial \delta n}{\partial t} + \mathbf{v}\frac{\partial \delta n}{\partial \mathbf{r}} - \frac{\partial n_0}{\partial \varepsilon}\mathbf{v}\frac{\partial \delta\varepsilon}{\partial \mathbf{r}} + e\mathbf{E}\mathbf{v}\frac{\partial n_0}{\partial \varepsilon} = 0. \qquad (2)$$

We have neglected in (2) the collision integral, assuming henceforth that the range of the excitations, due at low temperatures to the scattering of the electrons by the impurities or by the phonons, is large compared with the wavelength of the zero-sound oscillations.

The term with $\partial\delta\epsilon/\partial\mathbf{r}$, first introduced by Landau, where $\delta\epsilon$ is given by (1), is the result of the interaction of the excitations in the Fermi liquid. For simplicity we assume for the time being that the function $f(\mathbf{p}, \mathbf{p}')$ does not contain a dependence on the electron spins.

E is the electric field, which accompanies the oscillations of the electron liquid in a metal. As already mentioned, the variation of the electron density should be equal to zero, for otherwise the oscillation frequency would be of the order of the plasma frequency ω_0 ($\omega_0 \sim 1$ eV in metals). In this region of frequencies, (2) is no longer valid. We shall determine the longitudinal part of the field E in (2) from the condition that the charge-density oscillations be equal to zero. The transverse part of the field **E** is connected with the transverse-current oscillations, which generally speaking cannot vanish, as will be shown presently. Let us change over to Fourier components in (2) and represent δn in the form

$$\delta n(\mathbf{p}) = \nu \partial n_0/\partial \varepsilon,$$

where ν depends only on the position of p directly on the Fermi surface itself. For ν we have

$$\omega\nu - \mathbf{kv}\hat{L}(\nu) + ie\mathbf{Ev} = 0, \tag{3}$$

where $\hat{L}(\nu) = \nu + \int f(\mathbf{p},\mathbf{p}')\nu(\mathbf{p}')dS$, and dS is the Fermi "surface element":

$$2d^3p/(2\pi)^3 = d\varepsilon\, dS. \tag{4}$$

The expressions for the charge density ρ and for the current **j** are obviously

$$\rho = -e\int \nu dS, \qquad \mathbf{j} = -e\int \mathbf{v}\hat{L}(\nu)\, dS. \tag{5}$$

Let us eliminate **E** from (3), expressing the transverse field $\mathbf{E}_\perp$ in terms of the current with the aid of Maxwell's equations, and using the condition $\rho = 0$ for the longitudinal field $\mathbf{E}_\parallel$. For this purpose we introduce

$$\omega\nu + ie\mathbf{Ev} = \omega X. \tag{6}$$

We then have in place of (3)

$$\omega X = \mathbf{kv}\hat{L}(\nu). \tag{7}$$

We apply to (6) the operator $\hat{L}$, multiply by **v**, and integrate over the Fermi surface. As a result we obtain

$$-\omega j_i + ie^2E_k\int (v_i\hat{L}(v_k))\, dS = \omega e\int v_i\hat{L}(X)\, dS.$$

For the longitudinal field $\mathbf{E}_\parallel$ we obtain directly from the condition of vanishing of the longitudinal current

$$ieE_\parallel = \omega\int v_\parallel\hat{L}(X)\, dS \Big/ \int v_\parallel\hat{L}(v_\parallel)\, dS$$
$$= \omega\int v_\parallel\hat{L}(X)\, dS \Big/ \frac{1}{3}\int \mathbf{v}\hat{L}(\mathbf{v})\, dS. \tag{8}$$

(To be specific, we confine ourselves to a crystal of cubic symmetry.)

From Maxwell's equations it follows that

$$\mathbf{j}_\perp = -i(c^2k^2/4\pi\omega)\mathbf{E}_\perp, \tag{9}$$

hence

$$ie\mathbf{E}_\perp\left[\frac{c^2k^2}{4\pi} + \frac{e^2}{3}\int(\mathbf{v}\hat{L}(\mathbf{v}))\, dS\right] = \omega e^2\int \mathbf{v}_\perp\hat{L}(X)\, dS. \tag{10}$$

Thus

$$\nu = X - \frac{v_\parallel\int v_\parallel\hat{L}(X)\, dS}{1/3\int \mathbf{v}\hat{L}(\mathbf{v})\, dS} - \frac{4\pi e^2\mathbf{v}_\perp\int \mathbf{v}_\perp\hat{L}(X)\, dS}{\omega_0^2 + c^2k^2}, \tag{11}$$

where

$$\omega_0^2 = \frac{4\pi e^2}{3}\int(\mathbf{v}\hat{L}(\mathbf{v}))\, dS$$

is of the order of the square of the plasma frequency.

Substituting (11) in (7), we arrive ultimately at the following equation:

$$\omega X - \mathbf{kv}\hat{L}(X) + \mathbf{kv}\left[\frac{\hat{L}(v_\parallel)\int v_\parallel\hat{L}(X)\, dS}{1/3\int(\mathbf{v}\hat{L}(\mathbf{v}))\, dS}\right.$$
$$\left. + \frac{4\pi e^2(\hat{L}(\mathbf{v}_\perp)\int \mathbf{v}_\perp)\hat{L}(X)\, dS}{\omega_0^2 + c^2k^2}\right] = 0. \tag{12}$$

It is seen from (12) that if the equation has a solution, then the spectrum $\omega = \omega(k)$ has two linear parts at low and high frequencies, respectively. In fact, if $c^2k^2 \ll \omega_0^2$, (12) is transformed into

$$\omega X - \mathbf{kv}\hat{L}(X) + (\mathbf{kv})\left(\hat{L}(\mathbf{v})\int \mathbf{v}\right)\hat{L}(X)\, dS \Big/ \frac{1}{3}\int(\mathbf{v}\hat{L}(\mathbf{v}))\, dS$$
$$= 0, \tag{13}$$

that is, it has the form of an ordinary equation for zero-sound oscillations in a Fermi liquid:

$$\omega X - \mathbf{kv}X - \mathbf{kv}\int F(\mathbf{p},\mathbf{p}')X(\mathbf{p}')\, dS = 0,$$

where, however, the function $F(\mathbf{p},\mathbf{p}')$ is connected with $f(\mathbf{p},\mathbf{p}')$ in the following manner:

$$F(\mathbf{p},\mathbf{p}') = f(\mathbf{p},\mathbf{p}') - (\hat{L}(\mathbf{v})\hat{L}'(\mathbf{v})) \Big/ \frac{1}{3}\int(\mathbf{v}\hat{L}(\mathbf{v}))\, dS. \tag{14}$$

In this case, according to (9) and (10), $\mathbf{j}_\perp$ is proportional to k^2, and consequently, as $k \to 0$ the transverse current vanishes, like the longitudinal one. The condition $c^2k^2 \ll \omega_0^2$ can be written in the form

$$\omega^2 \ll v^2\omega_0^2/c^2.$$

Together with the condition $kl \gg 1$ ($\omega\tau \gg 1$), it determines the region of radio frequencies in which the anomalous skin effect occurs in metals [5,6].

In the opposite limiting case $c^2k^2 \gg \omega_0^2$, the transverse current is not small, but, in accordance with (9) and (10), the transverse field is small. We can neglect in (12) the last term

$$\omega X - \mathbf{kv}\hat{L}(X) + (\mathbf{kv})\hat{L}(v_\parallel)\int v_\parallel\hat{L}(X)\, dS \Big/ \frac{1}{3}\int(\mathbf{v}\hat{L}(\mathbf{v}))\, dS$$
$$= 0. \tag{15}$$

This region of frequencies $\omega_0 v/c \ll \omega \ll \omega_0$ corresponds to the infrared part of the spectrum [6].

So far we have simply assumed that Eqs. (13) and (15) have a solution. Let us now discuss the requirements for making this possible. Even in an isotropic Fermi liquid (see, for example, [3]) it is impossible to obtain a solution for zero-sound oscillations in general form. One can see, however, that this solution must exist if the function $f(p, p')$ is sufficiently large and positive. The same holds true in the anisotropic case, too.

In the absence of anisotropy, $f(p, p')$ depends only on the angle between the vectors p and p' and can be expanded in Legendre polynomials. In the general case we can write instead

$$f(\mathbf{p}, \mathbf{p}') = \sum_i \lambda_i f_i(\mathbf{p}, \mathbf{p}'), \tag{16}$$

where $f_i(p, p')$ are functions which transform in each variable in accord with the i-th irreducible representation of the crystal symmetry group. To be specific, we assume that we are dealing with the O_h group. Let one of the components in (16) be large, $\lambda_{i_0} \to \infty$. For simplicity let $i_0 \neq F_{1u}$ (see [7]), that is, $f_{i_0}(p, p')$ and $v(p)$ transform in accordance with different representations. Then Eq. (13), for example, assumes the form

$$(\omega - \mathbf{kv}) X - (\mathbf{kv}) \Big\{\lambda_{i_0} \int f(\mathbf{p}, \mathbf{p}') X dS - \Big(\mathbf{v}\int \mathbf{v}'\Big) X dS \Big/ \frac{1}{3}\int v^2 dS \Big\} = 0. \tag{17}$$

Making the substitution $X = (\mathbf{kv})\chi/(\omega - \mathbf{kv})$, we see immediately that χ has the form

$$\chi = \chi_{i_0} - \mathbf{vc},$$

where the unknown function χ_{i_0} transforms in accordance with the representation i_0, and $\mathbf{c}$ is

$$\mathbf{c} = \int \mathbf{v} X dS \Big/ \frac{1}{3}\int v^2 dS. \tag{18}$$

If $\lambda_{i_0} \to \infty$, then it is obvious from (17) that $\omega \gg \mathbf{kv}$. Therefore

$$\frac{\mathbf{c}}{3}\int v^2 dS = \int \frac{\mathbf{v}(\mathbf{kv})}{\omega - (\mathbf{kv})} \chi_{i_0} dS. \tag{19}$$

For our purposes it is convenient to choose for i_0 a representation such that when $\mathbf{v}(\mathbf{kv})/(\omega - (\mathbf{kv}))$ is expanded in $\mathbf{kv}/\omega$, it appears first only in terms of sufficiently high order in $\mathbf{kv}/\omega$. Then $\mathbf{c}$ will be small. For example, the representation of F_{1g} in the expansion of $\mathbf{v}(\mathbf{kv})/(\omega - (\mathbf{kv}))$ enters, as can be readily verified, with a factor $(\mathbf{kv}/\omega)^3$. Then $c \sim v^{-1}(\mathbf{kv}/\omega)^3 \chi_{F_{1g}}$ and we can neglect the term with $\mathbf{c}$ in (17).

We obtain for the determination of χ_{i_0} an integral equation of the Fredholm type:

$$\chi_{i_0} = \frac{\lambda_{i_0}}{3}\frac{k^2}{\omega^2}\int f_{i_0} v^2 \chi_{i_0} dS, \tag{20}$$

which, according to the general theory, has positive eigenvalues $\omega^2 \sim k^2/\lambda_{i_0}$. In fact, when λ_{i_0} is sufficiently large, the variation in energy upon deviation from equilibrium is determined by a term quadratic in the energy

$$\delta E \backsim \lambda_{i_0} \int f_{i_0}(\mathbf{p}, \mathbf{p}')\, \nu(\mathbf{p})\, \nu(\mathbf{p}')\, dS.$$

The δE should be positive (from stability considerations), from which it follows that the kernel of the integral equation (20) is positive definite. Of course, this analysis applies also to Eq. (15).

Thus, if the function f is sufficiently large, the equation for the zero-sound oscillations must have a solution. The question therefore reduces to whether this quantity is sufficiently large in the given specific metal. Unfortunately, however, no information whatever is available on the function f in metals; it is clear from general considerations that the value of this function is of the order unity.

The foregoing pertains to zero sound with arbitrary direction of propagation. If we choose as the propagation direction one of the preferred directions in a crystal, then zero sound, subject to some limitations on the Fermi surface, can exist also for a positive function $f(p, p')$ which is as small as desired. To prove this we turn to the equation for zero-sound oscillations in the form (3):

$$(\omega - \mathbf{kv})\, \nu - \mathbf{kv}\int f(\mathbf{p}, \mathbf{p}')\nu dS + \mathbf{cv} = 0. \tag{21}$$

We consider first the simpler case of frequencies in the infrared region $v\omega_0/c \ll \omega \ll \omega_0$. As was already noted above, in this limit the vector $\mathbf{c}$, which is proportional to the electric field, has only one longitudinal component $c_{\parallel}$, which should be determined from the condition that the density oscillations be equal to zero. The transverse current is in this case different from zero.

If f is very small, then, as can be seen from (20), ν can be large only at those points, at which $\omega - \mathbf{kv}$ is small. The latter denotes that the propagation velocity $\mathbf{u}$ of the wave is close to the velocity of the electrons at some point on the Fermi surface [1]). In the vicinity of this point, $\mathbf{kv}$ has a maximum

$$\mathbf{kv} = (\mathbf{kv})_0 \{1 - [a_{11}(\theta - \theta_0)^2 + 2a_{12}(\theta - \theta_0)(\varphi - \varphi_0) + a_{22}(\varphi - \varphi_0)^2]\},$$

where $(\mathbf{kv})_0$ is the value at the point p_0, while the

[1]) Obviously, $|\mathbf{u}|$ is always larger than $|\mathbf{v}|$. Otherwise the excitation corresponding to zero sound would attenuate rapidly because of decay into an electron and a hole.

quantity in the square brackets is a positive-definite quadratic form, with the function ν having near $\mathbf{p}_0$ the form

$$\nu(\mathbf{p}) \approx A/(\omega - \mathbf{kv}). \quad (22)$$

If the vector $\mathbf{k}$ has an arbitrary direction and does not lie on any of the symmetry elements of the crystal, then, generally speaking, only one maximum of the quantity $\mathbf{kv}$ exists. The condition for the absence of density oscillations immediately yields, with logarithmic accuracy,

$$\delta\rho = -\frac{e\alpha}{h_0 k v_0} L A(\mathbf{p}_0) = 0, \quad L = \ln\frac{\text{const}}{\omega - k v_0}, \quad (22a)$$

where $(\mathbf{kv})_0 = kv_0$; h_0 is the Gaussian curvature of the Fermi surface at the point $\mathbf{p}_0$: $dS = h^{-1}d\Omega$; α is a numerical coefficient that depends on the logarithmic integration.

From (21) it follows, with the same accuracy, that

$$A(\mathbf{p}_0) = \alpha L h_0^{-1} f(\mathbf{p}_0, \mathbf{p}_0) A(\mathbf{p}_0) - c v_0.$$

We thus obtain $A(\mathbf{p}_0) = c = 0$, and there are no oscillations for arbitrarily small f.

Oscillations with arbitrarily small f are possible only if $\mathbf{k}$ lies on some one of the symmetry elements of the crystal (or close to it). Then Eq. (21) itself becomes invariant relative to the symmetry group of the vector $\mathbf{k}$. In this case, as is well known, the equation has several solutions, each of which transforms in accordance with some representations of its symmetry group. Let us consider for simplicity the symmetry plane (C_s group). It has two representations, A′ and A″ (see [7] for the notation). The components of the vector lying in the plane transform in accordance with the first representation, and the perpendicular component transforms in accordance with the second.

Let us denote the two solutions by $\nu_{A'}$ and $\nu_{A''}$. The equation for these solutions has the form

$$(\omega - \mathbf{kv})\nu_{A'} - \mathbf{kv}\int f_{A'}\nu_{A'} dS = c v_{\parallel},$$

$$(\omega - \mathbf{kv})\nu_{A''} - \mathbf{kv}\int f_{A''}\nu_{A''} dS = 0.$$

The first equation must be solved with the condition $\delta\rho = 0$, while the solution of the second equation satisfies this condition by definition. However, neither equation has solutions with arbitrarily small f if the maximum of $\mathbf{kv}$ is itself in the symmetry plane. In the former case this is connected with the need for satisfying the additional condition (22a), and in the latter case with the fact, which can be readily verified, that any function f_i which transforms in accordance with a non-unitary representation vanishes at a point lying on the symmetry element.

Solutions will exist if the maximum of $\mathbf{kv}$ does not lie on the symmetry plane. Then $\mathbf{kv}$ has two identical maximum values at the points $\mathbf{p}_1$ and $\mathbf{p}_2$. In the vicinity of these points $\nu_{A''}$ has the form

$$\nu_{A''}(\mathbf{p}_1) = \frac{A_1}{\omega - \mathbf{kv}}, \qquad \nu_{A''}(\mathbf{p}_2) = \frac{A_2}{\omega - \mathbf{kv}},$$

and $A_2 = -A_1$ by virtue of the symmetry. Substituting this in (21), we obtain with logarithmic accuracy

$$A_1 = \alpha h_0^{-1} L (f_{11}^{A''} - f_{12}^{A''}) A_1,$$

where $\hat{f}_{ik} = f(\mathbf{p}_i, \mathbf{p}_k)$. (We recall that by virtue of the symmetry $f_{22} = f_{11}$ and $f_{12} = f_{21}$.) A solution of this equation exists when $f_{11}^{A''} - f_{12}^{A''} > 0$, and has the form

$$\omega = kv_0 + \text{const}\cdot k \exp\{-h_0/\alpha (f_{11}^{A''} - f_{12}^{A''})\}.$$

It is easy to verify in exactly the same way that a solution of the type A′ with the additional condition $\delta\rho = 0$ exists when $f_{11}^{A'} - f_{12}^{A'} > 0$.

An analogous situation remains at infrared frequencies for the vectors $\mathbf{k}$ that lie on any symmetry element. Thus, in a crystal with arbitrarily small f the zero-sound of infrared frequency can propagate along directions that are close to all the symmetry elements, provided that the maximum of $\mathbf{kv}$ does not lie on these elements.

In the case of radio frequencies it is necessary to take into account not one but three supplementary conditions ($\mathbf{j} = 0$). Without repeating the perfectly similar considerations, we cite the results. Oscillations at arbitrarily small f are possible in this case, too, if $\mathbf{k}$ lies near one of the symmetry elements, and the maximum of $\mathbf{kv}$ does not lie on any of the symmetry elements. In this case oscillations of all types are possible along the three-fold and four-fold axes (corresponding to the representations of the C_{3v} and C_{4v} groups). Only oscillations of type A_2 are possible along the two-fold axis (group C_{2v}).

The foregoing analysis shows that zero sound must be sought predominantly along directions lying on the symmetry elements of the crystal. The probability of finding it is, roughly speaking, 50 per cent.

So far we have left out everywhere the spin-dependent terms. Assume now that we have in place of (1)

$$\delta\hat{\varepsilon} = \int \hat{f}(\mathbf{p}, \mathbf{p}')\,\delta\hat{n}\,\frac{d^3p'}{(2\pi)^3}, \quad (1')$$

where the function

$$\hat{f}(\mathbf{p}, \mathbf{p}') = f(\mathbf{p}, \mathbf{p}') + \zeta(\mathbf{p}, \mathbf{p}')\,\sigma\sigma' \quad (23)$$

has an exchange part $\zeta(\mathbf{p},\mathbf{p}')$. (Here σ are Pauli matrices, while $\delta\hat{\epsilon}$ and $\delta\hat{n}$ in (1) are two-row matrices with spin variables.)

In place of equation (3) we obtain

$$\omega\hat{\nu} + ie\mathbf{E}\mathbf{v} - \gamma(\mathbf{k}\mathbf{v})(\mathbf{H}\hat{\sigma})$$
$$- \mathbf{k}\mathbf{v}\left\{\hat{\nu} + \tfrac{1}{2}\mathrm{Sp}_{\sigma'}\int f_{\sigma\sigma'}(\mathbf{p},\mathbf{p}')\,\hat{\nu}' dS\right\} = 0 \qquad (3')$$

(in a magnetic field $\delta\hat{\epsilon} = \gamma\mathbf{H}\sigma$), which can be written in the form of two equations for the two components of the function $\hat{\nu}$:

$$\hat{\nu} = \nu_0 + \nu_1\sigma.$$

We have

$$\omega\nu_0 + ie\mathbf{E}\mathbf{v} - \mathbf{k}\mathbf{v}\left[\nu_0 + \int f(\mathbf{p},\mathbf{p}')\nu_0' dS\right] = 0,$$
$$\omega\nu_1 - \gamma(\mathbf{k}\mathbf{v})\,\mathbf{H} - (\mathbf{k}\mathbf{v})\left[\nu_1 + \int\zeta(\mathbf{p},\mathbf{p}')\,\nu_1' dS\right] = 0. \qquad (24)$$

The oscillations ν_0 and ν_1 are generally speaking coupled with each other, since the fields $\mathbf{E}$ and $\mathbf{H}$, in accordance with Maxwell's equations, are expressed in terms of a current $\mathbf{j}$, equal to

$$\mathbf{j} = -e\int\mathbf{v}\hat{L}(\nu_0)\,dS - i\mu c\int[\mathbf{k}\nu_1]\,dS. \qquad (25)^*$$

(Here and below we use the symbols $L(\nu_0)$ and $\hat{\hat{L}}(\nu_1)$ for the operators in the square brackets in the first and second equations of (24), respectively.) We shall now show that this coupling is very weak and therefore zero sound not accompanied by spin oscillations, and spin waves of a special kind, can propagate in a metal independently[1,3].

For this purpose we transform (24) in the same manner as in the derivation of (12) above. We put

$$\omega\nu_0 + ie\mathbf{v}\mathbf{E} = \omega X, \qquad \omega\nu_1 - \gamma(\mathbf{k}\mathbf{v})\,\mathbf{H} = \omega\mathbf{Y}. \qquad (26)$$

Then we have in place of (24)

$$\omega X = \mathbf{k}\mathbf{v}\hat{L}(\nu_0), \qquad \omega\mathbf{Y} = (\mathbf{k}\mathbf{v})\,\hat{\hat{L}}(\nu_1). \qquad (27)$$

With the aid of (26), using the definition (25) for the current and Maxwell's equations, we can express the intensities of the fields $\mathbf{E}$ and $\mathbf{H}$ in terms of integrals of X and $\mathbf{Y}$. For the longitudinal electric field $E_{\|}$ the previous expression (8) holds true.

In place of (10) we obtain

$$ie\mathbf{E}_{\perp} = \frac{4\pi\omega}{\omega_0^2 + c^2k^2}\int\{e^2\mathbf{v}_{\perp}\hat{L}(X) + i\mu ce\,[\mathbf{k}\mathbf{Y}]\}\,dS. \qquad (28)$$

In addition

$$\mathbf{H} = (c/\omega)\,[\mathbf{k}\mathbf{E}]. \qquad (29)$$

After substituting (29), (28), and (8) in (26) we obtain

$$\nu_0 = X - v_{\|}\int v_{\|}\hat{L}(X)dS \Big/ \tfrac{1}{3}\int v\hat{L}(v)\,dS$$
$$- \frac{4\pi}{\omega_0^2 + c^2k^2}\left\{e^2(\mathbf{v}_{\perp}\int\mathbf{v}_{\perp}')\,\hat{L}(X)\,dS + i\mu ec\,(\mathbf{v}_{\perp}\int[\mathbf{k}\mathbf{Y}]\,dS\right\},$$
$$\nu_1 = \mathbf{Y} - i\frac{4\pi c\gamma(\mathbf{k}\mathbf{v})}{\omega(\omega_0^2 + c^2k^2)}\int\{e\,[\mathbf{k}\mathbf{v}]\,\hat{L}(X) + i\mu c\,[\mathbf{k}\,[\mathbf{k}\mathbf{Y}]]\}\,dS. \qquad (30)$$

Estimates show immediately that $\nu_1 \approx \mathbf{Y}$, and that the cross terms can be neglected, since their order is $k\omega_0^2/mv(\omega_0^2 + c^2k^2)$. Thus, the equation for the spin oscillations has in the entire frequency interval the usual form:

$$\omega\nu_1 - (\mathbf{k}\mathbf{v})\,\hat{\hat{L}}(\nu_1) = 0. \qquad (31)$$

An analysis similar to that of the spinless oscillations above, shows that (31) has a solution both for very large values of $\zeta(\mathbf{p},\mathbf{p}')$ and for small ones. Moreover, in the latter case the solutions exist also for arbitrary direction of $\mathbf{k}$. Indeed, since unlike (21) no additional conditions are imposed on the solution of (31), in order to obtain a solution for small ζ it is sufficient to have only one point $\mathbf{p}_0$ at which the function $\mathbf{k}\mathbf{v}$ has a maximum.

Substituting in (31) the function ν_1 in the form (22), we obtain, with logarithmic accuracy,

$$\omega - \mathbf{k}\mathbf{v}_0 \sim kv_0\exp\{-\alpha/\zeta_0\},$$

where $\zeta_0 = \zeta(\mathbf{p}_0,\mathbf{p}_0) > 0$ should have the antiferromagnetic sign. Therefore, in our opinion, spin zero-sound oscillations should exist in practically any metal.

Without citing the corresponding formulas, we note that in a weak external magnetic field the law governing the dispersion of both spin and spinless oscillations begins with a certain constant frequency, proportional to the magnetic field and dependent, generally speaking, in a rather complicated manner on the function f and on the direction of the vector $\mathbf{k}$. Unlike the isotropic case[4], it is impossible to obtain simple formulas here.

Let us proceed now to the question of the possibility of observing zero sound. These oscillations are connected with the appearance of electromagnetic waves propagating in a metal, so that it is most natural to attempt to excite them with the aid of electromagnetic radiation. The excitation mechanism will be essentially different for the radio and for the infrared frequency intervals. The case of low frequencies is less convenient in this sense. We have already mentioned that the current in a zero-sound wave is very small in the case of long waves, and therefore these waves are weakly excited by the electromagnetic field. If we deal with the excitation

---
$^*[\mathbf{k}\,\nu_1] = \mathbf{k}\times\nu_1$

of a spinless wave, then in the radio-frequency region the current in such a wave is proportional to the square of the frequency.

In a spin wave (31), the current is proportional to k for small $k \to 0$. It must be borne in mind that in this case a contribution is made to the current (25) not only by the part with ν_1, but also by the spinless part ν_0, which accompanies, in accordance with (30), the ν_1 oscillations. Indeed, although ν_0 is small in such a wave,

$$\nu_0 \sim \frac{k}{mv}\frac{\omega_0^2}{\omega_0^2+c^2k^2}\nu_1$$

it makes in the region $\omega_0^2 \gg c^2k^2$ a contribution of the same order of magnitude to the current.

Let us consider normal incidence of radiation on a semi-infinite metal. In order to determine exactly the amplitude of the zero-sound wave penetrating into the metal, it is necessary to solve a rather difficult electromagnetic problem, which is further complicated by the presence of anisotropy. One can, however, estimate rather crudely the order of magnitude of the transmitted radiation. For this purpose we note that the zero-sound wave is formed at distances on the order of the wavelength $\lambda \sim v/\omega$. In the radio-frequency region, reflection of radiation from the surface of a pure metal is accompanied by anomalous skin-effect, where the field inside the metal decreases rapidly with distance from the surface. The characteristic depth of penetration is $\delta \ll \lambda \sim v/\omega$. The amplitude of the transmitted wave can be estimated from the condition that the field in the wave becomes comparable with the field of the skin layer at distances on the order of λ. Inasmuch as at distances $r \gg \delta$ the field in the skin layer decreases as $H \sim H_0(\delta/r)$ where H_0 is the field on the surface, we have

$$H_\lambda \sim H_0(\delta/\lambda)^2 \sim H_0(c/v)^2(\delta/\lambda_0)^2,$$

where λ is the wavelength of zero sound and λ_0 is the length of the corresponding electromagnetic wave in vacuum. This, of course, is a very small quantity.

It is therefore much more convenient to excite zero sound in the region of the infrared spectrum. Even from the preceding estimate we see that if $\lambda \sim \delta$ the field in the wave becomes of the order of the field on the surface. Let us discuss this case in greater detail. At these frequencies the field in the zero-sound wave is small in the sense that its influence on the particle-number oscillations can be neglected. In this case, however, the current does not vanish. For example, in the spin wave,*

$$\mathbf{j} = -\mu c \operatorname{rot}\int \nu_1 dS, \quad \mathbf{H} = -4\pi\mu\int \nu_1 dS.$$

ν_1 itself satisfies the equation

$$-i\omega\nu_1 + v_z d\hat{\hat{L}}(\nu_1)/dz = 0. \tag{32}$$

In infinite space this equation has sinusoidal solutions with a dispersion law $\omega = uk$. The solution of (32) in a bounded metal depends on the boundary conditions. In the case of specular electron reflection from the metal surface the sinusoidal solution can be used directly, after which it is easy to determine the transmission coefficient

$$D = E_0/H_0 = \omega/ck \sim v/c.$$

Of course, reflection from a metal surface is always close to diffuse[5,6]. The solution of (32) goes over therefore into a sinusoidal wave only at distances on the order of the wavelength, but the transmission coefficient, naturally, has the same order of magnitude. The same pertains to the spinless oscillations described by equation (15).

We note, however, that with zero-sound oscillations impossible in the infrared region the field attenuates at distances on the order of $\delta \sim c/\omega_0 \gg v/\omega$ ($\delta \sim 10^{-5}$) and the reflection coefficient R is very close to unity[6], with

$$1 - R \sim v/c.$$

To verify the existence of zero-sound waves, it would be most convenient to study indeed the passage of radiation through a plate of thickness d $\gg c/\omega_0$. In the infrared region, zero-sound waves would lead to transparency of such plates. The fraction of the transmitted radiation would be of the order of $v^2/c^2 \sim 10^{-3}$—10^{-4}, and would be independent of the thickness of the plate. Unfortunately, this is true, of course, only so long as the dimension of the plate d is small compared with the mean free path of the electrons in the metal. Actually, two mean free paths enter into the problem: the damping of the spinless oscillations is characterized by the ordinary electron mean free path l, which is involved in the conductivity in a constant field; in addition there is the mean free path l_s needed to flip the electron spin.

At infrared frequencies $l \sim v/\Theta \sim 10^{-5}$—$10^{-6}$, where Θ is the Debye temperature. To observe spinless zero sound it is therefore necessary to satisfy the condition $v/\Theta \gtrsim d \gg c/\omega_0$. This is a very difficult condition even for a substance with a relatively low Debye temperature.

For spin zero sound, the conditions of observation are much more favorable. One can expect that in this case the non-spin mean free path $l_s = \alpha l$

*rot = curl.

(with $\alpha \gg 1$) would be important here, with α determined by the smaller of the two quantities, Z (the charge of the impurity) or $(c/v)^2$ (the latter is connected with the fact that the spin flip can be caused by two factors: exchange interaction, which is of the order of Z^{-1} compared with the non-exchange interaction, and the spin-orbit interaction, which is of order $(v/c)^2$). The corresponding limitation on the plate thickness are then weaker

$$\alpha v/\Theta \gtrsim d \gg c/\omega_0.$$

In the region of lower frequencies, the limitations connected with the mean free path are much weaker, but the transmission coefficient decreases much more significantly. From the foregoing we can estimate that the coefficient of transmission through the plate has in the radio-frequency region an order of[2)] $(v/c)^2(\delta/\lambda)^4 \sim (c/v)^2(\delta/\lambda_0)^4$. At the same time the mean free path in this region is proportional to $l \sim v\Theta/\omega^3$ [8], if the scattering by the impurities can be neglected.

[2)]It might appear that, on the other side of the plate, the fraction of the transmitted radiation would be reduced by another factor $(\delta/\lambda)^4$. This does not occur, since there is no solution to the problem of the anomalous skin effect for field that increases with increasing depth in the metal.

Zero-sound oscillations could also manifest themselves in many other effects, such as the line width in electron diffraction, the characteristic losses of charged particles upon passage through a metal plate in a magnetic field, etc. It seems to us that it would be very interesting to observe this phenomenon in metals.

[1] L. D. Landau, JETP **30**, 1058 (1956), Soviet Phys. JETP **3**, 920 (1956).

[2] L. D. Landau, JETP **35**, 97 (1958), Soviet Phys. JETP **8**, 70 (1959).

[3] A. A. Abrikosov and I. M. Khalatnikov, UFN **66**, 177 (1958), Soviet Phys. Uspekhi **1**, 68 (1959).

[4] V. P. Silin, JETP **35**, 1243 (1958), Soviet Phys. JETP **8**, 870 (1959).

[5] G. Reiter and E. Sondheimer, Proc. Roy. Soc. **A195**, 336 (1948).

[6] V. L. Ginzburg and G. P. Motulevich, UFN **55**, 469 (1955).

[7] L. D. Landau and E. M. Lifshitz, Quantum Mechanics, Pergamon, 1958.

[8] A. B. Migdal, JETP **34**, 1438 (1958), Soviet Phys. JETP **7**, 996 (1958).

Translated by J. G. Adashko
264

SOVIET PHYSICS JETP VOLUME 16, NUMBER 6 JUNE, 1963

ON THE NATURE OF IMPURITY FERROMAGNETISM

A. A. ABRIKOSOV and L. P. GOR'KOV

Institute for Physics Problems, Academy of Science, U.S.S.R.

Submitted to JETP editor July 3, 1962

J. Exptl. Theoret. Phys. (U.S.S.R.) **43**, 2230–2233 (December, 1962)

The ferromagnetism arising upon introduction of a small amount of magnetic atoms into a nonmagnetic metal is considered. It is shown that the objection raised in Yosida's paper [3] against explaining this effect by exchange interaction between the admixture atoms and conductivity electrons is unfounded. The main magnetic characteristics of impurity ferromagnets are determined.

MATTHIAS, Suhl, and Corenzwit [1] have observed that ferromagnetism exists in a non-magnetic metal in which a small amount of paramagnetic atoms is dissolved. The idea that this ferromagnetism may be the consequence of exchange interaction between the momenta of the impurity atoms and the conduction electrons was advanced in the literature many times [2]. However, this idea was discarded after the publication of the paper by Yosida [3], where it was stated that such an interaction did not lead to the appearance of homogeneous polarization of the electron spin. Instead, according to Yosida, the polarization of the electron spin takes place near each impurity atom, and this polarization oscillates rapidly and decreases in amplitude with increasing distance from the impurity atom.

Actually, Yosida's statement is in error. Although the decrease in the spin density does indeed take place, it is nevertheless not very fast [proportional to $(p_0R)^{-3} \cos p_0R - (p_0R)^{-4} \sin p_0R$, where R is the distance from the impurity atom]. In view of this, in considering the polarization of the electron spin at a given point, it is necessary to take into account the influence of all the impurity atoms, not only the nearest one. This leads to an entirely different deduction than obtained by Yosida.

Indeed, let us take the Yosida formula for the density of the number of electrons with different spin orientations:

$$\rho_{\pm} = \frac{n}{V} \pm \frac{3}{4}\frac{n}{\epsilon_F NV} \sum_{q} J(\mathbf{q}) f(\mathbf{q}) \sum_{n} \mathbf{S}_n^z \cos[\mathbf{q}(\mathbf{r} - \mathbf{R}_n)], \quad (1)$$

where 2n is the total electron number density, ϵ_F the Fermi limiting energy, N the number of crystal cells, $J(\mathbf{q})$ the Fourier component of the exchange potential, $\mathbf{S}_n$ the spin of the n-th atom of the impurity, and

$$f(\mathbf{q}) = 1 + \frac{4p_0^2 - q^2}{4p_0 q} \ln\left|\frac{2p_0 + q}{2p_0 - q}\right|$$

(p_0 is the limiting Fermi momentum).

We assume that the spin of the impurity atoms averaged over all atoms is different from zero, $\overline{\mathbf{S}}_n \neq 0$. Since we are interested in homogeneous polarization, we can immediately confine ourselves to an examination of small values of **q** in (1). We also assume that the impurity atoms are distributed randomly in the lattice. It follows therefore that that sum over the impurity atoms can be replaced by an integral over $\mathbf{R}_n$; as a result we obtain

$$\rho_{\pm} = \frac{n}{V}\left(1 \pm \frac{3}{2} c_i \frac{J(0)}{\varepsilon_F} \overline{S}_z\right), \quad (2)$$

where c_i is the number of impurity atoms per crystal cell. The impurity spin polarization thus gives rise to homogeneous polarization of the electron spin.

It is seen from the foregoing argument that Yosida's objections to attributing the impurity ferromagnetism to exchange interaction of the impurity spins and of the conduction electrons is unfounded. We investigate below the thermodynamic properties of such a model.

The interaction between the electrons and the impurity atoms can be written in the form

$$U = -\int \mathbf{s}(\mathbf{r}) \sum_i \mathbf{S}_i J(\mathbf{r} - \mathbf{R}_i)\, d^3\mathbf{r}, \quad (3)$$

where $\mathbf{S}_i$ and $\mathbf{R}_i$ are the spin and coordinate of the i-th impurity atom, $J(\mathbf{R})$ some interaction function, and $\mathbf{s}(\mathbf{r})$ the electron spin density at the point **r**. The interaction (3) can be regarded as the result of the action of some fictitious magnetic field on the electron spin. We assume that the electrons become "magnetized" in this field, i.e., at each point of space some average electron moment is produced. In view of the random distribution of the impurity atoms, which in the mean is homogeneous in space, we can assume the average density of the electron spin to be independent of the coordinate. In this case U can be rewritten in the form

$$U = - as \sum_i \mathbf{S}_i, \tag{4}$$

where the constant is $a = \int J(\mathbf{r})\, d^3\mathbf{r}$.

Formula (4) corresponds fully to the energy of the independent moments in the external field and it is possible to obtain from it in the usual fashion the impurity part of the free energy of the system (per unit volume):

$$F_i = - N_i T \left[\ln \mathrm{sh} \frac{as(S + {}^1/_2)}{T} - \ln \mathrm{sh} \frac{as}{2T} \right], \tag{5*}$$

where N_i is the number of impurity atoms per unit volume.

The addition to the electron free energy for a specified magnetization (i.e., s) can be written in the form

$$F_e = (2\mu_0 s)^2 / 2\chi_0, \tag{6}$$

where μ_0 is the Bohr magneton and χ_0 the paramagnetic susceptibility of the electrons in the absence of impurities.

The value of s is obtained by minimizing the free energy $F_i + F_e$. The equation obtained is transcendental

$$4\mu_0^2 s/\chi_0 = N_i a S B_S (asS/T). \tag{7}$$

Here B_S is the known Brillouin function. It can be shown that for this value of s the sum $F_i + F_e < 0$ over the entire range of the temperatures where $s \neq 0$.

Let us consider Eq. (7) in the high-temperature region $T \gg asS$. Here

$$B_S(\alpha) \approx \frac{\alpha}{3}\frac{S+1}{S} - \frac{\alpha^3}{45}\frac{(S+{}^1/_2)^4 - ({}^1/_2)^4}{S^4}.$$

Substituting this in (7) we obtain in first approximation

$$s\left(\frac{4\mu_0^2}{\chi_0} - \frac{N_i a^2 S(S+1)}{3T}\right) = 0.$$

This equation determines the Curie temperature, starting with which a nonvanishing value of s can appear:

$$T_c = \frac{N_i S(S+1) a^2 \chi_0}{12\mu_0^2}. \tag{8}$$

Thus, the Curie temperature is proportional to the impurity concentration.

The next approximation of (7) in asS/T gives the temperature variation of s near T_c:

$$s^2 = \frac{5}{4\mu_0^2}\frac{S(S+1) N_i \chi_0}{S^2 + S + {}^1/_2}(T_c - T). \tag{9}$$

In the opposite limiting case $T \to 0$, $B_S \to 1$ and we obtain

$$s = N_i a S \chi_0 / 4\mu_0^2. \tag{10}$$

The total moment comprises the electronic moment and the impurity moment, equal to $N_i \mu_0 g_i S B_S(asS/T)$, where g_i is the gyromagnetic ratio for the impurity atom. Thus, according to (7), the total magnetic moment per unit volume is

$$M = 2\mu_0 (1 + 2g_i \mu_0^2 / a\chi_0)\, s. \tag{11}$$

From (9) and (11) we see that near T_c we have $M \sim (T_c - T)^{1/2}$. When T = 0 we obtain the saturation moment

$$M = \mu_0 N_i S (g_i + a\chi_0 / 2\mu_0^2). \tag{12}$$

Depending on the sign of the constant a, which can be different in different cases, the moment of the impurity and the moment of the electrons can be parallel or antiparallel. Note also that in this model we have assumed that there is no noticeable coupling between the impurity spin and the lattice. This means that $g_i \approx 2$ and thus only the new constant a appears in all the formulas.

To take into account the influence of the external magnetic field we must add to the free energy a term $-2\mu_0 sH$, and in addition, replace as in the impurity term of F_i [Eq. (5)] by $as + g_i\mu_0 H$.

As a result we obtain in place of (7)

$$\frac{4\mu_0^2}{\chi_0} s - 2\mu_0 H = N_i a S B_S \left[\frac{(as + g_i \mu_0 H) S}{T}\right]. \tag{13}$$

From this, in particular, we obtain the paramagnetic susceptibility at $T > T_c$:

$$\chi = \chi_0 \left\{1 + \frac{T_c}{T - T_c}\left(1 + \frac{2\mu_0^2 g_i}{a\chi_0}\right)^2\right\}. \tag{14}$$

The formulas obtained make it also possible to obtain the additional specific heat. Of course, as in all cases of second-order phase transitions, these formulas do not fit the experimentally observed singularity near the transition point as $T \to T_c$.

*sh = sinh.

At temperatures $T \ll T_c$ formula (7) leads to an exponential temperature variation of all the quantities. This conclusion, apparently, cannot be correct for arbitrarily low temperatures, in view of the fact that the collective excitations of the spin systems are not taken into account here.

In conclusion it must be noted that in accordance with the experimental data the ferromagnetism arises not in all weak solutions of magnetic atoms in non-magnetic metals. For example in the case of a weak solution of Mn in Cu we obtain antiferromagnetism [4]. The nature of this phenomenon is unclear at present. The attempt by Overhauser [5] to construct a theory for the antiferromagnetism of weak solutions appears unconvincing to us.

[1] Mathias, Suhl, and Corenzwit, Phys. Rev. **115**, 1597 (1959). Phys. Rev. Lett. **1**, 44, 92 (1958).

[2] H. Frohlich and F. R. N. Nabarro, Proc. Roy. Soc. **A175**, 382 (1940). C. Zener, Phys. Rev. **81**, 440 (1951). Owen, Browne, Knight, and Kittel, Phys. Rev. **102**, 1501 (1956).

[3] K. Yosida, Phys. Rev. **106**, 893 (1957).

[4] Owen, Browne, Arp, and Kip, Phys. Chem. Solids **3**, 324 (1957).

[5] A. W. Overhauser, Phys. Rev. Lett. **3**, 414 (1959); Phys. Chem. Solids **13**, 71 (1960).

Translated by J. G. Adashko
382

5B. Superconductivity

5B.1 Applications of the microscopic theory

Collected in this Section are papers on several aspects of superconductivity.

5B.1.1 *Knight shift and spin-orbit interaction.* The Knight shift is a shift of the resonance frequency of the nuclei spins caused by the additional magnetic field produced on the part of metallic electrons magnetized in the applied field. In the fully gapped BCS superconductors, the magnetic susceptibility at low temperatures exponentially decreases to zero and it was expected that the value of the Knight shift at $T = 0$ must be zero too. Experimentally, the Knight shift at zero temperature in Ref. 1 was finite and in some cases large. Paper [5Ba] addresses this issue.

NMR experiments are usually performed on samples with sizes smaller than the penetration depth to achieve better homogeneity of the field inside the sample. Such samples are imperfect. In heavy enough metals the spin-orbit interaction (SO) is quite sizable and tends to destroy the anti-parallel alignment of spins of the two electrons in the Cooper pair as the latter scatter on imperfections. Our detailed calculations [5Ba] of the paramagnetic susceptibility in the superconducting phase have shown that, indeed, in the presence of SO interactions the susceptibility remains finite at $T = 0$.

Data on thin films of tin in Ref. 1 were in good agreement with predictions [5Ba]. Troubling were the observations of the large Knight shift in *light elements* such as V and Al in which SO is weak. This controversy was finally resolved only in 1967 after R. H. Hammond and G. M. Kelly (Ref. 2) have shown that the previous data for Al were incorrect and, in effect, the Knight shift in superconducting aluminum vanishes at $T = 0$.

As far as the paramagnetic susceptibility remains finite at low temperatures, this also suggests the explanation to experimental fact of the ferromagnetism of the paramagnetic centers in some *superconducting* alloys. In the normal state, the mechanism consists in the mutual magnetization of spins of the paramagnetic centers and of the electronic sub-system (See **5A.3** and Ref. 3). Previously, it was usually maintained in literature that the s-wave BCS superconductivity and ferromagnetism never coexist because of the mismatch between the superconducting gap and unequal populations for spins-up and spins-down on one and the same Fermi surface.

5B.1.2 *Paramagnetic limit.* Paramagnetic effects may impose severe limitations H_{para} on values of the critical magnetic field. Loosely speaking, the mechanism is that the Zeeman magnetic field destroys the BCS singlet state by breaking the singlet state of the pair if $\mu_B H_{para} = \Delta(0)$ (here μ_B is the Bohr magneton and $\Delta(0)$ is the BCS gap at $T = 0$). The argument is known as the Shandrasekhar-Clogston criterion (Ref. 4). In [5Bb] it was shown that the paramagnetic limit can be achieved in the so-called "hard" superconductors where the mean free path is short and of the atomic order.

5B.1.3 *Time-dependent GL.* The paper [5Bc] concerns the non-stationary processes in superconductors. Behavior of a superconductor in the presence of variable magnetic fields is sometimes discussed in the literature with help of the so-called "time-dependent" GL equations (TDGL). The simplicity of the stationary GL theory provokes adding of the *time derivative* into the equations for the order parameter; that is, to write for the GL function $\Psi(t;\vec{r})$ the equation: $\gamma \frac{d\Psi}{dt} = \alpha\Psi + \beta|\Psi|^2\Psi + \frac{1}{2m^*}(-i\hbar\vec{\partial} - \frac{e^*}{c}\vec{A})^2\Psi$, with γ as a new phenomenological parameter.

Conclusions of [5Bc] in that regard are negative, for it is shown that the TDGL equations never follow from the microscopic theory with the only exception of the extreme limit of the "gapless" superconductivity. In the general case the non-zero gap in the energy spectrum directly below T_c leads to the equations for non-stationary processes in a BCS-like superconductor that are *non-local* in the time variable, even in the vicinity of the transition temperature $|T - T_c| \ll T_c$.

In the first part of [5Bc] the general rules for the analytical continuation for responses of the system to arbitrary time-dependent perturbations (see Eqs. (2′,4)) are formulated. From here we derive, as stated above, TDGL equations for superconducting alloys with the paramagnetic impurities in the "gapless" regime. Notice conductivity of the *normal* phase in the dissipative component of the current. (Non-linear TDGL equations are solved in a few instances to demonstrate the complexity of the behavior of a superconductor in variable fields).

5B.1.4 *Vortex moving in electric fields.* TDGL equations of the previous paper were applied to the motion of the vortices in electrical fields [5Bd]. In my way of thinking, there is a beauty in the paper to the extent that it presents the exact solution of the hydrodynamic problem of flow of the superfluid current past an isolated vortex line.

On the phenomenological level, Bardeen and Stephen (Ref. 5) treated viscosity at motion of the vortices as due to dissipation by the normal excitations *inside* the vortex cores. In solving the microscopic equations for the alloys with paramagnetic impurities, we were dealing with somewhat simpler physics because the gapless limit dissipation has the same origin as in the normal phase.

In our formulation of the problem, given is the current. The vortex moves under the action of the Magnus force. That is, the electric charge drops out. The moving vortices have the velocity component perpendicular to the current direction. Since each single vortex carries the unit of the magnetic flux, at crossing lines of the current an electric field is generated.

Neither TDGL equations nor the phenomenological models can fully account for all details of the relaxation mechanisms for the motion of vortices in alloys. For instance, in the ordinary alloys the relaxation becomes especially slow near T_c resulting in strong increase of the viscosity [5Be].

5B.2 "Exotic" superconductivity

What was called "exotic" at one time became the mainstream of the current research in superconductivity and I would like at first to briefly bring a few main steps of the story to recollections.

The field started as far back as the 1960s with the so-called low temperature resistivity minimum in alloys containing paramagnetic centers. In 1964, Kondo revealed a $\ln T$-term in the expression of the scattering matrix element in the resistivity of the third order in the exchange interaction $J(\vec{S} \cdot \vec{\sigma})$ between spins of electrons and the magnetic center.

The $\ln T$-contribution accounts (at the proper sign of the exchange) for the resistivity increase at temperatures below the temperature of the minimum, but the singular character of the $\ln T$-dependence has signaled a non-trivial physics there. In loose terms, the phenomenon could be termed as the screening of the local magnetic moments by the conduction electrons at low temperatures.

The problem became known as the "Kondo problem." Sharing the spin degrees between the localized spin and the conduction electrons results in the high local density of states (DOS) on the center proportional to $1/T_K$ where the Kondo temperature $T_K \propto J exp(-const \cdot J/E_F)$ is small at $J \ll E_F$. (The exact Bethe-*Ansatz* solution of the Kondo problem was obtained in 1980 independently by P. B. Wiegmann and N. Andrei for several models (Ref. 6)).

From one isolated center the Kondo problem evolved into the notion of the "Kondo lattice." Most related to the "exotic" theme are so-called "heavy fermions", the metallic compounds containing ions with the unfilled $4f$- or $5f$-orbits. Here low T_K (and, hence, a high local DOS) for a single Kondo center intertwines with narrowness of the conduction band formed by the weakly overlapping f-wave functions on the neighboring centers. As the result, DOS on the Fermi level can be extremely high. Indeed, the values of the Sommerfeld coefficient in linear electronic specific heat reach equivalent of the effective masses of the order of $m^* \sim (100-1000)m_e$. Masses of same order of magnitude follow from data on quantum oscillation as, for example, in the de Haas–van Alphen effect (dHvA). Hence the name: heavy fermions (HF).

In 1979 the Steglich group (Ref. 7) discovered superconductivity of CeCu2Si2. It seemed next to impossible to reconcile the fact with the BCS phonon mechanism, for interactions with the concentrated magnetic f-moments in the lattice should be detrimental for the s-wave pairing.

Observation in 1984 of the T^3-law for the electronic specific heat of superconductor $UBe_{13}, T_c = 0.85K$ (Ref. 8) immediately prompted speculations whether properties of the superconducting phase in UBe_{13} may have something in common with the A-phase of the superfluid helium-three; in the latter, as it is known, the energy gap in the spectrum has zeroes on the Fermi surface. The Cooper pairing in ^3He takes place with the angular momentum $l = 1$, but ^3He is neutral and has the isotropic spectrum, in contrast with the anisotropic liquid of HFs in the cubic UBe_{13}.

The question then is how to characterize pairing in the crystalline systems. The lattice is invariant with respect to transformations of the total symmetry group of a crystal; it is enough to confine oneself to the *point group* of rotations and reflections because periodicity of the lattice does not change in the superconducting transition. In short, in place of the angular momentum in an isotropic system, in a crystal the superconducting pairing can be classified in accordance with the irreducible representations of the point group.

The procedure is reduced to finding the *superconducting* classes. That is, instead of enumerating how the order parameter *changes* at one or other symmetry transformations; the classes are built of the symmetry elements of the total group as *subgroups* of the transformations with respect to which the order parameter is *invariant*.

In its most general form, the superconducting order parameter is given by the "gaps" $\hat{\Delta}_{\alpha\beta}(\vec{k})$ and $\hat{\Delta}^+_{\alpha\beta}(\vec{k})$; thus, we have $\hat{\Delta}_{\alpha\beta}(\vec{k}) \equiv -\sum_{\vec{k}',\omega} V_{\beta\alpha,\mu\lambda}(\vec{k},\vec{k}')\hat{F}_{\lambda\mu}(\vec{k}';\omega)$. Here $\hat{V}$ is the interaction and $\hat{F}$ the anomalous Gorkov Green function, both written in the general form with the spin indices present. As it admits the multiplication by a phase factor $e^{i\varphi}$, the order parameter breaks the $U(1)$ symmetry. From this it follows that as the total symmetry group of the *normal* phase must be taken the direct product $G \times R \times U(1)$(G–the point group; R the time reversal ($t \to -t$)). The procedure of constructing the superconducting classes for all the point groups G is described in [5Bg]. Properties of UBe_{13} were analyzed first in [5Bf] (independently, in Ref. 9).

5B.2.1 *Few remarks and summary of the main results.*

1) In HF the spin-orbit interactions are strong. Spins being tightly bound to the lattice rotate with the coordinates at the symmetry transformations.

2) For HF, all point symmetry groups possess the center of inversion. Although the weak coupling classification in terms of the singlet or the triplet pairing states is not applicable, classification is possible according to the parity; it seemed convenient to preserve the symbols $S = 0$ for odd parity and $S = 1$ for even parity.

3) The gap in the spectrum may have zeroes on the Fermi surface. Considering that, if $S = 0$, zeroes may occur on the *lines* of symmetry, thus contributing the T^2 term into the specific heat; at $S = 1$ the gap may have zeroes only at the symmetry *points*. That leads to the T^3 dependence for the specific heat at low temperatures.

4) For few high symmetry phases the superconductor may possess the magnetic moment and, hence, reveal the domain structure.

5) Classification in [5Bg] applies only to the phases that grow immediately below the temperature of the second order phase transition T_c. At further lowering of the temperature the first order phase transitions are possible.

5B.2.2 *Anisotropy of upper critical field.* For the ordinary GL theory anisotropy of the upper critical field is fixed by anisotropy of the lattice. In "exotic" superconductors the upper field H_{c2} at T_c can be anisotropic in the planes perpendicular to the high symmetry directions for the phases built upon two- and three-dimensional representations of the cubic group and two-dimensional representations in the tetragonal case [5Bh]. It would be of the especial interest to observe this effect in UBe_{13} for which $H_{c2}(0)$ is very high and the T^3-contribution in the low temperature specific heat suggest strong arguments in favor of one of the degenerate representations. However, the GL-interval $\Delta T = |T - T_c| \ll T_c$ in UBe_{13} seems to be too narrow for doing the experiment.

5B.2.3 *Stability to presence of defects.* In a clean $S = 0$ superconductor (i.e. at the "singlet" pairing) zeroes along the symmetry lines on the Fermi surface are unstable due to the presence of defects: small but non-zero density of states develop on the *lines* of zeros at *arbitrarily small* concentrations of the ordinary impurities, thus, leading to "gapless" superconductivity at a higher defects' concentration [5Bi].

5B.2.4 *Exotic phases without zeroes?* Both papers [5Bj, k] are relevant only to the multi-band superconductors.

Consider several Fermi surfaces centered each at the *non-equivalent* points of the Brillouin Zone. At transformations of the symmetry group of the crystal superconducting gaps on the different Fermi surfaces transform through each other in accordance to one or other representations of the group. In particular, it can be a non-identical representation that coincides with the superconducting class corresponding to one of the exotic phases discussed in [5Bg], although gaps on each of the FS are non-zero. Even the phonon mechanisms can realize such scenario. (Of course, if there were a Fermi surface centered at the Γ-point, the gap on *this* FS must possess zeroes. Similar situations are possible also in quasi-two-dimensional Fe-based superconductors).

Materials with multiple Fermi surfaces are indeed well known and few examples are listed in the papers. Stability of the "exotic" solution with respect to the ordinary impurities was investigated [5Bk]; "exotic" solutions are sensitive mainly to inter-band scattering.

References

1. G. M. Androes and W. D. Knight, *Phys. Rev.* **121**, 779 (1961)
2. R. H. Hammond and G. M. Kelly, *Phys. Rev. Lett.* **18**, 156 (1967)
3. L. P. Gor'kov and A. I. Rusinov, *Sov. Phys. JETP* **19**, 922 (1964)
4. A. M. Clogston, *Phys. Rev. Lett.* 9, 266 (1962); B. S. Shandrasekhar, *Appl. Phys. Lett.* **1**, 7 (1962)
5. J. Bardeen and M. J. Stephen, *Phys. Rev.* **140**, A1197 (1965)
6. N. Andrei, *Phys. Rev. Lett. Rev.* **45**, 379 (1980); P. B. Wiegmann, *Sov. Phys. JETP Lett.* **31**, 392 (1980)
7. F. Steglich *et al.*, *Phys. Rev. Lett.* **43**, 1892 (1979)
8. H. R. Ott *et al.*, *Phys. Rev. Lett.* **52**, 1915 (1984)
9. P. W. Anderson, *Phys. Rev. B* 30, 1549 (1984)

List of original papers

[5Ba] A. A. Abrikosov and L. P. Gor'kov, *Spin-orbit interaction and the Knight shift in superconductors*, Sov. Phys. JETP, 15, 752 (1962)

[5Bb] L. P. Gor'kov, *An estimate of the limiting values of the critical fields for hard superconductors*, Sov. Phys. JETP, 17, 518 (1963)

[5Bc] L. P. Gor'kov and G. M. Eliashberg, *Generalization of the Ginsburg-Landau equations for non-stationary problems in the case of alloys with paramagnetic impurities*, Sov. Phys. JETP, 27, 328 (1968)

[5Bd] L. P. Gor'kov and N. B. Kopnin, *Viscous vortex flows in superconductors with paramagnetic impurities*, Sov. Phys. JETP **33**, 1251 (1971)

[5Be] L. P. Gor'kov and N. B. Kopnin, *Some feature of viscous flow of vortices in superconducting alloys near critical temperature,* Sov. Phys. JETP, **37**, 183 (1973)

[5Bf] G. E. Volovik and L. P. Gor'kov, *An unusual superconductivity in* UBe_{13}, Sov. Phys. JETP Lett. **39**, 674 (1984)

[5Bg] G. E. Volovik and L. P. Gor'kov, *Superconducting classes in heavy-fermions systems*, Sov. Phys. JETP **61**, 843 (1985)

[5Bh] L. P. Gor'kov, *Anisotropy of the upper critical field in exotic superconductors*, JETP Lett. **40**, 1155 (1985)

[5Bi] L. P. Gor'kov and P. A. Kalugin, *Defects and unusual superconductivity*, JETP Lett. **41**, 253 (1985)

[5Bj] D. F. Agterberg, V. V. Barzykin and L. P. Gor'kov, *Conventional mechanisms for exotic superconductors*, Phys. Rev. B **60**, 14868 (1999)

[5Bk] D. F. Agterberg, V. V. Barzykin and L. P. Gor'kov, *Exotic ground states and impurities in multi-band superconductors*, Europhys. Lett. **48**, 449 (1999)

SOVIET PHYSICS JETP VOLUME 15, NUMBER 4 OCTOBER, 1962

SPIN-ORBIT INTERACTION AND THE KNIGHT SHIFT IN SUPERCONDUCTORS

A. A. ABRIKOSOV and L. P. GOR'KOV

Institute for Physical Problems, Academy of Sciences, U.S.S.R.

Submitted to JETP editor November 4, 1961

J. Exptl. Theoret. Phys. (U.S.S.R.) **42**, 1088-1096 (April, 1962)

A theory of the Knight shift in superconductors is presented. It is shown that the effect can be completely explained if the spin-orbit part of the interaction in the scattering of electrons on the crystal boundaries is taken into account. The agreement between the theory and the experimental data [8] is found to be good.

1. INTRODUCTION

MANY recent papers [1-7] have been devoted to an explanation of the nonvanishing shift in the frequency of the nuclear magnetic resonance in a superconductor at zero absolute temperature. As is known, this so-called Knight shift is proportional to the paramagnetic susceptibility of the conduction electrons, χ:

$$\Delta\omega/\omega = (8\pi/3N_{at})\,|\psi(0)|^2\,\chi$$

where $|\psi(0)|^2$ is the probability density for finding an electron at the position of the nucleus and N_{at} is the number of atoms in a unit volume.

According to the microscopic theory of superconductivity, the paramagnetic susceptibility of the electrons in the superconducting state vanishes for $T = 0$. This is in contradiction with the nonvanishing Knight shift observed experimentally. The vanishing of χ for $T = 0$ is connected with the fact that the states of a superconductor are classified according to the eigenvalues of the total spin. In the ground state all electrons are paired in Cooper pairs, and the total spin of the system is zero. Excitation of the system requires a finite energy 2Δ, where Δ is the gap in the energy spectrum of the superconductor.

The most natural explanation of the finite Knight shift in superconductors is based on the assumption of Ferrell [5] and Anderson [6] that the spin-orbit interaction plays an important role for electrons in small specimens. Owing to this interaction the electron can change its spin direction when it is scattered at the boundary. The states of such a system can therefore no longer be characterized by the eigenvalues of the spin, and the above-mentioned conclusion is not correct.

A finite shift at $T = 0$ in tin has definitely been observed by Androes and Knight in recent work. [8] The earlier results of Reif [9] on the Knight shift in colloidal mercury are difficult to interpret, since, as noted by the authors, [4] the uncertainty in the size of the particles was so large that an appreciable number of large particles may have gone over into the normal state in fields of the order of a few kilooersted thus simulating a finite shift in the superconducting state.

The experiments of Androes and Knight [8] were performed with stacks of thin films with thickness of 40 Å. The fields used in this work (from 1 to 8 kilooersted) were weaker than the critical fields, which reach a strength of 25 kilooersted. The ratio χ_s/χ_n, extrapolated to $T = 0$, was equal to ~ 0.77.

In the present paper we shall show that the above-mentioned spin-orbit interaction provides a correct quantitative explanation of the phenomenon. For the mathematical formulation of the problem it is important to take into consideration that the thin films as well as all small-sized specimens, have a polycrystalline structure. At the boundaries of the crystallites the electrons are scattered, which leads to a destruction of the correlations between a pair of electrons which were introduced in the transition to the superconducting state. These correlations are thus restricted to distances of the order of the dimensions of the crystallites. The dimensions of the crystallites, therefore, play the role of a mean free path for the electrons. Hence we shall below consider the equivalent problem of the scattering of an electron by impurities in the superconductor, where spin flip can occur on account of the spin-orbit interaction. As will be shown below, the final result will involve only the mean free path of the electrons with respect to spin flip, $l_{s.o.}$.

The amplitude for the scattering of an electron by an atom of the impurity can be written in the form

$$\hat{f}(\mathbf{p}, \mathbf{p}') = a(\mathbf{p}, \mathbf{p}') + ib(\mathbf{p}, \mathbf{p}')\, p_0^{-2}([\mathbf{p}, \mathbf{p}']\, \hat{\sigma}), \qquad (1)^*$$

where $\mathbf{p}$ and $\mathbf{p}'$ are the momenta of the electron before and after the collision, $\hat{\sigma}$ are the Pauli spin matrices, and p_0 is the Fermi momentum. The second term in (1), corresponding to the spin-orbit interaction, is of order $b/a \sim (Ze^2/\hbar c)^2$ and may, in general, be appreciable for heavy elements. To keep the calculations simple, we shall, however, assume that $|b/a|^2 \ll 1$. In other words, we shall assume that the mean free path $l_{s.o.}$ is much larger than the mean free path between collisions of the electrons with the impurities, l, which is determined essentially by the amplitude $a(\mathbf{p}, \mathbf{p}')$ in (1). It was shown in [4] that the scattering does not change the paramagnetic susceptibility of the superconductor in the absence of the spin-orbit interaction, so that there will be a Knight shift of the order of unity at T = 0 if

$$l_{s.o.} \sim \xi_0 \sim \hbar v/T_c \sim 10^{-4} - 10^{-5} \text{ cm}.$$

2. CALCULATION OF THE PARAMAGNETIC SUSCEPTIBILITY

The spin magnetic moment of a system of electrons in a uniform magnetic field is equal to [4]

$$\mathbf{M} = -\mu_0^2 \lim_{\substack{\tau' \to \tau + 0 \\ \mathbf{r}' \to \mathbf{r}}} \int_0^{1/T} d\tau_y \int d^3y \, \mathrm{Sp}\Big[\exp\Big(\frac{\Omega + \mu N - H_0}{T}\Big)$$

$$\times\, T_\tau (\psi_\alpha(x)\, \psi_\gamma^+(y)\, \psi_\delta(y)\, \psi_\beta^+(x'))\, \sigma_{\beta\alpha}(\sigma_{\gamma\delta}\mathbf{H})\Big]. \qquad (2)$$

Here μ_0 is the Bohr magneton. In the following we shall use the thermodynamic formulation [10,11] of perturbation theory, in which the field operators depend on the time parameter τ $(0 < \tau < 1/T)$.

Denoting the average over the Gibbs ensemble by $\langle \ldots \rangle$, we consider the average over the product of four operators appearing in (2):

$$\langle T(\psi_\alpha(x)\, \psi_\gamma^+(y)\psi_\delta(y)\, \psi_\beta^+(x'))\rangle.$$

In a superconductor this average can be written in the form

$$\langle T(\psi_\alpha(x)\, \psi_\gamma^+(y)\, \psi_\delta(y)\, \psi_\beta^+(x'))\rangle$$

$$= \mathfrak{G}_{\alpha\gamma}(x, y)\, \mathfrak{G}_{\delta\beta}(y, x') - \mathfrak{F}_{\alpha\delta}(x, y)\, \mathfrak{F}_{\gamma\beta}^+(y, x'), \qquad (3)$$

where

$$\mathfrak{G}_{\alpha\beta}(x, y) = -\langle T(\psi_\alpha(x)\, \psi_\beta^+(y))\rangle;$$

$$\mathfrak{F}_{\alpha\beta}(x, y) = \langle T(\psi_\alpha(x)\, \psi_\beta(y))\rangle;$$

$$\mathfrak{F}_{\alpha\beta}^+(x, y) = \langle T(\psi_\alpha^+(x)\, \psi_\beta^+(y))\rangle.$$

The diagram corresponding to expression (3) is shown in Fig. 1. The line with both arrows pointing in the same direction represents the function $\mathfrak{G}$ and the lines with the arrows pointing both either in or out represent $\mathfrak{F}$ and $\mathfrak{F}^+$.

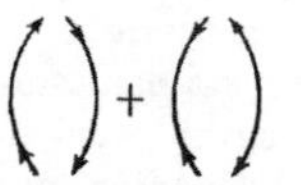

FIG. 1

In our case all quantities in (2) and (3) contain the interaction with the atoms of the impurities. In zeroth order in this interaction we must substitute the known expressions for the Green's functions of the pure superconductor in (3):

$$\mathfrak{G}_{\alpha\beta}(x - y) = \delta_{\alpha\beta}\, T\, (2\pi)^{-3} \sum_\omega \int d^3\mathbf{p}\, \mathfrak{G}(p) \exp\{i\mathbf{p}(\mathbf{x} - \mathbf{y})$$

$$- i\omega(\tau_x - \tau_y)\},$$

$$\mathfrak{F}_{\alpha\beta}^+(x - y) = -\mathfrak{F}_{\alpha\beta}(x - y) = \hat{g}_{\alpha\beta}\, T\, (2\pi)^{-3} \sum_\omega \int d^3\mathbf{p}\, \mathfrak{F}(p)$$

$$\times \exp\{i\mathbf{p}(\mathbf{x} - \mathbf{y}) - i\omega(\tau_x - \tau_y)\}, \qquad (4)$$

where

$$\mathfrak{G}(p) = -\frac{i\omega + \xi}{\omega^2 + \xi^2 + \Delta^2}, \qquad \mathfrak{F}(p) = \frac{\Delta}{\omega^2 + \xi^2 + \Delta^2}, \qquad (4')$$

g is the spin metric tensor ($g_{11} = g_{22} = 0$, $g_{12} = -g_{21} = 1$), Δ is the energy gap in the spectrum of the pure superconductor, and $\xi = v(p - p_0)$. The frequencies ω appearing in the Fourier series in τ take the values $\omega = (2n + 1)\pi T$ $(n = 0, \pm 1, \ldots)$, and $\sum_\omega \varphi(\omega)$ means $\sum_n \varphi(2n + 1)\pi T$.

Taking account of the interaction with the impurities changes expressions (4) and (4′). In Fig. 2 we show diagrams for expression (3) in which the crosses indicate the act of scattering of the electron with the separate atoms of the impurity. The scattering of the electron by an atom located at the point $\mathbf{r}_a$ gives rise to a factor $f_{\alpha\beta}(\mathbf{p}, \mathbf{p}') \times \exp\{i(\mathbf{p}' - \mathbf{p}) \cdot \mathbf{r}_a\}$ in the matrix element in the momentum representation.

FIG. 2

$^*([\mathbf{p}, \mathbf{p}']\, \hat{\sigma}) = [(\mathbf{p} \times \mathbf{p}') \cdot \hat{\sigma}]$

In the following we shall be interested in the values of all quantities, including (2), averaged over the positions of all atoms of the impurity. The relevant momenta in our problem are all of the order $p \sim p_0 \sim 1/a$, where a is the interatomic distance. Therefore the average of the factor $\exp\{i(\mathbf{p}'-\mathbf{p})\cdot\mathbf{r}_a\}$ over $\mathbf{r}_a$ is zero and so is every matrix element unless it contains yet another factor $\exp\{-i(\mathbf{p}'-\mathbf{p})\cdot\mathbf{r}\ \}$ corresponding to the scattering on the same atom of the impurity with the initial and final states interchanged. In taking the average we must therefore sum over pairs of scatterings on the same atom.

The whole averaging procedure can be illustrated by a special diagram technique which has been developed by the authors.[12] The pairwise averaging is represented by a dotted line in the diagram, which joins two identical crosses (scattering by the same atoms). In the matrix element the dotted line corresponds to the factor

$$n(2\pi)^{-3} f_{\alpha\beta}(\mathbf{p},\mathbf{p}')\, f_{\gamma\delta}(\mathbf{p}',\mathbf{p})$$

where n is the number density of the atoms of the impurity. With an accuracy up to small terms of order a/l it is sufficient to consider only nonintersecting dotted lines.

Let us introduce the notation

$$\Pi^{(1)}_{\alpha\beta}(x-y,\ y-x') = \overline{\langle T(\psi_\alpha(x)\,\psi^+_\delta(y)\,\sigma_{\delta\lambda}\psi_\lambda(y)\,\psi^+_\beta(x'))\rangle}. \tag{5}$$

(The bar denotes the average over the positions of the atoms of the impurity). Let us write

$$\Pi^{(1)}_{\alpha\beta}(x-y,\ y-x') = T^2(2\pi)^{-6}\sum_{\omega_+\omega_-}\iint d^3\mathbf{p}_+\, d^3\mathbf{p}_-\, \Pi^{(1)}_{\alpha\beta}(p_+,\ p_-)$$
$$\times \exp\{ip_+(\mathbf{x}-\mathbf{y}) + ip_-(\mathbf{y}-\mathbf{x}') - i\omega_+(\tau_x-\tau_y) - i\omega_-(\tau_y-\tau_{x'})\}. \tag{5'}$$

The magnetic moment is given in terms of $\Pi^{(1)}_{\alpha\beta}$ by the relation

$$\mathbf{M} = -\mu_0^2\sigma_{\alpha\beta}T\,(2\pi)^{-3}\sum_\omega\int(\mathbf{H}\Pi^{(1)}_{\alpha\beta}(p,\ p))\,d^3\mathbf{p}. \tag{6}$$

Some lowest order diagrams in the perturbation expansion of (5) are shown schematically (without indicating the direction of the arrows) in Fig. 3. In accordance with the general recipe we leave out all diagrams with intersecting dotted lines, as well as diagrams in which the dotted line goes into a vertex with a large ($\sim p_0$) momentum transfer. In these diagrams part of the integration is over momenta far away from the Fermi surface, which reduces the contribution from these diagrams by a factor of order $1/p_0 l \sim a/l \ll 1$ as compared to diagrams of the "ladder" type. In the remaining

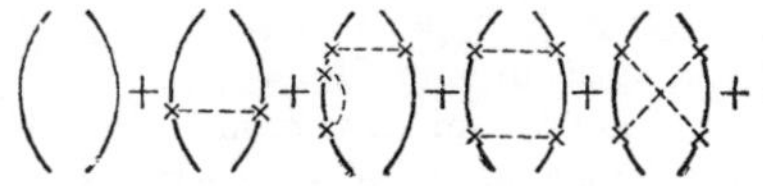

FIG. 3

diagrams the dotted line either joins two different electron lines or begins and ends on the same line (but without intersecting the "steps" of the ladder). In the latter case all such dotted lines evidently represent corrections from the interaction with the impurities to the Green's functions themselves. In order to derive an equation for $\Pi^{(1)}_{\alpha\beta}(p_+,\ p_-)$ it is therefore sufficient to consider only the "ladder" diagrams in which the dotted lines join different electron lines. The different sections of the electron lines in these diagrams represent the exact (averaged over the positions of the impurities) Green's functions $\mathfrak{G}$, $\mathfrak{F}$, and $\mathfrak{F}^+$ in the matrix element.

The equations for the Green's functions themselves in a superconductor with impurities coincide in this case exactly with the equations of [12]. Solving these equations, we find again that the averaged Green's functions for a superconductor with impurities can be obtained from (4) and (4') by the replacement

$$\{\omega,\ \Delta\} \to \{\omega\eta,\ \Delta\eta\}; \tag{7}$$

$$\eta = 1 + 1/2\tau\sqrt{\omega^2+\Delta^2}, \tag{8}$$

$$\frac{1}{\tau} = \frac{nmp_0}{4\pi^2}\int d\Omega\,(|a|^2 + |b|^2\sin^2\theta) = \frac{1}{\tau_0} + \frac{1}{\tau_1}. \tag{9}$$

Here we differ from [12] only in that $1/\tau$ now contains a contribution from the part of the scattering amplitude arising from the spin-orbit interaction:

$$\frac{1}{\tau_1} = \frac{nmp_0}{4\pi^2}\int|b|^2\sin^2\theta\, d\Omega. \tag{10}$$

As was shown in [12], formulas (4'), (7) to (9) imply that in the coordinate representation, the Green's function for a pure superconductor is to be multiplied by the factor

$$\exp\{-|r-r'|/2l\},$$

to account for the contamination by impurities, where $l = v\tau$ is the mean free path.* It follows from this that the introduction of impurities does not affect the thermodynamics of a superconductor, and, in particular, the transition temperature at which the system goes into the superconducting state remains unchanged. In the following all expressions will contain only the Green's functions for a superconductor with impurities, for which

*We note that $l_{s.o.} \sim v\tau_1$.

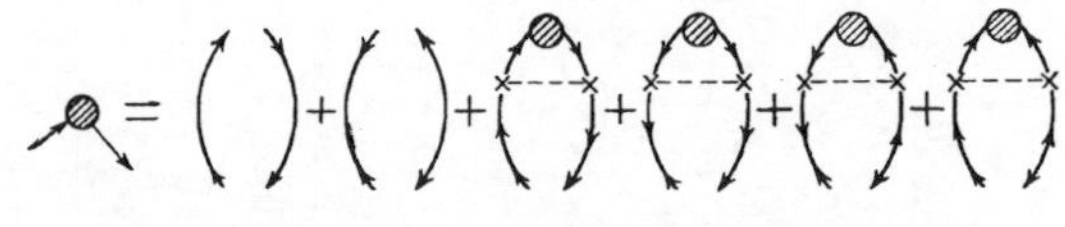

FIG. 4

we can, therefore, use the old symbols $\mathfrak{G}$, $\mathfrak{F}$, and $\mathfrak{F}^+$.

Figure 4 shows a graphical representation of the equation for $\Pi^{(1)}_{\alpha\beta}(p_+, p_-)$. We see that the determination of $\Pi^{(1)}_{\alpha\beta}(p_+, p_-)$ requires, in general, the knowledge of three new quantities which differ graphically from one another by the directions of the arrows. We define these quantities as Fourier transforms of the following expressions:

$$\Pi^{(2)}_{\alpha\beta}(x-y, y-x') = \overline{\langle T((\hat{g}\psi^+(x))_\alpha \psi^+_\delta(y)\sigma_{\delta\lambda}\psi_\lambda(y)\psi^+_\beta(x'))\rangle},$$

$$\Pi^{(3)}_{\alpha\beta}(x-y, y-x') = \overline{\langle T((\hat{g}\psi^+(x))_\alpha \psi^+_\delta(y)\sigma_{\delta\lambda}\psi_\lambda(y)(\psi(x')\hat{g})_\beta)\rangle},$$

$$\Pi^{(4)}_{\alpha\beta}(x-y, y-x') = \overline{\langle T(\psi_\alpha(x)\psi^-_\delta(y)\sigma_{\delta\lambda}\psi_\lambda(y)(\psi^+(x')\hat{g})_\beta)\rangle}.$$

Expression (6) involves only the component $\Pi^{(1)}_{\alpha\beta}(p, p)$. Therefore we shall not write down the general equations for $\Pi^{(i)}_{\alpha\beta}(p_+, p_-)$, but remark only that for $p_+ = p_- = p$ the equations imply

$$\Pi^{(3)}_{\alpha\beta}(p, p) = -\Pi^{(1)}_{\alpha\beta}(p, p), \qquad \Pi^{(4)}_{\alpha\beta}(p, p) = \Pi^{(2)}_{\alpha\beta}(p, p).$$

Using these results, we can obtain from the equation Fig. 4 the following expression for $\Pi^{(1)}_{\alpha\beta}(p, p)$:

$$\Pi^{(1)}_{\alpha\beta}(p, p) = [\mathfrak{G}^2(p) + \mathfrak{F}^2(p)](\sigma_{\alpha\beta} + \Lambda^{(1)}_{\alpha\beta}(p)) + 2\mathfrak{F}(p)\mathfrak{G}(p)\Lambda^{(2)}_{\alpha\beta}(p), \tag{11}$$

where we have introduced the notation

$$\Lambda^{(1)}_{\alpha\beta}(p) = \frac{n}{(2\pi)^3}\int \hat{f}_{\alpha\delta}(\mathbf{p}, \mathbf{p}')\,\Pi^{(1)}_{\delta\lambda}(p', p')\,\hat{f}_{\lambda\beta}(\mathbf{p}', \mathbf{p})\,d^3\mathbf{p}',$$

$$\Lambda^{(2)}_{\alpha\beta}(p) = \frac{n}{(2\pi)^3}\int \hat{f}_{\alpha\delta}(\mathbf{p}, \mathbf{p}')\,\Pi^{(2)}_{\delta\lambda}(p', p')\,\hat{f}_{\lambda\beta}(\mathbf{p}', \mathbf{p})\,d^3\mathbf{p}'. \tag{12}$$

The analogous expression for $\Pi^{(2)}_{\alpha\beta}$ is

$$\Pi^{(2)}_{\alpha\beta}(p, p) = -[\mathfrak{G}(p)\mathfrak{F}(p) - \mathfrak{F}(p)\mathfrak{G}(-p)](\sigma_{\alpha\beta} + \Lambda^{(1)}_{\alpha\beta}(p)) - (\mathfrak{F}^2(p) - \mathfrak{G}(p)\mathfrak{G}(-p))\Lambda^{(2)}_{\alpha\beta}(p). \tag{13}$$

The vector $\Lambda^{(i)}$ can be directed either along σ or $\mathbf{p}$. Accordingly it is convenient to write $\Lambda^{(i)}$ as the sum of two parts:

$$\Lambda^{(i)}_{\alpha\beta}(p) = \sigma_{\alpha\beta}\Lambda^{(i)}_\sigma(p) + (\sigma\mathbf{p})_{\alpha\beta}\mathbf{p}p_0^{-2}\Lambda^{(i)}_p(p). \tag{14}$$

Substituting (11) and (13) in (12), we obtain a system of four equations for the quantities $\Lambda^{(1,2)}_\sigma(p)$ and $\Lambda^{(1,2)}_p(p)$. Using the expressions (4′) and (7) for the Green's functions, we can easily show that all the quantities $\Lambda^{(1,2)}_{(\sigma,p)}(p)$ depend only on the fourth component of p, i.e., on the frequency ω. Owing to this circumstance the system of equations (11) to (14) can be reduced to algebraic form:

$$\left[(\omega^2+\Delta^2)^{3/2}\eta - \frac{\Delta^2}{2\tau_0}\right](1+\Lambda^{(1)}_\sigma) + \frac{i\Delta\omega}{2\tau_0} = (\omega^2+\Delta^2)^{3/2}\eta + \frac{1}{2\tau_0'}(\Delta^2\Lambda^{(1)}_p - i\Delta\omega\Lambda^{(2)}_p),$$

$$\left[(\omega^2+\Delta^2)^{3/2}\eta - \frac{\omega^2}{2\tau_0}\right]\Lambda^{(2)}_\sigma - \frac{i\Delta\omega}{2\tau_0}(1+\Lambda^{(1)}_\sigma) = \frac{1}{2\tau_0'}(i\Delta\omega\Lambda^{(1)}_p + \omega^2\Lambda^{(2)}_p),$$

$$\left[(\omega^2+\Delta^2)^{3/2}\eta - \frac{\Delta^2}{2\tau_0''}\right]\Lambda^{(1)}_p + \frac{i\Delta\omega}{2\tau_0''}\Lambda^{(2)}_p = -\frac{\Delta^2}{2\tau_1}(1+\Lambda^{(1)}_\sigma) + \frac{i\Delta\omega}{2\tau_1}\Lambda^{(2)}_\sigma,$$

$$\left[(\omega^2+\Delta^2)^{3/2}\eta - \frac{\omega^2}{2\tau_0''}\right]\Lambda^{(2)}_p - \frac{i\Delta\omega}{2\tau_0''}\Lambda^{(1)}_p = -\frac{i\Delta\omega}{2\tau_1}(1+\Lambda^{(1)}_\sigma) - \frac{\omega^2}{2\tau_1}\Lambda^{(2)}_\sigma. \tag{15}$$

Here

$$\frac{1}{\tau_0''} = \frac{1}{\tau_0} - \frac{3}{\tau_0'}; \qquad \frac{1}{\tau_0'} = \frac{nmp_0}{8\pi^2}\int |a|^2 \sin^2\theta\, d\Omega.$$

The last two equations are convenient for expressing $\Lambda^{(1)}_p$ and $\Lambda^{(2)}_p$ in terms of $\Lambda^{(1)}_\sigma$ and $\Lambda^{(2)}_\sigma$:

$$\Lambda^{(1)}_p = -\frac{1}{3}\frac{\tau_0'}{\tau_1}\frac{\Delta^2(1+\Lambda^{(1)}_\sigma) - i\Delta\omega\Lambda^{(2)}_\sigma}{\omega^2+\Delta^2},$$

$$\Lambda^{(2)}_p = -\frac{1}{3}\frac{\tau_0'}{\tau_1}\frac{\omega^2(1+\Lambda^{(1)}_\sigma) + i\Delta\omega\Lambda^{(2)}_\sigma}{\omega^2+\Delta^2}. \tag{16}$$

Since $\tau_0'/\tau_1 \ll 1$, we have, in the derivation of the equations above, neglected interference terms and terms quadratic in the amplitude of the spin-orbit interaction in all expressions containing $\Lambda^{(1)}_p$ and $\Lambda^{(2)}_p$. The last-mentioned terms appear as a result of the multiplication of two scattering amplitudes (1) in the definition of the quantities $\Lambda^{(i)}$ [formula (12)].

With the help of (16) we find from the first two equations (15)

$$1+\Lambda^{(1)}_\sigma = \frac{(\omega^2+\Delta^2)^{3/2}\eta - \omega^2(1/\tau_0 - 1/3\tau_1)/2}{(\omega^2+\Delta^2)[(\omega^2+\Delta^2)^{1/2}\eta - 1/2\tau_0 + 1/6\tau_1]},$$

$$\Lambda^{(2)}_\sigma = \frac{i\Delta\omega}{2}\frac{1/\tau_0 - 1/3\tau_1}{(\omega^2+\Delta^2)[(\omega^2+\Delta^2)^{1/2}\eta - 1/2\tau_0 + 1/6\tau_1]}. \tag{17}$$

We now have to calculate the susceptibility χ_S according to (6). Substituting (17) in (11), we obtain after integration finally

$$\frac{\chi_s}{\chi_n} = 1 - \Delta^2\pi T \sum_{\omega=-\infty}^{\infty} \frac{1}{(\omega^2+\Delta^2)\,[\sqrt{\omega^2+\Delta^2}+2/3\tau_1]}. \quad (18)$$

The contributions from $\Lambda_p^{(1)}$ and $\Lambda_p^{(2)}$ to χ can be neglected, since, according to (16), they are of the order $l/l_{s.o.}$, whereas the contributions from $\Lambda_\sigma^{(1)}$ and $\Lambda_\sigma^{(2)}$ are of the order ξ_0/l.*

In conclusion we write down the final formulas. If the spin-orbit interaction is very small ($l_{s.o.}$ large in comparison with ξ_0), then

$$\frac{\chi_s}{\chi_n} = \frac{N_n(T)}{N} - \frac{\pi\Delta}{6\tau_1}\frac{d}{d\Delta}\left(\frac{1}{\Delta}\,\mathrm{th}\,\frac{\Delta}{2T}\right). \quad (19)\dagger$$

The first term represents the ratio of the number of "normal" electrons over the total number of electrons and agrees with the results of Yosida and the earlier work of the authors.[1,4] For $T = 0$ we have, of course, $N_n(0) = 0$.

In the opposite limiting case of small mean free paths $l_{s.o.}$, we find that $\chi_s/\chi_n \to 1$:

$$\chi_s/\chi_n = 1 - \tfrac{3}{4}\pi\tau_1\Delta\,\mathrm{th}\,(\Delta/2T). \quad (20)$$

The symbol Δ in formulas (18) to (20) denotes the equilibrium gap in the spectrum of the superconductor at the given temperature.

In the general case, when the mean free path $l_{s.o.}$ is comparable with the correlation length ξ_0, closed expressions can be written down for χ_s in the cases $T = 0$ and $|T_c - T| \ll T_c$. For $T = 0$ we have ($\rho_0 = \tfrac{2}{3}\tau_1\Delta_0$):

$$\frac{\chi_s}{\chi_n} = \begin{cases} 1 - \dfrac{1}{\rho_0}\left[\dfrac{\pi}{2} - \dfrac{\mathrm{arch}\,\rho_0}{\sqrt{\rho_0^2-1}}\right], & \rho_0 > 1 \\ 1 - \dfrac{1}{\rho_0}\left[\dfrac{\pi}{2} - \dfrac{\arccos\rho_0}{\sqrt{1-\rho_0^2}}\right], & \rho_0 < 1 \end{cases} \quad (21)\ddagger$$

This function is shown in Fig. 5.

For $|T - T_c| \ll T_c$ ($\Delta \to 0$) we find

$$\frac{\chi_s}{\chi_n} = 1 - \frac{2\Delta^2}{(\pi T)^2\rho_c}\left\{\frac{\pi^2}{8} + \frac{1}{2\rho_c}\left[\psi\left(\frac{1}{2}\right) - \psi\left(\frac{1}{2}+\frac{\rho_c}{2}\right)\right]\right\}. \quad (22)$$

Here $\rho_c = \tfrac{2}{3}\tau_1\pi T_c = \rho_0/\gamma$, where γ is the Euler constant, $\gamma = 1.78$; $\psi(x)$ is the logarithmic derivative of the Γ function.

In the following section we shall compare our formulas with the experimental data.

3. DISCUSSION OF THE EXPERIMENTAL DATA

It has already been noted in [4] that such a comparison cannot be based on data which have been obtained with too large specimens and too strong fields. This limitation is due not only to the circumstance that the field is inhomogeneous in specimens with dimensions larger than the penetration depth. Another factor is important, too. It is known that if the dimensions of the specimen are smaller than the penetration depth the phase transition in the magnetic field is a second order transition and is connected with a gradual vanishing of the quantity Δ, the gap in the electron spectrum. The vanishing of the gap is, at the given temperature, accompanied by an increase in the number of "unpaired" electrons, i.e., an increase in the Knight shift. This last effect shows up, in particular, in the experiments of Androes and Knight (see Figs. 3 and 5 in [8]). In order to make this effect unimportant, the field must be weaker than the critical field of a specimen with the given dimensions d. Since $H_c \propto d^{-1}$, the effect of the change in the energy gap due to the external field is the smaller, the smaller the dimensions of the specimen.

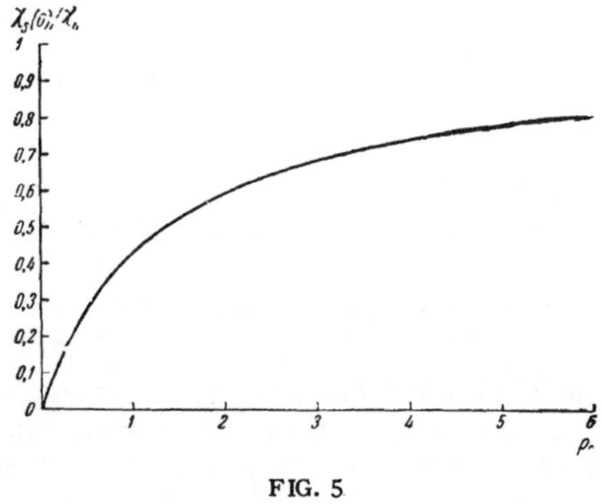

FIG. 5

For these reasons it is difficult to interpret the data of Reif[9] for colloidal mercury, for which the uncertainty in the size of the particles is large. This applies also to a considerable degree to the results of Androes and Knight[8] obtained with fields above 4 kilooersted. The critical field for the specimens used in [8] was 25 kilooersted at $T = 0$, and a field of 4 kilooersted could give a noticeable effect, in any case at temperatures which are close to the critical temperature. The attempt in [8] to interpret these data by taking account of the effect of the field through a change in the temperature scale does not seem convincing to us. A theory describing the effect of the field on the Knight shift does not exist as yet, and there are no indications whatever that a change in the temperature scale, and in particular, the one proposed in [8], has anything to do with the

*In the cases of interest to us the specimens have dimensions of the order 10^{-6} cm, i.e., $l \approx 10^{-6}$, so that $l \ll \xi_0 \approx 10^{-4}$ to 10^{-5} cm.

†th = tanh.

‡arch = $\cosh^{-1}$

problem. In particular, Androes and Knight [8] assume that near T_c one has $H_c \approx 2H(0)(1 - T/T_c)$, whereas it is known that for films $H_c \sim H_{cm}\delta/d$, where $H_{cm} \sim 1 - T/T_c$ is the field for a bulky specimen and $\delta \sim 1/\sqrt{1 - T/T_c}$ is the penetration depth, i.e., $H_c \sim \sqrt{1 - T/T_c}$.

Therefore, we have compared our results only with the data obtained in [8] with a field of 1.2 kilooersted (Fig. 6). The theoretical curves correspond to different values of ρ_0. The lower dashed curve refers to the case $\rho_0 = 0$, i.e., shows the dependence of the paramagnetic susceptibility of a superconductor in the absence of spin-orbit coupling. [1]

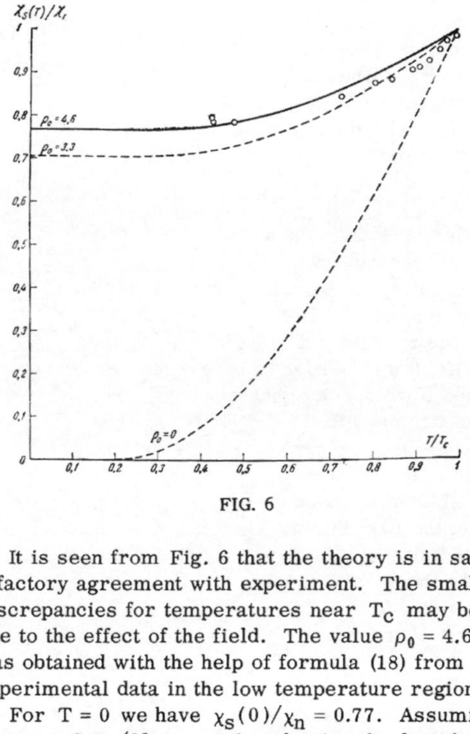

FIG. 6

It is seen from Fig. 6 that the theory is in satisfactory agreement with experiment. The small discrepancies for temperatures near T_c may be due to the effect of the field. The value $\rho_0 = 4.6$ was obtained with the help of formula (18) from experimental data in the low temperature region.

For T = 0 we have $\chi_s(0)/\chi_n = 0.77$. Assuming that $\rho_0 = 2\pi\xi_0/3l_{s.o.} = 4.6$ and using the fact that $\xi_0 = 2 \times 10^{-5}$ for tin, [7] we obtain $l_{s.o.} = 1.0 \times 10^{-5}$ cm. Assuming further that the mean free path l is of the order of the dimensions of the specimen, i.e., $l \sim 10^{-6}$, we find that $l \sim 0.1\, l_{s.o.}$, which implies that the spin-orbit interaction is somewhat weaker than usual.

In conclusion we make one further remark. Androes and Knight [8] have expressed doubt that the Knight shift can be explained with the help of the spin-orbit interaction, since they did not observe any appreciable dependence on the dimensions of the specimen in their experiments. Measurements on specimens with dimensions of ~ 1000 Å gave the same order of magnitude for the Knight shift as films with a thickness of 40 Å. In our opinion, these results are not sufficient to make the above-mentioned conclusion convincing. If the critical field for specimens with dimensions of 40 Å was 25 kilooersted, then specimens with dimensions of 1000 Å have a critical field of 1 kilooersted, and for this reason alone the Knight shift may have a magnitude of order unity.

[1] K. Yosida, Phys. Rev. **110**, 769 (1958).

[2] P. C. Martin and L. P. Kadanoff, Phys. Rev. Lett. **3**, 322 (1959).

[3] J. R. Schrieffer, Phys. Rev. Lett. **3**, 323 (1959).

[4] A. A. Abrikosov and L. P. Gor'kov, JETP **39**, 480 (1960), Soviet Phys. JETP **12**, 337 (1961).

[5] R. A. Ferrell, Phys. Rev. Lett. **3**, 262 (1959).

[6] P. W. Anderson, Phys. Rev. Lett. **3**, 325 (1959).

[7] J. Bardeen and J. R. Schrieffer, Progress in Low Temperature Physics **3**, Amsterdam (1961).

[8] G. M. Androes and W. D. Knight, Phys. Rev. **121**, 779 (1961).

[9] F. Reif, Phys. Rev. **106**, 208 (1957).

[10] T. Matsubara, Progr. Theor. Phys. **14**, 351 (1955).

[11] Abrikosov, Gor'kov, and Dzyaloshinskiĭ, JETP **36**, 900 (1959), Soviet Phys. JETP **9**, 636 (1959).

[12] A. A. Abrikosov and L. P. Gor'kov, JETP **35**, 1558 (1958) and **36**, 319 (1959), Soviet Phys. JETP **8**, 1090 (1959) and **9**, 220 (1959).

Translated by R. Lipperheide
179

AN ESTIMATE OF THE LIMITING VALUES OF THE CRITICAL FIELDS FOR HARD SUPERCONDUCTORS

L. P. GOR'KOV

Institute for Physics Problems, Academy of Sciences, U.S.S.R.

Submitted to JETP editor November 28, 1962

J. Exptl. Theoret. Phys. (U.S.S.R.) **44**, 767-769 (February, 1963)

VERY high critical fields ($\approx 10^5$ G) have recently been reported for several superconductors.[1] In the present note we give an estimate of the order of magnitude of the upper limit of the critical magnetic field in the case of weak current. A similar estimate was obtained by Clogston.[2] Clogston assumed that the maximum field is governed by the condition that the energy in the magnetic field of electron spins forming a Cooper pair is comparable with the binding energy of the pair

$$\mu H \sim T_c, \tag{1}$$

where μ is Bohr magneton. This mechanism does not allow for the fact that the superconductors which we are discussing here are always superconducting alloys. The high value of the critical field for such alloys is possible only due to the short mean free path of electrons. A field of the order of that given in Eq. (1) is obtained if we assume that the mean free path becomes comparable with the interatomic distances.

We shall consider first the situation in pure superconductors. The majority of known superconductors undergoes a transition of the first kind to the normal state at some critical value of the magnetic field equal, according to the theory of Bardeen, Cooper, and Schrieffer[3] (at T = 0), to

$$H_c = (T_c/\hbar\gamma)\sqrt{2\pi m p_F/\hbar} \quad (\gamma = 1.78). \tag{2}$$

Then μH_c is of the order

$$\mu H_c \sim T_c \sqrt{\frac{v_F}{c}\frac{e^2}{\hbar c}} \ll T_c.$$

As shown by the present author,[4] this situation occurs only in metals of the "Pippard" type for which the depth of penetration in a weak field is $\delta_P \ll \xi_0 = \hbar v_F / T_c$. Metals of "London" type transform from the normal to the superconducting state by a phase transition of the second kind in a field $H_{c2} > H_c$, where

$$H_{c2} \sim H_c \delta_L / \xi_0, \qquad \delta_L^{-1} = \sqrt{4\pi N e^2 / mc^2},$$

δ_L is the "London" depth of penetration ($\delta_L > \xi_0$). (At the moment of the transition there is no Meissner effect. For metals of "Pippard" type the field is $H_{c2} < H_c$ and has the meaning of the supercooling field.)

Substituting Eq. (2) we find the critical field for pure "London" superconductors (among which are obviously[4] such materials as pure La, V, Nb, Ta and others, which do not obey the Rutgers formula):

$$\mu H_{c2} \sim T_c (T_c / \varepsilon_F) \ll T_c.$$

Thus also in these pure superconductors the critical field never exceeds 10^2–10^3 G.

The situation is different in superconducting alloys. The present author[5] showed that for alloys with an electron mean free path $l \ll \delta_P$ (i.e., a path which is small compared with the depth of penetration in a weak field) we have:

$$H_{c2} \sim H_c \delta_L / l.$$

The transition in a magnetic field in alloys is always of the second kind.

To make our estimate we shall use the expression for the field H_{c2} at $T = 0$ obtained by Shapoval[6]:

$$H_{c2} = 1.5\, cT_c / elv_F, \qquad \mu H_{c2} = 1.5 T_c (\hbar / p_F l). \qquad (3)$$

The validity of the above formulas is limited to the region $p_F l \gg \hbar$, i.e., to defect concentrations for which the mean free path l is large compared with the interatomic distances $a \approx \hbar / p_F$. To estimate the upper limit of the critical field in alloys we shall assume that $\hbar / p_F l \approx 1$, whence we obtain, using Eq. (1), $H_{c\,max} \approx 10^4 T^0$ G. We emphasize once more that such high fields ($\approx 10^5$ G) can in principle occur only in alloys. The less careful the preparation of these alloys the higher the critical field. The electron mean free path or the residual resistance of a sample can be used as a measure of its critical field. In particular for relatively low defect concentrations Eq. (3) can be rewritten in the form[6]

$$H_{c2} = \frac{3}{2\pi}\frac{ec\gamma T_c}{\sigma k},$$

where γ is the coefficient in the linear law for the electronic specific heat of a unit volume and σ is the conductivity. When the mean free path becomes comparable with the atomic distances we reach the upper limit of the critical field for alloys. As far as the author is aware the conductivity of such alloys has not been measured. It would be interesting to investigate in what region the law of proportionality of the critical field and the residual resistance is valid. If we had a graph of the dependence of H_c on $\rho = \sigma^{-1}$, which should be a curve with saturation, we could quite accurately predict the upper limit of the critical magnetic field for superconducting alloys.

[1] J. E. Kunzler, J. Appl. Phys. **33**, 1042 (1962); Berlincourt, Hake, and Leslie, Phys. Rev. Lett. **6**, 671 (1961).

[2] A. M. Clogston, Phys. Rev. Lett. **9**, 266 (1962).

[3] Bardeen, Cooper, and Schrieffer, Phys. Rev. **106**, 162 (1957).

[4] L. P. Gor'kov, JETP **37**, 833 (1959), Soviet Phys. JETP **10**, 593 (1960).

[5] L. P. Gor'kov, JETP **37**, 1407 (1959), Soviet Phys. JETP **10**, 998 (1960).

[6] E. A. Shapoval, JETP **41**, 877 (1961), Soviet Phys. JETP **14**, 628 (1962).

Translated by A. Tybulewicz
127

SOVIET PHYSICS JETP VOLUME 27, NUMBER 2 AUGUST, 1968

GENERALIZATION OF THE GINZBURG-LANDAU EQUATIONS FOR NON-STATIONARY PROBLEMS IN THE CASE OF ALLOYS WITH PARAMAGNETIC IMPURITIES

L. P. GOR'KOV and G. M. ÉLIASHBERG

Institute of Theoretical Physics, Academy of Sciences of the U.S.S.R.

Submitted August 22, 1967

Zh. Eksp. Teor. Fiz. 54, 612–626 (February, 1968)

We develop a simple diagrammatic method, which allows us to consider non-linear, non-stationary problems in the theory of superconductivity; we operate directly with physical quantities depending on the real "physical" time. By applying this method to the case of alloys with paramagnetic impurities we were able to obtain the non-stationary generalization of the Ginzburg-Landau equations valid for all important frequencies. Using the equations obtained we solve some non-linear problems.

1. INTRODUCTION

IN the last two or three years many papers have appeared in the literature which were devoted to the kinetics of superconductors (see, e.g.,[1,2]). However, even though there have been obvious successes by the theory of superconductivity in explaining the thermodynamic properties of superconductors, the position is much more complicated regarding problems about their behavior in non-stationary fields. If we forget about phenomenological models referring to the motion of vortices, and so on, the microscopic theory was constructed only for the interaction of weak electromagnetic radiation with superconductors. The cause of difficulties lies first of all in the complicated apparatus of the BCS theory which in the general case has the form of equations for two Green functions formulated on the "imaginary" time axis. Therefore it is necessary to solve for a non-stationary problem these complicated equations in the general form and then to continue these solutions onto the physical time axis.

The form of the equations for the Green functions in the case of static problems of superconductors enables us to simplify considerably in the range of temperatures close to the critical temperature where these equations turn out to be equivalent to the set of Ginzburg-Landau equations. Formally this is achieved by expanding all quantities in powers of the ratio of Δ, the energy gap in the spectrum of the superconductor, to the critical temperature T_c. One may therefore hope that in that temperature range one may simplify the general scheme of superconductivity theory also for non-stationary problems, and thus obtain some generalized Ginzburg-Landau equations containing only the electromagnetic fields and the parameter Δ. Unfortunately, we did not succeed so far in proceeding further than the linear equations for the non-stationary correction to Δ. For instance, although in the paper by Abrahams and Tsuneto[1] some non-linear equations were written down for the gap, the region of applicability in the physically interesting cases is the region where the non-linear terms are in fact small. In the papers by Kemoklidze and Pitaevskiĭ[2] non-linear equations for Δ were given, which were obtained for a pure superconductor under the assumption that the change in all quantities takes place sufficiently slowly. In pure superconductors, however, the anomalous skin effect is very important, so that the assumption about the spatial change of the gap for a given temporal behavior of the external fields is practically always violated in physical problems.

Summarizing we can say that so far one has not succeeded in writing down a generalization of the Ginzburg-Landau equations with a range of applicability which covered the whole range of frequencies of physical interest. In the following we shall do this for superconducting alloys with paramagnetic impurities. This problem is of immediate interest and also, in our opinion, extremely important as it enables us to comprehend the situation as a whole.

Before we proceed, we note that the statement of the problem of non-linear equations for the gap in variable fields by itself is possible only in the region of temperatures close to the critical one. Indeed, to obtain equations that are independent of the experimental arrangement it is necessary that the Joule losses be small. These can be estimated as follows. The quantity $cH^2\zeta/8\pi\omega\delta$, where ζ is the impedance and δ the penetration depth is of the order of the radiative energy dissipated per unit volume in one period. This quantity must be small compared with the thermal energy of the electrons per unit volume, $\gamma T^2/2$. However $c\zeta/\omega\delta \sim 1$, and since the critical field $H_c \sim \sqrt{\gamma}(T_c - T)$ for $T \sim T_c$, we get

$$\left(\frac{H}{H_c}\right)^2 \ll \left(\frac{T_c}{T_c - T}\right)^2, \tag{1}$$

which can be satisfied when $H \sim H_c$ if $T_c - T \ll T_c$. For superconductors with paramagnetic impurities[3] the corresponding condition has the form

$$\left(\frac{H}{H_c}\right)^2 \ll \frac{T^2}{T_c^2 - T^2}\frac{1}{(\tau_s T_c)^2}. \tag{1'}$$

We shall further assume that $\tau_s T_c \ll 1$, so that in the stationary case the equations would have the form of the Ginzburg-Landau equations.[3]

We bear in mind that the derivation of the local Ginzburg-Landau equations was based upon the possibility of expanding all quantities in a series in the coordinates since the kernels of the integral equations which determine, e.g., the parameter Δ, decrease exponentially near T_c over distances of order $\xi_0 \sim kv/T_c$. An analysis of the corresponding equations for the non-stationary case (see below) shows that in the whole temperature range these kernels have terms which decrease slowly or which oscillate for large values of the difference in the time variables with a frequency $2\Delta(T)$. The temporal equations for $\Delta(r, t)$ are thus in general not local. Physically this means that as in a pure superconductor or in a normal alloy there is a gap in the energy spectrum, there will be a singularity for all quantities in their frequency dependence corresponding to an absorption threshold $\omega = 2\Delta$. Because of this it turns out that it is impossible to expand the various quantities in powers of Δ. Paramagnetic alloys have no such threshold when $\tau_s T_c \ll 1$, since there is no energy gap in that range of concentrations.

Finally, we note that to describe alloys with paramagnetic impurities we shall apply the same approach as in a paper by Abrikosov and one of the authors.[3] In other words, we shall not take the Kondo effect[4] (the formation of a complex consisting of an electron and a paramagnetic impurity) into account of the appearance of ferromagnetic ordering in such alloys at low temperatures. As far as the former is concerned, the Kondo effect depends on the sign of the exchange interaction and in general only changes the magnitude of the mean free flight time τ_s for a spin flip. Ferromagnetism widens the region where gapless superconductivity occurs. In that case one should take into account the interaction with spin waves. We reserve that problem for another paper and in the following we shall assume that we are above the line of ferromagnetic ordering. Formally such a model is not self-contradictory.[5]

2. DERIVATION OF THE GENERAL EQUATIONS OF THE NON-STATIONARY PROBLEM

There is no simple variant of the Feynman technique for the Green functions at finite temperatures. The further calculations are based upon a direct analytical continuation of the thermodynamic quantities from the imaginary axis to the real frequency axis. Let $\Delta_\omega(k)$ and $\Delta^*_\omega(k)$ be the Fourier components of the superconductivity parameter in the thermodynamic technique ($\omega = 2n\pi iT$). As is done usually[6] we introduce two Green functions $G_{\epsilon_1\epsilon_2}(k_1, k_2)$ and $F^+_{\epsilon_1\epsilon_2}(k_1, k_2)$ which because of the spatial and temporal non-uniformity depend on two frequencies and two momenta. We have

$$\Delta_\omega^*(k) = |g| T \sum_\varepsilon \int \frac{d^3\mathbf{p}}{(2\pi)^3} F^+_{\varepsilon\varepsilon-\varphi}(\mathbf{p}, \mathbf{p}-\mathbf{k}), \quad \varepsilon = i(2n+1)\pi T. \qquad (2)$$

The idea of the calculations is to write F^+ as a power series in Δ_ω and Δ^*_ω considering the coefficients $B^{(2k+1)}$ in the corresponding series for Δ^*,

$$\Delta_\omega^* = B^{(1)}(\omega)\Delta_\omega^* + \sum_{\omega'+\omega''+\omega'''=\omega} B^{(3)}(\omega'\omega''\omega''')\Delta_{\omega'}^*\Delta_{\omega''}\Delta_{\omega'''}^* + \ldots$$

as "response" of the system to the perturbations Δ and Δ^*. A similar expansion could also be made in respect to the potentials of the electromagnetic field, but for the sake of simplicity we shall not yet do that. We must continue the expressions for the coefficients B which are given initially in discrete points $2n\pi iT$ for each frequency onto the real axis. The way to continue is determined by the condition of causality: the right-hand side must contain Δ and Δ^* at a time which precedes the time on the left-hand side. The causality condition determines the coefficients $B(\omega', \omega'', \ldots)$ as functions which are analytic in the upper half-plane for each frequency ω^i. This method is a simple generalization of the approach used in the case of linear response.[6] It is equivalent to Keldysh's method[7] with a specially chosen Hamiltonian.

The equations for the Green functions

$$\left(-\frac{\partial}{\partial\tau_1} - \hat{h}_1 + \mu\right)G(12) + \Delta(1)F^+(12) = \delta(1-2),$$

$$\left(\frac{\partial}{\partial\tau_1} - \hat{h}_1 + \mu\right)F^+(12) - \Delta^*(1)G(12) = 0,$$

in which Δ and the field entering in $\hat{h}_1$ and $\hat{h}_1$ depend on the "imaginary" time, give for $G_{\epsilon\,\epsilon-\omega}$ and $F^+_{\epsilon\,\epsilon-\omega}$ the series

$$G_{\varepsilon\varepsilon-\omega}(\mathbf{p}, \mathbf{p}-\mathbf{k}) = \frac{1}{\varepsilon - \xi_p}$$

$$+ \frac{1}{\varepsilon - \xi_p}\Delta_{\omega'}(k')\frac{1}{\varepsilon - \omega' + \xi_{\mathbf{p}-\mathbf{k}'}}\Delta_{\omega''}^*(k'')\frac{1}{\varepsilon - \omega' - \omega'' - \xi_{\mathbf{p}-\mathbf{k}'-\mathbf{k}''}} + \ldots,$$

$$F^+_{\varepsilon\varepsilon-\omega}(\mathbf{p}, \mathbf{p}-\mathbf{k}) = -\left\{\frac{1}{\varepsilon + \xi_p}\Delta_{\omega'}^*(k')\frac{1}{\varepsilon - \omega' - \xi_{\mathbf{p}-\mathbf{k}'}} + \ldots\right\},$$

$$\varepsilon = i(2n+1)\pi T; \quad \omega, \omega' = i2n\pi T. \qquad (2')$$

In each term of the series summation over the frequencies ω^i and integration over the momenta k^i with conditions $\Sigma\omega^i = \omega$ and $\Sigma k^i = k$ is assumed. It is clear that the general term in the series contains a product of simple factors $G^{\pm}(\epsilon) = 1/(\epsilon \pm \xi)$ with alternating G^- and G^+ after each Δ and Δ^*.

In the sum over the frequencies in Eq. (2) each term can be considered as the residue of an integral along a contour around the point $z = \epsilon_n$:

$$T\sum_\varepsilon \to \frac{1}{4\pi i}\oint_C \operatorname{th}\frac{z}{2T}\,dz.$$

For instance,

$$T\sum G^+(\varepsilon)G^-(\varepsilon-\omega')G^+(\varepsilon-\omega'-\omega'')\ldots$$

$$= \frac{1}{4\pi i}\oint_C dz\operatorname{th}\frac{z}{2T}G^+(z)G^-(z-\omega')G^+(z-\omega'-\omega'')\ldots$$

Deforming the contour C along the lines $z = \omega'$, $z = \omega' + \omega''$, ... (Fig. 1) and bearing in mind that ω', ω'', ... at the moment are fixed in the imaginary points $2n\pi iT$ (all $n > 0$ since we shall be interested in the continuation from the upper half-plane for each of the ω^i) we put the integral in the form

$$\frac{1}{4\pi i}\int_{-\infty}^{\infty} d\varepsilon\operatorname{th}\frac{\varepsilon}{2T}\{[G^{+R}(\varepsilon) - G^{+A}(\varepsilon)]G^{-A}(\varepsilon-\omega')$$
$$\times G^{+A}(\varepsilon-\omega'-\omega'')\ldots + G^{+R}(\varepsilon+\omega')[G^{-R}(\varepsilon)$$
$$- G^{-A}(\varepsilon)]G^{+A}(\varepsilon-\omega'')\ldots + \ldots\}. \qquad (3)$$

Here G^R and G^A are the retarded and advanced functions which have no singularities in ϵ in the upper and

FIG. 1

lower half-planes, respectively. So far, therefore, while all ω', ω'', ... lie in the upper half-plane, Eq. (3) determines a function which is analytic in each of these variables. For the required continuation onto the real axis all ω^i must be replaced by $\omega^i + i\delta$, where now the ω^i are real. Performing then a simple substitution of integration variable in each term of Eq. (3) starting with the second one, we get

$$\frac{1}{4\pi i}\int_{-\infty}^{\infty} d\varepsilon\left\{-\operatorname{th}\frac{\varepsilon}{2T}G^{+A}(\varepsilon)G^{-A}(\varepsilon-\omega')G^{+A}(\varepsilon-\omega'-\omega'')\ldots\right.$$
$$\left.+\operatorname{th}\frac{\varepsilon-\omega}{2T}G^{+R}(\varepsilon)G^{-R}(\varepsilon-\omega')G^{+R}(\varepsilon-\omega'-\omega'')\ldots\right\}$$
$$+\frac{1}{4\pi i}\int_{-\infty}^{\infty} d\varepsilon G^{+R}(\varepsilon)\left(\operatorname{th}\frac{\varepsilon}{2T}-\operatorname{th}\frac{\varepsilon-\omega'}{2T}\right)G^{-A}(\varepsilon-\omega')G^{+A}(\varepsilon-\omega'-\omega'')\ldots$$
$$+\frac{1}{4\pi i}\int_{-\infty}^{\infty} d\varepsilon G^{+R}(\varepsilon)G^{-R}(\varepsilon-\omega')\left(\operatorname{th}\frac{\varepsilon-\omega'}{2T}-\operatorname{th}\frac{\varepsilon-\omega'-\omega''}{2T}\right)$$
$$\times G^{+A}(\varepsilon-\omega'-\omega'')\ldots+\ldots \tag{4}$$

In Fig. 2 we have illustrated the rule to construct Eq. (4). It consists in that the sum over frequencies of the product of G-functions as the result of the analytical continuation changes to an integral over a sum of several terms. The first two contain the product of only advanced and only retarded functions. The remaining terms contain all such products, in each of which in only one place a change from retarded to advanced functions occurs. The factors corresponding to each of the terms are clear from (4).

Using these results we get easily for Eq. (2) the following expression:

$$T\sum_{\varepsilon}F^{+}_{\varepsilon\,\varepsilon-\omega}\to-\frac{1}{4\pi i}\int_{-\infty}^{\infty}d\varepsilon\left[\operatorname{th}\frac{\varepsilon}{2T}F^{+A}_{\varepsilon\,\varepsilon-\omega}-\operatorname{th}\frac{\varepsilon-\omega}{2T}F^{+R}_{\varepsilon\,\varepsilon-\omega}\right]$$
$$-\frac{1}{4\pi i}\int_{-\infty}^{\infty}d\varepsilon\int_{-\infty}^{\infty}d\varepsilon_1\left\{[G^{+R}_{\varepsilon\varepsilon_1}\Delta^{*}_{\omega'}G^{-A}_{\varepsilon_1-\omega'\,\varepsilon-\omega}+F^{+R}_{\varepsilon\varepsilon_1}\Delta_{\omega'}F^{A}_{\varepsilon_1-\omega'\,\varepsilon-\omega}]\right.$$
$$\left.+[G^{+R}_{\varepsilon\varepsilon_1}F^{A}_{\varepsilon_1-\omega'\,\varepsilon-\omega}+F^{+R}_{\varepsilon\varepsilon_1}G^{-A}_{\varepsilon_1-\omega'\,\varepsilon-\omega}]\left[\frac{e}{c}(\mathbf{v}\mathbf{A}_{\omega'})-e\varphi_{\omega'}\right]\right\}$$
$$\times\left(\operatorname{th}\frac{\varepsilon_1}{2T}-\operatorname{th}\frac{\varepsilon_1-\omega'}{2T}\right) \tag{5}$$

Here A and φ are the electromagnetic potentials. The functions $G^{-R(A)}$ and $F^{+R(A)}$ are formally defined by the series (2′) in which we must make the substitution $\epsilon \to \epsilon + i\delta$ ($\epsilon \to \epsilon - i\delta$) for the retarded (advanced) functions and assume all ω^i to be real. The functions G^+ and F^- are obtained from G^- and F^+ by changing the sign of ξ in the denominators in (2′). The functions $G^{-R(A)}(12)$ and $F^{+R(A)}(12)$ satisfy the set of equations

$$\left(-i\frac{\partial}{\partial t_1}-\hat{h}_1+\mu\right)G^{-}(12)+\Delta(1)F^{+}(12)=\delta(1-2),$$
$$\left(-i\frac{\partial}{\partial t_1}+\hat{h}_1-\mu\right)F^{+}(12)+\Delta^{*}(1)G^{-}(12)=0. \tag{6}$$

The choice of solutions is determined by the fact that the Fourier components $G^{-R(A)}_{\epsilon\,\epsilon-\omega}$ and $F^{+R(A)}_{\epsilon\,\epsilon-\omega}$ are analytical in ϵ in the upper (lower) half-plane. Hence,

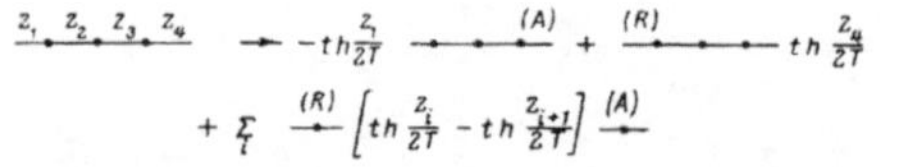

FIG. 2

$G^{-R(A)}(12)$ and $F^{+R(A)}(12)$ are non-vanishing for $t_1 > t_2$ ($t_1 < t_2$).

We turn to Eq. (5). The first integral contains separately the functions F^{+R} and F^{+A} which are analytical in ϵ in the appropriate half-planes. Deforming the contour of integration passing along the real axis in the upper and lower half-plane, respectively, we reduce this term to the sum of the residues in the poles of the tangents:

$$T\sum_{n>0}F^{+}_{\varepsilon-\omega\,\varepsilon}+T\sum_{n<0}F^{+}_{\varepsilon\,\varepsilon+\omega}+4\pi imp_0e\tau_s(\Delta^{*}\psi)_{\omega}. \tag{7}$$

where $\epsilon = (2n+1)\pi iT$ and ω is a real frequency. The third term corresponding to the second integral in Eq. (5) contains the definition of the new function ψ. The appropriateness of such a definition will become clear from what follows. The first two terms in (7) are analogous to the corresponding expressions in the static case and when $T_c - T \ll T_c$, $\omega \ll T_c$ they can be expanded in the ratios of Δ, Δ^*, and ω to T_c. In the case of pure superconductors and normal alloys one can check that if $\Delta \ll T$, $\omega' \ll T$, also ϵ, $\epsilon_1 \ll T$. Therefore

$$\operatorname{th}\frac{\varepsilon_1}{2T}-\operatorname{th}\frac{\varepsilon_1-\omega'}{2T}\approx\frac{\omega'}{2T}\operatorname{ch}^{-2}\frac{\varepsilon_1}{2T}\approx\frac{\omega'}{2T}. \tag{8}$$

In the coordinate representation the anomalous terms in (5) reduce in this case to integrals of the kind

$$\int F^{+R}(12)F^{A}(21)\dot{\Delta}(2)\,dt_2\,d^3r_2.$$

As the retarded and advanced Green functions of an ordinary superconductor do not decrease exponentially for large $t_1 - t_2$ but oscillate with frequency $|\Delta|$, it becomes clear why it is not possible to transfer the simple procedure for deriving the static Ginzburg-Landau equations to the non-stationary case.

The situation is different for alloys with paramagnetic impurities. When $\tau_s T_c \ll 1$, the important values are here $\epsilon_1 \gg T$ since the Green functions decrease exponentially over a time $t_1 - t_2 \sim \tau_s$.[1)] We have thus instead of (8)

$$\operatorname{th}\frac{\varepsilon_1}{2T}-\operatorname{th}\frac{\varepsilon_1-\omega'}{2T}\approx 2\omega'\delta(\varepsilon_1). \tag{8′}$$

The method presented in the present section enables us thus also for the non-stationary case to evaluate as before all physical quantities (such as the current, the density, the superconductivity parameters Δ and Δ^*) using Feynman diagrams. Equations (5) to (7) and Fig. 2 are a summary of the rules which do not need additional explanations. It is useful to note that in this scheme impurities are included automatically. Since scattering by static impurities does not change the frequencies of the Green function we can apply here all methods for averaging over the impurity positions using the standard diagram technique developed for superconducting alloys.[6]

3. DERIVATION OF THE EQUATIONS FOR ALLOYS WITH PARAMAGNETIC IMPURITIES

We turn to the case of paramagnetic impurities. For the sake of simplicity we shall assume the scattering

---

[1)] One can immediately check that this is correct by studying the expressions for the Green functions [3]. This will also be clear from the subsequent calculations.

to be isotropic. We see that the answer can be expressed in terms of the diffusion coefficient. The assumption made is thus inessential but it appreciably simplifies the calculations. In the case of paramagnetic impurities two times between collisions τ_1 and τ_2 occur. The difference $\tau_1^{-1} - \tau_2^{-1} = 2\tau_s^{-1}$ determines the parameter τ_s which has the meaning of a free flight time between collisions associated with the electron spin flip. The rule obtained in [3] consists in that when averaging over the positions of the impurities to a dotted line involving an even number of vertices in which Δ or Δ^* appear the quantity τ_1^{-1} is associated, while for an odd number of vertices, we get τ_2^{-1}.

The first two terms in Eq. (7) can easily be evaluated by an expansion in $\Delta\tau_s$ and $\omega\tau_s$ which are assumed to be small. Substituting (7) into (2) we get

$$\frac{\partial\Delta^*}{\partial t} + \frac{\tau_s}{3}\left\{\left[-\pi^2(T_c^2 - T^2) + \frac{|\Delta|^2}{2}\right]\Delta^* - \frac{v^2\tau_1}{\tau_s}\left(\nabla + \frac{2ie}{c}\mathbf{A}\right)^2\Delta^*\right\} - 2ie\Delta^*\psi = 0 \tag{9}$$

It is clear that Eq. (9) contains not only parts corresponding to the static equation[2)] but also the first derivative of Δ with respect to the time and the unknown function ψ. It follows from (9) that we must evaluate this function assuming that the characteristic frequencies ω and the spatial scale k^{-1} over which Δ changes are found from the following relations

$$\omega \sim \tau_s|\Delta|^2 \sim k^2 vl \ll \tau_s^{-1}.$$

According to (8′) the anomalous terms on the right-hand side of (5), i.e., ψ, contain time derivatives of Δ, Δ^*, and the field. In the diagrams we shall indicate these terms by a vertex with wavy lines. An arbitrary term in the expansion of ψ in powers of Δ and the field can then be depicted by the diagram of Fig. 3a (the dotted line indicates averaging). In accordance with what we have said in the preceding section all $G^{\pm}$ - functions in the upper line are retarded and in the lower line they are advanced.

The rectangle in Fig. 3a encloses the sum of diagrams which have a singular value. We denote it by $I(\omega, k)$. A ladder-type equation for this quantity is depicted in Fig. 3b. Assigning a factor $(8\pi^2\tau_1 mp_0)^{-1}$ to a dotted line (we bear in mind that the scattering is assumed to be isotropic) we get

$$I(\omega, k) = \frac{1}{8\pi^2\tau_1 mp_0}\left\{1 + \int d^3\mathbf{p}\, G_\varepsilon^{+R}(\mathbf{p})\, G_{\varepsilon-\omega}^{+A}(\mathbf{p}-\mathbf{k})\, I(\omega, k)\right\}.$$

Substituting here the Green functions of the normal metal averaged over the impurities,

$$G^{+R} = \frac{1}{\varepsilon + \xi + i/2\tau_1}, \quad G^{+A} = \frac{1}{\varepsilon + \xi - i/2\tau_1},$$

we find easily

$$I(\omega, k) = \frac{1}{8\pi^2\tau_1^2 mp_0}\,\frac{1}{-i\omega + v^2k^2\tau_1/3}. \tag{10}$$

$I(\omega, k)$ thus describes the slow diffusion behavior of the non-stationary perturbations. For low frequencies the quantity I can be arbitrarily large and we must therefore before everything else sum all diagrams which contain I in some power. We emphasize that the dotted lines which form the dangerous quantity I necessarily connect opposite sides of the diagram of Fig. 3a. Therefore, in turn, the diagram containing an arbitrary number of I is a diagram of the ladder type in which I occurs as "rung". We see, moreover, that the expansion occurs in powers of the dimensionless quantity

$$\frac{\tau_s|\Delta|^2}{-i\omega + v^2k^2\tau_1/3}, \tag{11}$$

where ω and k are the frequency and wavevector characterizing the change in the parameter Δ.

Bearing in mind the discussions given above we express the function ψ in terms of the quantities $\Gamma_1(\omega, k)$ and $\Gamma_2(\omega, k)$ the diagrammatic definition of which is clear from Fig. 4. In the momentum representation Γ_1 and Γ_2 depend only on the difference of the frequencies and momenta on the upper and the lower lines. The exact equations which these quantities satisfy are depicted in Fig. 5. The heavy rungs correspond to the function I, and the external blocks contain retarded and advanced Green functions and also dotted lines corresponding to averaging. From these blocks one must, of course, exclude the diagrams included in I (Fig. 3b). An approximate evaluation of the blocks α, β, γ, and π is based upon the fact that they can be expanded in a power series in $\tau_s\Delta$ and also in the frequency and wavevector characterizing the change in Δ and that one can restrict oneself to the first terms in the expansion. The possibility of such an expansion is connected with the fact that the upper and lower lines of the blocks contain, respectively, only retarded and only advanced Green functions, each of which does not contain anomalous terms and therefore allows the above-mentioned expansion. In each term of the expansion of the blocks one must perform in the proper way the averaging over the impurities.

It is clear from Eq. (9) that $e\psi$ is of the order of the quantity $\tau_s|\Delta|^2$. As ψ contains a time derivative while the characteristic frequencies are of the same

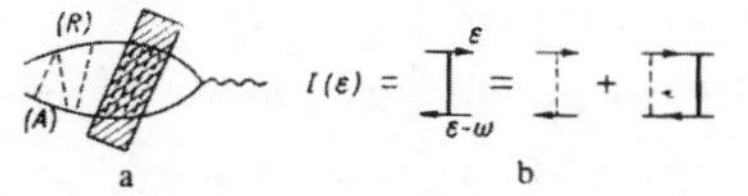

FIG. 3

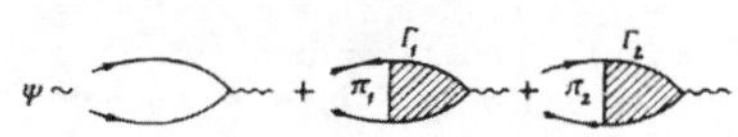

FIG. 4

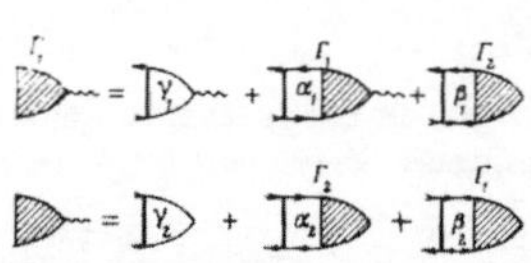

FIG. 5

FIG. 6

2)In this connection we draw attention to some misprints occurring in [3].

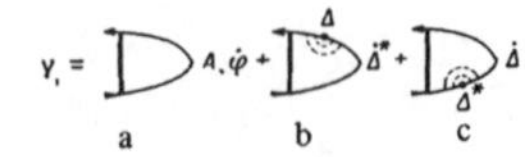

FIG. 7

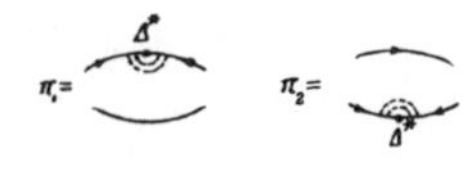

FIG. 8

FIG. 9

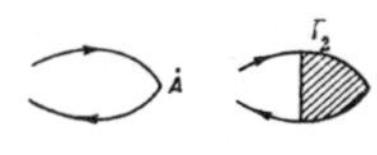

FIG. 10

order as $\tau_s|\Delta|^2$ the first term in ψ in Fig. 4 reduces to the diagram of Fig. 6 where the lines are the Green functions of the normal metal. This diagram vanishes when integrated over the electron energy ξ. In Fig. 7 we show diagrams which give a contribution to γ_1. The arrow in Fig. 7a shows that in this diagram one could insert the external field, but such terms vanish. The diagrams 7b and 7c contain dotted ladders about the vertices Δ and Δ^*. Summation of these ladders gives for small frequencies (see [3])

$$\Delta_\omega \to \frac{\tau_s}{2\tau_1}\Delta_\omega; \quad \Delta_\omega^* \to \frac{\tau_s}{2\tau_1}\Delta_\omega^*. \tag{12}$$

After some calculations we get

$$\gamma_1 = -\gamma_2 = \frac{i}{-i\omega + v^2k^2\tau_1/3}\left\{\frac{2}{3}ev^2 \operatorname{div}\dot{A} + \frac{2e}{\tau_1}\dot{\varphi} + \frac{i\tau_s}{\tau_1}(\Delta^*\dot{\Delta} - \Delta\dot{\Delta}^*)\right\}_{\omega k}. \tag{13}$$

The blocks π_1 and π_2 reduce to the diagrams of Fig. 8:

$$\pi_1 = -\pi_2 = -i4\pi^2 m p_0 \tau_s \tau_1 \Delta_\omega^*(k). \tag{14}$$

Calculation shows that the blocks α and β do not depend on the electromagnetic field and are proportional to $|\Delta|^2$. As an example we give in Fig. 9 the diagrams for β_1. The second and third diagrams in that figure contain one dotted line connecting the upper with the lower line. These diagrams play a very important role as they lead to the substitution of one of the factors τ_1 by τ_S in the expression for the first diagram of Fig. 9.

The results obtained enable us to write down the equation of Fig. 5 explicitly. It then turns out that $\Gamma_2 = -\Gamma_1$ and we have

$$\Gamma_1(\omega, k) = \frac{1}{-i\omega + v^2k^2\tau_1/3}\left\{i\frac{2}{3}ev^2 \operatorname{div}\dot{A} + i\frac{2e}{\tau_1}\dot{\varphi} - \frac{\tau_s}{\tau_1}(\Delta^*\dot{\Delta} - \Delta\dot{\Delta}^*) - 2\tau_s(|\Delta|^2\Gamma_1)\right\}_{\omega, k}. \tag{15}$$

Using the definition of ψ in (7) and the connection between ψ and Γ (Fig. 4) and Eq. (14) we find

$$2e\psi = -i\tau_1\Gamma_1. \tag{16}$$

The function ψ then satisfies the equation

$$\dot{\psi} = D\nabla^2\psi - 2\tau_s|\Delta|^2\psi + D\frac{1}{c}\operatorname{div}\dot{A} + \dot{\varphi} + \frac{i\tau_s}{2}(\Delta^*\dot{\Delta} - \Delta\dot{\Delta}^*), \tag{17}$$

in which $D = v^2\tau_1/3$ is the diffusion coefficient.

Up to now we have not taken into account the condition of electrical neutrality, which is connected with the characteristics of superconductors such as metals. It is well known that this condition means that the electrical current must be transverse:

$$\operatorname{div} \mathbf{j} = 0. \tag{18}$$

However, we shall verify now that in the isotropic model considered the equation of continuity

$$\partial\rho/\partial t + \operatorname{div}\mathbf{j} = 0. \tag{18'}$$

is automatically satisfied. To show this we evaluate the charge and current densities. For the current density we have in the thermodynamic technique the following expression:

$$\mathbf{j}(\omega, k) = -\frac{2e}{m}T\sum_\varepsilon\int\frac{d^3\mathbf{p}}{(2\pi)^3}\mathbf{p}G_{\varepsilon\,\varepsilon-\omega}(\mathbf{p}, \mathbf{p}-\mathbf{k}) - \frac{Ne^2}{mc}\mathbf{A}_\omega(k).$$

Repeating the procedure with the analytical continuation which in the case of the F^+ function led us to Eq. (5) we find that the expression for the current consists of two parts. The first part does not contain time derivatives and leads to the static expression for the current.[4] The second part, depicted in Fig. 10 contains two diagrams one of which is connected with the time derivative of the vector potential **A** while the other is expressed in terms of ψ. We note that the part which directly contains the scalar potential φ is small since mixed spatial and temporal derivatives of φ occur in it. As a result we get

$$\mathbf{j} = -\sigma\left(\frac{1}{c}\dot{\mathbf{A}} + \nabla\psi\right) + \frac{N\tau_s\tau_1}{2m}\left[ie(\Delta\nabla\Delta^* - \Delta^*\nabla\Delta) - \frac{4e^2}{c}\mathbf{A}\right]. \tag{19}$$

One can similarly obtain an expression for the change in the charge density

$$\rho = \frac{3e^2N}{mv^2}(\psi - \varphi) \tag{19'}$$

(the term containing φ is the quasistatic part of ρ and arises from the sum over frequencies in an expression such as (5) for ρ). The equation is then obtained directly as a consequence of Eqs. (9) for Δ and Δ^* and Eq. (17) for ψ. (The equation for Δ is the one which is the complex conjugate of (9) while it is clear from (17) that ψ is real.)

Because of (18′) the condition $\operatorname{div}\mathbf{j} = 0$ is thus equivalent to the condition

$$\dot{\psi} = \dot{\varphi}. \tag{20}$$

If we neglect the interaction with sound waves we can require that the stronger condition

$$\psi = \varphi, \tag{20'}$$

which by virtue of (19′) corresponds to $\rho = 0$, is satisfied.

We see thus that the Coulomb condition appreciably simplifies the equation. We now rewrite the set of equations, introducing the notation

$$\Delta = \Delta' e^{2ie\chi/c}, \quad \mathbf{Q} = \mathbf{A} - \nabla\chi. \tag{21}$$

In this notation we have

$$\begin{gathered}\dot{\Delta}' + \frac{\tau_s}{3}\left[-\pi^2(T_c^2 - T^2) + \frac{\Delta'^2}{2}\right]\Delta' - D\nabla^2\Delta' + \frac{4e^2}{c^2}DQ^2\Delta' = 0,\\ \Delta'^2(\dot{\chi} + c\varphi) + D\operatorname{div}[\Delta'^2\mathbf{Q}] = 0,\\ \mathbf{j} = \sigma\mathbf{E} - \frac{2\sigma\tau_s}{c}\Delta'^2\mathbf{Q},\\ \operatorname{div}\mathbf{j} = 0.\end{gathered} \tag{22}$$

To those equations we must add the Maxwell equations

$$\operatorname{div}\mathbf{H} = 0, \quad \operatorname{rot}\mathbf{H} = \frac{4\pi}{c}\mathbf{j}, \quad \operatorname{rot}\mathbf{E} = -\frac{1}{c}\dot{\mathbf{H}} \tag{23}$$

and the relations

$$\mathrm{H} = \mathrm{rot}\,\mathrm{Q}, \quad \mathrm{E} = -\frac{1}{c}(\dot{\mathrm{Q}} + \nabla\dot{\chi}) - \nabla\varphi. \tag{23'}$$

To conclude this section we must say something about the boundary conditions. If the sample is finite, but its dimensions large compared to $\sqrt{(\xi_0 l)}$, where $\xi_0 \sim hv/T_c$ and l the mean free path, the presence of a surface is reflected in the expressions for the functions F^+ in (7). Expanding F^+ in powers of Δ and Δ^* and bearing in mind that Δ is a slowly changing function, we see that in Eqs. (9) additional terms arise of the form $n(-i\nabla + 2eA/c)\Delta^*$, where n is a vector which is normal to the surface. This is connected with the fact that the volume of integration in the integral kernels is finite. Since these terms are anomalously large we must require that

$$\mathbf{n}\left(-i\nabla + \frac{2e}{c}\mathbf{A}\right)\Delta^* = 0 \tag{24}$$

and a similar condition for the conjugate quantity. These conditions are the same as the boundary conditions for Δ and Δ^* in the case of the static Ginzburg-Landau equations. In the notation (21)

$$\partial\Delta'/\partial n = 0, \quad Q_n = 0. \tag{24'}$$

Combining this with the condition that the normal component of the current must vanish, we get $E_n = 0$ which together with the conditions of continuity of the magnetic field H and of the tangential components of E (see [8], Sec. 45) gives us the complete set of boundary conditions.

4. PARAMAGNETIC ALLOYS IN A STRONG VARIABLE FIELD

As an example of applications of Eqs. (22) we solve some plane problems. Let, e.g., there be a weak magnetic field $H = H_0 e^{i\omega t}$ parallel to the boundary of the half-space occupies by the superconductor ($z < 0$). We choose the gauge in which the scalar potential φ vanishes. Assuming all quantities to depend on z only, we get, as $E_z = 0$, $\dot{\chi} = 0$ and $Q_z = 0$. From Maxwell's equations (23) for the transverse part of Q,

$$\frac{\partial^2 Q}{\partial z^2} = \frac{4\pi\sigma}{c^2}[\dot{Q} + 2\tau_s\Delta'^2 Q] \tag{25}$$

we see that if the field is weak, it is damped into the superconductor according to the law

$$Q = Q_0 e^{kz - i\omega t},$$

where k is connected with ω and Δ' by the simple relation

$$k^2 = -\frac{4\pi\sigma}{c^2}(i\omega - 2\tau_s\Delta'^2) = -2i\delta_{\mathrm{skin}}^{-2} + \delta_L^{-2}. \tag{26}$$

Taking H along the x-axis ($H_x = -\partial Q_y/\partial z$) we get for the surface impedance ζ [8] ($\mathbf{E}(0) = \zeta\mathbf{H} \times \mathbf{n}$)

$$\zeta = -ikc/\omega. \tag{27}$$

Above we introduced in (26) the notation δ_{skin} and δ_L, respectively, for the penetration depth in the case of a normal skin effect in the normal metal and the penetration depth of a static field in the superconductor.

We now turn to the first of Eqs. (22). The magnetic field causes changes in Δ' and Δ_1 of second order in the magnitude of the field. Linearizing the equation for Δ'

$$\dot{\Delta}_1 + \frac{\tau_s\Delta'^2}{2}\Delta_1 - D\nabla^2\Delta_1 = -\frac{4e^2}{c^2}DQ^2\Delta' \tag{28}$$

and using the fact that when $\kappa \gg 1$ where

$$\varkappa = c/4\sqrt{3\pi D}\sigma, \tag{29}$$

we can neglect the term $D\nabla^2\Delta_1$, we get

$$\dot{\Delta}_t + \frac{\tau_s\Delta'^2}{2}\Delta_t = -\frac{4e^2}{c^2}DQ^2\Delta'$$

or

$$\Delta_{1\omega} = \frac{4e^2 D\Delta'(Q^2)_\omega}{c^2(i\omega - \tau_s\Delta'^2/2)}.$$

This expression is valid only for distances $|z| \gg (\kappa k)^{-1}$ since it does not satisfy the condition $(\partial\Delta_1/\partial z)_{z=0} = 0$. However, for electromagnetic problems this region is the basic one. The non-linear effects lead thus to the appearance of combination frequencies in an electromagnetic field:

$$\frac{\partial^2 Q_{\omega+\omega'}}{\partial z^2} = \frac{4\pi\sigma}{c^2}[\dot{Q}_{\omega+\omega'} + 2\tau_s\Delta'^2 Q_{\omega+\omega'} + 4\tau_s\Delta'\Delta_{1\omega'}Q_\omega]$$

or

$$Q_{\omega+\omega'} = \frac{16\pi\sigma\tau_s\Delta'^2}{c^2}\,\frac{4e^2 DQ_\omega(Q^2)_{\omega'}}{c^2(i\omega' - \tau_s\Delta'^2/2)\{[k_1(\omega') + k(\omega)]^2 + 4\pi\sigma i(\omega+\omega')/c^2 - \delta_L^{-2}\}} + Ae^{k(\omega+\omega')z}.$$

The quantity A is determined by the boundary conditions in the resonator. A weak field can thus generate radiation with a tripled frequency. If the resonator possesses such oscillations we must substitute the condition that the magnetic field has an antinode: $E(0) = 0$. In that case $k_1(\omega') + k(\omega) = 3k(\omega)$. For the amplitude of the third harmonic we find

$$\frac{H_{3\omega}}{H_\omega} = \frac{1}{4}\left(\frac{H_\omega}{H_c}\right)^2 \mathrm{Re}\left\{\left(1 - \frac{4i\omega}{\tau_s\Delta^2}\right)\left(3 + \frac{k(3\omega)}{k(\omega)}\right)(k(\omega)\delta_L)^4\right\}^{-1},$$

where H_c is the thermodynamic critical field of the superconductor. In the low frequency region $k\delta_L \approx 1$ this ratio does, however, not exceed the magnitude κ^{-2} since for fields $H_\omega \sim H_{c1} \sim H_c/\kappa$ for type II superconductors the formation of vortices is possible and this steeply increases losses. If the normal skin penetration depth becomes comparable with δ_L the formation of vortices is, clearly made more difficult even for a large amplitude of the field.

In conclusion, we discuss an example where one can completely consider a problem about the destruction of superconductivity by a high-frequency field. Let a thin film of a paramagnetic alloy with dimensions 2d which are less than the characteristic dimension of the vortices be placed in a variable magnetic field, which is the same on both sides of the plate and parallel to it. We can then assume that Δ' is in first approximation constant over the thickness of the film. The condition that the magnetic field distribution is symmetric imposes, by virtue of (23), the requirement that the total current through the film vanishes. Neglecting the induced electrical field under the conditions when $\delta \gg d$ and the change of Δ' across the film we get after averaging (22) over the z-coordinate

$$\int_{-d}^{d} j\,dz = \bar{Q}_\perp = 0,$$

so that $Q_x = -H_0(t)z$ where the coordinate origin is chosen in the center of the film. The first of Eqs. (22) gives, after averaging over z with the condition

$(d\Delta/dz)_{z=\pm d} = 0$,

$$\dot{\Delta} + \frac{\tau_s}{3}\left[-\pi^2(T_c^2 - T^2) + \frac{\Delta^2}{2}\right]\Delta + \frac{4e^2}{c^2} DH^2(t)\frac{d^2}{3}\Delta = 0.$$

Writing H_c for the static critical field of a thin film

$$\frac{\tau_s}{3}\pi^2(T_c^2 - T^2) = \frac{4e^2}{c^2} H_c^2 \frac{d^2}{3},$$

we get

$$\dot{\Delta} + \frac{\tau_s \Delta^3}{6} - \beta[1 - h^2(t)]\Delta = 0,$$

where

$$\beta = \frac{4e^2 D d^2}{3c^2} H_c^2, \quad h^2 = \frac{H^2}{H_c^2}.$$

This equation can be solved in its general form. Introducing a new function η:

$$\Delta = \eta \exp\left\{\beta \int_0^t [1 - h^2(\tau)]\, d\tau\right\},$$

we find

$$-\frac{\dot{\eta}}{\eta^3} = \frac{\tau_s}{6}\exp\left\{2\beta \int_0^t [1 - h^2(\tau)]\, d\tau\right\},$$

whence

$$\Delta = \frac{\exp\left\{\beta \int_0^t [1 - h^2(\tau)]\, d\tau\right\}}{\left(C + \frac{\tau_s}{3}\int_0^t \exp\left\{2\beta \int_0^\tau [1 - h^2(\tau_1)]\, d\tau_1\right\} d\tau\right)^{1/2}} \tag{30}$$

Here C is some initial value. The behavior of the expression at large values of the time is determined by the law $1 - \overline{h^2}$ where the bar indicates the average value of $h^2(\tau)$. If $1 - \overline{h^2} > 0$,

$$\int_0^t \exp\left\{2\beta \int_0^\tau [1 - h^2(\tau_1)]\, d\tau_1\right\} d\tau = \frac{\exp\{\beta t(1 - \overline{h^2})\}}{2\beta(1 - \overline{h^2})}$$

and we can neglect the initial condition. In the opposite case Δ from (30) decreases exponentially with time.

If the external field changes harmonically, $H = H_0 \sin \omega t$,

$$\Delta = \frac{\exp\left\{\beta(1 - \overline{h^2})t + \frac{\beta}{2\omega}\overline{h^2}\sin 2\omega t\right\}}{\left(C + \frac{\tau_s}{3}\int_0^t \exp 2\left\{\beta(1 - \overline{h^2})\tau + \frac{\beta}{2\omega}\overline{h^2}\sin 2\omega\tau\right\} d\tau\right)^{1/2}}. \tag{31}$$

We transform the integral under the radical sign to the form

$$e^{2\beta(1-\overline{h^2})t}\int_0^t \exp 2\left\{-\beta(1 - \overline{h^2})u + \frac{\beta}{2\omega}\overline{h^2}\sin 2\omega(t - u)\right\} du.$$

If $\omega t \gg 1$ it is convenient to rewrite this last expression as

$$\frac{\exp\{2\beta(1 - \overline{h^2})t\}}{1 - \exp[-2\beta(1 - \overline{h^2})/\omega]}\frac{1}{2\omega}\int_0^{2\pi} \exp\frac{\beta}{\omega}[-(1 - \overline{h^2})\lambda + \overline{h^2}\sin(2\omega t - \lambda)]\, d\lambda.$$

Thus, when $1 - \overline{h^2} > 0$,

$$\Delta(t) = \frac{\exp\left[\frac{\beta}{2\omega}\overline{h^2}\sin 2\omega t\right]\left\{1 - \exp\left[-\frac{2\beta}{\omega}(1 - \overline{h^2})\right]\right\}^{1/2}}{\left(\frac{\tau_s\pi}{6\omega}\int_0^{2\pi} \exp\frac{\beta}{\omega}[-(1 - \overline{h^2})\lambda + \overline{h^2}\sin(2\omega t - \lambda)]\, d\lambda\right)^{1/2}}. \tag{31'}$$

We consider two limiting cases of high and low frequencies. When $\beta/\omega \ll 1$

$$\Delta(t) = \Delta_0\sqrt{1 - \overline{h^2}}, \tag{32}$$

where Δ_0 is the thermodynamic equilibrium value of Δ. In the opposite limiting case the values $\lambda \sim \omega/\beta \ll 1$ are important in the integral in (31′) and therefore

$$\Delta(t) = \left(\frac{6\omega}{\tau_s \int_0^{2\pi} \exp\left[-\frac{\beta}{\omega}(1 - h^2(t))\lambda\right] d\lambda}\right)^{1/2} = \Delta_0\left(\frac{1 - h^2(t)}{1 - \exp\left[-\frac{2\pi\beta}{\omega}(1 - h^2(t))\right]}\right)^{1/2},$$

where $h^2(t)$ is the instantaneous value of the field. As long as $1 - h^2(t) > 0$,

$$\Delta(t) = \Delta_0\sqrt{1 - h^2(t)}, \tag{33}$$

i.e., $\Delta(t)$ follows the field adiabatically. If $1 - h^2(t) < 0$,

$$\Delta(t) = \Delta_0\sqrt{h^2(t) - 1}\exp\left[-\frac{\pi\tau_s\Delta_0^2}{6\omega}(h^2(t) - 1)\right] \tag{33'}$$

is exponentially small.

It is of interest to note that according to (32) for high frequencies $\overline{H^2(t)}$ plays the role of an effective constant field. As a result superconductivity with respect to a constant current will, in particular, be destroyed in thin films when $\overline{H^2(t)} = H_c^2$.

In conclusion we note that we are talking about frequencies and fields of the following order of magnitude:[3]

$$\omega / T_c \sim (\Delta / T_{c0})^2, \quad H \sim H_{c0}(\Delta / T_{c0})^2,$$

where H_{c0} and T_{c0} are the critical field and the transition temperature of a pure superconductor. These conditions are realistic ones.

The authors are grateful to L. P. Pitaevskiĭ and A. I. Larkin for a number of useful discussions.

[1] E. Abrahams and T. Tsuneto, Phys. Rev. 152, 416 (1966).

[2] M. P. Kemoklidze and L. P. Pitaevskiĭ, Zh. Eksp. Teor. Fiz. 50, 243 (1966); 52, 1556 (1967) [Soviet Phys.-JETP 23, 160 (1966); 25, 1036 (1967)].

[3] A. A. Abrikosov and L. P. Gor'kov, Zh. Eksp. Teor. Fiz. 39, 1781 (1960) [Sov. Phys.-JETP 12, 1243 (1961)].

[4] J. Kondo, Prog. Theoret. Phys. 32, 37 (1964).

[5] L. P. Gor'kov and A. I. Rusinov, Zh. Eksp. Teor. Fiz. 46, 1363 (1964) [Sov. Phys.-JETP 19, 922 (1964)].

[6] A. A. Abrikosov, L. P. Gor'kov, and I. È. Dzyaloshinskiĭ, Metody kvantovoĭ teorii polya v statisticheskoĭ fizike (Quantum Field Theoretical Methods in Statistical Physics) Fizmatgiz, 1962 [English translation published by Pergamon Press, 1965].

[7] L. V. Keldysh, Zh. Eksp. Teor. Fiz., 47, 1515 (1964) [Sov. Phys.-JETP 20, 1018 (1965)].

[8] L. D. Landau and E. M. Lifshitz, Elektrodinamika sploshnykh sred (Electrodynamics of Continuous Media) Gostekhizdat, 1957 [English translation published by Pergamon Press, 1960].

Translated by D. ter Haar
74

SOVIET PHYSICS JETP VOLUME 33, NUMBER 6 DECEMBER, 1971

VISCOUS VORTEX FLOW IN SUPERCONDUCTORS WITH PARAMAGNETIC IMPURITIES

L. P. GOR'KOV and N. B. KOPNIN

L. D. Landau Institute of Theoretical Physics, USSR Academy of Sciences

Submitted December 31, 1970

Zh. Eksp. Teor. Fiz. 60, 2331–2343 (June, 1971)

An expression for the conductivity of a vortex lattice in fields $H \ll H_{c2}$ is obtained on the basis of the nonstationary equations for superconductors with paramagnetic impurities. The case of high temperatures (small paramagnetic impurity concentrations) leads effectively to a conductivity greater than in the low temperature region. This agrees qualitatively with the experiments with ordinary alloys. Macroscopic electrodynamics equations are written down for vortex-structure superconductors.

1. INTRODUCTION AND FORMULATION OF PROBLEM

As demonstrated by a number of experiments (see, e.g.,[1]), superconductors of the second kind (alloys) have a finite resistance in a magnetic field exceeding the lower critical value H_{c1}. The nature of these dissipative effects is due to motion of the Abrikosov vortex structure[2] under the influence of the Lorentz force resulting from the flow of current[3-5]. Attempts to study this very interesting question were based mainly on phenomenological models (see[3-8]). With the exception of the work of Bardeen and Stephen[5], an analogy is usually suggested with the model of an ideal charged liquid, which apparently has little bearing even on the case of a pure superconductor of the second kind, owing to the microscopic dissipative effects. The latter are due to the existence of gapless excitations in the center of the vortex filament[9]. In particular, an ideal model would give a large Hall angle, in direct contradiction to an experiment[10] in which the Hall effect turned out to be of the same order as in the normal metal. Bardeen and Stephen[5] attempted to take phenomenological account of these spectrum singularities and were thus able to obtain the correct value of the resistance.

The motion of the vortex structure was investigated on the basis of the BCS microscopic theory of superconductivity only in the vicinity of the upper critical field H_{c2} (see[11]), where the resistance due to this motion constitutes a small correction. The general case calls for the use of equations that describe the dynamic properties of the superconductors. A theory that includes nonstationary processes within the framework of the BCS microscopic theory of superconductivity was constructed in[12]. In the present article we confine ourselves to a study of superconducting alloys with paramagnetic impurities, where it turns out that the problem of vortex motion admits of an analytic solution. Ordinary alloys or a pure superconductor of the second kind can be handled in the same manner, but additional difficulties arise, connected with the complex picture of the excitation spectrum in the core of the vortex[9,13].

An attempt to solve the problem in alloys with paramagnetic impurities was made by Baba and Maki[14]. The results obtained below for the resistance in fields $H \ll H_{c2}$ do not agree with the results of Baba and Maki. The point is that in[14] all the expressions for the physical quantities, for example for the current density, are expanded simultaneously in terms of the electric field intensity and the magnitude of the superconducting parameter Δ. For $H \ll H_{c2}$ this, of course, is incorrect, since the dissipation of the current (the resistance) is still small and is due only to the motion of the structure if one deals with small currents. Finally, Baramidze[15] considered the spectrum of the "oscillations" of an individual filament in the same model (the spectrum has here a diffuse character). Some inaccuracies in Baramidze's result will be pointed out in Sec. 2.

We assume throughout that the parameter of the Ginzburg-Landau theory is $\kappa \gg 1$. Because of this, the distribution of all the quantities (fields and currents) can be investigated separately far from the vortex filaments and close to them. The matching of the resultant expressions gives the solution of our problem. Such an approach is valid so long as the distance d between individual filaments is large compared with the dimension of the core of the vortex. In other words, the induction $B = n\phi_0 \ll H_{c2}$ (n is the density of the filaments and ϕ_0 is the flux quantum).

The equations have the simplest form in the case of a superconductor with large paramagnetic-impurity concentration, $\tau_s T_c \ll 1$[12]. Putting $\Delta = |\Delta| e^{i\theta}$, $\mathbf{Q} = \mathbf{A} - (c/2e)\nabla\theta$ and $\mu = \dot{\theta} + 2e\varphi$, where ω and $\mathbf{A}$ are the scalar and vector potentials and θ is the phase of the ordering parameter, we write the equations in the form

$$|\dot{\Delta}| + \frac{\tau_s}{3}\left[-\pi^2(T_c^2 - T^2) + \frac{|\Delta|^2}{2}\right]|\Delta| - D\nabla^2|\Delta| + \frac{4e^2}{c^2}DQ^2|\Delta| = 0,$$

$$\frac{c}{2e}|\Delta|^2\mu + D\,\mathrm{div}[|\Delta|^2\mathbf{Q}] = 0, \qquad \mathbf{j} = \sigma\mathbf{E} - \frac{2\sigma\tau_s}{c}|\Delta|^2\mathbf{Q}.$$

Here $D = v_F l/3$ is the diffusion coefficient and τ_s the time between collisions of the electron with the impurity atoms relative to the spin flip. In addition, we have the relations

$$\mathrm{div}\,\mathbf{j} = 0, \quad \mathrm{rot}\,\mathbf{Q} = \mathbf{H}, \quad \mathrm{rot}\,\mathbf{H} = \frac{4\pi}{c}\mathbf{j}, \quad \mathbf{E} = -\frac{1}{c}\dot{\mathbf{Q}} - \frac{1}{2e}\nabla\mu.$$

The convenience of this notation lies in the fact that the combinations of $\mathbf{Q}$ and μ are gauge-invariant.

We introduce the following dimensionless variables:

$$|\Delta| = \Delta_\infty \Delta', \quad \omega = 2\tau_s \Delta_\infty^2 \omega', \quad r = \delta r',$$

$$Q = \frac{\phi_0}{2\pi\xi} Q', \quad j = \frac{c\phi_0}{8\pi^2\delta^2\xi} j', \quad \mu = 2\tau_s \Delta_\infty^2 \mu'.$$

Here $\phi_0 = \pi\hbar c/e$ is the flux quantum, $\Delta_\infty^2 = 2\pi^2(T_c^2 - T^2)$, $\delta = (8\pi\sigma\tau_s\Delta_\infty^2)^{-1/2}c$—the depth of penetration, and $\xi = (6D/\tau_s\Delta_\infty^2)^{1/2}$—the coherence radius, with $\kappa = \delta/\xi = c(48\pi D\sigma)^{-1/2} \gg 1$ as the parameter.

In terms of this notation, the equations take the form (the primes designating the reduced quantities will henceforth be omitted):

$$12\dot{\Delta} + (\Delta^2 - 1 + Q^2)\Delta - \frac{1}{\varkappa^2}\nabla^2\Delta = 0, \tag{1}$$

$$12\Delta^2\mu + \frac{1}{\varkappa}\operatorname{div}[\Delta^2 Q] = 0, \tag{2}$$

$$j = \operatorname{rot}\operatorname{rot} Q = -\dot{Q} - \frac{1}{\varkappa}\nabla\mu - \Delta^2 Q. \tag{3}$$

The condition div j = 0 jointly with (2) yields

$$12\Delta^2\mu - \frac{1}{\varkappa^2}\nabla^2\mu = \frac{1}{\varkappa}\operatorname{div}\dot{Q}. \tag{4}$$

We write out also the connection between the fields and the potentials:

$$H = \operatorname{rot} Q, \quad E = -Q - \frac{1}{\varkappa}\nabla\mu.$$

In the case of an immobile lattice of vortex filaments, $\dot{Q} = \dot{\Delta} = \mu = 0$ and Eqs. (1)–(3) turn into the ordinary Ginzburg-Landau equations. The solutions of the static equations for an isolated vortex were obtained in[2]. In cylindrical coordinates with z axis directed parallel to the magnetic field H_0 we have, for a vortex containing one flux quantum, $\theta = \varphi$ (φ is the polar angle) and $\Delta = f(r)$, where f satisfies the relation

$$\frac{1}{\varkappa^2}\left(\frac{d^2 f}{dr^2} + \frac{1}{r}\frac{df}{dr}\right) + (1 - f^2)f - Q_0^2 f = 0, \tag{5}$$

Q has only the component $Q_\varphi \equiv Q_0$, satisfying the equation

$$\frac{d}{dr}\frac{1}{r}\frac{d}{dr}(rQ_0) - f^2 Q_0 = 0. \tag{6}$$

Since the scale of variation of f is of the order of $1/\kappa$, which is much smaller than the scale of variation of Q_0 (the latter is of the order of unity), we get from (5) and (6) $Q_0 = -\kappa^{-1}K_1(r)$; f is not expressed in analytic form, but when $r \gg 1/\kappa$ we have $f = 1 - 1/2\kappa^2 r^2$, and when $r \ll 1/\kappa$ we have $f \backsim \kappa r$.

Equations (2) and (4) are satisfied identically.

2. FIELD IN THE VICINITY OF AN INDIVIDUAL MOVING FILAMENT

In the presence of a current, the vortex lattice begins to move and becomes deformed. So long as the current is small, the rates of motion and of deformation are also small. We choose on one of the filaments a point with which we connect the cylindrical coordinate system (ρ, φ, z), directing the z axis along the undeformed filament. Then the coordinates of the remaining points along the filament are characterized by a two-dimensional deformation vector $u(z, t)$. The coordinates along the remaining filaments $\rho_{0i} + u_i(z, t)$, where ρ_{0i} is the equilibrium position of the i-th filament in the absence of deformations.

Let us consider first the picture of motion in the vicinity of one vortex. For an individual filament, the zeroth approximation $(\tilde{f}_0, \tilde{Q}_0)$ corresponds to a displacement and rotation of the unperturbed solutions

$$\tilde{f}_0(r) = f(\rho) - (u(z,t)\,\nabla)f(\rho),$$
$$\tilde{Q}_0 = Q_0(\rho) - (u(z,t)\nabla)Q_0 + [\delta\varphi Q_0], \tag{7}$$

where

$$\delta\varphi = \left[n\frac{\partial u}{\partial z}\right]. \tag{8*}$$

Here f and $Q_0 = (0, Q_0, 0)$ are the solutions of Eqs. (5) and (6), and the vector n is directed along the z axis. The solutions (7) satisfy identically the system of equations (1)–(4) if we neglect the dependence of u on t and $\partial u/\partial z$ on z. We shall seek corrections to solutions (7) and (8) in the form

$$\Delta = \tilde{f}_0 + \chi, \quad Q = \tilde{Q}_0 + q, \tag{9}$$

where the quantities χ and q are proportional, as we shall show below, to $\partial u/\partial t$ and to $\kappa^{-2}\partial^2 u/\partial z^2$. We note immediately that from (4) it follows that μ is small in this approximation, since the zeroth approximation $\tilde{Q}_0$ satisfies the condition div $\tilde{Q}_0 = 0$. We can similarly verify the smallness of the component q_z. The term $\dot{\Delta}$ in (1) is transformed into $(\dot{u}\cdot\nabla)f$ and give the right-hand side, from the form of which we conclude that the dependence of the solution (χ, q) on the angles φ is conveniently sought in a form proportional to $e^{i\varphi}$. Omitting the intermediate steps, we present the equations obtained for the components (χ, q) from (1)–(3) after substituting in them (7)–(9) (we put henceforth $u = u_x - iu_y$):

$$\frac{1}{\varkappa^2}\left[\frac{1}{\rho}\frac{d}{d\rho}\left(\rho\frac{d\chi}{d\rho}\right) - \frac{\chi}{\rho^2}\right] - (3f^2 - 1)\chi - Q_0^2\chi - 2fQ_0 q_\varphi$$
$$= \left(12\dot{u} - \frac{\partial^2 u}{\partial z^2}\right)\psi, \tag{10a}$$

$$2\psi q_\rho + f\left[\frac{q_\rho}{\rho} + \frac{dq_\rho}{d\rho}\right] + \frac{if}{\rho}q_\varphi + \frac{2iQ_0\chi}{\rho} = i\frac{\partial^2 u}{\partial z^2}Q_0, \tag{10b}$$

$$\left(f^2 + \frac{1}{\rho^2}\right)q_\rho + \frac{i}{\rho}\left(\frac{dq_\varphi}{d\rho} + \frac{q_\varphi}{\rho}\right) = i\frac{\partial^2 u}{\partial z^2}\frac{1}{\rho}\frac{d}{d\rho}(\rho Q_0). \tag{10c}$$

We have left out of the last of these equations the term resulting from $\dot{Q}$ in (3), which is small when $\kappa \gg 1$. By putting $\chi = \psi w$ we transform Eq. (10a):

$$2Q_0Q_0'fw + \frac{1}{\varkappa^2}\left[\left(2\psi' + \frac{\psi}{\rho}\right)w' + \psi w''\right] - 2fQ_0q_\varphi = \left(12\dot{u} - \frac{1}{\varkappa^2}\frac{\partial^2 u}{\partial z^2}\right)\psi. \tag{11}$$

We have used here the following relation for the function $\psi = df/d\rho$:

$$\frac{1}{\varkappa^2}\frac{1}{\rho}\frac{d}{d\rho}\left(\rho\frac{d\psi}{d\rho}\right) - \frac{1}{\varkappa^2\rho^2}\psi - (3f^2 - 1)\psi - Q_0^2\psi = 2Q_0Q_0'f, \tag{12}$$

which is derived from (5) by differentiation. Multiplying now (11) by $\rho\psi$, (10b) by $i\rho^2 fQ_0'$ and then adding, we obtain the integral of the system for a solution that is finite at zero:

$$-\left[12\dot{u} - \frac{1}{\varkappa^2}\frac{\partial^2 u}{\partial z^2}\right]\int_0^\rho \rho\psi^2 d\rho - \frac{\partial^2 u}{\partial z^2}\int_0^\rho \rho f^2(\rho Q_0^2)'d\rho$$
$$= \rho\psi^2 w' + i\rho^2 f^2 Q_0' q_\rho - \rho Q_0 f^2 q_\varphi.$$

Recognizing that the function ψ decreases rapidly when

*$[n\dot{u}] \equiv n \times \dot{u}$.

$\rho \gg \kappa^{-1}$, and $f \to 1$, we can write in the region $\kappa^{-1} \ll \rho \ll 1$ 1)

$$-12\dot{u}\int_0^\infty \rho\psi^2 d\rho + \frac{\partial^2 u}{\partial z^2}\left[\frac{1}{\varkappa^2}\int_0^\infty \rho\psi^2 d\rho - \int_0^\rho \rho f^2(\rho Q_0^2)' d\rho\right] = \frac{1}{\varkappa}(q_\varphi + iq_\rho), \tag{13'}$$

where $\int_0^\infty \rho\psi^2 d\rho = \gamma \approx 0.247$. The term in the square brackets can be rewritten in the form

$$\int_0^\infty\left[\frac{1}{\varkappa^2}\left(\frac{df}{d\rho}\right)^2 - f^2\frac{d}{d\rho}(\rho Q_0^2)\right]\rho d\rho + \int_\rho^\infty \rho(\rho Q_0^2)' d\rho = \frac{\varepsilon}{2\pi} + \int_\rho^\infty \rho(\rho Q_0^2)' d\rho,$$

where, according to[2], $\epsilon \approx 2\pi\kappa^{-2}(\ln\kappa + 0.08)$ is the energy per unit length of the vortex. The obtained relation can be written in vector form. Since $j_1 = -q$ when $\rho \gg \kappa^{-1}$, we find that

$$j_1 = -q = 6\varkappa\gamma[n\dot{u}] - \frac{\varkappa\varepsilon}{4\pi}\left[n\frac{\partial^2 u}{\partial z^2}\right] - \frac{\varkappa}{2}\int_\rho^\infty (\rho Q_0^2)'\rho d\rho\left[n\frac{\partial^2 u}{\partial z^2}\right]. \tag{14*}$$

When $\rho \ll 1$ we have, accurate to terms $\sim\rho^2$,

$$\int_\rho^\infty \rho(\rho Q_0^2)' d\rho \approx -\frac{1}{\varkappa^2}K_0(\rho) - \frac{1}{2\varkappa^2}.$$

Formula (14) gives an expression for the correction to the current near the vortex; this correction is necessitated by the motion and deformation of the vortex.

3. FIELD AT LARGE DISTANCES FROM THE CORE OF THE VORTEX FILAMENT

To find the current produced by all vortices in the vicinity $\kappa^{-1} \ll r \ll d$ of a given vortex, we proceed as follows. The microscopic magnetic field h outside the core ($r \gg \kappa^{-1}$) is described by the Londons' generalized equation (cf., e.g.,[7]). In the usual units

$$\delta^2 \operatorname{rot}\operatorname{rot} \mathbf{h} + \mathbf{h} = \Phi_0 \sum_j \int d\mathbf{S}_j \delta(\mathbf{R} - \mathbf{R}_j).$$

Here $d\mathbf{S}_j$ is an element of length along the j-th filament, $\delta(\mathbf{R} - \mathbf{R}_j)$ a three-dimensional δ function, and $\mathbf{R}_j$ a three-dimensional vector drawn to a point on the j-th filament.

From this we easily obtain an expression for the magnetic field

$$\mathbf{h}(\mathbf{R}) = \frac{\Phi_0}{4\pi\delta^2}\sum_j \int d\mathbf{S}_j \frac{1}{|\mathbf{R} - \mathbf{R}_j|}\exp\left[-\frac{|\mathbf{R} - \mathbf{R}_j|}{\delta}\right]$$

and for the current

$$\mathbf{j}(\mathbf{R}) = -\frac{c\Phi_0}{16\pi^2\delta^2}\sum_j \int\left[d\mathbf{S}_j \operatorname{grad}\left(\frac{1}{|\mathbf{R} - \mathbf{R}_j|}\exp\left\{-\frac{|\mathbf{R} - \mathbf{R}_j|}{\delta}\right\}\right)\right]$$

where $\mathbf{R}_j = (\rho_{0j} + u_j; z_j)$. We put $\Phi[(\mathbf{R} - \mathbf{R}_j)^2] = |\mathbf{R} - \mathbf{R}_j|^{-1}\exp(-|\mathbf{R} - \mathbf{R}_j|/\delta]$. We expand (15) in powers of a small displacement u_j, after first writing $j(\mathbf{R})$ in the form

$$\mathbf{j}(\mathbf{R}) = -\frac{c\Phi_0}{8\pi^2\delta^2}\sum_j \int [d\mathbf{S}_j(\mathbf{R} - \mathbf{R}_j)]\Phi'[(\mathbf{R} - \mathbf{R}_j)^2].$$

Since $d\mathbf{S}_j = (\mathbf{n} + \partial\mathbf{u}_j/\partial z)dz_j$, where $\mathbf{n} \parallel \mathbf{H}_0$, we get

$$\mathbf{j}(\mathbf{R}) = -\frac{c\Phi_0}{8\pi^2\delta^2}\sum_j\Big\{\int[\mathbf{n}\rho_{0j}]\Phi'[(\mathbf{R} - \mathbf{R}_{0j})^2]dz_j - \int[\mathbf{n}\mathbf{u}_j]\Phi'[(\mathbf{R} - \mathbf{R}_{0j})^2]dz_j - 2\int[\mathbf{n}\rho_{0j}](\mathbf{u}_j\rho_{0j})\Phi''[(\mathbf{R} - \mathbf{R}_{0j})^2]dz_j$$
$$+ \int\left[\frac{\partial\mathbf{u}_j}{\partial z}\mathbf{n}\right]z_j\Phi'[(\mathbf{R} - \mathbf{R}_{0j})^2]dz_j + \int\left[\frac{\partial\mathbf{u}_j}{\partial z}\rho_{0j}\right]\Phi'[(\mathbf{R} - \mathbf{R}_{0j})^2]dz_j. \tag{16}$$

Assuming that u varies slowly over distances on the order of δ (for a loose lattice, $d \gg \delta$, it is necessary to stipulate that **u** vary over distances on the order of d in a plane perpendicular to the constant magnetic field), we represent u_j in the form

$$\mathbf{u}_j = \mathbf{u}^0 + (\rho_{0j\alpha}\nabla_\alpha)\mathbf{u} + z_j\frac{d\mathbf{u}}{dz} + \frac{1}{2}(\rho_{0j\alpha}\rho_{0j\beta}\nabla_\alpha\nabla_\beta)\mathbf{u} + \left(\rho_{0j\alpha}\nabla_\alpha z_j\frac{\partial}{\partial z}\right)\mathbf{u} + \frac{1}{2}z_j^2\frac{\partial^2\mathbf{u}}{\partial z^2}$$

(α and β are two-dimensional vector indices). For the selected vortex we have

$$\mathbf{u} = z\frac{\partial\mathbf{u}}{\partial z} + \frac{1}{2}z^2\frac{\partial^2\mathbf{u}}{\partial z^2}.$$

Substituting in (16) the expansion for u and integrating with respect to z from $-\infty$ to $+\infty$ and summing over a lattice with hexagonal symmetry, we obtain near the selected vortex (we can put $\rho = 0$ in the regular terms)

$$\mathbf{j}_1 = \frac{c\Phi_0}{16\pi^2\delta^2}\Big\{[\mathbf{n}\nabla\operatorname{div}\mathbf{u}]\sum_j{}'\frac{1}{2}\rho_{0j}^4\int\Phi''(\mathbf{R}_{0j}^2)dz + [\mathbf{n}\nabla^2\mathbf{u}]\sum_j{}'\left[\frac{1}{4}\rho_{0j}^4\int\Phi''(\mathbf{R}_{0j}^2)dz + \frac{1}{2}\rho_{0j}^2\int\Phi'(\mathbf{R}_{0j}^2)dz\right]$$
$$+ \left[\mathbf{n}\frac{\partial^2\mathbf{u}}{\partial z^2}\right]\sum_j{}'\left[\frac{1}{2}\int\Phi(\mathbf{R}_{0j}^2)dz - \frac{1}{2}\rho_{0j}^2\int\Phi'(\mathbf{R}_{0j}^2)dz\right] \tag{17}$$
$$- \frac{1}{2}\operatorname{rot}\frac{\partial\mathbf{u}}{\partial z}\sum_j{}'\left[\rho_{0j}^2\int\Phi'(\mathbf{R}_{0j}^2)dz\right] + \left[\mathbf{n}\frac{\partial^2\mathbf{u}}{\partial z^2}\right]\left[\frac{1}{2}\int\Phi(\mathbf{R}^2)dz - \frac{1}{2}\rho^2\int\Phi'(\mathbf{R}^2)dz\right]\Big\}.$$

Here j_1, as in (14), in a correction to the distribution of the current of the undeformed lattice. Since the coordinate system was chosen with its origin coinciding at any given instant with the undeformed vortex, there is no current corresponding to the shift in (17). The selected vortex drops out of the sum. The last term in (17) corresponds precisely to the increments to the velocity field of the selected filament. The sums in (17) are expressed in terms of the elastic constants of a triangular lattice of vortex filaments, obtained by Labusch[17]. We ultimately obtain

$$\mathbf{j}_1 = \frac{c}{n\Phi_0}\Big\{(c_{11} - c_{66})[\mathbf{n}\nabla\operatorname{div}\mathbf{u}] + c_{66}[\mathbf{n}\nabla^2\mathbf{u}] + c_{44}\left[\mathbf{n}\frac{\partial^2\mathbf{u}}{\partial z^2}\right] + (c_{12} - c_{66})\frac{\partial}{\partial z}\operatorname{rot}\mathbf{u}\Big\} -$$
$$- \frac{c\Phi_0}{16\pi^2\xi^2}\frac{\varepsilon}{2\pi}\left[\mathbf{n}\frac{\partial^2\mathbf{u}}{\partial z^2}\right] + \frac{c\Phi_0}{16\pi^2\delta^2}\left[\mathbf{n}\frac{\partial^2\mathbf{u}}{\partial z^2}\right]\left\{K_0\left(\frac{\rho}{\delta}\right) + \frac{\rho}{2\delta}K_1\left(\frac{\rho}{\delta}\right)\right\}, \tag{18}$$

1)Baramidze [15] investigated the oscillations of an isolated filament in the same model, using the condition of translational invariance of the initial equations. He found that the dispersion law is given by $\omega \sim -i(k^2/\kappa^2)\ln(k/\kappa)$ for all k. It is seen from (13) that by setting q_ρ, q_φ, and w′ approach zero for an isolated filament as $\rho \to \infty$, we obtain $\omega \sim -i(k^2/\kappa^2)\ln\kappa$. Baramidze's result is valid only if $\kappa \gg k \gg 1$. The condition of Galilean invariance was used earlier by Pitaevskii in a paper on oscillations of a vortex in a weakly-nonideal Bose gas [16].

*$[\mathbf{n}\dot{\mathbf{u}}] \equiv \mathbf{n} \times \dot{\mathbf{u}}$.

where the constants c are given by

$$c_{11}=\frac{B^2}{4\pi}\frac{\partial H}{\partial B}+\frac{1}{8\pi}\int_0^B x^2H''(x)dx,$$

$$c_{44}=\frac{BH(B)}{4\pi},\quad c_{66}=\frac{1}{8\pi}\int_0^B x^2H''(x)dx.$$

In dimensionless units ($c = \frac{1}{2}\kappa(\phi_0/2\pi\delta^2)^2c'$) we have

$$j_1'=\frac{1}{B'}\Big\{(c_{11}'-c_{66}')[n\nabla\,\mathrm{div}\,u]+c_{66}'[n\nabla^2u]$$

$$+c_{44}'\Big[n\frac{\partial^2u}{\partial z^2}\Big]+(c_{12}'-c_{66}')\frac{\partial}{\partial z}\,\mathrm{rot}\,u\Big\}$$

$$+\frac{1}{2\kappa}\Big[n\frac{\partial^2u}{\partial z^2}\Big]\Big\{K_0(\rho)+\frac{\rho}{2}K_1(\rho)-\frac{\kappa^2\varepsilon}{2\pi}\Big\}. \qquad (18')$$

Expression (18′) must be matched to (14) in the region $\kappa^{-1}\ll\rho\ll 1$. As a result we obtain an equation describing the motion of the vortex structure:

$$\frac{1}{B'}\Big\{(c_{11}'-c_{66}')[n\nabla\,\mathrm{div}\,u]+c_{66}'[n\nabla^2u]$$

$$+c_{44}'\Big[n\frac{\partial^2u}{\partial z^2}\Big]\Big\}=6\kappa\gamma[n\dot{u}]. \qquad (19)$$

In the usual units we have

$$6\gamma H_{c2}\frac{\sigma}{c^2}B[n\dot{u}]=(c_{11}-c_{66})[n\nabla\,\mathrm{div}\,u]+c_{66}[n\nabla^2u]+c_{44}[n\frac{\partial^2u}{\partial z^2}]. \qquad (20)$$

The component along the z axis in (18′) does not enter in the condition (14).

4. RESISTANCE OF A SAMPLE WITH VORTEX FILAMENTS

For macroscopic problems it is necessary to relate the fields **E** and **H** as well as the macroscopic current j_e flowing through the sample with the lattice deformation. To this end we write down the thermodynamic Gibbs potential (see[18]) accurate to terms of order $(\nabla_i u_k)^2$ and $(\partial u_i/\partial z)^2$:

$$F=F_0+\int\Big\{\Big[\frac{1}{2}(c_{11}-c_{66})\Big(\frac{\partial u_i}{\partial x_i}\Big)^2+c_{66}\Big(\frac{\partial u_k}{\partial x_i}\Big)^2$$

$$+c_{44}\Big(\frac{\partial u_i}{\partial z}\Big)^2\Big]-\frac{H-H_0}{4\pi}B\Big\}dV;$$

here F_0 corresponds to the undeformed lattice, and i, k = 1, 2. Deformation changes the induction **B** by an amount

$$\Delta B=B_0\frac{\partial u}{\partial z}-B_0 n\frac{\partial u_i}{\partial x_i},\quad B_0=n\phi_0 \qquad (21)$$

By varying the potential over the deformation vector **u**, we obtain

$$\delta F=\int\Big\{-\Big[(c_{11}-c_{66})\nabla\,\mathrm{div}\,u+c_{66}\nabla^2u+c_{44}\frac{\partial^2u}{\partial z^2}\Big]$$

$$+\frac{B_0}{4\pi}\Big[\frac{\partial H}{\partial z}-\nabla(H_z-H_0)\Big]\Big\}\delta u\,dV$$

$$+\int\Big(c_{44}\frac{\partial u}{\partial z}-\frac{B_0}{4\pi}H\Big)\delta u\,ds+\int\Big[(c_{11}-c_{66})\mathrm{div}\,u\delta_{ik}$$

$$+c_{66}\frac{\partial u_k}{\partial x_i}+\frac{B_0}{4\pi}(H_z-H_0)\delta_{ik}\Big]\delta u_k d\sigma_i,$$

where ds is a surface element perpendicular to the magnetic field H_0, and $d\sigma$ is parallel to H_0.

The condition that the potential be a minimum yields

$$(c_{11}-c_{66})\nabla\,\mathrm{div}\,u+c_{66}\nabla^2u+c_{44}\frac{\partial^2u}{\partial z^2}=\frac{B_0}{4\pi}\Big[\frac{\partial H}{\partial z}-\nabla(H_z-H_0)\Big] \qquad (22)$$

(all the vectors and the vector differentiations are two-dimensional). The boundary conditions supplementing (2) are

$$\nu_k(c_{11}-c_{66})\mathrm{div}\,u+c_{66}\nu_i\frac{\partial u_k}{\partial x_i}+c_{44}\nu_z\frac{\partial u_k}{\partial z}$$

$$=\nu_z\frac{B_0H_k}{4\pi}-\nu_k\frac{B_0(H_z-H_0)}{4\pi} \qquad (23)$$

($\nu = (\nu_i; \nu_z)$ is the vector normal to the surface).

Calculating curl **H**, we readily see that

$$\Big[n\frac{\partial H}{\partial z}\Big]-[n\nabla(H_z-H_0)]=\frac{4\pi}{c}j_e,$$

therefore

$$(c_{11}-c_{66})[n\nabla\,\mathrm{div}\,n]+c_{66}[n\nabla^2u]+c_{44}\Big[n\frac{\partial^2u}{\partial z^2}\Big]=\frac{B_0}{c}j_e. \qquad (24)$$

Comparing (24) with (20) we see that

$$j_e=6\gamma H_{c2}\sigma[n\dot{v}]/c.$$

The vortex motion velocity $v = \dot{u}$ is perpendicular to the current, and consequently there is no Hall effect in this approximation.

Motion of the vortex lattice gives rise to an electric field **E**, which can be determined from Maxwell's equation curl $E = -c^{-1}\partial B/\partial t$, where **B** is expressed in terms of **u** by means of (21):

$$E=B_0[nv]/c. \qquad (25)$$

Thus, the superconductor in a field $B \ll H_{c2}$ has a finite conductivity

$$\sigma_{\mathrm{eff}}=6\gamma\sigma H_{c2}/B_0. \qquad (26)$$

Experiments[1] yielded $\sigma_{\mathrm{eff}}/\sigma \approx H_{c2}/B_0$.

Equation (22) has the form of the equilibrium conditions of an elastically deformed lattice $\partial\sigma_{ik}/\partial x_k = F_i$, where the role of the external force is played by the Lorentz force $F = c^{-1}j_e \times B_0$. Taking the force acting on one vortex, we find that $F = \eta v$, where the "viscosity" coefficient is

$$\eta=6\gamma\sigma\phi_0 c^{-2}H_{c2}.$$

The derived equations make it possible to describe the reflection of an electromagnetic wave from the surface of a superconductor with a vortex structure. Let us consider a plane wave incident normally on the surface of the superconductor. The intensity of the alternating electric and magnetic fields will be denoted by $E_\sim$ and $H_\sim$. Two geometries are possible: $H_\sim \parallel H_0$ and $H_\sim \perp H_0$. These are shown in Figs. 1 and 2. The z axis is always parallel to H_0.

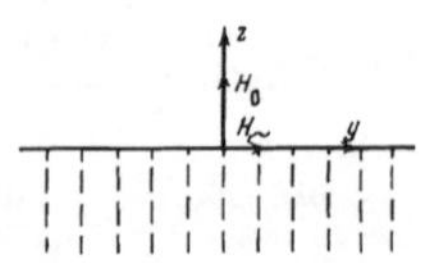

FIG. 1. Alternating magnetic field perpendicular to the vortex filaments (the dashed lines represent the vortex filaments).

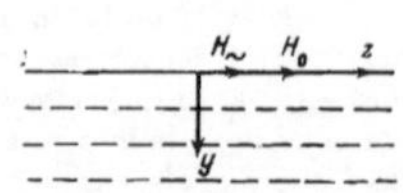

FIG. 2. Alternating magnetic field parallel to the vortex filaments.

Let first $\mathbf{H}_\sim \perp \mathbf{H}_0$, and then $H_{\sim z} = 0$. In terms of (22) and (23) this means that $H_z - H_0 = 0$. It follows from the boundary conditions (23) that on the surface $z = 0$ (the superconductor fills the half-space $z < 0$) we have

$$c_{44}\frac{\partial \mathbf{u}}{\partial z} = \frac{B_0 \mathbf{H}_\sim}{4\pi}.$$

By virtue of the homogeneity in x and y, all the quantities depend only on z. From (22) it follows that

$$c_{44}\frac{\partial^2 \mathbf{u}}{\partial z^2} = \frac{B_0}{4\pi}\frac{\partial \mathbf{H}_\sim}{\partial z}.$$

Since $\mathbf{H}_\sim$ is continuous on the boundary, the magnetic field inside the superconductor is

$$H_\sim = \frac{4\pi c_{44}}{B_0}\frac{\partial \mathbf{u}}{\partial z}.$$

Expression (21) yields

$$\mathbf{B}_\sim = B_0 \partial \mathbf{u} / \partial z,$$

here $\mathbf{B}_\sim$ is the induction produced by the alternating field. Comparing the last two formulas, we see that $\mathbf{B}_\sim = \mu_\perp \mathbf{H}_\sim$, where $\mu_\perp = B_0^2/4\pi c_{44} = B_0/H_0$.

If $\mathbf{H}_\sim \parallel \mathbf{H}_0$, then $H_z - H_0 = H_{\sim z}$ in (22) and (23). From the boundary conditions we find that when $y = 0$

$$\frac{\partial u_x}{\partial y} = \frac{\partial u_x}{\partial z} = \frac{\partial u_y}{\partial z} = 0;$$

$$c_{66}\frac{\partial u_y}{\partial y} + (c_{11} - c_{66})\operatorname{div}\mathbf{u} = -\frac{B_0}{4\pi}H_{\sim z}.$$

By virtue of the homogeneity in x and z, all the quantities depend only on y. From (22) it follows that $u_x = 0$ and

$$c_{11}\frac{\partial^2 u_y}{\partial y^2} = -\frac{B_0}{4\pi}\frac{\partial H_{\sim z}}{\partial y}.$$

This means that inside the superconductor

$$-\frac{B_0}{4\pi}H_{\sim z} = c_{11}\frac{\partial u_y}{\partial y}.$$

The induction is $B_{\sim z} = -B_0\partial u_y/\partial y = H_{\sim z}B_0^2/4\pi c_{11}$. Thus, the "magnetic permeability" is $\mu_\parallel = B_0^2/4\pi c_{11}$.

We can now write our equations in the form of Maxwell's equations

$$\operatorname{rot}\mathbf{E}_\sim = -\frac{1}{c}\frac{\partial B_\sim}{\partial t}, \quad \operatorname{rot}\mathbf{H}_\sim = \frac{4\pi}{c}\mathbf{j}_\sim,$$

and $\mathbf{j}_\sim = \sigma_{\text{eff}}\mathbf{E}_\sim$ and $\mathbf{B}_\sim = \mu\mathbf{H}_\sim$, where μ is equal to $\mu_\perp$ or $\mu_\parallel$, and σ_{eff} is determined by (26). Proceeding in the usual manner, we can obtain the "skin" depth of penetration of the alternating field $\lambda_{\perp,\parallel}$, where

$$\lambda_\perp = (1+i)\frac{c}{\sqrt{8\pi\mu_\perp\sigma_{\text{eff}}\omega}} = (1+i)\frac{c}{\sqrt{48\pi\gamma\sigma\omega}}\left(\frac{H_0}{H_{c2}}\right)^{1/2},$$

$$\lambda_\parallel = (1+i)\frac{c}{\sqrt{8\pi\mu_\parallel\sigma_{\text{eff}}\omega}} = (1+i)\frac{c}{\sqrt{12\gamma\sigma\omega}}\left(\frac{c_{11}}{B_0 H_{c2}}\right)^{1/2}.$$

The impedance of the superconductor $\zeta = -\omega\lambda/c$ is, respectively

$$\zeta_\perp = (1-i)\sqrt{\frac{\omega}{48\pi\gamma\sigma}}\left(\frac{H_0}{H_{c2}}\right)^{1/2}, \quad \zeta_\parallel = (1-i)\sqrt{\frac{\omega}{12\gamma\sigma}}\left(\frac{c_{11}}{B_0 H_{c2}}\right)^{1/2}.$$

The last formulas are valid if $\lambda \gg \delta$. In addition, the depth of penetration should be large compared with the distances between filaments.

5. MOTION OF VORTICES IN A SUPERCONDUCTOR WITH SMALL PARAMAGNETIC-IMPURITY CONCENTRATION

So far we have considered the case of large paramagnetic-impurity concentrations, corresponding to a near-zero critical temperature. At finite temperatures, as is well known, the so-called anomalous terms in the nonstationary equations for the superconductors become significant. To ascertain their role, let us consider the limiting case of low paramagnetic-impurity concentrations $\tau_s T_c \gg 1$, when the anomalous terms are large. We assume for simplicity that the concentration of the nonmagnetic impurities is large ($\kappa \gg 1$). The corresponding equations were obtained by Éliashberg[12]. They can be written in the form

$$\frac{\pi}{8T_c}\{-|\dot\Delta| + D\nabla^2|\Delta| - DQ^2|\Delta|\} + \left[\frac{T_c - T}{T_c} - \frac{7\zeta(3)|\Delta|^2}{8(\pi T_c)^2}\right]|\Delta| + |\Delta| U_1 = 0$$

$$\dot U_1 - D\nabla^2 U_1 = -\frac{\pi\tau_s}{8T_c}|\Delta||\dot\Delta|.$$

$$\frac{\pi}{8T_c}\{-|\Delta|^2\mu - D\operatorname{div}[|\Delta|^2\mathbf{Q}]\} + |\Delta|^2\overline{U}_\gamma = 0$$

$$\overline{U}_\gamma = -\frac{i\tau_1}{4\tau_s}\int\frac{\gamma_\varepsilon d\varepsilon}{\varepsilon^2 + \tau_s^{-2}},$$

$$\gamma_\varepsilon - D\nabla^2\gamma_\varepsilon = \frac{1}{\tau_1}\frac{1}{2T_c}\operatorname{ch}^{-2}\left(\frac{\varepsilon}{2T_c}\right)\frac{2i|\Delta|^2\mu}{\tau_s(\varepsilon^2 + \tau_s^{-2})} - \frac{i}{\tau_1}\frac{1}{2T_c}\operatorname{ch}^{-2}\left(\frac{\varepsilon}{2T_c}\right)(D\operatorname{div}\dot{\mathbf{Q}} + D\nabla^2\mu) - \frac{2|\Delta|^2}{\tau_s(\varepsilon^2 + \tau_s^{-2})}\gamma_\varepsilon,$$

$$\mathbf{j} = \sigma\mathbf{E} - \frac{mp_0 e}{4\pi c}\frac{|\Delta|^2}{T_c}D\mathbf{Q}.$$

A simplifying circumstance in this rather complicated system is the fact that owing to the electroneutrality condition $\operatorname{div}\mathbf{j} = 0$ the equations for γ_e and μ contain in the right-hand side terms proportional to $\operatorname{div}\dot{\mathbf{Q}}$. It can therefore be readily shown, as was already noted in Sec. 2, that μ, γ_e, and $\overline{U}_\gamma$ are small. An important role is therefore played only by the anomalous part of $\dot U_1$. Changing over to the dimensionless variables of the Ginzburg-Landau theory, and also introducing the time scale $t = \omega_0^{-1}t'$, where $\omega_0^{-1} = 14\zeta(3)/\pi\Delta_\infty^2$, and putting $U_1 = (7\zeta(3)\Delta_\infty^2/8\pi^2T_c^2)U_1'$, we obtain a dimensionless system in which Eq. (1) is replaced by the following two equations (the primes have been omitted, just as in the system (1)–(4)):

$$\frac{1}{\varkappa^2}\nabla^2\Delta - (\Delta^2 - 1 + Q^2)\Delta = a\dot\Delta - U_1\Delta, \tag{27}$$

$$a\dot U_1 - \frac{1}{\varkappa^2}\nabla^2 U_1 = -a\nu(\dot{\Delta^2}). \tag{28}$$

Here $a = \pi^4/14\zeta(3) \approx 5.8$ and

$$\nu = \frac{\pi^3}{14\zeta(3)}\tau_s T_c \gg 1. \tag{29}$$

The reasoning now follows that of Sec. 2. Replacing $\dot\Delta$ by $(\dot{\mathbf{u}}\cdot\nabla)\Delta$, we solve Eqs. (27) and (28) jointly with Eq. (3) and with the condition $\operatorname{div}[\Delta^2\mathbf{Q}] = 0$ by perturbation theory. For the component $U_1 \propto e^{i\varphi}$ we have from (28)

$$\frac{1}{\varkappa^2}\frac{d}{d\rho}\left(\frac{1}{\rho}\frac{d}{d\rho}(\rho U_1)\right) = 2a\nu\dot u f\psi,$$

whence, stipulating that U_1 decrease as $\rho \to \infty$ and be finite at zero, we readily get

$$U_1 = \frac{\varkappa^2}{\rho} a\nu\dot{u} \int_0^\rho (f^2 - 1)\rho\, d\rho.$$

This expression is contained in the right-hand side of (27) together with the term $a\dot{u}\psi$. We put

$$\gamma_a = -\varkappa^2 \int_0^\infty f \frac{df}{d\rho}\left[\int_0^\rho (f^2 - 1)\rho'\, d\rho'\right] d\rho$$

(the constant γ_a is positive and is of the order of unity).

We see thus that all the formulas obtained in the preceding sections can be used also for a superconductor with low paramagnetic-impurity concentration, by replacing 12γ with $a(\gamma + \nu\gamma_a) \approx a\nu\gamma_a$. By the same token, the vortex velocity decreases in order of magnitude by a factor $\nu \gg 1$: the anomalous terms suppress strongly the motion of the vortex filaments and the associated energy dissipation. In particular, the effective conductivity of a superconductor with a vortex structure increases:

$$\sigma_{\text{eff}} = \frac{a\gamma_a}{2} \nu \frac{H_{c2}}{B_0} \sigma = \frac{a^2\gamma_a}{2\pi} (\tau_s T_c) \frac{H_{c2}}{B_0} \sigma. \qquad (30)$$

It is interesting to note in this connection that in[1], within the limits of the experimental accuracy, the ratio $\sigma_{\text{eff}}/\sigma = H_{c2}(0)/H$ had no temperature dependence. Equation (3) contains $H_{c2}(T)$ at a temperature close to T_c, but the smallness of $H_{c2}(T)$ is essentially compensated by the factor $\tau_s T_c \gg 1$.

In conclusion we note that all our results are valid only at sufficiently low vortex velocities. In order for perturbation theory to be valid in the case of a superconductor with a large paramagnetic-impurity concentration it is necessary to have $(v \cdot \nabla) f \ll f$, i.e., $v \ll 1/\kappa$. For a superconductor with low paramagnetic-impurity concentration it is necessary to have $U_1 f \ll f$, i.e., $v \ll 1/\kappa\nu$. This imposes the condition $j_e \ll H_{c2}c/4\pi\delta\kappa$ on the magnitude of the current density.

In the case when the anomalous terms are large, perturbation theory is initially invalid for the function Δ with increasing current. We performed no calculations in this region, but qualitatively the picture is such as if the large viscosity were to cause the core of the vortex to become strongly deformed in the direction of motion.

The authors are grateful to A. I. Larkin, L. P. Pitaevskiĭ, and G. M. Éliashberg for useful discussions and to Yu. L. Rodin for help with the numerical calculations.

[1] Y. B. Kim, C. F. Hempstead, and A. R. Strnad, Phys. Rev., 139, A1163 (1965).
[2] A. A. Abrikosov, Zh. Eksp. Teor. Fiz. 32, 1442 (1957) [Sov. Phys.-JETP 5, 1174 (1957)].
[3] P. G. de Gennes and J. Matricon, Rev. Mod. Phys., 36, 45 (1964).
[4] W. F. Vinen and P. Nozieres, Phil. Mag., 14, 667 (1966).
[5] J. Bardeen and M. J. Stephen, Phys. Rev., 140, A1197 (1965).
[6] A. L. Fetter, P. C. Hohenberg, and P. Pincus, Phys. Rev., 147, 140 (1966).
[7] A. L. Fetter, Phys. Rev., 163, 390 (1967).
[8] A. A. Abrikosov, M. P. Kemoklidze, and I. M. Khalatnikov, Zh. Eksp. Teor. Fiz. 48, 765 (1965) [Sov. Phys.-JETP 21, 506 (1965)].
[9] C. Caroli, P. G. de Gennes, and J. Matricon, Phys. Lett., 9, 307 (1964).
[10] W. A. Reed, E. Fawcett, and Y. B. Kim, Phys. Rev. Lett., 14, 790 (1965).
[11] C. Caroli, and K. Maki, Phys. Rev., 164, 591 (1967). K. Maki, J. of Low Temp. Phys. 1, 54 (1969).
[12] L. P. Gor'kov, and G. M. Éliashberg, Zh. Eksp. Teor. Fiz. 54, 612 (1968); 55, 2430 (1968); 56, 1297 (1969) [Sov. Phys.-JETP 27, 328 (1968); 28, 1291 (1969); 29, 698 (1969)]. G. M. Éliashberg, ibid. 55, 2443 (1968) [28, 1298 (1969)].
[13] J. Bardeen, R. Kümmel, A. E. Jacobs, and L. Tewordt, Phys. Rev., 187, 556 (1969).
[14] Ya. Baba and K. Maki, (France), 1970.
[15] G. A. Baramidze, Zh. Eksp. Teor. Fiz. 48, 765 (1965) [sic!].
[16] L. P. Pitaevskiĭ, Zh. Eksp. Teor. Fiz. 40, 646 (1961) [Sov. Phys.-JETP 13, 451 (1961)].
[17] R. Labusch, Phys. Stat. Solidi, 19, 715 (1967).
[18] A. I. Larkin, Zh. Eksp. Teor. Fiz. 58, 1959 (1970) [Sov. Phys.-JETP 31, 1056 (1970)].

Translated by J. G. Adashko
251

Some features of viscous flow of vortices in superconducting alloys near the critical temperature

L. P. Gor'kov and N. B. Kopnin

(L. D. Landau Institute of Theoretical Physics, USSR Academy of Sciences

(Submitted July 10, 1972)

Zh. Eksp. Teor. Fiz. **64**, 356-370 (January 1973)

Time equations for slow ($\omega \ll \Delta$) motion of the vortex structure in alloys with $l \ll \xi(T)$ near T_c are derived on the basis of the microscopic superconductivity theory. The major role is played by the term responsible for the slow diffusion mechanism of relaxation. The motion of vortex lines is considered for the case of a magnetic field $H \ll H_{c2}$. This relaxation mechanism leads to a large value of the viscosity coefficient.

1. INTRODUCTION

It is known that the finite resistance of a type-II superconductor in a magnetic field exceeding the lower critical field H_{c1} is due to energy dissipation that occurs when the vortex structure moves in the superconductor. This question has been under experimental study for quite some time [1], whereas the theoretical study of the motion of vortex filaments was based until recently only on phenomenological models [2]. These turned out (see [3-6]) not to cover all the significant kinetic processes that occur in superconductors. A consistent approach to this problem obviously calls for the use of microscopic time-dependent equations of the theory of superconductivity [3].

The motion of a vortex structure in strong magnetic fields (near H_{c2}), where the superconductivity is greatly suppressed, was investigated on the basis of the microscopic theory, for example, by Caroli and Maki [7], and by Baba and Maki [8]. Great interest, however, attaches to the wide range of fields $H \ll H_{c2}$. In an earlier paper [9], using alloys with paramagnetic impurities as an example, we have proposed a scheme by which to describe the vortex motion within the framework of microscopic differential time-dependent equations of the Ginzburg-Landau type.

Our result [9] is valid for the case of large paramagnetic-impurity concentrations, strictly speaking, only for an uncharged Fermi gas, since we took no account of the term due to the normal energy loss in the center of the vortex. This circumstance was pointed out by a number of authors [10,11].

It follows from [9] that at finite temperatures, when the anomalous terms in the equations for the superconductors become important, the viscosity experienced by the moving vortex increases strongly, i.e., the vortex speed is low. This is caused by a circumstance already noted by Éliashberg and one of the authors [4], that the anomalous term becomes larger in order of magnitude than the remaining terms with time derivatives when the damping of the excitations is determined by slow diffusion processes. The reason for this increase in viscosity is that to ensure stationary motion of the vortex it is essential that the perturbations produced around the moving vortex have time to attenuate within a time on the order of $\xi(T)/u$, where u is the vortex speed and $\xi(T)$ is the scale of variation of the ordering parameter. In superconductors, the times of homogeneous energy relaxation are quite long (Θ_D^2/T^3 and E_F/T^2 for electron-phonon and electron-electron interactions, respectively). The damping of the excitations is therefore determined by a diffusion mechanism which, nonetheless, is still slow enough at temperatures $T_c - T \ll T_c$.

This article is devoted to a study of these features of relaxation for vortex filaments moving at $T_c - T \ll T$ under the more favorable experimental conditions of superconducting alloys with ordinary impurities. It will be shown that the relaxation mechanism mentioned above plays the principal role in this case, so that an analytic solution of the resultant equations can be obtained.

2. DERIVATION OF EQUATIONS FOR THE ANOMALOUS GREEN'S FUNCTIONS IN ALLOYS AT A TEMPERATURE CLOSE TO T_c

The general scheme for deriving the time-dependent equations for superconductors was developed by Éliashberg and one of the authors [3]. It was shown that when the expressions for the ordering parameter Δ and for the current are analytically continued to the real frequency axis, the result consists of terms that are regular in the upper or lower half of the complex ϵ plane, as well as the so-called "anomalous terms," which do not possess this property. We write down the expressions derived in [3] and choose immediately for Δ a gauge whereby we obtain throughout only the gauge-invariant combinations

$$|\Delta|, \quad \mathbf{Q} = \mathbf{A} - \frac{c}{2e}\nabla\chi, \quad \mu = \varphi + \frac{1}{2e}\dot{\chi},$$

where χ is the phase shift of the ordering parameter (we shall henceforth omit the absolute-magnitude symbols of Δ):

$$\frac{\Delta(\mathbf{r})}{|g|} = \int \frac{d\varepsilon}{4\pi i}\left[\operatorname{th}\frac{\varepsilon-\omega}{2T}F^{R}_{\varepsilon,\varepsilon-\omega}(\mathbf{r},\mathbf{r}) - \operatorname{th}\frac{\varepsilon}{2T}F^{A}_{\varepsilon,\varepsilon-\omega}(\mathbf{r},\mathbf{r})\right] + \int \frac{d\varepsilon}{4\pi i}F^{(a)}_{\varepsilon,\varepsilon-\omega}(\mathbf{r},\mathbf{r}),$$

$$\mathbf{j} = -\frac{e}{m}\int\frac{d\varepsilon}{4\pi i}\left\{(\hat{\mathbf{p}}-\hat{\mathbf{p}}')\left[\operatorname{th}\frac{\varepsilon-\omega}{2T}G^{R}_{\varepsilon,\varepsilon-\omega}(\mathbf{r},\mathbf{r}') - \operatorname{th}\frac{\varepsilon}{2T}G^{A}_{\varepsilon,\varepsilon-\omega}(\mathbf{r},\mathbf{r}') + G^{(a)}_{\varepsilon,\varepsilon-\omega}(\mathbf{r},\mathbf{r}')\right]\right\}_{\mathbf{r}'=\mathbf{r}} - \frac{Ne^2}{mc}\mathbf{Q}_{\omega}(\mathbf{r}). \quad (1)$$

Here $G^{R(A)}$ and $F^{R(A)}$ are retarded and advanced Green's functions, $\hat{\mathbf{p}} = -i\nabla$, and $\hat{\mathbf{p}}' = -i\nabla'$.

It is convenient to write down the definitions of the anomalous functions directly in matrix form. To this end, we introduce the matrices

$$\mathscr{G}^{R(A)}_{\varepsilon,\varepsilon_1}(\mathbf{r},\mathbf{r}') = \begin{pmatrix} G^{R(A)}_{\varepsilon,\varepsilon_1}(\mathbf{r},\mathbf{r}') & F^{R(A)}_{\varepsilon,\varepsilon_1}(\mathbf{r},\mathbf{r}') \\ -F^{+R(A)}_{\varepsilon,\varepsilon_1}(\mathbf{r},\mathbf{r}') & \bar{G}^{R(A)}_{\varepsilon,\varepsilon_1}(\mathbf{r},\mathbf{r}') \end{pmatrix},$$

$$\mathscr{G}^{(a)}_{\varepsilon,\varepsilon-\omega}(\mathbf{r},\mathbf{r}') = \begin{pmatrix} G^{(a)}_{\varepsilon,\varepsilon-\omega}(\mathbf{r},\mathbf{r}') & F^{(a)}_{\varepsilon,\varepsilon-\omega}(\mathbf{r},\mathbf{r}') \\ -F^{+(a)}_{\varepsilon,\varepsilon-\omega}(\mathbf{r},\mathbf{r}') & \bar{G}^{(a)}_{\varepsilon,\varepsilon-\omega}(\mathbf{r},\mathbf{r}') \end{pmatrix},$$

$$\hat{H}_{\omega}(\mathbf{r}) = \begin{pmatrix} -e\mathbf{Q}_{\omega}\hat{\mathbf{p}}/mc + e\mu_{\omega} & -\Delta_{\omega} \\ \Delta_{\omega} & e\mathbf{Q}_{\omega}\hat{\mathbf{p}}/mc + e\mu_{\omega} \end{pmatrix}.$$

The equation for the anomalous functions $\mathcal{G}^{(a)}_{\epsilon,\epsilon-\omega}$ contain retarded and advanced functions $\mathcal{G}^{R(A)}_{\epsilon,\epsilon_1}$ that depend on two frequencies (see [3]). In view of the slowness of the considered processes ($\omega \ll \Delta$), the regular functions can be replaced by the equilibrium functions $\mathcal{G}^{R(A)}_{\epsilon} 2\pi\delta(\epsilon - \epsilon_1)$, and we can omit from $\hat{H}_\omega$ the μ-containing terms, which are of the order of ω. Ultimately we get

$$\mathcal{G}^{(a)}_{\varepsilon,\varepsilon-\omega}(\mathbf{r},\mathbf{r}') = -\frac{\omega}{2T}\operatorname{ch}^{-2}\frac{\varepsilon}{2T}\int \mathcal{G}^R_\varepsilon(\mathbf{r},\mathbf{r}_1)\hat{H}_\omega(\mathbf{r}_1)\mathcal{G}^A_{\varepsilon-\omega}(\mathbf{r}_1,\mathbf{r}')\,d^3\mathbf{r}_1$$
$$+(2\pi\tau)^{-1}\int \mathcal{G}^R_\varepsilon(\mathbf{r},\mathbf{r}_1)\frac{2\pi^2}{mp_0}\mathcal{G}^{(a)}_{\varepsilon,\varepsilon-\omega}(\mathbf{r}_1,\mathbf{r}_1)\mathcal{G}^A_{\varepsilon-\omega}(\mathbf{r}_1,\mathbf{r}')\,d^3\mathbf{r}_1. \quad (2)$$

This formula corresponds to the diagram equation shown in the figure. The dashed lines denoting averaging over the impurities corresponds, as usual, to the factor $(2\pi\tau)^{-1}$, where τ is the free path time.

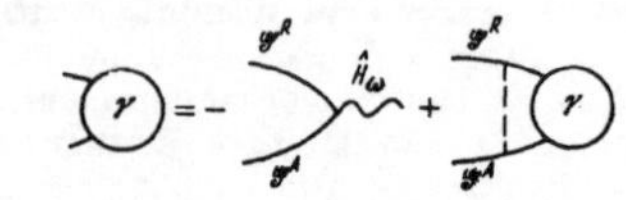

We introduce also the notation

$$\hat{\gamma}(\mathbf{r}) = \begin{pmatrix} \gamma_1 & \gamma_2 \\ -\gamma_2^+ & \bar{\gamma}_1 \end{pmatrix} = \frac{2\pi^2}{mp_0}\mathcal{G}^{(a)}_{\varepsilon,\varepsilon-\omega}(\mathbf{r},\mathbf{r}),$$

$$\hat{\boldsymbol{\gamma}}(\mathbf{r}) = \begin{pmatrix} \boldsymbol{\gamma}_1 & \boldsymbol{\gamma}_2 \\ -\boldsymbol{\gamma}_2^+ & \bar{\boldsymbol{\gamma}}_1 \end{pmatrix} = \frac{2\pi^2}{mp_0}\left[\frac{\hat{\mathbf{p}}-\hat{\mathbf{p}}'}{2p_0}\mathcal{G}^{(a)}_{\varepsilon,\varepsilon-\omega}(\mathbf{r},\mathbf{r}')\right]_{\mathbf{r}'=\mathbf{r}} \quad (3)$$

p_0 is the Fermi momentum.

Expansions of the regular parts of (1) in powers of Δ/T are well known [6]. We can use them to write down expressions for Δ and for the current:

$$\frac{\pi}{8T_c}\left[-\frac{\partial}{\partial t}+D_1\left(\nabla^2-\frac{4e^2}{c^2}Q^2\right)\right]\Delta+\left[\frac{T_c-T}{T_c}-\frac{7\zeta(3)}{8\pi^2T_c^2}\Delta^2\right]\Delta$$
$$+\int_{-\infty}^{\infty}\frac{d\varepsilon}{4\pi i}\frac{(\gamma_2+\gamma_2^+)}{2}=0, \quad (4a)$$

$$\frac{i\pi}{8T_c}D_1\operatorname{div}\left[\frac{2e}{c}Q\Delta^2\right]=\int_{-\infty}^{\infty}\frac{d\varepsilon}{4\pi i}\frac{(\gamma_2-\gamma_2^+)\Delta}{2}, \quad (4b)$$

$$\mathbf{j}=\mathbf{j}^{(a)}-\frac{mp_0e^2}{2\pi c}\frac{\Delta^2}{T_c}D_1Q, \quad (4c)$$

$$\mathbf{j}^{(a)}=-\frac{ep_0^2}{2\pi^2}\int_{-\infty}^{+\infty}\frac{d\varepsilon}{4\pi i}(\boldsymbol{\gamma}_1-\bar{\boldsymbol{\gamma}}_1). \quad (4d)$$

Here

$$D_1=\frac{v_0^2\tau}{3}y(\tau T),\qquad y(x)=\frac{8}{\pi^2}\sum_{n=1}^{\infty}\frac{1}{(2n+1)^2[(2n+1)2\pi x+1]}.$$

At $\tau T \ll 1$, we have $y(\tau T) = 1$ and at $\tau T \gg 1$ we get $y(\tau T) = 7\zeta(3)/2\pi^3\tau T$.

We mention immediately (and will demonstrate later on) that the condition that $|\Delta|$ be real (Eq. (4b)) follows from the expression for the current with allowance for the electroneutrality of the superconductor, $\operatorname{div}\mathbf{j} = 0$. The problem thus reduces to a solution of (2) for the anomalous functions.

We confine ourselves to alloys in which $l \ll \xi(T)$, where $l = v_0\tau$ is the mean free path. In this region we can draw the following conclusions concerning the spatial behavior of the functions $\mathcal{G}^{R(A)}_{\epsilon}(\mathbf{r},\mathbf{r}')$. We represent them in the form

$$\mathcal{G}^{R(A)}_\varepsilon(\mathbf{r},\mathbf{r}')=\frac{m\exp(-|\mathbf{r}-\mathbf{r}'|/2l)}{2\pi|\mathbf{r}-\mathbf{r}'|}\{\hat{g}^{R(A)}_+(\mathbf{r},\mathbf{r}')\exp(ip_0|\mathbf{r}-\mathbf{r}'|)$$
$$+\hat{g}^{R(A)}_-(\mathbf{r},\mathbf{r}')\exp(-ip_0|\mathbf{r}-\mathbf{r}'|)\}. \quad (5)$$

Applying to the right and to the left of (5) the operator

$$\mathcal{G}^{-1R(A)}_\varepsilon(\mathbf{r})=\begin{pmatrix}-\varepsilon+\frac{\hat{\mathbf{p}}^2}{2m}-\frac{e}{mc}Q\hat{\mathbf{p}}-E_F & -\Delta \\ \Delta & \varepsilon+\frac{\hat{\mathbf{p}}^2}{2m}+\frac{e}{mc}Q\hat{\mathbf{p}}-E_F\end{pmatrix}-\Sigma^{R(A)}_\varepsilon(\mathbf{r}),$$

where

$$\Sigma^{R(A)}_\varepsilon(\mathbf{r})=\frac{1}{2\pi\tau}\frac{2\pi^2}{mp_0}\mathcal{G}^{R(A)}_\varepsilon(\mathbf{r},\mathbf{r})$$

is the impurity self-energy part, we can easily obtain the conditions satisfied by the function $\hat{g}^{R(A)}_{\pm}(\mathbf{r},\mathbf{r}')$. We introduce the vector $\mathbf{n} = (\mathbf{r}-\mathbf{r}')/|\mathbf{r}-\mathbf{r}'|$; then

$$\int\{\hat{g}_+(\mathbf{r},\mathbf{r}')+\hat{g}_-(\mathbf{r},\mathbf{r}')\}^{R(A)}_{\mathbf{r}'=\mathbf{r}}\frac{do_{\mathbf{n}}}{4\pi}=1, \quad (6)$$

$$\left[\mp iv_0\mathbf{n}\nabla-\Sigma^{R(A)}_\varepsilon(\mathbf{r})\pm\frac{i}{2\tau}\right]\hat{g}^{R(A)}_\pm(\mathbf{r},\mathbf{r}')+\begin{pmatrix}-\varepsilon_\pm & -\Delta\\ \Delta & \varepsilon_\pm\end{pmatrix}\hat{g}^{R(A)}_\pm(\mathbf{r},\mathbf{r}')=0,$$

$$\hat{g}^{R(A)}_\pm(\mathbf{r},\mathbf{r}')\left[\pm iv_0\mathbf{n}\nabla'-\Sigma^{R(A)}_\varepsilon(\mathbf{r}')\pm\frac{i}{2\tau}\right]+\hat{g}^{R(A)}_\pm(\mathbf{r},\mathbf{r}')\begin{pmatrix}-\varepsilon_\pm & -\Delta\\ \Delta & \varepsilon_\pm\end{pmatrix}=0,$$

where $\epsilon_\pm = \epsilon \pm v_0 e\mathbf{n}\cdot\mathbf{Q}/c$ and the operator ∇' acts on $\mathbf{r}'$.

Combining the corresponding matrix elements of these equations, we easily find that the functions $\hat{g}^{R(A)}_{\pm}(\mathbf{r},\mathbf{r}')$ with coinciding points ($\mathbf{r}' \to \mathbf{r}$) are connected with the functions

$$\hat{g}^{R(A)}_\varepsilon(\mathbf{v}_0,\mathbf{k})=\int\mathcal{G}^{R(A)}_\varepsilon(\mathbf{p},\mathbf{p}-\mathbf{k})\frac{d\xi}{\pi i},$$

which are integrated with respect to the energy variable $\xi = v_0(p - p_0)$, by the relations

$$[\hat{g}^{R(A)}_\pm(\mathbf{r},\mathbf{r}')]_{\mathbf{r}'\to\mathbf{r}}={}^1\!/_2\{\hat{1}\pm\hat{g}^{R(A)}_\varepsilon(\pm\mathbf{v}_0,\mathbf{r})\},$$
$$[\hat{g}^{R(A)}_\pm(\mathbf{r}',\mathbf{r})]_{\mathbf{r}'\to\mathbf{r}}={}^1\!/_2\{\hat{1}\pm\hat{g}^{R(A)}_\varepsilon(\mp\mathbf{v}_0,\mathbf{r})\}, \quad (7)$$

where $\mathbf{v}_0 = v_0\mathbf{n}$. Indeed, the functions

$$\hat{g}^{R(A)}_\varepsilon(\mathbf{v}_0,\mathbf{r})=\begin{pmatrix}g_\varepsilon(\mathbf{v}_0,\mathbf{r}) & f_\varepsilon(\mathbf{v}_0,\mathbf{r})\\ -f^+_\varepsilon(\mathbf{v}_0,\mathbf{r}) & \bar{g}_\varepsilon(\mathbf{v}_0,\mathbf{r})\end{pmatrix}^{R(A)}$$

satisfy the equations derived by Eilenberger [12] and the corresponding boundary conditions at infinity:

$$-i\mathbf{v}_0\nabla g_\varepsilon+\Delta(f^+_\varepsilon-f_\varepsilon)+\frac{i}{2\tau}(f_{0\varepsilon}f^+_\varepsilon-f_\varepsilon f^+_{0\varepsilon})=0,$$

$$-i\mathbf{v}_0\nabla\bar{g}_\varepsilon-\Delta(f_\varepsilon-f^+_\varepsilon)+\frac{i}{2\tau}(f^+_{0\varepsilon}f_\varepsilon-f^+_\varepsilon f_{0\varepsilon})=0,$$

$$\left[-2\varepsilon-i\mathbf{v}_0\left(\nabla-i\frac{2e}{c}Q\right)\right]f_\varepsilon+\Delta(g_\varepsilon-\bar{g}_\varepsilon)$$
$$+\frac{i}{2\tau}(g_\varepsilon-\bar{g}_\varepsilon)f_{0\varepsilon}+\frac{i}{2\tau}(\bar{g}_{0\varepsilon}-g_{0\varepsilon})f_\varepsilon=0, \quad (8)$$
$$\left[-2\varepsilon+i\mathbf{v}_0\left(\nabla+i\frac{2e}{c}Q\right)\right]f^+_\varepsilon+\Delta(g_\varepsilon-\bar{g}_\varepsilon)$$
$$+\frac{i}{2\tau}(g_\varepsilon-\bar{g}_\varepsilon)f^+_{0\varepsilon}+\frac{i}{2\tau}(\bar{g}_{0\varepsilon}-g_{0\varepsilon})f^+_\varepsilon=0,$$

The subscript 0 designates throughout functions averaged over the direction of $\mathbf{v}_0$.

It follows from (8) that

$$g^{R(A)}_\varepsilon(\mathbf{v}_0,\mathbf{r})+\bar{g}^{R(A)}_\varepsilon(\mathbf{v}_0,\mathbf{r})=0,$$
$$[g^{R(A)}_\varepsilon(\mathbf{v}_0,\mathbf{r})]^2-f^{R(A)}_\varepsilon(\mathbf{v}_0,\mathbf{r})f^{+R(A)}_\varepsilon(\mathbf{v}_0,\mathbf{r})=1. \quad (9)$$

We have used here the fact that as $\mathbf{r} \to \infty$ we get

$$g^{R(A)}_\varepsilon(\mathbf{v}_0,\mathbf{r})=\frac{\varepsilon}{\lambda^{R(A)}_\infty},\quad f^{R(A)}_\varepsilon(\mathbf{v}_0,\mathbf{r})=\frac{\Delta_\infty}{\lambda^{R(A)}_\infty},\quad \lambda_\infty=\sqrt{\varepsilon^2-\Delta_\infty^2},$$

where $\Delta = \Delta_\infty$ = const; $\lambda^{R(A)}$ are so defined that when $\epsilon = \epsilon_0 + i\delta$, $\epsilon_0 > \Delta > 0$, and $\delta \to +0$ we have $\lambda^{R(A)} = \pm\lambda$.

In the limit when $l \ll \xi(T)$, the functions $\hat{g}^{R(A)}(\mathbf{r}, \mathbf{r}')$, which vary over distances on the order of $\xi(T)$, can be regarded as slow in comparison with $\exp[-|\mathbf{r} - \mathbf{r}'|/2l]$, while the functions $g_\epsilon^{R(A)}(\mathbf{v}_0, \mathbf{r})$ can be regarded in first approximation as isotropic with respect to the directions of the vector $\mathbf{v}_0$ (cf.[13]), i.e., they can be represented in the form

$$\hat{g}_\epsilon^{R(A)}(\mathbf{v}_0, \mathbf{r}) = \hat{g}_{0\epsilon}^{R(A)}(\mathbf{r}) + \frac{\mathbf{v}_0}{v}\hat{\mathbf{g}}_\epsilon^{R(A)}(\mathbf{r}),$$

where $|\mathbf{g}| \ll g_0$. We see from (6) and (7) that the last assumption is the consequence of the slowness of the variation of $g_\pm$. Indeed, the functions $\hat{g}_{0\epsilon}$ drop out from the large terms $\pm i(2\tau)^{-1}\hat{g}_\pm - \Sigma\hat{g}_\pm$ of (6) as $r \to r'$, and it turns out that $|\mathbf{g}| \sim (l\Delta/\xi_0 T_c)g_0$ (an important factor in this estimate is that the principal role will henceforth be assumed by frequencies $\epsilon \sim \Delta$).

Using the representation (5), we could calculate directly the kernel $\mathscr{G}_\epsilon^R(\mathbf{r}, \mathbf{r}')\mathscr{G}_{\epsilon-\omega}^A(\mathbf{r}', \mathbf{r})$ in (2) and obtain equations for the functions $\hat{\gamma}(\mathbf{r})$. The determinant of the matrix equation for $\hat{\gamma}$, taken accurate only to zeroth order terms in l, vanishes in this case. It becomes necessary to expand the kernel in (2) up to terms of second order in l, but this entails extremely cumbersome calculations. We therefore choose a somewhat different method, which is also valid so long as it can be assumed that $l \ll \xi(T)$, an assumption that can always be made in a sufficiently small vicinity of T_c. Our results in this region are valid not only for "dirty" alloys with $l \ll \xi_0$, but also for relatively pure superconductors, where

$$v_0/T_c \ll l \ll \frac{v_0}{T_c}(1 - T/T_c)^{-1/2}.$$

We write down the equations that are satisfied by the functions

$$\hat{g}_{\epsilon,\epsilon-\omega}^{(a)}(\mathbf{v}_0, \mathbf{k}) = \int \mathscr{G}_{\epsilon,\epsilon-\omega}^{(a)}(\mathbf{p}, \mathbf{p} - \mathbf{k})\,d\xi, \quad \xi = v_0(p - p_0),$$

using the same procedure as Eilenberg[12] (see above) and Eliashberg[14]. We apply to Eq. (2) the operator $\mathscr{G}_\epsilon^{-1R}$ from the left, the operator $\mathscr{G}_\epsilon^{-1A}$ from the right, and subtract one resultant equation from the other. After integrating with respect to ξ we obtain

$$\begin{aligned}\begin{pmatrix} A_1 & A_2 \\ -A_2^+ & \bar{A}_1 \end{pmatrix} &= \Sigma^R(\mathbf{r})\hat{g}^{(a)}(\mathbf{v}_0, \mathbf{r}) - \hat{g}^{(a)}(\mathbf{v}_0, \mathbf{r})\Sigma^A(\mathbf{r}) \\ &+ i(2\tau)^{-1}[\hat{\gamma}(\mathbf{r})\hat{g}^A(\mathbf{v}_0, \mathbf{r}) - \hat{g}^R(\mathbf{v}_0, \mathbf{r})\hat{\gamma}(\mathbf{r})] \\ &- \pi i\frac{\omega}{2T}\mathrm{ch}^{-2}\frac{\varepsilon}{2T}[\hat{H}_\omega\hat{g}^A(\mathbf{v}_0, \mathbf{r}) - \hat{g}^R(\mathbf{v}_0, \mathbf{r})\hat{H}_\omega],\end{aligned} \tag{10}$$

where

$$\begin{Bmatrix} A_1 \\ \bar{A}_1 \end{Bmatrix} = (-i\mathbf{v}_0\nabla \mp \omega)\begin{Bmatrix} g^{(a)}(\mathbf{v}_0, \mathbf{r}) \\ \bar{g}^{(a)}(\mathbf{v}_0, \mathbf{r}) \end{Bmatrix} \pm \Delta(f^{+(a)}(\mathbf{v}_0, \mathbf{r}) - f^{(a)}(\mathbf{v}_0, \mathbf{r})),$$

$$\begin{Bmatrix} A_2 \\ A_2^+ \end{Bmatrix} = \left[\mp(2\varepsilon - \omega) - i\mathbf{v}_0\left(\nabla \mp i\frac{2e}{c}\mathbf{Q}\right)\right]\begin{Bmatrix} f^{(a)}(\mathbf{v}_0, \mathbf{r}) \\ f^{+(a)}(\mathbf{v}_0, \mathbf{r}) \end{Bmatrix} \pm \Delta(g^{(a)}(\mathbf{v}_0, \mathbf{r}) - \bar{g}^{(a)}(\mathbf{v}_0, \mathbf{r})).$$

For the sake of brevity, we shall henceforth drop the subscripts ϵ and $\epsilon - \omega$, bearing in mind that $\hat{g}^R$ refers to the frequency ϵ, g^A to the frequency $\epsilon - \omega$, and $\hat{g}^{(a)}$ to both frequencies.

Averaging (10) over the directions of the vector $\mathbf{v}_0$, we obtain

$$\begin{aligned}&-iv_0\mathbf{d}\hat{\gamma} + \begin{pmatrix} \Delta(\gamma_2^+ - \gamma_2) - \omega\gamma_1 & -2\varepsilon\gamma_2 + \Delta(\gamma_1 - \bar{\gamma}_1) \\ -2\varepsilon\gamma_2^+ + \Delta(\gamma_1 - \bar{\gamma}_1) & \Delta(\gamma_2 - \gamma_2^+) + \omega\bar{\gamma}_1 \end{pmatrix} \\ &= \pi i\frac{\omega\Delta_\omega}{2T}\mathrm{ch}^{-2}\frac{\varepsilon}{2T}\begin{pmatrix} f_0^R - f_0^{+A} & -g_0^R + \bar{g}_0^A \\ \bar{g}_0^R - g_0^A & f_0^{+R} - f_0^A \end{pmatrix} \\ &+ \pi i\frac{\omega}{2T}\mathrm{ch}^{-2}\frac{\varepsilon}{2T}\frac{v_0}{3}\frac{e}{c}\mathbf{Q}\begin{pmatrix} -\mathbf{g}^R + \mathbf{g}^A & \mathbf{f}^R + \mathbf{f}^A \\ \mathbf{f}^{+R} + \mathbf{f}^{+A} & \bar{\mathbf{g}}^R - \bar{\mathbf{g}}^A \end{pmatrix}.\end{aligned} \tag{11}$$

We denote by $\mathbf{d}\hat{\gamma}$ the gauge-invariant differentiation

$$\begin{pmatrix} \nabla\gamma_1 & \left(\nabla - i\frac{2e}{c}\mathbf{Q}\right)\gamma_2 \\ -\left(\nabla + i\frac{2e}{c}\mathbf{Q}\right)\gamma_2^+ & \nabla\bar{\gamma}_1 \end{pmatrix}$$

The terms containing τ^{-1} drop out after averaging.

To derive the equations satisfied by the quantities $\hat{\gamma}$, we proceed as follows. With the aid of the representation (5), using the fact that the functions $\hat{g}_\pm^{R(A)}$ vary slowly over distances on the order of l, we calculate the vector $\hat{\boldsymbol{\gamma}}$ directly from the definition (3). It suffices to do so accurate to first order in l. We then obtain the sought equations by substituting the values of $\hat{\boldsymbol{\gamma}}$ in (11).

We note that the method used by Usadel[13] to obtain the equations for the functions f and g averaged over the angles $\mathbf{v}_0$ are not valid here. Indeed, multiplying (10) by $\mathbf{v}_0$ and averaging over the directions of $\mathbf{v}_0$ we would obtain one more system of relations between $\hat{\gamma}$ and $\hat{\boldsymbol{\gamma}}$. Since $\Sigma^{R(A)} = i(2\tau)^{-1}\hat{g}_0^{R(A)}$, the principal terms with $\hat{\gamma}$ are contained in the combination $i(2\tau)^{-1}[\hat{g}_0^R\hat{\boldsymbol{\gamma}} - \hat{\boldsymbol{\gamma}}\hat{g}_0^A]$. Were it possible to express the quantities $\boldsymbol{\gamma}_1$, $\bar{\boldsymbol{\gamma}}_1$, $\boldsymbol{\gamma}_2$, $\boldsymbol{\gamma}_2^+$ with the aid of this system in terms of $\hat{g}_0^{R(A)}$ and $\hat{\mathbf{g}}^{R(A)}$, then, by substituting them in (11), we would obtain equations for $\hat{\gamma}$. However, the determinant of the four-row matrix for the quantities $\boldsymbol{\gamma}_1$, $\bar{\boldsymbol{\gamma}}_1$, $\boldsymbol{\gamma}_2$, $\boldsymbol{\gamma}_2^+$ is equal to zero.

Substituting (2) and (5) in (3), we get

$$\begin{aligned}\frac{mp_0}{2\pi^2}\hat{\boldsymbol{\gamma}} &= -\frac{\omega}{2T}\mathrm{ch}^{-2}\frac{\varepsilon}{2T}\int\left(\frac{m}{2\pi R}\right)^2 e^{-R/l}\mathbf{n}\{\hat{g}_+^R(\mathbf{r}, \mathbf{r}')\hat{h}_-(\mathbf{r}')\hat{g}_-^A(\mathbf{r}', \mathbf{r}) \\ &\qquad - \hat{g}_-^R(\mathbf{r}, \mathbf{r}')\hat{h}_+(\mathbf{r}')\hat{g}_+^A(\mathbf{r}', \mathbf{r})\}d^3\mathbf{r}' \\ &+ (2\pi\tau)^{-1}\int\left(\frac{m}{2\pi R}\right)^2 e^{-R/l}\mathbf{n}\{\hat{g}_+^R(\mathbf{r}, \mathbf{r}')\hat{\gamma}(\mathbf{r}')\hat{g}_-^A(\mathbf{r}', \mathbf{r}) \\ &\qquad - \hat{g}_-^R(\mathbf{r}, \mathbf{r}')\hat{\gamma}(\mathbf{r}')\hat{g}_+^A(\mathbf{r}', \mathbf{r})\}d^3\mathbf{r}',\end{aligned} \tag{12}$$

$$\hat{h}_\pm(\mathbf{r}') = \begin{pmatrix} \pm v_0 e\mathbf{n}\mathbf{Q}_\omega(\mathbf{r}')/c & -\Delta_\omega(\mathbf{r}') \\ \Delta_\omega(\mathbf{r}') & \mp v_0 e\mathbf{n}\mathbf{Q}_\omega(\mathbf{r}')/c \end{pmatrix}, \quad \mathbf{R} = \mathbf{r} - \mathbf{r}'.$$

We now expand the quantities $\hat{g}_\pm^R(\mathbf{r}, \mathbf{r}')$, $\hat{\mathbf{g}}^A(\mathbf{r}', \mathbf{r})$, $\hat{\gamma}(\mathbf{r}')$ and $\hat{h}_\pm(\mathbf{r}')$ in powers of $\mathbf{R}$ about the point $\mathbf{r}$:

$$\hat{g}_\pm^R(\mathbf{r}, \mathbf{r}') = [\hat{g}_\pm^R(\mathbf{r}, \mathbf{r}')]_{\mathbf{r}'=\mathbf{r}} - R[\mathbf{n}\nabla'\hat{g}_\pm^R(\mathbf{r}, \mathbf{r}')]_{\mathbf{r}'=\mathbf{r}}$$

etc. The derivatives $\mathbf{n}\nabla'\hat{g}_\pm(\mathbf{r}, \mathbf{r}')$, $\mathbf{r}' \to \mathbf{r}$, can be expressed with the aid of (6) in terms of $\hat{\mathbf{g}}_\pm(\mathbf{r}, \mathbf{r})$. Integrating in (12) with respect to R and averaging over the directions of $\mathbf{n}$, we obtain with the aid of (7) after rather laborious calculations

$$\begin{aligned}\hat{\boldsymbol{\gamma}} &= -{}^1/_3\{[\hat{g}_0^R\hat{\boldsymbol{\gamma}}\hat{\mathbf{g}}^A + \hat{\mathbf{g}}^R\hat{\boldsymbol{\gamma}}\hat{g}_0^A] + {}^1/_4[\hat{\mathbf{g}}^R(\hat{g}_0^R\hat{\gamma} - \hat{\gamma}\hat{g}_0^A) \\ &- (\hat{g}_0^R\hat{\gamma} - \hat{\gamma}\hat{g}_0^A)\hat{\mathbf{g}}^A] + {}^1/_2 l[\hat{g}_0^R\mathbf{d}\hat{\gamma} - (\mathbf{d}\hat{\gamma})\hat{g}_0^A]\} \\ &+ \frac{\pi l}{3}\mathrm{ch}^{-2}\frac{\varepsilon}{2T}\frac{\omega}{2T}\frac{e}{c}\mathbf{Q}_\omega\{\sigma_z - \hat{g}_0^R\sigma_z\hat{g}_0^A\},\end{aligned} \tag{13}$$

where σ_z is a Pauli matrix.

Before we continue our conclusion, we turn to the determination of the regular functions $\hat{g}^{R(A)}(\mathbf{v}_0, \mathbf{r})$. We can eliminate from (8) the vectors $\mathbf{g}^{R(A)}$, $\mathbf{f}^{R(A)}$, and $\mathbf{f}^{+R(A)}$

with the aid of the conditions (8) (see[13]):

$$f = l\left[\frac{f_0}{2g_0}\nabla(f_0f_0^+) - g_0\left(\nabla - i\frac{2e}{c}Q\right)f_0\right],$$

$$f^+ = l\left[-\frac{f_0^+}{2g_0}\nabla(f_0f_0^+) + g_0\left(\nabla + i\frac{2e}{c}Q\right)f_0^+\right], \qquad (14)$$

$$g = {}^1/_2 l\left[f_0\left(\nabla + i\frac{2e}{c}Q\right)f_0^+ - f_0^+\left(\nabla - i\frac{2e}{c}Q\right)f_0\right],$$

and obtain equations for $f_0^{R(A)}$ and $f_0^{+R(A)}$:

$$-2\varepsilon f_0 + 2\Delta g_0 + iD\left(\nabla - i\frac{2e}{c}Q\right)\left[g_0\left(\nabla - i\frac{2e}{c}Q\right)f_0 - \frac{f_0}{2g_0}\nabla(f_0f_0^+)\right] = 0,$$

$$-2\varepsilon f_0^+ + 2\Delta g_0 + iD\left(\nabla + i\frac{2e}{c}Q\right)\left[g_0\left(\nabla + i\frac{2e}{c}Q\right)f_0^+ - \frac{f_0^+}{2g_0}\nabla(f_0f_0^+)\right] = 0, \qquad (15)$$

where $D = v_0 l/3$ is the diffusion coefficients. For the sake of brevity, the superscripts R and A to the functions f_0 and g_0 are omitted. The function g_0 is so defined that

$$g_0^2 - f_0f_0^+ = 1.$$

The rather complicated equations in (15) become simpler in the considered temperature interval $T_c - T \ll T_c$ because the terms with the spatial derivatives and with the field are small. Indeed, as is seen from the static equations (4) ($\partial\Delta/\partial t$ and $\gamma_2 + \gamma_2^+$ are equal to zero), $\nabla^2\Delta$ and $(2eQ/c)^2\Delta$ are of the order of $\Delta\xi^{-2}(T) \sim \Delta^3/D_1T_c$. The combinations $D\nabla^2\Delta$ and $D(2eQ/c)^2\Delta$ in (15) are therefore of order $D\Delta^2/D_1T_c$ relative to Δ. At $\tau T_c \ll 1$ (i.e., $l \ll \xi_0$) this reduces to Δ^2/T_c, which is much less than Δ at $3.3(1 - T/T_c)^{1/2} \ll 1$. On the other hand, if $\tau T_c \gg 1$, then we have

$$\frac{D}{D_1}\frac{\Delta^2}{T_c} \sim \frac{l}{\xi(T)}\Delta.$$

The ratio

$$\frac{l}{\xi(T)} = 1{,}3\frac{l}{\xi_0}\left(1 - \frac{T}{T_c}\right)^{1/2}$$

becomes small in the vicinity of $(1 - T/T_c)^{1/2} \ll \xi_0/l$.

Thus, at $l \ll \xi(T)$ we can neglect in (15) the terms with the derivative and field in comparison with Δ and ϵ. We see that at $l \ll \xi(T)$ the functions f_0 and g_0 depend adiabatically on the distance r:

$$f_0^{R(A)} = f_0^{+R(A)} = \frac{\Delta}{\lambda^{R(A)}}, \quad g_0^{R(A)} = \frac{\varepsilon}{\lambda^{R(A)}}, \quad \lambda = \sqrt{\varepsilon^2 - \Delta^2}, \qquad (16)$$

the $\lambda^{R(A)}$ are so defined that we have $\lambda^{R(A)} = \pm\lambda$ at $\epsilon = \epsilon_0 + i\delta$, $\epsilon_0 > \Delta > 0$, and $\delta \to +0$. The matching of the expressions at $\epsilon < \Delta(r)$ and $\epsilon > \Delta(r)$ should occur, as follows from (15), near the point $\mathbf{r}_c$, where $\Delta(\mathbf{r}_c) = \epsilon$, at distances on the order of $\xi(D/\xi^2\Delta)^{2/3}$, which is much less than $\xi(T)$. The retarded and advanced functions determined in this manner are equal on the real axis in the interval $-\Delta < \epsilon < \Delta$, apart from small terms containing D.

We now return to Eq. (11). Since, as seen from (13), $\hat{\gamma} \propto l$, we have with the required accuracy

$$-\varepsilon(\gamma_2 + \gamma_2^+) + \Delta(\gamma_1 - \bar{\gamma}_1) = -\pi i\frac{\omega\Delta_\omega}{2T}(g_0^R + g_0^A)\operatorname{ch}^{-2}\frac{\varepsilon}{2T}. \qquad (17)$$

In addition

$$-iv_0\nabla(\gamma_1 + \bar{\gamma}_1) - \omega(\gamma - \bar{\gamma}_1) = \pi i\frac{\omega\Delta_\omega}{T}\operatorname{ch}^{-2}\frac{\varepsilon}{2T}(f_0^R - f_0^A). \qquad (18)$$

On the other hand, $\gamma_2 - \gamma_2^+$ is proportional to l, and therefore, substituting in (18) $\gamma_1 + \bar{\gamma}_1$ from (13), we obtain with the aid of (14) and (16), retaining the principal terms in the right-hand side

$$D\nabla\left\{\frac{1}{g_0^R}[(f_0^R - f_0^A)(\nabla f_0^R)(\gamma_1 - \bar{\gamma}_1) - \nabla(f_0^R - f_0^A)g_0^R(\gamma_2 + \gamma_2^+)]\right.$$
$$\left. + \frac{g_0^R - g_0^A}{2}\nabla(\gamma - \bar{\gamma}_1) - \frac{f_0^R - f_0^A}{2}\nabla(\gamma_2 + \gamma_2^+)\right\} + i\omega(\gamma_1 - \bar{\gamma}_1) \qquad (19)$$
$$= \pi\frac{\omega\Delta_\omega}{T}\operatorname{ch}^{-2}\frac{\varepsilon}{2T}(f_0^R - f_0^A).$$

In the derivation of (19) we have left out of the left-hand sides of the advanced functions the frequency ω, which is small in comparison with ϵ, since terms containing ω can be neglected in comparison with $i\omega(\gamma_1 - \bar{\gamma}_1)$ in view of the smallness of $D\nabla^2$.

As already mentioned, there are two different regions, $|\epsilon| > \Delta$ and $|\epsilon| < \Delta$. At $|\epsilon| > \Delta$ Eq. (17) yields

$$\varepsilon(\gamma_2 + \gamma_2^+) = \Delta(\gamma_1 - \bar{\gamma}_1)$$

and we readily obtain from (19)

$$-\frac{\partial}{\partial t}\left[\frac{\varepsilon(\gamma_2 + \gamma_2^+)}{\Delta}\right] + D\nabla^2\left[\frac{\gamma_2 + \gamma_2^+}{f}\right] = \frac{2\pi i}{T}\operatorname{ch}^{-2}\frac{\varepsilon}{2T}f\frac{\partial\Delta}{\partial t}. \qquad (20)$$

We have put

$$f = \begin{cases}\Delta/(\varepsilon^2 - \Delta^2)^{1/2}, & \varepsilon > \Delta\\ -\Delta/(\varepsilon^2 - \Delta^2)^{1/2}, & \varepsilon < -\Delta\end{cases}; \qquad g = \begin{cases}\varepsilon/(\varepsilon^2 - \Delta^2)^{1/2}, & \varepsilon > \Delta,\\ -\varepsilon/(\varepsilon^2 - \Delta^2)^{1/2}, & \varepsilon < -\Delta.\end{cases}$$

We see that owing to the smallness of $D\nabla^2$ (see above) the quantity $(\gamma_2 + \gamma_2^+)$ is anomalously large in the region $|\epsilon| > \Delta$.

We note here the fact that at $|\epsilon| > \Delta$ the integral

$$\int(\gamma_2 + \gamma_2^+)d\varepsilon/4\pi i$$

is determined by the frequencies $\epsilon \sim \Delta \ll T$.

At $|\epsilon| < \Delta$, the terms of first order in ω in the right and left sides of (19) vanish. In this case it is necessary to take into account in the right-hand side the difference between $f_{0\epsilon}^R$ and $f_{0\epsilon-\omega}^A$, which is due to ω. As a result we get

$$\gamma_2 + \gamma_2^+ = -\frac{\pi i}{T}\operatorname{ch}^{-2}\frac{\varepsilon}{2T}\frac{\varepsilon^2}{(\Delta^2 - \varepsilon^2)^{3/2}}\frac{\partial\Delta}{\partial t}.$$

This expression coincides with Eq. (8) of[4] at $|\epsilon| < \Delta$. Unlike (20), it does not contain the small factor $D\nabla^2$ in the left-hand side, and therefore in the region $|\epsilon| < \Delta$ the quantity $(\gamma_2 + \gamma_2^+)$ turns out to be small in comparison with its value at $|\epsilon| > \Delta$.

As to Eq. (4b), we shall now show that it is not independent, but follows from the expressions (4c, d) for the current and the continuity equation

$$\operatorname{div}\mathbf{j} + \partial\rho/\partial t = 0,$$

where ρ is the charge density. The continuity equation in a superconductor, in turn, just as in any metal, means $\operatorname{div}\mathbf{j} = 0$, since no charges accumulate, owing to the strong Coulomb interaction, and $\partial\rho/\partial t = 0$[3] (the quasi-neutrality condition).

We note that the last equation in our model is satisfied identically; namely, in slow nonstationary processes the deviation of the charge density from the equilibrium value is proportional to the frequency, so that $\partial\rho/\partial t$ is of second order of smallness and should be taken equal to zero. In other words, Eq. (4b) is the consequence of Maxwell's equation

$$\operatorname{rot}\mathbf{H} = \frac{4\pi}{c}\mathbf{j}$$

and of the equation for the current. Indeed, it is seen from (11) that, accurate to terms of first order in the frequency

$$v_0\nabla(\gamma_1-\bar{\gamma}_1)=2i\Delta(\gamma_2-\gamma_2^+).$$

Multiplying this equation by $(emp_0/2\pi^2)(d\epsilon/4\pi i)$ and integrating with respect to ϵ, we obtain with the aid of (4d)

$$\operatorname{div}\mathbf{j}^{(a)}-\frac{mp_0e^2}{2\pi c}D_1\operatorname{div}\left[\frac{\Delta^2}{T_c}\mathbf{Q}\right]=\frac{2iemp_0}{\pi^2}\left\{\frac{i\pi}{8T_c}D_1\operatorname{div}\left[\frac{2e}{c}\mathbf{Q}\Delta^2\right]-\right.$$
$$\left.-\int\frac{d\varepsilon}{4\pi i}\frac{\Delta(\gamma_2-\gamma_2^+)}{2}\right\}.$$

Comparing this with (4c) we obtain immediately (4b).

Thus, it suffices to write only the equations that determine $\mathbf{j}^{(a)}$. To find $\gamma_1-\bar{\gamma}_1$, it is necessary in general to solve the entire system (11). We can verify, however, that the contribution made to the energy dissipation in the vortex by the anomalous part of the current $\mathbf{j}^{(a)}$ is small in comparison with the contribution of $\gamma_2+\gamma_2^+$. Indeed, the expression for $\gamma_1-\bar{\gamma}_1$ at $|\epsilon|>\Delta$ is

$$(\gamma_1-\bar{\gamma}_1)=\frac{l}{3}\left\{-g^2\nabla\left(\frac{\gamma_1+\bar{\gamma}_1}{g}\right)+\frac{2\pi i}{T}\operatorname{ch}^{-2}\frac{\varepsilon}{2T}g^2\frac{e}{c}\frac{\partial\mathbf{Q}}{\partial t}\right\}.$$

For $\gamma_1+\bar{\gamma}_1$ we have in turn from (11) and (13)

$$\nabla^2(\gamma_1+\bar{\gamma}_1)-\left[\frac{(\nabla f)^2}{g^2}+\left(\frac{2e}{c}\mathbf{Q}f\right)^2\right](\gamma_1+\bar{\gamma}_1)$$
$$=\frac{2\pi i}{T}\operatorname{ch}^{-2}\frac{\varepsilon}{2T}\nabla\left(g^3\frac{e}{c}\frac{\partial\mathbf{Q}}{\partial t}\right)g^{-1}.$$

Retaining in the integral

$$\int(\gamma_1-\bar{\gamma}_1)\,d\varepsilon/4\pi i$$

only the principal terms in Δ/T, we obtain

$$\mathbf{j}^{(a)}=\sigma\left\{-\frac{1}{c}\frac{\partial\mathbf{Q}}{\partial t}+\frac{1}{2e}\nabla\int(\gamma_1+\bar{\gamma}_1)\frac{d\varepsilon}{4\pi i}\right\},$$
$$\nabla^2(\gamma_1+\bar{\gamma}_1)=\frac{2\pi i}{T}\operatorname{ch}^{-2}\frac{\varepsilon}{2T}\frac{e}{c}\operatorname{div}\left(\frac{\partial\mathbf{Q}}{\partial t}\right),$$

where $\sigma=p_0^3e^3\tau/3\pi^2m$ is the conductivity of the normal metal. Thus, the divergence of the normal current is equal to zero. As applied to the motion of vortex filaments, the last equation gives a stronger result. Namely, the vector potential $\mathbf{Q}$ at the center of the filament varies like ρ^{-1}, where ρ is the distance to the filament axis[15]. Stipulating that the normal current be finite in the center of the vortex $\rho\ll\xi$, we obtain for the principal terms in Δ/T

$$\nabla(\gamma_1+\bar{\gamma}_1)=\frac{2\pi i}{T}\operatorname{ch}^{-2}\frac{\varepsilon}{2T}\frac{e}{c}\frac{\partial\mathbf{Q}}{\partial t}.$$

That is to say, the normal current is equal to zero in this case. The corrections to $\mathbf{j}^{(a)}$ in the next higher approximation in Δ/T, as seen from the equation for $\gamma_1+\bar{\gamma}_1$, are of the order of Δ/T relative to $c^{-1}\sigma\partial\mathbf{Q}/\partial t$ and, consequently, make a small contribution to the energy dissipated by the moving vortex.

3. MOTION OF VORTEX FILAMENTS IN AN ALLOY AT $T_c-T\ll T_c$

Just as in the earlier work[9], we assume a magnetic field $H_0\ll H_{c2}$. This means that the vortex-filament density B_0/Φ_0 is small ($\Phi_0=\pi c/e$ is the flux quantum and B_0 is the induction), i.e., the distance d between filaments is much larger than the dimension $\xi(T)$ of the core of the filament. This condition enables us to reduce the problem of the motion of a lattice of vortices to the problem of the single filament. Namely, assume that a transport current $\mathbf{j}_{tr}$ perpendicular to the magnetic field H_0 flows through the sample.

As shown by us in[9], the current produced by an individual vortex at distances ρ, $\xi\ll\rho\ll d$ from the center is equal to the transport current. It is thus necessary to establish the connection between the velocity of a single filament and the current at large distances from the filament center. Knowing the vortex velocity, we can determine from the formula $\operatorname{curl}\bar{\mathbf{E}}=-c^{-1}\partial B/\partial t$ the intensity of the average electric field

$$\bar{\mathbf{E}}=\frac{B_0}{c}[\mathbf{n}_H\times\mathbf{u}] \qquad (21)$$

($\mathbf{n}_H$ is a unit vector in the magnetic-field direction and $\mathbf{u}$ is the velocity of the vortex filament), and consequently determine the resistance of the sample. We shall henceforth assume that the Ginzburg-Landau parameter is $\kappa=\delta/\xi\gg1$.[1)]

The function $\Delta(\rho)$ for an immobile vortex was obtained by Abrikosov[15]. We shall label the corresponding functions Δ and $\mathbf{Q}$ by the index 0. At low velocities it can be assumed that in first approximation the vortex moves as a unit, i.e., Δ and $\mathbf{Q}$ are the functions Δ_0 and $\mathbf{Q}_0$ of $\mathbf{r}-\mathbf{u}t$, and the deviations due to the "deceleration" are small corrections, Δ_1 and $\mathbf{Q}_1$. We choose also a cylindrical coordinate system (ρ,φ,z) in which the z axis coincides at the initial instant of time with the filament axis. We have

$$\Delta=\Delta_0(\mathbf{r}-\mathbf{u}t)+\Delta_1,\quad \mathbf{Q}=\mathbf{Q}_0(\mathbf{r}-\mathbf{u}t)+\mathbf{Q}_1. \qquad (22)$$

We recall that Eqs. (4) without the anomalous term and the time derivative are static equations that are invariant relative to any displacement of the origin. The functions $\Delta_0(\mathbf{r}+\mathbf{a})$ and $\mathbf{Q}_0(\mathbf{r}+\mathbf{a})$, where $\mathbf{a}$ is an arbitrary constant vector, are therefore also solutions of (4). We put $\Delta_1^0=(\mathbf{a}\cdot\nabla)\Delta_0$ and $\mathbf{Q}_1^0=(\mathbf{a}\cdot\nabla)\mathbf{Q}_0$. Then Δ_1^0 and $\mathbf{Q}_1^0$ also satisfy the static equations linearized relative to the small deviations Δ_1 and $\mathbf{Q}_1$.

We substitute the expressions of (22) in the equations of (4). Linearizing them and recalling that $\Delta_0=t(\mathbf{u}\cdot\nabla)\Delta_0$ and $\mathbf{Q}_0-t(\mathbf{u}\cdot\nabla)\mathbf{Q}_0$ are static solutions, we obtain

$$\frac{\pi}{8T_c}D_1\left[\nabla^2\Delta_1-\frac{4e^2}{c^2}Q_0^2\Delta_1-\frac{8e^2}{c^2}\mathbf{Q}_0\mathbf{Q}_1\Delta_0\right]$$
$$+\left[\frac{T_c-T}{T_c}-3\Delta^2\frac{7\zeta(3)}{8\pi^2T_c^2}\right]\Delta_1=-\Delta^{(a)}, \qquad (23')$$

$$\mathbf{j}_1+\frac{mp_0e^2D_1}{2\pi cT}[\Delta_0^2\mathbf{Q}_1+2\Delta_0\mathbf{Q}_0\Delta_1]=0;$$

$$\Delta^{(a)}=\int_0^\infty\frac{d\varepsilon}{4\pi i}(\gamma_2+\gamma_2^+). \qquad (23'')$$

The quantity $\gamma_2+\gamma_2^+$ is determined from (20) where $\partial\Delta/\partial t$ is replaced by $-(\mathbf{u}\cdot\nabla)\Delta_0$ and $\Delta=\Delta_0$.

As already noted, $\Delta^{(a)}$ is large, and we can therefore omit from (23′) the term $(\pi/8T_c\partial\Delta/\partial t$ in the right hand-side. As to $\mathbf{j}^{(a)}$, this quantity, as already mentioned, is of the order of $\sigma c^{-1}\partial\mathbf{Q}/\partial t(\Delta/T)$.

For a moving vortex we have therefore

$$j^{(a)}\sim\frac{\sigma\Delta}{Te\xi^2}u\sim\frac{\Delta}{T}\frac{\sigma H_{c2}u}{c}.$$

We shall show below (see (28)) that this makes a smaller contribution (by a factor Δ^2/T^2) to the conductivity than the anomalous term $\Delta^{(a)}$, so that $\mathbf{j}^{(a)}$ can be left out from the right-hand side of (23″)[2)].

To determine the current $\mathbf{j}_1$ at large distances $\rho\gg\xi(T)$ from the center of the filament, we use the procedure employed by us earlier[9] to find the integral of the system of equations for Δ_1 and $\mathbf{Q}_1$. We need also the equation stemming from the superconductor electroneutrality condition:

$$\operatorname{div}[\Delta_0^2 \mathbf{Q}_1 + 2\Delta_0 \mathbf{Q}_0 \Delta_1] = 0. \tag{24}$$

We are not interested here in the bending of the vortex filaments, and therefore leave out the corresponding terms in (23) and (24). The results can be easily generalized to this case.

Let the velocity of the vortex $\mathbf{u}$ be parallel to the x axis. Then

$$\frac{\partial\Delta}{\partial t} = -(\mathbf{u}\nabla)\Delta_0 = -u\frac{\partial\Delta}{\partial\rho}\cos\varphi.$$

It is seen from (20) that $\Delta^{(a)}$ is also proportional to $\cos\varphi$. If we express (23) and (24) in cylindrical coordinates and separate the dependence on the angle, then we can obtain, using the same procedure as in [9],

$$[\mathbf{n}_H \mathbf{a}]\mathbf{j}_{tr} = [\mathbf{n}_H \times \mathbf{a}]\mathbf{j}_t = \frac{mp_0 e}{\pi^3}\int \Delta^{(a)}\Delta_1^0 \rho\, d\rho\, a\varphi. \tag{25}$$

We now calculate $\Delta^{(a)}$. We can neglect the term with $\partial/\partial t$ in the left-hand side of (20), since the vortex velocity is low. We express (20) in cylindrical coordinates, putting

$$\gamma_2 + \gamma_2^+ = uf(\rho)w(\rho)\cos\varphi.$$

Since $\Delta_1^0 = (\mathbf{a}\cdot\nabla)\Delta_0$, the integral in (25) is proportional to the scalar product $\mathbf{u}\cdot\mathbf{a}$:

$$[\mathbf{n}_H \times \mathbf{a}]\mathbf{j}_{tr} = \frac{emp_0}{\pi^2}(\mathbf{ua})\int_0^\infty \rho\, d\rho\frac{d\Delta}{d\rho}\int_{\varepsilon>0}\frac{d\varepsilon}{4\pi i}w\frac{\Delta}{(\varepsilon^2-\Delta^2)^{1/2}}, \tag{26}$$

and w is obtained from the equation

$$D\frac{a}{d\rho}\left[\frac{1}{\rho}\frac{d}{d\rho}(\rho w)\right] = \frac{2\pi i}{T}\frac{d}{d\rho}(\varepsilon^2-\Delta^2)^{1/2}, \quad |\varepsilon| > \Delta(\rho), \tag{27}$$

with $w = 0$ at $|\epsilon| < \Delta$.

Equation (27) has a solution that ensures that the functions $\gamma_1 - \bar{\gamma}_1$ and $\gamma_2 + \gamma_2^+$ are finite at $\rho = 0$. This solution is

$$w = \frac{2\pi i}{DT}\frac{1}{\rho}\int_0^\rho [(\varepsilon^2-\Delta^2)^{1/2} - C]\rho'\, d\rho'.$$

To determine the constant C, we recall that in our approximation $\gamma_2 + \gamma_2^+$ should be regarded as different from zero only if $|\epsilon| > \Delta(\rho)$. The condition that $\gamma_2 + \gamma_2^+$ decrease at large distances therefore determines C only when $\epsilon > \Delta_\infty$, then there exists a certain value $\rho = \rho_\epsilon$ at which $\epsilon = \Delta(\rho)$. Thus, $\gamma_2 + \gamma_2^+$ differs from zero when $\rho < \rho_\epsilon$ and equals zero when $\rho > \rho_\epsilon$.

In the approximation in which the retarded and advanced functions $f_0^{R(A)}$ and $g_0^{R(A)}$ (16) vary slowly, the rigorous boundary condition for the values of γ at $\rho = \rho_\epsilon$ necessitates that the solutions be matched together in this region. We consider it natural, however, to impose on the real quantities $(\gamma_2 + \gamma_2)/4\pi i$ etc. the requirement that they be continuous at the point ρ_ϵ. Putting $w(\rho_\epsilon) = 0$, we get

$$C = C_\varepsilon = \frac{2}{\rho_\varepsilon^2}\int_0^{\rho_\varepsilon}(\varepsilon^2-\Delta^2)^{1/2}\rho\, d\rho, \quad \varepsilon < \Delta_\infty.$$

Thus

$$C = \begin{cases}(\varepsilon^2-\Delta_\infty^2)^{1/2}, & \varepsilon > \Delta_\infty\\ C_\varepsilon, & \varepsilon < \Delta_\infty\end{cases}$$

To calculate the integral in (26), it is convenient to integrate first with respect to ρ by parts, and then again interchange the order of integration. We introduce the dimensionless quantities

$$\rho = \xi\tilde{\rho}, \quad \Delta = \Delta_\infty\tilde{\Delta}, \quad \varepsilon = \Delta_\infty\tilde{\varepsilon},$$

where, as seen from (4)

$$\xi = (\pi^3 T D_t / 7\zeta(3)\Delta_\infty^2)^{1/2}, \quad \Delta_\infty = (8\pi^2 T_c(T_c - T)/7\zeta(3))^{1/2},$$

and $\tilde{\Delta}$ satisfies the equation (we recall that $\rho \ll \delta$, where δ is the field penetration depth):

$$\left(\frac{d^2}{d\tilde{\rho}^2} + \frac{1}{\tilde{\rho}}\frac{d}{d\tilde{\rho}} - \frac{1}{\tilde{\rho}^2}\right)\tilde{\Delta} + (1-\tilde{\Delta}^2)\tilde{\Delta} = 0$$

with boundary conditions $\tilde{\Delta} \to 1$, $\tilde{\rho} \to \infty$ and $\tilde{\Delta}(0) = 0$). From (26) we have

$$\mathbf{j}_{tr} = \alpha\left(\frac{\pi^3}{7\zeta(3)}\right)^2\frac{\sigma H_{c2}}{c}\frac{T}{\Delta_\infty}y^2(\tau T)[\mathbf{n}_H \times \mathbf{u}], \tag{28}$$

$$\alpha = \int_0^\infty \tilde{\rho}\, d\tilde{\rho}\left[\int_{\tilde{\Delta}}^1(\tilde{\varepsilon}^2-\tilde{\Delta}^2)\, d\tilde{\varepsilon} + \int_1^\infty[(\tilde{\varepsilon}^2-\tilde{\Delta}^2)^{1/2} - (\tilde{\varepsilon}^2-1)^{1/2}]^2\, d\tilde{\varepsilon}\right] - \frac{1}{2}\int_0^1 \tilde{C}_\varepsilon^2\tilde{\rho}_\varepsilon^2\, d\varepsilon,$$

$$C_\varepsilon = \frac{2}{\tilde{\rho}_\varepsilon^2}\int_0^{\tilde{\rho}_\varepsilon}(\tilde{\varepsilon}^2-\tilde{\Delta}^2)^{1/2}\tilde{\rho}\, d\tilde{\rho}.$$

To calculate the constant α we have approximated $\tilde{\Delta}(\tilde{\rho})$ by the function $\tilde{\rho}(1+\tilde{\rho}^2)^{-1/2}$, which has the correct asymptotic form as $\tilde{\rho} \to \infty$ and duplicates $\tilde{\Delta}(\tilde{\rho})$ quite well in the remaining range of $\tilde{\rho}$. All the integrals can then be easily calculated and yield $\alpha = 0.26$.

We see from (28) that in our case there is no velocity perpendicular to the current, i.e., there is no Hall effect in our approximation. Comparing (28) with expression (21) for the electric field, we obtain the effective conductivity of the superconductor with vortex filaments:

$$\sigma_{eff} = \alpha\left(\frac{\pi^3}{7\zeta(3)}\right)^2\frac{T}{\Delta_\infty}\sigma\frac{H_{c2}}{B_0}y^2(\tau T) = 1{,}1\left(1-\frac{T}{T_c}\right)^{-1/2}\sigma\frac{H_{c2}}{B_0}y^2(\tau T).$$

We can also write down an expression for the viscosity coefficient η. The energy dissipation per unit volume and per vortex of unit length is $\sigma_{eff}E^2\Phi_0/B_0$. Equating it to ηu^2 we obtain

$$\eta = \alpha\left(\frac{\pi^3}{7\zeta(3)}\right)^2\frac{\sigma\Phi_0 H_{c2}}{c^2}\frac{T}{\Delta_\infty}y^2(\tau T).$$

For alloys with $l \ll \xi_0$ we have, in particular $(y(\tau T) = 1)$:

$$\sigma_{eff} = \alpha\left(\frac{\pi^3}{7\zeta(3)}\right)^2\frac{T}{\Delta_\infty}\frac{\sigma H_{c2}}{B_0} = 1{,}1\left(1-\frac{T}{T_c}\right)^{-1/2}\frac{\sigma H_{c2}}{B_0}.$$

We call attention to the fact that the temperature dependence of the conductivity in ordinary alloys differs from the dependence of σ_{eff} in alloys with paramagnetic impurities, obtained earlier in [9], owing to the presence of the additional factor T/Δ_∞.[3)] The last circumstance reflects the slowness of the relaxation processes near T_c. The existence of such a temperature dependence is indicated by recent measurements of σ_{eff} near T_c [16].

In conclusion, the authors thank K. K. Likharev and G. M. Éliashberg for useful discussions.

[1)]Since $\kappa \sim 1$ in real pure superconductors ($\tau T_c \gg 1$), the condition $\kappa \gg 1$ is actually satisfied in experiments only in the limit of "dirty" alloys, where $l \ll \xi_0$.

[2)]The vanishing of the normal current evidently allows us to disregard the normal energy loss at the center of the vortex [10,11].

[3)]A temperature dependence of the same kind can be obtained also from our results [9], by cutting off, in order of magnitude, the large parameter $\tau_S T_C$ as $\tau_S \to \infty$ at the value T/Δ (see [10]).

[1] Y. B. Kim, C. F. Hempstead and A. R. Strnad, Phys. Rev. **139**, A1163, 1965.

[2] P. G. de Gennes and J. Matricon, Rev. Mod. Phys. **36**, 45, 1964. J. Bardeen and M. J. Stephen, Phys. Rev. **140**, A1197, 1965. A. Schmid, Phys. Kond. Materie, **5**, 302, 1966. W. E. Vinen and P. Nozieres, Phil. Mag. **14**, 667, 1966. A. L. Fetter, Phys. Rev. **163**, 390, 1967. A. L. Fetter, P. C. Hohenberg and P. Pincus, Phys. Rev. **147**, 140, 1966. A. A. Abrikosov, M. P. Kemoklidze and I. M. Khalatnikov, Zh. Eksp. Teor. Fiz. **48**, 765 (1965) [Sov. Phys.-JETP **21**, 506 (1965)].

[3] L. P. Gor'kov and G. M. Éliashberg, Zh. Eksp. Teor. Fiz. **54**, 612 (1968) [Sov. Phys.-JETP **27**, 328 (1968)].

[4] L. P. Gor'kov and G. M. Éliashberg, ibid. **56**, 1297 (1969) [**29**, 698 (1969)].

[5] L. P. Gor'kov and G. M. Éliashberg, J. Low Temp. Phys. **2**, 161 (1970).

[6] G. M. Éliashberg, Zh. Eksp. Teor. Fiz. **55**, 2443 (1968) [Sov. Phys.-JETP **28**, 1298 (1969)].

[7] C. Caroli and K. Maki, Phys. Rev. **164**, 591, 1967. K. Maki, J. Low. Temp. Phys. **1**, 54, 1969.

[8] Y. Baba and K. Maki, Progr. Theor. Phys. **44**, 1431, 1970.

[9] L. P. Gor'kov and N. B. Kopnin, Zh. Eksp. Teor. Fiz. **60**, 2331 (1971) [Sov. Phys.-JETP **33**, 1251 (1971)].

[10] M. Yu. Kupriyanov and K. K. Likharev, ZhETF Pis. Red. **15**, 349 (1972) [JETP Lett. **15**, 247 (1972)].

[11] C. R. Hu and R. S. Thompson, Phys. Rev. **B6**, 110, 1972.

[12] G. Eilenberger, Z. Phys. **214**, 195, 1968.

[13] K. D. Usadel, Phys. Rev. Lett. **25**, 507, 1970.

[14] G. M. Éliashberg, Zh. Eksp. Teor. Fiz. **61**, 1254 (1971) [Sov. Phys.-JETP **34**, 668 (1972)].

[15] A. A. Abrikosov, ibid. **32**, 1442 (1957) [**5**, 1174 (1957)].

[16] V. N. Gubankov, Fiz. Tverd. Tela **14**, 2618 (1972) [Sov. Phys.-Solid State **14**, 2264 (1973)].

Translated by J. G. Adashko
38

An unusual superconductivity in UBe_{13}

G. E. Volovik and L. P. Gor'kov
L. D. Landau Institute of Theoretical Physics, Academy of Sciences of the USSR

(Submitted 10 May 1984)
Pis'ma Zh. Eksp. Teor. Fiz. **39**, No. 12, 550–553 (25 June 1984)

The most symmetric superconducting phases in cubic crystals are found. Some of their thermodynamic and magnetic properties are discussed.

The compound UBe_{13}, which goes superconducting at $T_c = 0.85$ K, exhibits several exotic properties.[1] Above 100 K its spin susceptibility obeys a Curie law, becoming independent of the temperature only at $T_F \sim 10$ K. The state density extracted from this behavior and from data on the electron specific heat corresponds to an effective mass $m^* \simeq 200 m_e$. A similar inference comes from the anomalously high critical fields [more precisely, $(dH_{c2}/dT)_{T_c} = -257$ kOe/K; Ref. 1]. It has therefore been suggested that for this compound [and also $CeCu_2Si_2$ (Ref. 2) and U_6Fe (Ref. 3)] it would be better to speak in terms of a Fermi liquid of heavy fermions ($5f$ electrons) with a degeneracy temperature $T_F \sim 10$ K. Correspondingly, we would expect that the superconducting pairing would not be of a phonon nature but would instead result from magnetic mechanisms (the exchange of paramagnons) and would correspond to a triplet pairing. More briefly, it is expected that in these compounds the superconductivity might in many respects be more reminiscent of the superfluidity in ^{3}He than the ordinary Bardeen-Cooper-Schrieffer superconductivity. Recent measurements[4] of the electron specific heat of UBe_{13} below T_c have shown that its low-temperature behavior corresponds to $C_{se} \simeq \gamma T (T/T_c)^2$, instead of the exponential dependence $\exp(-\Delta/T)$ for BCS superconductivity with a gap. Such a dependence is characteristic of ^{3}He-A, where the gap in the quasiparticle spectrum varies along the Fermi surface, vanishing at two points on this surface.

Nevertheless, a direct analogy with the superfluid phases of ^{3}He (in particular, the triplet pairing mechanism) does not follow from these results. The primary distinction between a superconductor and ^{3}He stems from its discrete crystal symmetry. For example, the compound UBe_{13} belongs to the cubic group.[5] In the present paper we are basically listing the types of superconducting states (the superconducting symmetry cases) which are possible for cubic crystals. According to Ott *et al.*,[4] the UBe_{13} ground state differs from the ordinary type, but these results do not rule out the possibility of a singlet pairing. We will also briefly mention certain phenomena which might help resolve the question.

To list the substantially different symmetry classes for an equilibrium superconducting state is only slightly more complicated than to construct the magnetic classes.[6] This assertion also applies to the case of triplet pairing, provided that the spin-orbit effects are sufficiently strong (i.e., provided that rotations of the lattice simultaneously rotate the spins pinned to it). Treating the singlet and triplet states separately, we can split off an inversion transformation. The complete group whose discrete subgroups are to be found contains the spatial rotation group O, time reversal R, and Abelian gauge-rotation group U(1). In order to construct expanded subgroups [whose elements contain transformations from O in combination with elements from U(1) and R] we need to construct factor groups on one of the invariant subgroups (there are two in O: $D_2 \equiv V$ and T). From the elements $(e^{2\pi i/3}, e^{4\pi i/3})$ or $(e^{\pi i/2}, e^{\pi i}, e^{-\pi i/2})$, and an element of R we in turn construct a group which is isomorphic to the factor group. Invariant combinations can be constructed on any basis of representations of the group O whose characters are the same as those for the factor group.

The superconducting order parameter is[7]

$$\hat{\Delta}(\mathbf{k}) = \psi(\mathbf{k}) i\hat{\sigma}_y \quad (S = 0);$$
$$\hat{\Delta}(\mathbf{k}) = i(\hat{\sigma}\mathbf{d}(\mathbf{k}))\hat{\sigma}_y \quad (S = 1).$$

The coordinate part [$\psi(\mathbf{k})$ and $\mathbf{d}(\mathbf{k})$] is correspondingly either even or odd under the replacement $\mathbf{k} \to -\mathbf{k}$. Table I lists only those superconducting phases in a crystal of group O_h (UBe_{13}) which have the highest symmetries (in the theory of the superfluidity of ^{3}He, these phases are called "inert").

The superconducting class $O \times R$ has an analog either an ordinary superconductor ($S=0$) or the B phase of ^{3}He ($S=1$). In other cases the rotation elements are linked up with elements of the gauge group (this is a violation of relative gauge-rotation invariance). As a result, at the points at which the Fermi surface intersects certain axes, or on lines where this surface intersects certain symmetry planes, the superconducting gap in the excitation spectrum vanishes [if this condition is written as $\det\hat{\Delta}(\mathbf{k}) = 0$, it would correspond in the case $S=1$ to $\mathbf{d}^2(\mathbf{k}) = 0$]. The behavior $C_{se} \propto T^3$ in UBe_{13} (Ref. 4) is satisfied by four phases: a singlet phase $O(D_2)$ and three triplet phases : $O(D_2)$, $O(T) \times R$, and $D_4(E)$.

Let us assume[8] $\det\hat{\Delta}(\mathbf{k}) = |\det\hat{\Delta}(\mathbf{k})| \exp(i\varphi(\mathbf{k}))$. Near a point where we have $\det\hat{\Delta}(\mathbf{k}) = 0$, either the condition $(\mathbf{k}) = 0$ holds [e.g., $O(T) \times R$], or the phase $\varphi(\mathbf{k})$ acquires an increment of $2\pi N$ when this point is circumvented (a boojum). In the A phase of ^{3}He, this singularity is associated with an orbital angular momentum. In a super-

TABLE I. The most symmetric superconducting phases in a cubic crystal

Superconducting class	Degeneracy	Positions of zeros in spectral gap Type of order parameter	Magnetic moment	Isotropic analog
$O \times R$ ($S = 0$)	1	None $\psi(\mathbf{k}) = f(\mathbf{k}) = f(-\mathbf{k})$, real cubic function	—	Superconducting in S state
$O \times R$ ($S = 1$)	1	None $\mathbf{d} = (\tilde{x}k_x + \tilde{y}k_y + \tilde{z}k_z)f(\mathbf{k})$	—	B-^{3}He
$O(D_2)$ ($S = 0$)	2	8 intersections of Fermi surface with threefold axes $\psi = (k_z^2 + k_x^2 e^{2\pi i/3} + k_y^2 e^{-2\pi i/3})f(\mathbf{k})$	—	—
$O(D_2)$ ($S = 1$)	2	The same zeros $\mathbf{d} = (\tilde{z}k_z + \tilde{x}k_x e^{2\pi i/3} + \tilde{y}k_y e^{-2\pi i/3})f(\mathbf{k})$	—	α-phase ^{3}He, $S=1$
$O(T) \times R$ ($S = 0$)	1	Lines where the Fermi surface intersects diagonal planes of the cube $\psi = (k_x^2 - k_y^2)(k_y^2 - k_z^2)(k_z^2 - k_x^2)f(\mathbf{k})$	—	—
$O(T) \times R$ ($S = 1$)	1	Points at which the Fermi surface intersects fourfold axes $\mathbf{d} = \{\tilde{x}k_x(k_y^2 - k_z^2) + \tilde{y}k_y(k_z^2 - k_x^2) + \tilde{z}k_z(k_x^2 - k_y^2)\}f(\mathbf{k})$	—	—
$D_4(E)$ ($S = 1$)	6	2 points at which the Fermi surface intersects a fourfold axis $\mathbf{d} = \{a\tilde{z}(k_x + ik_y) + bk_z(\tilde{x} + i\tilde{y})\}\ f(\mathbf{k})$	Along fourfold axis	A-^{3}He
$D_4(E)$ ($S = 0$)	6	At the same places and also at the intersection of the Fermi surface with the perpendicular symmetry plane $\psi = k_z(k_x + ik_y)f(\mathbf{k})$	Along fourfold axis	—

Here $\tilde{x}, \tilde{y}, \tilde{z}$ are unit vectors along the coordinates. Here the groups are $O(T) = (E, 8C_3, 3C_2, 6C_2e^{i\pi}, 6C_4e^{i\pi})$; $D_4(E) = (E, C_2e^{i\pi}, C_4e^{i\pi/2}, C_4^3e^{-i\pi/2}, U_{2x}e^{i\pi}R, U_{2y}R, 2U_2'e^{\pm \pi i/2}R)$; $O(D_2) = (E, 3C_2, 2C_4^xR, 2C_2^{x\perp}R, 2C_4^ye^{2\pi i/3}R, 2C_2^{y\perp}e^{2\pi i/3}R, 2C_4^ze^{-2\pi i/3}R, 2C_2^{z\perp}e^{-2\pi i/3}R, 4C_3e^{-2\pi i/3}, 4C_3^2e^{2\pi i/3})$.

conductor, it would correspond to an orbital magnetic moment (and also a spin moment in the triplet phase). It is easy to show that the corresponding moments in $O(D_2)$ are ordered antiferromagnetically (along threefold axes) because of the cubic symmetry. A slight uniaxial deformation, however, gives rise to a magnetic moment piezoelectric magnetism).

The question of the orbital angular momentum and its magnitude are most important for the $D_4(E)$ $(S=1)$ phase, which is a superconductor with an orbital and spin ferromagnetism. Even in the ground state, a current should flow along the surface of the sample and cancel the field of the current inside it. (The field can be estimated to be $H \sim \mu_B n \sim 1$ kOe.) A discrete (sixfold) degeneracy allows the formation of domain structures which offset the loss of magnetic energy.

Let us assume that the spin-orbit interaction is nonvanishing but still weak enough that the angle (θ) through which the spins rotate around a fourfold axis could vary over space. In this case the charging current would acquire an additional term[9]:

$$\mathbf{j} = \rho_s (\vec{\nabla} \varphi - \frac{2e}{c} \mathbf{A}) + \rho_1 \vec{\nabla} \theta \, .$$

From this term we would expect a nontrivial flux-quantization condition: $\Phi = \varphi_0[N_1 + N_2(\rho_1/\rho_s)]$. Here we have $\rho_1/\rho_s \sim 1$, and this ratio is otherwise arbitrary. The finite spin-orbit coupling determines how advantageous the formation of an additional spin structure will be. This structure might form, for example, around a vortex associated with a change in θ upon a circumvention of 2π $(N_2 = 1,2,...)$.

The magnitude of the spin-orbit coupling is also important for distinguishing the singlet pairing from the triplet pairing in the most symmetric phase, $O(D_2)$. Spin degrees of freedom might be excited at frequencies smaller than the energy gap. We have in mind the possibility that modes of this type would be related to the spin modes (ferromagnetic or antiferromagnetic resonances) in magnetic materials. Spin oscillations associated with the rotation of the system of spins with respect to the lattice of course exist in any phase with $S = 1$. For a charged system, however, their description requires appealing to a model.

We wish to thank V. P. Mineev for useful discussions.

[1]H. Ott, H. Rudigier, Z. Fisk, and J. L. Smith, Phys. Rev. Lett. **50**, 1595 (1983).
[2]W. Lieke, U. Rauchschwalbe, C. Bredl, F. Steglich, J. Aarts, and F. DeBoer, J. Appl. Phys. **53**, 2111 (1982).
[3]L. DeLong, J. Huber, K. N. Yang, and M. B. Maple, Phys. Rev. Lett. **51**, 312 (1983).
[4]H. R. Ott, H. Rudigier, T. M. Rice, k. Ueaa, Z. Fisk, and J. L. Smith, Preprint, 1984.
[5]E. Bucher, J. P. Maita, G. W. Hull, R. C. Fulton, and A. S. Cooper, Phys. Rev. B **112**, 440 (1975).
[6]L. D. Landau and E. M. Lifshitz, Élektrodinamika sploshnykh sred, § 38, Nauka, Moscow, 1982 (Electrodynamics of Continuous Media, Pergamon, Oxford).
[7]V. P. Mineev, Usp. Fiz. Nauk **139**, 303 (1983) [Sov. Phys. Usp. **26**, 160 (1983)].
[8]G. E. Volovik and V. P. Mineev, Zh. Eksp. Teor. Fiz. **83**, 1025 (1982) [Sov. Phys. JETP **56**, 579 (1982)].
[9]G. E. Volovik, Usp. Fiz. Nauk **143**, 73 (1984) [Sov. Phys. Usp. (to be published)].

Translated by Dave Parsons
Edited by S. J. Amoretty

Superconducting classes in heavy-fermion systems

G. E. Volovik and L. P. Gor'kov

L. D. Landau Institute of Theoretical Physics, USSR Academy of Sciences
(Submitted 18 October 1984)
Zh. Eksp. Teor. Fiz. **88**, 1412–1428 (April 1985)

A mathematical method for constructing the superconducting classes for nontrivial superconductors is described, and all the phases that can be produced directly in a transition from the normal state are enumerated for the cubic, hexagonal and tetragonal symmetries. It is shown that in the triplet case the type of zeros in the energy gap always corresponds to points on the Fermi surface, whereas whole lines of zeros are possible in singlet pairing. For phases having zeros on lines or at points, the low-temperature heat capacity is proportional to T^2 and T^3, respectively. Superconducting phases that stem from non-one-dimensional representations can have a magnetic moment that generates currents on the surface of a single-domain sample even in the absence of an external magnetic field. A specific example of a domain wall is considered and it is shown that large magnetic currents flow in it.

1. INTRODUCTION

The development of the BCS microscopic theory of superconductivity raised immediately the question of feasibility of a nontrivial superconductivity corresponding (in isotropic Fermi-liquid terms) to Cooper pairing with nonzero angular momentum (Ref. 1).[1)] In Ref. 2 it was shown for the first time that a state with spontaneous breaking of rotational symmetry is possible in such a system. The system in which nontrivial pairing was achieved turned out to be ^{3}He, for which this fact was theoretically predicted[3] as a consequence of the role of Van der Waals attraction forces. The main properties of ^{3}He superfluidity are reported in recent reviews.[4,5] We mention only that in ^{3}He was observed the so-called *A*-phase (the Anderson-Brinkman-Morel state), the occurrence of which is difficult to explain without invoking an additional interaction mechanism via spin fluctuations of "paramagnons,"[6] Special attention was therefore paid in searches for nontrivial superconductivity (see Ref. 7) to compounds with so-called heavy fermions that have paramagnetic behavior (the Curie law) in a wide temperature range. Without listing for now these specific compounds, we indicate only that their properties are somehow connected with strongly localized 4*f* shells of Ce or 5*f* of U. This is evident from the anomalously large values of the density of states, as deduced from data on the electronic heat capacity and on the Pauli susceptibility at low temperatures. The narrowness of the effective band is conveniently characterized by an effective mass m^* that reaches in these materials values $m^* \sim (10^2–10^{-3})m_e$.

The paramagnetic mechanism favors for ^{3}He, in a simple model, triplet pairing into an *A* phase. Owing to the nontrivial breaking of the gauge symmetry,[5] the latter has, in particular, the remarkable property that the energy gap in its spectrum vanishes at two points on the Fermi surface. Were there no phase transition in ^{3}He into the isotropic *B* phase with decreasing temperature, the heat capacity of the *A* phase would have a T^3 dependence at the lowest temperatures. Calorimetric measurements[8] indeed yielded for UBe_{13} a relation $C_e \sim \gamma(T(T/T_c)^2$ that differs from the BCS-theory activation dependence. This result, however, is as yet not unambiguous. For many heavy-fermion superconductors ($CeCu_2Si_2$, Ref. 9; U_6Fe, Ref. 10; UPt_3, Ref. 11), the results available for the ultrasound-absorption coefficient (UPt_3, Ref. 12) and the reciprocal spin-relaxation time ($CeCu_2Si_2$ Ref. 13; Ube_{13}, Ref. 14) offer evidence in favor of a temperature dependence that would correspond to solid lines of zeros in the energy gap on the Fermi surface. Similar properties would be possessed in ^{3}He by the polar phase, but the latter does not correspond at all to the free-energy minimum and appears in ^{3}He-A only in vertex cores.[15]

The microscopic mechanism of the superconductivity in these compounds, the competition between superconductivity and magnetism, and the mutual role of 5*f* electrons of U and *sp* electrons of Be (e.g., in UBe_{13}) are all questions that remain far from understood. Yet the answers to many physical questions depend only on the symmetry properties of the superconducting order parameter $\Delta_{\alpha\beta}(\mathbf{k}) = \langle a_{\mathbf{k}\alpha} a_{-\mathbf{k}\beta} \rangle$. One such question is: are we dealing with triplet or singlet pairing? (It would be more correct to speak of parity, as shown below.)

In an earlier paper[16] we have already indicated some of the most symmetric superconducting classes, and have shown that $S = 1$ always corresponded to zeros only at points (not on lines) on the Fermi surface. Our present purpose is a complete analysis of the possible superconducting classes, not only for the cubic group but also for some other symmetries (hexagonal, tetragonal, rhombohedral). Anticipating somewhat, we indicate that the foregoing result remains in force in the general case. In Sec. 2 we describe briefly a method of enumerating all the superconducting classes, with the cubic groups as the example. This classification does not depend at all on the microscopic model. In the sections that follow we show that it is more convenient in practice to employ a certain generalization of the methods of phase-transition theory.[17] The region near T_c is experimentally of great interest, and a symmetry-theory analysis will enable us to identify in Secs. 3–5 those superconducting states into which a system can go over directly from the normal phase by a continuous transition (of second order). We find the types of zeros, the magnetic properties, and the degree of degeneracy for each of these states. It is shown in Sec. 6 that in some superconducting phases a magnetic moment is induced by the superconducting transition, and the current-carrying domain walls that separate domains with different magnetic-moment orientations are described.

2. CONSTRUCTION OF SUPERCONDUCTING CLASSES

The order parameter in the superconducting state is the already mentioned quantity $\hat{\Delta}_{\alpha\beta}(\mathbf{k})$, which has the meaning of the Cooper-pair wave function in the condensate. It is precisely its transformation properties under the influence of various symmetry transformations which determine the singularities of the superconducting phase. Since, in accord with the definition of the Bose condensate, there is only one such function, it should be transformed into itself under symmetry transformations that do not alter the superconducting state. The symmetric states of the system can therefore be enumerated by indicating all those possible subgroups, from among the total group, with respect to which the order parameter remains invariant. In the considered crystal-symmetry case, the total group consists of the crystal point group G, of the time-reversal operation R, and the gauge-transformation group $U(1)$. Superconductors are subdivided into superconductivity classes in a manner that is somewhat similar to the construction of magnetic classes (see, e.g., Ref. 18), i.e., by enumerating the subgroups of the expanded group $G\times R\times U(1)$ with account taken, of course, of the fact that the operation R (complex conjugation) does not commute with the transformation from U(1).

The classes of the first type are characterized by the subgroups $H\times R$ from $G\times R$. These are ordinary superconductors or else superconductors whose transition from the normal phase is accompanied by breaking of spatial symmetry (we know of no example of transitions of the latter type).

The second, no longer trivial, type includes classes whose symmetry groups contain combined elements made up of elements of the $G\times R$ group and the $U(1)$ group. The geometric locus of the zeros (if they exist) in the energy gap on the Fermi surface should be determined by the symmetry relations, just as the linkage of the gauge and spatial groups leads in ^{3}He to zeros in the gap for the A phase and for the polar phase. Thus, the main difference from ordinary superconductivity lies in the internal symmetry of the state, and this question can be investigated independently of the microscopic pairing mechanism. It can apparently now be stated that UBe_{13}, UPt_3, and possible $CeCu_2Si_2$ belong to one of the nontrivial superconductivity types.

It has already been noted in Refs. 16 and 19 that although the spin–orbit interactions in the compounds considered are large, their point groups contain inversion. By definition, the order parameter is antisymmetric relative to simultaneous permutation of all the variables: $\hat{\Delta}_{\alpha\beta}(\mathbf{k}) = -\hat{\Delta}_{\beta\alpha}(-\mathbf{k})$. We can therefore immediately distinguish between the parities of the two ansatzes

$$\hat{\Delta}(\mathbf{k})=\psi(\mathbf{k})\, i\hat{\sigma}^{y}\quad (S=0),$$
$$\hat{\Delta}(\mathbf{k})=(\hat{\sigma}\mathbf{d}(\mathbf{k}))\, i\hat{\sigma}^{y}\quad (S=1), \tag{1}$$

where $\psi(\mathbf{k})$ is a scalar even function and $\mathbf{d}(\mathbf{k}) = -\mathbf{d}(-\mathbf{k})$ is an odd vector function. Following the already established terminology, we shall call these two possibilities, as before, singlet and triplet states, respectively. In the broad sense in which the nontrivial cases were defined above, the triplet pairing is always of the nontrivial type: the vector $\mathbf{d}(\mathbf{k})$ is invariant not to the subgroup C_i (inversion) but to the group $C_i \exp(\pi i)$. We shall find it more convenient, however, to consider the two cases (1) separately. Then the remaining complete group is $G'\times R\times U(1)$, where G' is the point group of all the rotations. Since the spins, as mentioned, are "frozen" in the lattice, each rotation element A from G' acts in the triplet case as follows: $\hat{A}\,\mathbf{d}\equiv A\,\mathbf{d}(A\,\mathbf{k})$. We shall see later on that the properties of triplet and singlet nontrivial classes are very similar. The differences lie in the types of the zeros and are caused precisely by the different behaviors of the scalar and vector functions on the symmetry-allowed geometric locus of the points under the action of the transformations A from G'.

We shall describe here briefly the general method of listing all the expanded groups. The number of superconducting classes is very large, and we confine ourselves to the cubic system as an example. In accordance with the statements made above concerning parity, the group G' is in this case the group O of all the rotation axes of the cube. The group O contains the subgroups $T, D_4, D_3, D_2, C_4, C_3, C_2$ (we use throughout the notation of Ref. 20). The meaning of the required procedure is quite clear—the order parameter for nontrivial superconductors is not invariant to certain transformations from O, but is multiplied by phase factors or is transformed into its complex conjugate. These factors must be compensated for by combining the operations of group O with elements of subgroups from $U(1)$, namely $(1, e^{\pm i\pi/2}, e^{i\pi})$, $(1,\varepsilon,\varepsilon^2)$ and $(1, e^{i\pi})$, which are isomorphic to the rotation subgroups of fourth, third ($\varepsilon = e^{2\pi i/3}$) and second order, respectively, and of time reversal (complex conjugation). The implementation of this procedure is facilitated by the fact that the representations of an arbitrary group contain representation of factor groups made up of all its invariant subgroups.[16] For group O, the invariant subgroups are the tetrahedron group T and the group of rotations D_2 about three mutually perpendicular twofold axes that are perpendicular to the faces of the cube. The factor group O/T has the index 2, is isomorphic to the subgroup $(1, e^{i\pi})$ from $U(1)$, and its representations are one-dimensional. Therefore the combined group $O(T)$ has the elements $(E, 8C_3, 3C_2, 6U_2e^{i\pi}, 6C_4e^{i\pi})$.

The factor-group O/D_2 has the index 6 and is isomorphic to D_3. It corresponds to the two-dimensional representation E of group O. Since the operation R does not commute with operations from $U(1)$, it is possible to construct from their combinations a group isomorphic to O/D_2. The elements of the combined $O(D_2)$ group are

$$(E, 3C_2, 2C_4^x R, 2C_4^y \varepsilon R, 2C_4^z \varepsilon^2 R, 4C_3 \varepsilon^2, 4C_3^2 \varepsilon,$$
$$2U_2^{\perp x} R, 2U_2^{\perp y} \varepsilon R, 2U_2^{\perp z} \varepsilon^2 R).$$

From the elements of $U(1)$ and R we can construct a group that has representations with dimensionality higher than two. It is therefore impossible to split the representations of O further without lowering the crystal symmetry. The tetrahedron subgroup admits of the combined group $T(D_3) = (E, 3C_2, 4C_3\varepsilon, 4C_3^2\varepsilon^2)$. This group, however, is a subgroup of $O(D_2)$ and, as will be shown below, cannot stem directly from the cubic normal phase (T-subgroups of lower symmetry (threefold axis) are included in the system D_3).

The next subgroup of O is D_4. There exists a maximal combined group

$$D_4(E) = (E, C_2 e^{i\pi}, C_4 e^{i\pi/2}, C_4^3 e^{-i\pi/2}, U_{2x} e^{i\pi} R, U_{2y} R, 2U_2' e^{\pm i\pi/2} R).$$

The invariant subgroups in the group D_4 are C_4 and D_2 (the factor-groups D_4/C_4 and D_4/D_2 have the index 2). Three other combined groups are also possible:

$$D_4(C_4)\times R = (E, C_2, 2C_4, 2e^{i\pi}U_2, 2e^{i\pi}U_2')\times R,$$
$$D_4^{(1)}(D_2)\times R = (E, C_2, 2U_2, 2C_4 e^{i\pi}, 2U_2' e^{i\pi})\times R,$$
$$D_4^{(2)}(D_2)\times R = (E, C_2, 2U_2', 2C_4 e^{i\pi}, 2U_2 e^{i\pi})\times R.$$

For C_4, all the possibilities correspond to subgroups of the just-considered groups from D_4.

The same procedure is used to construct

$$D_3(E) = (E, C_3\varepsilon, C_3^2\varepsilon^2, U_2' R, U_2'' \varepsilon^2 R, U_2''' \varepsilon R),$$
$$D_3(C_3)\times R = (E, 2C_3, 3U_2 e^{i\pi})\times R.$$

In analogy with the case of C_4, the subgroups of these groups exhaust all the possibilities for C_3.

In the subgroup D_2, finally, there are three possibilities:

$$D_2(C_{2x})\times R = (E, C_{2x}, C_{2y} e^{i\pi}, C_{2z} e^{i\pi})\times R,\ D_2(C_{2y})\times R,\ D_2(C_{2z})\times R.$$

Their subgroups also exhaust all the possibilities of the subgroup C_2.

We did not mention above the subgroups T, D_4, C_4, D_3, C_3, D_2, and C_2 themselves. For the singlet case they would correspond to the aforementioned usual superconducting classes. The triplet case, as already mentioned, would correspond to linkage of the gauge group for the inversion transformation $C_i e^{i\pi}$.

The combined groups listed above for D_4, D_3, and D_2 indicate also all the possibilities for the tetragonal, rhombohedral, rhombic, and monoclinic systems (for those classes that have an inversion center). The hexagonal system will be considered by us later. We have likewise not written out the basis functions. Some are contained in Ref. 16, where the most symmetric classes were considered, and some are given in the next sections.

A subordination scheme in the Landau theory, for all the possible second-order phase transitions without change of the number of atoms per unit cell, was constructed in Ref. 21. For the superconducting classes this analysis must be carried out anew. Thus, for example, the restrictions that follow from the requirement that the expansion for the energy functional contain no terms of third order in the order parameter are entirely lifted for the complex order parameter of a superconducting phase by gauge invariance. (The terms linear in the gradient—the Lifshitz criterion—do not appear in the systems investigated by us, which have inversion centers.) We consider below only those states to which can be reached directly from the normal metal; this requirement limits the number of classes listed.

The last requirement allows us to confine ourselves to the vicinity of T_c. Of importance to the order parameter $\hat{\Delta}(\mathbf{k})$ near T_c is only one of the representations of the group G. On a specified representation there can be realized not any arbitrary symmetry group (the superconducting class in this case), but only a definite set of classes, from among which we choose only those that can effect an absolute minimum of the Ginzburg-Landau functional in a certain region of the parameters of the latter. Examining all the representations of a given group and finding, in the manner indicated, for each of them the possible symmetry subgroups, we obtain all the symmetry classes that can be realized also at low temperatures, provided, of course, that no additional phase transition takes place when the temperature is lowered, since the symmetry cannot vary continuously.

It was noted in Ref. 22 that, owing to the strong spin-orbit coupling in these systems, the Ginzburg-Landau functional takes the same form for triplet and singlet pairing, inasmuch as all the possibilities are exhausted, at any choice of the order parameter from (1), by enumerating the representations of the rotation group G'. However, as indicated above, the types of zeros on the Fermi surface (as well as the basis functions) are generally speaking different for the two types of pairing.

3. SUPERCONDUCTIVITY IN THE CUBIC GROUP

The symmetry O_h is possessed by UBe_{13}. Separating in the O_h group the rotation subgroup $O_h = O\times C_i$, we have for O five representations: two one-dimensional A_1 and A_2, one two-dimensional E, and two three-dimensional F_1 and F_2 (see, e.g., Ref. 20).

a) One-dimensional representations. A_1 is the only representation of the O group. At $S = 0$ it describes the state A_{1g}, which has the symmetry of the complete cubic group $O_h\times R$ and corresponds to ordinary superconductivity. Parity is violated for $S = 1$, therefore the state is of the nontrivial type and coincides with the odd representation A_{1u} of the O_h group (total group $O\times R\times C_1 e^{i\pi}$). Whereas in the former case ($S = 0$) the wave function $\psi(\mathbf{k})$ can be chosen to be a real function of $f(\mathbf{k})$ and invariant to all cube-symmetry transformations, for $S = 1$ the basis function is of the form

$$\mathbf{d}(\mathbf{k}) \propto (\tilde{\mathbf{x}}k_x + \tilde{\mathbf{y}}k_y + \tilde{\mathbf{z}}k_z) f(\mathbf{k}), \tag{2}$$

where $\tilde{\mathbf{x}}$, $\tilde{\mathbf{y}}$, $\tilde{\mathbf{z}}$ are the unit vectors of the principal cubic axes. The properties of this phase are similar to the B phase of ^{3}He, with the natural exception that the state (2) is not degenerate on account of crystal symmetry. The question of the zeros of the gap in the excitation spectrum of the triplet phase is equivalent to the possibility of vanishing of the scalar $\mathbf{d}^2(\mathbf{k})$

$=k^2 f^2(\mathbf{k})$ on the Fermi surface. There are no symmetry-based grounds for the apperans of zeros for either $S=0$ or $S=1$, and the heat capacity decreases exponentially at low temperatures. The state (2) is not magnetic, since time reversal was not violated.

The representation A_2 is not unique and corresponds, as can be easily verified from the table of characters, to the nontrivial class $O(T)\times R$. The basis functions are

$$\psi(\mathbf{k}) \propto (k_x^2-k_y^2)(k_y^2-k_z^2)(k_z^2-k_x^2)f(\mathbf{k}), \tag{3}$$

$$\mathbf{d}(\mathbf{k}) \propto \{\tilde{\mathbf{x}}k_x(k_z^2-k_y^2)+\tilde{\mathbf{y}}k_y(k_x^2-k_z^2)+\tilde{\mathbf{z}}k_z(k_y^2-k_x^2)\}f(\mathbf{k}) \tag{4}$$

for $S=0$ and $S=1$, respectively. Wave-function zeros exist in (3) on the lines of intersection of the Fermi surface (FS) with all the diagonal planes of the cube, whereas in the triplet state the zeros appear on the intersection of the FS with threefold and fourfold axes (the last fact was not noted in Ref. 16).

We show now that the appearance of these zeros is connected with nontrivial elements of the $O(T)\times R$ group. For $S=0$ we have $C_2\psi=\psi$, $C_i\psi=\psi$ and $U_2e^{i\pi}\psi=\psi$. We take an arbitrary point $\mathbf{k}_0=(a,a,k_z)$ (i.e., $k_x=k_y=a$) on a diagonal plane of the cube and use the equations

$$\psi(\mathbf{k}_0)=-\psi(U_2\mathbf{k}_0)=-C_iC_2^z\psi(a,\hat{a},-k_z)$$
$$=-\psi(a,a,k_z)=-\psi(\mathbf{k}_0)=0. \tag{5}$$

Thus, in the singlet phase the zeros appear in the gap on a whole line and the heat capacity at low temperature varies like T^2.

In the triplet phase $O(T)\times R$ the form of the basis function (4) indicates that a similar analysis need be carried out only for points lying on threefold and fourfold axes. Let $\mathbf{k}_0$ lie on a threefold axis. Since rotation of C_3 around this axis is an invariant operation, it follows that $\mathbf{d}(\mathbf{k}_0)$ is parallel to $\mathbf{k}_0$. On the other hand, we should have $\hat{U}_2e^{i\pi}\mathbf{d}(\mathbf{k}_0)=\mathbf{d}(\mathbf{k}_0)$ (the rotation axis U_2 perpendicular to the chosen threefold axis). We have then

$$\mathbf{d}(\mathbf{k}_0)=\hat{U}_2e^{i\pi}\mathbf{d}(\mathbf{k}_0)\equiv -U_2\mathbf{d}(U_2\mathbf{k}_0)=\mathbf{d}(-\mathbf{k}_0)=-\mathbf{d}(\mathbf{k}_0)=0 \tag{5'}$$

when account is taken of the fact that the representation is odd. Let now the point $\mathbf{k}_0$ lie on a fourfold axis. Again $\mathbf{d}(\mathbf{k}_0)\|\mathbf{k}_0$ (C_2 transformation). Applying the transformation $C_4e^{i\pi}$, we have

$$\mathbf{d}(\mathbf{k}_0)=\hat{C}_4e^{i\pi}\mathbf{d}(\mathbf{k}_0)=-\mathbf{d}(\mathbf{k}_0)=0.$$

The gap thus vanishes in such a triplet superconducting state at 14 points on the FS, so that at $T\ll T_c$ we have for the heat capacity $C_e\propto T^3$.

The properties of superconducting phases on one-dimensional representations are gathered in Table I. Since the representations are one-dimensional, there is no discrete degeneracy (a domain structure is impossible) and the phases are not magnetic (the time reversal R is not violated).

b. Two-dimensional representation. The order parameter for $S=0$ and $S=1$ can be written in like form as a sum over the basis functions:

$$\hat{\Delta}(\mathbf{k})=\eta_1\Phi^{(1)}(\mathbf{k})+\eta_2\Phi^{(2)}(\mathbf{k}), \tag{6}$$

where the basis functions for $S=0$ can be represented in the form

$$\Phi^{(1)}(\mathbf{k})\propto k_x^2+\varepsilon k_y^2+\varepsilon^2k_z^2,\quad \Phi^{(2)}(\mathbf{k})\propto k_x^2+\varepsilon^2k_y^2+\varepsilon k_z^2, \tag{7}$$

and for $S=1$,

$$\boldsymbol{\Phi}^{(1)}(\mathbf{k})\propto\tilde{\mathbf{x}}k_x+\varepsilon\tilde{\mathbf{y}}k_y+\varepsilon^2\tilde{\mathbf{z}}k_z,\quad \boldsymbol{\Phi}^{(2)}(\mathbf{k})\propto\tilde{\mathbf{x}}k_x+\varepsilon^2\tilde{\mathbf{y}}k_y+\varepsilon\tilde{\mathbf{z}}k_z \tag{8}$$

[here $\varepsilon=\exp(2\pi i/3)$], and the remaining $\mathbf{k}$-dependences corresponding to cubic symmetry have been left out. As usual, it is convenient to refer the transformation properties of the order parameter under symmetry transformations to two complex components (η_1,η_2).

The Ginzburg-Landau functional takes the form

$$F=-\alpha(|\eta_1|^2+|\eta_2|^2)+\beta_1(|\eta_1|^4+|\eta_2|^4)$$
$$+2\beta_2|\eta_1|^2|\eta_2|^2+\gamma(\eta_1^3\eta_2^{*3}+\eta_1^{*3}\eta_2^3). \tag{9}$$

We have added here a sixth-order term which will be needed later. If we neglect this term initially, the minimum of the functional corresponds at $\beta_2>\beta_1$ to a state with $|\eta_1|^2=\alpha/2\beta_1$, $\eta_2=0$, and at $\beta_2<\beta_1$ to a state with $|\eta_1|^2=|\eta_2|^2=\alpha/2(\beta_1+\beta_2)$. In the latter case there is additional degeneracy

TABLE I. Superconductivity classes for one-dimensional representations of the cubic group.

Representation		Class	Heat capacity $C_e(T)$
A_1	$S=0$, A_{1g}	$O\times R$	$\exp(-\Delta/T)$
	$S=1$, A_{1u}		$\exp(-\Delta/T)$
A_2	$S=0$, A_{2g}	$O(T)\times R$	T^2
	$S=1$, A_{2u}		T^3

For the type of the basis functions see expressions (2), (3), and (4) in the text. The superconducting phases with symmetry $O\times R$ have no zeros in the gap of the Fermi-excitation spectrum. The heat capacity decreases therefore in the usual exponential manner with decreasing temperature. The triplet phase from the class $O(T)\times R$ must contain zeros at the points where the FS intersects the three- and fourfold axes, whereas the zeros of the singlet phase with the same symmetry $O(T)\times R$ lie on the lines of intersection of the FS with the diagonal planes of the cube. As a result, the heat capacities of the triplet and singlet phases behave differently at low temperatures, like T^3 and T^2, respectively.

with respect to the phase difference $\varphi_1 - \varphi_2$ between the components $\eta_1 = |\eta_1| \exp i\varphi_1$, $\eta_2 = |\eta_2| \exp i\varphi_2$. The additional symmetry of functional that contains only fourth-order terms is lifted by the sixth-order terms which take the form

$$\frac{\gamma}{4}\left(\frac{\alpha}{\beta_1+\beta_2}\right)^3 \cos 3(\varphi_1-\varphi_2). \tag{10}$$

As a result it follows from (10) that $\varphi_1 - \varphi_2 = (\pi + 2\pi n)/3$, at $\gamma > 0$ and $\varphi_1 - \varphi_2 = 2\pi n/3$ at $\gamma < 0$.

The two-dimensional representation offers therefore three possibilities. At $\beta_2 > \beta_1$, as can be seen from (6)–(8), the state corresponds to the superconducting class $O(D_2)$, in which the symmetry calls for the vanishing of the gap in the intersections of all the threefold axes with the Fermi surface, for both singlet and triplet pairing.[16] In fact, applying a combined-symmetry element, say to $\psi(\mathbf{k}_0)$ ($\mathbf{k}_0$ on a threefold axis), we obtain

$$\psi(\mathbf{k}_0) = C_3 e^{2\pi i/3} \psi(\mathbf{k}_0) = \varepsilon\psi(C_3\mathbf{k}_0) \equiv \varepsilon\psi(\mathbf{k}_0) = 0. \tag{11}$$

By analogous reasoning we verify that in the case of triplet pairing $\mathbf{d}^2(\mathbf{k}_0)$ vanishes at the point $\mathbf{k}_0$. The types of the zeros in the gap likewise coincides for this class in both cases of (1). The heat capacity always varies asymptotically like T^3.

The eight points where the FS intersects the threefold axes constitute the so-called boojums on the Fermi surface, viz., topologically stable point vortices in $\mathbf{k}$ space, and when these are bypassed the phase of the function $\psi(\mathbf{k})$ or $\mathbf{d}^2(\mathbf{k})$ changes by $2\pi N$.[23] In this case N on the values ± 1. The boojum arrangement is shown in Fig. 1, where the points and crosses mark boojums with $N = 1$ and $N = -1$, respectively. It is easily seen that both superconducting states are doubly degenerate, so that domain walls are possible. We defer the discussion of the domain wall and of the magnetic nature of the superconducting state (which is ferromagnetic) to Sec. 6.

To establish the symmetry of the states produced at $\beta_2 < \beta_1$ and $\gamma > 0$, we write one of the solutions:

$$\psi(\mathbf{k}) \infty 2k_x^2 - k_z^2 - k_y^2 \quad (S=0), \tag{12}$$

$$\mathbf{d}(\mathbf{k}) \infty 2\tilde{x}k_x - \tilde{z}k_z - \tilde{y}k_y \quad (S=1). \tag{13}$$

Obviously the symmetry is $D_4 \times R$, i.e., in the singlet state we have a transition to a trivial type of superconductivity (with tetragonal symmetry of the real order parameter). The transition to the triplet state is essentially of the same type, if we disregard the linking of gauge-invariance element $\exp(i\pi)$ with the inversion $C_i \exp(i\pi)$. In (12), $\psi(\mathbf{k})$ has lines of zeros.

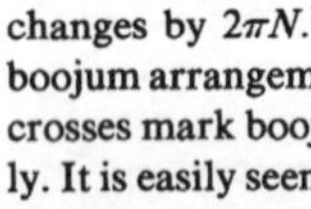

FIG. 1. Arrangement of zeros in class $O(D_2)$. The points and crosses show boojums with $N = 1$ and $N = -1$, respectively.

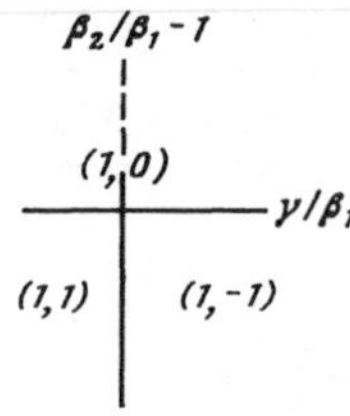

FIG. 2. Phase diagram of superconducting states near T_c for two-dimensional representation of the cubic group.

There are no combined symmetry elements in the $D_4 \times R$ group, and the zeros in (12) have a random character (a property of the functions of the representation E itself). In the nonlinear problem, i.e., with decrease of temperature, these zeros are smeared out because of the mixing of several representations of like symmetry. From the viewpoint of symmetry theory, the heat capacity should decrease exponentially at the very lowest temperatures. Within the framework of a theory of the BCS type, the admixture of solutions corresponding to other eigenvalues in the integral equation for the gap is small to the extent that the dimensionless interaction constant is small.

Finally, the states corresponding to $\beta_2 < \beta_1$, $\gamma > 0$, have basis functions of the type

$$\psi(\mathbf{k}) \infty k_x^2 - k_y^2 \quad (S=0), \tag{14}$$

$$\mathbf{d}(\mathbf{k}) \infty \tilde{x}k_x - \tilde{y}k_y \quad (S=1). \tag{15}$$

The symmetry of these solutions corresponds to the combined group $D_4^{(1)}(D_2) \times R$. The or the zeros in this group stem from nontrivial combined $C_4 \exp(i\pi)$ and $U_2 \exp(i\pi)$ symmetry elements. In the singlet state the zeros lie on the lines where the FS intersects the diagonal planes of the cube (heat capacity $\propto T^2$), and in the triplet states the zeros correspond to the points of intersection of the FS with one of the fourfold axes. The results for the two-dimensional representation E are gathered in Table II. Figure 2 shows the phase diagram for the functional (9).

c. Three-dimensional representations. The rotation group O has two irreducible three-dimensional representations: the vector representation F_1 and a representation F_2 that corresponds to transformations of a symmetric tensor of the form (xy, yz, xz). In the expansion of the order parameter

$$\hat{\Delta}(\mathbf{k}) = \sum_{i=1}^{3} \eta_i \Phi^{(i)}(\mathbf{k}) \tag{16}$$

the three basis functions $\Phi^{(i)}(\mathbf{k})$ can be chosen such that the Ginzburg-Landau functional takes the form

$$F = -\alpha(\eta\eta^*) + \beta_1(\eta\eta^*)^2 + \beta_2|\eta^2|^2 + \beta_3(|\eta_x|^4 + |\eta_y|^4 + |\eta_z|^4), \tag{17}$$

which is the same for the representations F_1 and F_2 (and, of course, for $S = 0$ and 1). The three coefficients η_i are regarded hereafter as a complex three-dimensional vector. This choice for F_1 can be:

TABLE II. Superconductivity classes from two-dimensional representations E of the cubic group.

(η_1, η_2)		Class	$C_e(T)$	Degeneracy	Magn. properties
(1, 0)	$S=0, E_g$	$O(D_2)$	T^3	2	A
	$S=1, E_u$		T^3	2	A
(1, −1)	$S=0, E_g$	$D_4^{(1)}(D_2)\times R$	T^2	3	–
	$S=1, E_u$		T^3	3	–
(1, 1)	$S=0, E_g$	$D_4\times R$	$\exp(-\Delta/T)$	3	–
	$S=1, E_u$		$\exp(-\Delta/T)$	3	–

For the types of basis function see expressions (7), (8), and (12)–(15) in the text; A—"antiferromagnetism." The triplet and singlet phases from the class $O(D_2)$ have zeros at the points where the FS intersects the threefold axes, the phase $D_4^{(1)}(D_2)\times R$ ($S=1$) has zeros at the intersectionof the FF with the fourfold axis, and the phase $D_4^{(1)}(D_2)\times R$ ($s=1$) on the lines of intersection of the FS with two mutually perpendicular vertical symmetry planes.

$$F_{1g}(S=0):\quad k_yk_z(k_y^2-k_z^2),\ k_zk_x(k_z^2-k_x^2),\ k_yk_x(k_x^2-k_y^2),\tag{18}$$

$$F_{1u}(S=1):\quad \tilde{y}k_z-\tilde{z}k_y,\ \tilde{z}k_x-\tilde{x}k_z,\ \tilde{x}k_y-\tilde{y}k_x,\tag{19}$$

and for F_2,

$$F_{2g}(S=0):\quad k_yk_z,\ k_zk_x,\ k_xk_y,\tag{20}$$

$$F_{2u}(S=1):\quad \tilde{y}k_z+\tilde{z}k_y,\ \tilde{z}k_x+\tilde{x}k_z,\ \tilde{x}k_y+\tilde{y}k_x\tag{21}$$

(the subscript g or u indicates that the representation is constructed on an even or odd representation of the O_h group, depending on the choice of S in (1).

The cubic anisotropy is represented in (17) only by the term with β_3. Omitting this term for a while, we see that the first term is invariant in the SO(6) group, and the second introduces a linkage between the real and imaginary components of the three-dimensional complex vector

$$\boldsymbol{\eta}=\boldsymbol{\eta}'+i\boldsymbol{\eta}'',\tag{22}$$

leaving the SO(3) symmetry. Elementary calculations show that in the isotropic case the minimum of (17) at $\beta_2<0$ corresponds to the choice of one real vector ($\boldsymbol{\eta}=\boldsymbol{\eta}'$), whereas at $\beta_2>0$ the vector is substantially complex, $|\boldsymbol{\eta}'|=|\boldsymbol{\eta}''|$, $\boldsymbol{\eta}'\perp\boldsymbol{\eta}''$, (i.e., $\boldsymbol{\eta}^2=0$) and the state is characterized by a triad of vectors $\boldsymbol{\eta}'\perp\boldsymbol{\eta}''$ and $\mathbf{l}$, where $\mathbf{l}=\boldsymbol{\eta}'\times\boldsymbol{\eta}''$. The continuous degeneracy is lifted by the cubic anisotropy.

The case of a real vector ($\beta_2<0$) is simpler. The expected symmetric solutions for $\boldsymbol{\eta}$ correspond to fourfold, threefold, and twofold anisotropy axes $\boldsymbol{\eta}=(1,0,0)$, $(1,1,1)$, and $(1,1,0)$, respectively. The third solution does not correspond to a minimum of the functional (17) at arbitrary β_3 and is therefore discarded. The symmetry group of the remaining solutions can be easily determined from the form of the basis functions. These groups are, of course, different for the representations F_1 and F_2. The solution (1,1,1) corresponds to $D_3(C_3)\times R$ in F_1 and to $D_2\times R$ in F_2. The solution (1,0,0), similarly, corresponds to $D_4(C_4)\times R$ for F_1 and to $D_4^{(2)}(D_2)\times R$ for F_2.

If a complex vector is chosen ($\beta_2>0$) the symmetry is lowered primarily via the choice of the direction of the vector $\mathbf{l}$, whose physical meaning is the direction of the magnetic moment (see below). The remaining degeneracy is connected with the choice of $\boldsymbol{\eta}'$ and $\boldsymbol{\eta}''$. The crystal anisotropy can, in particular, lift the condition that the vectors be equal. This

TABLE III a. Superconductivity classes from three-dimensional representation F_1 of the cubic group.

(η_x, η_y, η_z)		Class	$C_e(T)$	Degeneracy	Magn. properties
$(1, \varepsilon, \varepsilon^2)$	$S=0, F_{1g}$	$D_3(E)$	T^3	8	F
	$S=1, F_{1u}$		T^3	8	F
(1, 1, 1)	$S=0, F_{1g}$	$D_3(C_3)\times R$	T^2	4	–
	$S=1, F_{1u}$		T^3	4	–
$(1, i, 0)$	$S=0, F_{1g}$	$D_4(E)$	T^2	6	F
	$S=1, F_{1u}$		T^3	6	F
(1, 0, 0)	$S=0, F_{1g}$	$D_4(C_4)\times R$	T^2	3	–
	$S=1, F_{1u}$		T^3	3	–

See expressions (18) and (19) of the text for the type of basis functions; F stands for "ferromagnetism." The phases $D_3(E)$ ($S=0,1$) and $(D_3(C_3)\times R$ ($S=1$) have zeros at the points of intersection of the FS with spontaneous-anisotropy axis (threefold axis). The phases $D_4(C_4)\times R$ ($S=1$), $D_4(E)$ ($S=1,0$) have zeros at the points of intersection of the FS with a fourfold axis, and the last of them $D_4(E)$ $S=0$) has in addition a line of zeros on the intersection of the FS with the horizontal symmetry plane. The singlet phases $D_3(C_3)\times R$ and $D_4(C_4)\times R$ have lines of zeros on the intersection of the FS with the diagonal planes of the cube and with the vertical symmetry planes, respectively.

TABLE IIIb. Superconductivity classes from three-dimensional representations F_2 of cubic group.

(η_x, η_y, η_z)		Class	$C_e(T)$	Degeneracy	Magn. properties
$(1, \varepsilon, \varepsilon^2)$	$S=0$, F_{2g}	$D_3(E)$	T^3	8	F
	$S=1$, F_{2u}		T^3	8	F
$(1, 1, 1)$	$S=0$, F_{2g}	$D_3 \times R$	$\exp(-\Delta/T)$	4	–
	$S=1$, F_{2u}		$\exp(-\Delta/T)$	4	–
$(1, i, 0)$	$S=0$, F_{2g}	$D_4(E)$	T^2	6	F
	$S=1$, F_{2u}		T^3	6	F
$(1, 0, 0)$	$S=0$, F_{2g}	$D_4^{(2)}(D_2) \times R$	T^2	3	–
	$S=1$, F_{2u}		T^3	3	–

See Eqs. (20) and (21) of the text for the type of basis functions. The zeros phases are at the points of intersection of the FS with a threefold axis in both $D_3(E)$ phases and at the points of intersection of the FS with a fourfold axis in the phases $D_4^{(2)}(D_2) \times R$ $(S=1)$, $D_4(E)$ $(S=0,1)$. In addition, the phase $D_4(E)$ $(S=0)$ has a line of zeros on the intersection of the FS with the horizontal symmetry plane. The singlet $D_4^{(2)}(D^2) \times R$ phase has zeros on the lines of intersection of the FS with two mutually perpendicular vertical symmetry planes.

takes place, for example, if the vector **l** is directed along a twofold axis.[2)] This choice can again be rejected, since it does not minimize the functional (17) in any range of values of the parameter β_3.

A vector **l** directed along a threefold axis corresponds to the solutions $(1,\varepsilon,\varepsilon^2)$. The superconductivity class is in this case $D_3(E)$ for both representations F_1 and F_2.

A direction of **l** along a fourfold axis corresponds to the solution $(1,i,0)$. The superconductivity class for both representations F_1 and F_2 is $D_4(E)$.

Figure 3 shows the regions of existence of the different phases for three-dimensional representations (we did not determine the boundary where the degeneracy just discussed is lifted). The properties of all four solutions and of the phases corresponding to them are listed in Tables IIIa and IIIb. All the complex solutions should be magnetic, since time reversal is violated in them. The magnetism (spontaneous moment) is caused by the boojums that stem from the nontrivial combined symmetry $C_3e^{2\pi i/3}$ or $C_4e^{i\pi/2}$. We see that the disposition of the boojums in these superconducting classes is "ferromagnetic." The magnetism due to boojums will be discussed in Sec. 6. Table III illustrates also the low-temperature behavior of the heat capacity. We note again that in the triplet case the energy gap can have zeros only at points on the FS. In evaluating the procedural aspect of our subject, we see that the use of basis functions is advantageous because it frequently yields the symmetry of the state. The disposition of the zeros in the nonlinear problem, however, is governed just by symmetry and generally speaking does not coincide with that of the basis-function zeros.

4. SUPERCONDUCTING PHASES IN CRYSTALS OF THE $D_{6h} \times R$ GROUP

The superconductivity classes in hexagonal crystals (such as UPt_3) are obtained similarly. Since $D_{6h} = D_6 \times C_i$, we shall need hereafter representations of only the rotation

TABLE IV. Superconductivity classes from one-dimensional representations of group D_6.

Representation		Type of basis functions	Class	$C_e(T)$
A_1	$S=0$, A_{1g}	Symm. function	$D_6 \times R$	$\exp(-\Delta/T)$
	$S=1$, A_{1u}	$a\tilde{z}k_z + b(\tilde{x}k_x + \tilde{y}k_y)$		$\exp(-\Delta/T)$
A_2	$S=0$, A_{2g}	$(k_x^3 - 3k_xk_y^2)(k_y^3 - 3k_yk_x^2)$	$D_6(C_6) \times R$	T^2
	$S=1$, A_{2u}	$\tilde{z}k_z(k_x^3 - 3k_xk_y^2)(k_y^3 - 3k_yk_x^2)$		T^3
B_1	$S=0$, B_{1g}	$k_z(k_x^3 - 3k_xk_y^2)$	$D_6(D_3) \times R$	T^2
	$S=1$, B_{1u}	$\tilde{z}(k_x^3 - 3k_xk_y^2)$		T^3
B_2	$S=0$, B_{2g}	$k_z(k_y^3 - 3k_yk_x^2)$	$D_6(D_3') \times R$	T^2
	$S=1$, B_{2u}	$\tilde{z}(k_y^3 - 3k_yk_x^2)$		T^3

Elements of combined groups $D_6(C_6) = (E, C_2, 2C_3, 2C_6, 3U_2e^{i\pi}, 3U_2'e^{i\pi})$; $D_6(D_3) = (E, 2C_3, 3U_2, C_2e^{i\pi}, 2C_6e^{i\pi}, 3U_2'e^{i\pi})$; $D_6(D_3') \to D_6(D_3)(U_2 \to U_2')$. The triplet phases from classes $D_6(C_6) \times R$, $D_6(D_3) \times R$, $D_6(D_3') \times R$ have zeros at the points of intersection of the FS with the sixfold axis. The relative positions of the zero lines in the singlet phases are on the intersections of the FS with all vertical symmetry planes in the $D_6(C_6) \times R$ phase, on the intersection of the FS with the horizontal plane and with three vertical ones in the $D_6(D_3) \times R$ phase, and on the intersections with the horizontal and with three other vertical planes in the $D_6(D_3') \times R$ phase.

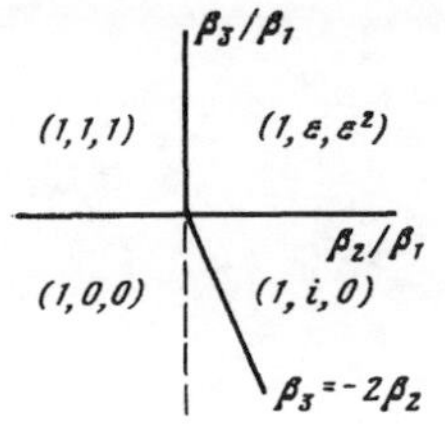

FIG. 3. Regions of existence of various solutions for three-dimensional representations of the cubic group.

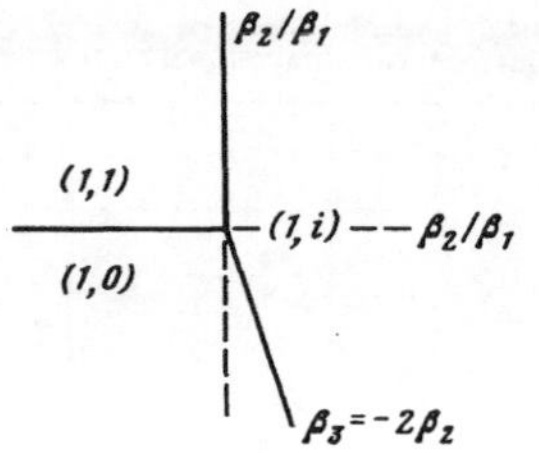

FIG. 4. Regions of existence of solutions for the two-dimensional representation E_1 of the hexagonal group.

group D_6 which contains four one-dimensional and two two-dimensional representations.

a. One-dimensional representations. The superconducting phases in these representations are not degenerate (there are no domain walls) and have no spontaneous ferro- or antiferromagnetism. The order parameter is always a complex scalar, and the corresponding Ginzburg-Landau functional is of standard form. The superconducting classes (nontrivial subgroups of the $D_6 \times R \times U(1)$ group) can be constructed either directly from the characteristics of the representations, as explained in Sec. 2, or from the form of the basis functions of these representations. All these data (including the location of the zeros, which generally speaking does not coincide with that of the zeros of the simplest basis functions) are gathered in Table IV.

b. The two-dimensional representation E_2. The order parameter contains, in analogy with (6) two complex components (η_1, η_2). The basis functions can be chosen in the form

$$S=0: \quad (k_x+ik_y)^2, \quad (k_x-ik_y)^2, \tag{23}$$

$$S=1: \quad (\tilde{x}+i\tilde{y})(k_x+ik_y), \quad (\tilde{x}-i\tilde{y})(k_x-ik_y). \tag{24}$$

The functional invariant to the total group $D_6 \times R \times U(1)$ is, as usual, the same for the triplet and singlet pairings in the case of strong spin-orbit coupling and coincides with the already investigated expression (9) for the two-dimensional representation of the cubic group. For (η_1, η_2) we have again the following solutions: (1,0) at $\beta_2 > \beta_1$; (1, — 1) at $\beta_2 < \beta_1$, $\gamma > 0$ and (1.1) at $\beta_2 < \beta_1$, $\gamma < 0$. The corresponding function symmetries, gap zeros, heat capacity, and magnetic properties are indicated in Table V.

c. Two-dimensional representation E_1. In contrast to the representation E_2, the two-component order parameter (η_1, η_2) is transformed in the E_1 representation as a two-dimensional vector located in a plane perpendicular to the axis:

$$\psi(\mathbf{k}) \propto \eta_x k_z k_x + \eta_y k_z k_y \quad (S=0), \tag{25}$$

$$\mathbf{d}(\mathbf{k}) \propto \eta_x \tilde{z} k_x + \eta_y \tilde{z} k_y \quad (S=1). \tag{26}$$

The Ginzburg-Landau functional, naturally, takes the same form as for the vector order parameter in the cubic group (17):

$$F = -\alpha(\boldsymbol{\eta}\boldsymbol{\eta}^*) + \beta_1(\boldsymbol{\eta}\boldsymbol{\eta}^*)^2 + \beta_2|\boldsymbol{\eta}^2|^2 + \beta_3(|\eta_x|^4 + |\eta_y|^4). \tag{27}$$

The solutions are obtained in the same way. There are three of them: (1,0), (1,1), and (1,*i*). (The statements made above concerning the hidden symmetry of the functional (17) are valid also for (27) in the last case of the solution with $\boldsymbol{\eta}' \perp \boldsymbol{\eta}''$). For each of the phases there is on the (β_2, β_3) plane a region where this phase has a lower energy that the others (see Fig. 4). The properties of the singlet and triplet phases are gathered in Table VI.

TABLE V. Superconductivity classes from representations E_2 of group D_6.

(η_1, η_2)	Class	$C_e(T)$	Degeneracy	Magn. properties
(1, 0) $S=0$, E_{2g}	$D_6(C_2)$	T^3	2	F
(1, 0) $S=1$, E_{2u}		T^3	2	F
(1, −1) $S=0$, E_{2g}	$D_2(C_2) \times R$	T^2	3	–
(1, −1) $S=1$, E_{2u}		T^3	3	–
(1, 1) $S=0$, E_{2g}	$D_2 \times R$	$\exp(-\Delta/T)$	3	–
(1, 1) $S=1$, E_{2u}		$\exp(-\Delta/T)$	3	–

See Eqs. (23) and (24) for the basis functions. The elements of the combined groups are $D_6(C_2) = (E, C_2, \varepsilon^2 C_3, \varepsilon C_3^2, \varepsilon C_6, \varepsilon^2 C_6^5, U_2 R, U_2' \varepsilon R, \ldots)$; $D_2(C_2) = (E, C_2, U_2 e^{i\pi}, U_2' e^{i\pi})$. Both phases from Class $D_6(C_2)$ and the triplet phase from $D_2(C_2) \times R$ have zeros at the point of intersection of the FS with a sixfold axis. The singlet phase from $D_2(C_2) \times R$ has a line of zeros on the intersection of the FS with two mutually perpendicular vertical symmetry planes.

TABLE VI. Superconductivity classes from two-dimensional representations E_1 of group D_6.

(η_1, η_2)		Class	$C_e(T)$	Degeneracy	Magn. properties
(1, *i*)	$S=0$, E_{1g}	$D_6(E)$	T^2	2	*F*
	$S=1$, E_{1u}		T^3	2	*F*
(1, 0)	$S=0$, E_{1g}	$D_2(C_2)\times R$	T^2	3	–
	$S=1$, E_{1u}		T^3	3	–
(1, 1)	$S=0$, E_{1g}	$C_2(E)\times R$	T^2	3	–
	$S=1$, E_{1u}		$\exp(-\Delta/T)$	3	–

The triplet phases $D_4(C_4)\times R$, $D_4^{(1)}(D_2)\times R$, $D_4^{(2)}(D_2)\times R$ have zeros at the points of intersection of the FS with a fourfold axis. The singlet phases have zeros on the following lines of intersection of the FS: for $D_4(C_4)\times R$—with all the vertical symmetry planes, for $D_4^{(1)}(D_2)\times R$—with two mutually perpendicular planes, and for $D_4^{(2)}(D_2)\times R$—with the two remaining vertical symmetry planes.

5. SUPERCONDUCTIVITY CLASSES IN TETRAGONAL SYMMETRY

This symmetry is possessed, e.g., by $CeCu_2Si_2$ and U_6Fe. The rotation subgroup D in the group $D_{4h}=D_4\times C_i$ four one-dimensional and one two-dimensional representation. The properties of the nondegenerate nonmagnetic superconducting phases that are possible on the basis of the one-dimensional representations are given in Table VII. In a two-dimensional representation the order parameter transforms again as a planar vector (η_x, η_y). The functional is of the same form (27) and has the same solutions. The expansion in the basis functions coincides with (25) and (26). The properties of the corresponding phases obtained from the two-dimensional representations are shown in Table VIII.

Since no rhombohedral-system compounds with heavy fermion are known so far, we shall not discuss this system in detail (see Sec. 2).

6. SUPERCONDUCTING STATES WITH SPONTANEOUS MAGNETISM

It was shown above that, on going to certain superconducting states belonging to a nontrivial class, invariance to time reversal is violated and consequently one should expect the appearance of magnetic properties in these states. These are the classes $D_3(E)$, $D_4(E)$, $D_6(E)$, $D_6(C_2)$ as well as the class $O(D_2)$, which are realized both in singlet and in triplet pairing. Moreover, a property of these classes is that they contain vectors whose transformation properties are analogous to the magnetic moment. These are vectors that emerge from boojums with negative "charge" $N=-1$ on the Fermi surface and enter the point of the boojum with $N=1$ (crosses and dots in Fig. 1). Time reversal (complex conjugation) reverses the signs of the charges, the points become crosses, and the system goes into another degenerate state.

The structures of these states are similar in many respects to the *A* phase of ^{3}He, where the Cooper pair is in a state with orbital angular momentum $L=1$ and with projection $L_z=1$ on the axis $\mathbf{l}\|\mathbf{z}$, and has consequently an angular momentum $\hbar\mathbf{l}$. It is intuitively clear that, say at $T=0$, when the system is totally in a coherent state, its orbital angular momentum should be $\hbar N\mathbf{l}/2$, where N is the total number of particles in the volume. Yet calculations show that the local angular momentum in ^{3}He-*A* is small and finite only to the extent that the asymmetry of the particles and hole on the FS is small. The angular momentum is only of the order of $\hbar(T_c/$

TABLE VII. Superconductivity classes from one-dimensional representations of group D_4.

Representation		Type of basis function	Class	$C_e(T)$
A_1	$S=0$, A_{1g}	Symmetr. Function	$D_4\times R$	$\exp(-\Delta/T)$
	$S=1$, A_{1u}	$a\tilde{z}k_z+b(\tilde{x}k_x+\tilde{y}k_y)$		$\exp(-\Delta/T)$
A_2	$S=0$, A_{2g}	$k_xk_y(k_x^2-k_y^2)$	$D_4(C_4)\times R$	T^2
	$S=1$, A_{2u}	$(\tilde{x}k_y+\tilde{y}k_x)(k_x^2-k_y^2)$		T^3
B_1	$S=0$, B_{1g}	$k_x^2-k_y^2$	$D_4^{(1)}(D_2)\times R$	T^2
	$S=1$, B_{1u}	$\tilde{x}k_x-\tilde{y}k_y$		T^3
B_2	$S=0$, B_{2g}	k_xk_y	$D_4^{(2)}(D_2)\times R$	T^2
	$S=1$, B_{2u}	$\tilde{x}k_y+\tilde{y}k_x$		T^3

The triplet phases $D_4(C_4)\times R$, $D_4^{(1)}(D_2)\times R$, $D_4^{(2)}(D_2)\times R$ have zeros at the points of intersection of the FS with a fourfold axis. The singlet phases have zeros on the following lines of intersection of the FS: for $D_4(C_4)\times R$—with all the vertical symmetry planes, for $D_4^{(1)}(D_2)\times R$—with two mutually perpendicular planes, and for $D_4^{(2)}(D_2)\times R$—with the two remaining vertical symmetry planes.

TABLE VIII. Superconductivity classes from two-dimensional representations of group D_4.

(η_1, η_2)	Class	$C_e(T)$	Degeneracy	Magn. properties
$(1, i)$ $S=0,\ E_g$	$D_4(E)$	T^2	2	F
$(1, i)$ $S=1,\ E_u$	$D_4(E)$	T^3	2	F
$(1, 0)$ $S=0,\ E_g$	$D_2(C_2)\times R$	T^2	2	–
$(1, 0)$ $S=1,\ E_u$	$D_2(C_2)\times R$	T^3	2	–
$(1, 1)$ $S=0,\ E_g$	$D_2(C_2')\times R$	T^2	2	–
$(1, 1)$ $S=1,\ E_u$	$D_2(C_2')\times R$	T^3	2	–

See Eqs. (25) and (26) of the text for the type of basis functions. The phases from the class $D_4(E)$ have zeros at the points of intersection of the FS with a fourfold axis, while the singlet phase has furthermore a line of zeros on the intersection of the FS with the horizontal symmetry plane. The triplet phases $D_2(C_2)\times R$ and $D_2(C_2')\times R$ have zeros at the points of intersection of the FS with the corresponding twofold axis, and the singlet phases from these classes have zeros on the lines of intersection of the FS with the horizontal plane and one of the vertical symmetry planes.

$T_F)^2$ per atom of the liquid. According to an explanation proposed for this paradox in Refs. 23 and 24, the angular momentum should be defined not in an infinite system, but with account taken of the geometry of the vessel holding the liquid. The internal angular momentum of the pair produces a superfluid flow along the boundary of the A-phase with the vessel, and this flow imparts to the fluid a total angular momentum that is no longer small and is of the order of $\hbar$ per ^{3}He atom (to our knowledge, there is no proof that this angular momentum is identically equal to $\hbar/2$ per ^{3}He atom, since at first glance the flows in the texture could depend on the boundary conditions.

The question of the value of the angular momentum in a superconductor is even more important from the viewpoint of its physical manifestions, since the carriers are charged and a finite angular momentum would mean existence of an orbital (and spin) magnetic moment. At the same time, the local moment in the interior of the superconductor must be exactly zero because of the Meissner effect. Owing to the crystal-lattice field, the states of a Cooper pair cannot be classified at all in accordance with the value of the orbital angular momentum (as well of the spin, in view of the spin-orbit coupling). Nonetheless, the indicated close analogy between the aforementioned nontrivial superconductivity classes and ^{3}He-A suggests that here, too, circulating superconducting currents can flow over the surface of a single-domain superconductor even in the absence of an external magnetic field. The analysis of the structure of such a surface current is quite complicated, since the entire question is closely connected with the problem of choosing the boundary conditions. As for the latter, there are as yet none in general form. Obviously, the boundary conditions are very sensitive to the quality of the surface.

To answer the fundamental question of the existence of surface magnetic current, we shall circumvent the complications with the boundary conditions by studying the current distribution on the interface between two degenerate superconducting states (domains) with different orientations of the angular momentum (more accurately, of the boojums). In contrast to the usual domain wall in a magnet, the boundary between two superconducting domains that appear only during a transition into the superconducting state is described (near T_c) in terms of the very same Ginzburg-Landau functional as for the superconductivity itself, without invoking additional terms for the anisotropy. By way of a very simple example, we choose the class $D_4(E)$ (for either the singlet or triplet case) from the two-dimensional representation of the group D_4. Other examples are more difficult to calculate, but should lead to the same conclusions.

To find the structure of the wall separating domains with different order-parameter orientations, we must add to the corresponding Ginzburg-Landau functional of the last section, which describes a two-dimensional representation in the D_4 group, the gradient terms

$$F=-\alpha(\eta\eta^*)+\beta_1(\eta\eta^*)^2+\beta_2|\eta^2|^2+\beta_3[|\eta_x|^4+|\eta_y|^4] +\frac{1}{2m_1'}(|\partial_x\eta_x|^2+|\partial_y\eta_y|^2)+\frac{1}{2m_1''}(|\partial_x\eta_y|^2+|\partial_y\eta_x|^2) +\frac{1}{2m_2}(|\partial_z\eta_x|^2+|\partial_z\eta_y|^2)+\frac{1}{4m_3'}[(\partial_x\eta_y)(\partial_y\eta_x^*) +(\partial_x\eta_y^*)(\partial_y\eta_x)]+\frac{1}{4m_3''}[(\partial_y\eta_y)(\partial_x\eta_x^*)+(\partial_y\eta_y^*)(\partial_x\eta_x)]. \tag{28}$$

Here ∂_k is the covariant derivative

$$\partial_k\eta=\nabla_k\eta-(2ie/c)A_k\eta. \tag{29}$$

The answer given in Ref. 22 contains $m_1^{-1}\equiv(m_1')^{-1}+(m_1'')^{-1}$, while $m_3'=m_3''$. In the enumeration of all the possible invariants in the real functional (28), m_3' and m_3'' are independent. Their equality (accurate to terms of order T_c^2/T_F^2) is due to the approximate electron-hole symmetry on the FS. We assume also that the penetration depth δ of the magnetic field exceeds substantially the coherence length ξ. The condition $\delta\gg\xi$ is apparently well satisfied in superconductors with heavy fermions. The vector potential can then be left out of the Ginzburg-Landau equations.

We consider the following geometry. The domain wall lies in the yz plane, so that the order parameter η depends only on the x coordinate, and the "magnetic moment" is directed along the z axis as $x\to\infty$ and is oppositely directed as $x\to-\infty$. In this geometry the current should flow along the y axis. Varying (28) with respect to $\delta\mathbf{A}$ and taking (29) into account, we get

$$j_y = \frac{ie}{2m_3}[\eta_x\eta_y^{*\prime} - \eta_x^*\eta_y' + \eta_y\eta_x^{*\prime} - \eta_y^*\eta_x']. \tag{30}$$

We consider for simplicity the case of strong anisotropy, $\beta_3 \gg \beta_1, \beta_2$. In the initial approximation η_x and η_y have equal moduli but different phases:

$$\eta_x = (\alpha/2\beta_3)^{1/2}e^{i\varphi_1}, \quad \eta_y = (\alpha/2\beta_3)^{1/2}e^{i\varphi_2}. \tag{31}$$

The small term

$$\beta_2|\eta^2|^2 = \frac{\beta_2\alpha^2}{2\beta_3^2}[1+\cos 2(\varphi_1-\varphi_2)]$$

determines the phase difference outside the domain wall, with $\beta_2 > 0$ to ensure just the $D_4(E)$ phase far from the wall. Different orientations of the angular momentum on the right and left correspond to the conditions

$$\varphi_1 - \varphi_2 = \begin{cases} \pi/2, & x \to \infty \\ -\pi/2, & x \to -\infty \end{cases}. \tag{32}$$

The equations for φ_1 and φ_2 follow from variations of (28):

$$\frac{1}{2m_1'}\varphi_1'' + \frac{2\beta_2\alpha}{\beta_3}\sin 2(\varphi_1-\varphi_2) = 0, \tag{33}$$

$$\frac{1}{2m_1''}\varphi_2'' + \frac{2\beta_2\alpha}{\beta_3}\sin 2(\varphi_2-\varphi_1) = 0.$$

The condition that the current component perpendicular to the wall be zero yields

$$\varphi_1'/m_1' + \varphi_2'/m_1'' = 0. \tag{34}$$

With the aid of this condition we can rewrite the system (33) in the form of the standard sine-Gordon equation for the difference $u = \varphi_2 - \varphi_1$:

$$\lambda^2 u'' + \sin 2u = 0, \tag{35}$$

where $\lambda^2 = \beta_3/4\beta_2\alpha(m_1' + m_1'')$.

A solution of (35) satisfying the boundary conditions (32) is

$$\sin u(x) = \text{th}\,(\sqrt{2}x/\lambda). \tag{36}$$

Returning to Eq. (30) for the current density j_y and using (31) and (34), we obtain

$$j_y(x) = \frac{\alpha e}{2m_3\beta_3}\frac{m_1''-m_1'}{m_1''+m_1'}\cos u\, u'. \tag{37}$$

Integration of this expression with respect to x with the aid of (36) yields directly the total current in the wall:

$$i = \int_{-\infty}^{+\infty} j_y(x)\,dx = \frac{\alpha e}{m_3\beta_3}\frac{m_1''-m_1'}{m_1''+m_1'}, \tag{38}$$

which differs from zero, since there are no special conditions whatever on the relation between m_1' and m_1''. This current is small only to the extent that $T_c - T$ is small (at $\beta_3 \sim \beta_2$), and this proves the validity of the arguments advanced at the beginning of this section.

According to (38), the magnetic moment is $M_n \sim i/2c$. Comparing this with the estimate for the thermodynamic critical field $H_c^2/8\pi \sim \alpha^2/\beta$, we get

$$M_n \sim (\xi/\delta)H_c \sim (1/\varkappa)H_c \sim H_{c1}. \tag{39}$$

The moment produced at the domain wall in the region ξ is screened at a distance δ by the superconducting currents. Unlike for a ferromagnet, for a superconductor there are thus no large magnetic-energy terms that prevent the existence of large single-domain samples. The exact relation in (39) contains a complicated combination of parameters of the Ginzburg-Landau functional. In principle, M_n could exceed the lower critical field H_{c1} at which vortex formation becomes favored. The situation wherein a magnetic-moment field screened in the interior of a superconductor exists on its very surface could turn out by the same token to be unstable to vortex formation in the screening region.

7. CONCLUSION

We have considered above all the possible superconducting states into which a system can go directly from the normal metallic state in the case of nontrivial pairing. These states are determined for the most part by that symmetry-group representation which is responsible for the onset of superconducting instability at T_c in a given specific substance. This representation can be determined, as indicated in Ref. 22, by measuring the anisotropy of the upper critical field H_{c2} near T_c. The properties of the corresponding phases below T_c are described in practically the same way in both the singlet and the triplet cases, except for important differences in the types of the zeros. In triplet pairing zeros appear in the energy gap only at points on the FS, whereas in singlet pairing the geometric loci of the zeros can be whole lines on the FS. This is so far the only method that permits identification of the pairing with some degree of assurance. If, say, the heat capacity depends on temperature like T^2, triplet pairing is excluded, but a T^3 dependence is possible for both types of pairing. The experimental situation is in this respect not yet clear. According to Ref. 8, in UBe_{13} this law corresponds to a T^3 dependence, whereas the data on UPt_3 (Ref. 12) and $CeCu_2Si_2$ (Ref. 13) favor more readily singlet superconductivity, as already mentioned in Sec. 1.

An interesting distinguishing feature of the new superconductors is that they have magnetic properties. The presence of a magnetic moment, meaning hence of a magnetic field near the sample surface even in the absence of an external field, can undoubtedly be investigated by nuclear-magnetic-resonance methods.

[1]In §20 of Ref. 1 the discussion of pairing with nonzero angular momentum is based on unpublished results obtained in 1958 by A. A. Abrikosov, L. P. Gor'kov, L. D. Landau, and I. M. Khalatnikov.

[2]This was pointed out to us by N. Konyshev.

[1]A. A. Abrikosov, L. P. Gor'kov, and I. E. Dzyaloshinskii, Quantum Field-Theoretical Methods in Statistical Physics, Pergamon, 1965, §20.
[2]P. W. Anderson and P. Morel, Phys. Rev. **123**, 1911 (1961).
[3]L. P. Pitaevskiĭ, Zh. Eksp. Teor. Fiz. **37**, 1794 (1959) [Sov. Phys. JETP **10**, 1267 (1960)].
[4]V. P. Mineev, Usp. Fiz. Nauk **139**, 303 (1983) [Sov. Phys. Usp. **26**, 160 (1983)].
[5]G. E. Volovik, *ibid.* **143**, 73 (1984) [**27**, 363 (1984)].
[6]P. W. Anderson and W. F. Brinkman, Phys. Rev. Lett. **30**, 1108 (1973).
[7]C. M. Varma, Bull. Am. Phys. Soc. **29**, 404 (1984).
[8]H. R. Ott, H. Rudiger, T. M. Rice, K. Ueda, Z. Fisk, and J. L. Smith, Phys. Rev. Lett. **52**, 1915 (1984).
[9]W. Lieke, U. Rauchschwalbe, C. Bredl, P. Steglich, J. Aarts, and F. de Boer, J. Appl. Phys. **53**, 2111 (1982).
[10]L. E. de Long, J. G. Huber, K. N. Yang, and M. B. Maple, Phys. Rev. Lett. **51**, 312 (1983).
[11]G. R. Stewart, Z. Fisk, J. O. Willis, and J. L. Smith, *ibid.* **52**, 679 (1984).
[12]D. J. Bishop, C. M. Varma, B. Batlogg, E. Bucher, Z. Fisk, and J. L. Smith, *ibid.* **53**, 1009 (1984).
[13]D. E. MacLaughlin, Cheng Tien, L. C. Gupta, J. Aarts, F. R. de Boer, and Z. Fisk, Phys. Rev. **B30**, 1577 (1984).
[14]W. G. Clark, Z. Fisk, K. Glover, M. D. Lan, D. E. MacLaughlin, J. L. Smith, and C. Tien, Proc. LT-17 Conference, North-Holland, 1984, p. 227.
[15]P. Muzikar, J. de Phys. **39**, C6-53 (1978).
[16]G. E. Volovik and L. P. Gor'kov, Pis'ma Zh. Eksp. Teor. Fiz. **39**, 550 (1984) [JETP Lett. **39**, 674 (1984)].
[17]L. D. Landau and E. M. Lifshitz, Statistical Physics, Part 1, Pergamon, 1980.
[18]L. D. Landau and E. M. LIfshitz, Electrodynamics of Continuous Media, Pergamon, 1983, §38.
[19]P. W. Anderson, Phys. Rev. **B30**, 1549 (1984).
[20]L. D. Landau and E. M. Lifshitz, Quantum Mechanics, Nonrelativistic Theory, Pergamon, 1977.
[21]V. L. Indenbom, Kristallografiya **5**, 115 (1960) [Sov. Phys. Crystallogr. **5**, 106 (1960)].
[22]L. P. Gor'kov, Pis'ma Zh. Eksp. Teor. Fiz. **40**, 351 (1984) [JETP Lett. **40**, 1155 (1984)].
[23]G. E. Volovik and V. P. Mineev, Zh. Eksp. Teor. Fiz. **83**, 1025 (1982) [Sov. Phys. JETP **56**, 579 (1982)].
[24]G. E. Volovik and V. P. Mineev, *ibid.* **81**, 989 (1981); **86**, 1667 (1984) [**54**, 524 (1981); **59**, 972 (1984)].

Translated by J. G. Adashko

Anisotropy of the upper critical field in exotic superconductors

L. P. Gor'kov
L. D. Landau Institute of Theoretical Physics, Academy of Sciences of the USSR

(Submitted 17 September 1984)
Pis'ma Zh. Eksp. Teor. Fiz. **40**, No. 8, 351–353 (25 October 1984)

While the superconducting order parameter is characterized by a multidimensional representation of the rotation group (a state of the type ^{3}He-A), the upper critical field near T_c is anisotropic even in a crystal with a high degree of symmetry. This circumstance can be used to describe the nature of the observed superconductivity. The degeneracy can in principle be lifted by elastic stress.

Recent experiments[1] on the measurement of the specific heat in the superconductor UBe_{13} at low temperatures provided the first direct proof that an analog of the superfluid A phase of ^{3}He is realized in this compound. The dependence $C_e \sim T^3$ (instead of the usual activation dependence) indicates the presence of zeroes in the order parameter on the Fermi surface, just as in the Anderson–Morell–Brinkman–Rice state. It has already been pointed out (see, for example, Ref. 2) that an unusual superconductivity can exist in several compounds with the so-called "heavy fermions" ($CeSi_2Cu_2$[3],U_6Fe[4],UPt_3[5] and others), which have many exotic properties.

The mechanisms leading to nontrivial pairing are not discussed below. In the present letter we show that in the case of such pairing one of the main characteristics of superconductivity—the upper critical field H_{c2}—exhibits a considerable anisotropy near T_c even in a crystal with a high degree of symmetry (a cubic symmetry, for

 0021-3640/84/201155-04$01.00

example). This anisotropy distinguishes the observed superconductivity from the usual superconductivity and imposes some restrictions on the ground-state symmetry. This is important, since the presence of zeroes in the energy gap does not determine this symmetry uniquely.[6]

Near the transition point the Ginzburg–Landau expansion is applicable for an anisotropic metal.[7,8] For the problem of the upper critical field H_{c2}, it is sufficient to restrict the analysis to only the gradient terms in the equation for the order parameter $\hat{\Delta}(\mathbf{p})$. The transition point is determined by a condition of the form

$$\hat{\Delta}(\mathbf{p}) = \ln \frac{\bar{\omega}}{T_c} \int \hat{K}(\mathbf{p}, \mathbf{p}') \hat{\Delta}(\mathbf{p}') d\Omega_{\mathbf{p}'} . \tag{1}$$

In (1) $\hat{\Delta}$ and the interaction $\hat{K}$ depend on spinor indices. By slightly changing the normalization $\hat{\Delta}$, we can put the kernel $\hat{K}(\mathbf{p},\mathbf{p}')$ in a Hermitian form. This kernel is invariant under all transformations of the crystal-symmetry group and under time reversal. It is appropriate to note here the spin nature of the order parameter.

The systems under discussion contain atoms of heavy elements, and the spin-orbit interaction in them is large. If we assume that the order parameter can be classified by assigning parity to spatial inversion, we will have two alternatives[9]:

$$\begin{aligned} \hat{\Delta}(\mathbf{p}) &= i\hat{\sigma}_y \psi(\mathbf{p}); \qquad (S=0), \\ \hat{\Delta}(\mathbf{p}) &= (\hat{\sigma}\mathbf{d}(\mathbf{p}))i\hat{\sigma}_y; \qquad (S=1). \end{aligned} \tag{2}$$

Substituting them separately into (1), we note that in the first case we obtain a scalar integral equation for $\psi(\mathbf{p})$, while in the second case we will obtain a vector equation for $\mathbf{d}(\mathbf{p})$. The strong spin-orbit coupling (the spins are "frozen" into the lattice and rotate together with it under the symmetry transformations) accounts for the fact that the solutions of Eq. (1) in both cases realize the irreducible representations of a subgroup of rotations of the corresponding complete point group of the crystal. As far as the basis functions are concerned, the complete set of basis functions for $\psi(\mathbf{p})$ is described, for UBe_{13}, for example, by even representations of the group O_h, while the vector basis functions of the corresponding representations for $\mathbf{d}(\mathbf{p})$ are constructed from the direct products of the spin unit vectors $(\tilde{x}, \tilde{y}, \tilde{z})$ and odd scalar representations of O_h. Since $S=0$ and $S=1$ can, according to (2), be decoupled, it is sufficient that the corresponding Ginzburg–Landau functional be invariant with respect to the rotation groups. Denoting the basis functions of the irreducible representation by ϕ^i, we can write the order parameter in each case in (2) as

$$\hat{\Delta}(\mathbf{p}) = \Sigma\, \eta_i \phi^i(\mathbf{p}) . \tag{3}$$

The group O has five representations (in the notation of Ref. 10): two one-dimensional representations, A_1 and A_2; one two-dimensional representation, E; and two three-dimensional representations, F_1 and F_2. As usual, the coordinate transformations are assumed to operate on the coefficients η_i and not on $\phi^i(\mathbf{p})$.

In selecting a unitary representation for the order parameter, the anisotropy of the crystal is taken into account near T_c only by the mass tensor.[7,8] The field H_{c2} near

T_c is thus isotropic in a cubic crystal and independent of the angle in the plane perpendicular to the principal axis for the tetragonal or hexagonal symmetry. For the one-dimensional (nonunitary) representation of any group, the problem, of course, reduces to the mass tensor (see the derivation in Ref. 8).

We now consider the degenerate solutions of (1) in the cubic group. In the two-dimensional representation E, we can select the basis functions in such a way that they would transform as

$$\phi^1 = x^2 + \epsilon y^2 + \epsilon^{-1} z^2; \quad \phi^2 = \phi^{1*}; \quad \epsilon = \exp(2\pi i/3) . \tag{4}$$

The product of gradient-invariant derivatives $\partial_i \partial_k$, $(\partial_i = \partial/\partial x_i - ieA_i/c)$ appears in the free-energy functional only with $i = k$. It is convenient to write the following combinations:

$$\Delta = \partial^2_{ii} ; \quad \Delta_1 = \partial^2_{xx} + \epsilon \partial^2_{yy} + \epsilon^{-1} \partial^2_{zz} ; \quad \Delta_2 = \partial^2_{xx} + \epsilon^{-1} \partial^2_{yy} + \epsilon \partial^2_{zz} . \tag{5}$$

The gradient terms in the functional for the E representation are

$$\frac{1}{2m} (\eta^*_1 \Delta \eta_1 + \eta^*_2 \Delta \eta_2) + \frac{1}{2m'} (\eta^*_1 \Delta_2 \eta_2 + \eta^*_2 \Delta_1 \eta_1). \tag{6}$$

The basis functions of the three-dimensional representation F_2 transform like the symmetric tensor $\xi_{ij} = (xy, yz, zx)$, i.e., with $i = j\ \xi_{ii} = 0$. A classification based on the irreducible representations shows that there are three invariants, which can be chosen, for example, as follows:

$$\xi^*_{ji} \partial^2_{il} \xi_{lj} ; \quad \xi^*_{ij} \partial^2_{ll} \xi_{ji} ; \quad (\xi^*_{xy} \partial^2_{yz} \xi_{zx} + \ldots). \tag{7}$$

In the vector basis $\eta = (\eta_x, \eta_y, \eta_z)$ of the representation F_1 we have the following cubic invariants:

$$\eta^*_i \partial^2_{jj} \eta_i ; \quad \eta^*_i \partial^2_{ij} \eta_j ; \quad \eta^*_i \partial^2_{ii} \eta_i . \tag{8}$$

The linear equation of stability, which determines H_{c2}, can be obtained through appropriate variation of the functional and in all cases (5)–(8) leads to an unexpected (for a cubic group) anisotropy of the upper field, which should be sought near T_c in UBe_{13}. Analysis of the angular dependence of H_{c2} in cubic and hexagonal structures is rather complicated and will be carried out separately. The simplest manifestation of the effect is in tetragonal symmetry (D_4), to which the $CeSi_2Cu_2$ and U_6Fe superconductors belong.

The group D_4 has only one two-dimensional representation (η_x, η_y), and the equations determining the field H_{c2} can be written [compare with (8)]

$$\frac{1}{2m_1} \partial^2_{ii} \eta_x + \frac{1}{2m_2} \partial^2_{zz} \eta_x + \frac{1}{2m_3} \partial^2_{xy} \eta_y = (T - T_c)\eta_x \tag{9}$$

(in a corresponding way, they can be written for η_y). After selecting the field in the basal plane ($H\cos\varphi$, $H\sin\varphi$), and after simple calculations, we find from (9)

$$H^{\perp}_{c2}(\varphi) \propto (m_1 m_2)^{1/2} \left(1 - \frac{m_1}{m_3} | \sin\varphi \cos\varphi |\right)^{-1/2} . \tag{10}$$

Observation of the aforementioned singularities in the anisotropy of the field H_{c2} is thus a convenient method for determining the degenerate representation of the order parameter in (1) (the spin of the pair, however, is not determined). We recall that most high-symmetry classes, which have points or lines where the gap vanishes, are associated precisely with such representations.[6] Acoustic deformations, we might note, also lift the degeneracy of the order parameter, which leads to a splitting of the transition temperature in (1) by the elastic field. The effect, however, has a small value T_c/T_F because of the electron-hole symmetry near the Fermi level.

I thank G. E. Volovik and D. E. Khmel'nitskiĭ for a useful discussion of the results.

[1]H. R. Ott, H. Rudiger, T. M. Rice, K. Ueda, Z. Fisk, and J. L. Smith, Phys. Rev. Lett. **52**, 1915 (1984).
[2]C. M. Varma, Bull. Am. Phys. Soc. **29**, 404 (1984).
[3]W. Lieke, U. Rauchschwalbe, C. Bredl, F. Steglich, J. Aarts, and F. De Boer, J. Appl. Phys. **53**, 2111 (1982).
[4]L. E. De Long, J. G. Huber, K. N. Yang, and M. B. Maple, Phys. Rev. Lett. **51**, 312 (1983).
[5]G. R. Stewart, Z. Fisk, J. O. Willis, and J. L. Smith, Phys. Rev. Lett. **52**, 679 (1984).
[6]G. E. Volovik and L. P. Gor'kov, Pis'ma Zh. Eksp. Teor. Fiz. **39**, 550 (1984) [JETP Lett. **39**, 674 (1984)].
[7]V. L. Ginzburg, Zh. Eksp. Teor. Fiz. **23**, 236 (1952).
[8]L. P. Gor'kov and T. K. Melik-Barkhudarov, Zh. Eksp. Teor. Fiz. **45**, 1493 (1963) [Sov. Phys. JETP **18**, 1031 (1964)].
[9]V. P. Mineev, Usp. Fiz. Nauk **139**, 303 (1983) [Sov. Phys. Usp. **26**, 160 (1983)].
[10]L. D. Landau and E. M. Lifshitz, Kvantovaya mekhanika (Quantum Mechanics), Nauka, Moscow, 1974, Sec. 95.

Translated by M. E. Alferieff
Edited by S. J. Amoretty

Defects and an unusual superconductivity

L. P. Gor'kov and P. A. Kalugin
L. D. Landau Institute of Theoretical Physics, Academy of Sciences of the USSR

(Submitted 31 January 1985)
Pis'ma Zh. Eksp. Teor. Fiz. **41**, No. 5, 208–210 (10 March 1985)

If the superconducting gap in a pure material has zeros on an entire line along the Fermi surface, the state density is finite at $\epsilon = 0$, beginning with the lowest defect concentrations. The splitting of the superconducting transition by impurities, $U_{1-x}Th_xBe_{13}$ is discussed.

Arguments in favor of a nontrivial nature of the superconductivity in heavy-fermion systems are based on the power-law behavior of several measured properties at low temperatures (the specific heat,[1] the ultrasonic absorption coefficient,[2] and NMR line widths,[3] among others). All the superconducting classes that are possible for these systems were listed in Refs. 4 and 5. It turns out that in the triplet class ($S = 1$) the superconducting gap can vanish only at isolated points on the Fermi surface, while in the singlet case ($S = 0$) it can vanish on entire lines along the Fermi surface. Depending on the particular features of the gap, the specific heat (for example) would vary in proportion to T^3 and T^2, respectively ($T \ll T_c$).

Let us examine the stability of these results with respect to impurities. We first assume $S = 0$; then the order parameter[4] is a scalar, $\psi(\mathbf{k})$. To determine the state density for excitations, we use a generalization of the expression in Ref. 6 for the entropy:

$$S_e = -\frac{1}{(2\pi)^3 T^2} \int_0^\infty \frac{\epsilon\, d\epsilon}{\mathrm{ch}^2(\epsilon/2T)} \frac{dS_{\mathbf{k}}}{v(\mathbf{k})} \,\mathrm{Im} \sqrt{|\tilde{\psi}(\mathbf{k},\epsilon)|^2 - \tilde{\epsilon}^2(\mathbf{k},\epsilon)}\,. \tag{1}$$

The functions $\tilde{\epsilon}(\mathbf{k},\epsilon)$ and $\tilde{\psi}(\mathbf{k},\epsilon)$ are the result of an analytic continuation from the upper thermodynamic frequency axis ($\epsilon = i\epsilon_n$) of the solution of the equations

$$\tilde{\epsilon}_n = i\epsilon_n + \frac{1}{2}\langle \tau^{-1}(\mathbf{k},\mathbf{k}') \frac{\tilde{\epsilon}}{\sqrt{|\tilde{\psi}|^2 - \tilde{\epsilon}^2}} \rangle_{\mathbf{k}'}\,, \tag{2}$$

$$\tilde{\psi}(\mathbf{k}, i\epsilon_n) = \psi(\mathbf{k}) + \frac{1}{2}\langle \tau^{-1}(\mathbf{k},\mathbf{k}') \frac{\tilde{\psi}}{\sqrt{|\tilde{\psi}|^2 - \tilde{\epsilon}^2}} \rangle_{\mathbf{k}'}.$$

Here $\langle \dots \rangle_{\mathbf{k}'} = \int \dots dS_{\mathbf{k}'} / \int dS_{\mathbf{k}}$.

Equations (2) determine the self-energy parts of the Green's functions averaged over impurities. For a uniform impurity distribution, the kernel $\tau^{-1}(\mathbf{k},\mathbf{k}')$ has the initial crystal symmetry. The symmetry of $\tilde{\psi}(\mathbf{k},\epsilon)$ is thus the same as $\psi(\mathbf{k})$; in particular, $\tilde{\psi}(\mathbf{k},\epsilon)$ has zeros at the same locus of points.

For simplicity, we assume an isotropic impurity $[(\tau^{-1}(\mathbf{k},\mathbf{k}') \equiv \tau^{-1}]$. The second of

Eqs. (2) gives us $\tilde{\psi} = \psi$, while the first becomes

$$\tilde{\epsilon} = i\epsilon_n + \frac{1\tilde{\epsilon}}{2\tau}\langle (|\psi|^2 - \tilde{\epsilon}^2)^{-1/2}\rangle .$$

If $\psi(\mathbf{k})$ vanishes on lines, the integral here has a logarithmic singularity,

$$\tilde{\epsilon}\left(1 - \frac{1}{2\tau\Delta_0}\ln\frac{2\Delta_0}{-i\tilde{\epsilon}}\right) = \epsilon \qquad (3)$$

$[T\to 0, |\epsilon|, |\tilde{\epsilon}| \ll \Delta_0$, where Δ_0 is the scale dimension of $|\psi(\mathbf{k})|]$. The value $\epsilon = 0$ corresponds to

$$\tilde{\epsilon}_0 = i2\Delta_0\exp(-2\tau\Delta_0). \qquad (4)$$

Expanding (3) in the neighborhood of (4), we find from (1) a finite state density in a narrow region $\epsilon \simeq 0$:

$$\nu_S/\nu_N = 4\tau^2\Delta_0^2\exp(-2\tau\Delta_0). \qquad (5)$$

The general form of the kernel $\hat{\tau}^{-1}(\mathbf{k},\mathbf{k}')$ is

$$\hat{\tau}^{-1}(\mathbf{k},\mathbf{k}') = \sum_{i,\lambda}\tau_i^{-1}\hat{\theta}^{i}_{\lambda}(\mathbf{k})\hat{\theta}^{i\lambda}(\mathbf{k}'), \qquad (6)$$

where the operator symbol incorporates the spin variables, the index i specifies all the eigenvalues τ_i^{-1}, $\hat{\theta}^{i,\lambda}$ are the corresponding eigenfunctions, which always belong to one of five representations of the cubic group, and λ specifies the basis functions of the representation if it is degenerate. The sign of τ_i^{-1} is, in general, arbitrary, except for the principal value, $\tau_0^{-1} > 0$, which belongs to the identity representation. We now see that (4) and (5) are also valid in the general case, since the corrections to $\tilde{\psi}(\mathbf{k},\epsilon)$ do not contain logarithmic terms. Finally, if the zeros of the gap correspond to isolated points on the Fermi surface, a perturbation theory will be valid at low concentrations. A "gap-free superconductivity" is possible only at a finite defect concentration, $\tau\Delta_0 \sim 1$.

In the case of a nontrivial superconductivity, impurities will lower T_c. It was found in Ref. 7, however, that the superconducting transition in $U_{1-x}Th_xBe_{13}$ also splits in two. An explanation for this effect[8] was proposed on the basis of the suggestion that the superconductivity in UBe_{13} is anisotropic and is characterized by a vector whose direction is pinned by defects in a random way. The second transition (at $T_{c2} < T_{c1}$) was identified with an orientational transition. The estimates in Ref. 8, however, contain some parameters which were literally small. In this connection, it was mentioned in Ref. 8 that in the fourth-degree terms in the Ginzburg-Landau functional

$$\beta_1|\vec{\eta}|^4 + \beta_2|\vec{\eta}^2|^2 + \beta_3(|\vec{\eta}_x|^4 + |\vec{\eta}_y|^4 + |\vec{\eta}_z|^4) \qquad (7)$$

($\vec{\eta}_i$ are the coefficients in the expansion of the order parameter in the basis functions of the three-dimensional representations of the O group) the relations between the coefficients β_i depend on the impurities. The derivations of the coefficients β_i in the weak-coupling model, which is completely analogous to that of Ref. 9, shows that all the β_i vary over a rather broad range, depending on the assumptions regarding the anisotropy of the basis functions and the signs of τ_i^{-1} [several terms corresponding to different representations must be retained in (6)]. Of these results we will mention two: As the impurity concentration x is increased, the superconducting transition may become a first-order transition; i.e., (7) may lose its positive definite form. Second, in the weak-coupling model and for $x = 0$ we have $\beta_2 = \frac{1}{2}\beta_1 > 0$ for $S = 0$, and, according to Ref. 5, the superconducting phase near T_c corresponds to a symmetry $D_4(E)$ or $D_3(E)$. In other words, it has a magnetic moment. For triplet pairing, $S = 1$, we have $\beta_2 < 0$ (the latter result was also derived in Ref. 10). Correspondingly, there are classes[5] $D_3 \times R$, $D_3(C_3) \times R$, $D_4(C_4) \times R$, $D_4^{(2)}(D_2) \times R$, in which the order parameter is real.

Returning to the possibility of describing the splitting of the transition in $U_{1-x}Th_xBe_{13}$ by this approach (we are assuming that the second transition is also a superconducting transition), we are confronted by the circumstances that these classes exhaust all the extrema of the Ginzburg-Landau functional[5] (7). Continuous transitions between them are not possible within the framework of a single representation (this also applies to two-dimensional representations). Measurements of NMR line widths[3] indicate that the second transition is continuous.

The splitting of the temperature of the superconducting transition can be understood easily by assuming that there is a preferred orientation of the impurities and that this orientation lowers the symmetry of the cube. That this is indeed the case in the UBe_{13} system was brought to our attention by N. E. Alekseevskiĭ, whom the authors thank. We also thank G. E. Volovik and D. E. Khmel'nitskiĭ for many useful discussions.

[1]H. R. Ott, H. Rudigier, T. M. Rice, K. Ueda, Z. Fisk, and J. L. Smith, Phys. Rev. Lett. **52**, 1915 (1984).
[2]D. J. Bishop, C. M. Varma, B. Batlogg, E. Bucher, Z. Fisk, and J. L. Smith, Phys. Rev. Lett. **53**, 1009 (1984).
[3]D. E. Maclaughlin, Cheng Tien, W. G. Clark, M. D. Lan, Z. Fisk, J. L. Smith, and H. R. Ott, Phys. Rev. Lett. **53**, 1833 (1984).
[4]G. E. Volovik and L. P. Gor'kov, Pis'ma Zh. Eksp. Teor. Fiz. **39**, 550 (1984) [JETP Lett. **39**, 674 (1984)].
[5]G. E. Volovik and L. P. Gor'kov, Preprint 1984.
[6]G. M. Éliashberg, Zh. Eksp. Teor. Fiz. **43**, 1105 (1962) [Sov. Phys. JETP **16**, 780 (1963)].
[7]H. R. Ott, H. Rudigier, Z. Fisk, and J. L. Smith, Preprint, 1984.
[8]G. E. Volovik and D. E. Khmel'nitskiĭ, Pis'ma Zh. Eksp. Teor. Fiz. **40**, 469 (1984) [JETP Lett. **40**, 1299 (1984)].
[9]L. P. Gor'kov, Zh. Eksp. Teor. Fiz. **37**, 1407 (1959) [Sov. Phys. JETP **10**, 998 (1960)].
[10]K. Ueda and T. M. Rice, Preprint, 1985.

Translated by Dave Parsons

PHYSICAL REVIEW B VOLUME 60, NUMBER 21 1 DECEMBER 1999-I

Conventional mechanisms for exotic superconductivity

D. F. Agterberg and Victor Barzykin
National High Magnetic Field Laboratory, Florida State University, Tallahassee, Florida 32310

Lev P. Gor'kov
National High Magnetic Field Laboratory, Florida State University, Tallahassee, Florida 32310 and L.D. Landau Institute for Theoretical Physics, Chernogolovka, 142432, Russia

(Received 19 May 1999)

We consider the pairing state due to the usual BCS mechanism in substances of cubic and hexagonal symmetry where the Fermi surface forms pockets around several points of high symmetry. We find that the symmetry imposed on the multiple pocket positions could give rise to a multidimensional nontrivial superconducting order parameter. The time-reversal symmetry in the pairing state is broken. We suggest several candidate substances where such ordering may appear, and discuss means by which such a phase may be identified. [S0163-1829(99)00942-X]

Most conventional superconductors are described very well by the BCS theory.[1] The electron-phonon interaction mediates an attraction between electrons that is stronger than the Coulomb repulsion. This gives rise to the Cooper instability of the normal state leading to the appearance of a condensate of pairs. The order parameter (the anomalous function $\mathcal{F}$; Ref. 2) in this case belongs to the "s-wave" type, *i.e.*, it is invariant with respect to the transformations of $G \otimes R$, where G is the crystal point group and R is time-reversal operation. As a result the quasiparticle spectrum has a gap, which leads to well-known experimental consequences. A variety of materials: ^{3}He,[3] UBe_{13},[4] UPt_3,[5] high-T_c materials,[6] and Sr_2RuO_4,[7] have been discovered that potentially break the $G\otimes R$ symmetry of the normal state. A well-known example is the A phase of ^{3}He,[3] which is not rotationally or time-reversal invariant (note that the B phase of ^{3}He is both rotation and time-reversal invariant). Such non-s-wave superconductors are usually expected to have a gapless excitation spectrum and arise when the interaction itself depends upon the superconducting ground state (in the case of ^{3}He the BCS ground state is the B phase and the spin-fluctuation feedback effect is required to stabilize the A phase[8]). All the possible symmetry classes of the superconducting state in crystalline materials were enumerated in Ref. 9 (for a review, see Refs. 10 and 11).

We show below that exotic superconductivity can be a much more common phenomenon and does not require unusual mechanisms. The electron-phonon and Coulomb interactions are enough to give rise to a multidimensional order parameter which would have lower symmetry than the normal state — including the breaking of time-reversal invariance. The effects we consider are possible in metals with several pockets which are centered at or around some symmetry points of the Brillouin zone (BZ). A BCS approximation generalized to the multiband case (see, e.g., Ref. 12) will be used. The point here is that since the form of the interaction parameters describing the two electron scattering on and between the different pockets of the Fermi surface (FS) is fixed by symmetry, the resulting superconducting state need not be s wave. Below we consider three cases in detail: (a) three FS pockets centered about the X-points of a simple cubic lattice; (b) three FS pockets at the M points of the hexagonal lattice; (c) four FS pockets at the L points in the face-centered-cubic lattice. A complete analysis of all other high-symmetry points is possible, and will be published elsewhere.[13]

We emphasize that this FS structure is not unusual. Indeed, superconductivity with pockets as in case (a) and $T_c \sim 0.1$ K is found in LaB_6.[14] Another example is given by the superconducting semiconductors such as PbTe, SnTe, or $SrTiO_3$.[15] Many materials exist where such FS sheets coexist with other nonsymmetry related FS sheets and some of these materials have anomalous superconducting properties. One example is $CeCo_2$;[16,17] Fig. 1 shows some of the FS sheets of $CeCo_2$($T_c = 1.6$ K). We will return to this later.[13]

We will use the generalized Ginsburg-Landau (GL) functional to identify possible nontrivial superconducting phases. The Hamiltonian for several separate pieces of the FS can be written in the following form:

$$H=\sum_{\alpha\sigma\mathbf{p}} \epsilon(\mathbf{p}) a^{\dagger}_{\alpha\sigma}(\mathbf{p}) a_{\alpha\sigma}(\mathbf{p}) + \frac{1}{2}\sum_{\mathbf{k},\mathbf{k}',\mathbf{q}}\sum_{\alpha\beta\sigma\sigma'} \times \lambda_{\alpha\beta}(\mathbf{q}) a^{\dagger}_{\alpha\sigma}(\mathbf{k}+\mathbf{q}) a^{\dagger}_{\beta\sigma'}(\mathbf{k}'-\mathbf{q}) a_{\alpha\sigma'}(\mathbf{k}') a_{\beta\sigma}(\mathbf{k}), \tag{1}$$

where σ and σ' are spin indices, $\lambda_{\alpha\beta}(\mathbf{q})$ includes the interaction for scattering two electrons from the pocket α into pocket β which is due to both Coulomb and electron-phonon terms. Introducing the anomalous Green's function $\hat{\mathcal{F}}_\alpha(x - x')$ for each FS sheet α, the corresponding Gor'kov equations[18] can be used to obtain the following solution at finite temperatures for the case of singlet pairing:

$$\mathcal{F}^{\dagger}_{\alpha}(\omega_n,\mathbf{p}) = \frac{\Delta^*_\alpha(\mathbf{p})}{\omega_n^2+\xi^2+|\Delta_\alpha(\mathbf{p})|^2}, \tag{2}$$

where

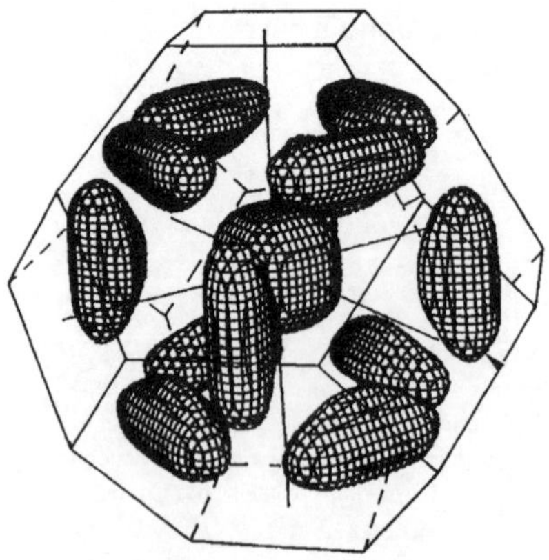

FIG. 1. Some FS sheets of $CeCo_2$ (from Ref. 17).

$$\Delta_\alpha^*(\mathbf{p}) = -\frac{T}{(2\pi)^3}\sum_\beta\sum_n\int d\mathbf{k}\lambda_{\beta\alpha}(\mathbf{p}-\mathbf{k})F_\beta^\dagger(\omega_n,\mathbf{k}). \tag{3}$$

Equation (2) being expanded in $|\Delta_\alpha|$ up to the third order, Eq. (3) becomes a variation of the GL functional with respect to the vector order parameter $\Delta_\alpha(\mathbf{p})$.[19,20] For simplicity we assume that each $\Delta_\alpha(\mathbf{p})$ is constant along the corresponding FS, and to fourth order in Δ_α we can write

$$F_s - F_n = -\Delta_\alpha^*\left[(\hat{\lambda}^{-1})_{\alpha\beta} - \delta_{\alpha\beta}\frac{mp_0}{2\pi^2}\ln\left(\frac{2\gamma\omega_D}{\pi T}\right)\right]\Delta_\beta + \frac{7\zeta(3)mp_0}{32\pi^4T^2}\sum_\beta|\Delta_\beta|^4, \tag{4}$$

where $\hat{\lambda}^{-1}$ is the matrix inverse to the interaction $\lambda_{\alpha\beta}$, ω_D is the cutoff (Debye) frequency.

We now analyze three different cases for multiple FS sheets.

(*a*) *Three X points in a cubic lattice.* The interaction matrix $\hat{\lambda}$ for three X points takes the following general form

$$\lambda_{\alpha\beta} = \lambda\delta_{\alpha\beta} + \mu(1-\delta_{\alpha\beta}). \tag{5}$$

Here λ is the interaction on the same pocket, μ couples any two different pockets. Consider first the linearized gap equation Eqs. (2) and (3) to determine T_c:

$$\Delta_\alpha^*\frac{2\pi^2}{mp_0} = -\sum_\beta\lambda_{\alpha\beta}\Delta_\beta^*\ln\left(\frac{2\gamma\omega_D}{\pi T_c}\right). \tag{6}$$

The three Δ_α transform among each other at cubic symmetry transformations forming a three-dimensional (3D) reducible representation of the cubic group O_h, which *is split* into a 1D A_{1g} and a 2D E_g irreducible representation. These two representations correspond to different order parameters with two critical temperatures:

$$T_{c,E} = \frac{2\gamma\omega_D}{\pi}\exp\left(\frac{2\pi^2}{mp_0(\lambda-\mu)}\right) \quad (2\text{D}), \tag{7}$$

$$T_{c,A} = \frac{2\gamma\omega_D}{\pi}\exp\left(\frac{2\pi^2}{mp_0(\lambda+2\mu)}\right) \quad (1\text{D}) \tag{8}$$

(the terms in the exponents must be negative for the Cooper effect to take place). The basis wave function for 1D identical representation is

$$l = (\Delta_1+\Delta_2+\Delta_3)/\sqrt{3} \tag{9}$$

and the basis wave functions for the 2D representation can be chosen as

$$\eta_1 = (\Delta_1+\epsilon\Delta_2+\epsilon^2\Delta_3)/\sqrt{3},$$
$$\eta_2 = (\Delta_1+\epsilon^2\Delta_2+\epsilon\Delta_3)/\sqrt{3}, \tag{10}$$

where $\epsilon=\exp(2\pi i/3)$. From Eq. (8) if $\lambda-\mu<0$ and $\mu>0$ then superconductivity will belong to the nontrivial 2D E_g representation, i.e., if the interaction *between two* different FS pockets is dominated by Coulomb repulsion. Let us consider the latter case in detail. Rewriting the Landau functional Eq. (4) in terms of l, η_1, and η_2, we obtain, for temperatures T *near* $T_{c,E}$:

$$\frac{2\pi^2}{mp_0}\delta F = \frac{T-T_{c,E}}{T_{c,E}}(|\eta_1|^2+|\eta_2|^2) + \ln(T_{c,E}/T_{c,A})|l|^2 + \frac{7\zeta(3)}{48\pi^2T_{c,E}^2}(|\eta_1|^4+|\eta_2|^4+4|\eta_1|^2|\eta_2|^2+F_{l\eta}^{(4)}), \tag{11}$$

where $F_{l\eta}^{(4)}$ is the fourth-order term in the GL functional which may admix the 1D representation l:

$$F_{l\eta}^{(4)} = 2l(\eta_1^*)^2\eta_2 + 2l\eta_1(\eta_2^*)^2 + \text{H.c.} \tag{12}$$

With $T_{c,A}<T_{c,E}$ the superconducting instability will correspond to the 2D representation E_g of the cubic point group. The fourth-order coefficients in Eq. (11) indicate that the class $O(D_2)$ is the most preferable energetically.[9] This class corresponds to a phase with $\eta_2=0$, $\eta_1\neq 0$ in Eq. (11). The symmetry properties of this class are known[9]. Time-reversal symmetry is broken and allows for antiferromagnetic domains and for fractional vortices to appear (see, e.g., (11)). In principle point nodes should appear where the FS intersects the cube diagonals. This would lead to the *electronic* contribution in the T^3 behavior for the heat capacity at low temperatures. In our case, however, there is no FS along the diagonals of the cube and the low-temperature thermodynamic properties will be determined by the gap of the same magnitude for all three FS sheets. In the presence of another FS, for example, at the Γ point, the nontrivial order $O(D_2)$ will be induced on it. In this case the point nodes will exist and power laws in thermodynamic properties due to the superconductivity should be seen experimentally. The anisotropy of the upper critical field (H_{c2}) (Ref. 21) *near* T_c for this class also requires that (at least) two vortex lattice phases (with a second-order transition between them) exists when the magnetic field is applied along the (1,1,0) and equivalent directions. Note from Eq. (12) that the terms linear in l identically disappear for this class, i.e., there will be no admixture of the s-wave component.

(*b*) *Three X points in a hexagonal lattice*. Calculations in this case are the same as in the previous case, i.e., Eqs. (5)–(12) apply. We only have to specify the symmetry properties and the superconducting class in this case since the symmetry of the lattice is different. The three-dimensional representation is split by the hexagonal group $D_{6h}\otimes R$ into 1D (A_{1g}) and 2D (E_{2g}). Note that the basis functions for E_{2g} can be once again chosen as given by Eq. (10). The phase with $\eta_1\neq 0, \eta_2=0$ has the lowest free energy and in this case corresponds to the nontrivial class $D_6(C_2)$. Time-reversal symmetry for this class is broken,[9] and *ferromagnetism* is allowed. Point nodes (at two points of intersection of an additional FS at the Γ point with the sixfold axis) can be seen in thermodynamic properties but again these nodes are only present if such a FS exists. The upper critical field is isotropic near T_c for this superconductivity representation. Nevertheless, it can be shown that there will also exist (at least) two distinct vortex lattice phases with a second-order transition between them for the magnetic field applied in the basal plane.[13]

(*c*) *Four L points in the fcc lattice*. The interaction and the linearized gap equation for the four L points again take the form Eqs. (5) and (6). This time the 4D representation Δ_α is split into the 1D (A_{1g}) and 3D (F_{2g}) irreducible representations. The critical temperatures for the two representations are now given by

$$T_{c,F}=\frac{2\gamma\omega_D}{\pi}\exp\left(\frac{2\pi^2}{mp_0(\lambda-\mu)}\right) \quad (3\mathrm{D}), \tag{13}$$

$$T_{c,A}=\frac{2\gamma\omega_D}{\pi}\exp\left(\frac{2\pi^2}{mp_0(\lambda+3\mu)}\right) \quad (1\mathrm{D}). \tag{14}$$

The basis functions for the 1D and 3D representations are

$$l=(\Delta_1+\Delta_2+\Delta_3+\Delta_4)/2 \quad (1\mathrm{D}) \tag{15}$$

and

$$\begin{aligned}\eta_x&=(\Delta_1-\Delta_2-\Delta_3+\Delta_4)/2,\\ \eta_y&=(\Delta_1+\Delta_2-\Delta_3-\Delta_4)/2 \quad (3\mathrm{D})\\ \eta_z&=(\Delta_1-\Delta_2+\Delta_3-\Delta_4)/2.\end{aligned} \tag{16}$$

The nontrivial 3D representation is stable if $\lambda-\mu<0$ and $\mu>0$, i.e., if the interaction is *attractive* for each pocket alone, while it is *repulsive* between two different pockets. As above, we can expand Eq. (4) in terms of l and $\vec{\eta}$. Dropping the 1D identical representation, we get

$$\frac{2\pi^2}{mp_0}\delta F=\frac{T-T_{c,F}}{T_{c,F}}(\vec{\eta}\vec{\eta}^*)+\frac{7\zeta(3)}{64\pi^2T_{c,F}^2}[2(\vec{\eta}\vec{\eta}^*)^2+|\vec{\eta}^2|^2-2(|\eta_x|^4+|\eta_y|^4+|\eta_z|^4)]. \tag{17}$$

The GL coefficients in Eq. (17) places the system right on the boundary of two phases, superconducting classes $D_4^{(2)}(D_2)\otimes R$ and $D_4(E)$ (see Fig. 2). This degeneracy is an artifact of the BCS theory, it is *not* lifted by higher-order terms in the GL functional. The presence of a FS at the Γ point lifts this degeneracy. As a result, the magnetic super-

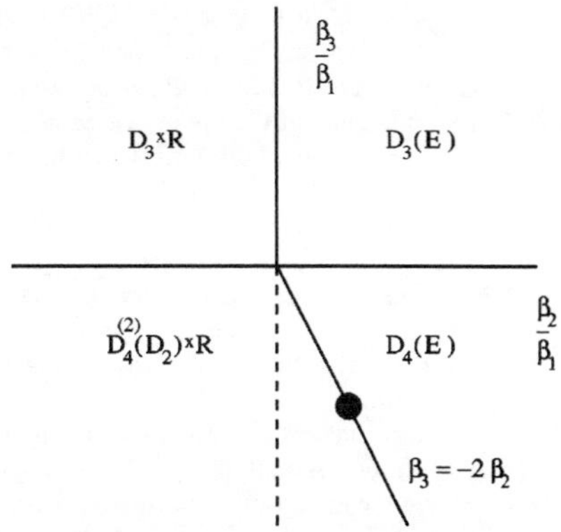

FIG. 2. Regions of existence of different superconducting phases on the basis of the three-dimensional representations of the cubic group (from Ref. 9). A point on the boundary of two phases corresponds to our BCS solution.

conducting class $D_4(E)$ is likely to appear. This class also allows ferromagnetism. Note that it has a line of nodes (at the intersection of the FS with the horizontal plane of symmetry) i.e., $C\propto T^2$ at low temperatures.[9] There are also point nodes at the intersection of the FS with the fourfold symmetry axis. This representation also exhibits an anisotropy of H_{c2} near T_c. In principle multiple vortex lattice phases can also exist for this class but they are not required by symmetry as they are for the E representations discussed above.

In the above cases, other than the standard isotropic order parameter, only multidimensional order parameters appeared. This leads to the possibility of domain walls between different equivalent superconducting states and as a possible consequence the existence of inhomogeneous magnetic order[9] (see also Refs. 10 and 11). Also in all the above cases the resulting superconducting states had gaps of equal magnitude on each of the FS sheets. In such a case a Hebel-Slichter peak in $1/T_1$ measurements may be present. The nontrivial representations were stable when the pair interaction between the different sheets was repulsive (independent of the intrasheet interaction). In many materials the Coulomb repulsion can be comparable to the attraction between electrons due to electron-phonon interactions. This is illustrated through the reduction of isotope effect due to Coulomb repulsion. It is well known that for a number of metals $T_c\propto M^{-\alpha}$, where not only $\alpha\neq 0.5$ but it may even have the opposite sign ($\alpha\simeq -2$ for $\alpha-U$; Ref. 22). The sensitivity of these exotic superconducting phases to impurities needs a more detailed analysis. However, provided the interpocket defect scattering amplitudes are much smaller than the intrapocket amplitudes, these phases will survive the presence of a considerable amount of defects (due to the ordinary BCS pairing on each sheet). This will be studied in more detail in Ref. 13.

In summary, we have shown that exotic superconductivity can appear merely as a competition of the phonon and Coulomb interactions if the FS consists of several pockets located at some symmetry points. Time-reversal symmetry is broken for the nontrivial order, meaning that the superconducting transition should be accompanied by some kind of magnetic order. The simplest methods to identify exotic order parameters are, apart from the phase-sensitive measurements, the power-law dependence of the heat capacity (due to the nodes), measurements of the upper critical field anisotropy $H_{c2}(\theta)$ at T_c,[21] or the observation of transitions between different vortex lattice phases. Note that if the FS pockets are fully isolated then the nodes are absent since the order parameter is then constant on each FS pocket and changes phase as one moves from one pocket to another. Nodes could appear, however, if there are "necks" connecting different sheets[14] or if superconductivity is induced on a FS centered, for example, around the Γ point. The upper critical-field anisotropy near T_c does not work as a test of nontrivial order in the hexagonal group.[21] The magnetic order, on the other hand, can be observed in μSR (muon spin-rotation) measurements or magnetization measurements in small enough samples (where the dimensions are on the order of the penetration depth). A general classification of all cases of nontrivial superconductivity of the type considered above is possible. These FS sheets are not always centered on the BZ boundary, as, for example, in some doped semiconductors[15] and $CeCo_2$.[17] We postpone a detailed analysis of the various possibilities to a future work.[13]

We would like to thank Z. Fisk, D. Khokhlov, J. R. Schrieffer, and the members of the NHMFL condensed-matter theory group seminar for useful discussions and comments. This work was supported by the National High Magnetic Field Laboratory through NSF cooperative agreement No. DMR-9527035 and the State of Florida.

[1] J. Bardeen, L.N. Cooper, and J.R. Schrieffer, Phys. Rev. **108**, 1175 (1957).
[2] L.P. Gor'kov, Zh. Éksp. Teor. Fiz. **34**, 735 (1958) [Sov. Phys. JETP **7**, 505 (1958)].
[3] D.D. Osheroff, R.C. Richardson, and D.M. Lee, Phys. Rev. Lett. **28**, 885 (1972).
[4] H.R. Ott, H. Rudigier, P. Desling, and Z. Fisk, Phys. Rev. Lett. **52**, 1551 (1984).
[5] G.R. Stewart, Z. Fisk, J.O. Willis, and J.L. Smith, Phys. Rev. Lett. **52**, 679 (1984).
[6] J.G. Bednorz and K.A. Müller, Z. Phys. B **64**, 189 (1986).
[7] Y. Maeno, H. Hashimoto, K. Yoshida, S. Nishizaki, T. Fujita, J.G. Bednorz, and F. Lichtenberg, Nature (London) **372**, 532 (1994).
[8] P.W. Anderson and W.F. Brinkmann, Phys. Rev. Lett. **30**, 1108 (1973).
[9] G.E. Volovik and L.P. Gor'kov, Zh. Éksp. Teor. Fiz **88**, 1412 (1985) [Sov. Phys. JETP **61**, 843 (1985)].
[10] L.P. Gor'kov, Sov. Sci. Rev., Sect. A **9**, 1 (1987).
[11] M. Sigrist and K. Ueda, Rev. Mod. Phys. **63**, 239 (1991).
[12] H. Suhl, B.T. Matthias, and L.R. Walker, Phys. Rev. Lett. **3**, 552 (1959).
[13] D. F. Agterberg, V. Barzykin, and L. P. Gor'kov (unpublished).
[14] A.J. Arko, G. Grabtree, J.B. Ketterson, F.M. Mueller, P.F. Walch, L.R. Windmiller, Z. Fisk, H.F. Hoyt, A.C. Mota, R. Viswanathan, D.E. Ellis, A.J. Freeman, and J. Rath, Int. J. Quantum Chem. **S9**, 569 (1975).
[15] Marvin L. Cohen, in *Superconductivity*, edited by R. D. Parks (Marcel Dekker, New York, 1969), Vol. 1.
[16] H. Sugawara, O. Inoue, Y. Kobayashi, H. Sato, T. Nishigaki, Y. Aoki, H. Sato, R. Setta, and Y. Onuki, J. Phys. Soc. Jpn. **64**, 3639 (1995).
[17] H. Sugawara, M. Higuchi, O. Inoue, T. Nishigaki, Y. Aoki, H. Sato, R. Settai, Y. Onuki, and A. Hasegawa, J. Phys. Soc. Jpn. **65**, 1744 (1996).
[18] A. A. Abrikosov, L. P. Gor'kov, and I. E. Dzyaloshinskii, *Methods of Quantum Field Theory in Statistical Physics* (Prentice-Hall, Inc., Englewood Cliffs, NJ, 1963).
[19] V.L. Ginsburg and L.D. Landau, Zh. Eksp. Teor. Fiz. **20**, 1064 (1950).
[20] L.P. Gor'kov, Zh. Éksp. Teor. Fiz. **36**, 1918 (1959) [Sov. Phys. JETP **9**, 1364 (1959)].
[21] L.P. Gor'kov, Pis'ma Zh. Éksp. Teor. Fiz. **40**, 351 (1984) [JETP Lett. **40**, 1155 (1984)].
[22] G.H. Lander, E.S. Fisher, and S.D. Bader, Adv. Phys. **43**, 1 (1994).

EUROPHYSICS LETTERS 15 November 1999

Europhys. Lett., **48** (4), pp. 449-454 (1999)

Exotic ground states and impurities in multiband superconductors

D. F. AGTERBERG[1], V. BARZYKIN[1] and L. P. GOR'KOV[1,2]

[1] *National High Magnetic Field Laboratory, Florida State University*
1800 E. Paul Dirac Dr., Tallahassee, Florida 32310, USA
[2] *L. D. Landau Institute for Theoretical Physics - Chernogolovka, 142432, Russia*

(received 28 July 1999; accepted 9 September 1999)

PACS. 74.20De – Phenomenological theories (two-fluid, Ginzburg-Landau, etc.).
PACS. 74.25Dw – Superconductivity phase diagrams.

Abstract. – We consider the effect of isotropic impurity scattering on the exotic superconducting states that arise from the usual BCS mechanism in substances of cubic and hexagonal symmetry where the Fermi surface contains inequivalent but degenerate pockets (*e.g.* around several points of high symmetry). As examples we look at $CeCo_2$, $CeRu_2$, and LaB_6; all of which have such Fermi surface topologies and the former exhibits unconventional superconducting behavior. We find that while these non *s*-wave states are suppressed by non-magnetic impurities, the suppression is much weaker than would be expected for unconventional superconductors with isotropic non-magnetic impurity scattering.

Recently it was shown that in substances with cubic or hexagonal symmetry, exotic superconducting states can arise from conventional BCS mechanisms [1]. This can occur when these metals have Fermi surface (FS) pockets that are centered at or around some three-fold or higher degenerate symmetry points of the Brillouin zone (BZ). The resulting superconducting state is either a conventional *s*-wave state or corresponds to a multidimensional superconducting representation for which the ground state breaks time reversal symmetry. Such FS topologies exist in many superconductors. Examples are $CeRu_2$ [2], LaB_6 [3], and $CeCo_2$ [4]; the latter exhibits a low temperature behavior specific heat that appears to vary as a power law with temperature [4,5] which is consistent with unconventional superconductivity. Here we investigate the stability of these exotic superconducting states in the presence of non-magnetic isotropic impurity scattering. In particular we examine in detail a model that applies both to $CeCo_2$ and to the FS pockets centered at the six N-points of a body-centered cubic (BCC) lattice and discuss the results of the analogous theory for $CeRu_2$ and LaB_6. We show that the exotic superconducting states are less susceptible to non-magnetic impurities than might be expected from earlier theories of isotropic impurity scattering in single band unconventional superconductors [6, 7].

A BCS approximation for the multi-band case will be used (see, *e.g.*, [8]). The form of the interaction parameters describing the two electron scattering on and between the different degenerate pockets of the FS is fixed by symmetry. Consequently, the resulting superconducting state need not be s-wave when the interaction does not depend on the direction of the Fermi momenta of an individual pocket. In our earlier work [1] we considered three cases in detail: a) three FS pockets centered about the X-points of a simple cubic lattice; b) three FS pockets at the M-points of the hexagonal lattice; c) four FS pockets at the L-points in the face-centered cubic (FCC) lattice. Case a) may describe the superconducting states in LaB_6 and in $CeRu_2$ since these materials both have FS pockets centered at the X-point.

It has recently been demonstrated that the specific heat exhibits a power law temperature dependence at low temperatures in $CeCo_2$ [4,5]. For this reason we consider here FS pockets centered at the six N-points of the BCC lattice which is formally identical to that in FCC $CeCo_2$ for the FS pockets which are located along the $(1,1,0)$ and equivalent directions [4,9] (note that in this material there also exist FS sheets centered at the Γ-point).

The Hamiltonian for several separate pieces of the FS can be written in the following form:

$$H = \sum_{\alpha\sigma\mathbf{p}} \epsilon(\mathbf{p}) a^{\dagger}_{\alpha\sigma}(\mathbf{p}) a_{\alpha\sigma}(\mathbf{p}) + \frac{1}{2} \sum_{\mathbf{k},\mathbf{k}',\mathbf{q}} \sum_{\alpha\beta\sigma\sigma'} \lambda_{\alpha\beta}(\mathbf{q}) a^{\dagger}_{\alpha\sigma}(\mathbf{k}+\mathbf{q}) a^{\dagger}_{\beta\sigma'}(\mathbf{k}'-\mathbf{q}) a_{\alpha\sigma'}(\mathbf{k}') a_{\beta\sigma}(\mathbf{k}), \quad (1)$$

where σ and σ' are spin indices, $\lambda_{\alpha\beta}(\mathbf{q})$ includes the interaction for scattering two electrons from the pocket α into pocket β which is due to both Coulomb and electron-phonon terms. For simplicity we will take $\lambda_{\alpha,\beta}(\mathbf{q})$ to be independent of the direction of $\mathbf{q}$. Scattering by impurities is described by the Hamiltonian

$$H_{\mathrm{imp}} = \sum_{\sigma} \sum_{k,q} \sum_{\alpha,\beta} \sum_{j} \exp[i(\mathbf{q}-\mathbf{k})\mathbf{R}_j] V_{\alpha,\beta}(\mathbf{k},\mathbf{q}) a^{\dagger}_{\beta\sigma}(\mathbf{k}) a_{\alpha\sigma}(\mathbf{q}). \quad (2)$$

Introducing the anomalous Green's function $\hat{\mathcal{F}}_{\alpha}(x-x')$ for each FS sheet α, the corresponding Gor'kov equations can be found for the case of singlet pairing in exactly the same manner as in ref. [10]. Including the isotropic impurity scattering within the Born approximation the following Gor'kov equations are found:

$$[i\omega_n - \epsilon_{\alpha}(\mathbf{k}) - \bar{\mathcal{G}}_{\alpha}(\omega)]\mathcal{G}_{\alpha}(\mathbf{k}, i\omega_n) - [\Delta^*_{\alpha} + \bar{\mathcal{F}}^{\dagger}_{\alpha}(\omega)]\mathcal{F}_{\alpha}(\mathbf{k}, i\omega_n) = 1\,,$$

$$[i\omega_n + \epsilon_{\alpha}(\mathbf{k}) + \bar{\mathcal{G}}_{\alpha}(-\omega)]\mathcal{F}^{\dagger}_{\alpha}(\mathbf{k}, i\omega_n) - [\Delta^*_{\alpha} + \bar{\mathcal{F}}^{\dagger}_{\alpha}(\omega)]\mathcal{G}_{\alpha}(\mathbf{k}, i\omega_n) = 0\,,$$

where

$$\bar{\mathcal{G}}_{\alpha}(\omega) = n_i \sum_{\mathbf{k},\beta} |u_{\alpha,\beta}|^2 \mathcal{G}_{\beta}(\mathbf{k}, i\omega_n)\,,$$

$$\bar{\mathcal{F}}_{\alpha}(\omega) = n_i \sum_{\mathbf{k},\beta} |u_{\alpha,\beta}|^2 \mathcal{F}_{\beta}(\mathbf{k}, i\omega_n)\,,$$

$$\Delta^*_{\alpha} = -T \sum_{\mathbf{k},\beta} \sum_{n} \lambda_{\beta\alpha} \mathcal{F}^{\dagger}_{\beta}(\omega_n, \mathbf{k})\,,$$

$|u_{\alpha,\beta}|^2 = \int \frac{\mathrm{d}\Omega}{4\pi} \int \frac{\mathrm{d}\Omega'}{4\pi} |V_{\alpha,\beta}(\mathbf{k}_{\mathrm{F}}, \mathbf{k}'_{\mathrm{F}})|^2$ and n_i is the concentration of impurities. Introducing $\tilde{\Delta}_{\alpha,n} = \Delta_{\alpha} + \bar{\mathcal{F}}^{\dagger}_{\alpha}(\omega)$ and $i\tilde{\omega}_{\alpha,n} = i\omega_n - \bar{\mathcal{G}}_{\alpha}(\omega)$, the gap and self-energy equations can be expressed as

$$\Delta_{\alpha} = T\pi \sum_{\beta,n} \frac{N_0 \lambda_{\alpha,\beta} \tilde{\Delta}_{\beta,n}}{\sqrt{\tilde{w}^2_{n,\beta} + |\tilde{\Delta}_{\beta,n}|^2}}, \quad (3)$$

$$\tilde{w}_{n,\alpha} = w_n + \sum_{\beta} \frac{\Gamma_{\alpha,\beta}\tilde{w}_{n,\beta}}{\sqrt{\tilde{w}^2_{n,\beta} + |\tilde{\Delta}_\beta|^2}}, \tag{4}$$

and

$$\tilde{\Delta}_{n,\alpha} = \Delta_\alpha + \sum_{\beta} \frac{\Gamma_{\alpha,\beta}\tilde{\Delta}_{n,\beta}}{\sqrt{\tilde{w}^2_{n,\beta} + |\tilde{\Delta}_{\beta,n}|^2}}, \tag{5}$$

where $\Gamma_{\alpha,\beta} = \pi N_0 |u_{\alpha,\beta}|^2$, and N_0 is the normal density of states on a single pocket. Equations (3), (4), and (5) have been studied in the context of multi-band superconductivity by a variety of authors (see, *e.g.*, [11-13]).

We consider here FS pockets centered at the six N-points of the BCC lattice. This situation is formally identical to that describing the FS pockets in FCC $CeCo_2$ located along the (110) and equivalent directions. The N-points of the BCC lattice lie at $\frac{1}{2}\mathbf{b}_i$ and $\frac{1}{2}(\mathbf{b}_i - \mathbf{b}_j)$ with $i > j$ where the $\mathbf{b}_i$ are the reciprocal lattice basis vectors $[\mathbf{b}_1 = \frac{2\pi}{a}(0,1,1)$, $\mathbf{b}_2 = \frac{2\pi}{a}(1,0,1)$, and $\mathbf{b}_3 = \frac{2\pi}{a}(1,1,0)]$. The interaction between the electrons forming a Cooper pair on Fermi pockets centered at these points takes the form

$$V = \begin{pmatrix} \lambda & \nu & \mu & \nu & \nu & \nu \\ \nu & \lambda & \nu & \mu & \nu & \nu \\ \mu & \nu & \lambda & \nu & \nu & \nu \\ \nu & \mu & \nu & \lambda & \nu & \nu \\ \nu & \nu & \nu & \nu & \lambda & \mu \\ \nu & \nu & \nu & \nu & \mu & \lambda \end{pmatrix}, \tag{6}$$

where λ is the interaction on the same pocket, μ couples two pockets connected by a $\frac{4\pi}{a}(1,0,0)$, $\frac{4\pi}{a}(0,1,0)$, or a $\frac{4\pi}{a}(0,0,1)$ translation, and ν characterizes the remaining couplings between nearest-neighbor N-points. The matrix for the impurity scattering (with elements $\Gamma_{\alpha,\beta}$) has the same form and will be described by elements Γ_0, Γ_1, and Γ_2 replacing λ, ν, and μ, respectively.

Consider the clean limit, the linearized gap equation is

$$\Delta^*_\alpha = -\sum_\beta N_0 \lambda_{\alpha\beta} \Delta^*_\beta \ln\left(\frac{2\gamma\omega_D}{\pi T_c}\right). \tag{7}$$

The six Δ_α transform among each other under cubic symmetry transformations forming a 6D reducible representation of the cubic group O_h, which is split into a 1D A_{1g}, a 2D E_g, and a 3D F_{2g} irreducible representation. These three representations correspond to different order parameters with three critical temperatures:

$$T_{c0,A} = \frac{2\gamma\omega_D}{\pi} \exp\left[\frac{1}{N_0(\lambda + \mu + 4\nu)}\right], \tag{8}$$

$$T_{c0,E} = \frac{2\gamma\omega_D}{\pi} \exp\left[\frac{1}{N_0(\lambda + \mu - 2\nu)}\right], \tag{9}$$

$$T_{c0,F} = \frac{2\gamma\omega_D}{\pi} \exp\left[\frac{1}{N_0(\lambda - \mu)}\right], \tag{10}$$

where the factors in the exponentials must be negative for a non-zero transition temperature. When $\nu > 0$ or $2\nu + \mu > 0$ (*i.e.* for repulsive inter-pocket interactions) the higher-dimensional representations have the higher T_c. The basis wave function for 1D A_{1g} identical representation is

$$l = (\Delta_1 + \Delta_2 + \Delta_3 + \Delta_4 + \Delta_5 + \Delta_6)/\sqrt{6}, \tag{11}$$

the basis wave functions for the 2D E_g representation can be chosen as

$$\eta_1 = (\Delta_1 + \epsilon\Delta_2 + \Delta_3 + \epsilon\Delta_4 + \epsilon^2\Delta_5 + \epsilon^2\Delta_6)/\sqrt{6}\,,$$
$$\eta_2 = (\Delta_1 + \epsilon^2\Delta_2 + \Delta_3 + \epsilon^2\Delta_4 + \epsilon\Delta_5 + \epsilon\Delta_6)/\sqrt{6}\,, \tag{12}$$

where $\epsilon = \exp[2\pi i/3]$, and the basis wave functions for the 3D F_{2g} representation can be chosen as

$$\eta_x = (\Delta_5 - \Delta_6)/\sqrt{2}\,,$$
$$\eta_y = (\Delta_1 - \Delta_3)/\sqrt{2}\,, \tag{13}$$
$$\eta_z = (\Delta_2 - \Delta_4)/\sqrt{2}\,.$$

Following refs. [14, 1], the Ginzburg Landau functional for the 2D and 3D representations are found to be

$$\delta F_E/N_0 = \frac{T - T_{c0,E}}{T_{c0,E}}(|\eta_1|^2 + |\eta_2|^2) + \frac{7\zeta(3)}{96\pi^2 T_{c0,E}^2}(|\eta_1|^4 + |\eta_2|^4 + 4|\eta_1|^2|\eta_2|^2)\,, \tag{14}$$

$$\delta F_F/N_0 = \frac{T - T_{c0,F}}{T_{c0,F}}(\vec{\eta}\vec{\eta}^*) + \frac{7\zeta(3)}{32\pi^2 T_{c0,F}^2}(|\eta_x|^4 + |\eta_y|^4 + |\eta_z|^4). \tag{15}$$

Equation (14) for the E_g representation implies $(\eta_1, \eta_2) = (1, 0)$ is a stable ground state. Following the notation of ref. [15] this corresponds to the superconducting class $O(D_2)$. The properties of such a ground state are the same as those found for the analogous calculation for FS pockets centered at the three X-points in a simple cubic lattice and this is also the situation that applies to both LaB_6 and $CeRu_2$ (we refer to ref. [1] for more details). Equation (15) for the F_{2g} representation implies the ground state solution $(\eta_x, \eta_y, \eta_z) = [1, \exp[i\phi_1], \exp[i\phi_2]]$ where ϕ_1 and ϕ_2 are arbitrary phase factors. The degeneracy arising from the phases ϕ_1 and ϕ_2 is an artifact of the BCS theory and is not lifted by higher-order terms in free energy functional. In the notation of ref. [15] the free energy of eq. (15) places the system on the boundary of two phases with superconducting classes $D_3 \times R$ and $D_3(E)$. A similar situation arose for FS pockets centered at the four L-points of a face-centered cubic lattice where a degeneracy was found between solutions corresponding to the classes $D_4^{(2)}(D_2) \times R$ and $D_4(E)$ of the F_{2g} representation [1]. The presence of a FS centered at the Γ-point lifts this degeneracy. As a result the solution $(\eta_x, \eta_y, \eta_z) = (1, \epsilon, \epsilon^2)$ corresponding to the magnetic class $D_3(E)$ is likely. This class allows ferromagnetism and it will have point nodes on the Γ-centered FS (the N-point centered FS pockets will have no nodes) [15, 16].

Now consider the effect of including impurity scattering on $T_{c,A}, T_{c,F}, T_{c,E}$. It is convenient to define the following relaxation times:

$$(2\tau_A)^{-1} = \Gamma_0 + 4\Gamma_1 + \Gamma_2\,,$$
$$(2\tau_E)^{-1} = \Gamma_0 + \Gamma_2 - 2\Gamma_1\,, \tag{16}$$
$$(2\tau_F)^{-1} = \Gamma_0 - \Gamma_2\,. \tag{17}$$

The lifetime τ_A determines the elastic mean free path while the lifetimes τ_E and τ_F cannot be easily measured. Solution of the linearized equations indicate that the transition temperatures $T_{c,i}$ are given by

$$\ln \frac{T_{c0,i}}{T_{c,i}} = \psi\left(\frac{1}{2} + \frac{\tau_A^{-1} - \tau_i^{-1}}{4\pi T_{c,i}}\right) - \psi\left(\frac{1}{2}\right), \tag{18}$$

where $\psi(x)$ is the digamma function. From this expression it is clear that $T_{c,A}$ is not suppressed by non-magnetic impurities while $T_{c,E}$ and $T_{c,F}$ are suppressed. Note that if $\Gamma_1 = \Gamma_2 = 0$ (there is only intra-pocket impurity scattering) then all three transition temperatures are not suppressed by impurities. It is only inter-pocket scattering that reduces $T_{c,E}$ and $T_{c,F}$ from $T_{c0,E}$ and $T_{c0,F}$. Consequently, the suppression of T_c as the impurity concentration is increased is not as rapid as might be expected from previous analysis of unconventional single band superconductors [6, 7]. Furthermore, an impurity-induced transition is possible from the non s-wave superconducting states to the s-wave superconducting states when $\lambda < -4\nu - \mu$ where negative λ corresponds to an attractive intra-pocket interaction and either $\nu > 0$ or $2\nu + \mu > 0$. The observation of an abrupt change in the T_c-dependence on impurity concentration corresponding to this phase transition (*i.e.* T_c initially decreasing with impurity concentration in the exotic state and then becoming impurity independent in the s-wave state) may help to identify such exotic superconducting states.

The results of the above analysis may be directly applied to $CeCo_2$ and predicts that due to the presence of the the Γ-centered FS sheet, the low temperature specific heat will vary as T^3 for the exotic states considered (with superconducting classes $O(D_2)$ or $D_3(E)$). Note that initial measurements appear more consistent with a T^2 behavior, however, these measurements only exist down to $T/T_c \approx 0.13$ [5] which may not yet be sufficiently low to extract the low temperature behavior reliably. For the materials LaB_6 and $CeRu_2$ the FS pockets are centered at the X-points for which the exotic superconducting state belongs to the 2D E_g representation which is described in detail in ref. [1]. Impurity scattering will have qualitatively the same effect on this state as on the exotic states examined above.

In conclusion we have considered the effect of isotropic non-magnetic impurity scattering on the exotic superconducting states that arise in a BCC lattice when the FS forms pockets centered at the six N-points. We have found that increasing the impurity concentration does not suppress these non s-wave states as rapidly as would be expected from earlier studies of impurity scattering in unconventional superconductors. These results are not specific to FS pockets centered at the reciprocal lattice N-points of a BCC lattice but will apply to any degenerate set of FS pockets.

***

We would like to thank Z. Fisk, D. Khokhlov, J. R. Schrieffer, and the members of the NHMFL condensed matter theory group seminar for useful discussions and comments. This work was supported by the National High Magnetic Field Laboratory through NSF cooperative agreement No. DMR-9527035 and the State of Florida.

REFERENCES

[1] Agterberg D. F., Barzykin V. and Gor'kov L. P., Preprint, cond-mat/9906394 (1999).
[2] Hedo M., Inada Y., Ishida T., Yamamoto E., Haga Y., Onuki Y., Higuchi M. and Hasegawa A., *J. Phys. Soc. Jpn.*, **64** (1995) 4535.
[3] Arko A. J., Grabtree G., Ketterson J. B., Mueller F. M., Walch P. F., Windmiller L. R., Fisk Z., Hoyt H. F., Mota A. C., Viswanathan R., Ellis D. E., Freeman A. J. and Rath J., *Int. J. Quantum Chem.*, **S9** (1975) 569.

[4] SUGAWARA H., INOUE O., KOBAYASHI Y., SATO H., NISHIGAKI T., AOKI Y., SATO H., SETTA R. and ONUKI Y., *J. Phys. Soc. Jpn.*, **64** (1995) 3639.

[5] AOKI Y., NISHIGAKI T., SUGAWARA H. and SATO H., *Phys. Rev. B*, **55** (1997) 2768.

[6] GOR'KOV L. P. and KALUGIN P. A., *JEPT Lett.*, **41** (1985) 253.

[7] UEDA K. and RICE T. M., in *Theory of Heavy Fermions and Valence Fluctuations*, edited by T. KASUYA and T. SASO, (Springer, Berlin) 1985.

[8] SUHL H., MATTHIAS B. T. and WALKER L. R., *Phys. Rev. Lett.*, **3** (1959) 552.

[9] SUGAWARA H., HIGUCHI M., INOUE O., NISHIGAKI T., AOKI Y., SATO H., SETTA R., ONUKI Y. and HASEGAWA A., *J. Phys. Soc. Jpn.*, **65** (1996) 1744.

[10] ABRIKOSOV A. A., GOR'KOV L. P. and DZYALOSHINSKII I. E., *Methods of Quantum Field Theory in Statistical Physics* (Prentice-Hall, Inc., Engelewood Cliffs) 1963.

[11] MOSKALENKO V. A. and PALISTRANT M. E., *Sov. Phys. JEPT*, **22** (1966) 536.

[12] SCHOPOHL N. and SCARNBERG K., *Solid State Commun.*, **22** (1977) 371.

[13] GOLUBOV A. A. and MAZIN I. I., *Phys. Rev. B*, **55** (1997) 15146.

[14] GOR'KOV L. P., *Sov. Phys. JETP*, **9** (1959) 1364.

[15] GOR'KOV L. P., *Sov. Sci. Rev. A Phys.*, **9** (1987) 1.

[16] SIGRIST M. and UEDA K., *Rev. Mod. Phys.*, **63** (1991) 239.

Chapter 6

Organic Conductors

6.1 Superconductivity in one dimension?

In 1964 W. Little set forward a proposal for superconductivity in the *linear "metallic" chains* such as in conducting polymers (Ref. 1). Within the mechanism in which electrons on a conducting chain interact via an exchange, instead of phonons, by excitations on the neutral complexes adjacent to the chain he ended with the estimate for the transition temperature T_c of several hundred degrees. So large T_c in his model originated from the factor in the BCS formula: $T_c \simeq \omega_{el} \exp[-1/\lambda]$ where, instead of a typical phonon frequency $\omega_D \approx 100{-}300K$, would stand $\omega_{el} \sim 1eV$, a characteristic energy of the *electronic* excitations on the neutral complex. In hindsight, it was a childish idea, but the paper created unprecedented excitement and prompted the active experimental and theoretical studies around the world in the field of organic conductors.

While the possibility of superconductivity in one dimension regardless of the details of the Little's mechanism was immediately criticized on theoretical grounds; predictions of such high values for T_c attracted much attention and even caused a tumult and not only among the scientific circles, but even among some officials, at least some Russian officials. In my case it was Academician N. N. Semenov, Head of the Academy Center in Chernogolovka and Director of the Institute for Chemical Physics of the Soviet Academy. He was the Nobel Laureate in Chemistry, 1956, for research in chemical chain reactions (with Sir Cyril Hinshelwood). The Institute of Theoretical Physics at the Noginskii Center was yet in early stage, but by that time I have already left the Kapitza Institute, worked at the Branch of the Institute for Chemical Physics in Chernogolovka and lived there. Semenov immediately demanded more information concerning details of the Little's claim and what may the prospects be in the area.

Our paper [6a] was such first attempt to clarify the situation. Objections at the time against having superconductivity in 1D were two-fold. On one side, few authors e.g. R. Ferrell (Ref. 2) argued that there could be no long-range phase correlations in one dimension. The arguments seemed not to be directly relevant to us, for we were focusing on the problem of stability of the *metallic* ground state at $T = 0$. To us the main danger seemed to be in the structural Peierls instability; the famous Peierls dimerization in the chain polymers such as polyacetylene, $[CH]x$ doubles the periodicity of the metallic $[-C-C-]$ network imposing the insulating energy gap at the 1D Fermi "surface" (at the Fermi *points*: $\pm p_F \equiv \pm(\pi/a)$).

Conclusions from [6a] were non-trivial and encouraging: at the proper (negative) effective sign of the electron-electron interactions the Peierls instability would indeed occur, but it would develop *simultaneously* and at the same rate as instability in the Cooper channel. If only the phase transition took place in point of fact, the *two* gaps would coexist. (At the opposite sign of interactions the ground state remains metallic). As we know better now, no long-range ordering is possible in 1D. However our results supported the notion of superconductivity phenomena in organic conductors, at least on the level of fluctuations. (For discussion and more details, especially regarding the mathematics related to solving the parquet equations, see Ref. 3).

Later on I. Dzyaloshinskii and A. Larkin (Ref. 4) pointed out that *if* the conducting chain contains strictly one electron per chain period and the matrix element for the *umklapp* electron-electron scattering is large enough so that there is room for instability in the antiferromagnetic, i.e. the spin density wave (SDW) channel.

Thanks to the Little's affair, in 1965 I visited the USA for the first time. The adventure is illustrative of that epoch. I mentioned somewhere that traveling abroad for anyone from our group was, to say the least, difficult. The Academy bureaucrats just disliked Landau and his circle. Any foreign trip in those years was considered by the same Academy authorities as a political venture for which the permission was necessary from "High Level" to avert any possibility of any "ideological inconsistency", if I may say so. It just so happened, however, that Semenov had an invitation to one of the Gordon Chemistry Conferences at the time. As a Nobel Laureate, he was extremely influential in the Soviet hierarchy and I was delegated instead of him to that Chemistry conference, much to the organizers surprise. For a couple of days I was entertained by two people of athletic build until my scientific identity was established.

6.2 Role of three-dimensional effects

Further efforts both in theory and in experiment were concentrated on ways and means of converting 1D instabilities into the real phase transitions with help of different three-dimensional (3D) factors. The latter are inherent in all real situations because chains of the conducting molecules must be packed together into a crystal.

The most important 3D effects were investigated first in [6b]. Subsequent theoretical papers elaborated further details and particular models, but in essence the conclusions remained the same. In simplified models to restrain fluctuations at a putative phase transition, the authors deliberately coupled the diverging one-dimensional responses to 3D mechanisms in a mean field fashion (Refs. 5, 6).

In [6b] we started with the role of interactions in the crystal between electrons on different chains. The possibility of long-range order hinges on the solutions of the generalized parquet equations that we called the "*moving poles*". That is, position of the pole (i.e. position of the zero in denominator of an electron-electron vertex) must depend on the transverse components of the momentum. Then it is possible to describe the transition and the new phase by a theory of the GL type. At the same time it was clear that such solutions (as well as in the case with electrons interacting via dispersive phonon modes) cannot describe the *superconductivity* transitions. The latter is possible *only* at a non-zero inter-chain electron tunneling.

The complete list of the ground states that can emerge at zero temperature out of the parquet instabilities in the presence of 3D features is given in the concluding part of [6b]. Note the interesting solution corresponding to a regime of long-range correlations that we dubbed as the *standing pole*. Such solutions were presumed not to be rare as they may arise from the broad range of initial conditions. I am more inclined to think that in reality the "standing pole" solutions correspond to the 1D fluctuations synchronized on the different chains via interactions. How such fluctuations could be seen experimentally was an interesting direction, but we left it unfinished. Of course any actual phase transition occurs via 3D mechanisms.

As the on-chain interactions are also weak, even minor deviation from the strictly one-dimensional picture ensures for a 1D instability to outgrow into the second order phase transition at *low enough* temperatures. The implications were very significant for the experiment because, in reality, all new compounds are "quasi-one-dimensional" (Q1D) materials. In particular, as the conducting molecules chains are packed together in the crystalline matrix rather loosely, most critical for superconductivity are the values of tunneling integrals in perpendicular to the chains directions. Tunneling imparts the three-dimensional (3D) "corrugations" to the two initially *planar* Fermi surface sheets.

Hence, the necessary precondition for stabilizing superconductivity in Q1D conductors was to enhance the 3D features in the *electronic spectrum*. In 1980 the Danish-French group observed for the first time the superconductivity transition *under pressure* in $(TMTSF)_2PF_6$, in the "metal" that contains *no* atoms of the metallic elements (Ref. 7).

Compounds $(TMTSF)_2X$ form the whole class of Q1D conductors named the Bechgaard salts after the Danish chemist Klaus Bechgaard. The abbreviation TMTSF stands for the organic *tetramethyl-tetraselena fulvalene* molecules; as a matter of fact, for a theorist — in most cases — it is more than enough to know that the conducting chains consist of stacks of planar donor TMTSF molecules packed into crystal; X stands for a whole variety of inorganic anions, PF6, ClO4, NO3, etc. The conduction network in these materials is extremely anisotropic. By applying pressure the Jerome group managed to adequately increase the overlap between the neighboring chains.

6.3 Superconducting fluctuations and conductivity

The first known organic *metal* was the so called TTF-TCNQ. (TTF-TCNQ is built of the two parallel stacks of the planar organic molecules). Early in the 70's, much attention was directed to this material after the Pennsylvania team made the claim (Ref. 8) that before transition into the semiconducting Peierls state the TTF-TCNQ-conduction shows a huge metallic increase down to 60 K. Some made haste to interpret the peak as coming due to superconducting fluctuation.

The claim Ref. 8 seemed dubious. Indeed, on the theoretical side [6c], even enhanced by one-dimensional fluctuations the Q1D conductivity seemed not to agree with so pronounced anomaly. More rigorously the expression for the fluctuation conductivity was recalculated in [6d]; only weak singularity was found near temperature of the 3D structural transition. In fact, the notorious conductivity peak at ~60K in TTF-TCNQ has never been confirmed by any other experimental group.

6.4 General properties of $(TMTSF)_2PF_6$

6.4.1 *Why SDW as the ground state?* Simplicity of the crystalline structure of $(TMTSF)_2PF_6$ is deceptive. Although, in general, our understanding concerning the low temperature properties of the Bechgaard salts may seem good, the material in many respects is uncommon [6e].

The prevailing feature in the (T, p)-phase diagram (p is the pressure) is the SDW transition ($T_{SDW} \simeq 12K$ at ambient pressure); with the increase in pressure T_{SDW} decreases; superconductivity begins to grow *from inside* the area in the phase diagram occupied by the SDW state.

The SDW phase and dependence of the transition temperature T_{SDW} on pressure can be interpreted, even quantitatively in frameworks of the so-called "nesting" instability. In this scenario there are the two open Fermi sheets that can be superimposed on top of each other by the shift on vector $\vec{Q} = (2p_F; \pi/b)$; electron-electron interactions tend to close the metallic spectrum with a gap. At a low temperature the instability resolves into a phase transition with $\vec{Q}$ as the structural vector.

The theoretical analysis of how a *one-dimensional* regime at higher energies or *high* temperatures scales down and merges with the low temperature Q1D nesting regime was performed in [6e]. At the bandwidth in the Bechgaard salts of the order of ~0.8eV the smallness of the SDW energy scale (~10K) is consistent with the assumed weakness of electron-electron interactions. Values of the interchain tunneling integrals in the Bechgaard salts never exceeding one-two hundred Kelvin give further support to the model.

At that, however, unresolved remains the question why the nesting instability occurs namely *in the SDW*- and not in the CDW-channel. To answer this question, it seems necessary to reconsider the role of some subtle details of the lattice structure. To understand what the problem is, we return again to the high energy regime. In this limit the system behaves as a one-dimensional one and most essential are the *on-chain* interactions. In the 1D limit, the SDW instability prevails against instability in the CDW channel *only* in case of the *strong Umklapp* coupling (Ref. 4). *Formally*, indeed, there are *two* TMTSF molecules in $(TMTSF)_2PF_6$ and *one* hole per period as is the case of the half-filled reciprocal unit cell; so the Umklapp interactions could be possible, only that the difference in distances between two TMTSF molecules in the unit cell is so small that such dimerization is always deemed negligible. In that last case the Brillouin Zone would be quarter-filled and processes with the momentum transfer $4p_F$ at the particle-particle scattering possible in the second approximation in interactions only. As umklapp interactions seem to be necessary for the condensation namely into the SDW phase, importance of the dimerization must be revised anew.

6.4.2 *Superconductivity.* The nature of the superconducting pairing in $(TMTSF)_2PF_6$ (that is, the mechanism and the type of pairing) is not established reliably even until now. In [6f] it was pointed out that the experimental values of the upper critical fields are very high and exceed considerably the limitation imposed by the Clogston criterion, thus supporting the notion of the triplet pairing $S = 1$. More recently, in Ref. 9 we arrived to same conclusion that superconductivity in $(TMTSF)_2PF_6$ must bear the triplet pairing, at least in that part of the (T, p)-phase diagram inside the SDW phase where the superconducting phase sets in for the first time.

6.5 Field induced SDW

Congruency of the two Fermi sheets in $(TMTSF)_2$ X is not perfect and in the energy spectrum are present the so-called "anti-nesting" terms. The latter are small, but $T_{SDW} \approx 12K$in $(TMTSF)_2$ PF_6 is small too. Application of hydrostatic pressure increases the overlap between neighboring chains and by that the deviations from the perfect nesting, hence, decreasing T_{SDW} (a moderate pressure is capable of fully destroying the gapped SDW state). At higher pressure $(TMTSF)_2PF_6$ fully returns back in the metallic state with the two open Fermi surface sheets. Metallic energy spectrum in the normal phase is usually chosen as: $\varepsilon_\pm(\vec{p}) = \pm v_F(p_a \mp p_F) + 2t_b \cos(p_b b) + 2t_c \cos(p_c c) + t'_b \cos(2p_b b)$ (b, c are the lattice periods perpendicular to the chains; t'_b give the amplitude of the "anti-nesting" term. The tunneling integral t_c is small and usually can be omitted).

In spite of such a simple form for the energy spectrum, the metallic phase in high magnetic fields behaved rather strange. In fact, early experiments have revealed oscillations almost periodic in $1/H$ thereby reminiscent of the quantum oscillations in the de Haas–van Alphen (dHvA) effect in ordinary metals. That seemed impossible in a Q1D metal with the two open Fermi surfaces only [6g].

The observations were explained in [6g] as due to the "one-dimensionalization" of carriers in the magnetic field. That is, the magnetic field produces the *qualitative* changes in how electrons in Q1D conductors, such as $(TMTSF)_2PF_6$, can move in the *real space*. To understand, consider first a trajectory of one quasiclassical electron moving freely with the velocity $\vec{u}(\vec{p}) = \partial\varepsilon_\pm(\vec{p})/\partial\vec{p}$. Along the chains' direction velocity is $u_a = \pm v_F$; along the b-axis: $u_b(p_b) = -2bt_b \sin(p_b b)$, i.e, the trajectory is *unbounded* as $x(t) = \pm v_F t$, $y(p_b; t) = u_b(p_b)t$ (t is the time).

Now switch on the magnetic field H perpendicular to the (a, b)-plane. Choose the vector potential as $\vec{A}(\vec{r}) = (0; Hx)$ and substitute $p_b \to p_b - (e/c)A_b$ in the equations of motion. Along the chain, as before, $x(t) = \pm v_F t$ but the motion in the transverse direction $y_b(p_b; t) = 2(c/eHb)t_b \cos[p_b b - (e/c)v_F Ht]$ is now periodic. Increase of the magnetic field *confines* further amplitude of the oscillations in perpendicular to the chains directions thereby *improving* the nesting conditions. Analysis in [6g] has shown that the phenomenon unfolds as a *cascade* of the Field Induced SDW transitions (FISDW).

6.6 Cyclotron resonance on open orbits

Periodicity of the electronic motion in the magnetic field perpendicular to the (a, b)-plane along b-axis can be detected directly in the real space via the cyclotron resonance. Expressions for the surface impedance were derived and analyzed in [6h].

6.7 Other effects

In the above I concentrated on $(TMTSF)_2PF_6$ mainly because experimentally, this material was most studied, especially in the early years. Other compounds demonstrate the same physics as well. Still, $(TMTSF)_2PF_6$ remains the cleanest material even in the whole class of the Bechgaard salts, so the most delicate features could be investigated here best. This concerns, e.g. observation of the cyclotron resonance and of the FISDW cascade. The experimental details and a good summary of the recent experimental results can be found in Ref. 10.

One-dimensional physics leads to other interesting phenomena in organic conductors, such as the appearance of domains. Consider first the strictly one-dimensional Peierls *insulator*. (It can be as the SDW — as the CDW — phase). Unlike the electron-hole pairs in semiconductors, excitations in the 1D Peierls state are the so-called *solitons*. Solitons are the non-linear formations built upon the electronic states from both the lower and the upper bands. Energy of a single soliton lies in the middle of the gap. At a non-zero transverse tunneling one soliton can jump from chain to chain. However, as the tunneling integrals start to exceed some threshold, it becomes energetically favorable for solitons to spread over many chains into the *metallic* domain wall with the Fermi spectrum for the excitations (Ref. 11). Related to this effect is the spontaneous formation of such domain walls at some conditions. The regime of the periodic *soliton walls* as the ground state of $(TMTSF)_2PF_6$ was suggested in Ref. 12. Recent experiments (Ref. 13) seem to confirm the existence of a domain structure in part of the phase diagram of $(TMTSF)_2PF_6$, although not all details were yet resolved.

6.8 Field-induced superconductivity

Another interesting low-dimensional phenomenon is superconductivity in the $\lambda - (BETS)_2FeCl_4$ that can be observed only in strong enough magnetic fields.

Organic conductor $\lambda - (BETS)_2FeCl_4$ is antiferromagnetic at low temperatures. (For interested readers, in the organic chemistry, *BETS* has the name: *bis(ethylenedithio)tetraselenafulvalene*). Molecules of *BETS* form the conduction layers prone, as it will be clear from what follows, to superconducting pairing. As the conduction planes are separated by layers of $FeCl_4$, coupling to the magnetic iron ions tends to destroy the s-wave superconductivity.

When the magnetic field is imposed it also acts on the spins of the Fe^{3+} ions ($S = 5/2$). It can be shown that layers of the irons' spins being oriented by the *in-plane* field decouple the conducting *BETS* planes thus making the problem a two dimensional (2D) one. A strong magnetic field by itself would, of course, also destroy the singlet superconductivity by breaking the Cooper pairs. But the Fe^{3+}-magnetic moments oriented in the magnetic field exert the *exchange* field $J\langle S\rangle$ on spins of the conduction electrons via the exchange interaction $J(\vec{S} \cdot \vec{s})$. The sum of two fields H_{eff} is: $\mu_B H_{eff} = \mu_B B + J\langle S\rangle$ and at $J < 0$ the field H_{eff} is below the Clogston limit in the whole interval of *high B*.

References

1. W. A. Little, *Phys. Rev.* **134**, A1416 (1964)
2. R. A. Ferrell, *Phys. Rev. Lett.* **13**, 330 (1964)
3. Yu. A. Bychkov, L. P. Gor'kov, and I. E. Dzyaloshinskii, *Sov. Phys. JETP* **23**, 489 (1966)
4. I. E. Dzyaloshiskii and A. I. Larkin, *Sov. Phys. JETP* **34**, 422 (1972)
5. P. A. Lee, T. M. Rice and R. A. Klemm, *Phys. Rev B.* **15**, 2984 (1977)
6. J. Solyom, *Adv. Phys.* **28**, 201 (1979)
7. D. Jerome *et al.*, *J. Phys. France Lett.* **41**, L-95 (1980)
8. L. B. Coleman *et al.*, *Solid State Comm.* **12**, 1125 (1973)
9. L. P. Gor'kov and P. D. Grigoriev, *Phys. Rev. B* **75**, 020507(R) (2007)
10. A.V. Kornilov and V. M. Pudalov (p. 487), in *The Physics of Organic Superconductors and Conductors*, ed. A. G. Lebed; Springer, Berlin, 2008
11. S. A. Brazovskii, L. P. Gor'kov and J. R. Schrieffer, *Physica Scr.* **25**, 425 (1981)
12. L. P. Gor'kov and P. D. Grigoriev, *Europhys. Lett.* **71**, 425 (2005)
13. N. Kang *et al.*, *Phys. Rev. B* **81**, 100509 (2010)

List of original papers

[6a] Yu. A. Bychkov, L. P. Gor'kov and I. E. Dzyaloshinskii, *Concerning one-dimensional superconductivity*, Sov. Phys. JETP Lett. 2, 92 (1965)

[6b] L. P. Gor'kov and I. E. Dzyaloshinskii, *Possible phase transitions in systems of interacting metallic filaments (quasi-one-dimensional metals)*, Sov. Phys. JETP 40, 198 (1975)

[6c] L. P. Gor'kov and I. E. Dzyaloshinskii, *Conductivity of quasi-one-dimensional crystals in the metallic phase*, Sov, Phys. JETP Lett. **18**, 401 (1973)

[6d] L. P. Gor'kov and I. E. Dzyaloshinskii, *Fluctuation increment to the conductivity of quasi-one-dimensional system near a structure transition*, Sov, Phys. JETP Lett. **20**, 16 (1974)

[6e] L. P. Gor'kov, *Quasi-One-Dimensional Organic Metals: Theory and Experiment*, J. Phys. I (France) **6**, 1697 (1996)

[6f] L. P. Gor'kov and D. Jerome, *Back to the problem of the upper critical field in organic superconductors*, J. Physique Lett, **46**, L-642 (1985)

[6g] L. P. Gor'kov and A. G Lebed', *On the stability of the quasi-one-dimensional metallic phase in magnetic field against the spin density wave formation*, J. Phys. Lett. **45**, L-433 (1984)

[6h] L. P. Gor'kov and A. G Lebed', *Cyclotron Resonance on Open Orbits in Organic Materials*, Phys. Rev. Lett. **71**, 3874 (1993)

[6i] L. Balicas, J.S. Brooks, K. Storr, S. Uji, M. Tokumoto, H. Tanaka, H. Kobayashi, A. Kobayashi, V. Barzykin, and L. P. Gor'kov, *Superconductivity in an Organic Insulator at Very High Magnetic Fields*, Phys. Rev. Lett. **87**, 067002 (2001)

CONCERNING ONE-DIMENSIONAL SUPERCONDUCTIVITY

Yu. A. Bychkov, L. P. Gor'kov, and I. E. Dzyaloshinskii
Institute of Physics Problems, USSR Academy of Sciences; Institute of Theoretical Physics, USSR Academy of Sciences
Submitted 15 June 1965

The publication of Little's paper [1] has evoked great interest in one-dimensional conducting systems, such as the molecules of certain linear polymers. We have therefore investigated the model of a monovalent metal, taking into account both the interaction between the electrons, due to Coulomb forces and polarization [1], and the interaction between electrons and ions (phonons). It is important here that our system, being one-dimensional, is at the same time sufficiently "thick" with respect to the Coulomb forces. In this sense it is similar to the molecules of a polymer, in which the transverse dimension of the main chain always amounts to several Angstroms, while the Debye radius does not exceed 1 Å [1]. The plasma frequency ω_p of such a system does not differ in order of magnitude from ω_p of a three-dimensional metal. Low-frequency oscillations of the electron density are thus forbidden, and this enables us to disregard completely electron interactions with small momentum transfer.

We chose an interaction with large momentum transfer in the form

$$\Gamma^{(0)}_{\alpha\beta,\gamma\delta}(p_1p_2,p_3p_4) = g(\delta_{\alpha\gamma}\delta_{\beta\delta} - \delta_{\alpha\delta}\delta_{\beta\gamma}) - \lambda^2\delta_{\alpha\gamma}\delta_{\beta\delta} \qquad (1)$$

The constant g describes the Coulomb interaction and the Little interaction [1], and is effective only at interatomic distances, so that the corresponding integrals are cut off at atomic energies ϵ_0. λ is the constant of electron-phonon interaction which, being long-range, is cut off at the Debye frequency ω_D. The constants λ and g were assumed to be small, and we retained in the calculations only terms of order $g^{n+1}\ln^n \epsilon_0/T_c$ (or $\lambda^{2n+2}\ln^n \omega_D/T_c$), where T_c is the transition temperature. A detailed analysis, the reasoning of which is briefly described below, has shown the following:

1) In the presence of effective attraction, the system always goes over into the superconducting state. The transition temperature (or the energy gap Δ_{eff}, which coincides with it in our approximation) is equal to

$$T_c = \epsilon_0 \exp(\pi v/g) \qquad \text{if } (g/\pi v)\ln(\epsilon_0/\omega_D) < -1$$

$$T_c = \omega_D \exp[\pi v/(\bar{g} - \lambda^2)] \qquad \text{if } (g/\pi v)\ln(\epsilon_0/\omega_D) > -1 \qquad (2)$$

$$\bar{g} = g/[1 + (g/\pi v)\ln(\epsilon_0/\omega_D)$$

where v is the velocity on the Fermi surface.

2) The superconducting transition is of necessity accompanied by doubling of the lattice period, i.e., by a real shift of the ions. Therefore the effective gap is given by $\Delta^2_{eff} = |\Delta|^2 + |\delta|^2$, where Δ is due to Cooper pairing and δ to the doubling of the period. The superconducting current is in this case proportional to

$$I_{super} \sim |\Delta|^2/\Delta^2_{eff}$$

and thus vanishes in the absence of pairing.

3) If the interaction is an effective repulsion ($g > 0$, $\bar{g} - \lambda^2 > 0$), the system remains a normal metal at all temperatures.

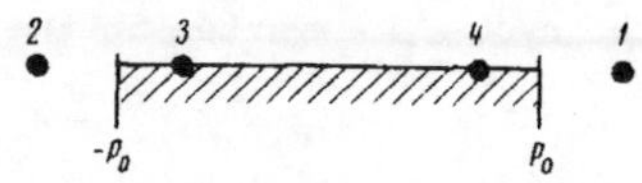

Fig. 1

Doubling of the period in the superconducting transition is a distinctive feature of one-dimensional systems. Whereas in the three-dimensional case it is essential, in order for superconductivity to occur, to have only interaction of two electrons (or two holes) lying in the Fermi surface, in the one-dimensional case it is just as important to have interaction between electron 1 and hole 3 (electron 2 and hole 4) (Fig. 1). Thus, a specific feature of one-dimensional superconductivity will be not pairs, but groups of four. In the diagram language this means that along with the Cooper diagram of Fig. 2a, the order of magnitude of which at small $p_1 + p_2$ is $g^2 \ln \epsilon_0/(p_1 + p_2)$, it is necessary to take into account also a diagram which has at $p_1 - p_3 \approx 2p_0$ the same order $[g^2 \ln \epsilon_0/(p_1 - p_3 - 2p_0)]$. It is easy to understand that in our approximation when $g \ln \epsilon_0/(p_1 + p_2) \sim v$, all the diagrams of the type of Fig. 2c will also be important. A distinguishing feature of this class of diagrams is that any of them can be divided into two parts by cutting two internal lines.

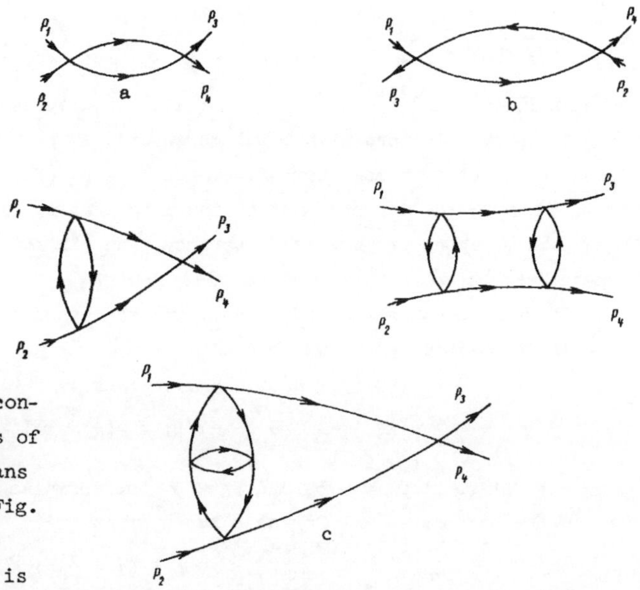

Fig. 2

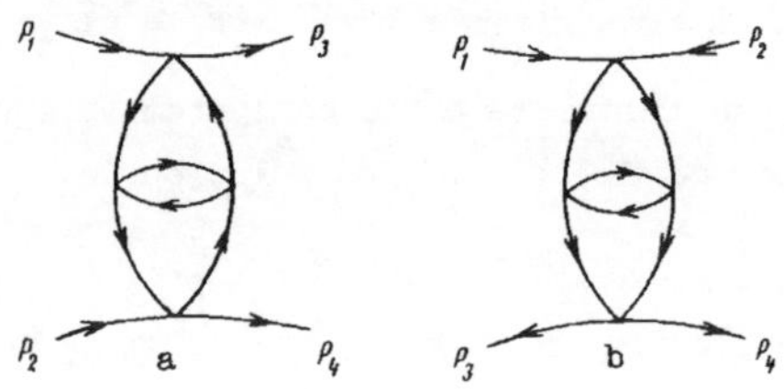

Fig. 3

The determination of the sum of all diagrams of this type, the so-called "parquet," reduces to a solution of the system of nonlinear integral equations derived by Dyatlov, Sudakov, and Ter-Martirosyan [2]. The concrete form of the system depends on the spin structure of the interaction. In our case, when the spin dependence of the total vertex part at $p_1p_4 \sim -p_2p_3 \sim p_0$ is determined by the formula

$$\Gamma_{\alpha\beta,\gamma\delta}(p_1p_2,p_3p_4) = \gamma_1(\delta_{\alpha\gamma}\delta_{\beta\delta} - \delta_{\alpha\delta}\delta_{\beta\gamma}) + \gamma_2\delta_{\alpha\delta}\delta_{\beta\gamma},$$

the system is written in the form

$$\gamma_{2,3}(\eta,\xi) = s_{2,3}(\eta) + s_{2,3}(\eta)\int_{\eta}^{\xi}\gamma_{2,3}(\zeta,\xi)d\zeta + \int_{0}^{\eta}s_{2,3}(\zeta)\gamma_{2,3}(\zeta,\xi)d\zeta, (\eta < \xi);$$

$$\gamma_{\pm}(\xi,\eta) = \sigma_{\pm}(\xi) \mp \sigma_{\pm}\int_{\xi}^{\eta}\gamma_{\pm}(\zeta,\eta)d\zeta \mp \int_{0}^{\xi}\sigma_{\pm}(\zeta)\gamma_{\pm}(\zeta,\eta)d\zeta, \qquad (\eta > \xi);$$

$$\gamma_1(\xi,\xi) = s_1(\xi) + \sigma_1(\xi) - \gamma_1^{(0)}, \quad \gamma_2(\xi,\xi) = s_2(\xi) + \sigma_2(\xi) + \sigma_2(\xi) - \gamma_2^{(0)} \quad (\eta = \xi)$$

Here

$$\gamma_3 = -2\gamma_1 + \gamma_2, \qquad \gamma_{\pm} = \gamma_1 \pm \gamma_2;$$

$$\xi = (1/2\pi v)\ln \epsilon_0/v|p_1 + p_2|, \qquad \eta = (1/2\pi v)\ln \epsilon_0/v|p_1 - p_3 - 2p_0|$$

The quantities s and σ are the sums of all the diagrams which can not be cut respectively along two parallel lines with total momentum $p_1 + p_2$ (for example, Fig. 3a) and along two antiparallel lines with momentum difference $p_1 - p_3$ (Fig. 3b). In the definition of the quantities ξ and η, the momenta must be taken to mean the largest of the three following quantities: the momentum proper, the frequency, and (in the case of diagram technique at $T \neq 0$) the temperature.

The system can be solved exactly by the method proposed by Ansel'm [3]. At temperatures above the transition point we obtain

$$\gamma_1(\xi,\xi) = g/(1 + 2g\xi), \quad \gamma_2(\xi,\xi) = (1/2)[\gamma_1(\xi,\xi) - g], \quad \xi < \xi_0$$

$$\gamma_1(\xi,\xi) = (\bar{g} - \lambda^2)/[1 + 2(\bar{g} - \lambda^2)(\xi - \xi_0)], \quad \xi > \xi_0 = (1/2\pi v)\ln \epsilon_0/\omega_D \tag{4}$$

From this follow at $p_1 = -p_2$ and $p_3 = p_0$ the formulas given above for T_c. The Green's function of the phonons at $k = 2p_0$ also has a singularity at the point $T = T_c$

$$D \sim (I + 2g\xi)^{-3/4} \tag{5}$$

At absolute zero, the superconductor is described by three Green's functions:

$$G = -i\langle Ta_p a_p^+\rangle, \qquad F^+ = \langle Ta_p^+ a_p^+\rangle, \qquad \bar{G} = -i\langle Ta_p a^+_{p-2p_0}\rangle \tag{6}$$

for which we can write a closed system of equations analogous to the three-dimensional case (see, for example, [4]). Its solution is

$$G = (\omega + \epsilon_{0p})/(\omega^2 + E_p^2), \quad F^+ = [-i\Delta^*(p)]/(\omega^2 + E_p^2), \quad \bar{G} = [i\delta(p)]/(\omega^2 + E_p^2);$$

$$E_p^2 = \epsilon_{0p}^2 + |\Delta(p)|^2 + |\delta(p)|^2 \tag{7}$$

The functions Δ and δ satisfy the integral equations ($\xi_{eff} = (1/2\pi v)\ln(\epsilon_0/\Delta_{eff})$)

$$\Delta[\eta = (1/2\pi v)\ln(\epsilon_0/v|p - p_0|)] = s(\eta)\int_{\eta}^{\xi_{eff}}\Delta(\zeta)d\zeta + \int_{0}^{\eta}s_3(\zeta)\Delta(\zeta)d\zeta,$$

$$\delta[\eta = (1/2\pi v)\ln(\epsilon_0/v|p - p_0|)] = \sigma_+(\eta)\int_{\eta}^{\xi_{eff}}\delta(\zeta)d\zeta - \int_{0}^{\eta}\sigma_+(\zeta)\delta(\zeta)d\zeta \tag{8}$$

which have nonvanishing solutions only when $1 + 2g\xi_{eff} = 0$.

The system of equations for G, F^+, and $\bar{G}$ enables us to find the current flowing in the superconductor as a result of application of an alternating electric field E. Calculations with account of the specific nature of the Coulomb forces (see [4], Sec. 37) yield

$$I(\omega) = i(2e^2v/\pi\omega)(|\Delta|^2/\Delta^2_{eff})\ E(\omega) \tag{9}$$

We note, finally, that Ferrell's arguments [5] against the existence of one-dimensional superconductors do not seem convincing to us. It can be shown (this will be done in a detailed article) that the presence of ion oscillations (there are no electron oscillations in our system) will merely cause the one-dimensional superconductor to be always similar to an alloy.

[1] W. A. Little, Phys. Rev. 134, A1416 (1964).

[2] Dyatlov, Sudakov, and Ter-Martirosyan, JETP 32, 767 (1957), Soviet Phys. JETP 5, 631 (1957).

[3] A. A. Ansel'm, JETP 38, 1887 (1960), Soviet Phys. JETP 11, 1356 (1960).

[4] Abrikosov, Gor'kov, and Dzyaloshinskii, Metody kvantovoi teorii polya v staticheskoi fizike (Methods of Quantum Field Theory in Statistical Physics), Fizmatgiz, 1962.

[5] R. A. Ferrell, Phys. Rev. Lett. 13, 330 (1964).

ERRATUM

Figure 1 of the article by Yu. Kagan and V. Sobakin, "Anisotropy of Odd Photomagnetic effect in Semiconductors of Cubic Symmetry," in JETP Letters 2, 46 (1965), was erroneously identified in the caption as pertaining to the even photomagnetic effect, in place of the odd effect.

Possible phase transitions in systems of interacting metallic filaments (quasiunidimensional metals)

L. P. Gor'kov and I. E. Dzyaloshinskiĭ

L. D. Landau Institute of Theoretical Physics, USSR Academy of Sciences

(Submitted March 11, 1974)
Zh. Eksp. Teor. Fiz. 67, 397–417 (July 1974)

The effects of interactions between filaments on the properties of quasiunidimensional metals are discussed. It is shown that, together with the state considered previously, [9,10] in which dielectric and superconductive pairing coexist on a single filament, the interaction makes possible either a purely dielectric state of the Peierls type or a mixed state in which the superconducting and dielectric gaps Δ^{ik} are essential functions of the filament numbers i and k. The influence of transverse motion on the transition temperature and the possibility of superconducting states are taken into account.

1. INTRODUCTION AND CHOICE OF MODEL

Experimental studies in recent years have reawakened interest in the properties of one-dimensional systems. The most relevant of these are experiments with complexes based on TCNQ and variable-valency complexes based on Pt and Ir (see, for instance, the review paper by Shchegolev [1]). It has been found that for the set of filamentary-structure substances Qn-TCNQ [1], NMP-TCNQ [2], and TTF-TCNQ [3] the temperature dependence of the conductivity is metallic in character. At room temperature the conductivity reaches a value of about 10^2–10^3 $\text{ohm}^{-1}\cdot\text{cm}^{-1}$ with a very high anisotropy ($\sim 10^2$–10^3). In addition, these substances display an unusual "one-dimensional" metal-dielectric transition, accompanied by anomalous damping of the phonon mode [4].

The existence of dynamic instabilities in a one-dimensional metal has been discussed many times. Instability with respect to ion shifts and damping of the phonon mode was discussed by Peierls [5] and Froehlich [6] (see also Afanas'ev and Kagan [7]), and antiferromagnetic instability (giant spin waves) by Overhauser [8]. It was shown by Bychkov and the present authors [9] that the Peierls instability is always accompanied by an instability of the BCS type, with the formation of Cooper pairs. It was explained in [9] that the formulas of Froehlich and Overhauser are incorrect by themselves. They correspond to the Hartree-Fock approximation (or, what is the same thing, the "ladder" approximation), which is inapplicable in the one-dimensional case and leads to a transition with the incorrect sign of the interaction. From [9] (the so-called "parquet" approximation) there follows the simultaneous appearance of dielectric (Peierls) and superconducting instabilities. It was later shown by Larkin and one of the present authors [10] that the antiferromagnetic instability can appear only in the case of a single electron in the unit cell, when Umklapp processes are significant in the one-dimensional case.

However, the even more rigorous parquet approximation [9,10] has extremely limited accuracy in the one-dimensional case. As a phenomenological theory of phase transitions, it corresponds to the molecular field approximation. When fluctuations are included in the one-dimensional case, the transition becomes badly smeared out. The formulas of [9,10] have numerical meaning in the region $T - T_{c0} \gtrsim T_{c0}$ (T_{c0} is the temperature "according to molecular field theory").

In the present work we intend to take into account the various three-dimensional effects that, on the one hand, make the "true" transition temperature finite, and on the other permit us to determine the region $|T - T_c| \lesssim T_c$ in which theories of the Ginzburg-Landau type are applicable.

We consider a system of parallel metallic filaments packed into a plane lattice with distance a between filaments. In the first approximation, when in principle it should not lead to an incorrect answer, we may neglect electron hopping from one filament to another[1)].

For the Green's function of an electron in the temperature method, and for all the correlation functions in the following, it is convenient to use the mixed representation that indicates the momentum p along the filament and the number i, k, l, ... of the corresponding filament. In this representation

$$\mathfrak{G}^{ik}_{\alpha\beta}(p) = \frac{\delta_{ik}\delta_{\alpha\beta}}{i\varepsilon_n - v_F(|p| - p_F)};$$

here α and β are spin indices, and $\epsilon_n = 2\pi(n + ½)T$. On each filament the region between $-p_F$ and p_F in the one-dimensional reciprocal cell $(-\pi/c, \pi/c)$ is filled by electrons (Fig. 1), so that the total three-dimensional electron density is $n = 2p_F/a^2\pi$. In the case when there is exactly one electron per period along the filament, $p_F = \pi/2c$.

If the electron hopping is significant we must transform from the "node" representation i, k, l, ... to the transverse quasimomentum p. Then we can write

$$\mathfrak{G}_{\alpha\beta}(p, p_\perp) = \frac{\delta_{\alpha\beta}}{i\varepsilon_n - v_F(|p| - p_F) - w(p_x, p_y)},$$

where w is the "hopping energy," which is small compared to the Fermi energy: $\epsilon_F = v_F p_F \gg w$. In the three-dimensional reciprocal cell the absence of hopping corresponds to a plane Fermi surface, and the inclusion of hopping leads to a weak "corrugation" of the Fermi plane.

Even when the electrons move only along their respective filaments, their interactions can still be significant (cf. [1]). These include both the direct Coulomb interaction and the indirect interaction through phonons, since the filaments are densely bound ($a \sim c$) in an elastic matrix and the phonon spectrum remains, in principle, three-dimensional. Non-phonon exchange mechanisms of a different type also lead to interactions between separate filaments [11,12]. In our model "phonon interactions" will be those that are cut off at energies higher than the Debye energy ω_D. Non-phonon interactions are retained up to an energy $\bar{\omega}$ on the order of the Fermi energy ϵ_F.

2. BASIC EQUATIONS (NON-PHONON INTERACTIONS)

From the mathematical point of view, the principal difference between the one-dimensional problem of the electron-spectrum instability and the BCS theory of superconductivity is the existence of two special scattering channels. The first (Cooper) channel (Fig. 2a) corresponds to the coupling of two electrons at opposite ends of the Fermi surface (Fig. 1), and contributes a logarithmic integral of order $\ln(\overline{\omega}/T)$. The same logarithm is given by the second (Peierls) channel (Fig. 2b), which corresponds to the pairing of electrons with a hole at the opposite end of the Fermi surface. The anomalous scattering in this channel is characterized by momentum $2p_F$, and corresponds to Peierls doubling of the period[5,6] and a strong Kohn anomaly in the phonon spectrum[7].

In the model with a weak interaction the summation of terms of the form $(g\ln(\overline{\omega}/T))^n$ leads, owing to the mutual insertion of diagrams 2a and 2b into one another, to a system of parquet equations. We shall select the bare interactions as shown in Fig. 3. The difference from [10] consists in the fact that the constants g_1^{ik} and g_2^{ik} are now essential functions of the distance between filaments.

In the case of exactly one electron per (one-dimensional) cell, $p_F = \pi/2c$, as in [10], logarithmic terms are given not only by the diagrams of Figs. 2a and 2b, but also by the diagrams related to the Umklapp processes $p_1 + p_2 = p_3 + p_4 \pm 2\pi/c$. These processes are described by the additional constant g_3^{ik}.

When hopping is neglected, we can explicitly separate the dependence of the vertex parts Γ on the spins $(\alpha\beta\gamma\delta)$ and the filament numbers $(ikl m)$:

$$\Gamma_{\alpha\beta\gamma\delta}^{iklm}\ (+--+)=\gamma_1^{ik}\delta_{\alpha\gamma}\delta_{\beta\delta}\delta_{il}\delta_{km}-\gamma_2^{ik}\delta_{\alpha\delta}\delta_{\beta\gamma}\delta_{im}\delta_{kl}, \tag{1}$$

where the $\pm$ signs correspond to electrons lying at one or another of the "Fermi points" $\pm p_F$; the functions γ_1^{ik} and γ_2^{ik} depend only on the longitudinal momenta and the frequencies. If scattering with Umklapp is possible, then together with the amplitude Γ in (1), which conserves the total momentum, we must also, as in [10], take into account the vertex part

$$\Gamma_{\alpha\beta\gamma\delta}^{iklm}\ (++--)=\gamma_3^{ik}(\delta_{\alpha\gamma}\delta_{\beta\delta}\delta_{il}\delta_{km}-\delta_{\alpha\delta}\delta_{\beta\gamma}\delta_{im}\delta_{kl}). \tag{2}$$

The fundamental scheme of the parquet equations is well known and has been described many times. In addition to [9], a detailed discussion is given in the first work by Dyatlov et al. [13] and in the article by Nozieres et al. [14], where the procedure suggested by Sudakov [13] is analyzed in detail.

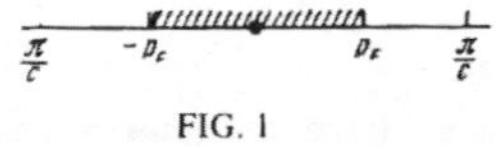

FIG. 1

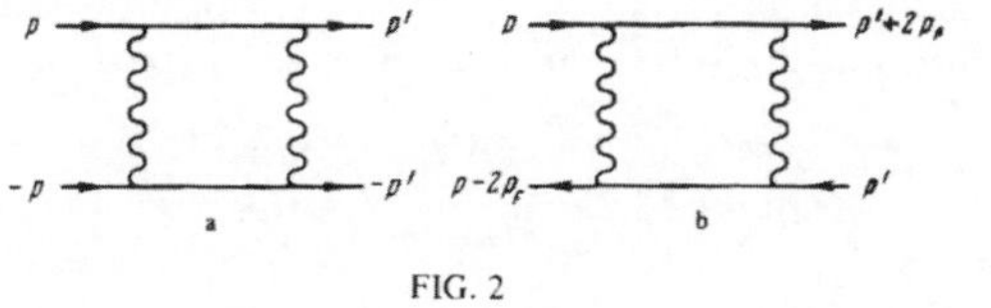

FIG. 2

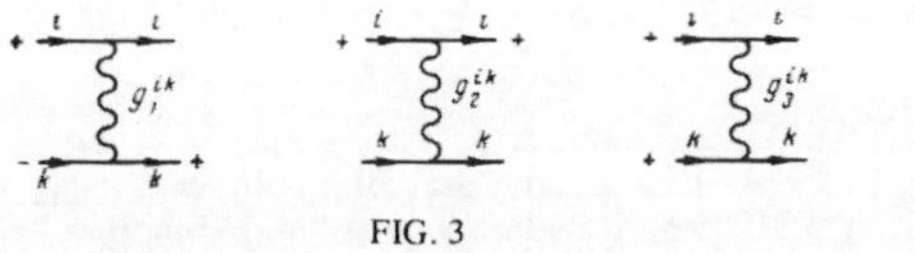

FIG. 3

Assuming that all the external momenta (in the sense of their distance $\Delta p = |p| - p_F$ from the corresponding Fermi point) and the frequencies ϵ_n are of the same order of magnitude, $v_F\Delta p \sim \epsilon_n$, we can verify that all the diagrams in the logarithmic approximation depend only on the single variable

$$\xi = \ln\frac{\omega}{\max\{v_F\Delta p, T\}}.$$

The total vertex parts $\gamma_1(\xi)$ and $\gamma_2(\xi)$ are the sums of the bare amplitudes $g_{1,2}$ and the amplitudes that describe multiple scattering in the Cooper (C) and Peierls (P) channels:

$$\gamma_{1,2}^{ik}=g_{1,2}^{ik}+C_{1,2}^{ik}+P_{1,2}^{ik}. \tag{3}$$

The Sudakov method (see [10,13,14]) gives expressions for C and P in terms of γ:

$$C_1^{ik}(\xi)=-\int_0^\xi d\zeta\,\gamma_1^{ik}(\zeta)\gamma_2^{ik}(\zeta),$$

$$C_2^{ik}(\xi)=-\frac{1}{2}\int_0^\xi[(\gamma_1^{ik}(\zeta))^2+(\gamma_2^{ik}(\zeta))^2]d\zeta,$$

$$P_1^{ik}(\xi)=\int_0^\xi d\zeta[\gamma_2^{ii}(\zeta)\gamma_1^{ik}(\zeta)-\sum_l\gamma_1^{il}(\zeta)\gamma_1^{lk}(\zeta)], \tag{4}$$

$$P_2^{ik}(\xi)=\frac{1}{2}\int_0^\xi(\gamma_2^{ik}(\zeta))^2d\zeta.$$

In the case of a single electron per cell ($g_3 \neq 0$) the equations become even more cumbersome. We therefore write down at once the equations for γ:

$$\gamma_1^{ik}=g_1^{ik}-\int_0^\xi d\zeta\Big[\gamma_1^{ik}\gamma_2^{ik}-\gamma_1^{ik}\gamma_2^{ii}+\sum_l\gamma_1^{il}\gamma_1^{lk}+(\gamma_3^{ik})^2-\sum_l\gamma_3^{il}\gamma_3^{lk}\Big],$$

$$\gamma_2^{ik}=g_2^{ik}-\frac{1}{2}\int_0^\xi d\zeta[(\gamma_1^{ik})^2-(\gamma_3^{ik})^2], \tag{5}$$

$$\gamma_3^{ik}=g_3^{ik}+\int_0^\xi d\zeta\Big[\gamma_2^{ik}\gamma_3^{ik}+\gamma_2^{ii}\gamma_3^{ik}+\gamma_1^{ik}\gamma_3^{ii}-\sum_l(\gamma_1^{il}\gamma_3^{lk}+\gamma_3^{il}\gamma_1^{lk})\Big].$$

The new equations (3), (4), and (5) are very complex and cannot be solved in general form. In the absence of interactions between the filaments ($g^{ik} = 0$ for $i \neq k$) they transform into the solutions that we have already discussed[9,10]. We recall that in the absence of Umklapp ($g_3 = 0$) these solutions are of the form

$$\gamma_1=\frac{g_1}{1+g_1\xi},\quad \gamma_2=g_2-\frac{1}{2}\frac{g_1^2\xi}{1+g_1\xi}. \tag{6}$$

In the case of attraction ($g_1 < 0$) both vertices have pole singularities[2)]: $\gamma_1 \approx 2\gamma_2 \approx 1/(\xi - \xi_0)$ with $\xi_0 = 1/|g_1|$, which corresponds to a transition temperature $T_{c0} = \overline{\omega}\exp(-1/|g_1|)$.

An analysis of the system (3), (4) shows that the solution (6), which is exact in the absence of interactions between the filaments, becomes unstable even with the inclusion of an arbitrarily weak interaction.

The system of integral equations (3) and (4) is in fact equivalent to the following system of differential equations:

$$\frac{d\gamma_1^{ik}}{d\xi} = -\gamma_1^{ik}\gamma_2^{ik} + \gamma_1^{ik}\gamma_2^{ii} - \sum_l \gamma_1^{il}\gamma_1^{lk},$$
$$\frac{d\gamma_2^{ik}}{d\xi} = -\frac{1}{2}(\gamma_1^{ik})^2 \tag{7}$$

with the "initial" conditions

$$\gamma_1^{ik}(0) = g_1^{ik}, \quad \gamma_2^{ik}(0) = g_2^{ik}. \tag{8}$$

Since the right side of the system (7) does not depend explicitly on ξ and is a homogeneous quadratic function of γ, it can, in principle, have stationary singular solutions of the form

$$\gamma_1^{ik} = \frac{A_{ik}}{\xi - \xi_0}, \quad \gamma_2^{ik} = \frac{B_{ik}}{\xi - \xi_0}, \tag{9}$$

with arbitrary ξ_0, among which is Eq. (6) with

$$A_{ik} = 2B_{ik} = \delta_{ik}. \tag{10}$$

The purely one-dimensional[9] solution (6) corresponds to the special initial conditions $g_1^{ik} = g_1\delta^{ik}$ and $g_2^{ik} = g_2\delta^{ik}$.

Let $g^{ik} \neq 0$ and $|g^{ik}| \ll |g^{ii}|$. In the approximation linear in γ^{ik} ($i \neq k$) the equations for γ^{ii} on a single filament do not change form:

$$\frac{d\gamma_1^{ii}}{d\xi} = -(\gamma_1^{ii})^2, \quad \frac{d\gamma_2^{ii}}{d\xi} = -\frac{1}{2}(\gamma_1^{ii})^2,$$

so that near (10) we again have

$$\gamma_1^{ii} = 2\gamma_2^{ii} = \frac{1}{\xi - \bar{\xi}_0} \equiv \gamma$$

with a shifted pole $\xi_0 \to \bar{\xi}_0$. For small values of γ^{ik} ($i \neq k$) we have

$$\frac{d\gamma_1^{ik}}{d\gamma} = \frac{3}{2}\frac{\gamma_1^{ik}}{\gamma}, \quad \gamma_2^{ik} = 0.$$

Accordingly, the equation of a typical trajectory is $\gamma_1^{ik} = \text{const} \cdot \gamma^{3/2}$. A plot of the trajectories using the variables γ_1^{ik}/γ and γ^{-1} is given in Fig. 4. The one-dimensional system corresponds to the abscissa axis in Fig. 4. It is apparent that the system can reach the "one-dimensional" singular point (9), (10) only from this axis. Physically this means that the singularities in the vertex part on separate filaments are not independent.

It is not hard to show that the "one-dimensional" solutions of the system (5) are also unstable in the presence of Umklapp[10].

In addition to special solutions of the type (9) with a pole ξ_0 which does not depend on the coordinates i and k of the filaments (in the following we shall call such poles "stationary"), the system (7) has solutions of a different form. These are the so-called "moving" poles discussed earlier by one of the present authors with Katz and Brazovskiĭ[15]. In seeking them it is more convenient to rewrite the system (7) in the two-dimensional momentum

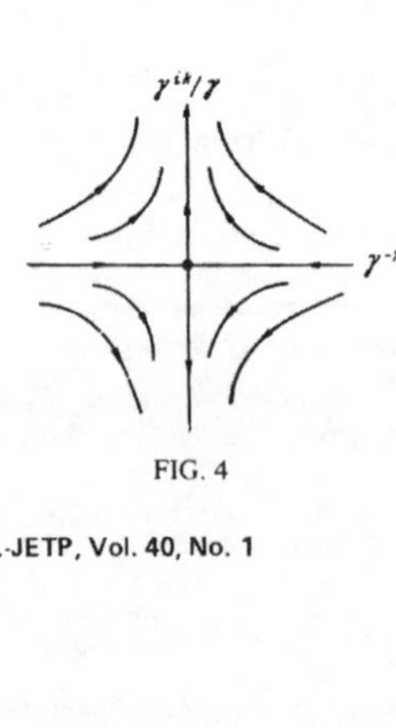

FIG. 4

representation q with respect to the differences ρ_{ik} of the two-dimensional coordinates of the filaments:

$$\gamma(q) = \sum_k \gamma^{ik}\exp(iq\rho_{ik}), \quad \gamma^{ik} = \int_S \gamma(q)\exp(-iq\rho_{ik})\frac{d^2q}{S},$$

where S is the area of the reciprocal cell (for a square, $S = 4\pi^2/a^2$). We have

$$\frac{d\gamma_1(q)}{d\xi} = -\gamma_1^2(q) + \gamma_1(q)\int\gamma_2(q_1)\frac{d^2q_1}{S} - \int\gamma_1(q-q_1)\gamma_2(q_1)\frac{d^2q_1}{S},$$
$$\frac{d\gamma_2(q)}{d\xi} = -\frac{1}{2}\int\gamma_1(q-q_1)\gamma_1(q_1)\frac{d^2q_1}{S}. \tag{11}$$

System (11) has a special solution with a polar singularity

$$\gamma_1(q) \approx \frac{1}{\xi - \xi_0(q)}, \quad \gamma_2 \approx 0, \tag{12}$$

where $\xi_0(q)$ is an essential function of the momentum and is determined by matching with the initial conditions[3)] (8). The transition occurs at the point $\xi = \ln(\bar{\omega}/T_{c0}) = \xi_0(q_0)$, $\xi_0(q_0) = \min \xi_0(q)$. Assuming that near $\xi_0(q_0)$ the "trajectory" of the pole $\xi_0(q)$ is of the form

$$\xi_0(q) = \xi_0(q_0) + b(q - q_0)^2, \quad b > 0$$

(see footnote 3), we can verify that the increments of γ_1 and γ_2 from Eq. (12) near $\xi = \xi_0(q_0)$ have weaker, nonpole singularities (for additional details see[15]).

The character of the transition determined by the moving pole can be assessed from the fact that Eq. (12) corresponds to a simple pole (with residue 1!) in the Peierls channel P_1 in Eq. (4) and to weaker singularities in the channel P_2 and in both Cooper channels. In accordance with[9,10], this means that the transition has the purely dielectric character of a Peierls instability. In this case the phonon frequency ω also vanishes ($q_{\parallel} = 2p_F$, $q_{\perp} = q_0$) (see[9,10] and Sec. 4). However, this does not mean that the result could have been obtained by a simple summation of ladder diagrams; although the pole is only in the P channel and its residue is 1, its formation and the trajectory of $\xi_0(q)$, and in particular the transition temperature T_{c0}, are determined by all the parquet diagrams[4)].

Let us now turn to the special solutions of the "standing pole" type (9). We write the equations for the residues A_{ik} and B_{ik}:

$$A_{ik} = A_{ik}B_{ik} - A_{ik}B_{ii} + \sum_l A_{il}A_{lk}, \quad B_{ik} = \frac{1}{2}A_{ik}^2 \tag{13}$$

or

$$A_{ik} = \frac{1}{2}A_{ik}^3 - \frac{1}{2}A_{ik}A_{ii}^2 + \sum_l A_{il}A_{lk}.$$

System (13) does not have any arbitrary parameters of the problem. However, it can be shown[5)] that it has a two-parameter family of solutions. For this purpose we change in Eq. (13) to the momentum representation:

$$A(q) = A^2(q) - \frac{1}{2}A(q)\left[\int A(q_1)\frac{d^2q_1}{S}\right]^2$$
$$+ \frac{1}{2}\iint A(q_1)A(q_2)A(q-q_1-q_2)\frac{d^2q_1}{S}\frac{d^2q_2}{S} \tag{14}$$

and as the zeroth approximation we use the solutions of the equation

$$A_0(q) = A_0^2(q). \tag{15}$$

Besides the two obvious solutions $A_0(q) \equiv 1$ (corresponding to a one-dimensional standing pole) and $A_0(q) \equiv 0$, Eq. (14) also has a wide class of "step-function" solutions in which $A_0(q)$ equals 1 in some portions of the reciprocal cell and zero in the remainder. The "elementary" solution of this type is the "unit function"

$$A_0(q) = \theta_\lambda(\mathbf{q}), \quad \theta_\lambda(\mathbf{q}) = \theta_{\lambda x}(q_x)\theta_{\lambda y}(q_y);$$

$$\theta_\lambda(q) = \begin{cases} 1, & |q| < \lambda/2 \\ 0, & |q| > \lambda/2 \end{cases}. \tag{16}$$

It is easy to see that if the width λ is small compared to the cell dimension $2\pi/a$, then, as in the case of a moving pole, the neglected terms in Eq. (14) will only provide a small correction of order $(\lambda a)^4$ to solution (16). The situation remains the same for any superposition of the elementary solutions:

$$\theta_{\lambda_0}(\mathbf{q}) + \theta_{\lambda_1}(\mathbf{q} \pm \mathbf{q}_{01}) + \theta_{\lambda_2}(\mathbf{q} \pm \mathbf{q}_{02}) + \ldots, \tag{16'}$$

when the condition $\lambda_i \ll |q_{0i} - q_{0k}| \lesssim 2\pi/a$ is satisfied, i.e., when the function (17) consists of a few rather narrow spikes[6]; see Fig. 5.

The "elementary" solution (16) corresponds to the establishment of long-range correlations, accompanied by slow oscillations of period $\sim 1/\lambda$. The corresponding special solution is of the form

$$\gamma_{1\lambda}^{ik} = \frac{1}{\xi - \xi_0} \frac{4 \sin(\frac{1}{2}\lambda_x x_{ik}) \sin(\frac{1}{2}\lambda_y y_{ik})}{S x_{ik} y_{ik}}. \tag{17}$$

Although solution (16) is of interest primarily as a proof of the existence of solutions to the general system (13), it is also interesting to consider the nature of the transition to which it corresponds. According to Eq. (4), at small distances there exists a pole $\sim\lambda^4/(\xi - \xi_0)$ in the Cooper channel, and at large distances, where $|x_{ik}|$, $|y_{ik}| \gtrsim 1/\lambda$, a pole with residue 1 exists in the P channel. Thus, in this extremely peculiar state there exists, against the background of a large dielectric gap (Peierls instability), a superconducting correlation that is small in magnitude but drops off slowly according to a power law.

In addition to these solutions which are "close" to the moving dielectric pole, there exists also a two-parameter set of solutions close to the "one-dimensional standing pole" (10). By making the change of variable $A(q) = 1 - \frac{1}{2}C(q)$ in Eq. (15) we can show that the "elementary" solution is of the form

$$A_0(q) = 1 - \tfrac{1}{2}\theta_\lambda(\mathbf{q}). \tag{18}$$

The difference now is that the small increments to the solution are of another order of magnitude, $\sim\lambda^2 a^2$. The solutions of type (16′) are constructed similarly. The elementary solution (18) corresponds to pairing in both the Cooper and dielectric channels, when all four particles lie on a single filament. In addition, according to (18), (17), and (16), long-range Cooper and Peierls correlations of type (17) can be superimposed on these "one-dimensional" pairings.

As we would naturally expect, states such as (18) are unstable in the same sense as the purely one-dimensional states (10) (see Appendix I). However, since the number of such solutions is "continuous," there exist in principle "continuous" regions of the initial data (8) from which these special points can be attained[7].

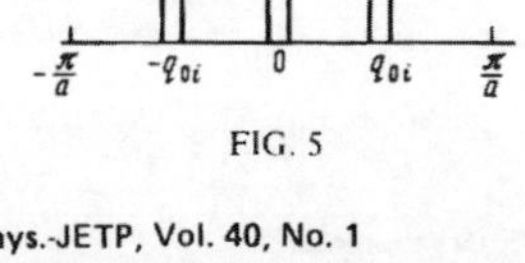

FIG. 5

Investigating the stability of solutions (16) and (16′) is a more complex problem. For a two-dimensional system of filaments such an investigation can only be carried out numerically. We shall therefore consider only the case of a one-dimensional filament system, for which analytical methods are completely sufficient (see Appendix I). Only the elementary solution $\theta_\lambda(\mathbf{q})$ in Eq. (16) is found to be stable; all the combined solutions of type (16′) are unstable, although, as in the case of the nearly one-dimensional solutions (18), they can be reached from a "continuum" of initial conditions (see, however, [7]). The structure of the problem concerning the stability of solutions (16) and (16′) leaves practically no doubt that the elementary solution (16) also remains stable in the case of a real two-dimensional system of filaments.

3. EFFECTS OF HOPPING. THE MOVING COOPER POLE

We now discuss the role of electron hopping. Neglecting the overlap integrals between the filaments, whose characteristic value we denote by V_{12}, the three-dimensional electron spectrum is flat. The inclusion of hopping distorts the Fermi surface of the electrons, leading to a finite effective mass in the transverse direction. The amount of this distortion, in energy units, is of order $w \approx V_{12}^2/\epsilon_F$. At low temperatures the deviation of the Fermi surface from a plane affects the values of the logarithmic integrals in the Peierls channel in Fig. 2b. Actually, the logarithmic cutoff at the lower limit in the Cooper channel at $\mathbf{k} = \mathbf{p}_1 + \mathbf{p}_2 = 0$ always occurs at $v_F \Delta p = T$ because of the invariance of the spectrum under time inversion, $\epsilon(-\mathbf{p}) = \epsilon(\mathbf{p})$. In the Peierls channel the appearance of logarithmic integrals is due to the degenerate form of the electron dispersion law. Thus at small w the overlap between filaments enters into the parquet equations only as the cutoff limit of the Peierls cross sections; it becomes significant when w is comparable (on the logarithmic scale) with the transition temperature specified by the pole in Eq. (6), (9), or (12). We may therefore assume that w is exponentially small, so that, as before, we will neglect electron transitions from one filament to another in the initial inter-electronic interactions. The transitions enter the problem only because the electron Green's function $\mathfrak{G}^{ik}(p)$ is nondiagonal in i and k.

Consider, for example, the diagram of Fig. 6 for the vertex part with longitudinal momentum transfer $2p_F$. In principle, here $l \neq i$ and $m \neq s$. However, if we take the Fourier components with respect to the transverse momenta t, the internal part of the diagram is proportional to

$$\int U(t_1 - t')U(t' - t_4)\frac{d^2t'}{v_F} \ln\frac{\bar{\omega}}{w(t')},$$

where U corresponds to the interaction between the filaments, the integration is carried out over the three-dimensional Fermi surface (which effectively coincides with a plane), and $v_F \approx v_{Fz}$. Neglecting, with logarithmic accuracy, the dependence of $w(t')$ on the point t', we see that the vertex in Fig. 6 is nonzero only when $l = i$ and

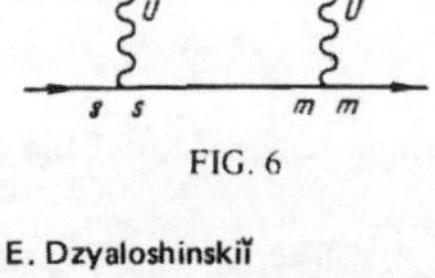

FIG. 6

$m = s$. In other words, in this approximation we still retain the picture of separate filaments that was used to derive the system (3)—(5). The only difference is that the cutoff in the Peierls channel is limited by the value of w. If $\nu = \ln(\bar{\omega}/w) > \xi_0$, where $\xi_0 = \ln(\bar{\omega}/T_0)$ is the pole of Eq. (6), (9), or (12), the jumps affect only the fluctuation region below T_0. Now let $\nu < \xi_0$. In this case the Peierls channel has no pole and there is no dielectric transition.

We introduce the Cooper vertex part $\gamma^{ik}(\nu, \xi)$, where the momenta p_1, p_2, p_3, and p_4, measured from the Fermi surface, are of order w/v_F, but $p_1 + p_2 \sim T/v_F \ll w/v_F$. Employing the usual procedure for separating the cross section in which the variable of integration is smallest, we obtain the equations

$$\gamma_1^{ik}(\nu,\xi) = \gamma_1^{ik}(\nu) - \int_\nu^\xi \gamma_1^{ik}(\nu,\zeta)\gamma_2^{ik}(\nu,\zeta)\,d\zeta,$$

$$\gamma_2^{ik}(\nu,\xi) = \gamma_2^{ik}(\nu) - \frac{1}{2}\int_\nu^\xi [(\gamma_1^{ik}(\nu,\zeta))^2 + (\gamma_2^{ik}(\nu,\zeta))^2]\,d\zeta. \tag{19}$$

By solving these we find

$$\gamma_\pm^{ik}(\nu,\xi) = [\pm{}^1/_2(\xi-\nu) + 1/\gamma_\pm^{ik}(\nu)]^{-1}, \tag{20}$$

where $\gamma_\pm = \gamma_1 \pm \gamma_2$. When $\xi < \nu$, $\gamma^{ik} \equiv \gamma^{ik}(\xi)$. The pole $\tilde{\xi}_0$ in Eq. (20) specifies the Cooper singularity if $\tilde{\xi}_0 > \nu$, i.e., if $\tilde{T}_0 < w$. In the opposite case the pole in Eq. (20) has no meaning.

There are two possible cases with respect to the transition temperature T_0 obtained by neglecting the overlap integral between filaments: $\tilde{T}_0 > T_0$ and $\tilde{T}_0 < T_0$. Thus, if ν is close to ξ_0 (on the logarithmic scale), we obtain formally from Eq. (15)

$$\tilde{\xi}_{0\pm}^{ik} = \nu \pm 2(\xi_0 - \nu)/(A_{ik} \pm {}^1/_2 A_{ik}^2). \tag{21}$$

For example, in the solution for a single filament (with the interaction between the filaments small compared to w) $A_{ik} = \delta_{ik}$:

$$\tilde{\xi}_{0+} = \nu + {}^4/_3(\xi_0 - \nu), \quad T_{c+} \sim T_{c0}(T_{c0}/w)^{1/3}.$$

The second pole in Eq. (21), which would correspond to a solution with $T_{c-} > w$, does not exist. For the general form of A_{ik}, obviously, both cases are possible. The poles $\tilde{\xi}^{ik}_{0+}$ and $\tilde{\xi}^{ik}_{0-}$ correspond to singlet and triplet pairing, respectively, where $i \neq k$ implies the combining of electrons on different filaments. The transition is specified by the largest of the temperatures in Eq. (21).

To establish a definite correspondence with the three-dimensional formulation of the BCS theory, we write in place of Eq. (20) the three-dimensional equations in the momentum representation in the following schematic form, corresponding to the ladder approximation ($\mathbf{p}_1 = -\mathbf{p}_2$):

$$\Gamma(\mathbf{p}_1, \mathbf{p}_3) = \Gamma^\nu(\mathbf{p}_1, \mathbf{p}_3) + \int_\nu^\xi \overline{\Gamma(\mathbf{p}_1, \mathbf{l})\Gamma(\mathbf{l}, \mathbf{p}_3)}\,d\zeta, \tag{22}$$

where the momenta $\mathbf{p}$ and $\mathbf{l}$ lie on the three-dimensional (nearly planar) Fermi surface, the logarithmic integration is taken along the normals to the Fermi surface, and the bar denotes the integral over this surface. Γ^ν has the same meaning as in Eq. (20). The pole part in the vertex part $\Gamma(\mathbf{p}_1, \mathbf{p}_3)$ has the usual form

$$\frac{\Delta(\mathbf{p}_1)\Delta(\mathbf{p}_3)}{\xi - \tilde{\xi}_0}$$

where $\Delta(\mathbf{p})$ determines (when $\xi > \nu$) the dependence of the resultant energy gap on the point of the Fermi surface. By establishing the correspondence, in our approximation, between Eq. (22) and the representation that uses the filament numbers, we easily find that the pole in Eq. (20) at a given $|\boldsymbol{\rho}_{ik}| = \rho(i-k)$ corresponds to the following form of $\Delta(\mathbf{p})$:

$$\Delta(\mathbf{q}) \propto \sum_{|i-k|} \exp(i\mathbf{q}\boldsymbol{\rho}_{ik}),$$

where the sum is taken over all neighbors in the two-dimensional lattice. In particular, the pole with $i = k$ corresponds to a gap that does not depend on the vector $\mathbf{q}$. Thus the standing pole of Eqs. (6) and (9), with hopping between the filaments taken into account, becomes at $w > T_{c0}$ the Cooper pole of the three-dimensional problem. If $\mathbf{p}_1 + \mathbf{p}_2 = \mathbf{q} \neq 0$ in the vertex part, then with the inclusion of the nonlogarithmic terms the pole in the Cooper channel becomes a moving pole:

$$\tilde{\xi}_0(\mathbf{q}) = \tilde{\xi}_0 + (v_F/T_c)^2(\alpha q_\parallel^2 + \beta w^2 q_\perp^2/\varepsilon_F),$$

where α and β are numerical coefficients. The usual estimates indicate that at $w \gg T_c$ the region of the essential singularity around the transition point extends over a temperature interval $\Delta T/T_c \sim (T_c/w)^4$.

4. PHONONS

The problem of the interaction with phonons requires separate consideration, because their energies are limited by frequencies on the order of the Debye frequency $\omega_D \ll \epsilon_F$. Thus for $T > \omega_D$ the interaction via phonons provides no contribution to the principal logarithmic terms in the vertex part, since in all the logarithmic integrals in the diagrams of Fig. 7 (where the broken line denotes the phonon D-function) the phonon line enters at energies above the Debye energy. Accordingly, at high temperatures the vertex part is given by the old equations (3), (4), and (5) with the bare interaction g specified by the non-phonon mechanism. However, as one of the present authors has shown[16], an electronic phase transition with $T_c > \omega_D$ cannot occur at temperatures corresponding to the pole in Γ. In this case (a pole in Γ at $T_{c0} > \omega_D$!) it is always preceded by a purely structural dielectric transition in the ion system.

Actually, when $T > \omega_D$ it is more convenient to write the usual equation for the D-function in terms of the polarization operator (with $q_\parallel \approx 2p_F$, $\omega_n = 0$):

$$D^{-1}(0, \mathbf{q}) = -\omega_0^2(\mathbf{q}) - \Pi, \tag{23}$$

$\Pi =$ 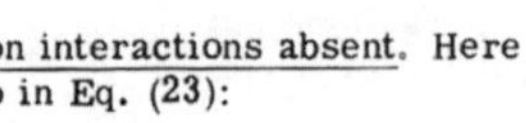

The vertex part of the polarization operator in Eq. (23) depends only on the nonphonon mechanisms. Accordingly, the logarithmic integral in the loop in Eq. (23) is also cut off at the nonphonon energy $\bar{\omega} \sim \epsilon_F$. The bare phonon frequency $\omega_0(\mathbf{q})$ is, as we have already indicated, an essential three-dimensional function of the momentum $\mathbf{q}$.

We cite here the expressions for Π in several special cases.

1. Nonphonon interactions absent. Here Π is given by the simple loop in Eq. (23):

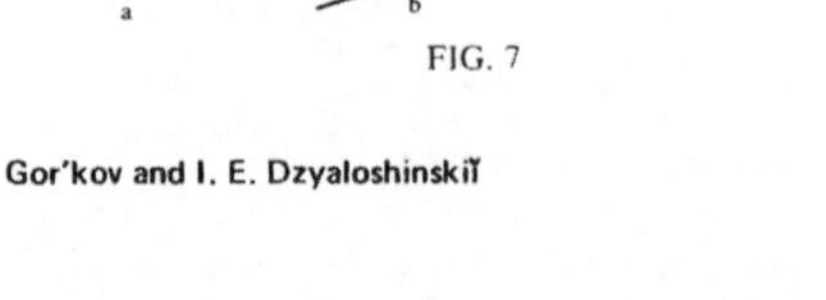

FIG. 7

$$\Pi(\mathbf{q}) = -\frac{1}{2}\,\omega_0^2(\mathbf{q})\,g_{ph}^2(\mathbf{q})\ln\frac{\bar{\omega}}{T}, \tag{24}$$

where g_{ph} is the electron-phonon constant. The transition occurs at the temperature where $D^{-1} = 0$, i.e.,

$$1 - \tfrac{1}{2} g_{ph}^2 \ln(\bar{\omega}/T_c) = 0, \tag{25}$$

if the value of T_{c0} given by this equation is actually larger than the Debye energy[8] ω_D.

2. Nonphonon mechanism. In this case the dielectric transition is "amplified." For instance, in the case of a "one-dimensional" pole (6) with $g_1 < 0$, the polarization operator itself becomes infinite at $\xi = \xi_0 = 1/|g_1|$. Let $T_0 = \bar{\omega}\exp(-\xi_0) > \omega_D$. Near T_0 a singular term[10] appears in Eq. (25):

$$1 - \frac{1}{2} g_{ph}^2 \frac{1}{(\xi_0 - \xi)^{1/2}} \frac{\exp(1/2 - g_2/g_1)}{|g_1|^{1/2}} = 0 \tag{26}$$

and the dielectric transition must occur before the "one-dimensional" transition, for any value (even a low value) of the constant g_{ph}.

Since it is effectively impossible to determine the non-unidimensional solutions to the parquet equations (3), (4), and (5) in analytic form, we shall cite here only the expression for the singular part Π near the "moving" pole (12). In accordance with its meaning as a pole in the P channel, we also find for Π the following simple pole[9]:

$$\Pi(q_\parallel \approx 2p_F, \mathbf{q}_\perp) \sim -\frac{g_{ph}^2(\mathbf{q}_\perp)}{\xi_0(\mathbf{q}_\perp) - \xi}, \tag{27}$$

where $\xi_0(\mathbf{q}_\perp)$ is the trajectory of the "moving" pole.

Thus, at $T > \omega_D$ there is no purely electronic transition due to nonphonon mechanisms. It is forestalled, in quasi-unidimensional metals, by a structural Peierls transition[5,6].

One further point should be made with respect to Eqs. (24)–(27). In the phonon D-function there appears the "bare" phonon spectrum $\omega_0(\mathbf{q})$, which has no particular meaning. The real spectrum is specified by the poles of the exact phonon Green's function. Therefore, in the expressions for the polarization operator it is the temperature dependence of the phonon frequencies that has physical meaning. The remaining terms represent renormalizations, which can be included, for instance, in the determination of the phonon spectra near the melting temperature.

When $T < \omega_D$ the phonon mechanism does not differ essentially from the nonphonon mechanism. Here the diagrams of Fig. 7 also contribute to Γ. Their inclusion leads only to the "renormalization" of the interaction constants and the limits of the integration in Eqs. (3)–(5):

$$\xi = \ln(\bar{\omega}/T) \to \xi = \ln(\omega_D/T),$$
$$g_1 \to \gamma_1(\xi_D) - g_{ph}^2, \quad g_2 \to \gamma_2(\xi_D), \quad g_3 \to \gamma_3(\xi_D), \tag{28}$$

where $\gamma_1(\xi_D)$, $\gamma_2(\xi_D)$, and $\gamma_3(\xi_D)$ are solutions of the system (3)–(5) at the point $\xi_D = \ln(\bar{\omega}/\omega_D)$. In particular, for the one-dimensional solution (6)[9] we have

$$g_1 \to \frac{g_1}{1 + g_1\xi_D} - g_{ph}^2, \quad g_2 \to g_2 - \frac{1}{2}\frac{g_1^2\xi_D}{1 + g_1\xi_D}.$$

The Peierls ion shift now occurs only simultaneously with the transition in the electron system[9], and Eq. (23) for D is replaced by

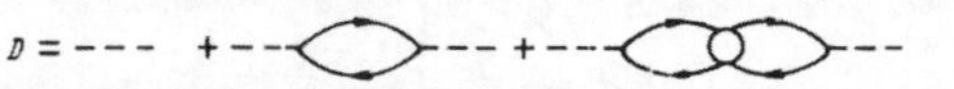
(29)

where the broken line is the free-phonon function $D_0 = -1/\omega_0^2(\mathbf{q})$, and the vertex in the third term of Eq. (29) includes diagrams such as that in Fig. 7c. At the transition point D approaches infinity in the same manner as Π in Eqs. (27) and (26) (cf.[9,10]). If there are several phonon branches that interact with the electrons, D becomes infinite in Eq. (29) when one of the phonon frequencies becomes zero at the transition point.

5. APPLICABILITY OF OUR RESULTS

The basic equations (4), (5) or (11) were obtained with the logarithmic accuracy $g\ln(\bar{\omega}/T) \sim 1$. Thus, for instance, in solution (6) for a single filament[9] the polar singularity point T_0 was determined only apart from a factor. By making use of the expression obtained for the gap at zero temperature in the exact solution for one of the one-dimensional models[18],

$$\Delta \sim g^{1/2}\bar{\omega}e^{-1/|g|}$$

and assuming that $\Delta \sim T_0$, we find that additional terms appear in the denominators of Eq. (6):

$$1 + g\left(\ln\frac{\bar{\omega}}{T} + \frac{1}{2}\ln g + \text{const}\right). \tag{30}$$

Therefore, in the preceding analysis we have essentially assumed that the higher-order corrections with respect to the logarithmic approximation in Eq. (30)

$$1 + g\ln\frac{\bar{\omega}}{T} + \Sigma\left(g, \frac{\omega}{T}\right) \tag{31}$$

have a structure which allows us (near $1 + g\ln(\bar{\omega}/T) \ll 1$) to write Eq. (31) in the form

$$g(\xi - \xi_0). \tag{32}$$

In the one-dimensional case there is no phase transition and, strictly speaking, the vertex part can become infinite only at $T = 0$. However, since the terms $g\ln(\bar{\omega}/T)$ in Eq. (31) represent the basic approximation, expression (32) for the vertex should certainly be valid over some range of temperatures. To estimate this range, i.e., the region of applicability of Eq. (32) within the region where the vertex part is large, becomes an extremely difficult problem in the purely one-dimensional case[9]. In particular, we cannot exclude the possibility that the inclusion of non-parquet diagrams will separate the singularities (6) in the Cooper and Peierls channels. This uncertainty also relates to the special standing-pole solution (9).

Although we have no way to resolve this question exactly, we can still make the following comments. In solution (9) dielectric pairing of electrons and holes is possible, of course, because the filament system is essentially three-dimensional. The familiar proof by Hohenberg[19] that superconducting averages $\Delta = \langle a_p a_{-p}\rangle$ are impossible in the one-dimensional case is based on the continuity equation

$$\frac{\partial\rho}{\partial t} + \operatorname{div}\mathbf{j} = 0,$$

which, in the absence of hopping from one filament to another, holds true also in our model. We have not extended the proof of[19] to the case $\Delta_{ik} \neq 0$ at $i \neq k$, but if $A_{ii} \neq 0$ in Eq. (9), then the temperature of a real transition with the formation of a superconducting gap in the spectrum contains an additional order of smallness due to the hopping energy w.

At the same time it should be noted that the existence of a finite superconducting gap is not necessary in order for a superconducting current to flow. All that is needed for superconductivity is a change in the density of current states. Unfortunately, the logarithmic approximation does not permit us to determine exactly the structure of the states below the transition point. An approximate qualitative determination is given in Appendix II. It appears, however, that the standing-pole solutions represent many new possibilities.

For the moving pole the situation is much simpler. In this case it is possible to use the molecular-field approximation everywhere except for a small region of true singularity near the phase-transition point. It is easiest to understand this from an example. Consider the dependence of the polarization operator $\Pi(q_{\parallel}, \mathbf{q}_{\perp})$ on the longitudinal and transverse momenta in Eq. (23) for the D-function. Then in Eq. (24), because of the flat shape of the spectrum, the integration in the electron loop leads to a term that depends only on the longitudinal momentum $q_{\parallel}$ ($v_F q_{\parallel} \ll T$):

$$\ln\frac{\tilde{\omega}}{T} \to \ln\frac{\tilde{\omega}}{T_0} - \alpha\left(\frac{v_F q_{\parallel}}{T_0}\right)^2 - \frac{T-T_0}{T_0}$$

(where α is a number). Both $\omega_0(\mathbf{q})$ and $g^2_{ph}(\mathbf{q})$ depend on the transverse momentum $\mathbf{q}_{\perp}$, but this dependence is weak:

$$\omega_0(\mathbf{q}) = \omega_0\left(1 + C_1\left(\frac{q_{\perp}}{q_0}\right)^2\right), \quad g_{ph}^2(\mathbf{q}) = g_{ph}^2\left(1 + C_2\left(\frac{q_{\perp}}{q_0}\right)^2\right), \tag{33}$$

where the expansion is about the reciprocal-lattice point at which Eq. (23) becomes zero. The order of magnitude of q_0 is atomic ($q_0 \sim 2\pi/a$). Thus, near the transition point T_0:

$$1 - {}^1\!/_2 g_{ph}^2 \ln(\tilde{\omega}/T_0) = 0,$$

and we have

$$D(\mathbf{q}, T) = \frac{(g_{ph}\omega_0)^{-2}}{\tau + \alpha(v_F q_{\parallel}/T_0)^2 + g_{ph}^{-2}\beta(q_{\perp}/g_0)^2}$$

$$\tau = (T - T_0)/T_0.$$

This result also applies directly to the most general case, in which all the interactions involved in the problem show a dependence on the transverse momentum (33). Finally, in the case of a moving pole we have for the vertex part near the transition point

$$\gamma(\tau, \mathbf{q}) = [\tau + \alpha(v_F q_{\parallel}/T_c)^2 + \beta g^{-1}(q_{\perp}/q_0)^2]^{-1}, \tag{34}$$

where α and β are numerical coefficients, and g represents the order of smallness of the interaction.

From the viewpoint of the theory of phase transitions, Eq. (34) behaves like the correlation function near the transition point [20]. From the familiar formulas for the fluctuations of the order parameter (see, e.g., [20]) we find that outside a narrow neighborhood of the transition temperature

$$\tau > g^2, \; g_{ph}^4 \tag{35}$$

the fluctuations are small. At the end of Sec. 3 we have noted that when the number of jumps is sufficient to cause the formation of a moving Cooper pole, the analogous estimate is of the form

$$\tau > (T_c/w)^4. \tag{36}$$

Outside the temperature range specified by Eq. (35) or (36), the corresponding structural or superconducting transition can be described by a theory of the Ginzburg-Landau type.

It is not without interest to show that the estimate (35) coincides with the region of applicability of the parquet equations (within the accuracy of terms including g ln g and g). In fact, we can estimate the contribution from the first of the non-parquet diagrams (the "envelope") shown in Fig. 8. At high temperatures its contribution is $g^4 \ln(\tilde{\omega}/T)$ (relative magnitude $\sim g^2$). Near $\tau \to 0$ the principal terms in Eq. (34) are those that contain sums over the filaments, since in the solution with the moving pole the γ^{ik} themselves are small.

The order of magnitude of the diagram in Fig. 8 is

$$\frac{1}{T^2}\sum_i \int dq_{\parallel 1}\, dq_{\parallel 2} (\gamma_1^{ii}(q_{\parallel 1}))^2 (\gamma_1^{ik}(q_{\parallel 2}))^2. \tag{37}$$

For the Fourier component with respect to the difference i − k with $\mathbf{q}_{\perp} \neq 0$ we have

$$\frac{1}{T^2}\int \gamma_1(\mathbf{t}_1, q_{\parallel 1})\gamma_1(\mathbf{q}_{\perp} - \mathbf{t}_1, q_{\parallel 1})\gamma_1(\mathbf{t}_2, q_{\parallel 2})\gamma_1(\mathbf{q}_{\perp} - \mathbf{t}_2, q_{\parallel 2})\, d^2t_1\, d^2t_2\, dq_{\parallel 1}\, dq_{\parallel 2}.$$

By substituting Eq. (34) in this expression we obtain the estimate g^2/τ. The condition for the smallness of this quantity coincides with Eq. (35).

6. CONCLUSION

As we have seen, the fact of three-dimensionality has significant effects on the given model of quasi-unidimensional filaments. The purely one-dimensional solution for a single filament [9,10] is, in general, unstable, and can exist only if the interaction between the filaments is so small that the actual transition is due solely to hopping effects. We have shown, however, that although the interaction between filaments can lead to a dielectric structural transition, there is also the possibility that it will lead, as before, to a transition with simultaneous dielectric and superconducting pairing. In the latter case the actual transition point clearly is dependent on the hopping processes from one filament to another. In the opposite case, when the probability of an electron transition from one filament to another, while still small, becomes predominant (the quasi-three-dimensional spectrum), the system becomes capable of superconducting pairing alone. Here, however, in contrast to the BCS theory, the system of parquet equations must be used to determine the transition temperature.

Above, for lack of space, we did not dwell in detail upon the case of only one electron in the reciprocal cell ($2p_F = \pi/c$). In this case there is an additional scattering corresponding to Umklapp processes, which in the one-dimensional case creates the possibility of independent antiferromagnetic ordering [10]. An analysis of the general equations (5) similar to that given above shows that the moving pole can arise either in γ_1 separately or in γ_1 and γ_3 simultaneously. Similarly, the standing pole is possible either in γ_1 and γ_2 only, or in all three amplitudes at once. Since the singularity in γ_3 is related to the dielectric scattering channel, sufficiently large hopping will suppress both dielectric and antiferromagnetic pairing.

We can now list all the possibilities for the states in the given system at low temperatures.

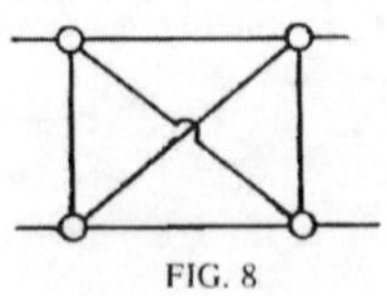

FIG. 8

1. Dielectric state: a structural transition due to the moving pole in γ_1.

2. Dielectric state with antiferromagnetic ordering: a moving pole in γ_1 and γ_3.

3. Superconducting state with changed lattice structure: a standing pole in γ_1 and γ_2.

4. Superconducting state with antiferromagnetic ordering and changed structure: a standing pole simultaneously in γ_1, γ_2, and γ_3.

5. Superconducting state: jumps predominant.

6. Normal metallic state.

The details of these states can be explained only outside the framework of the logarithmic approximation, although, as we saw above, there are many possibilities here also.

We have discussed here only the thermodynamic picture of the transitions. Therefore, in speaking of the superconducting transition we have in mind the formation of an energy gap in the spectrum due to Cooper pairing. The current state, the conductivity and the paraconductivity, and similar questions, which are of primary interest in experimental work [1-4], have been left out of consideration. While leaving these problems to a later article, we may nevertheless point out a rather fundamental difficulty, namely, that any defects and impurities, in the Mott sense [21], will lead to the localization of the electron states in the one-dimensional case. Thus, at zero temperature the metallic state in a single filament possesses zero conductivity, as Berezinskiĭ recently confirmed by direct calculations [23].

In light of all the above we would like to point out that experiments [3] on TTF-TCNQ have indicated the presence of a dielectric transition in this system. Measurements of the phonon spectrum in Magnus salts [4] also indicate a damping of the phonon mode. The above arguments tend to be confirmed by the observed narrowness of the region of dip in the phonon frequencies $\Delta q_{\parallel}/q_{\parallel}$. From Eq. (33), the corresponding value is $\Delta q_{\parallel}/q_{\parallel} \sim T_0/\epsilon_F$.

In conclusion the authors wish to thank B. S. Mityagin for his valuable advice, and A. I. Larkin, S. P. Novikov, and Ya. G. Sinaĭ for many useful discussions.

APPENDIX I

Let us consider the stability of the "standing" poles (9). Writing

$$\gamma_1(q)=\frac{A(q)}{\xi}+\frac{\delta_1(q)}{\xi^{\Lambda}},\quad \gamma_2(q)=\frac{B(q)}{\xi}+\frac{\delta_2(q)}{\xi^{\Lambda}},$$

where A(q) and B(q) are solutions of Eqs. (13) and (14), δ_1 and δ_2 are small increments, and Λ are characteristic numbers that determine the stability, we obtain the linearized equations (11) for δ:

$$\Lambda\delta_1=2A(q)\delta_1-A(q)\int\delta_2(q_1)\frac{d^2q_1}{S}-\delta_1(q)\int B(q_1)\frac{d^2q_1}{S}$$
$$+\int B(q-q_1)\delta_1(q_1)\frac{d^2q_1}{S}+\int A(q-q_1)\delta_2(q_1)\frac{d^2q_1}{S}, \qquad \text{(A.I.1)}$$
$$\Lambda\delta_2=\int A(q-q_1)\delta_1(q_1)\frac{d^2q_1}{S}.$$

The system (A.I.1) specifies the eigenfunctions δ_1 and δ_2 and the corresponding eigenvalues Λ. If among the latter there are any $\Lambda > 1$, the corresponding "standing" pole will be unstable.

Let us now consider the "elementary" solution of the type (16). It is apparent at once that in this case the dangerous eigenvalues (larger than 1) lie near $\Lambda = 2$ (with accuracy $\sim\lambda^4$). It is also clear that in the earlier approximation the value of the function δ in the outer region $|q| > \lambda/2$ can be expressed, using Eq. (A.I.1), in terms of their values in the internal region. After we eliminate the external regions and express δ_2 in terms of δ_1, we obtain a closed integral equation for δ_1 in the internal region, into which we must substitute the expressions for A and B, including the terms that follow from (16).

In the following calculations we shall use a one-dimensional momentum q, assuming also, for simplicity, that the distance between filaments is a = 1. We have:

$$A(q)=\left[1+\frac{1}{8\pi^2}\left(\frac{\lambda^2}{4}+q^2\right)\right]\theta_\lambda(q)+\frac{1}{16\pi^2}\left(\frac{3\lambda}{2}-|q|\right)^2(\theta_{3\lambda}(q)-\theta_\lambda(q)),$$
$$B(q)=\frac{1}{4\pi^2}(\lambda-|q|)\theta_{2\lambda}(q) \qquad \text{(A.I.2)}$$

and the equation for δ_1 in the internal region $|q| < \lambda/2$:

$$\left[2-\Lambda-\frac{\lambda^2}{16\pi^2}+\frac{q^2}{4\pi^2}\right]\delta_1-\frac{\lambda}{8\pi^2}\int_{-\lambda/2}^{\lambda/2}\delta_1(q_1)dq_1$$
$$+\frac{1}{4\pi^2}\int_{-\lambda/2}^{\lambda/2}\delta_1(q_1)(\lambda-|q-q_1|)dq_1=0. \qquad \text{(A.I.3)}$$

Equation (A.I.3) is equivalent to a differential equation with boundary conditions that correspond to the two possible solutions of (A.I.3): symmetric ($\delta_s(q) = \delta_s(-q)$) and antisymmetric ($\delta_a(q) = -\delta_a(-q)$). Introducing the notation

$$\delta_{s,a}=\frac{Q_{s,a}}{\nu+q^2},\quad \nu=(2-\Lambda)4\pi^2-\frac{\lambda^2}{4},$$

we obtain a second-order equation for $Q_{s,a}$:

$$\frac{d^2Q_{s,a}}{dq^2}-\frac{2Q_{s,a}}{\nu+q^2}=0$$

with the boundary conditions

$$\frac{dQ_s(0)}{dq}=Q_s\left(\frac{\lambda}{2}\right)=Q_a(0)=\frac{dQ_a(\lambda/2)}{dq}=0.$$

The equation can be solved by quadratures:

$$Q=C_1(\nu+q^2)+C_2(\nu+q^2)\int_0^q\frac{dq_1}{(\nu+q_1^2)^2}.$$

For the even solutions Q_s the constant $C_2 = 0$:

$$Q_s=C(\nu_s+q^2)$$

and the condition $Q_s(\lambda/2) = 0$ then clearly gives

$$\nu_s=-\lambda^2/4 \text{ or } \Lambda_s=2.$$

It can be shown that the eigenvalue $\Lambda_s = 2$ corresponds to a simple shift of the pole position by a constant amount, and does not mean a loss of stability.

An investigation of the antisymmetric case shows that it corresponds to the eigenvalue

$$\Lambda_a=2+\frac{\lambda^2}{16\pi^2}(y^2-1),$$

where $y \approx 1.20$ is the root of the equation

$$\ln\frac{y+1}{y-1}=2y.$$

However, the presence of an antisymmetric perturbation does not imply the instability of solution (16), because when $\gamma(q)$ is antisymmetric in momentum space the vertices are also asymmetric, $\gamma^{ik} = -\gamma^{ki}$ (because they are real in coordinate space). But any reasonable interaction g^{ik} between the filaments will naturally depend symmetrically on the distance between i and k, and this symmetry will be conserved in all solutions of Eqs. (3)—(5) for γ^{ik}. Thus the unstable antisymmetric perturbation is not physical.

The situation changes when we consider the stability of solutions in the more general form of (16). Calculations (which we shall not cite here) show that in this case, together with $\Lambda = 2$, there appear perturbations that are nontrivial and unstable, but symmetric on the whole. We can similarly demonstrate the instability of solutions of type (19).

APPENDIX II

We showed above that a theory of the Ginzburg-Landau type is valid outside the region (35) below the transition point. In all essentially parquet-type problems we are unable to formulate the nonlinear equations for the gap Δ, and are forced to consider only their linearized version.

The problem of determining the gap for a pure one-dimensional pole (6) has already been solved by the present authors [9]. We shall therefore present only the final results.

It was shown in [9] that each gap Δ (in [9] there were two: the Cooper and dielectric gaps) is an essential function of the two logarithmic variables, $\Delta(\eta, \xi)$:

$$\xi = \frac{1}{2}\ln\frac{\bar{\omega}^2}{\Delta_1^2+\Delta_2^2+\ldots}, \tag{A.II.1}$$

which contains the sum $\Delta_1^2 + \Delta_2^2 + \ldots$ of all the anomalous nonzero pairings, and the variable

$$\eta(p) = \ln\frac{\bar{\omega}}{v_F|p\mp p_F|}, \tag{A.II.2}$$

which determines the dependence of the gap on the distance from the Fermi surface when $\eta < \xi$, i.e., when $v_F^2(\Delta p)^2 > \Delta_1^2 + \Delta_2^2 + \ldots$. The function $\Delta(\eta, \xi)$ satisfies the same equations as the vertex parts, but in contrast to γ in Eqs. (3), (4), and (5) it depends on two momenta with different orders of magnitude (the variables ξ and η). There are two types of such vertices: $\gamma_C(\eta, \xi)$ and $\gamma_P(\eta, \xi)$. They correspond respectively to the situations of Figs. 2a and 2b with

$$\xi = \ln\frac{\bar{\omega}}{v_F|p\mp p_F|} > \eta = \ln\frac{\bar{\omega}}{v_F|p'\mp p_F|}.$$

In our problem all four forms of anomalous pairings are possible:

$$\langle a_{p\alpha}a_{-p\beta}\rangle_{s,t}, \quad \langle a_{p\alpha}{}^{+}a_{2p_F-p\beta}\rangle_{s,t},$$

where the indices s and t denote singlet (s) or triplet (t) pairing with respect to the spin variables. We denote the corresponding gaps (or anomalous mass operators [9]) by $\Delta_{s,t}(\eta, \xi)$ and $\kappa^k_{s,t}(\eta, \xi)$; of course, these are also functions of the filament numbers. In [9] we established the following relationships[10)] between $\gamma_{C,P}$ and Δ, κ:

$$\Delta_s{}^{ik} \sim \gamma_{C+}{}^{ik} \equiv \gamma_{C1}{}^{ik} + \gamma_{C2}{}^{ik}, \quad \Delta_t{}^{ik} \sim \gamma_{C-}{}^{ik} \equiv \gamma_{C1}{}^{ik} - \gamma_{C2}{}^{ik},$$
$$\kappa_s{}^{ik} \sim \gamma_{P1}{}^{ik}, \quad \kappa_t{}^{ik} \sim \gamma_{P2}{}^{ik}.$$

The equations for $\gamma_{P,C}$ have the following form (cf. [9,14]):

$$\gamma_{C\pm}{}^{ik}(\eta,\xi) = \gamma_\pm{}^{ik}(\eta) \mp \frac{1}{2}\int_\eta^\xi d\zeta\,\gamma_{C\pm}{}^{ik}(\eta,\zeta)\gamma_\pm{}^{ik}(\zeta),$$

$$\gamma_{P1}{}^{ik}(\eta,\xi) = \gamma_1{}^{ik}(\eta) - \int_\eta^\xi d\zeta\,\Big[\sum_l \gamma_{P1}{}^{il}(\eta,\zeta)\gamma_1{}^{lk}(\zeta)$$
$$-\frac{1}{2}\gamma_{P1}{}^{ik}(\eta,\zeta)\gamma_2{}^{ii}(\zeta) - \frac{1}{2}\gamma_{P2}{}^{ii}(\eta,\zeta)\gamma_1{}^{ik}(\zeta)\Big], \tag{A.II.3}$$
$$\gamma_{P2}{}^{ik}(\eta,\xi) = \gamma_2{}^{ik}(\eta) + \frac{1}{2}\int_\eta^\xi d\zeta\,\gamma_{P2}{}^{ik}(\eta,\zeta)\gamma_2{}^{ik}(\zeta).$$

It was shown in [9] that the linear homogeneous equations for the gaps Δ and κ have solutions at the same values of ξ at which the corresponding $\gamma_{C,P}$ become infinite. The dependence on η of the term in $\gamma_{C,P}$ that tends to infinity then determines the dependence of the gaps on the momentum (i.e., on η).

For the moving pole (12), none of the vertices, except γ_{P1}, becomes infinite. According to Eq. (A.II.3) the latter vertex has the same form of pole,

$$\gamma_{P1}(\eta, \xi; q) = \frac{1}{\xi - \xi_0(q)},$$

so that

$$\kappa_s(\eta, q) = \text{const}, \quad \kappa_t = \Delta_s = \Delta_t = 0$$

and $\kappa_s = \bar{\omega}\exp(-\min\xi_0(q))$, as in the BCS theory.

For an arbitrary standing pole, the system (A.II.3) does not have a simple form. If we take a standing pole of type (16), then by neglecting terms of smallness A_{ik}^2 where possible we obtain

$$\gamma_{C\pm}{}^{ik}(\eta,\xi) = \gamma_\pm{}^{ik}(\eta)\exp\Big\{\mp\frac{1}{2}\int_\eta^\xi d\zeta\,\gamma_\pm{}^{ik}(\zeta)\Big\}$$

$$\gamma_{P1}(\eta,\xi;q) = \gamma_1(\eta;q)\exp\Big\{-\int_\eta^\xi d\zeta\,\gamma_1(\zeta;q)\Big\}.$$

$$\gamma_{P2}{}^{ik}(\eta,\xi) = \gamma_2{}^{ik}(\eta)\exp\Big\{\frac{1}{2}\int_\eta^\xi d\zeta\,\gamma_2{}^{ik}(\zeta)\Big\}.$$

Near the poles (9) with A_{ik} and B_{ik} specified by Eqs. (13), (16), and (17) we have, in the old approximation,

$$\gamma_{P1}(\eta,\xi;q) = \frac{\theta_\lambda(q)}{\xi-\xi_0},$$

$$\gamma_{P2}{}^{ik}(\eta,\xi) = \frac{A_{ik}{}^2}{2(\eta-\xi_0)}\left(\frac{\xi-\xi_0}{\eta-\xi_0}\right)^{A_{ik}{}^2/4}, \tag{A.II.4}$$

$$\gamma_{C\pm}{}^{ik}(\eta,\xi) = \frac{A_{ik}}{\eta-\xi_0}\left(\frac{\eta-\xi_0}{\xi-\xi_0}\right)^{\pm A_{ik}/2};$$

$$A_{ik} = \frac{4\sin(\tfrac{1}{2}\lambda x_{ik})\sin(\tfrac{1}{2}\lambda y_{ik})}{S x_{ik} y_{ik}}. \tag{A.II.5}$$

The formulas (A.II.4) show that in such a transition:

1) singlet Peierls pairing depends on the distance between the filaments,

$$\kappa_s{}^{ik}(\eta) \sim A_{ik};$$

2) triplet Peierls pairing is absent;

3) singlet Cooper pairs occur only at distances between the filaments such that $A_{ik} > 0$, and here

$$\Delta_s{}^{ik}(\eta) \sim A_{ik}(\eta-\xi_0)^{-1+A_{ik}/2};$$

4) triplet Cooper pairing occurs at distances such that $A_{ik} < 0$:

$$\Delta_t{}^{ik}(\eta) \sim A_{ik}(\eta-\xi_0)^{-1-A_{ik}/2}.$$

For small distances between the filaments, $\rho_{ik}\lambda \lesssim 1$, the A_{ik} are constant and positive ($A_{ik} \sim \lambda^2/S$), so that, as in the case of a one-dimensional pole, dielectric and singlet Cooper pairing will exist at these distances. At large distances A_{ik} begins to oscillate, which, according to the preceding formulas, leads to alternating singlet Δ_s and triplet Δ_t Cooper pairs.

[1)]In fact, a need not be too large compared to the period c along the filaments, since in TCNQ complexes the strong directionality of the electron orbitals tends to suppress the hopping.

[2)]We recall, to avoid misunderstandings, that in our normalization of Eqs. (3)–(5) and choice of the variable ξ the constants g and the γ themselves are measured in units of πv_F. To transform to the usual notation of the BCS theory we must make the substitutions $g \to g/\pi v_F$ and $\gamma \to \gamma/\pi v_F$.

[3)]In reality, this matching and the determination of $\xi_0(q)$ can be accomplished only by the numerical integration of Eq. (11). In [15] these calculations were carried out for some particular model and it was found that $\xi_0(q)$ is of the form $\xi_0(q) = \xi_{min} + bq^2$.

[4)]Note also that the moving pole is a "stable" singularity of system (7), since, on the one hand, it contains the arbitrary function $\xi_0(q)$, and thus can be joined to the entire region of the initial conditions, and on the other hand the corrections to solution (12) have a weaker singularity.

[5)]The authors are deeply grateful to B. S. Mityagin for having pointed out to them solutions of the type (16) and (18).

[6)]The only condition is that solution (16) must be an even function, since both A^{ik} and its Fourier transform A(q) are real.

[7)]There are strong arguments to indicate, however, that these initial conditions themselves must be of the long-range type (17).

[8)]The condition $T_{c0} > \omega_D$ was the model criterion for the validity of the 1954 theory of Froehlich [6] (see also [7]). Metal-dielectric transitions in semiconductors under similar conditions ($T_c > \omega_D$) were also discussed by Kopaev and Timerov [17].

[9)]For the procedure by which Π is calculated in the parquet approximation, see [9,10,14].

[10)]In making comparisons we must keep in mind that the dependence of γ on the spin indices was introduced in [9] in a different manner than in Eq. (1). To reconcile Eq. (1) with [9] we must make the substitutions $\gamma_1 \to \gamma_1$ and $\gamma_1 \to \gamma_1 - \gamma_2$.

[1] I. F. Shchegolev, Phys. Stat. Sol. (a) **12**, 9 (1972).

[2] A. J. Epstein, S. Etemad, A. F. Garito, and A. J. Heeger, Phys. Rev. **B5**, 952 (1972).

[3] L. B. Coleman, M. J. Cohen, D. J. Sandman, F. G. Yamagishi, A. F. Garito, and A. J. Heeger, Sol. St. Comm. **12**, 1125 (1973).

[4] B. Renker, H. Rietschel, L. Pintschovius, W. Glaeser, P. Bruesch, D. Kuse, and M. J. Rice, Phys. Rev. Lett. **30**, 1144 (1973).

[5] R. Peierls, Quantum Theory of Solids, Clarendon Press, 1955 (Russ. Transl., IIL, 1956).

[6] H. Froehlich, Proc. Roy. Soc. **A223**, 296 (1954).

[7] A. M. Afanas'ev and Yu. Kagan, Zh. Eksp. Teor. Fiz. **43**, 1456 (1962) [Sov. Phys.-JETP **16**, 1030 (1963)].

[8] A. W. Overhauser, Phys. Rev. Lett. **4**, 462 (1960).

[9] Yu. A. Bychkov, L. P. Gor'kov, and I. E. Dzyaloshinskiĭ, ZhETF Pis. Red. **2**, 146 (1965) [JETP Lett. **2**, 92 (1965)]; Zh. Eksp. Teor. Fiz. **50**, 738 (1966) [Sov. Phys.-JETP **23**, 489 (1966)].

[10] I. E. Dzyaloshinskiĭ and A. I. Larkin, Zh. Eksp. Teor. Fiz. **61**, 791 (1971) [Sov. Phys.-JETP **34**, 422 (1972)].

[11] W. A. Little, Phys. Rev. **A134**, 1416 (1964).

[12] V. L. Ginzburg, Usp. Fiz. Nauk **101**, 185 (1970) [Sov. Phys.-Uspekhi **13**, 335 (1970)].

[13] I. T. Dyatlov, V. V. Sudakov, and K. A. Ter-Martirosyan, Zh. Eksp. Teor. Fiz. **32**, 767 (1957) [Sov. Phys.-JETP **5**, 631 (1957)].

[14] P. Nozieres, J. Gavoret, and B. Roulet, Phys. Rev. **178**, 1072, 1084 (1969).

[15] I. E. Dzyaloshinskiĭ and E. I. Kats, Zh. Eksp. Teor. Fiz. **62**, 1104 (1972) [Sov. Phys.-JETP **35**, 584 (1972)]; S. A. Brazovskii, Zh. Eksp. Teor. Fiz. **61**, 2401 (1971) [Sov. Phys.-JETP **34**, 1286 (1972)].

[16] L. P. Gor'kov, Zh. Eksp. Teor. Fiz. **65**, 1658 (1973) [Sov. Phys.-JETP **38**, 830 (1974)].

[17] Yu. V. Kopaev and R. Kh. Timerov, Zh. Eksp. Teor. Fiz. **63**, 290 (1972) [Sov. Phys.-JETP **36**, 153 (1973)].

[18] E. H. Lieb and F. Y. Wu, Phys. Rev. Lett. **20**, 1445 (1968).

[19] P. C. Hohenberg, Phys. Rev. **158**, 383 (1967).

[20] L. D. Landau and E. M. Lifshitz, Statisticheskaya fizika (Statistical Physics), Nauka, 1964, Sec. 119 (Addison-Wesley, 1969).

[21] N. F. Mott, Adv. Phys. **16**, 49 (1967).

[22] V. L. Berezinskiĭ, Zh. Eksp. Teor. Fiz. **65**, 1251 (1973) [Sov. Phys.-JETP **38**, 620 (1974)].

Translated by R. Rutherford III
46

CONDUCTIVITY OF QUASI-ONE-DIMENSIONAL CRYSTALS IN THE METALLIC PHASE

L. P. Gor'kov and I. E. Dzyaloshinskii
L. D. Landau Institute of Theoretical Physics, USSR Academy of Sciences
Submitted 30 October 1973
ZhETF Pis. Red. 18, No. 11, 686 — 689 (5 December 1973)

Features of the conductivity of quasi-one-dimensional crystals in the metallic phase are considered.

Detailed experimental reports on organic complexes based on TCNQ have recently been published [1 - 3]. The most important result is the observation of so-called "phase transitions" into an antiferromagnetic state [1, 2] or of a Peierls transition [3]. The latter is of particular interest in view of the connection previously noted [4, 5] between the structural instability and the possible appearance of phenomena of the superconductivity type. There is still no theory of fluctuations in a quasi-one-dimensional metal (see [4, 5]). We confine ourselves here only to sufficiently high temperatures, when the fluctuations do not play an essential role, but nevertheless the behavior of all physical quantities exhibits certain singularities inherent in one-dimensional systems (see also [5, 6]).

We consider the conductivity of such systems in the temperature region where the kinetic equation is still valid, i.e., at $\hbar \ll \tau T$. In our case this condition means weakness of the interaction of the electrons with the phonons and of the electrons with one another, or else smallness of the defect concentration.

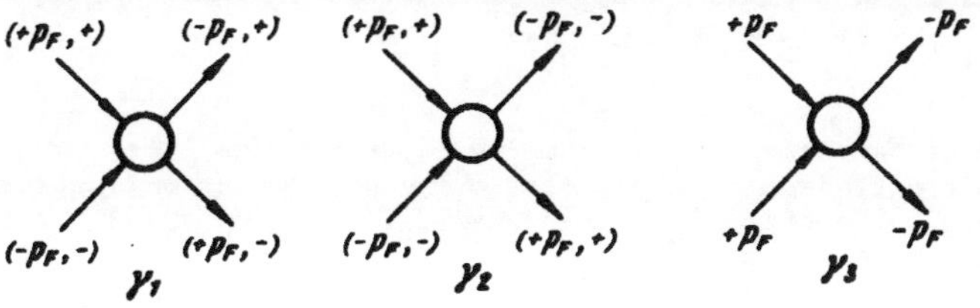

Fig. 1

The conductivity is determined, as always, by three mechanisms — impurity, electron-phonon, and electron-electron. An important distinguishing feature of the last two mechanisms is that the Fermi surface reduces to two points in the reciprocal lattice $\{-p_F, +p_F\}$. There is therefore no diffusion relaxation of the momentum, and the resistance is always determined by a direct transition of the electron from the point $-p_F$ to $+p_F$. The electron-phonon time at low temperatures is exponentially large

$$\tau_{eph}^{-1} \sim \exp\{-\omega_0(2p_F)/T\},$$

where $\omega_0(k)$ is the phonon spectrum. For the same reason, an exponential factor appears also in τ_{ee}^{-1}, provided that the number of electrons per cell is not exactly equal to unity, when $2p_F = \pi/a$ (a is the lattice period) and the transition from p_F to $-p_F$ is made possible by Umklapp processes without a change of energy.

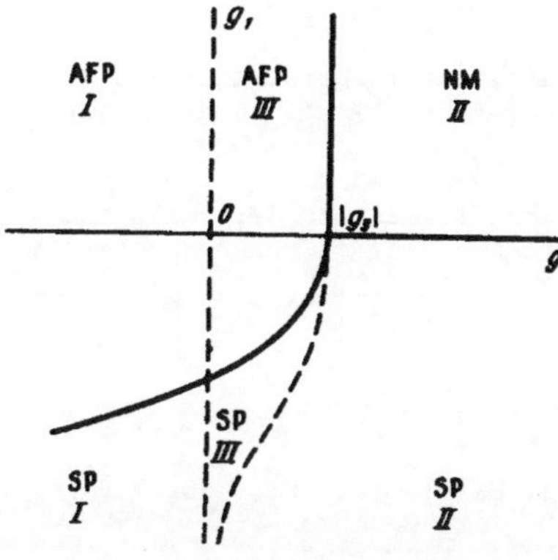

Fig. 2

The non-exponential behavior of the relaxation time takes place in the considered simple model at temperatures higher than the Debye temperature ($T \gg \omega_D$), and in the case of exactly one electron per cell it occurs also at $T \ll \omega_D$. The latter takes place apparently in the experimentally investigated NMP - TCNQ [2] and TTF - TCNQ [3]. Let $T \ll \omega_D$, and let the effective mechanism be electron-electron scattering or scattering by defects. To avoid misunderstanding, we emphasize that in a one-dimensional system the repulsion integral for the first of these mechanisms, without allowance for the interaction effects described below, yields precisely $\tau_{ee}^{-1} \propto T$, and not $\tau_{ee}^{-1} \propto T^2$, as in the two- and three-dimensional cases. However, neither τ_{imp} nor $\tau_{ee}T$ remains constant at $T \ll \varepsilon_F$. The reason is that in the one-dimensional case all the scattering amplitudes γ are slowly-varying functions of the logarithm of the temperature (or energy) $\xi \equiv \ln(\varepsilon_F/T)$.

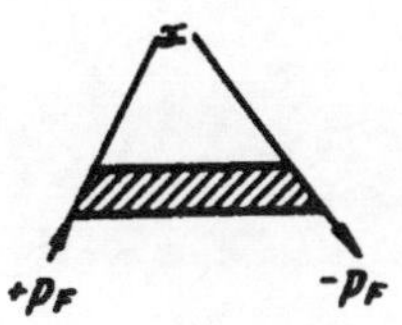

Fig. 3

As shown in [5], the properties of the state are determined by three amplitudes, $\gamma_1(\xi)$, $\gamma_2(\xi)$, and the Umklapp amplitude $\gamma_3(\xi)$, which are shown in Fig. 1 (the ± signs designate the spin projections). Depending on the relation between the initial constants g_1, g_2, and g_3, the system can remain a normal metal (NM), or some type of instability can develop in it, such as a combination of superconducting and Peierls (SP) instability or a combination of antiferromagnetic and Peierls (AFP) instability. The thick solid lines of Fig. 2 show schematically the boundaries between the regions on the plane $g_4 = g_1 - 2g_2$, g_1.

The electron-electron time is

$$\tau_{ee}^{-1} \sim T|\gamma_3(\xi)|^2 . \tag{1}$$

Formulas for γ_3 are given in [5]. An investigation of these formulas shows that $\tau_{ee}T$ can behave in three ways with increasing $\ln T$: monotonic growth (I), decrease (II), and a curve with a maximum (III). The boundaries between regions I, II, and III are shown in Fig. 2 by dashed lines. In the Hubbard model, when $g_1 = g_2 = g_3 = g_4 = g$, we have (see [5])

$$T\tau_{ee} \sim (1 - g\xi)^2 \tag{1'}$$

For impurity relaxation, similar calculations [5, 6] yield

$$\tau_{imp}^{-1} \sim \exp\left\{ -\int_0^{\xi}\left[\frac{3}{2}\gamma_1(\eta) + \frac{1}{2}\gamma_4(\eta) + \gamma_3(\eta)\right] d\eta\right\} . \tag{2}$$

The nature of the resultant renormalization factors is clear from Fig. 3, which shows the inserts corresponding to the electron interactions in impurity scattering with momentum transfer $2p_F$. At different ratios of g_1, g_2, and g_3, the time τ_{imp} as well as $\tau_{ee}T$ can exhibit all three types of dependences on lnT. In Hubbard's model we have

$$\tau_{imp} \sim (1 + g\xi)^{3/2}(1 - g\xi)^{-1/2}, \tag{3}$$

i.e., τ_{imp} decreases when the superconducting instability region is approached (this behavior is typical of the Kondo effect), and increases in the case of the antiferromagnetic instability.

The amplitude of the electron-phonon interaction, which turns out to be significant at $T \gg \omega_D$, is likewise renormalized in accordance with Fig. 3. Therefore $1/T\tau_{eph}$ is determined in this region by formula (2).

The conductivity data used in the experiments [1 - 3] is insufficient to determine the parameters in the logarithmic formulas (1) and (3). The metallic behavior of the conductivity in TTF - TCNQ [3] above the transition is in qualitative agreement with (1).

[1] F. Shchegolev, Phys. Stat. Sol. (a) 12, 9 (1972).
[2] A. J. Ejstein, S. Etemad, A. F. Garito, and A. J. Heeger, Phys. Rev. B5, 952 (1972).
[3] L. B. Coleman, M. J. Cohen, D. J. Sandman, F. G. Ygamagish, A. F. Garito, and A. J. Heeger, Sol. St. Comm. 12, 1125 (1973).
[4] Yu. A. Bychkov, L. P. Gor'kov, and I. E. Dzyaloshinskii, Zh. Eksp. Teor. Fiz. 50, 738 (1966) [Sov. Phys.-JETP 23, 489 (1966)].
[5] I. E. Dzyaloshinskii and A. I. Larkin, ibid. 61, 791 (1971) [34, 422 (1972)].
[6] L. P. Gor'kov, ibid. 65, 1658 (1973) [38, No. 4 (1974)].

Fluctuation increment to the conductivity in quasi-one-dimensional systems near a structure transition

L. P. Gor'kov and I. E. Dzyaloshinskiĭ

L. D. Landau Institute of Theoretical Physics, USSR Academy of Sciences
(Submitted May 27, 1974)
ZhETF Pis. Red. 20, 37–41 (July 5, 1974)

It is shown that the contribution made to the conductivity by fluctuations near a structure transition is determined by the kinetics of the electronic interactions. The derived expression does not differ from the earlier results. "Three-dimensionality" of the phonons annihilates the fluctuation growth in the conductivity.

A noticeable conductivity maximum was observed in TTF-TCNQ crystals[1] near $T_s = 60\,^\circ$K. It was confirmed in[2] that in a number of cases this maximum takes the form of a sharp peak. According to the explanation proposed for this effect in[3], a structure transition takes place at the point of conductivity maximum. In the vicinity of the transition, the fluctuation gap in the energy spectrum of the one-dimensional electrons attunes itself in the presence of the current to the motion of these electrons in such a way as to hinder the dissipative processes. Attempts were made in[4,5] to derive the mechanism of[3] from the microscopic theory. The results of[4] and[5] do not agree. Since the calculations of[4] have not been published,[1)] we present below the results of our calculations. These results do not agree with those of[4,5], although in the limiting case of one-dimensional phonons they lead to the same temperature dependence of the increment on the conductivity, $\Delta\sigma \propto (T - T_s)^{-1/2}$, as obtained in[3–5]. What is more important is that when the "three-dimensionality" of the phonons is taken into account the structure transition leads only to a singularity in the derivative of the conductivity.

In a one-dimensional metal, the structure and Cooper transitions cannot be separated in the principal approximation.[6,7] At the same time a phase transition is impossible in the one-dimensional case. An analysis of the role of effects of the three-dimensional structure, undertaken in[7], has shown that a structure transition is realized if the interaction between electrons on different filaments (without their jumping over from filament to filament) is large enough. If the jumping over of electrons prevails, then the one-dimensional singularities go over into Cooper pairing in this case. In the first case we deal also with a softening of the corresponding phonon mode. The fluctuation region of temperatures corresponds to

$$(T - T_s)/T_s > g^2, \tag{1}$$

where g^2 is a certain effective electron-electron interaction constant. We consider below for simplicity the model of a structure transition that takes place at a temperature T_s above the Debye temperature. In this case, in the region of applicability of the self-consistent-field approximation (1), the phonon D function takes the form

$$D^R(\omega, k) = \frac{\omega_0^2(k)}{\omega^2 - \omega_0^2 g_{ph}^2 [f - i\beta\omega]}, \tag{2}$$

where

$$f = \frac{T - T_s}{T_s} + \frac{7\zeta(3)}{16\pi^2 T_s^2}(v_F k_\parallel)^2 + B g_{ph}^{-2} S^{-1} k_\perp^2\,; \quad \beta = \pi/8T_s,$$

and g_{ph}^2 is the square of the electron-phonon interaction, S is the cross-section area of the reciprocal lattice, and $B \sim 1$ is a constant that characterizes the dispersion of the phonon frequency. The electron spectrum in the absence of the jumps remains flat ($v_F \equiv v_{zF}$). The fluctuation increments of the conductivity, resulting from the soft mode in (2), are due to the diagrams shown in the figure (see[4,5]), where a wavy line corresponds to the D function of phonons with a transfer $k_\parallel = \pm 2p_F$. For an analytic continuation in ω_m, it is necessary to represent all the sums over the frequencies in the matrix elements of these diagrams in terms of contour integrals. Starting, as in[5], from sums over the phonon frequencies, it is easy to verify that the main contribution to the current matrix element connected with small k and z in (2) is proportional to

$$T^2 \sum_{n,n'} \int_{-\infty}^{+\infty} \frac{dz}{4\pi i} \operatorname{cth}\frac{z}{2T} D^R_{i\omega_m - z}(D^R_z - D^A_z)$$
$$\times \{p_z G_{i\epsilon_n} G_{i\epsilon_n + i\omega_m}[G_{z + i\epsilon_n} + G_{i\epsilon_n + i\omega_m - z}]\}$$
$$\times \{[G_{i\epsilon_{n'} + z} + G_{i\epsilon_{n'} + i\omega_m - z}] G_{i\epsilon_{n'}} G_{i\epsilon_{n'} + i\omega_m} p'_z\}. \tag{3}$$

(We have left out terms in which the D function contains electron frequencies of the order of either T or $1/\tau$. Here $1/\tau$ is the electron-phonon collision frequency.) The succeeding calculations reduce to a determination of the vertices in (3), consisting of products of three electronic Green's functions. Since the D function (2) has a singularity at a longitudinal-momentum transfer $\pm 2p_F$, the two parts of the diagrams in the figure are not independent. If, say, the integration with respect to the electron momentum takes place in the left electronic loop of diagram (a) near $+p_F$, then the point $+p_F$ makes also a contribution in its right-hand loop. The opposite ends of the Fermi surface contribute to the diagram (b). A special case is $4p_F = 2\pi/a$, where a is

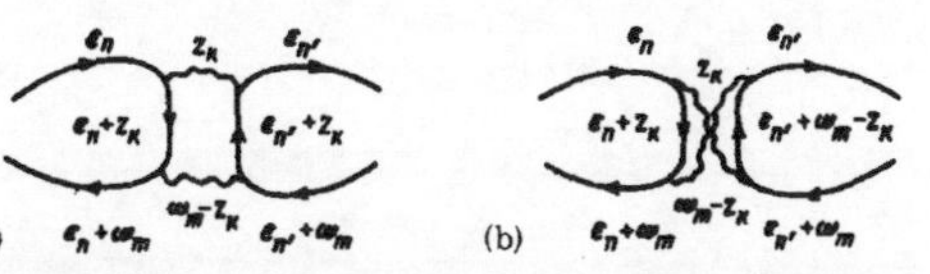

the period along the filament axis. This gives rise to new possibilities connected with Umklapp processes. It is easy to verify that in this case the matrix element should be multiplied by a factor $[1-g_{ph}^{*2}/g_{ph}^2]^2$, where g_{ph}^* corresponds to electron-phonon interaction with Umklapp. Generally speaking, there is no reason why this factor should be considered equal to zero, as was done in[4].

For an analytic continuation of (3) it is necessary to calculate the corresponding sums over the electron frequencies. Changing over to the integrals

$$\left(T\sum_n \ldots \to \frac{1}{4\pi i}\int \operatorname{th}(z/2T)\,dz\ldots\right),$$

we call attention to terms of the type

$$\frac{\omega}{2T}\int_{-\infty}^{+\infty}\operatorname{ch}^{-2}\frac{\epsilon}{2T}\,G_\epsilon^R G_{\epsilon+\omega}^A G_{z+\epsilon}^R\,d\epsilon,$$

which have a kinetic nature and therefore depend significantly on the relation between ω and $1/\tau$. These terms were omitted in[5], since the product of the electron loops (designated A^2 in[5]) was expanded in terms of the *Matsubara* frequencies (it appears that the same procedure was used in[4]). In fact the expression of[5] for $A^2(z)\propto z^2+2\omega z$ is replaced at $\omega\tau \ll 1$ by

$$z^2+\frac{2\pi^3}{7\zeta(3)}(\tau T_s)\omega z.$$

We finally obtain

$$\frac{\Delta\sigma}{\sigma_0}=\frac{v_F 7\zeta(3)}{\pi T_s S(2\pi)^3}\int d^3k\left\{\frac{T-T_s}{T_s}+\frac{7\zeta(3)(v_F k_\parallel)^2}{16\pi^2 T_s^2}+B\frac{k_\perp^2}{g_{ph}^2 S}\right\}^{-1}. \quad (4)$$

In the one-dimensional case this expression does not contain interaction constants, i.e., it does not depend on the range (cf.[4,5]). The main contribution to the integral (4) is made by large k, where the D function does not take the form (2), and the contribution from the fluctuations constitutes a small increment to the main conductivity. Thus, taking into account the three-dimensional character of the phonons, a square-root singularity exists only for the derivative

$$(1/\sigma_0)\frac{d\sigma}{dT}\sim\frac{1}{\sqrt{T-T_c}}.$$

The one-dimensional case corresponds formally to $B=0$, but in this case there is no phase transition, and (2) and (4) cease to be valid in the strong-interaction region[6] where, in principle, $\Delta\sigma/\sigma_0$ can become of the order of unity. Smallness of B would mean the absence of dispersion of the corresponding optical frequencies, for which no special reasons can be seen. It seems to us that the result is not limited to the considered phonon model of the structure transition, and has a more general character. Returning therefore to the experiments,[3] we point out that the weak maxima of $\sigma(T_s)/\sigma(300)\sim 10-20$ can be understood in principle in accordance with the remark made by the authors of[9], that the relaxation times increase as a result of the electron-electron processes like $T^{-1}f(\ln T)$. It is impossible to explain in this manner an increase of the conductivity by two orders of magnitude.

[1]All that is made is a qualitative statement that the fluctuations are important only if the number of electrons per period is not equal to unity.

[1]L. B. Coleman, M. J. Cohen, D. J. Sandman, F. G. Yamagishi, A. F. Garito, and A. J. Heeger, Sol. St. Comm. **12**, 1125 (1973).

[2]M. J. Cohen, L. B. Coleman, A. F. Garito, and A. J. Heeger, "The Electrical Conductivity of TTF-TCNQ," Preprint.

[3]J. W. Bray, J. Bardeen, and D. Allender, Phys. Rev. **9B**, 119 (1974).

[4]B. R. Patton and L. J. Sham, Phys. Rev. Lett. **31**, 631 (1973); Phys. Lett. **47A**, 133 (1974).

[5]S. Strassler and G. A. Toombs, Phys. Lett. **46A**, 321 (1974).

[6]Yu. A Bychkov, L. P. Gor'kov, and I. E. Dzyaloshinskiĭ, Zh. Eksp. Teor. Fiz. **50**, 738 (1966) [Sov. Phys.-JETP **23**, 489 (1966)].

[7]I. E. Dzyaloshinskiĭ and A. I. Larkin, *ibid.* **61**, 791 (1971) [**34**, 422 (1972)].

[8]L. P. Gor'kov and I. E. Dzyaloshinskiĭ, *ibid.* **67**, 397 (1974) [**39**, No. 1 (1974)].

[9]L. P. Gor'kov and I. E. Dzyaloshinskiĭ, ZhETF Pis. Red. **18**, 686 (1973) [JETP Lett. **18**, 401 (1973)].

J. Phys. I France 6 (1996) 1697–1710 DECEMBER 1996, PAGE 1697

Quasi-One-Dimensional Organic Metals: Theory and Experiment

L.P. Gor'kov (*)

National High Magnetic Field Laboratory, Florida State University, Tallahassee, Florida 32306-4005, USA
and
L.D.Landau Institute for Theoretical Physics, Russian Academy of Sciences. 117334 Moscow, Russia

(Received 25 June 1996, received in final form 5 August 1996, accepted 13 August 1996)

PACS.71.20.Rv – Polymers and organic compounds
PACS.71.27.+a – Strongly correlated electron systems; heavy fermions
PACS.75.30.Fv – Spin-density waves

Abstract. — Given the success of the weak coupling nesting model in explaining thermodynamical properties of the Bechgaard salts at low temperatures and in magnetic fields, we first concentrate on its implications to kinetics in the metallic phase. The model results in nonuniversal temperature dependencies of resistivity and magnetoresistance due to proximity of the metallic and the spin density wave phases, which are in a qualitative agreement with the available experimental data. We then analyze whether the phenomenological nesting model can be justified in frameworks of a more general model of electron-electron interactions in the one-dimensional system improved by three-dimensional effects of the interchain hopping. Properties of the Bechgaard salts look consistent with the Hubbard model with a weak repulsion. Considerable high temperature variation of the magnetic susceptibility is ascribed to localization of electrons by quasi-elastic scattering on thermal phonons. The fact that these materials correspond to the half-filled (hole) band was crucial for the analyses. Except for a new energy scale, introduced by the temperature of a spin density wave transition, no other electronic correlation effects stem from the analysis.

1. Introduction

The reader should not expect to find below a detailed comparison between theory and the experimental data. In our opinion, such a comparison would be a premature attempt, at least, if a more broad view is taken. Some experimental facts important for our discussion are not available yet, or the corresponding experiments still remain in a preliminary stage. In this presentation we discuss whether remarkable and diverse low temperature properties of quasi-one-dimensional (Q1D) conductors (especially, those of the Bechgaard's salts family) can be made consistent with predictions of the Fermi liquid theory. The issue is about the role played by electron-electron correlations in the system.

The class of materials has recently attracted again much of attention, both of experimentalists and theorists. There are important reasons for that. On the theory side, the compounds

(*) e-mail: gorkov@magnet.fsu.edu

realize the best known so far example of the so-called "quasi-one-dimensional" materials with a pronounced chain structure. As for the theory of electron-electron interactions in a one-dimensional (1D) system, it has been elaborated in great detail years ago [1, 2] (see also the book [3] for a few exact results). The strictly one-dimensional system would indeed possess numerous unusual interesting features, sometimes dramatically different from properties of electrons in ordinary metals described in frameworks of the Landau Fermi liquid theory. On the other hand, the discovery of high temperature superconductivity in cuprates, for instance, with all their unexpected properties, has produced the high demand for new ideas and theoretical approaches. Therefore, unusual features of the gas of interacting one-dimensional electrons are now very often viewed as an illustrative example, a "prototype" of nontrivial new physics, which may also be pertinent to many other systems, such as the 2D electron gas, heavy fermions, high T_c oxides, and so forth, remarkable for importance of electronic correlations (see, *e.g.* [4]).

In what follows, we concentrate our attention on properties of the selenium-based salts, $(TMTSF)_2X$. The detailed experimental information is available for these materials regarding their thermodynamic, transport and galvanomagnetic properties, especially, for the two of the most studied compounds, $(TMTSF)_2PF_6$ and $(TMTSF)_2ClO_4$. At very low temperature (~ 1 K) the materials may become superconductors (for reviews see [5]).

Taken as a whole, numerous existing low temperature experimental data are consistent with the idea of a strongly anisotropic three-dimensional metallic behavior in these systems, although with a few reservations. Namely, among many striking features characteristic of these materials, there is a proximity to some transition, a structural transition, such as the anion ordering (AO) in ClO_4, or a spin density wave (SDW) instability, which occur at rather low temperatures (10-20 K) in $(TMTSF)_2X$ with a symmetric anion, X. It is well established theoretically, that electron-electron interactions in the one-dimensional conductors reveal a tendency to produce various instabilities [1,2]; the instabilities then may wind up as a thermodynamical phase transition [6], if additional three-dimensional features, such as interchain electron tunneling, or Coulomb interactions between electrons on different chains, were added to the purely one-dimensional analysis. However, as the low temperature properties of the Bechgaard salts are concerned ($T < 50$ K), it seems, at least, at first sight, that for the selenium-based compounds, $(TMTSF)_2X$, there is no immediate necessity to invoke the one-dimensional physics at all. All phase transitions in the materials mentioned above are rather well defined mean field transitions with critical fluctuations important only in a narrow enough vicinity of the transition temperature. The mechanism for the SDW state itself, its suppression by external pressure, and the peculiar phenomenon of restoration of the SDW-like states in high magnetic fields, in particular, can be described even quantitatively in a simple and plausible model of weekly interacting electrons with the metallic Fermi surfaces possessing an approximate "nesting" property, as the starting point [7].

Some *sulfur* analogs of the Bechgaard salts, such as $(TMTTF)_2PF_6$, however, reveal somewhat more pronounced one-dimensional behavior (see discussion [5d]). A number of them show a tendency to instabilities at considerable higher temperature, sometimes exceeding estimated values for their transverse tunneling integrals. Therefore in these materials one may anticipate a more important role of the on-chain interactions, although, again, it seems that low temperature properties of both the sulfur- and selenium-based materials qualitatively are rather similar and can be mapped on the top of each other by applying external pressure.

As it is well known, the Landau Fermi liquid theory is nothing but a self-consistent approach based, in particular, on the assumption that in metals there is only one energy scale (typically, of an atomic order) which characterizes both the bare electronic spectrum itself and the strength of effective electron-electron interactions. The Q1D materials are remarkable for strongly anisotropic conducting properties. This fact alone means additional scales for the electron spectrum. As for electron-electron interactions, there are no special reasons for them to be strong, especially because high polarizabilities of the constituting organic molecules may essentially reduce the Coulomb forces. The spin density wave (SDW) transition, say, in $(TMTSF)_2PF_6$, occurs at $T_{SDW} \sim 12$ K. This scale is almost three order of magnitude less than the bandwidth $\sim$1 eV, as estimated for electronic motion along the chain direction. Again, as it was mentioned above, the superconducting transition occurs at $T \sim 1$ K. The weak coupling theory of the SDW-transition, based on the "nesting" of the two open Fermi surfaces, explains many experimental results at low temperatures surprisingly well (see [5]). There are, however, some facts which contradict the idea that interactions are so weak: the magnetic susceptibility at high enough temperatures (above $\sim$100 K) displays the considerable temperature change. This fact cannot be explained in terms of weakly interacting electrons (see [5d]). We will address this important issue.

The fact that many of the Bechgaard salts may exist in a non-metallic ground state at $T = 0$, implies that their low temperature properties are not the properties of a Fermi liquid in the strict sense. On the other hand, the SDW-phase can be removed by an external pressure, and the same material will now behave metallically down to lowest temperatures. The proximity to an insulating state, hence, casts some reasonable doubts regarding the Fermi liquid description in the adjacent metallic phase. There is a growing amount of evidences that, at least, *some* properties of these materials somehow differ from simple expectations of the theory of metals. Among these properties are: unusual temperature dependence of resistivity, large magnetoresistance in weak enough fields ($\sim$10 Tesla) and temperatures of order of 10 K, a non-Korringa temperature dependence of the NMR-relaxation rate (see [5d]). Note that in the three cases the new features appear in the kinetic characteristics.

In the next few Sections we show how the aforementioned simple model of two Fermi surface sheets with an approximate "nesting" properties could also explain some of the new findings in kinetics of the Q1D-organic conductors at low temperatures. (This model, however, does not rule out an involvement of different physics [4]). In Section 5, we will make an attempt, on a speculative level, though, to trace how interactions in Q1D conductor may renormalize themselves in a manner so that the weak coupling standard treatment of the SDW-phenomena remains correct. We will stress, however, that there are new facts which cannot be immediately explained by the theory, although it seems that the clue to understanding of these new findings lies in a 3D-low temperature physics of these compounds.

2. Model

Before going into more details, it is helpful to recall the model, some standard notations, and the main known facts used below. As usual, we approximate the monoclinic system of the Bechgaard salts by an orthorhombic symmetry with a as the lattice period along the chain direction, while b and c correspond to periodicity in directions transverse to the chain. For $(TMTSF)_2PF_6$ the values of the three main tunneling integrals along the corresponding axes are estimated as:

$$t_a \; : \; t_b \quad t_c{}' = 0.2 \text{ eV} \quad 200 \text{ K} \quad 10 \text{ K} \tag{1}$$

This anisotropy is consistent with data on the anisotropy of conductivity, as obtained at room temperatures. Such an anisotropy of conducting properties suggests that the "bare" electronic spectrum may be adequately described in terms of a tight binding model with only a few tunneling (hopping) matrix elements involved. (In view of the exceedingly large anisotropy between the b- and the c-directions, the materials are also called quasi-two-dimensional (Q2D) conductors). The tight binding spectrum of non-interacting electrons would be of a generic form:

$$E(\mathbf{p}) = v_{\mathrm{F}}(\pm \mathbf{p}_x - \mathbf{p}_{\mathrm{F}}) - 2t_b \cos \mathbf{p}_y b - 2t_c \cos \mathbf{p}_z c - 2t'_b \cos 2\mathbf{p}_y b \tag{2}$$

The first term describes an electron moving along the chain. It is linearized in the vicinity of each of the two open Fermi surfaces, for its right and left sides, respectively. Two other terms, t_b and t_c, correspond to hopping between nearest-neighboring chains along b and along c, respectively; as for the t'_b-term, it measures an effective hopping between the next nearest-neighboring chains. In absence of the latter term ($t'_b = 0$) the spectrum in equation (2) possesses the so-called "nesting" degeneracy:

$$E(\mathbf{p} + \mathbf{Q}) = -E(\mathbf{p}) \tag{3}$$

Here

$$\mathbf{Q} = \mathbf{Q}_0 \equiv (2\mathbf{p}_{\mathrm{F}};\ \frac{\pi}{b};\ \frac{\pi}{c}) \tag{2'}$$

At $t'_b \neq 0$ equation (3) takes place only approximately, provided that t'_b is small enough compared to t_b. (In what follows the tunneling integral t_c introduces no new physics and will be omitted for brevity).

The Bechgaard salts, $(\mathrm{TMTSF})_2\mathrm{X}$, are the charge transfer compounds with one electron transferred to each anion, X. Therefore, the salts are hole conductors with one hole per unit cell in the conducting TMTSF-network. The conduction band is half-filled; for a single metallic chain, $4\mathbf{p}_{\mathrm{F}} = 2\pi/a$.

For simplicity of the analysis, electron-electron interactions are usually chosen as short range (Coulomb) interactions. The effective screening in real materials is due to the fact that all chains are packed together into the lattice, so that even in absence of interchain hopping charges on neighboring chains may adjust themselves to screen a charge placed on a given chain. Such a screening will be rather anisotropic.

For electrons confined to one metallic chain all interactions at the Fermi surface are reduced to the following three interaction constants:

- g_1: the amplitude for the backward scattering of one electron by another;
- g_2: the forward scattering amplitude (each electron remains on its side of the Fermi surface after scattering);
- g_3: the Umklapp processes; the total momentum is not conserved.

(For the Hubbard model all three constants are equal: $g_1 = g_2 = g_3 = g^*$). The 1D analyses [1,2] proceeds assuming weak enough interactions, $g_i \ll 1$.

Interactions between electrons on a metallic one-dimensional chain may lead to a few instabilities which would compete with each other for opening of a gap at the Fermi level [1,2]. There are three competing channels: superconducting channel (SC); formation of charge density wave (CDW); the spin density wave channel (SDW). No thermodynamic phase transition is possible in one dimension.

To separate the above channels and to fix one of the above instabilities as the thermodynamical phase transition, it is necessary to account for three-dimensional features in the real materials. There are only two distinctly different three-dimensional effects [6]: 1) interactions between electrons on different chains and 2) interchain electron hopping. The type of the possible ground state essentially depends on which of these two effects prevails. However, in presence of any 3D-features the phase transition would happen at some low but finite temperature, T_c. The transition turns out to be a mean field transitions. The provision that the temperature interval for critical fluctuations is narrow, is given by smallness of all the interaction constants,

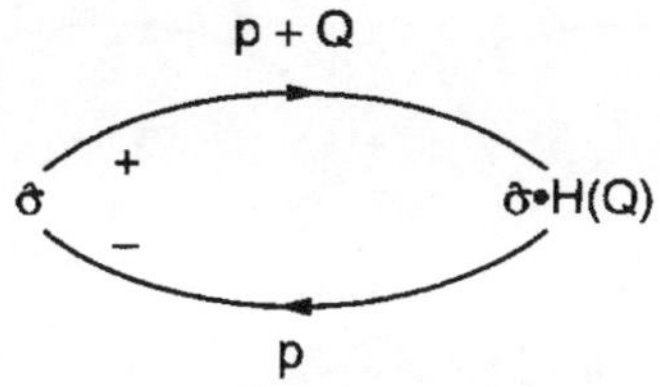

Fig. 1. — The generalized magnetic susceptibility for free electrons in its diagrammatic form: two lines correspond to the two Green functions in the matrix element of equation (5).

g_i, *i.e.*, $g_i \ll 1$ (In the $(TMTSF)_2X$-salts hopping prevails as a 3D effect; it is assumed below that $T_c \ll t_b$).

3. SDW-Instability

The nesting instability scenario has been extensively studied by many authors (*e.g.*, see [5] for review). We shall only outline below some major steps of the calculations. Thus, in case of the SDW-instability the analysis starts with calculation of the generalized linear magnetic response to a staggered field, $\mathbf{H}(\omega_0, \mathbf{Q})$:

$$\mathbf{M}(\omega_0; \mathbf{Q}) = \chi(\omega_0, \mathbf{Q})\mathbf{H}(\omega_0, \mathbf{Q}) \tag{4}$$

The free electron response function, $\chi(\omega_0, \mathbf{Q})$, is proportional to the matrix element constituted of two Green functions, shown in Figure 1:

$$\chi(\omega_0, \mathbf{Q}) \sim T\sum \int \mathrm{d}\mathbf{p}\, G(\omega_n; \mathbf{p})G(\omega_n - \omega_0; \mathbf{p} - \mathbf{Q}) \Rightarrow$$
$$\Rightarrow \int \mathrm{d}\mathbf{p}[i\omega_0 + E(\mathbf{p}) - E(\mathbf{p} - \mathbf{Q})]^{-1} \times [\tanh(E(\mathbf{p})/2T) - \tanh(E(\mathbf{p} - \mathbf{Q})/2T)] \tag{5}$$

At $\mathbf{Q} = \mathbf{Q}_0$ and $\omega_0 = 0$, the last integral becomes:

$$\nu(E_\mathrm{F}) \int \frac{\mathrm{d}E}{E} \tanh \frac{E}{2T} \sim 2\nu(E_\mathrm{F}) \ln(\bar{E}/T) \tag{5'}$$

where $\nu(E_\mathrm{F})$ is the electronic density of states (per spin), $\bar{E}$ is a cut-off energy. Note that when the vector $\mathbf{Q}$ is commensurate, the integral over momentum, $\mathbf{p}$, comprises of the two equal logarithmic contributions arising each from two sides of the Fermi surface (+ or −), because 2 $\mathbf{Q}$ is the reciprocal lattice vector.

With the above information concerning the single "bubble" diagram in Figure 1, consider corrections to $\chi(\omega_0, \mathbf{Q})$ due to electron-electron interactions. Some of those corrections are drawn schematically in Figure 2. Adding each new internal block into Figure 2 introduces a factor to the matrix element of exactly the same form as the integral calculated before, in equations (5, 5′). The integral is logarithmically divergent at low temperatures. This is the familiar logarithmic problem: at low enough temperatures the large logarithmic factors may compensate weakness of interactions, and the whole expression has to be summed up to take this fact into account. To find the effective interactions, Γ_2 and Γ_3, one is to solve the diagramatic equations for the two vertices shown in Figure 3. Their form is self-explanatory.

With the bare interactions, g_2 and g_3, independent on transverse momentum, the two equations become a system of the two algebraic equations, and at low temperatures each vertex

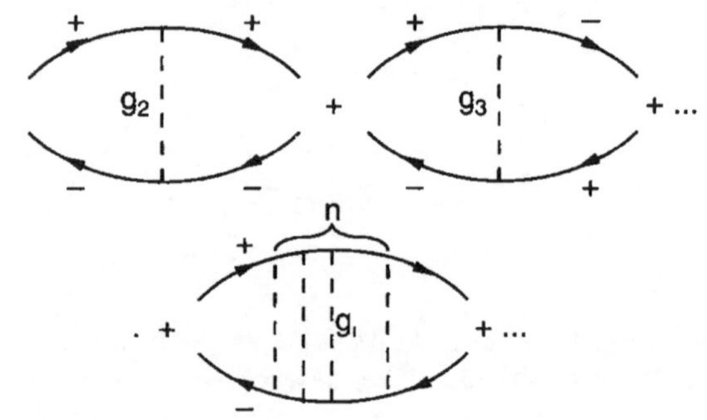

Fig. 2. — Electron-electron interactions (the dashed lines) form the "ladder" corrections to $\chi(\omega_0, \mathbf{Q})$; the two Green functions inside each block produce the logarithmic term of the form in equations (5, 5′)

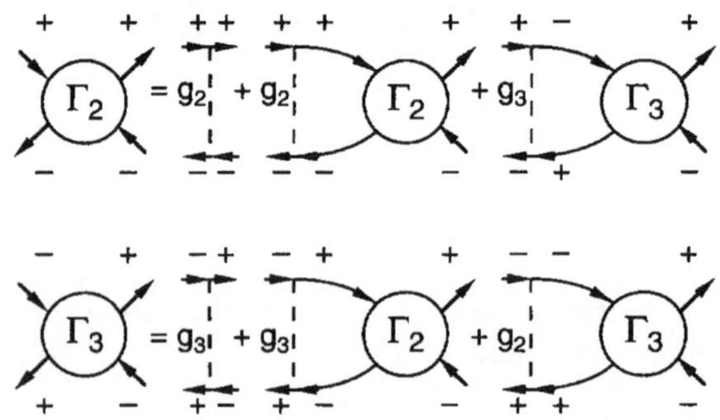

Fig. 3. — Equations for the renormalized vertices, Γ_2, Γ_3; their structure reflects the structure of diagramatic corrections in Figure 2.

acquires the form:

$$\Gamma^* = g^* \left[1 - 2g^* \ln \frac{\bar{E}}{T}\right]^{-1} \equiv \frac{1}{2} \ln \frac{T}{T_0}, \quad \Gamma^* \equiv \Gamma_2 = \Gamma_3 \tag{6}$$

(at $g_2 \equiv g_3$, denoted as g^*). Here T_0, as defined by the standard relation:

$$1 = 2g^* \ln \frac{\bar{E}}{T_0} \tag{7}$$

determines the temperature for the onset of the SDW-phase. It is easy to verify that the mean field character of the transition is guaranteed at $T_0 \ll t_b \ll E_F$.

The main contribution into integral (5), which provides the pole in Γ's, is due to the logarithmic contribution from the extended interval $T_0 \ll E \ll \bar{E} \sim E_F$. Deviations from the exact nesting condition, for instance, due to non-zero t'_b in equation (2), only weakly influence the contribution from that main integration interval, and may not destroy formation of a low temperature SDW-phase.

4. Precursor Effects in Kinetics

As it has already been emphasized above, outside a narrow temperature interval around T_0, $\Delta T/T_0 \ll 1$, fluctuations introduce only negligible corrections to all physical quantities. Therefore the critical fluctuations cannot account for any considerable effect in a wide temperature interval (10-40 K) where unexpected features have been observed in the temperature dependence of resistivity (and in the magnetoresistance) for $(TMTFT)_2X$, where $X = PF_6$ and ClO_4 (see [5d, 8, 9]).

Consider properties of two vertices, Γ_2, Γ_3, in more details. Each of them diverges at $T = T_0$ for $\mathbf{Q} = \mathbf{Q}_0$ and $\omega_0 = 0$. At $T \sim T_0$ and $\mathbf{Q} = \mathbf{Q}_0 + \mathbf{q}$, where $\mathbf{q}$ is small:

$$2t_b \mathbf{q}_y b \sim T_0 \tag{8}$$

the Γ's are finite and are of the order of unity. In other words, in the above domain of the transverse momentum transfer the effective interactions become large compared to the weak bare interactions, $g_i \ll 1$. This interval however, constitutes only a small fraction of the Fermi surface area of order of $T_0/t_b \ll 1$. Here lies the main distinction between the 1D and the 2D- and 3D-cases. In case of uncoupled linear chains each chain exercises the independent renormalization of interactions. If a vertex becomes large as the result of this, the corresponding enhancement takes place along the whole Fermi surface in the Brillouin zone. On the contrary, if three-dimensional effects cut-off the 1D-renormalization at $E \sim t_b$, the phase space becomes restricted by the narrow interval (8), so that corrections due to the proximity to a phase transition usually do not contribute significantly to the Fermi liquid characteristics. We show now that in the model under discussion, the above interval (8), however, plays an important role in the kinetics. Thus, for electron relaxation times the above interval in the momentum space *is singled out by the conservation laws of energy and momentum*, which have to be fulfilled at a real process of electron-electron scattering [10].

To elaborate on the idea, consider in some more details the temperature dependence of resistivity, $\rho(T)$. When electron-electron scattering is the dominant mechanism of resistivity, as is probably the case for these rather clean materials in the temperature range under consideration, the dissipation rate for momentum, $1/\tau_U$, is proportional to the matrix element of the Umklapp scattering, g_3. Omitting all nonessential factors, one can write $1/\tau_U$ in the following familiar form:

$$\frac{1}{\tau_U} \propto |g_3|^2 \frac{1}{T} \int n_{\mathbf{p}_1} n_{\mathbf{p}_2} (1 - n_{\mathbf{p}_3})(1 - n_{\mathbf{p}_4}) \times \delta(\mathbf{p}_1 + \mathbf{p}_2 - \mathbf{p}_3 - \mathbf{p}_4 - \mathbf{K}) \delta(E_1 + E_2 - E_3 - E_4)\, d\mathbf{p}_1 \ldots d\mathbf{p}_4 \tag{9}$$

Here the Fermi functions, $n_{\mathbf{p}}$, together with the δ-functions expressing the conservation laws of energy and momentum, would lead to the well-known T^2-dependence for $\rho(T)$ (more exactly, in 2D-case $\rho(T) \sim T^2 \ln T$). Making use of the expression (2) for the total electron energy spectrum, as a sum of two components for the longitudinal and transverse dispersions, respectively:

$$E(\mathbf{p}) = \xi(\mathbf{p}_x) + \varepsilon(\mathbf{p}_\perp) \tag{10}$$

and choosing the Umklapp vector as $\mathbf{K} = (4\mathbf{p}_F; 0; 0)$, the conservation laws in (9) may be identically re-written in the following manner:

$$\delta(\mathbf{p}_{1\perp} + \mathbf{p}_{2\perp} - \mathbf{p}_{3\perp} - \mathbf{p}_{4\perp})\ \delta(E_1 + E_2 - \frac{\Omega}{2})\ \delta(E_3 + E_4 - \frac{\Omega}{2}) \tag{11}$$

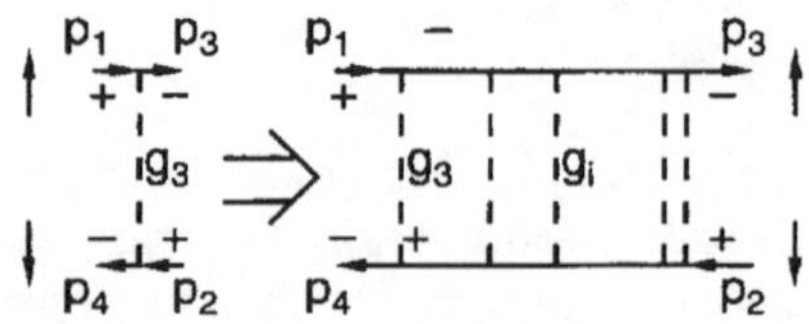

Fig. 4. — The bare Umklapp matrix element, g_3, renormalized by interactions along the SDW-channel.

Here $\Omega = \sum_i \varepsilon(\mathbf{p}_{i\perp})$). The resulting expression for (9) can be easily integrated over all energies, $E(\mathbf{p}_i)$, and for the relaxation rate $1/\tau_{\rm U}$ one obtains:

$$\frac{1}{\tau_{\rm U}} \propto \frac{|g_3|^2}{T} \int \Omega^2 \sinh^{-2}\left(\frac{\Omega}{2T}\right) \times \delta(\mathbf{p}_{1\perp} + \mathbf{p}_{2\perp} - \mathbf{p}_{3\perp} - \mathbf{p}_{4\perp})\, d\mathbf{p}_{1\perp} \ldots d\mathbf{p}_{4\perp} \qquad (12)$$

The merit of this transformation is that now one can explicitly see which part of the momentum space makes the main contribution into resistivity. At low enough temperatures, $T \sim T_0$, such a contribution comes from a vicinity of the phase space where $\Omega \sim T$, (at $T \ll t_b$), that is Ω in (12) is to be small. Without the t_b'-term, for the brevity sake, it is possible to present Ω in the explicit form:

$$\Omega = -8t_b \cos\left(\frac{(\mathbf{p}_{1y} + \mathbf{p}_{2y})b}{2}\right) \cdot \cos\left(\frac{(\mathbf{p}_{3y} - \mathbf{p}_{1y})b}{2}\right) \cdot \cos\left(\frac{(\mathbf{p}_{3y} + \mathbf{p}_{2y})b}{2}\right) \qquad (13)$$

Thus, there are three planes in the $(\mathbf{p}_{1y}; \mathbf{p}_{2y}; \mathbf{p}_{3y})$-phase space on which $\Omega = 0$; in the discussion of the SDW-transition we need to consider $\mathbf{p}_{3y} - \mathbf{p}_{2y} = \pi/b$ (We omitted the t_c-term.).

In fact, as shown in Figure 4, one should again consider the "ladder"-type corrections in Figure 2 instead of the bare matrix element g_3, because each "bubble" introduces the large logarithmic contributions, that has led to expressions (5′) and (6) (Note the opposite directions of two spins for the Γ's contributing into the SDW-channel.). The choice of $\mathbf{p}_{3y} - \mathbf{p}_{2y} = \pi/b$ together with the longitudinal component equal to $2\mathbf{p}_{\rm F}$ specifies the momentum transfer between $\mathbf{p}_3$ and $\mathbf{p}_2$ exactly equal to $\mathbf{Q}_0$, equation (2′), responsible for the SDW-instability itself. Finite temperatures in (12) spread essential $\mathbf{Q}$'s over interval (8). Therefore, the sum of diagrams shown in Figure 4, would provide an enhancement of the bare matrix element g_3. The corresponding equations are those in Figure 3. After solving the equations for Γ's, and substituting the results into equation (9) ($g_3 \to \Gamma_3$), one gets for $1/\tau_{\rm U}$ an expression which significantly modifies the resistivity temperature dependence on the temperature interval of the order of T_0 itself. We will not go into further details, and we will restrict ourselves to simple physical implications from the above considerations.

Thus, the relaxation rates for various scattering processes acquire an additional temperature dependence due to proximity of the metallic phase in the Bechgaard salts under pressure to a SDW-state. The scattering processes between electrons shown in Figure 4, depend on energies and momenta of interacting electrons. This fact can be taken into account by a modification of the effective matrix element of the form:

$$g_3 \Rightarrow \frac{1}{2} / \ln\left(\frac{T}{\alpha T_0}\right) \qquad (14)$$

(Here $\alpha < 1$ is a constant which can be determined only from the full solution.)

The new effective interaction is *of order of unity* in the temperature range $T \sim T_0$, even if the bare interactions were small. The dominant contribution into resistivity in this temperature range is now of the form:

$$\rho(T) \Rightarrow \frac{T^2}{\ln^2\left(\frac{T}{\alpha T_0}\right)} \tag{15}$$

Temperature dependence in equation (15) displays a new peculiarity in that it is not monotonous and has a minimum at

$$T_{\min} = e\alpha T_0 \; ; \; (e = 2.71..) \tag{16}$$

Note that resistivity remains finite at the transition temperature ($\alpha < 1$). The fast increase of resistivity starts below the transition, together with the development of the SDW-gap. (In the very vicinity of the transition all quantities, including resistivity, are corrected by critical fluctuations.)

So far we were omitting the so-called "anti-nesting" term in equation (2). Calculations at $t'_b \neq 0$ are more tedious. From the physical point of view, however, this term suppresses the SDW-phase and with the further increase above some critical value, t'_b, stabilizes the metallic state. (The term t'_b, therefore, mimics the role of the external pressure.) We incorporate these features by re-writing equation (15) in the following manner:

$$\rho(T) \Rightarrow \frac{T^2}{\ln^2\left(\frac{T + \beta t'_b}{\alpha T_0}\right)} \tag{17}$$

Here β is another constant. Remember that T_0 is T_{SDW} in absence of the t'_b term. To find the temperature, T_{SDW}, of the actual SDW-transition as a function of pressure (*i.e.*, t'_b), one has to return back to equations in Figure 3 in presence of $t'_b \neq 0$.

The temperature at which $\rho(T)$ may have a minimum, would now be determined from equation:

$$\ln\left(\frac{T + \beta t'_b}{\alpha T_0}\right) = \frac{T}{T + \beta t'_b} \tag{18}$$

Depending on the t'_b-value, the resistivity minimum may, or may not be seen in $\rho(T)$.

The temperature dependence of resistivity in the metallic phase of $(TMTSF)_2PF_6$ is a somewhat controversial issue. However, some main peculiarities in the resistivity behavior are known for a long time. The resistivity displays a few rather unusual features [5a]. The most remarkable are 1) considerable deviations of $\rho(T)$ at lower temperatures (10-30 K) from the T^2-dependence (resistivity decreases faster! [9]) and 2) large positive magnetoresistance in the same temperature range for rather low fields of order of 10 Tesla. It was first suggested in [5a] that these peculiarities may be a manifestation of strong superconducting fluctuations, which develop in these chain materials due to strong electron-electron correlation effects, specific for the one-dimensional physics. In view that the superconducting transition temperature in $(TMTSF)_2PF_6$ is about 1 K, and superconductivity itself has the well-defined three-dimensional character, such an explanation looks somewhat exotic.

A more detailed recent study [8] of magnetoresistance effects performed on $(TMTSF)_2ClO_4$, has revealed that *there is* a minimum in resistivity plotted as a function of temperature at a given magnetic field (in the field range 12-27 Tesla). The corresponding temperature, $T_{\min}(H)$, increases considerably with the field increase. The two materials, $(TMTSF)_2PF_6$ and $(TMTSF)_2ClO_4$, are known for similarity in many of their properties, and we will discuss whether the theoretical results above can help provide a qualitative explanation of the experimental findings [10].

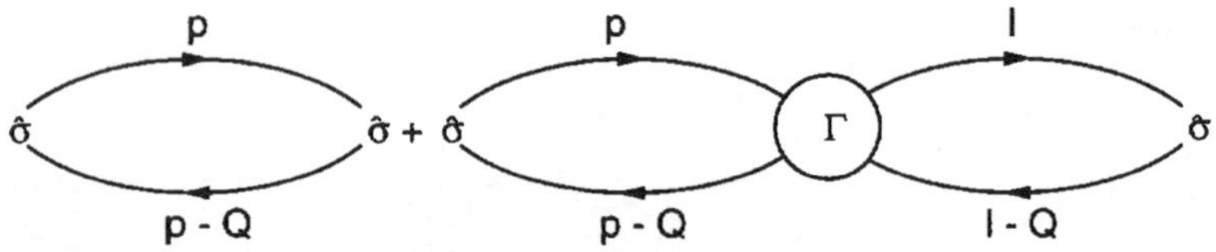

Fig. 5. — The generic form of the magnetic susceptibility response for interacting electrons. Unlike in Figure 1, the lines correspond to the exact Green functions.

As for temperature dependences of resistivity of $(TMTSF)_2PF_6$, note first that no minimum has been observed at zero field [9]. In accordance with equation (18), this fact would impose some restrictions (from below) on the actual value of t'_b. Assuming again a similar behavior for both materials, ClO_4 and PF_6 (at least, in higher fields), we conclude that for the minimum, $T_{\min}(H)$, to develop with the field increase, magnetic field itself must somehow diminish the effect of t'_b. Such a mechanism is well known [7a]. Taking the scale for the effect of the magnetic field on the "anti-nesting" term from [7a], we may assume in equations (17, 18):

$$t'_b \Rightarrow t'_b \cdot \frac{\lambda}{1+\lambda}, \quad \text{where } \lambda = \frac{2ct'_b}{v_F eHb} \equiv \frac{2t'_b}{\omega_c} \tag{19}$$

With the further increase of the cyclotron frequency, ω_c, in large fields $T_{\min}(H)$ would tend to its value from equation (16). Also if the interpretation suggested above, is correct, a plot of $\rho(T)\cdot T^{-2}$ *vs.* T would merely measure the temperature dependence of the effective relaxation time, in accordance with equations (15, 17).

5. Renormalized Interactions at Low Temperatures

While the precursor effects of the preceding section may successfully explain the non-Fermi liquid features in the temperature dependence of the single particle relaxation, some other facts need further discussion.

We have already mentioned that the NMR-relaxation rate, T_1^{-1}, both in $(TMTSF)_2ClO_4$ and in the metallic $(TMTSF)_2PF_6$ (under pressure) display pronounced deviations from the familiar Korringa behavior in the temperature interval 5-30 K [5d, 11]. For the free electron gas the Korringa law is obtained as the probability of the nuclear spin to flip due to a hyperfine interaction with conduction electrons, $V_{p,p'}$.

$$T_1^{-1}\alpha \int\int |V_{\mathbf{p},\mathbf{p}-\mathbf{Q}}|^2 n_{\mathbf{p}}[1-n(\mathbf{p}-\mathbf{Q}]\delta(E(\mathbf{p})-E(\mathbf{p}-\mathbf{Q}))d^3\mathbf{p}d^3\mathbf{Q} \propto T\nu(E_F)^2 \tag{20}$$

The momentum transfer, $\mathbf{Q}$, between the initial and final electronic states in equation (20), is arbitrary because relaxation of the nuclear spin has the local origin. T_1^{-1} can also be expressed in terms of the generalized magnetic susceptibility:

$$T_1^{-1} \propto T \sum_{\mathbf{Q}} \mathrm{Im}\chi(\mathbf{Q},\omega=0) \tag{20'}$$

The diagramatic expression for $\chi(\mathbf{Q})$ is shown in Figure 5.

The vertex $\Gamma(\mathbf{p},\mathbf{l}-\mathbf{Q};\ \mathbf{p}-\mathbf{Q},\mathbf{l})$ is of a general form and includes all electron-electron interactions in the system. For a *single chain* the vector $\mathbf{Q}$ has only two components: $\mathbf{Q}_{\|} \simeq 0, \pm 2\mathbf{p}_F$. The $2\mathbf{p}_F$-component is always strongly affected by interactions in the 1D case. On the other hand, the proximity to the three-dimensional SDW transition, as described above, results in the enhancement of Γ's in Figure 3 only in a narrow interval, as given by equation (8), and, hence, is not expected to drastically change the temperature behavior of T_1^{-1} Therefore, the low temperature NMR data [11] remain unexplained in terms of the nesting model. The interpretation of the T_1^{-1}-behavior in terms of one-dimensional physics, suggested by the Orsay group, seems to be self contradictory in view of so many other facts which reliably establish the 3D features in these compounds at low enough temperatures. In the simple interpretation of equation (20′), the NMR relaxation rate is a measure of the density of states. We conjecture that some undetected structural transition, probably related to a phonon mode dealing with anions, superimposes itself on the phenomena described in the preceding section.

There is one more troublesome question regarding the strength of electron-electron interactions in the $(TMTSF)_2X$ salts. From the point of view that $T_{SDW} \simeq 12$ K is so low compared to the electron bandwidth ~0.8 K [5d], it seems plausible to interpret this last fact as an argument in favor of a rather weak coupling in the system. On the other hand, it seems that for some other phenomena related to a higher energy scale, this conclusion may not be true. Thus, the optical data [12], photoemission experiments [13], and some others have shown deviations from a single particle description in the frequency range $\omega \sim 100 - 1000\ \mathrm{cm}^{-1}$.

We will stay in more details on the temperature dependence of the static magnetic susceptibility, $\chi(T)$, as an example. Instead of the temperature independent Pauli susceptibility, expected for a free electron gas, the magnetic susceptibility for many of the $(TMTSF)_2X$ salts displays a sizable (~50%) decrease from 300 K down to approximately 100 K (see [5d]). The contribution into $\chi(T)$ due to *interactions* is represented by the vertex Γ in Figure 5, and $(\mathbf{Q} \equiv 0)$ is proportional to the integral:

$$\delta\chi \propto \int\int \frac{1}{T}\mathrm{ch}^{-2}(\frac{E(\mathbf{p})}{2T})\Gamma(E(\mathbf{p}), E(\mathbf{l});\ E(\mathbf{p}), E(\mathbf{l}))\frac{1}{T}\mathrm{ch}^{-2}(\frac{E(\mathbf{l})}{2T})\mathrm{d}^3\mathbf{p}\mathrm{d}^3\mathbf{l} \qquad (21)$$

As it was shown in [2], in the logarithmical approximation, $g_i \ll 1$, this vertex has the form:

$$\Gamma \Rightarrow \gamma_1 = \frac{g_1}{1 + g_1 \ln \dfrac{\bar{E}}{T}} \qquad (21')$$

Equation (21′) is applicable at $T > t_b$ and may give an essential behavior of $\delta\chi$ at $g_1 < 0$ at temperatures of order of T^*:

$$T^* \sim \bar{E} \exp - \frac{1}{|g_1|} \qquad (22)$$

where coupling would become strong. One option is to suggest that T^*, the temperature below which purely one-dimensional interactions would manifest the strong coupling regime, is by the order of magnitude equal to the transverse tunneling integral, t_b. At lower temperatures the strong coupling behavior in (21, 21′) would become gradually saturated. By analogy with equations (4, 7), one could assume that the temperature behavior of Γ in (21) below T^* is of a form:

$$\Gamma \simeq - \frac{1}{\ln \dfrac{T^*}{T + \alpha t_b}} \qquad (23)$$

and that namely this last regime is experimentally seen in $\chi(T)$ before it finally saturates to the low temperature constant, χ_s. Expression (23) reflects the tendency in (21′) to a substantial increase of bare interactions as it is given by γ_1, which then is cut off by the value of the tunneling integrals. The temperature dependence in $\chi(T)$ below T^* would be on the scale of

T/t_b. For $|g_1|$ one would estimate from (22) a bare coupling as $|g_1| \sim \frac{1}{2} \div \frac{1}{3}$. (Our approach to evaluation of the bare $|g_1|$ differs from the one adopted in [5d]). From the above one would conclude that the Fermi liquid characteristics below $T^*(\sim t_b)$ are renormalized by factors of order of unity.

Such a scenario, although it may explain the high temperature susceptibility behavior, is difficult to reconcile with other low temperature properties which are in a general agreement with the nesting model. For instance, let us consider, as before, the Hubbard model, for simplicity sake. Then $g_1 = g_2 = g_3 = g^*$. Recall that to get the SDW-transition in terms of the nesting model, g^* had to be positive. Also, in the 1D-physics a negative $g_1 < 0$ would enhance the Cooper channel for expense of the SDW-channel at $T \sim t_b$ [2]. Therefore, we will temporarily put aside the discussion of the $\chi(T)$-behavior, and will seek first for a possibility of deriving such a weak coupling scheme for the three-dimensional SDW-transition which would smoothly merge with the one-dimensional physics at higher energies. In our opinion, the smallness of T_{SDW} is indeed a manifestation of the fact that interactions $g^* > 0$ are reasonably small. Consider now vertices Γ_2 and Γ_3 in Figure 3 and write down the corresponding solutions for the purely one-dimensional problem [2]:

$$\Gamma_2 \Rightarrow \gamma_2 = \frac{(1/2)}{\xi_0 - \xi}; \quad \Gamma_3 \Rightarrow \gamma_3 = \frac{1}{\xi_0 - \xi} \tag{24}$$

In equation (24):

$$\xi = \ln \frac{\bar{E}}{T}, \; \xi_0 = \frac{1}{g^*} = \ln \frac{\bar{E}}{T_0} \tag{24'}$$

and the cut-off, $\bar{E} \sim E_{\mathrm{F}}$. (At $g^* > 0$, γ_1, in (21′) has no pole).

In the case of the 1D problem, the logarithmic divergencies are inherent in all channels, while the logarithmic contributions in equation (5) are due to the "nesting channel" at $\mathbf{Q} = \mathbf{Q}_0$. Therefore, for the integrals in Figure 2 the logarithmic contribution from the integration interval $E < t_b$ is to be treated by taking, instead of the bare g_2 and g_3, their expressions corrected by all contributions from the higher energy intervals, $t_b < E < \bar{E}$. As for the interval $E > t_b$ in equation (5) and in the equations in Figure 3, it should be considered in the frameworks of a single chain 1D problem. The results of that treatment are expressions of equation (24). We now assume that T_0 (24′) is so small that the 3D-tunneling, t_b, significantly changes a strong coupling 1D regime in (24):

$$T_0 \ll t_b \tag{25}$$

so that the purely 1D expressions in (24) have to be merged with the 3D-physics of equation (2). The corresponding analysis can be done in the manner very close to the treatment of the low temperature logarithmic singularity in the Cooper channel done in [6]. After some calculations with the logarithmic accuracy, one gets the estimate for the SDW-transition temperature:

$$T_{\mathrm{SDW}} \sim t_b \left(\frac{T_0}{t_b} \right)^{2/3} \tag{26}$$

This expression is consistent with a rather small $g^*(\sim \frac{1}{4} \div \frac{1}{5})$. If the above view indeed describes the real situation in the $(\mathrm{TMTSF})_2\mathrm{X}$-salts, there are no essential renormalizations in the density of states or other characteristics of the electron spectrum. The results (26) can be somewhat modified if instead of the Hubbard model, a more general form of electron-electron interactions is chosen, *i.e.* $g_1 \neq g_2 \neq g_3$.

The previous discussion has shown that to justify the low temperature "nesting" model, one has to accept the assumptions which make the interpretation of the $\chi(T)$-dependence at higher temperatures in terms of electron-electron interactions not possible. Hence, one is to look for alternative mechanisms. We see one possibility in the role of the lattice degrees of freedom. If $\bar{\omega}$ is a phonon frequency for some mode, at $T \gg \bar{\omega}$ the lattice produces a quasi-static random potential for scattering of electrons. It has been noticed [14] that in the case of the half-filled band ($4\mathbf{p}_\mathrm{F} = 2\pi/a$) static defects not only localize the one-dimensional electrons, but they may also produce a change in the electronic *density of states* (DOS) leading to a pronounced singularity at the chemical potential. We argue that the change in DOS at high temperature due to the thermally activated random lattice potential could be a cause for the magnetic susceptibility temperature dependence: with the temperature decrease scattering of electrons becomes inelastic, chains get coupled, and the additional DOS gradually disappears.

6. Conclusions

The bare tight-binding electronic spectrum seems to be a good starting point for the Bechgaard salts. Restricting the electron tunneling only to the nearest- and next-nearest-neighboring chains results in the two open Fermi surfaces with the approximate nesting degeneracy. The experimental energy scale for all low temperature phenomena is of order of 10-30 K and is rather small compared with the total bandwidth ~0.8 eV. This suggests that the electron-electron interactions in the system may be rather weak. The weak coupling model utilizing the nesting degeneracy, is known to properly describe the most striking features in these salts, such as the SDW-transition itself, its pressure dependence, and the restoration of the SDW-state in magnetic fields. It was shown above that this very model leads to considerable deviations from the Fermi liquid temperature dependence for various single particle relaxation rates, which may account for unusual temperature dependencies of the resistivity and magnetoresistance. More experiments have to be done, as the two last phenomena are concerned.

We have demonstrated that the phenomenological model can be justified by the analysis which treats all the energy scales on a rigorous basis to merge the one-dimensional and three-dimensional regimes. The fact that the $(\mathrm{TMTSF})_2\mathrm{X}$-salts have exactly one hole per unit cell has been extensively used. First, it specifies the mechanism of resistivity at low temperatures *via* the Umklapp processes. Secondly, it is the commensurability, $4\mathbf{p}_\mathrm{F} = \pi/a$, that guarantees the tendency to the SDW-instability in the 1D-approximation for the repulsive interaction. In addition to that, the commensurability effects may also be responsible for the puzzle with the temperature dependence of the magnetic susceptibility, as indicated at the end of the previous section.

In the scenario described above, there is almost no room seen for strong electron-electron correlations. Interactions between electrons modify all properties by introducing a new energy scale, T_SDW. In the metallic phase, a proximity to the SDW-transition may result in a non-Fermi liquid temperature dependence for the resistivity and related properties. The quasi-particle relaxation rates acquire an additional dependence on temperature related to this energy scale. Experimental situation remains, nevertheless, somewhat unclear, because this mechanism may explain, at least, qualitatively, low temperature behavior of resistivity and magnetoresistance, but is not applicable for the NMR-relaxation rate anomaly. Interrelations between the SDW-transitions and the structural (anion ordering) transitions are not well understood yet, either experimentally or theoretically.

Acknowledgments

The article is dedicated to the respectful memory of Prof. I.F. Schegolev. In more than thirty years of our scientific and personal relationship there were not many problems in the field which have not been discussed together. The author acknowledges with gratitude numerous stimulating discussions with Dr. Jérome. This work was supported by NHMFL through NSF cooperative agreement #DMR-9016241 and the State of Florida.

References

[1] Bychkov Yu.A.. Gor'kov L.P. and Dzyaloshinskii I.E., *Sov. Phys. JETP* **23** (1966) 489.

[2] Dzyaloshinskii I.E. and Larkin A.I., *Sov. Phys. JETP* **34** (1972) 422.

[3] The Many-Body Problem: An Encyclopedia of Exactly Solved Models in One-Dimension, D.C. Mattis Ed. (World Scientific, Singapore, 1993).

[4] Clarke D.G., Strong S.P. and Anderson P.W., *Phys. Rev. Lett.* **72** (1994) 3218.

[5] a) Jérome D. and Schulz H.J., *Adv.Phys.* **31** (1982) 229; b) Gor'kov L.P., *Sov. Phys. Usp.* **27** (1984) 809; c) Ishiguro T. and Yamaji K., Organic Superconductors (Springer-Verlag, Berlin, 1990); d) Jérome D., in "Organic Conductors, J.P. Farges Ed. (M. Dekker, New York, 1994).

[6] Gor'kov L.P. and Dzyaloshinskii I.E., *Sov. Phys. JETP* **40** (1975) 198.

[7] a) Gor'kov L.P. and Lebed A.G., *J. Phys. Lett. France* **45** (1984) L-443; b) Chaikin P.M., *Phys. Rev.* **31** (1985) 4770; c) Heritier M., Montanbaux G., and Lederer P., *J. Phys. Lett. France* **45** (1984) L-943.

[8] Behnia K. *et al., Phys. Rev. Lett.* **74** (1995) 5272.

[9] Balicas L. and Jérome D., (unpublished).

[10] Gor'kov L.P., *Europhys. Lett.* **31** (1995) 49.

[11] Wzietek P. *et al., J. Phys. I France* **3** (1993) 171.

[12] Donovan S., Degiorgi L. and Grüner G., *Europhys. Lett.* **19** (1992) 443; Donovan S. *et al., Phys. Rev.* **B49** (1994) 3363; Cao N., Timusk T. and Bechgaard K., *Bull. Am. Phys. Soc.* **41** (1996) 118.

[13] Dardel B. *et al.. Europhys. Lett.* **24** (1993) 687.

[14] Bulayevsky L.N. *et al.*, *ZhETF* **62** (1972) 725; Gor'kov L.P. and Dorkhov O.N., *Solid State Commun.* **20** (1976) 789; *ibid* **21** (1977) N7; *Fiz. Niskich Temp.* **4** (1978) 332.

J. Physique Lett. **46** (1985) L-643 - L-646 15 JUILLET **1985**, PAGE L-643

Classification
Physics Abstracts
74.70

Back to the problem of the upper critical fields in organic superconductors

L. P. Gor'kov

L. D. Landau Institute for Theoretical Physics, The U.S.S.R. Academy of Sciences, 117334 Moscow V-334, U.S.S.R.

and D. Jérome

Laboratoire de Physique des Solides (*), Université Paris-Sud, 91405 Orsay Cedex, France

(Reçu le 8 avril 1985, accepté le 29 mai 1985)

Résumé. — Nous montrons que les valeurs des champs critiques supérieurs des supraconducteurs organiques peuvent être interprétées dans le cadre de la théorie des supraconducteurs propres de type II. L'observation de valeurs dépassant la limite de Clogston pourrait indiquer l'existence d'une supraconduction avec couplage triplet.

Abstract. — It is shown that the properties of the upper critical fields in organic superconductors can be understood in terms of pure type II superconductivity. The experimental observation of high critical fields violating Clogston's criterion could give support in favour of triplet-paired superconductivity.

In this short Letter we shall reinvestigate the theoretical interpretation of the experimental observed upper critical fields, H_{c_2}, in the Bechgaard salt superconductors $(TMTSF)_2X$. We want to comment as to whether evidence in favour of the Pauli limitations (the Clogston criterion) are really present in these materials. Owing to the rather large values of the critical fields H_{c_2} obtained in these superconductors [1], the data has been interpreted in terms of the so-called dirty limit of the BCS-like superconductors [2]. Meanwhile, there is strong evidence that these materials can be extremely clean, which is inconsistent with this approach. Thus, the concentration of defects, as estimated from the irradiation effect on the transition temperature T_c (and on the residual resistance) gives approximately 100 ppm of defects per molecule [3]. In other words, the mean free path along the chain (i.e. the a-direction) is $l > 10^4\, a$, where a is the lattice constant. Taking the tight binding estimate for the electron velocity, v_a :

$$v_a = \frac{a t_a}{\hbar \sqrt{2}} \tag{1}$$

(*) Laboratoire associé au CNRS.

and $t_a \approx 2\,000$ K [4], one immediately obtains

$$\hbar/\tau \lesssim 0.1 \text{ K}$$

while T_c is about 1 K.

Because of these reasons we adopt the point of view that these superconductors are of the IInd type already in the pure limit, and we shall try to use the microscopic expressions for the effective mass tensor in the Ginzburg-Landau equation (near T_c) derived in reference [5].

$$\frac{1}{m_{ik}} = \frac{7\,\zeta(3)}{12\,\pi^2\,T} \langle\, v_i\, v_k \mid \psi^2 \mid \rangle\,. \tag{2}$$

(Here the average is taken over the Fermi surface, $\psi(k)$ is the normalized superconductivity order parameter, $\langle \mid \psi^2 \mid \rangle = 1$.)

Using the orthorhombic approximation for the $(TMTSF)_2X$-lattice and the tight binding dispersion law for all three directions (which is useful for the purpose of semiquantitative estimations),

$$\varepsilon(\tilde{p}) = 2\,t_a \cos \tilde{p}\,\frac{\tilde{a}}{2} + 2\,t_b \cos \tilde{p}\tilde{b} + 2\,t_c \cos \tilde{p}\tilde{c}\,, \tag{3}$$

one gets the following « mass-tensor » :

$$\begin{aligned} \frac{1}{m_a} &= \frac{7\,\zeta(3)\,t_a^2\,a^2}{24\,\pi^2\,T_c\,\hbar^2} \\ \frac{1}{m_b} &= \frac{7\,\zeta(3)\,t_b^2\,b^2}{6\,\pi^2\,T_c\,\hbar^2} \\ \frac{1}{m_c} &= \frac{7\,\zeta(3)\,t_c^2\,c^2}{6\,\pi^2\,T_c\,\hbar^2}\,. \end{aligned} \tag{4}$$

A variation of the factor $\mid \psi \mid^2$ along the F.S. in equation (2) has been omitted in the evaluation of equation (4). Since $\mid \psi \mid^2$ varies along the F.S., including its variation would increase the resulting fields, H_{c2}.

As the Cooper pair charge equals $2\,e$, one would immediately get the well known result of the G.-L. theory :

$$\frac{e\hbar H_{c2}^i}{c^* \sqrt{m_k\, m_j}} = T_c - T \tag{5}$$

where c^* in (5) is the velocity of light, and $i \neq k \neq j$ correspond to the three mutually orthogonal directions. Substition of (4) into (5) leads to the results :

$$\begin{aligned} H_{c2}^a &= \frac{T_c - T}{T_c}\,\frac{6\,\pi^2\,c^*\,\hbar T_c^2}{7\,\zeta(3)\,e\,t_b\,t_c\,bc} \\ H_{c2}^b &= \frac{T_c - T}{T_c}\,\frac{12\,\pi^2\,c^*\,\hbar T_c^2}{7\,\zeta(3)\,e\,t_a\,t_c\,ac} \\ H_{c2}^c &= \frac{T_c - T}{T_c}\,\frac{12\,\pi^2\,c^*\,\hbar T_c^2}{7\,\zeta(3)\,e\,t_a\,t_b\,ab}\,. \end{aligned} \tag{6a}$$

Equations (6a) can also be rewritten in a more practical form

$$H_{c2}^{a} = \frac{4.84 \times 10^4}{t_b\, t_c}\, T_c(T_c - T)$$
$$H_{c2}^{b} = \frac{9.68 \times 10^4}{t_a\, t_c}\, T_c(T_c - T) \qquad (6b)$$
$$H_{c2}^{c} = \frac{19.3 \times 10^4}{t_a\, t_b}\, T_c(T_c - T)$$

where the magnetic fields are given in kOe and t_i in kelvin.

Taking the transverse critical field data of $(TMTSF)_2ClO_4$ (R-state, $T_c = 1.2$ K) reported by Ishiguro *et al.* [6], namely $dH_{c2}^{b}/dT = 12.5$ kOe/K and $dH_{c2}^{c}/dT = 0.8$ kOe/K, and using the following set of band parameters $t_a = 2\,000$ K, $t_b = 200$ K and $t_c = 5$ K [7], one would derive from equations (6b) $dH_{c2}^{b}/dT = 11.6$ kOe/K and $dH_{c2}^{c}/dT = 0.6$ kOe/K. The latter values lie, admittedly, within the accuracy of the central field experimental determination (considering the approximation in Eq. (3)).

With these parameters in mind we now turn to the values of H_{c2}^{a} which are possible at low temperatures in the pure limit. This would give $dH_{c2}^{a}/dT = 58$ kOe/K, and probably even more if $|\psi|^2$ in equation (2) is taken into account.

Such large values of H_{c2}^{a} are consistent with the little existing experimental data for $(TMTSF)_2ClO_4$; for example, an early experiment with $T_c = 0.86$ K [3] (the sample was not properly in a R-state) has led to $dH_{c2}^{a}/dT \approx 50$ kOe/K while an other study with $T_c = 1$ K [8] has provided $dH_{c2}^{a}/dT \approx 38$ kOe/K. Measurements of transverse fields have also been reported in $(TMTSF)_2AsF_6$ under pressure of 11 kbar [1] with slopes dH_{c2}/dT of 16.6 kOe/K and 1 kOe/K for *b* and *c*-directions respectively.

At this point we would like to emphasize that the precise determination of the critical fields along the *a*-direction meets several complications, especially within a high pressure vessel at very low temperature. The correct alignment of the sample along the magnetic field is only one of the problems. But there are also problems regarding the low symmetry of the $(TMTSF)_2X$ phases (the angle between *a* and *a** axis is about 19°) and the possibility of twinning among $(TMTSF)_2X$ crystals.

The quality of data for $(TMTSF)_2PF_6$ under pressure is too low to allow any definite conclusions. The T_c of 0.8 K under 11 kbar [3] is different from other reported results existing in the literature [2].

The values of H_{c2}^{a} mentioned above exceed the limitation imposed by the Pauli pair-breaking mechanism by a factor about two ($2\,\mu_B\, H_{Pauli}(0) = 3.55\, T_c$). The spin-orbit interactions violating the applicability of the Clogston criterion in its simplest form are probably not important due to a relatively weak spin-orbit coupling and the minor presence of impurities. In this respect, superconductivity compounds built around sulfur-based organic molecules seem to show similar behaviours [9], but the quality of samples is not as good as in the Bechgaard series.

Thus, we conclude that the orbital mechanism destroying superconductivity in the presence of a large magnetic field is able to provide the correct order of magnitude for H_{c2} along the *a*-direction, even in the pure limit, and in addition, the large values of fields are not suppressed by the Pauli mechanism. We claim that the critical fields in $(TMTSF)_2X$ superconductors can be explained in the present pure-limit picture, as well as in the dirty-limit which was adopted for previous interpretations. This last fact could be considered as an indication in favour of « triplet superconductivity ». Furthermore, as indicated by H_{c2} measurements close to the SDW state in $(TMTSF)_2AsF_6$ at 9.5 kbar [1], critical fields seem to be somewhat enhanced and T_c depressed by the proximity to the SDW state.

It seems worthwhile emphasizing once again that the previous arguments were based on the Fermi surface picture using a single-particle approach, say, to the conductivity. Meanwhile, there are indications that this may not be valid [2]. As an example, we mention the very unusual behaviour of the magnetoresistance at low temperature (at $T \approx 4$ K) and at low fields ($H \lesssim 10$ kG) which is very large, $\Delta\rho/\rho \sim 1$, and is strongly dependent on the field orientation.

Acknowledgments

One of the authors (L.P.G.) acknowledges his gratitude to the Laboratoire de Physique des Solides for the hospitality extended to him during his stay at Orsay.

References

[1] BRUSETTI, R., RIBAULT, M., JÉROME, D. and BECHGAARD, K., *J. Physique* **43** (1982) 801.
[2] JÉROME, D. and SCHULZ, H. J., *Adv. Phys.* **31** (1982) 299.
[3] GREENE, R. L., HAEN, P., HUANG, S. Z., ENGLER, E. M., CHOI, M. Y. and CHAIKIN, P. M., *Mol. Cryst. Liq. Cryst.* **79** (1982) 183.
[4] BECHGAARD, K., JACOBSEN, C. S., MORTENSEN, K., PEDERSEN, H. J. and THORUP, N., *Solid State Commun.* **33** (1980) 1119.
[5] GOR'KOV, L. P. and MELIK-BARKHUDAROV, T. K., *Sov. Phys. JETP* **18** (1964) 1031.
[6] ISHIGURO, T., MURATA, K., KAJIMURA, K., KINOSHITA, N., TOKUMOTO, H., TOKUMOTO, M., UKACHI, T., ANZAI, H. and SAITO, G., *J. Physique Colloq.* **44** (1983) C3-831.
[7] The values of the band parameters are fairly consistent with recent data of quantum oscillations in $(TMTSF)_2PF_6$, see for example [2] and
ULMET, J. P., AUBAN, P., KHMOU, A., ASKENAZY, S. and MORADPOUR, A., *J. Physique Lett.* **46** (1985) L-535.
[8] GUBSER, D. U., FULLER, W., POEHLER, T. O., STOKES, J., COWAN, D. O., LEE, M. and BLOCH, A. N., *Phys. Rev. B* **24** (1981) 478.
[9] YAGUBSKII, E. B., SCHEGOLEV, I. F., PESOTSKII, S. I., LAUKHIN, V. N., KONONOVICH, P. A., KARTSOVNIK, M. V., ZVARYKINA, A. V., *Pisma JETP* **39** (1984) 275.

J. Physique Lett. **45** (1984) L-433 - L-440 1er MAI **1984**, PAGE L-433

Classification
Physics Abstracts
75.50 — 71.30

On the stability of the quasi-one-dimensional metallic phase in magnetic fields against the spin density wave formation

L. P. Gor'kov and A. G. Lebed'

L.D. Landau Institute for Theoretical Physics, The USSR Academy of Sciences, 117334 Moscow V-334 ul. Kosygina 2, U.S.S.R.

(*Reçu le 24 janvier 1984, accepté le 16 mars 1984*)

Résumé. — Nous montrons que la stabilité d'un état à ondes de densité de spin d'un métal quasi unidimensionnel, à surface de Fermi ouverte, augmente sous champ magnétique. Ce phénomène peut conduire à une série de transitions sous champ. Nous suggérons que ce modèle peut être une interprétation du diagramme de phase des composés $(TMTSF)_2X$ sous champ magnétique.

Abstract. — For a simple anisotropic metal model with two open nested Fermi surfaces, it is shown that the tendency to spin density wave formation increases when a magnetic field is applied, which could lead to a series of phase transitions as functions of the field value. The results are suggested as an explanation of the phase diagram of the $(TMTSF)_2X$ compounds.

In this letter we report an unexpected result, predicting in the framework of a certain theoretical model, the instability of the quasi-onedimensional (Q1D) metal state with respect to spin density wave (SDW) formation in a moderate magnetic field. It is in direct correspondence with the recent experimental discovery of the remarkable phase diagram in the $(TMTSF)_2X$ compound family (see in Refs. [1-6]). A typical phase diagram in the (T, P, H)-space for these materials, as has been obtained for $X = PF_6$ and $X = ClO_4$, is shown in figure 1. The twice shaded part of the surface is the boundary between the anisotropic (Q1D) metal phase with quasiplanar open Fermi surfaces and the field-induced spin density wave (FISDW) state. This boundary is marked with the so-called « onset » or threshold field H_0 which is both temperature- and pressure-dependent [3, 4]. The magnetic nature of the FISDW state is tested in NMR measurements [5, 6], the Shubnikov-de Haas (SdH) oscillations, observed in the magnetic field H exceeding the onset field $H > H_0$ [3, 4], have enabled one to conclude the semi-metallic character of this state. However some unusual features of these oscillations [1, 4] and especially the most recent observation [7, 8] of the step-like Hall voltage behaviour with the field in $(TMTSF)_2ClO_4$, in particular, confirming the fact that the Fermi surface is open at $H < H_0$, have led to the idea [7] that these fields H_k at which the « SdH » oscillations occur, could actually correspond to some cascade of phase subtransitions in the FISDW state. (The alternative explanation [7] of the observed Hall steps in terms of the quantized Hall conductivity is not discussed here, since the longitudinal resistivity does not testify to any decrease at these fields.)

Low values for all typical parameters ($T_{SDW} \sim 10$ K, $P \sim 10$ kbar, $H \sim 50$ kOe), characterizing the phenomena reflected by the phase diagram in figure 1, demand an explanation which

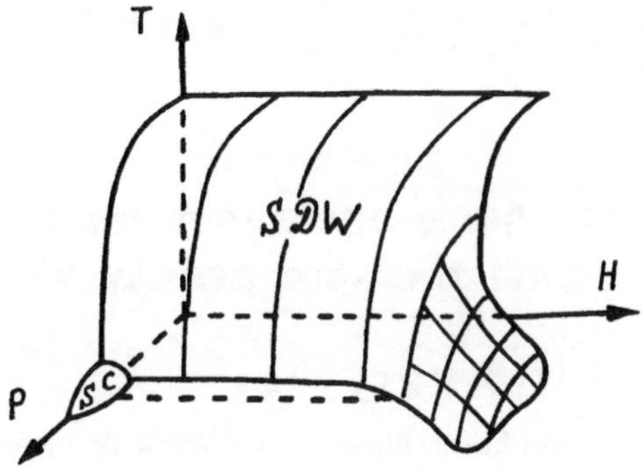

Fig. 1. — Phase diagram of the $(TMTSF)_2X$ salts in the (P, T, H)-space.

would permit one to forget high temperature properties of these compounds due to their original Q1D structure. If this point of view is accepted the simplest explanation of the appearance of the antiferromagnetic state in these materials lies in the Overhauser mechanism of SDW formation [1] based on the Keldysh-Kopaev model [9] with the electron spectrum property :

$$\varepsilon(\mathbf{p} + \mathbf{Q}) = -\varepsilon(\mathbf{p}) \tag{1}$$

(here $\mathbf{Q}$ is the nesting vector characterizing the spin wave periodicity).

This explanation has first been suggested in [10] where the tight binding model has been used for the energy spectrum :

$$\varepsilon(\mathbf{p}) = \pm v(p_{\parallel} \mp k_F) + t_{\perp}(\mathbf{p}_{\perp}) \tag{2}$$

where the plus and minus correspond to the spectrum near the right-hand and left-hand quasi-planar sheets of the Fermi surface, respectively. In the nearest-neighbour interaction approximation [10] :

$$t_{\perp}^{0}(\mathbf{p}_{\perp}) = t_{b^*} \cos(p_{b^*} b^*) + t_{c^*} \cos(p_{c^*} c^*) \tag{3}$$

perfect nesting is achieved if

$$\mathbf{Q} = (2 k_F, \pi/b^*, \pi/c^*) \tag{4}$$

which for the $(TMTSF)_2X$ family would correspond to the double lattice periods in all space directions. (In what follows we use the orthorhombic lattice model.) The 1D nesting vector

$$\mathbf{Q}' = (2 k_F, 0, 0) \tag{4'}$$

shall be discussed later, but at first glance this choice would not be compatible with the existence of the gap (dielectric state at $T = 0$) in $(TMTSF)_2PF_6$ and with a fairly low T_{SDW} at ambient pressure ($\sim$ 12 K) and the estimated values of $t_{b^*} \sim 100$ K [11]. An account of the next-neighbour interactions would slightly change the spectrum (2) :

$$t_{\perp}(\mathbf{p}_{\perp}) = t_{\perp}^{0}(\mathbf{p}_{\perp}) + t_{\perp}'(\mathbf{p}_{\perp}) \tag{5}$$

(1) We do not discuss here the reasons why the tendency to the SDW formation prevails over the CDW pairing in these materials taking merely the fact as given.

and $t'_\perp(\mathbf{p}_\perp)$ is now responsible for violations of perfect nesting of the Fermi surface sheets (3) and (4). As is known, the nesting pairing allows for some deviations from the perfect conditions if the value of $t'_\perp$ does not considerably exceed the magnitude of $T_0 \equiv T_{\text{SDW}}$ at $t'_\perp \equiv 0$.

The key idea for the subsequent discussion is that, when the pressure is applied, the value of $t'_\perp$ increases gradually suppressing the SDW state in accordance with figure 1. (The change of $t^0_\perp$ does not violate the nesting conditions, we shall also not consider the narrow field interval for the superconductivity on the phase diagram).

These comments are equally relevant to both the SDW and CDW types of pairing. A magnetic field introduces distinction between the two behaviours. Whereas the antiferromagnetic type of pairing corresponds to the nesting vector between the Fermi surfaces for spins of opposite directions (the vector is not affected by the field), the CDW pairing takes place for spins of the same directions. As a result, instead of (1)

$$\varepsilon(\mathbf{p} + \mathbf{Q}) + \varepsilon(\mathbf{p}) = 2\,\mu_{\text{B}}\,H \tag{1'}$$

which points to the mechanism by which the applied field, in principle, inhibits the CDW pairing.

To conclude this short discussion, let us also mention that in most experimental environments the phase diagram of figure 1 corresponds to the geometry when the field is oriented in a parallel direction to the c^*-axis of the sample ($H /\!/ c^*$). These materials at low temperatures are actually more like layered compounds with the TMTSF-conductivity sheets in the (a, b^*)-plane. Correspondingly, it is also manifested in very small values $t_{c^*} \ll t_{b^*}$.

Thus, it is necessary to find a mechanism *restoring* in high magnetic fields the tendency to SDW pairing in the metallic phase (the respective pressure region is shown by the dotted line in Fig. 1). The physical idea is that an electron moving in p-space along one of the open Fermi surface sheets in direct space performs motion along the quasiclassical trajectory, determined by the momentum $\pm k_{\text{F}}$ and this motion becomes limited and periodic in the perpendicular direction in the presence of the field due to the periodicity of the dispersion law $t_\perp(\mathbf{p}_\perp)$ in (2). This would increase the effective one-dimensionality of electron properties. To demonstrate this effect quantitatively, we shall consider the generalized susceptibility, $\chi(\mathbf{Q}, T, P, H) \equiv \chi(\mathbf{Q})$, describing the linear response of the system to the external staggered field :

$$\hat{h}(\mathbf{Q}) = (\hat{\sigma}_y)_{\alpha\beta}\exp(i\mathbf{Q}\mathbf{r})\,. \tag{6}$$

An expression for the susceptibility is easily obtained by summation of the standard « ladder » diagrams :

$$\chi(\mathbf{Q}) = \chi_0(\mathbf{Q})/[1 - \lambda\chi_0(\mathbf{Q})]\,. \tag{7}$$

(Here $\chi_0(\mathbf{Q})$ is the response for noninteracting electrons at given (T, P, H), λ is an appropriate interaction constant [10].) As for $\chi_0(\mathbf{Q})$, it is more convenient to write it now in the mixed $(\omega, \mathbf{p}_\perp, x)$ representation :

$$\chi_0(\mathbf{Q}) = T\sum_{\omega_n}\int\frac{d\mathbf{p}_\perp}{(2\,\pi)^2}\int dx'\, g_{++}(i\omega_n, \mathbf{p}_\perp; x, x')\, g_{--}(i\omega_n, \mathbf{p}_\perp - \mathbf{Q}; x', x) \tag{8}$$

where $g_{++}(g_{--})$ are slowly varying parts of the electron Green function near the right(left)-hand Fermi surface sheets. Thus

$$G_{++}(i\omega_n, \mathbf{p}_\perp; x, x') = \exp[ik_{\text{F}}(x - x')]\, g_{++}(i\omega_n, \mathbf{p}_\perp; x, x')\,. \tag{8'}$$

Taking the geometry into account ($H /\!/ c^*$), substituting $\mathbf{p} \to \mathbf{p} - \frac{e}{c}\mathbf{A}$ into (8) and choosing the vector potential, **A**, in the form $\mathbf{A} = (0, Hx, 0)$, we arrive at the equation defining g_{++}

$$\left\{ i\omega_n + iv \frac{\mathrm{d}}{\mathrm{d}x} - t_\perp\left(p_{b^*} - \frac{eHx}{c}, p_{c^*} \right) \right\} g_{++}(i\omega_n, \mathbf{p}_\perp; x, x') = \delta(x - x')$$

which is immediately integrated in the form

$$g_{++} = \frac{\operatorname{sign} \omega_n}{iv} \exp\left\{ -\frac{\omega_n(x - x')}{v} - \frac{i}{v} \int_{x'}^{x} t_\perp\left(p_{b^*} - \frac{eHu}{c}, p_{c^*} \right) \mathrm{d}u \right\} \tag{9}$$

at $\omega_n(x - x')/v > 0$ (and is zero in the opposite case). The substitution of (9) into (8) (the expression for g_{--} results from (9) due to the change of the sign of v) shows that only the terms $t'_\perp(\mathbf{p}_\perp)$ making the nesting imperfect are retained in equation (8). In what follows :

$$t'_\perp(\mathbf{p}_\perp) = 2\, t'_{b^*} \cos(2\, p_{b^*} b^*) + 2\, t'_{c^*} \cos(2\, p_{c^*} c^*) . \tag{3'}$$

Unlike (3), equation (3′) includes the next-neighbour interactions. By means of (3′), (8) can be rewritten explicitly. After averaging (8) over $\mathbf{p}_\perp$ and after subsequent simplifications the stability condition for the metallic phase, implying that the denominator in (7) is positive, finally becomes :

$$\frac{1}{\zeta} - \int_d^\infty J_0\left[\frac{4\, ct'_{b^*}}{veHb^*} \sin\left(\frac{b^*\, eH}{c} x \right) \right] J_0\left(\frac{4\, t'_{c^*}}{v} x \right) \frac{2\, \pi T\, \mathrm{d}x}{v \operatorname{sh}\left(\dfrac{2\, \pi T x}{v} \right)} \geqslant 0 . \tag{10}$$

(Here the dimensionless constant $\zeta = \dfrac{\lambda b^{*-1}\, c^{*-1}}{2\, \pi v}$ and the low cutoff, d, were introduced. As usual, the cutoff will drop out from equation (10), if the relation $\zeta^{-1} = -\ln \dfrac{\pi T_0\, d}{v}$ is used, where T_0 is the transition temperature at $t'_{b^*} = t'_{c^*} = 0$.)

The physical effect which has already been qualitatively explained above, is now evident from equations (9) and (10). In fact, at $H \neq 0$ the argument of the first Bessel function is now periodically dependent on x. If $t'_{c^*} \ll t'_{b^*}$, in the first approximation equation (10) transforms

$$\frac{1}{\zeta} - \int_d^\infty J_0\left[\frac{4\, ct'_{b^*}}{veHb^*} \sin\left(\frac{b^*\, eH}{c} x \right) \right] \frac{2\, \pi T\, \mathrm{d}x}{v \operatorname{sh}\left(\dfrac{2\, \pi T x}{v} \right)} \geqslant 0 . \tag{10'}$$

A new feature of this expression, compared to the case $H = 0$, is the appearance of the logarithmic divergence in the integral at large x ($x > c/eHb^*$). Averaging over x and using the relation

$$\frac{1}{\pi} \int_0^\pi J_0(z \sin \varphi)\, \mathrm{d}\varphi = J_0^2(z/2) \tag{11}$$

one sees from (10′) and (11) that (at $t'_{c^*} = 0$) the metallic phase would never be stable at zero temperature and finite H, except a discrete set, H_k, of certain values of the magnetic field, H, defined by the condition (11) :

$$J_0\left(\frac{2\, ct'_{b^*}}{veHb^*} \right) = 0 . \tag{11'}$$

Bearing in mind that $2\,t'_{b*} > \pi T_0/\gamma$ (i.e., at $H = 0$ the normal metallic phase corresponds to the given pressure) and substituting the first zero of the Bessel function, one finds that the first point of this set, H_1, from the right-hand side (i.e., from the side of large fields) lies at

$$H_1 \lesssim \frac{\pi c T_0}{2.405\ \gamma v e b^*}. \tag{12}$$

At $T_0 \simeq 10$ K, $v \sim 10^7$ cm/s, $b^* \simeq 7.7$ Å the right-hand side of (12) would yield $H_1 \lesssim 10^5$ Oe.

At larger H, according to (10′), the SDW phase is completely restored with $T_{SDW} = T_0$, which could be even higher than T_{SDW} at $P = 0$, since, of course, t'_{b*} is finite at ambient pressure. When the magnetic field decreases, the « points » of the metallic phase (at $T = 0$) are concentrated near $H = 0$ according to the asymptotic behaviour of the Bessel function zeroes (11′) :

$$2\,c t'_{b*}/veH_k\, b^* \simeq \pi\left(k - \frac{1}{4}\right), \qquad (k \gg 1). \tag{12′}$$

A schematic T-H phase diagram at fixed pressure (i.e. at fixed t'_{b*}) is shown in figure 2. The typical « transition temperature » $T_{SDW}(k)$, the value of the « subphase » k, exponentially decreases with an increase of the number k. With logarithmic accuracy the corresponding behaviour can be obtained directly from equation (10′) :

$$\ln \frac{\pi T_0}{2\,t'_{b*}\,\gamma} \simeq -\,J_0^2\left(\frac{2\,c t'_{b*}}{veHb^*}\right) \ln \frac{b^*\, eHv}{cT_{SDW}}. \tag{13}$$

This infinite sequence of transitions would resemble to some extent a sequence of exciton phases in semiconductors, the possibility of which was discussed in [12] many years ago. The phase diagram of the type shown in figure 2 is, of course, an artefact of the 2D dispersion law, (see below) and becomes unstable if some finite t'_{c*} is taken into account. In fact, according to (10) the presence of the factor $J_0\left(\frac{4\,t'_{c*}\,x}{v}\right)$ in the integral immediately makes again the whole integral convergent at large $x \sim v/t'_{c*}$. Comparing the value of T_{SDW} estimated at small H from equation (13) with the value of t'_{c*}, one gets that the oscillations shown in figure 2 do appear only at the onset field $H > H_0$ where

$$H_0 \simeq \frac{\pi c t'_{b*}}{veb^*} \ln \frac{\pi T_0}{2\,t'_{b*}\,\gamma} \Big/ \ln \frac{t'_{c*}}{t'_{b*}} \tag{13′}$$

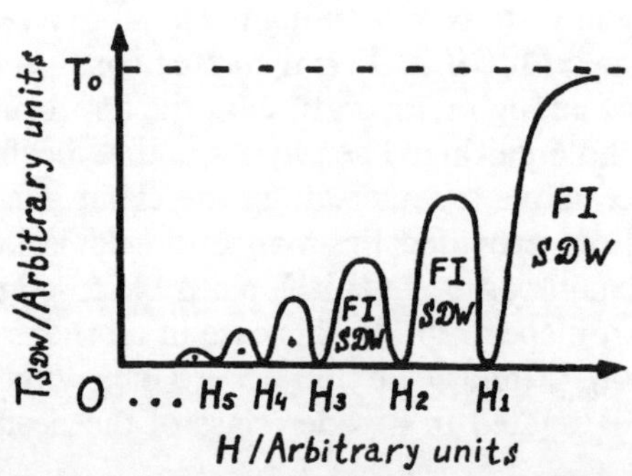

Fig. 2. — Infinite series of possible phase transitions in the presence of the magnetic field H.

and only a few oscillations can be realized in practice. Let us also add that the finite t'_{c^*} converts the « points » of the metallic phase stability, H_k, into small stability « islands », the width of which can again be estimated from the condition :

$$J_0^2\left(\frac{2\,ct'_{b^*}}{veHb^*}\right)\ln\frac{b^*\,eHv}{ct'_{c^*}}\sim 1$$

for the magnetic fields in the vicinity of the corresponding H_k from (11′).

At this stage, we shall not go into further details, since our stability condition (10) holds for both the thermodynamic stability of the phase and its metastability. Otherwise, the boundary line in figure 2, in principle, could describe the « supercooling » line for the metallic phase while the real transition into the SDW phase (or between two different SDW phases) would be of the first order. Some hysteresis features have been observed in [4] near some field values ascribed to the SdH oscillations. The sharp temperature-dependent steps in the Hall voltage in $(TMTSF)_2ClO_4$ [7] also point in the same direction. Note that the magnetic field appears in the discussed phenomenon approximately in the same combination as would be also expected in the SdH effect definition of the closed orbit areas for the hypothetic semimetallic SDW phase (cf. Eqs. (10) and (3′)). As for $t'_\perp(\mathbf{p}_\perp)$, it has been assumed in the form of equation (3′) only for the sake of convenience for the subsequent discussion. Its modification does not change the precedent results qualitatively.

We would also like to repeat that at the field $H > H_1$ from equation (12) the SDW finally becomes fixed. Since the above discussion shows that the dielectric state emerges in a large field (at $T = 0$), all the conducting mechanisms are expected to disappear. However a more thorough investigation of the SDW phases, restored by the magnetic field, is necessary.

Unlike the SDW phase, the CDW state in general is suppressed by a strong magnetic field, according to equation (1′). Nevertheless, the competition between this inhibiting mechanism and the phenomenon of the « one-dimensionalization » of the electron spectrum expressed by equation (10), could be important also for the CDW pairing at the intermediate fields because both mechanisms involve the same dimensionless combination of parameters.

So far we have preferred to explain the phenomenon in terms of the nesting vector from equation (4), since most of new effects observed, including the unusual « SDW » oscillations, may be described in this case by the appropriate choice of T_0 and $t'_\perp$. The main exception is that the sizes of the electron (hole) pockets, as defined by the Hall measurements [7, 8, 13] in $(TMTSF)_2ClO_4$, are too large. This gives rise again to the necessity of discussing the pairing with the « direct » nesting vector $(2\,k_F, 0, 0)$. There are two points which should be mentioned in this connection.

For the 3D (or 2D) electron dispersion law electron-electron interactions restore the Fermi liquid description, if the temperature is low enough. The above results (Eqs. (8, 9)) state rigorously that the 2D electron spectrum, if its Fermi surface consists of two separate open sheets, is absolutely unstable at a low enough temperature (at the appropriate sign of interaction) with respect to the SDW electron-hole pairing. The physics is that electron and hole from the opposite sides of the Fermi surface move together along the chain axis, their motion being limited in space in transverse direction, provided the magnetic field is imposed.

For the $(TMTSF)_2X$ one-chain-compounds it is plausible to consider equation (3) as a good approximation for their electron spectrum. In this case in all the expressions one should substitute t_{b^*,c^*} (and $b^*/2$, $c^*/2$) for t'_{b^*,c^*} (and b^*, c^*) in accordance with (3). It appears now that the instability mentioned above is related to a wider class of the nesting wave vectors

$$\mathbf{Q} = (2\,k_F + \delta, 0, 0)$$

and, instead of equation (10′), one has

$$\frac{1}{\zeta} - \int_d^\infty J_0\left[\frac{8\,ct_{b^*}}{eHvb^*}\sin\left(\frac{eHb^*}{2\,c}\,x\right)\right]\cos(\delta x)\,\frac{2\,\pi T\,\mathrm{d}x}{v\,\mathrm{sh}\,\frac{2\,\pi Tx}{v}} \geqslant 0\,.$$

Using the same procedure of the averaging of the oscillating factors over x at $T \to 0$, one obtains easily that the logarithmic divergence at large x is now connected with the coefficient :

$$J_m^2\left(\frac{4\,ct_{b^*}}{veHb^*}\right); \qquad m = \frac{\delta c}{eHb^*}\,.$$

This coefficient reaches its maximum at

$$m \simeq \frac{4\,ct_{b^*}}{veHb^*}; \qquad \delta = \frac{4\,t_{b^*}}{v}\,. \tag{14}$$

In other words, at low fields the instability would for the first time correspond to the biggest possible electron (or hole) pockets. The size of pockets given by (14) is in a good correspondence with the t_{b^*} estimates ($t_{b^*} \sim 150$ K) and the Hall effect data [7, 8]. The possibility to construct the ground states of a set of subphases characterized by δ which decreases with an increase in magnetic field, is now under investigation.

In conclusion, we have given theoretical arguments explaining the puzzling fact of restoration of the SDW phase in the $(TMTSF)_2X$ compounds by the magnetic field. The rôle of fluctuations of any kind has not been considered, since temperatures are assumed to be low enough to permit the Fermi liquid approach. Two models (i.e. the « perfect » nesting and the « direct », or one-dimensional pairing) could correspond to the physical situation in different compounds. The Hall data [7, 8] for ClO_4 showing comparatively large pocket volume in this material might be understood in terms of such a direct nesting. The expected phase diagram is shown schematically in figure 3.

Acknowledgments

One of us (L. G.) is grateful to D. Jérome who has first drawn his attention to the phase diagram of these materials and to S. Barisič for stimulating discussions.

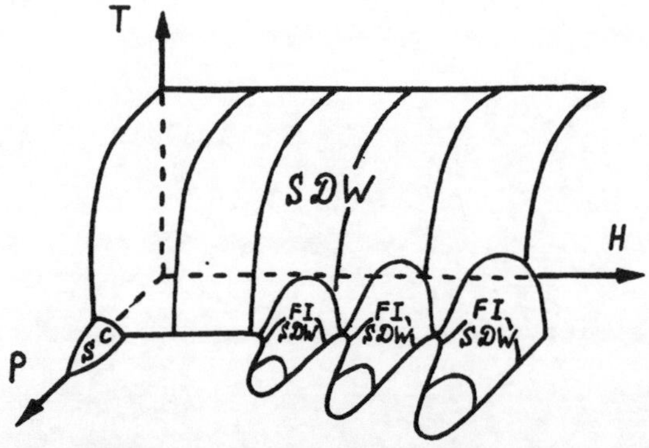

Fig. 3. — The phase diagram of Fig. 1 with the theoretical representations, developed in the text, taken into account.

References

[1] KWAK, J., *J. Physique Colloq.* **44** (1983) C3-839.
[2] BRUSETTI, R., GAROCHE, P. and BECHGAARD, K., *J. Physique Colloq.* (1983) C3-1051.
[3] KWAK, J., SCHIRBER, J., GREEN, R. and ENGLER, E., *Phys. Rev. Lett.* **46** (1981) 1296.
[4] KAJIMURA, K., TOKUMOTO, H., TOKUMOTO, M., MURATA, K., UKACHI, T., ANZAI, H., ISHIGURO, T. and SAITO, G., *J. Physique Colloq.* **44** (1983) C3-1059.
[5] AZEVEDO, L., SCHIRBER, J., GREEN, R. and ENGLER, E., *Physica B* **108** (1981) 1183.
[6] TAKAHASHI, T., JEROME, D. and BECHGAARD, K., *J. Physique Lett.* **43** (1982) L-565.
[7] RIBAULT, M., JÉROME, D., TUCHENDLER, J., WEYL, C. and BECHGAARD, K., *J. Physique Lett.* **44** (1983) L-953.
[8] CHAIKIN, P. M., MU-YONG CHOI, KWAK, J. F., BROOKS, J. S., MARTIN, K. P., NAUGHTON, M. J., ENGLER, E. M. and GREENE, R. L., *Phys. Rev. Lett.* **51** (1983) 2333.
[9] KELDYSH, L. and KOPAEV, Yu., *Fiz. Tverd. Tela* **6** (1964) 2791.
[10] HOROVITZ, B., GUTFREUND, H. and WEGER, M., *Phys. Rev. B* **12** (1975) 3174.
[11] JEROME, D., *J. Physique Colloq.* **44** (1983) C3-775.
[12] KOHN, W., *Phys. Rev. Lett.* **19** (1967) 439.
[13] KWAK, J. F., *Phys. Rev. B* **28** (1983) 3277.

VOLUME 71, NUMBER 23 PHYSICAL REVIEW LETTERS 6 DECEMBER 1993

Cyclotron Resonance on Open Orbits in Organic Materials

L. P. Gor'kov and A. G. Lebed'

National High Magnetic Field Laboratory, 1800 E. Paul Dirac Drive, P.O. Box 4005, Tallahassee, Florida 32306-4005 and L.D. Landau Institute for Theoretical Physics, Russian Academy of Sciences, 117334 Moscow, 2 Kosygina St., Russia

(Received 20 September 1993)

The cyclotron resonance on open orbits is suggested to demonstrate the one dimensional character of electron motion along the main conducting axis in the presence of a perpendicular magnetic field for organic materials with warped open Fermi surfaces. Singular behavior at resonant frequencies in the surface impedance and the helicon waves spectrum are obtained.

PACS numbers: 76.30.–v, 78.30.Jw

Initial studies of metallic organic compounds have been motivated by the search for superconductivity in these nontraditional materials. This goal has been achieved first in the Bechgaard salts, $(TMTSF)_2X$, and, later on, in a variety of the $(BEDT\text{-}TTF)_2X$ charge transfer salts. [The common abbreviation is $(ET)_2X$. In both cases X stands for various cations; for a review, see [1–4]]. From the structural point of view the main difference between the two classes is that while in the Bechgaard salts electronic properties may be described in terms of weakly coupled conducting chains, members of the $(ET)_2X$ family are layered compounds (this also makes them interesting objects for comparison with high-T_c oxides). For the $(TMTSF)_2X$ family the electronic spectrum corresponds to two open sheets of the Fermi surface and in the simplest tight binding approximation is of the form

$$\epsilon_\pm(\mathbf{p}) = \pm v_F(p_x \mp p_F) + 2t_b\cos(p_{b^*}b^*) + 2t_c\cos(p_{c^*}c^*)\,. \tag{1}$$

Here $\pm$ stands for the right $(+p_F)$ and left $(-p_F)$ Fermi surfaces, v_F is the Fermi velocity along the most conducting direction, b^* and c^* are the lattice constants in the corresponding directions ($b^* \approx 7.71$ Å, $c^* \approx 13.52$ Å); x is along the main axis a. From the resistivity anisotropy data the corresponding anisotropy of the electronic spectrum is estimated as $2t_a : 2t_b : 2t_c = 0.5 : 0.05 : 0.002$ eV (see [1,2]).

Therefore the Fermi surfaces are just two slightly warped planes perpendicular to the direction of the main conducting axis (assuming an orthorhombic Brillouin zone—the actual structure is monoclinic). On the other hand, the in-layer conductivity anisotropy for the $(ET)_2X$ family is usually less pronounced, though it may vary depending on the constituents. Correspondingly, in most cases the Fermi surface is a cylinder also slightly warped along the third direction, i.e., perpendicular to the layer. The warping may be quite weak ($t_b \sim 10$ K [4]), reflecting the genuine layered structure of these materials. However, it has recently been established that the $(BEDT\text{-}TTF)_2M\mathrm{Hg(SCN)_4}$ salts ($M = \mathrm{K, Tl, Rb, NH_4}$), for which the charge transfer mechanism would provide two conducting electrons per unit cell, are compensated metals where cylindrical Fermi surfaces (i.e., the closed orbits) may coexist with plane sheets warped slightly along both directions. In other words, such almost plane Fermi surfaces may occur both in the Bechgaard salts and in the new generation of organic conductors and superconductors, thus adding to their galvanomagnetic properties a number of new phenomena in sufficiently high magnetic fields.

Spectrum (1) possesses the "nesting" properties for the vector $Q = (2p_F, \pi/b^*, \pi/c^*)$; this feature is commonly believed to be an indication of some charge density wave (CDW) or spin density wave (SDW) instability to occur in the Bechgaard salts at low enough temperatures. The SDW transition takes place, indeed, in $(TMTSF)_2PF_6$ ($T_{SDW} \approx 12$ K) and some others. However, the most remarkable experimental feature of these materials is that if then the SDW transition is suppressed, say, by applied pressure, in high enough magnetic fields the instability reappears in the form of a cascade of the so-called field induced spin density wave (FISDW) transitions [2, 3]. The mechanism for the SDW state restoration in the presence of magnetic field has been proposed by the present authors [5] (see also [6]). While this theory successfully explains the experimental facts by proving that electron-electron interactions would always lead to a SDW instability in sufficiently high fields and low temperatures provided the electron interactions have a corresponding sign, its most curious general conclusion is that any arbitrary weak magnetic field perpendicular to the main axis transforms such an anisotropic 3D metal into the one dimensional system. Thus, for instance, if H_{b^*} and H_{c^*} are the field components along the two corresponding axes and the vector potential is chosen as $A_{b^*} = xH_{c^*}, A_{c^*} = -xH_{b^*}$, then the Schrödinger equation for the electron wave functions,

$$\Psi_\varepsilon^\pm(x, p_{b^*}, p_{c^*}) = \exp(\pm i p_F x)\, u_\varepsilon^\pm(x, p_{b^*}, p_{c^*})\,, \tag{2}$$

acquires the form

$$\left[\pm i v_F \frac{d}{dx} + 2t_{b^*}\cos\left(p_{b^*}b^* - \frac{e}{c}H_{c^*}b^*x\right) + 2t_{c^*}\cos\left(p_{c^*}c^* + \frac{e}{c}H_{b^*}c^*x\right)\right] u_\varepsilon^\pm(x) = \varepsilon u_\varepsilon^\pm(x)\,. \tag{3}$$

0031-9007/93/71(23)/3874(4)$06.00

One sees that electron motion in the perpendicular directions now needs a quantum approach even though Eq. (2) corresponds to the quasiclassical behavior along a. In classical terms, an electron moving, say, with velocity $(+v_F)$ along the a axis experiences a complicated oscillatory behavior in the (b^*, c^*) plane restricted by the amplitudes

$$X_{b^*} = \frac{2t_{b^*}c}{ev_F H_{c^*}}, \quad X_{c^*} = \frac{2t_{c^*}c}{ev_F H_{b^*}} \tag{4}$$

in directions b^* and c^*, respectively. This "one dimensionalization" of electronic motion by magnetic field possesses a number of intriguing problems regarding the properties of such a system in strong magnetic fields. The further development of the initial ideas [5] had shown that the approach based on equations (1)–(3) in truth explains remarkably well, at least qualitatively, most phenomena, including the phase diagram for FISDW transitions, the so-called "magic angles," the sharp angular oscillations of the magnetoresistance, and other properties [7–10].

On the other hand, reasons for such success remain not quite clear. Even the applicability of the Fermi liquid picture is an open question regarding the very small scale for the Fermi surface warping ($t_{c^*} \leq 10$ K) and not negligible electron-electron interactions as estimated from the resistivity data. It is, therefore, highly desirable to validate somehow the above picture for the electron spectrum by an unambiguous and direct experiment which would not even invoke any electron-electron interactions at all.

Below we suggest for this purpose new cyclotron resonance experiments to utilize the above sinusoidal-type motion of electrons in directions perpendicular to the main axis in the presence of magnetic field (to contrast it from the Azbel'-Kanner's resonance on closed orbits [11]). The effect is schematically shown in Fig. 1. The sample's surface is chosen in the (a, c^*) plane; the sample lies at $y < 0 (y \parallel b^*)$. To simplify the discussion, the magnetic field H is first chosen to be strictly parallel to the c^* axis while the ac electric field $E_{0\omega}$ is along the main axis a. An electron (with $\pm v_F$ along a) periodically enters the skin layer to pick up the energy from the alternating electric field. The resonance conditions are

$$\omega_n = n\omega_c = n\frac{eHv_F b^*}{c}, \tag{5}$$

as follows from Eq. (3). The cyclotron resonance on open orbits is, of course, known as the quasiclassical phenomenon for 3D metals [12]. The difference lies, first of all, in the fact that in the quasiclassical 3D case only the extremal orbits contribute significantly and all singularities in the surface impedance are correspondingly weaker. Second, the above effect of a magnetic field to restrict the electronic perpendicular motion is, generally speaking, of quantum origin and all the final results were obtained by making use of the expressions for the exact wave functions [Eq. (3)], obtained in [5].

We assume the skin effect conditions for the alternating

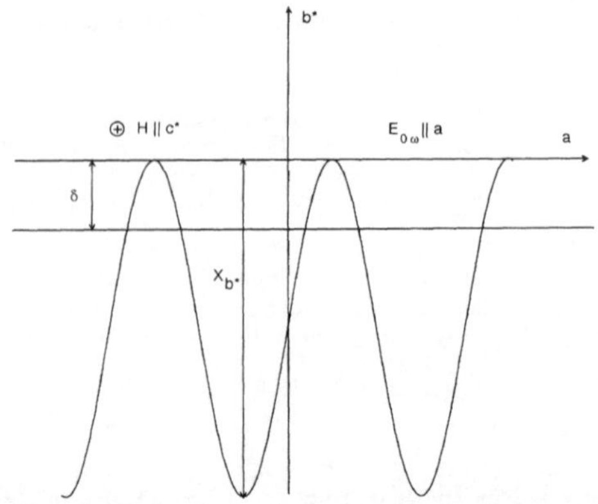

FIG. 1. The most favorable geometry for the cyclotron resonance on open sheets of the Fermi surface in the Bechgaard salts $(TMTSF)_2X$. The external field H is chosen along the c^* direction to make the amplitude for the transverse sinusoidal electron motion, $X_{b^*}(H)$, bigger to hold conditions for the cyclotron resonance $X_{b^*}(H) \geq \delta$. Here δ is the skin depth for an alternating field, $E_{0\omega}$ ($E_{0\omega}$ is parallel to the main axis, a).

field since there is no homogeneous cyclotron resonance in the absence of scattering mechanisms. Even though samples of high purity are now available, so that $\omega\tau \gg 1$ (at least for the Bechgaard salts), it is difficult to satisfy the conditions for the anomalous skin effect due to the large anisotropy ($v_{b^*} \sim 0.1 v_F$) in conducting properties. ω_c in Eq. (5) should not be too large—otherwise both scales (4) for the perpendicular electronic motion may become too short compared with the skin depth, δ. This point also explains our choice of the geometry in Fig. 1: $t_{b^*} \gg t_{c^*}$ and hence, $X_{b^*} \gg X_{c^*}$. As shown below, at values of $H \sim 1$ T (i.e., below the FISDW phases) one has

$$\delta \sim X_{b^*} . \tag{6}$$

We shall also assume specular boundary conditions on the sample surface (at $z = 0$). Then the expression for the surface impedance, $Z(\omega, H)$, acquires the standard form (see, e.g., in [13]):

$$Z(\omega, H) = -i\frac{8\pi}{c^2}\omega \int_0^\infty \frac{dK}{K^2 + (4\pi/c)Q(K, \omega, H)}, \tag{7}$$

where $Q(K, \omega, H)$ stands for the Fourier component of the kernel which relates the current $[j_x(y) \parallel a]$ and the applied alterating field, $E_x(y)$:

$$\mathbf{j}(\mathbf{r}, \omega, H) = -\int \hat{Q}(\mathbf{r} - \mathbf{r}', \omega, H)\mathbf{A}(\mathbf{r}')\, d\mathbf{r}' . \tag{8}$$

Making use of the explicit form for the Green functions obtained in [5], we get the following expression for $Q(K, \omega, H)$ in Eq. (7):

$$Q(K,\omega,H) = -\frac{i\omega c}{4\pi\delta^2}\int_0^\infty \exp[i(\omega+i/\tau)x/v_F] \times J_0\left[2X_{b^*}K\sin\left(\frac{x\omega_c}{2v_F}\right)\right]dx\,. \quad (9)$$

Here $\delta^2 = c^2b^*c^*/4e^2v_F$ and $J_0(Z)$ is the Bessel function.

To study peculiarities of the surface impedance at the resonance conditions, it is more convenient to expand the Bessel function into harmonics:

$$J_0\left[2X_{b^*}K\sin\left(\frac{x\omega_c}{2v_F}\right)\right] = \sum_{n=-\infty}^{+\infty} J_n^2(X_{b^*}K)\exp\left(\frac{ix\omega_c n}{v_F}\right) \quad (10)$$

so that $Q(K,\omega,H)$ can be easily calculated (near resonances).

Before proceeding further with the results for $Z(\omega,H)$ we shall discuss in short the spectrum of the so-called "extraordinary helicon waves" ($\mathbf{E}\perp\mathbf{H}$), which is to be determined from poles of the denominator in Eq. (7):

$$K^2 + \frac{4\pi}{c}Q(K,\omega,H) = 0\,. \quad (11)$$

Consider the very vicinity of the nth resonance, $\omega \gg |\, n\omega_c - \omega \,| \gg \frac{1}{\tau}$. By combining (9) and (10) with Eq. (11), one gets for the long wavelength helicons ($X_{b^*}K \ll 1$) the spectrum

$$\omega(K) = \omega_c\left[n - \frac{1}{(n!)^2}\left(\frac{KX_{b^*}}{2}\right)^{2n}\right]. \quad (12)$$

In addition, Eq. (11) always has short wavelength solutions at $X_{b^*}K \sim 1$. In fact, from (11) one gets near the resonance:

$$(\delta K)^2 = \frac{\omega}{n\omega_c - \omega}J_n^2(X_{b^*}K). \quad (13)$$

This equation has to have solutions near each of the zeros of the Bessel function in (13). In the analytical form solutions can be written down only making use of the asymptotic behavior for the mth zero of the Bessel function, $J_n(Z_m) = 0$, $Z_m \approx \pi(m + n/2 + 3/4)$, at large $m \gg n$:

$$(KX_{b^*} - Z_m)^2 = \frac{\pi}{2}\left(\frac{n\omega_c - \omega}{\omega}\right)\frac{\delta^2}{X_{b^*}^2}Z_m^3\,. \quad (14)$$

The result shows the existence of excitations in a narrow vicinity of each of the discrete values for the wave vector K_m, such that

$$K_m = \pi X_{b^*}^{-1}\left(m + \frac{n}{2} + \frac{3}{4}\right). \quad (15)$$

The dispersion of the "extraordinary helicon waves," as obtained above, differs from the results known in the 3D problem already in the long wavelength limit. Thus, the dispersion law of the form (12) exists in the case of the isotropic closed orbits for the "ordinary helicon waves" only (i.e., $\mathbf{E}\parallel\mathbf{H}$). All these results use the fact that in our model with the orthorhombic zone the Hall component for the conductivity tensor is identically zero. However, the most remarkable new feature in the helicon spectrum as compared to the isotropic case is now the appearance of waves around the discrete values of the wave vectors, K_m, at each nth resonance, as determined by zeros of the Bessel functions, $J_n(K_mX_{b^*}) = 0$, in Eq. (13). Such waves are closely connected with "peaks" of the electromagnetic field in the bulk on distances from the surface equal to the integer multiple of the orbit's "diameter" (i.e., in our case, mX_{b^*}). In the isotropic 3D case the orbit's diameter depends on its position on the Fermi surface and this would influence the helicon dispersion near the cyclotron resonance. We shall not stay any longer here on the possibility of observation of the helicon waves in organic materials. Instead, we just summarize the results of our calculations for the singular behavior of the surface impedance in vicinities of resonances, $\omega \gg |\omega - n\omega_c| \gg \frac{1}{\tau}$:

$$Z_1 = -\frac{16\pi^{7/3}\omega}{3\sqrt{3}c^2}\delta^{2/3}X_{b^*}^{1/3}|\,(\omega-\omega_c)/\omega_c\,|^{1/3} \times \begin{cases} i/2 - \sqrt{3}/2 & \text{for } \omega < \omega_c\,, \\ i & \text{for } \omega > \omega_c\,, \end{cases} \quad (16)$$

$$Z_{n>1} = -\frac{8\pi^2}{n}(n!)^{1/n}\frac{\omega\delta^2}{c^2X_{b^*}}|\,(\omega - n\omega_c)/\omega_c\,|^{1/2n} \times \begin{cases} -1 + i\cot(\pi/2n) & \text{for } \omega < n\omega_c\,, \\ i/\sin(\pi/2n) & \text{for } \omega > n\omega_c\,. \end{cases} \quad (17)$$

(As usual, a finite $1/\tau$ would not allow the observation of resonances for too large n.)

In accordance with Eqs. (11)–(13), the real part of the impedance is nonzero only at $n\omega_c > \omega$, because the only source of dissipation is due to excitation of helicon waves propagating into the bulk. At $n\omega_c < \omega$ the surface ideally reflects the electromagnetic radiation. One sees that the singular behavior in dZ/dH becomes more pronounced with the increase of the number, n. At all $n > 1$ the main contribution into this singularity comes from the long wavelength part of helicon spectrum, Eq. (13). At the main resonance, i.e., $\omega = \omega_c$, the contribution is due to the short wavelength spectrum, $KX_{b^*} \gg 1$. It is also seen from Eqs. (16) and (17) that dZ/dH has a more singular behavior at resonances than in the isotropic case for any $n > 1$.

In the above considerations we kept in mind mainly the Bechgaard salts. They indeed meet all the criteria, making observation of the effect possible. Thus, various estimates show that in these materials one may have

$\delta \approx 2 \times 10^{-5}$ cm $\approx X_{b^*}(H)$ for magnetic fields $H \sim 1$ T while $\frac{1}{\tau} \sim 0.1$ K. Therefore the expressions (16) and (17) for dZ/dH have to be applicable in $(TMTSF)_2X$ salts. At the same time, as we have already mentioned, there are also reasons to believe that in some of the new (ET)-based materials open trajectories and closed orbits may coexist and, for reasons which are not quite clear now, the open Fermi surfaces are very close to the picture of slightly warped planes used above. Even though we have not enough information so far concerning δ, X_{b^*} (or ω_c and $\frac{1}{\tau}$), the phenomenon of cyclotron resonance due to the one dimensionalization of electrons belonging to open Fermi surfaces in magnetic fields could give important information concerning electronic spectra in this new group of organic materials.

We shall now briefly mention a few points of experimental significance. Since $X_{b^*} \gg X_{c^*}$, $\delta \gg X_{c^*}$ and no resonances should be seen if the magnetic field is along the b^* axis. Therefore the only effect of an inclination of the magnetic field from the c^* direction in the case of the Bechgaard salts is to reduce the H_{c^*} component (i.e., the dependence on $H\cos\theta$ only in all the above expressions). This makes it possible to perform experiments on mosaics of single crystals aligned along the a axis but randomly oriented with respect to b^* and c^* direction: only samples with $c^* \parallel H$ contribute to Eqs. (16) and (17). In the case of the Bechgaard salts large enough platelike crystals are usually not available. In the ET-based materials, if they possess open Fermi surfaces, the cyclotron resonance due to open trajectories would coexist with the more familiar cyclotron resonances due to closed orbits.

To summarize, we suggest the simple experiment which can probe the one dimensionalization of the electron's motion along the main conduction axis by a perpendicular magnetic field—the concept which is crucial for understanding a number of phenomena in the Q1D organic Bechgaard salts. Similar phenomena may exist in other organic materials possessing open Fermi surfaces. The cyclotron resonance discussed above may become a useful tool to detect the presence of such slightly warped Fermi surfaces in the electron spectrum. Singularities in the surface impedance are investigated and shown to be more pronounced compared with the well-known results for 3D metals. The spectrum of helicon waves shows new features specific for open orbits only.

The authors acknowledge with gratitude stimulating discussions with J. S. Brooks, P. M. Chaikin, and S. Valfells. This work was supported by the National High Magnetic Field Laboratory through the NSF Cooperative Agreement No. DMR-9016241 and the State of Florida.

[1] D. Jerome and H.J. Schulz, Adv. Phys. **31**, 299 (1982).
[2] L.P. Gor'kov, Usp. Fiz. Nauk **144**, 381 (1984) [Sov. Phys. Usp. **27**, 809 (1984)].
[3] T. Ishiguro and K. Yamaji, *Organic Superconductors* (Springer-Verlag, Berlin, 1990).
[4] W. Kang *et al.*, Phys. Rev. Lett. **62**, 2559 (1989) [for review see J. Wosnitza (to be published)].
[5] L.P. Gor'kov and A.G. Lebed', J. Phys. (Paris), Lett. **45**, L433 (1984).
[6] P.M. Chaikin, Phys. Rev. B **31**, 4770 (1985).
[7] M. Heritier, G. Montambaux, and P. Lederer, J. Phys. (Paris), Lett. **45**, L943 (1984).
[8] A. Virosztek, L. Chen, and K. Maki, Phys. Rev. B **34**, 3371 (1986).
[9] A.G. Lebed', Pis'ma Zh. Eksp. Teor. Fiz. **43**, 137 (1986) [JETP Lett. **43**, 174 (1986)].
[10] W. Kang, S.T. Hannahs, and P.M. Chaikin, Phys. Rev. Lett. **69**, 2827 (1992).
[11] M.Ya. Azbel' and E.A. Kaner, Zh. Eksp. Teor. Fiz. **32**, 896 (1956) [Sov. Phys. JETP **5**, 730 (1957)].
[12] V.F. Gantmakher and E.A. Kaner, Zh. Eksp. Teor. Phys. **45**, 1430 (1963) [Sov. Phys. JETP **18**, 988 (1964)].
[13] E.A. Kaner and V.G. Skobov, Adv. Phys. **17**, 605 (1968).

VOLUME 87, NUMBER 6 PHYSICAL REVIEW LETTERS 6 AUGUST 2001

Superconductivity in an Organic Insulator at Very High Magnetic Fields

L. Balicas,[1] J. S. Brooks,[1] K. Storr,[1] S. Uji,[2] M. Tokumoto,[3] H. Tanaka,[4] H. Kobayashi,[4] A. Kobayashi,[5] V. Barzykin,[1] and L. P. Gor'kov[1]

[1]*National High Magnetic Field Laboratory, Florida State University, Tallahassee, Florida 32306*
[2]*National Research Institute for Metals, Tsukuba, Ibaraki 305-0003, Japan*
[3]*Nanotechnology Research Institute, National Institute of Advanced Industrial Science and Technology (AIST), Tsukuba, Ibaraki 305-8568, Japan*
[4]*Institute for Molecular Science, Okazaki, Aichi 444-8585, Japan*
[5]*Research Centre for Spectrochemistry, Graduate School of Science, The University of Tokyo, Bunkyo-ku, Tokyo 113-0033, Japan*
(Received 22 March 2001; published 19 July 2001)

We investigate by electrical transport the field-induced superconducting state (FISC) in the organic conductor λ-$(BETS)_2FeCl_4$. Below 4 K, antiferromagnetic-insulator, metallic, and eventually superconducting (FISC) ground states are observed with increasing in-plane magnetic field. The FISC state survives between 18 and 41 T and can be interpreted in terms of the Jaccarino-Peter effect, where the external magnetic field *compensates* the exchange field of aligned Fe^{3+} ions. We further argue that the Fe^{3+} moments are essential to stabilize the resulting singlet, two-dimensional superconducting state.

DOI: 10.1103/PhysRevLett.87.067002 PACS numbers: 74.70.Kn, 74.25.Dw, 74.25.Ha

Superconductivity is usually destroyed by diamagnetic currents induced in the presence of strong magnetic fields. This effect has orbital character and prevails in most conventional "*s*-wave" superconductors that involve the singlet state of the Cooper pairs. In addition, superconductivity can also be suppressed by the Pauli pair breaking mechanism: here the external field destroys the spin-singlet state of the Cooper pair, imposing the so-called Clogston-Chandrasekhar paramagnetic limit [1,2]. Nevertheless, and despite these well known physical limitations, Uji *et al.* [3] have recently reported the observation of a magnetic-field induced superconducting phase (FISC) in the quasi-two-dimensional organic conductor λ-$(BETS)_2FeCl_4$ [where BETS stands for Bis(ethylenedithio)tetraselenafulvalene] for fields exceeding 18 T, applied parallel to the conducting layers. This is particularly remarkable since this compound, at zero field, is an antiferromagnetic insulator (AI) below $T_p \cong 8.5$ K [4]. The AI state is suppressed by the application of magnetic fields above 10 T at low temperatures [5].

The present work was motivated by the apparent increase in the critical temperature of the FISC above 18 T with increasing magnetic field (Ref. [3]). Here, for instance, in the case of spin-triplet superconductivity, there would be, in principle, no limit on the upper critical field. The presence of Fe^{3+} magnetic moments, which coexist with the FISC state, adds further appeal to the triplet state model. To clarify the nature of the FISC, we have studied the λ-$(BETS)_2FeCl_4$ compound at low temperatures in steady, tilted magnetic fields up to 42 T. Our main result is the observation of reentrance towards the metallic state at a temperature-dependent critical field. We obtain a temperature-magnetic field phase diagram for the FISC state, which we interpret in terms of the Jaccarino-Peter (JP) field compensation effect [6], which was previously mentioned in Ref. [3] as one possible explanation for the observed phase transition. We argue further that the Fe^{3+} magnetic state is indeed necessary to stabilize the singlet superconducting state by suppression of interplane diamagnetic currents in the associated in-plane high magnetic fields.

λ-$(BETS)_2FeCl_4$ crystallizes in a triclinic unit cell. The BETS planar molecules are stacked along the crystallographic *a* axis and constitute conducting planes parallel to the *a*-*c* plane. These conducting layers alternate along the *b* axis with layers containing linear chains of $FeCl_4^-$ magnetic anions; hence the *b* axis is the least conducting direction. Spin interactions between localized Fe^{3+} $3d$ electrons and π conducting electrons are expected due to the short interatomic distance between the BETS molecules and the $FeCl_4$ anions.

Single crystals of λ-$(BETS)_2FeCl_4$ were obtained by electrocrystallization [7]. Annealed (low strain) gold wires ($\phi = 12.5$ μm) were attached with graphite paint in a four-terminal arrangement along the *c* axis. An ac current (10 to 100 μA) was used, and the voltage was measured by a conventional lock-in amplifier technique. Samples were mounted in a rotating sample holder in a ^{3}He refrigerator. The measurements were carried out in the Hybrid magnet at the DC Field Facility of the National High Magnetic Field Laboratory.

Our magnetic field dependent resistance of a λ-$(BETS)_2FeCl_4$ single crystal is shown in Fig. 1(a) for different temperatures. Here the magnetic field B is applied along the in-plane *c* axis. The main characteristic of the data is that between 18 and 41 T, the resistance of the material drops with decreasing temperature, reaching zero within experimental uncertainties below 2 K in a field range centered near 33 T. It is important to mention that in this part of the (B-T) phase diagram, the material behaves as a good metal. We find Shubnikov–de Haas oscillations (of order 700 T with effective mass $m^\star \sim 4m_0$) for the

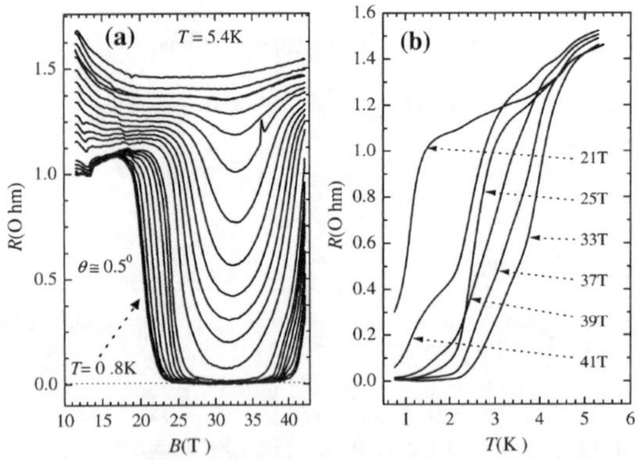

FIG. 1. (a) Resistance R as a function of magnetic field B, applied along the in-plane c axis ($\pm 0.3°$) of a λ-$(BETS)_2FeCl_4$ single crystal for temperature intervals of approximately 0.25 K, between 5.4 and 0.8 K. The superconducting state develops progressively with decreasing temperature but is suppressed for fields sufficiently away from (above or below) 33 T. (We note that since the Hybrid magnet is composed of a superconducting outsert coil in combination with a Bitter-type resistive insert coil, the field generated by the outsert is kept constant at approximately 11.5 T, while the field of the insert coil was ramped between 0 and and 31.5 T.) (b) Resistance as a function of temperature T for several values of B obtained from the field scans shown in (a). The FISC transition has a maximum transition temperature $T_c \simeq 4.2$ K near 33 T.

magnetic field perpendicular to the conducting planes [8], which for an isotropic model would give a Fermi energy $\varepsilon_F \sim 200$ K. The normal state resistivity is of order of $10^{-4}\ \Omega \cdot$ cm in the metallic state near 15 T. We estimate $k_F \ell \sim 20$ (where k_F is the Fermi wave vector and ℓ is the mean free path) and thus, despite the low scale of ε_F, the standard metallic conditions are fulfilled.

In the FISC state at higher fields, the resistivity drops typically by 2 to 4 orders of magnitude, putting it at or below the conductivity of copper, and beyond our ability to measure by standard ac lock-in methods. From the isothermal field scans we can extract the temperature dependence of the resistance at fixed values of the field; see Fig. 1(b). For fields between 18 and 37 T, the resistance shows a phase transition from the metallic phase (above 4.2 K) to the full superconducting state. Above a certain threshold field $B_{\rm th} = 18$ T, the onset of this transition *increases* with magnetic field, reaching a maximum $T_c \cong 4$ K at $B^\star = 33$ T. Above $B^\star$ the onset *decreases* in temperature with increasing field, and above 41 T the FISC is suppressed rapidly. We note that the experimental resistance does not fall immediately to zero below T_c. We interpret this as an evidence for complex magnetic flux penetration and trapping in our sample. Evidence for this is also provided by the hysteretic behavior observed in torque magnetization measurements shown in Ref. [3] and which was interpreted by the authors as evidence for flux pinning.

We next discuss a central question concerning the interpretation of the FISC state as "truly" superconducting, beyond the observation of zero resistance within experimental uncertainties. The Meissner effect—the standard test for the onset of superconductivity—where magnetic flux is excluded when a sample enters the superconducting state, may become a nontrivial experiment in the present case. This is due to the fact that the penetration of the magnetic field, which compensates the internal exchange field, is essential for the stabilization of the FISC state. Furthermore, the magnetic flux may penetrate and be trapped in between two-dimensional superconducting layers. And, in fact, torque magnetization measurements, which give thermodynamic evidence for a phase transition at $B_{\rm th}(T)$, indicate that the magnetic field remains pinned between the conduction layers producing a large hysteresis in the torque signal [3].

The present work provides an additional, independent piece of evidence that the FISC is superconducting: The FISC state is reentrant to a metallic state above 41 T, which excludes triplet pairing. This observation also rules out field-induced low-resistance models. When magnetic scattering or some other form of higher resistance state is removed by magnetic fields, restoration of disorder-related, inelastic processes at higher fields is very unlikely. Furthermore, and as is pointed out in Ref. [3], the application of a transverse magnetic field, i.e., perpendicular to the conducting layers, produces a dramatic effect on the FISC state. This behavior is quantitatively illustrated in Fig. 2(a), where we show results from a variation of the magnetic field away from the in-plane orientation at the lowest temperature of our investigation. The zero resistance state begins to vanish for a *transverse field* $B_{c\perp}$ greater than 3.5 T. This observation is elucidated in Fig. 2(b) by plotting the resistance for a *constant in-plane field* $B_{c\parallel} = B\sin(\theta)$ of about 33 T (i.e., $B_{c\parallel} = B^\star$) vs the transverse field $B_{c\perp} = B\cos(\theta)$. Hence the FISC state is removed when orbital components appear. Note that the critical field $B_{c\perp}$ for the FISC state is essentially identical to that of the nonmagnetic, isostructural material λ-$(BETS)_2GaCl_4$ [9]. The approximately linear increase in resistivity $\rho \propto (B_{c\perp} - B_{c\perp\rm out})$ may also be ascribed to trapped flux. A full explanation of this effect will require a further study (now being planned) involving a systematic experimental comparison of $B_{c\perp}$ for both Fe- and Ga-based compounds.

We next consider how the FISC state is stabilized. While the two anions (Fe^{3+} and Ga^{3+}) have different ground states at $T = 0$ in the low field range of the (B-T) phase diagram, alloying by Ga and external pressure restores superconductivity in the Fe-based material [10]. We expect that the superconducting states, in both cases, are close in energy. We therefore argue that the in-plane physics is similar for both materials, and the differences in the phase diagrams arise from correspondingly small energies related to, for instance, the interlayer coupling. Our model is as

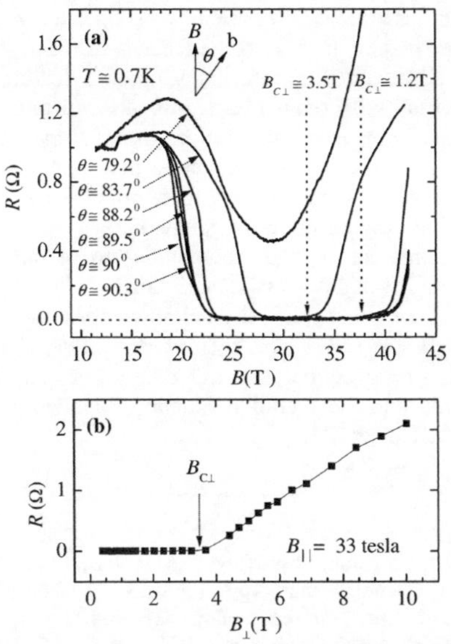

FIG. 2. (a) Resistance as a function of magnetic field at $T = 0.7$ K and for five different angles θ (indicated in the figure) between B and the interplane b axis. Notice that the interplane critical field $B_{c\perp}$, defining the orbital effect, decreases as θ approaches 90°. (b) Resistance for constant in-plane field $B_{c\|}$ vs transverse magnetic field $B_{c\perp}$ at $T = 0.7$ K.

follows. In-plane fields orient the $S = 5/2$ spin of the Fe^{3+} ions, and we assume that this decouples the BETS conducting layers. The problem then becomes two dimensional (2D), with no diamagnetic currents flowing between layers in the presence of a purely in-plane magnetic field (we return to this point later). For the 2D geometry, the in-plane field can destroy singlet superconductivity by means of paramagnetic effects only, i.e., breaking the Cooper pairs [1]. The $S = 5/2$ Fe^{3+} magnetic moments, oriented by magnetic field, exert the exchange field $J\langle S\rangle$ on the spins s of the conduction electrons via the exchange interaction, $J\vec{s} \cdot \vec{S}$. Thus, the effective field H_{eff} acting on the electron spin is

$$I(B) = \mu_B H_{\text{eff}} = \mu_B B + J\langle S\rangle. \qquad (1)$$

(For our experiments with $B \cong 20$–40 T and $T < 5$ K, the iron moments are saturated: $\langle S\rangle = 5/2$.) The Jaccarino-Peter effect is the compensation in Eq. (1) of the magnetic field, B, by the exchange field, $J\langle S\rangle$, at $J < 0$, so that the effective field, H_{eff}, may become below the Clogston limit [1,2] at higher B. According to [1], the first order phase transition from the paramagnetic normal metal into the superconducting state at $T = 0$ would occur at $I(B) = \Delta(0)/\sqrt{2}$, where $\Delta(0)$ is the superconducting gap at $T = 0$ (in the absence of the field). The actual situation is more complicated by the fact that the phase transition separating the normal and the superconducting states in the $(T, I(B))$-plane may be either first or second order [11–15]. Thus, at low enough temperatures the first order phase transition is preceded by the second order transition into the inhomogeneous LOFF (Larkin-Ovchinnikov-Fulde-Ferrell) phase [14,15] at $|I(B)| < \Delta(0)$. This phase is highly sensitive to the presence of defects. In the pure limit it would occupy a stripe in the $(T, I(B))$-plane as shown below in Fig. 3: at $T = 0$ $\Delta B \simeq 3.73$ T, which is beyond the resolution of our electrical transport measurements. Our experiments suggest $I(B^\star) = 0$ at $B^\star = 33$ T and $T_c \cong 4.2$ K. These parameters were used in the theoretical plots in Fig. 3. We see from Fig. 3 that the theoretical model discussed above reproduces the main features of the FISC phase diagram very well.

Let us now return to the question regarding the nature of the new FISC state in λ-(BETS)$_2$FeCl$_4$ and its relation to the "conventional" SC state in λ-(BETS)$_2$GaCl$_4$. Here we argue that remarkably, the Fe^{3+} magnetic ions may be essential to stabilize the two-dimensionality of the FISC state produced by the JP effect. Both compounds have similar, anisotropic, layered structure. Nevertheless, there is an inter-plane electronic coupling in the λ-(BETS)$_2$GaCl$_4$, as is evidenced in the finite upper critical field [9] for the in-plane direction,

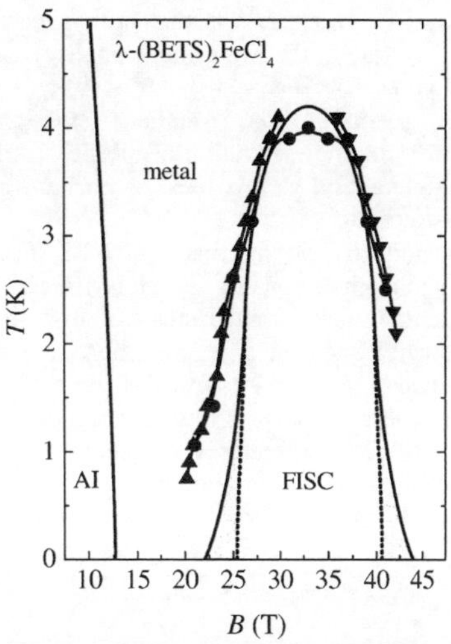

FIG. 3. Temperature-magnetic field phase diagram showing the AI, metallic, and FISC states for a λ-(BETS)$_2$FeCl$_4$ single crystal vs in-plane magnetic field. Solid triangles indicate the middle point of the resistive transition as a function of B [from Fig. 1(a)], while solid circles indicate the middle point of the resistive transition as a function of T [from Fig. 1(b)]. The solid line is a theoretical fit (see text) to a second order phase transition towards the FISC while the dashed line indicates a first order transition from the inhomogeneous LOFF state (shaded area) into the bulk SC state. Our electrical transport measurements do not have sufficient resolution to resolve both the second and the first order phase transitions.

$H_{C2}(0) = 12-15$ T, which is well below the field where the FISC is stable. To restore superconductivity in λ-$(BETS)_2FeCl_4$ within the field range of the JP compensation effect, one needs a mechanism to fully eliminate the diamagnetic interlayer currents in this compound. We suggest that the coupling between 2D BETS layers comes about through the bridging of MCl_4 tetrahedra, so that in the second order effective tunneling matrix element, the $(MCl_4)^0$ state shows up as the intermediate state in simple perturbation theory. While the p shell is empty for Ga^{3+}, all the "d levels" are occupied in Fe^{3+}, according to Hund's rule. Placing an additional electron at the Fe^{3+} site (with spin antiparallel to $\vec{S}$) is energetically unfavorable; i.e., it costs the Hund coupling energy. Furthermore, when the $S = 5/2$ spins of Fe^{3+} are aligned by the field, an electron with parallel spin has no accessible states on the d shell, and the energy cost inside the $FeCl_4$ complex is expected to increase even further. Therefore, tunneling across magnetically oriented Fe^{3+} tetrahedra becomes a spin selective process; consequently, transport of an s-wave Cooper pair between adjacent planes is excluded. Clearly, in λ-$(BETS)_2GaCl_4$ there is no spin selective process, and hence a correspondingly smaller upper critical field is observed [9].

Finally, we consider why $T_c^\star \cong 4.2$ K in λ-$(BETS)_2FeCl_4$ is less than $T_c^\star \cong 5.5$ K for λ-$(BETS)_2GaCl_4$. Our guess is that both superconducting ground states (due to the two dimensionality) are basically the same at $T = 0$ [i.e., $\Delta(0)_{Fe} \cong \Delta(0)_{Ga} = 1.76T_c^{Ga}$]. $T_c^\star = T_c^{Fe}$ seems to be smaller, since at higher temperatures, thermal activation of the Fe^{3+} spins via the exchange interaction J provides a mechanism for pair breaking scattering.

We conclude by noting that λ-$(BETS)_2FeCl_4$ [along with the nonmagnetic analog λ-$(BETS)_2GaCl_4$] has provided a rich new area for the study of low-dimensional superconductivity and magnetism, where the two mechanisms compete on a very low energy scale. In this Letter we have provided a simple theoretical picture that describes the broader features of the newly discovered high field induced superconducting state. We show experimentally that in-plane diamagnetic currents ($B_{c\perp}$) can destroy the FISC state. We have argued that magnetic ions are actually essential to suppress the coupling between planes in the presence of in-plane magnetic fields. However, there are many unusual features in this system that will require a significant, further level of both experimental and theoretical work. In particular, the mechanism by which the magnetic field penetrates in the bulk of our 2D superconducting samples remains unclear and may produce hysteretic behavior.

We thank the Hybrid Magnet Group at the NHMFL for their invaluable assistance during these measurements. One of us (J. S. B.) acknowledges support from NSF-DMR-99-71474 for this work. The NHMFL is supported by a cooperative agreement between the State of Florida and the National Science Foundation through NSF-DMR-0084173.

[1] A. M. Clogston, Phys. Rev. Lett. **9**, 266 (1962).
[2] B. S. Chandrasekhar, Appl. Phys. Lett. **1**, 7 (1962).
[3] S. Uji *et al.*, Nature (London) **410**, 908 (2001).
[4] A. Kobayashi, T. Udagawa, H. Tomita, T. Naito, and H. Kobayashi, Chem. Lett. **12**, 2179 (1993).
[5] L. Brossard *et al.*, Eur. Phys. J. B **1**, 439 (1998).
[6] V. Jaccarino and M. Peter, Phys. Rev. Lett. **9**, 290 (1962).
[7] H. Kobayashi, H. Tomita, T. Naito, A. Kobayashi, F. Sakai, T. Watanabe, and P. Cassoux, J. Am. Chem. Soc. **118**, 368 (1996).
[8] S. Uji *et al.*, Phys. Rev. B **64**, 024531 (2001); L. Balicas *et al.* (to be published).
[9] M. A. Tanatar, T. Ishiguro, H. Tanaka, A. Kobayashi, and H. Kobayashi, J. Supercond **12**, 511 (1999).
[10] A. Sato, E. Ojima, H. Akutsu, Y. Nakazawa, H. Kobayashi, H. Tanaka, A. Kobayashi, and P. Cassoux, Phys. Rev. B **61**, 111 (2000).
[11] L. P. Gor'kov and A. I. Rusinov, Sov. Phys. JETP **19**, 922 (1964).
[12] V. Barzykin and L. P. Gor'kov, Phys. Rev. Lett. **84**, 2207 (2000).
[13] L. Balents and C. M. Varma, Phys. Rev. Lett. **84**, 1264 (2000).
[14] A. I. Larkin and Yu. N. Ovchinnikov, Sov. Phys. JETP **20**, 762 (1962); P. Fulde and R. A. Ferrell, Phys. Rev. **135**, A550 (1964).
[15] L. N. Bulaevskii, Sov. Phys. JETP **38**, 634 (1974).

Chapter 7

Localization on Defects and Statistics of Levels in Low Dimension

7.1 Introduction

Because of the general interest in the problem, much attention has always been given to phenomena related to localization by the static disorder. For the onset of localization in 3D, known as the Anderson transition, the disorder must be strong. But in 1D conductivity falls with temperature already at a weak disorder. The tendency to localization finds its formal reflection in the difficulties of the "cross-technique" elaborated in Sect. 2 (see [2b]) as one can directly verify that the vertex corrections neglected in the 3D case are not small in 1D. (Below localization is studied for non-interacting particles).

In 1961 Mott and Twose argued that in the presence of a disorder *all* electronic states in one dimension are localized so that the *dc*-conductivity at zero temperature is zero *identically* (Ref. 1). In proof of their statement, Mott and Twose considered the model of centers localized at points of space and overlapping only weakly. That last assumption looks ungrounded because the wave functions of *free* quasiclassical electrons are plane waves with non-zero amplitude along the whole chain. Verifying Mott's assertion was one of the goals in the theoretical papers discussed below.

In two dimensions (2D) localization is a result of even more intricate interference between the wave functions of electrons scattered on many defects and happens on the distances exponentially larger than the mean free path (the latter being defined by strength of the backward scattering). Ways and means of treating the localization phenomenon are different in 1D and 2D and below are discussed separately.

7.1.1 *Disorder in one dimension.* I start with paper [7a] that with a simple example demonstrates the Schmidt's method (Ref. 2) based on counting zeros in the wave function of an electron along the chain. In particular, [7a] addresses thermodynamic properties of the three materials: NMP-TCNQ, Q-TCNQ, Ad-TCNQ, remarkable for the fact that the conducting TCNQ-chains have well-defined number of carriers in unit cells so that the period of the Peierls structure would correspond, respectively, to the doubling (for the first one) and quadrupling (for the rest two) periodicity of the lattice along the chain direction. The materials are metals at higher temperatures. Inherent to all three salts is a weak disorder related to asymmetry of cations.

Calculated in [7a] are the singularities in the electronic densities of states (DOS) at the Fermi level owed to the disorder. Such singularity is most pronounced in the case of NMP-TCNQ with the *half-filled* band where the Fermi "surface" consists of the two points $\pm p_F = \pm(\pi/2a)(2p_F = (\pi/a)$ corresponds to the doubled periodicity below the Peierls transition).

The effect is related to multiple scattering. That is, consider an electron, say at the Fermi point–p_F; scattering on imperfections with the transfer of the momentum $2p_F = (\pi/a)$ *twice* returns that electron back to the $-p_F$ point as $4p_F = (2\pi/a)$ is the inverse vector of the lattice. The phenomenon is akin to the parametric resonance and results in a singularity in DOS capable to account for the anomaly in the magnetic susceptibility of NMP-TCNQ.

Paper [7b] consists of two parts. The first part is general and formulates the localization criteria in a random field that are applicable in any dimension. The localization is shown to manifest itself via the singularity in expressions for several correlation functions.

In the second part we develop Schmidt's method further, thereby opening the possibility for calculating in one dimension the correlators between arbitrary numbers of the wave functions. (Schmidt's method was discussed by B. Halperin (Ref. 3); his paper of a methodological character contains no practical results related to the Mott-Twose problem).

Effectiveness of the method was demonstrated by calculating the conductivity: $\sigma(\omega) = (2l\nu^2/\pi) \ln^2 \nu$, at the low frequency $\nu = \omega l/v_F \ll 1$, where l is the kinetic mean free path for backward scattering. The expression formally coincides with the Mott-Twose result after redefinition of the localization length [7b].

Making use of [7b] in the next paper [7c] we calculated the *spatial* dependence for the correlator $F_{E,E+\omega}(x) = [\sum_{pq} \delta(E-\varepsilon_p)\delta(E+\omega-\varepsilon_q)\psi_p^2(0)\psi_q^2(x')]_{av}$. $F_{E,E+\omega}(x)$ is the density-of-states correlator for the two states close in energy ($\omega l/v_F \ll 1$). To some surprise, this correlator turned out not to be small on short distances $x \sim l$ where it is by the order of magnitude equal to $(\pi v_F)^{-2}$. As the typical localization length is l, the result may be interpreted as an *attraction* of the levels over distances of the order of the mean free path. (The levels become absolutely independent only on very large distances $|x| \gg l$ where one has for $F_{E,E+\omega}(x) \equiv (\pi v_F)^{-2}$).

Calculation of $F^1_{E,E+\omega}(x) = [\sum_{pq} \delta(E-\varepsilon_p)\delta(E+\omega-\varepsilon_q)\psi_p(0)\psi_q(0)\psi_p(x)\psi_q(x)]_{av}$ in [7d] revealed more interesting features. The two correlators $F_{E,E+\omega}(x)$ and $F^1_{E,E+\omega}(x)$ coincide as far as $x \sim l$. At larger distances both correlators rapidly drop in the value signifying strong *repulsion* between close levels on the quite *large spatial interval* (from $x \geq l$ and up to $x_0 = 2l \ln[8v_F/\omega l]$). At $x > x_0$ the first correlator gradually approaches its value $(\pi v_F)^{-2}$; $|F^1_{E,E+\omega}(x)|$ shows the Gaussian peak $\propto \exp[-(x-x_0)^2/4lx_0]$ and only then goes to zero. The last result confirmed that the wave functions are not correlated at $x \gg x_0$, but at the same time it reveals strong fluctuations in the interval $\approx (lx_0)^{1/2}$ around x_0. For further discussion and comparison with the statistics of levels for *small particles*, see [7d]. (More details are given in Ref. 4).

7.1.2 *Electron localization in the field of defects in 2D.* Our curiosity was aroused by the paper of the four authors dubbed in the literature as the "gang of four" (Ref. 5). Details of the calculations were not published in Ref. 5, so we decided to verify their claims by investigating the *frequency* dependence of conductivity in two dimensions. For the analysis of the fluctuation corrections we applied the renormalization group analysis (scaling) to the resistance.

The first result was the derivation of the diffusion-like pole in the Cooper channel. Diffusion poles in the Cooper and in the electron-hole (zero-sound) channels are now often described in the literature as "cooperon" and "diffuson", respectively. The cooperon pole owes its origin to the time reversal invariance $t \to -t$. As seen from our calculations, in two dimensions the localization is the result of the *exact* cancellation between contributions from the two channels. That is, any mechanisms violating the $t \to -t$ invariance such as the Zeeman Effect, scattering on magnetic centers and others effectively destroy localization. Non-zero spin-orbit interaction at the impurity mixes directions of the electronic spin; as shown in Ref. 6, it may change the sign in front of the logarithm in the Cooper channel.

References

1. N. F. Mott and W. D. Twose, *Adv. Phys.* **10**, 107 (1961)
2. H. Schmidt, *Phys. Rev.* **105**, 425 (1957)
3. B. I. Halperin, *Phys. Rev.* **139**, A104 (1965)
4. L. P. Gor'kov, O. N. Dorokhov and F. V. Prigara, *Sov. Phys. JETP* **58**, 852 (1984)
5. E. Abrahams, P. W. Anderson, D. C. Licciardello and T. V. Ramakrishnan, *Phys. Rev. Lett.* **42**, 673 (1979)
6. S. Hikami, A. I. Larkin and Y. Nagaoka, *Prog. Theor. Phys.* **63**, 707 (1980)

List of original papers

[7a] L. P. Gor'kov and O. N. Dorokhov, *Disorder and one-electron spectrum in a linear chain*, Sov. J. Low Temp. Phys. **4**, 160 (1978)

[7b] V. L. Berezinskii and L. P. Gor'kov, *On the theory of electrons localized in the field of defects*, Sov. Phys. JETP **50**, 1209 (1979)

[7c] L. P. Gor'kov, O. N. Dorokhov and F. V. Prigara, *Energy levels correlations in a disordered one-dimensional conductor*, Sov. Phys. JETP **57**, 838 (1983)

[7d] L. P. Gor'kov, O. N. Dorokhov and F. V. Prigara, *On the ac conductivity of slightly disordered one-dimensional conductors*, Sov. Phys. JETP Lett. **37**, 447 (1983)

[7e] L. P. Gor'kov, A. I. Larkin, and D. E. Khmel'nitski, *Particle conductivity in a two-dimensional random potential*, Sov. Phys. JETP Lett. **30**, 228 (1979)

Disorder and one-electron spectrum in a linear chain

L. P. Gor'kov and O. I. Dorokhov

Institute of Theoretical Physics, Academy of Sciences of the USSR

(Submitted July 25, 1977)

Fiz. Nizk. Temp. 4, 332–342 (March 1978)

Characteristic features of the electronic density of states of slightly disordered one-dimensional conducting systems are studied. It is shown that singularities in the density of states appear only in the case of half-filled bands and for specific types of the scattering potential. A similar problem is solved for systems with Peierls distortion superstructures. The relation of this problem to properties of several organic compounds is discussed.

PACS numbers: 63.20.Kn, 71.38.+i

INTRODUCTION

A large number of conducting organic compounds are now known, whose anisotropy is so large that the one-dimensional linear chain approximation can be used to describe them. Most experimental studies so far have been performed on the TCNQ base, particularly on the highly conducting TTF-TCNQ compound.[1] As to the theoretical description of the observed properties, the situation is much worse: Though one has qualitative understanding of effects possible in similar materials, the abundance of these effects, their mutual competition, and the complexity of the mathematical apparatus make a comparison between experimental data and theoretical conclusions rather difficult. In particular, a basic difficulty is that the complicated molecular structure of similar complexes contains a large number of unknown constants; therefore comparison of theoretical results with experimental data does not enable one to draw any unique conclusions.

Usually the description of the electron spectrum in terms of the band structure is a prerequisite. Without even mentioning that the possibility of such an approach is not at all obvious (this problem is touched upon Ref. 1; see also Refs. 2, 3), electron–electron interactions can lead in this case to a whole series of unusual collective effects (see, for example, Ref. 4). A simpler scheme is now adopted in the literature: the electron interaction is neglected, and only the electron–phonon interaction is taken into account, to the extent that it leads to appearance of Peierls (or Fröhlich) instability of active phonon modes or to generation of a kinetic momentum relaxation time.

Many experimental facts can be explained on this basis. In our opinion, the problem of quantitative comparison between theory and experiment is not decisive in the given case, since, for example, frequencies and electron–phonon interaction constants of numerous phonon branches appear in the theory for these compounds (see, for example, Ref. 5). Moreover, the one-electron approximation describes adequately even the temperature behavior of the dielectric constant and of the conductivity, with account taken of localized effects specific to the one-dimensional problem (this comparison with experiment for Q– and Adz–TCNQ was carried out in Ref. 6; see also the corresponding references).

We have pointed out, however,[7] that there exist experimental results[8] that, obviously, do not fit into the scheme described above, and are, therefore, most important for qualitative understanding of the general pattern. The low-temperature behavior of the magnetic susceptibility and of the heat capacity were measured[8] for the following groups of materials: NMP-TCNQ, Q-TCNQ, Ad-TCNQ. A sharp increase in the susceptibility in the region of very low temperatures (below 5-10°K) is characteristic of all these materials.

This group of materials possesses the following important features, distinguishing them, for example, from TTF-TCNQ. In the latter the appearance of carriers is related to overlap of electron energy bands of the TTF and TCNQ chains, due to which the charge redistribution between filaments corresponds in the band picture to some in no way special value of the Fermi momentum k_F. This situation leads, in particular, to a period of the Peierls structure incommensurate with the period of the original lattice. In the compounds mentioned a clearly defined number of carriers appears in the unit cells. In case of NMP-TCNQ the band is half-filled, while in the other two compounds one fourth of the band is filled. The period of the Peierls structure, $2\pi/2k_F$, would correspond, respectively, to doubling and quadrupling of the basic lattice periodicity along the direction of the chain. According to the results of Ref. 9, some phase transition occurs at least in NMP- and Q-TCNQ at substantially lower temperatures T_k than for TTF-TCNQ. For NMP-TCNQ $T_k \simeq 7$°K, while in the case[9] of Q-TCNQ $T_k \simeq 14$°K; the nature of the transition is nuclear, and, as far as we know no structural analysis was performed at low temperatures

In addition, an important feature of these materials is an inherent internal disorder, related to cation symmetry.[10]

The first attempt to explain the observed low-temperature features in the susceptibility and in the heat capacity, carried out in Ref. 8, was based on the formal similarity of the electron Hamiltonian in the strong-coupling approximation for random overlap integrals to the spectral problem of low-lying vibrational degrees of freedom of a disordered linear chain, a problem solved by Dyson.[11] The expression for the singularity in the density of one-particle states was borrowed directly from Ref. 11, and is related to the commensurability of the Fermi momenta in the materials mentioned.

The same problem was considered by us[7] in a more realistic model. As a result it was clarified that the Dy-

 0360-0335/78/0403-0160$02.20

son singularity in the density of states is indeed sometimes possible, but only under certain assumptions concerning the defect structure, and for bands exactly half-filled. Since the temperature dependence of the susceptibility referred to is approximately identical in all the compounds investigated,[8] the basic conclusion drawn[7] has been that the band scheme for describing electronic properties of these materials at low temperatures is not applicable, at least not in the absence of electron–electron interactions. The problem still remained, however, of whether the observed[9] low-temperature transitions in NMP- and Q-TCNQ are related to structural transformations, and how the results obtained in Ref. 7 are affected by gaps in the energy spectrum due to the appearance of additional superstructures in the system.

An attempt to find the density of electron states in a system of electrons with a semiconducting gap and disorder (defects) was recently undertaken in Ref. 12. The purpose of the present paper is to investigate the same problem. However, for systems of noninteracting electrons our conclusions differ substantially from the results of Ref. 12 in the presence of disorder, both in the absence of a Peierls gap and when the latter is taken into account. The reasons for this disagreement will be explained below.

1. IMPURITIES AND THE ELECTRON SPECTRUM IN THE METALLIC PHASE

The basic quantity in this approximation, used to find the thermodynamic characteristics of the system, is the one-particle density of states. As in Ref. 7, the method developed by Schmidt[13] for evaluating the number of zeros of a wave function is used in the calculation. This method enables one to retain more information concerning the characteristics of atomic lattice potentials or defects than does the Halperin equation, used in Ref. 12 (the latter is valid for a random potential with "white noise" correlation). Below we show that this fact is important for the results obtained, though the method of Ref. 13 is convenient in handling short-range atomic potentials (with a range of the order of the lattice constant). Since a trivial error[1)] was made in the original paper in deriving the basic equations, we briefly reproduce this derivation in the present section.

The mathematical problem of the motion of an electron in the field of a periodic lattice perturbed by defects (in the absence of a Peierls gap) can be modelled quite well by the condition that defects, even at low concentration, appear in rigorously defined chain sites:

$$z_n = an. \tag{1}$$

The unit cell number n is a random integer. The phase of the wave function is determined by the relation

$$\operatorname{ctg} \varphi(x) = \psi'(x)/k\psi(x) \tag{2}$$

(here $k^2 = 2mE$). To obtain results in analytic form, one usually assumes smallness (the Born approximation) of either the magnitude of the atomic defect potential or the concentration $c = N_D/N \ll 1$ (in what follows N_D is the total number of defects in the chain, and $Na = L$ is its length).

The method of Ref. 13 consists of determining the random Markov process for the change in phase $\varphi_j \equiv \varphi_j^+$ (for example, the phase to the right of the short-range atomic potential) with the use of the function $W(\varphi_j)$, the probability distribution function of the phase for each unit cell of the lattice $[aj, a(j+1)]$. The process remains Markovian as long as the phases in neighboring portions are related only by scattering from the j-th atom (generalizations are possible here, but they complicate the problem and do not lead to qualitatively new results).

If the potential is characterized by the parameters δ and γ, the first being the phase shift of a wave function of a passing electron, and the second determining its amplitude $|d|$ [the transmission coefficient is $d = \cos\gamma \cdot \exp(-i^{\delta})$], the relationship mentioned between the phases φ_j and φ_{j+1} is

$$\varphi_{j+1} = \varphi_j + ka - \delta + \frac{1}{2i} \ln \frac{1 + i\sin\gamma \exp[-i(2\varphi_j + 2ka - \delta)]}{1 - i\sin\gamma \exp[+i(2\varphi_j + 2ka - \delta)]}. \tag{3}$$

In the Born approximation we use the following expansion, accurate up to second-order terms:

$$\varphi_{j+1} = \varphi_j + ka - \delta_{j+1} + \gamma_{j+1}\cos(2\varphi_j + 2ka - \delta_{j+1}) - {}^1/_2\gamma_{j+1}^2 \sin(4\varphi_j + 4ka - 2\delta_{j+1}.) \tag{3'}$$

where δ_{j+1}, γ_{j+1} are characteristics of the potential in the (j + 1)-th lattice site. The equation[13] for the function $W(\varphi)$,

$$W(\varphi_{j+1}) = (1-c)W(\varphi_{j+1} - ka) + cW(\varphi_j(\varphi_{j+1}))\frac{d\varphi_j}{d\varphi_{j+1}}, \tag{4}$$

does not require explanation. The number of states n(k) with momentum less than k is

$$n(k) = \frac{k}{\pi} + \frac{c}{a\pi}\langle\Delta\varphi\rangle, \tag{5}$$

where $\langle\Delta\varphi\rangle$ is an average over the distribution $W(\varphi)$ of the phase "advance" (2) during motion from left to right.

We recall[7] that the appearance of singularities in the density of states for momentum values commensurate with the lattice is related to parametric resonance: Though the impurities are randomly located in the lattice, they are always fixed at lattice points z_n [Eq. (1)]. Effects of multiple electron scattering retain their "memory" concerning the latter circumstance. Therefore, as follows from Ref. 7, to obtain from the general relationship (4) equations describing resonance effects of interest to us it is convenient to use the Fourier representation

$$W(\varphi) = \sum_m e^{+2im\varphi} W_m, \quad W_m = \frac{1}{\pi}\int_0^\pi d\varphi W(\varphi) e^{-2im\varphi} \tag{6}$$

[the fact is here reflected that, according to (2), the function $W(\varphi)$ is defined in the interval $(0, \pi)$]. The significant harmonics in (6) are determined by terms in (4) not containing the concentration:

$$[1 - e^{-2imka}] W_m.$$

The requirement of smallness of the expression in the square brackets, necessary for the appearance of large corrections to $W(\varphi)$, also expresses the condition mentioned of parametric resonance

$$1 - e^{-2im\Delta ka} \approx 2im\Delta ka, \quad k = k_F + \Delta k \tag{7}$$

with $k_F a = \pi/n$.

The procedure of deriving the basic equations from relations (4), (3), (6), and (7) is extremely simple: In integrating the last term in the right-hand side of (4) with the factor $\exp(-2im\varphi_{j+1})$ the phase φ_{j+1} should be expressed in terms of the phase φ_j with the required accuracy. In the Born approximation, according to (3'), the exponent is expanded in a series up to second-order terms, which, generally speaking, leads to a second-order differential equation for the function $W(\varphi)$. This equation, however, reduces immediately to a first-order equation, since it follows from (7) and the form of Eq. (4) that in the Fourier components the factor m can be cancelled on both sides of the relation generated for the Fourier components. Recovering later on the form of the first-order differential equation, it is necessary to take into account that $m \neq 0$. Precisely this circumstance leads to a change in the right-hand side of Eq. (11) of Ref. 7.

In case of a half-filled band the equation is[2)]

$$[\lambda + \sin 2\psi]\, W(\psi) - [3s + \cos 2\psi]\, \frac{dW}{d\psi} = \frac{\lambda}{\pi} - \frac{1}{\pi}\int_0^{\pi} d\psi \sin 2\psi\, W(\psi),$$

where

$$\lambda = 2(\Delta ka - \delta)/c\gamma^2; \quad \psi = 2\varphi; \quad 3s = 1 + 2\delta^2/\gamma^2.$$

Unlike Ref. 7, the equations below are written down for the total function $W = W_0 + W_1$, normalized by the condition $\bar{W} = 1/\pi$.

Integrating both sides of this equation over ψ, it is easily shown that the function $W(\varphi)$ is periodic with the period π [i.e., $W(\varphi)$ has a period $\pi/2$]. Therefore the final system of equations for the function $W(\varphi)$ acquires the following form:

$$\begin{gathered} [\lambda + \sin 2\psi]\, W(\psi) - [3s + \cos 2\psi]\, W'(\psi) = A(\lambda), \\ W(0) = W(\pi), \quad \int_0^{\pi} W(\psi)\, d\psi = 1, \end{gathered} \tag{8}$$

and the density of states $\nu(\varepsilon)$ is represented by the derivative:

$$\nu(\varepsilon) = \partial A/\partial\lambda. \tag{9}$$

Solving the system of equations (8), we obtain

$$\begin{aligned} A(\lambda) = {} & \left\{1 - \exp\left(-\frac{\lambda\pi}{\sqrt{(3s)^2 - 1}}\right)\right\} \\ & \times \left[\int_0^{\pi} \frac{dx}{\sqrt{3s - \cos 2x}} \int_0^{\pi} dy\, \frac{\exp\left(-\lambda y/\sqrt{(3s)^2 - 1}\right)}{\sqrt{3s - \cos 2(x+y)}}\right]^{-1}. \end{aligned} \tag{10}$$

As in Ref. 7, a closed general expression can be obtained only for the density of states at the maximum point:

$$\nu|_{\lambda=0} = \frac{1}{\pi}\sqrt{\frac{3s+1}{3s-1}}\left(\pi/2K\sqrt{\frac{2}{3s-1}}\right)^2. \tag{11}$$

For $s = 1$ (corresponding to a δ-function potential) $\nu|_{\lambda=0} = 1.02\nu_0$ and for $s = 5/12$, $\nu|_{\lambda=0} = 1.6\nu_0$ ($\nu_0 = 1/\pi$). In other words, the numerical value of the peak height is not large. When the Fermi momentum corresponds to a band one-fourth filled, the height of the peak is of Born smallness. (The results of Ref. 7 are unchanged in this case.)

Thus, according to (10) and (11) a singularity in the density of states is possible only for a half-filled band and only in the limiting case $3s = 1$, corresponding to a scattering potential with no phase shift of the passing wave. It was pointed out[7] that in the class of NMP-TCNQ, Q-TCNQ, and Ad-TCNQ compounds disorder is related to the cation nonsymmetry, formed either by the CH_3^+ group in NMP or by the excess nitrogen atom in the remaining two compounds. The perturbation of the potential related to this nonsymmetry is odd with respect to the inversion transformation; therefore there is a special reason for using the limiting case $3s = 1$ at least in the Born approximation for these materials. The limiting transition in expression (10) to the case $3s = 1$ leads to the following expression for $A(\lambda)$:

$$A(\lambda) = \frac{4}{\pi^2}\left[J_0^2\left(\frac{\lambda}{2}\right) + N_0^2\left(\frac{\lambda}{2}\right)\right]^{-1} \tag{12'}$$

[$J_\nu(z)$, $N_\nu(z)$ are cylindrical functions]; for the singularity in the density of states (at $\lambda \ll 1$) we have

$$\nu \simeq \nu_0 \frac{2\pi}{\lambda\,|\ln^3\lambda|}. \tag{12''}$$

The expression for the magnetic susceptibility in the low-temperature region

$$\chi \simeq \frac{\pi\mu_B^2\nu_0}{2T\tau_-\ln^2 T\tau_-} \tag{12'''}$$

describes adequately the behavior of the susceptibility at temperatures near 10°K (see Ref. 8). We recall, however, that among the three materials mentioned, for which an anomalous increase in the paramagnetic susceptibility was observed at low temperatures, only the compound NMP-TCNQ corresponds to a half-filled band.

2. SINGULARITIES OF THE ELECTRON SPECTRUM IN A SYSTEM WITH PEIERLS DEFORMATION

In the present section we consider the same problem, but for a lattice having additional superstructure resulting from a Peierls instability, for example. The formation of the basic lattice parameters due to the latter effect is proportional to $\cos(2k_F x + \chi)$, i.e., has a period π/k_F. If this period is commensurate with the period of the original lattice, the phase in the cosine argument is determined by Umklapp processes. This fact is essential for period doubling, i.e., for a band half-filled (NMP-TCNQ). At higher multiplicity (p) of the newly generated period the corresponding "pinning" energies contain a high power of the ratio of the scale of deformation to the lattice period.[14]

Since the superstructure thus generated is quite small, so is the associated change in the potential of the field acting on the electron, and the respective gap in the electron spectrum. The potential due to the injected impurity or defect is, generally speaking, of the same order as the potential of the lattice atoms. Therefore, in the given case it is necessary to take into account the presence of a periodic field in the respective equations. This is particularly important for the results obtained. Below we use again the short-range potential model. We point out, however, that earlier we have simplified the problem somewhat, assuming that the scattering potential is symmetric in the replacement $x \to -x$. This fact is unimportant in the absence of a superstructure, but, as shown below, can qualitatively change the results for a lattice with a Peierls supersturcture. The scattering problem for a potential of general shape is, generally speaking, characterized by three parameters: the magnitude of the reflection coefficient γ, and the two phases δ and $\delta + \beta$ for the transmitted and reflected wave, respectively. As characteristics of the basic lattice atoms we introduce the following quantities: Γ_j, the amplitude of the reflection coefficient, and D_j and $D_j + B_j$, the phases of the transmitted and reflected wave. For defect atoms the corresponding quantities are denoted by γ, δ, $\delta + \beta$. As above, the Born approximation ($D, \Gamma, \delta, \gamma \ll 1$) is used, but the quantities Γ_j, D_j, and B_j are of the form

$$\Gamma_j = \Gamma + \Gamma_j^p;\ D_j = D + D_j^p;\ B_j = B + B_j^p, \tag{13}$$

where Γ_j^p, D_j^p, and B_j^p denote small modulations of these parameters due to the superstructure. Thus, if γ, δ, β are of the same order as Γ, D, and B, respectively, the quantities Γ_j^p, D_j^p, and B_j^p are small in comparison with them (the quantity $B \sim 1$ for an arbitrary shape of the potential).

The function $W(\varphi_j)$ should now be determined for the new lattice period. It is also necessary to take into account the increase in the period if one is interested in quantities (density of states, for example) referring to unit length of the chain.

If the multiplicity of the new period is p ($k_F = \pi/pa$), the equation replacing Eq. (4) of the preceding section acquires the form

$$W(\varphi_{l+p}) = (1 - pc) W(\varphi_l(\varphi_{l+p})) \frac{d\varphi_l}{d\varphi_{l+p}} + c \sum_{f=1}^{p} W(\varphi_l^{(f)}(\varphi_{l+p})) \frac{d\varphi_l^f}{d\varphi_{l+p}}, \tag{4'}$$

where the summation over the subscript f implies that the defect can appear at any site of the new lattice.

Within the Born approximation there is no particular difficulty in deriving the equation connecting φ_j and φ_{j+p}. This expression is too awkward to be written here. For p = 2, for example, we obtain

$$\begin{aligned}\varphi_{l+2} = {} & \varphi_l + \pi - D_{l+1} - D_{l+2} - (\Gamma_{l+1} - \Gamma_{l+2}) \cos 2\varphi_l \\ & - {}^1/_2 (\Gamma_{l+1}^2 + \Gamma_{l+2}^2) \sin 4\varphi_l - \Gamma_{l+1}(D_{l+1} + B_{l+1}) \sin 2\varphi_l \\ & + \Gamma_{l+2}(2D_{l+1} + D_{l+2} + B_{l+2} + 2\Gamma_{l+1} \cos 2\varphi_l) \sin 2\varphi_l.\end{aligned} \tag{14}$$

When there is a defect at a certain position the corresponding Γ, D, and B are replaced by the quantities γ, δ, and β. The following important facts are explained by studying this relationship. Terms due to the original lattice, i.e., independent of the subscript, lead only to energy renormalization:

$$k_F a \to k_F a - D.$$

This renormalization is trivial and will be omitted henceforth.

We compare the scale of the variation $\Delta k = k - k_F$ with the perturbation generating the superstructure, which is small and whose linear term only we retain in (14). Such expressions should be retained only in the term with unity in the right-hand side of (4'), since the cross terms related to defects are small either due to the low defect concentration $c \ll 1$, or as a result of the fact that at high defect concentrations[7] the perturbing defect potential is assumed to be small.

The terms in (14) due to the Peierls sublattice can be of different nature. The terms linear in Γ_j^p contribute to (14):

$$2\Gamma_l^p \cos 2\varphi_l. \tag{15}$$

Linear terms in D_j^p are absent, and the terms with B_j^p are[3)]

$$2\Gamma B_l^p \sin 2\varphi_l. \tag{15'}$$

[Terms of order $\Gamma\Gamma_j^p$ can be neglected in comparison with (15).] The magnitude of the perturbation amplitude β_j^p can be comparable with (15), but in this case it is still small due to the assumption of smallness of the original lattice deformation. Therefore it is necessary to take into account both possibilities, i.e., (15) and (15') can be written together as

$$2\Gamma^p \cos(2\varphi + \alpha). \tag{16}$$

Finally, it is easily verified that it is not the defect characteristics that appear in all equations, but their deviations from the same quantities for the original atoms $\tilde{\gamma} = \gamma - \Gamma$ and $\tilde{\delta} = \delta - D$. As to deriving the equations for the function $W(\varphi)$, it follows from the assumptions made above that all terms in these equations related to the presence of defects retain the same form as in Eqs. (8) and (9) of the preceding section with the replacement $(\gamma, \delta) \to (\tilde{\gamma}, \tilde{\delta})$.

It follows from the last statement that the case of a half-filled band should be considered separately, as above. If the bands are, say, one-third or one-fourth full, the terms responsible for generation of "parametric" resonance effects related to ordering of lattice defects in (1) contain additional small Born quantities. Within the same accuracy, when the characteristic energy scale (the Peierls gap) competes with the reciprocal kinetic time for electron scattering by defects, the equations obtained coincide with the equations for the so-called "incommensurate" case (arbitrary k_F). Below we provide without derivation the basic equations for these two cases, which are similar to Eq. (8), but obtained with account of the Peierls structure (13).[4)]

A. A half-filled band. For a general form of the superstructure (16) it is convenient to write the equation for the function $W(\psi)$ (where $\psi \equiv 2\varphi$):

$$\begin{gathered}[\lambda + \vartheta \cos(\psi + \alpha) + \sin 2\psi] W - [3s + \cos 2\psi] W' = A(\lambda), \\ W(0) = W(2\pi),\ \int_0^{2\pi} W(\psi)\, d\psi = 2.\end{gathered} \tag{17}$$

In this case these quantities can be expressed in terms of the energy gap in the electron spectrum, related to Peierls doubling and of kinetic relaxation times for electron scattering by a defect "forward" (τ_+) or "backward" (τ_-):

$$\lambda = 2\varepsilon\tau_-;\ \vartheta = 2\Delta\tau_-;\ 3s = 1 + 2\frac{\tau_-}{\tau_+}. \tag{17'}$$

We also provide the result of solving these equations for

the required quantity A(λ):

$$A(\lambda)=2\left\{1-\exp\left(-\frac{2\pi\lambda}{\sqrt{(3s)^2-1}}\right)\right\}$$
$$\times\left[\int_0^{2\pi}\frac{dx e^{f(x)}}{\sqrt{3s-\cos 2x}}\int_0^{2\pi}\frac{dy e^{-f(x+y)}}{\sqrt{3s-\cos 2(x+y)}}\right]^{-1}; \tag{18}$$
$$f(x)=\frac{\lambda x}{\sqrt{(3s)^2-1}}+\frac{\vartheta\cos\alpha}{\sqrt{2(3s+1)}}\,\mathrm{arth}\,\frac{\sqrt{2}\sin x}{\sqrt{3s-\cos 2x}}$$
$$+\frac{\vartheta\sin\alpha}{\sqrt{2(3s-1)}}\,\mathrm{arctg}\,\frac{\sqrt{2}\cos x}{\sqrt{3s-\cos 2x}}.$$

Due to the awkwardness of the last expression quantitative results can be obtained in the general case only numerically. The materials of interest to us, particularly NMP-TCNQ, possess, however, specific internal disorder, which by symmetry arguments corresponds to backward scattering only rather than the events discussed above. The limiting transition in (18) with account of all singularities at $3s-1\to 0$ gives

$$A(\lambda)=\frac{4}{\pi^2}\left[J^2_{1/2,\vartheta\cos\alpha}\left(\frac{1}{2}\sqrt{\lambda^2-\vartheta^2\sin^2\alpha}\right)\right.$$
$$\left.+N^2_{1/2,\vartheta\cos\alpha}\left(\frac{1}{2}\sqrt{\lambda^2-\vartheta^2\sin^2\alpha}\right)\right]^{-1}, \tag{19}$$
$$A(\lambda<\vartheta\sin\alpha)=0.$$

Choosing in (16) $\alpha=0$, the result coincides with Ref. 12 (only for a half-filled band!). The choice of α depends undoubtedly on the symmetry properties of the crystal and on the generated superstructure. The presence of a term of form (15') is quite important, for example, if the Peierls superstructure is due to interactions with the longitudinal acoustic branch of vibrations. In this case $\alpha\neq 0$, and, as follows from (19), a finite gap remains in the electron spectrum, despite the inherent disorder in the chain. It should be recalled that this statement refers only to the case of complete absence of forward scattering from the defect. Thus, the noted[12] singular behavior of the density of states inside the Peierls gap due to defects can be generated only at $\alpha=0$. Whether a structural transition occurs in the NMP-TCNQ compounds and what changes in the crystal symmetry accompany it are problems which are still open.

B. A superstructure period "incommensurate" with the basic lattice ($2k_Fa\neq\pi$). In this case the equations are simpler:

$$[\lambda+\vartheta\cos\psi]W-3sW'=A(\lambda),$$
$$W(0)=W(2\pi),\quad \int_0^{2\pi}W(\psi)d\psi=2. \tag{20}$$

Choosing (15) or (15') for the perturbed superstructure does not play any role, since the terms for either contain no explicit dependence on ψ. By the same reasoning no specifics of the defect are manifested. The solution of (20) is

$$A(\lambda)=\frac{3s}{2\pi}\left\{1-\exp\left(-\frac{2\pi\lambda}{3s}\right)\right\}\left[\int_0^{\pi}dx\exp\left(-\frac{2\lambda x}{3s}\right)I_0\left(\frac{2\vartheta}{3s}\sin x\right)\right]^{-1}, \tag{21}$$

where $I_0(y)$ is the Bessel function of an imaginary argument. As in (17'), the quantities λ, ϑ refer to defect characteristics:

$$\lambda=2\varepsilon\tau_-;\quad \vartheta=2\Delta\tau_-.$$

Performing in (21) the limiting transition $\lambda,\vartheta\to\infty$ in the absence of defects, we obtain the ordinary square-root singularity

$$\nu(\varepsilon)=\frac{1}{\pi}\frac{\varepsilon}{\sqrt{\varepsilon^2-\Delta^2}},\quad \varepsilon>\Delta. \tag{22}$$

The general expression (21) gives for the density of states at $\lambda=0$

$$\nu(0)=\frac{1}{\pi}I_0^{-2}\left(\frac{2\Delta\tau_-}{3s}\right), \tag{23}$$

i.e., a finite value. It is important to note that the assumption of the presence of a Peierls superstructure in the system at low temperatures does not lead to a singular density of states inside the gap due to defects, and, consequently, is not capable of describing the increase in magnetic susceptibility in Q-TCNQ and Ad-TCNQ, observed in Ref. 8.

In conclusion we explain the sharp disagreement between the results of our calculations and those of Ref. 12. In Ref. 12 a study was made of the system of equations for electron wave functions describing states near $+k_F(\psi_+)$ and $-k_F(\psi_-)$, respectively:

$$\left(\varepsilon+iv\frac{d}{dx}\right)\psi_+-(\Delta+\xi(x))\psi_-=0,$$
$$\left(\varepsilon-iv\frac{d}{dx}\right)\psi_--(\Delta+\tilde{\xi}(x))\psi_+=0, \tag{24}$$

where $\xi(x)$ are elements of a random potential due to impurities (without forward scattering). In these equations

$$\xi(x)=\sum_n v_{+-}(x-a_n)e^{-2ik_Fa_n-i\chi};\quad \tilde{\xi}(x)=\xi^*(x),$$

where V_{+-} are the parts of defect potentials responsible for backward scattering. It is assumed in Ref. 12 that $\tilde{\xi}=\xi$ and that the correlator of "random noise" $\langle\xi(x)\cdot\xi(x')\rangle=A\delta(x-x')$ is nonvanishing. In reality the correlator $\langle\xi\tilde{\xi}\rangle$ in (24) generally vanishes, since the quasimomentum $4k_F$ is not conserved in scattering processes (in this case the correlator $\langle\xi\tilde{\xi}\rangle$ does not vanish). This result remains true only for a half-filled band ($k_F=\pi/2a$), where the quantity $4k_F$ is a vector of the reciprocal lattice (in this case the phase $\chi=0$ is also fixed). We turn, finally, to more important physical remarks concerning the results obtained. It was assumed above that the defect substitutes for one of the lattice atoms. In reality, since electrons interact with defects and, in turn, participate in the formation of a Peierls structure, single defects cause also deformation of a Peierls superstructure.[16] This deformation is small in comparison with (13) (but only to the extent of assuming the presence of a Born potential for the defect!), however, it washes out over a distance $\xi_0\sim\hbar v/\Delta$ around the atomic defect. The method used by us "breaks down" for such "extended" defects, at least as long as we try to obtain results in analytical form. It can only be assumed that this fact could smooth out the singularities generated in the density of states, since $\xi_0\gg a$. A second effect, which was not taken into account in the preceding treatment and which can be particularly important in the so-called "incommensurate" case, is the fixing of an arbitrary phase of a Peierls superstructure on defects.[16]

NOTATION

Here, k_F is the Fermi momentum; a is the lattice

constant; ψ(x) and φ(x) are the wave function and its phase; c is the impurity concentration; N_D is the number of defects; W(φ) is the probability distribution; d is the transmission coefficient; γ, δ, and β are impurity atom characteristics; Γ, D, and B are host atom characteristics; Γ^p, D^p, and B^p are deviations of the respective characteristics due to Peierls distortion; n(k) is the number of electron states; $\nu(\varepsilon)$ is the density of states; τ_+ and τ_- are the relaxation times for forward and backward scattering; Δ is the Peierls gap; and ξ_0 is the coherence length.

1)See Erratum, Solid State Commun. 21, No. 7 (1977).

2)The authors are again grateful to A. Gogolin and V. Mel'nikov who, solving the same problem by diagram summation (Ref. 15), directed our attention to the difference in the final results.

3)For electrons moving along filaments in the crystalline lattice field the phase B vanishes (at least in the Born approximation) if the potential created by the molecule possesses a symmetry transformation changing the sign of x, such as inversion. Therefore for the TCNQ molecule B = 0, which can be incorrect for the perturbing potential related to a Peierls superstructure. Terms of type (15') are generated, for example, if the lattice deformation reduces to a displacement of the sublattice of the original atoms.

4)In case of commensurate bands the derivation of the equation for W(φ) can be greatly simplified. For this it is sufficient to recall that W(φ) is periodic with period π. Since, for example, for $k_F a = \pi/a$, Eq. (4') is written down for a doubled lattice, expanded in small Born quantities it reduces immediately to a second-order differential equation.

1A. F. Garito and A. Heeger, Low-dimensional cooperative phenomena (H. J. Keller, ed.), Plenum, New York (1975).

2V. Ya. Krivnov and A. A. Ovchinnikov, Zh. Eksp. Teor. Fiz. 67, 1568 (1974) [Sov. Phys. JETP 40, 781 (1974)].

3J. B. Torrance, I. Tomkiewicz, and B. D. Silverman, Phys. Rev. B (1977).

4Yu. A. Bychkov, L. P. Gor'kov, and I. E. Dzyaloshinskii, Zh. Eksp. Teor. Fiz. 50, 738 (1966) [Sov. Phys. JETP 23, 489 (1966)]; I. E. Dzyaloshinskii and A. I. Larkin, Zh. Eksp. Teor. Fiz. 61, 791 (1971) [Sov. Phys. JETP 34, 722 (1972)].

5M. J. Rice and N. O. Lipari, Phys. Rev. Lett. 38, 437 (1977).

6A. A. Gogolin, S. P. Zolotukhin, V. I. Mel'nikov, É. I. Rashba, and I. F. Shchegolev, Pis'ma Zh. Eksp. Teor. Fiz. 22, 564 (1975) [JETP Lett. 22, 278 (1975)].

7L. P. Gor'kov, and O. N. Dorokhov, Solid State Commun. 20, 789 (1976).

8L. N. Bulaevskii, A. V. Zvarykina, Yu. S. Karimov, R. B. Lyubovskii, and I. F. Shchegolev, Zh. Eksp. Teor. Fiz. 62, 725 (1972) [Sov. Phys. JETP 35, 384 (1972)].

9F. F. Igoshin, A. P. Kir'yanov, V. N. Topnikov, and I. F. Shchegolev, Zh. Eksp. Teor. Fiz. 68, 1203 (1975) [Sov. Phys. JETP 41, 597 (1975)]; V. N. Topnikov and I. F. Shchegolev, Pis'ma Zh. Eksp. Teor. Fiz. 20, 404 (1974) [JETP Lett. 20, 182 (1974)].

10A. N. Bloch, R. B. Weisman, and C. M. Varma, Phys. Rev. Lett. 28, 753 (1972).

11F. J. Dyson, Phys. Rev. 92, 1331 (1953).

12A. A. Ovchinnikov and N. S. Érikhman, Pis'ma Zh. Eksp. Teor. Fiz. 25, 197 (1977) [JETP Lett. 25, 180 (1977)].

13H. Schmidt, Phys. Rev. 105, 425 (1957); S. A. Gredeskul and L. A. Pastur, Fiz. Nizk. Temp. 1, 277 (1975) [Sov. J. Low Temp. Phys. 1, 135 (1975)].

14P. A. Lee, T. M. Rice, and P. W. Anderson, Solid State Commun. 14, 703 (1974).

15A. A. Gogolin and V. I. Mel'nikov, Zh. Eksp. Teor. Fiz. 73, 706 (1977).

16L. P. Gor'kov, Pis'ma Zh. Eksp. Teor. Fiz. 25, 384 (1977) [JETP Lett. 25, 358 (1977)].

Translated by Nathan Jacobi

On the theory of electrons localized in the field of defects

V. L. Berezinskiĭ and L. P. Gor'kov

L. D. Landau Institute of Theoretical Physics, Academy of Sciences of the USSR

(Submitted 2 July 1979)
Zh. Eksp. Teor. Fiz. 77, 2498–2517 (December 1979)

We formulate a mathematical criterion for the localization of single-particle wave functions in a random field. The localization is characterized by the appearance of singular terms in the expressions for the correlation functions of various physical quantities as functions as the energy (independent of the dimensionality of the problem). We propose for one-dimensional problems a method which, in principle, allows one to evaluate directly correlators which contain products of several exact wave functions of electrons in a random field, pertaining to different energies and taken at different points.

PACS numbers: 71.70.Ch

The present paper does not pretend to give new physical results, but has as its aim the solution of some methodological problems. We discuss in section 1 the problem of particle localization in the random field of randomly distributed defects and we propose a mathematical formulation of a criterion for the localization of the wave functions of eigenstates. The applicability of the results of this section is not limited to the case of one-dimensional problems. The criterion formulated there refers equally well to two- and three-dimensional cases.

The second part of the paper refers solely to a one-dimensional conductor. It contains a general method which allows us—at least in principle—to write down and evaluate an arbitrary correlator of the product of any number of exact wave functions with different energies of an electron in the random field of static defects, taken at different points in space.

The necessity to evaluate such correlators arises in a whole set of problems, such as, for instance, the determination of the temperature dependence of the conductivity. It is well known that the localization of electrons in the field of defects is complete in the sense that all states with arbitrary energies are localized.[1] The static conductivity therefore vanishes at zero temperature. Mott and Twose[2] were the first to note this fact and later on this result was obtained exactly by one of the present authors.[3] At non-zero temperatures the conductivity behaves like that of a semiconductor. The finite magnitude of the conductivity is caused by jumps of the electrons along the localized states due to inelastic interaction processes between them and the thermal phonons or between themselves. This effect was considered in Ref. 4 for the electron-phonon inter-

 0038-5646/79/121209-10$02.40

action mechanism under well-defined, rather strong assumptions about the temperature and the frequencies of the phonons participating in the inelastic process, and the diagram method of Ref. 3 was used. Afterwards the same result was obtained in Ref. 5 where, in turn, a somewhat different method (an S-matrix method) was proposed for the corresponding calculations.

So far it is unclear how the electron-electron and the other interactions affect the localization of the electrons by the static defects. The difficulty is here first and foremost a mathematical one. To study the role of the interactions it is necessary to be able to evaluate to any order of perturbation theory the matrix elements between the exact wave functions of the states of the non-interacting particles. The existing methods[3,5] are too complicated to be used as a basis to carry out this program. To a certain extent these difficulties reflect, of course, the complexity of the problem. From our point of view, however, their main defect is that the starting point of Refs. 3 and 5 is the approximation of free electrons in the absence of impurities. As we see it, the basic point which is characteristic only for localized electrons—the presence in the matrix elements of singular, δ-functionlike terms in the energy—is, as a result of this, not explicitly visible and can appear in the answer only after lengthy calculations. Therefore, a simplification of the mathematical apparatus remains as before very desirable.

In the present paper we develop a method which is, from our point of view, most suitable for the mathematical description of a one-dimensional problem. The new method stems from Schmidt's method[6] of calculating the number of zeroes of an electron wave function which, as has already been shown in a number of papers,[7] is able to simplify the calculations considerably.

1. LOCALIZATION CRITERION IN THE GENERAL CASE

In this section we make a few simple remarks about different possibilities of formulating mathematically the localization of wave functions by static defects. We remind ourselves that the electrons are assumed to be non-interacting. This practically reduces the problem to a single-particle one.

We introduce the simplest retarded density correlator:

$$K(x)=\langle\langle\hat\rho(\mathbf{r},t)-\bar\rho(\mathbf{r}),\hat\rho(0,0)-\bar\rho(0)\rangle\rangle,\quad t>0, \tag{1}$$

where the symbol $\langle\langle\ldots\rangle\rangle$ implies averaging both over a thermodynamic ensemble for a given realization of the random potential and subsequently over different configurations of that potential. Here

$$\hat\rho(\mathbf{r},t)=\sum_{\nu\nu',\sigma}\psi_\nu(x)\psi_{\nu'}(x)a^+_{\nu'\sigma}a_{\nu\sigma}$$

is the usual expression for the density operator in the second-quantization representation in terms of the wave functions $\psi_\nu(x)$ of the electron eigenstates with energies ε_ν. (We take the functions in a finite volume to be real.) An elementary evaluation of Eq. (1) gives

$$K(x)=2\left\langle\sum_{\nu\nu'}\exp\{-i(\varepsilon_{\nu'}-\varepsilon_\nu)t\}n_\nu(1-n_{\nu'})\psi_\nu(\mathbf{r})\psi_\nu(0)\psi_{\nu'}(\mathbf{r})\psi_{\nu'}(0)\right\rangle_U, \tag{1'}$$

where $\langle\ldots\rangle_U$ corresponds to the averaging over the realizations of the random potential. We assume in what follows that the latter is characterized by uniform correlation functions which are invariant under the rotation and reflection group in the space of the appropriate dimensionality d. This expression can be written in the form

$$K(x)=2\iint dE\,d\omega e^{-i\omega t}n_E(1-n_{E+\omega})\times\left\langle\sum_{\nu\nu'}\delta(E-\varepsilon_\nu)\delta(E+\omega-\varepsilon_{\nu'})\psi_\nu(\mathbf{r})\psi_{\nu'}(\mathbf{r})\psi_\nu(0)\psi_{\nu'}(0)\right\rangle_U. \tag{2}$$

Introducing the function

$$\langle\langle\rho_E(\mathbf{r})\rho_{E+\omega}(0)\rangle\rangle=\frac{1}{\nu(E)}\left\langle\sum_{\nu\nu'}\delta(E-\varepsilon_\nu)\delta(E+\omega-\varepsilon_{\nu'})\times\psi_\nu(\mathbf{r})\psi_{\nu'}(\mathbf{r})\psi_\nu(0)\psi_{\nu'}(0)\right\rangle_U \tag{3}$$

[$\nu(E)$ is the density of states for a single spin at an energy E] we write the Fourier component of (2) with respect to time in the form

$$K(\mathbf{r},\varepsilon)=2i\iint\frac{\nu(E)n_E(1-n_{E+\omega})\langle\langle\rho_E(\mathbf{r})\rho_{E+\omega}(0)\rangle\rangle dE\,d\omega}{\varepsilon-\omega+i\delta}. \tag{4}$$

The spectral density (3) has the following obvious properties:

$$\int_V d\mathbf{r}\langle\langle\rho_E(\mathbf{r})\rho_{E+\omega}(0)\rangle\rangle=\delta(\omega),\qquad \int d\omega\langle\langle\rho_E(\mathbf{r})\rho_{E+\omega}(0)\rangle\rangle=\delta(\mathbf{r}), \tag{5}$$

which for the spatial Fourier components $\langle\langle\rho_E\rho_{E+\omega}(\mathbf{q})\rangle\rangle$ means:

$$\langle\langle\rho_E\rho_{E+\omega}(\mathbf{q}=0)\rangle\rangle=\delta(\omega),\qquad \int d\omega\langle\langle\rho_E\rho_{E+\omega}(\mathbf{q})\rangle\rangle=1. \tag{5'}$$

Moreover, one shows easily, by using the assumptions made about the correlation functions of the random potential that $\langle\langle\rho_E\rho_{E+\omega}(\mathbf{q})\rangle\rangle\geq 0$.

It is now convenient to compare Eq. (1') for $K(x)$ and the properties expressed by Eq. (5'). Starting from (1') we see that there are present in it, in general, constant terms with $\varepsilon_\nu=\varepsilon_{\nu'}$. However, if the states are delocalized, their wave functions are normalized to the volume V and the corresponding contribution to (1') is thus of the order $\sim 1/V$. The situation is different if the states are localized. In that case the correlator $K(x)$ acquires a constant (i.e., independent of the time difference) term:

$$2\left\langle\sum_\nu n_\nu(1-n_\nu)\psi_\nu^2(\mathbf{r})\psi_\nu^2(0)\right\rangle_U, \tag{6}$$

which in terms of the spectral density is equivalent to a term of the form $A(\mathbf{q})\delta(\omega)$ in $\langle\langle\rho_E\rho_{E+\omega}(\mathbf{q})\rangle\rangle$. The wave functions are normalized to the localization length (mean free path) and an estimate of the term (6) gives

$$\sim(1/l)^d\int dE\nu(E)n_E(1-n_E).$$

We put it that one of the possible criteria for localization may be formulated as follows. We introduce the single-particle spectral densities:

$$\rho_E(\mathbf{r},\mathbf{r}')=\sum_\nu \delta(E-\varepsilon_\nu)\psi_\nu(\mathbf{r})\psi_\nu(\mathbf{r}').$$

In the average of the product of n quantities $\rho_{E_i}(\mathbf{r},\mathbf{r}')$ pertaining to energies $E_1,\ldots,E_n$ and arbitrary points $(\mathbf{r}_i,\mathbf{r}_i')$ there must be present δ-functionlike singularities corresponding to some of the energies E_i being equal. These energies correspond to localized states.[1)]

Turning to Eqs. (5′) we see that they, in turn, also express two possibilities. For the first of them, corresponding to the presence of localized states at an energy E, the δ-functionlike term is retained for finite $\mathbf{q}$ and for $q\neq 0$ we can write

$$\langle\langle\rho_E\rho_{E+\omega}(\mathbf{q})\rangle\rangle=A(\mathbf{q})\delta(\omega)+\rho_1(\mathbf{q},\omega), \tag{7}$$

and for small q

$$\langle\langle\rho_E\rho_{E+\omega}(\mathbf{q})\rangle\rangle=(1-\alpha q^2)\delta(\omega)+q^2\rho_1(\omega), \tag{7′}$$

where according to (5′)

$$\alpha=\int\rho_1(\omega)d\omega. \tag{7″}$$

Delocalization would correspond to a spreading out of the δ-functionlike singularities in (5′) at finite $\mathbf{q}$. The simplest variant corresponding to diffusive behavior means that the width $\Delta\omega$ of the singularity is quadratic in $\mathbf{q}$: $\Delta\omega\propto q^2$. We have

$$\langle\langle\rho_E\rho_{E+\omega}(\mathbf{q})\rangle\rangle=\frac{1}{\pi}\frac{D_E q^2}{\omega^2+(D_E q^2)^2}. \tag{8}$$

It follows already from the possibility of such a statement of the problem that the diffusive spreading out of the singularity is just one particular case of the behavior of $\langle\langle\rho_E\rho_{E+\omega}(\mathbf{q})\rangle\rangle$ for small $\mathbf{q}$ and there can be many other possibilities for its dependence on ω and $\mathbf{q}$ which satisfy the normalization condition (5′), among which there is also a singular behavior for small ω and $\mathbf{q}$. We deduce hence that delocalization does not necessarily imply a diffusive spreading out of the packet.

Comparing (7′) and (6) [i.e., evaluating the Fourier component of (6), the term q^2] we get

$$\alpha=\frac{1}{2d\nu(E)}\int r^2 dr\left\langle\sum_\nu\delta(E-\varepsilon_\nu)\psi_\nu^2(\mathbf{r})\psi_\nu^2(0)\right\rangle_U \tag{9}$$

(d is the dimensionality of the space). In other words, α can be expressed in terms of the mean square variance of the dipole moment.

Similar to the single-particle spectral density of the density-density correlator, we can introduce spectral densities for the current-current and current-density correlators. We can then use the equation of continuity to establish the following relations between these quantities:

$$\langle\langle\mathbf{j}_E\rho_{E+\omega}(\mathbf{q})\rangle\rangle=\langle\langle\rho_E\mathbf{j}_{E+\omega}(\mathbf{q})\rangle\rangle=\frac{\omega\mathbf{q}}{q^2}\langle\langle\rho_E\rho_{E+\omega}(\mathbf{q})\rangle\rangle,$$

$$\langle\langle j_{iE}j_{kE+\omega}(\mathbf{q})\rangle\rangle=\langle\langle\mathbf{j}_E\mathbf{j}^{\perp}_{E+\omega}(\mathbf{q})\rangle\rangle\left(\delta_{ik}-\frac{q_iq_k}{q^2}\right)+\langle\langle\mathbf{j}_E\mathbf{j}^{\parallel}_{E+\omega}(\mathbf{q})\rangle\rangle\frac{q_iq_k}{q^2}, \tag{10}$$

$$\langle\langle\mathbf{j}_E\mathbf{j}^{\parallel}_{E+\omega}(\mathbf{q})\rangle\rangle=\frac{\omega^2}{q^2}\langle\langle\rho_E\rho_{E+\omega}(\mathbf{q})\rangle\rangle.$$

The response in the current of a many-particle system to an external vector potential $\mathbf{A}_\varepsilon(\mathbf{r})$ is given by the formula[8]

$$j_i(\mathbf{r},\varepsilon)=-\frac{e^2\bar{\rho}}{m}A_{i\varepsilon}(\mathbf{r})+i\int d\mathbf{r}'\langle\langle[\hat{j}_i(x),\hat{j}_k(x')]\rangle\rangle_\varepsilon^{(R)}A_{k\varepsilon}(\mathbf{r}'),$$

where the integrand contains the retarded commutator of the current density operators. For the Fourier components with repsect to momenta we get[2)]

$$\mathbf{j}(\mathbf{q},\varepsilon)=-2e^2\iint\frac{\nu(E)dE\,d\omega[n_E-n_{E+\omega}]}{\varepsilon-\omega+i\delta}\frac{\omega^2}{q^2}\langle\langle\rho_E\rho_{E+\omega}(\mathbf{q})\rangle\rangle\mathbf{A}_\varepsilon(\mathbf{q}).$$

Using (8), letting $\mathbf{q}\to 0$ and substituting $\mathbf{A}_\varepsilon=\mathbf{E}/i\varepsilon$ we get in the diffusive case for the conductivity the well known expression:

$$\sigma=2e^2\nu(E_F)D.$$

In the localized case we have according to (7′)

$$\sigma(\varepsilon)=2e^2\nu(E_F)\varepsilon^2\rho_1(\varepsilon),$$

where[2,3] in the one-dimensional case

$$\rho_1(\varepsilon)\infty\ln^3(\varepsilon\tau).$$

Without using Eqs. (10) we can directly write down the following expression for the conductivity in terms of the wave functions of the system:

$$\sigma_{ik}(\varepsilon)=\delta_{ik}\sigma(\varepsilon)=\frac{2\pi e^2}{m^2}\int\left\langle\sum_{\nu\nu'}\delta(E-\varepsilon_\nu)\delta(\varepsilon_\nu-\varepsilon_{\nu'}+\varepsilon)\right.$$
$$\left.\times(\psi_{\varepsilon_\nu}(\mathbf{r})\hat{p}_{i,\mathbf{r}}\psi_{\varepsilon_{\nu'}}(\mathbf{r}))(\psi_{\varepsilon_\nu}(\mathbf{r}')\hat{p}_{k,\mathbf{r}'}\psi_{\varepsilon_{\nu'}}(\mathbf{r}'))\right\rangle_U d\mathbf{r}' \tag{11}$$

(here $\hat{p}_{i,\mathbf{r}}=-i\partial/\partial r_i$).

2. METHOD FOR EVALUATING AVERAGES OF WAVEFUNCTIONS IN THE ONE-DIMENSIONAL CASE

Equations (3), (6), (9), and (11) contain averages of the product of several electron wave functions with different or the same energies and at different points in space. It is possible to develop for the one-dimensional case a method for evaluating such averages which looks to us simpler and more general than the diagram method of one of the present authors[3] or the S-matrix method.[5] The present section is devoted to an exposition and discussion of that method.

One can in the one-dimensional case juxtapose the eigenvalue of the energy E_ν to the number of zeroes of its wave function $\psi_\nu(x)$ that satisfies the Schrödinger equation:

$$\{-(d^2/2m\,dx^2)+U(x)\}\psi_\nu(x)=E_\nu\psi_\nu(x).$$

This fact was used by Schmidt[6] as the basis for the evaluation of the single-particle density of states of a linear chain with a random distribution. The considerations given below are a further development of this approach. Schmidt's method[6] leading to the equations for a Markov process is valid for a potential $U(x)$ corresponding to a random set of short-range potentials and can be formulated both for a random Gaussian potential $U(x)$ (so-called white noise) and also for a lattice in whose points defect atoms are randomly dispersed and have a potential range less than the size of the selected cell. In the Born approximation both procedures are absolutely equivalent, but below we prefer to emphasize the lattice formulation, bearing in mind possible generalizations. In conclusion we give a derivation also for a white-noise kind of potential.

We thus consider a lattice of points with cell size a, while in each site there is a probability p that there is a defect situated with a potential with a range which for the sake of simplicity we assume to be negligibly small compared to a. When $p \ll 1$ this arrangement corresponds, in particular, to a low impurity concentration $c = p/a$. The phase φ_E of the wave function is given by the relation

$$\operatorname{ctg} \varphi_E = \psi'/k\psi, \quad k = (2mE)^{1/2} \tag{12}$$

and satisfies the equation

$$\varphi_E' = k - (U(x)/k) \sin^2 \varphi_E. \tag{13}$$

In the figure we have depicted the lattice potential where in agreement with the assumptions about the character of the potential in the whole cell (j), except at the lattice sites, the motion is free:

$$\psi_i(x) = A_j(x) + B_j(x) = R_j \sin \varphi_j(x); \quad \frac{d}{dx}\varphi_j(x) = k. \tag{14}$$

The connection between the amplitudes of the traveling waves $\{A_j B_j\}$ and $\{A_{j+1} B_{j+1}\}$ is determined by the scattering by the potential located at the site $j+1$. For the sake of argument, we shall match together all quantities A_j, B_j, φ_j taken immediately to the right of the site j, as shown by the arrow in the figure.[7]

One checks easily that for such a choice of the quantities in the points j and $j+1$ the connection between the coefficients A_j and B_j is given by the matrix

$$\begin{Bmatrix} A_{j+1} \\ B_{j+1} \end{Bmatrix} = \frac{1}{\cos\gamma} \begin{Bmatrix} e^{-i(ka-\delta)}, & i \sin\gamma e^{-ika} \\ -i \sin\gamma e^{ika}, & e^{-i(ka-\delta)} \end{Bmatrix} \begin{Bmatrix} A_j \\ B_j \end{Bmatrix}, \tag{15}$$

in which γ and δ are characterized by the problem of the scattering by the defect atom (for the sake of simplicity the potential is taken to be symmetrical under $x \to -x$) where δ is the advance in phase of the previous wave and $\sin^2\gamma$ the reflection coefficient.

We have

$$\begin{aligned} R_{j+1}^2 &= (R_j^2/\cos^2\gamma)[1 + \sin^2\gamma - 2\sin\gamma \sin(2\varphi_j + 2ak - \delta)], \\ R_j^2 &= (R_{j+1}^2/\cos^2\gamma)[1 + \sin^2\gamma + 2\sin\gamma \sin(2\varphi_{j+1} + \delta)] \end{aligned} \tag{16}$$

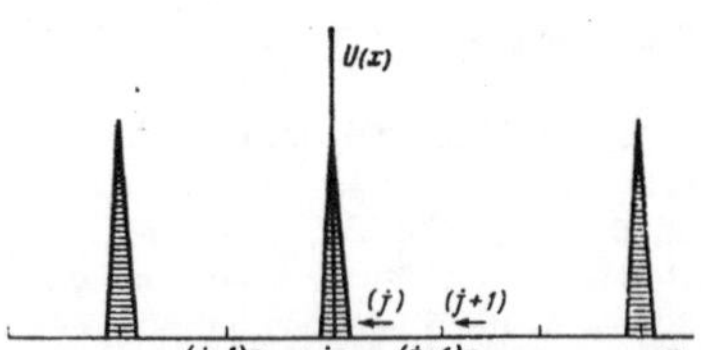

FIG. 1. Random potential $U(x)$ as potential of defects which are randomly positioned in lattice sites (lattice period a). The arrows indicate the positions where all quantities in the cell j and $j+1$ are matched.

and

$$\varphi_{j+1} = \varphi_j + ka - \delta + \frac{1}{2i} \ln\left\{\frac{1 - i\sin\gamma \exp[-i(2\varphi_i + 2ka - \delta)]}{1 + i\sin\gamma \exp[i(2\varphi_i + 2ka - \delta)]}\right\}, \tag{17}$$

$$\varphi_j = \varphi_{j+1} - ka + \delta + \frac{1}{2i} \ln\left\{\frac{1 + i\sin\gamma \exp[-i(2\varphi_{j+1} + \delta)]}{1 - i\sin\gamma \exp[i(2\varphi_{j+1} + \delta)]}\right\}. \tag{17'}$$

Equations (17) enable us to construct the phase φ_j as a monotonically increasing or decreasing function,[6] depending on whether we start the process from the left-hand or the right-hand end point. We shall mark the corresponding quantities with an index > or <.

We still need equations for the derivatives $\zeta^> = \partial\varphi^>/\partial k$ and $\zeta^< = \partial\varphi^</\partial k$. From (17) we get, for example,

$$\frac{\partial\varphi_{j+1}}{\partial k} = \left(\frac{\partial\varphi_j}{\partial k} + a\right) \frac{\cos^2\gamma}{1 + \sin^2\gamma - 2\sin\gamma \sin(2\varphi_j + 2ka - \delta)}. \tag{18}$$

In correspondence with the definition of the phase[6] the quantity $\zeta^>$ is always positive.

We write

$$\alpha(j|n) = R_j/R_n \tag{19}$$

and introduce as a definition

$$\int_{-L}^{x_j} \alpha^2(x|x_j)\,dx = \sum_{0<i<j} \alpha^2(i|j)a. \tag{20}$$

(The integral has here a symbolic meaning, but in what follows we shall often change to a continual description.) Using (16) and (18) one checks easily that

$$\zeta_j^> = \int_{-L}^{x_j} \alpha^2(x|x_j)\,dx, \tag{21}$$

as (20) satisfies Eq. (18) and the boundary condition $(\partial\varphi^>/\partial k)_{x=-L} = 0$. In turn, we have similarly

$$\zeta_j^< = -\int_{x_j}^{L} \alpha^2(x|x_j)\,dx = -\sum_{j<i<N} \alpha^2(i|j)a. \tag{21'}$$

In order not to encumber the paper with general formulae we choose first, to begin to demonsrate the idea of the method, the average of the electron Green function:

$$G_{i\omega_n}(x, x') = \sum_\nu \frac{\psi_\nu(x)\psi_\nu(x')}{i\omega_n - \varepsilon_\nu} \tag{22}$$

(here $\omega_n = (2n+1)\pi T$; we denote the energy reckoned from the level of the chemical potential by ε_ν). We write (22) in the form

$$G_{i\omega_n}(x, x') = \int d\xi \sum_\nu \delta(\xi - \varepsilon_\nu) \frac{\psi_\nu(x)\psi_\nu(x')}{i\omega_n - \xi} \tag{22'}$$

and, we choose, for instance, $x < x'$ in the mean value of the spectral density.

$$\langle\rho_\xi(x, x')\rangle = \left\langle \sum_\nu \delta(\xi - \varepsilon_\nu)\psi_\nu(x)\psi_\nu(x') \right\rangle_U. \tag{23}$$

The eigenwave functions $\psi_\nu(x)$ in (22) are normalized to unity:

$$\int \psi_\nu^2(x)\,dx = 1. \tag{24}$$

If the case is quasi-classical, $kl \gg 1$, one can write that condition as

$$\frac{1}{2}\sum_{0<i<N} R_i^2 a=1. \tag{24'}$$

Taking an arbitrary point j_0 we can rewrite (24') in terms of the quantity α from (19):

$$\frac{1}{2}R_{j_0}^2\int_{-L}^{L}\alpha^2(x|j_0)\,dx=1, \tag{25}$$

and the normalization constant $R_{j_0}^2$ is then according to (21) and (21') equal to

$$R_{j_0}^2=2\Big/\int\alpha^2(x|j_0)\,dx=\frac{2}{\zeta^>(j_0)-\zeta^<(j_0)}. \tag{26}$$

The presence of a δ-function in (23) expresses the fact that the energy ξ is the eigenvalue for a given configuration of defects. The same fact can be represented by the condition that the solutions for the phase, satisfying the boundary condition $\varphi|_{\pm L}=\varphi_0$, from the left $\varphi_j^>$ and from the right $\varphi_j^<$, differ by an integer times π in any arbitrarily chosen point j_0 where we match the solution. Taking therefore the phase in the range[3)] $(-\pi/2, \pi/2)$ we can write:

$$\delta(\xi-\varepsilon_\nu)=\delta(\varphi_{j_0}^>-\varphi_{j_0}^<)[(\partial\varphi^>/\partial\xi)_{j_0}-(\partial\varphi^</\partial\xi)_{j_0}]=(1/v)\delta(\varphi_{j_0}^>-\varphi_{j_0}^<)(\zeta_{j_0}^>-\zeta_{j_0}^<). \tag{27}$$

As a result Eq. (23), takes at the points $x=j$, $x'=j'$ [inside the cell, (23) is determined by the free-motion Eq. (14)], when we use (26), the form

$$(2/v)\langle\delta(\varphi_{j_0}^>-\varphi_{j_0}^<)\sin\varphi_j\alpha_\xi(j|j_0)\sin\varphi_{j'}\alpha_\xi(j'|j_0)\rangle_v.$$

We write it in the form of an integral:

$$2\int_{-\pi/2}^{+\pi/2}\frac{d\varphi}{v}\langle\delta(\varphi_{j_0}^>-\varphi)\delta(\varphi-\varphi_{j_0}^<)\sin\varphi_j\alpha_\xi(j|j_0)\sin\varphi_{j'}\alpha_\xi(j'|j_0)\rangle_v.$$

Let, for instance, $j_0=j$. We can then rewrite this average also differently:

$$\langle\rho_\xi(x,x')\rangle=\frac{2}{v}\iint_{-\pi/2}^{+\pi/2}d\varphi d\varphi'\langle\delta(\varphi_j^>-\varphi)\sin\varphi\times\delta(\varphi-\varphi^<(j|j',\varphi'))\alpha_\xi(j'|j)\sin\varphi'\delta(\varphi'-\varphi_{j'}^<)\rangle_v. \tag{28}$$

In other words, since this is a Markov process, we can perform the averaging in (28) over all defect configurations separately over the sections $-\infty=-L<j<j'<L=+\infty$. Accordingly we introduce the following quantities:

$$w_\xi(\varphi)\equiv w_\xi^>(\varphi)=\langle\delta(\varphi_{\xi,j}^>-\varphi)\rangle \tag{29}$$

the stationary phase distribution function, found going from the left. The analogous quantity obtained by going from the right is

$$w_\xi^<(\varphi)=\langle\delta(\varphi-\varphi_{\xi,j}^<)\rangle. \tag{29'}$$

The function for the transition from the point j' to the point j (from right to left) has the form

$$w_{jj',\xi}^{<(1)}(\varphi,\varphi')=\langle\alpha_\xi(j'|j)\delta(\varphi-\varphi_\xi^<(j|j',\varphi'))\rangle, \tag{30}$$

where the notation $\varphi^<(j|j',\varphi')$ indicates the result of solving the equation for the phase (from the right to the left) at the point j under the condition that the phase at the point j' equalled φ'. Thus

$$\langle\rho_\xi(x,x')\rangle=\frac{2}{v}\iint_{-\pi/2}^{+\pi/2}d\varphi d\varphi' w_\xi^>(\varphi)\sin\varphi\, w_{jj',\xi}^{<(1)}(\varphi,\varphi')\sin\varphi' w_\xi^<(\varphi'), \tag{31}$$

where one can use Schmidt's method[6] to obtain easily the equations for any of the probability functions which occur here, connecting through the transformations (16), (17) the neighboring points j and $j+1$. We thus have, according to Schmidt,[6] for the stationary probability distribution function $w(\varphi)$ the equation

$$w_\xi^>(\varphi_{j+1})=(1-p)w_\xi^>(\varphi_{j+1}-ka)+pw_\xi^>(\varphi_j(\varphi_{j+1}))(d\varphi_j(\varphi_{j+1})/d\varphi_{j+1}), \tag{32}$$

which expresses the fact that the site $j+1$ is unoccupied with a probability $1-p$ and occupied by a defect with a probability p. Schmidt's Eq. (32) is the basic one for deriving all further formulae.

We also note that for the symmetrical defect chosen by us Eq. (17) admits the substitution $\varphi^>\to-\varphi^<$. Since (17) defines a function $\varphi^>$ which increases monotonically from left to right, and (17') a function $\varphi^<$ which decreases monotonically from right to left, the stationary phase probability distribution functions satisfy the condition

$$w_\xi^<(\varphi)=w_\xi^>(-\varphi)\equiv w_\xi(-\varphi). \tag{33}$$

The generalization of the representation (31) to the case of correlators of products of a large number of $\rho_\xi(x,x')$ and the proof that the result of the averaging is independent of the choice of the point j_0 where we match the solutions are given by us in the Appendix.

So far in our considerations we had no need to resort to any simplifying assumptions regarding the potential, bar the assumption that it has a short range. However, we make here the following remark. Backward scattering of the electrons is important for the localization. At the same time the molecules in the experimentally studied organic conductors are rather large. The random potential then turns out to be relatively smooth which diminishes the amplitude for the backward scattering of an electron. (It is, of course, possible that for a number of substances a model in which the impurities split the chain into almost isolated sections is more suitable, but this case requires special considerations.) Starting here, therefore, and everywhere in what follows we use the Born approximation $\gamma\ll 1$.

The localization length

$$l=2a/p\gamma^2=2/c\gamma^2 \tag{34}$$

will be assumed to be large compared to "atomic" distances (while $ak\sim 1$) and also large compared to the average distance between defects

$$l\gg\bar{x}=a/p. \tag{35}$$

Using the fact that γ is small we simplify Eqs. (16), (17):

$$R_{j+1}^2=R_j^2[1+2\gamma^2-2\gamma\sin(2\varphi_j+2ak-\delta)], \tag{36}$$

$$R_j^2=R_{j+1}^2[1+2\gamma^2+2\gamma\sin(2\varphi_{j+1}+\delta)], \tag{36'}$$

$$\varphi_{j+1}=\varphi_j+ka-\delta-\gamma\cos(2\varphi_j+2ka-\delta), \quad (37)$$
$$\varphi_j=\varphi_{j+1}-ka+\delta+\gamma\cos(2\varphi_{j+1}+\delta). \quad (37')$$

We have dropped in (36), (37) terms of the form $\gamma^2\sin m\varphi$. From a formal point of view they are according to (34) of the necessary (second) order of magnitude in γ but in all equations below they would occur with a fast-oscillating factor $\sin m\varphi$ which drops out unless we make special assumptions about the magnitude of ka (to be more precise, require that k and $2\pi/a$ are commensurable). Indeed, in the vicinity of an arbitrarily positioned Fermi level all probabilities depend solely on differences in the energies and in the phases corresponding to them and do not contain in the main approximation a dependence on the phases φ_ξ themselves.

As an example we write down the equation which is satisfied by the stationary distribution function of several phases, $w(\varphi_{\xi_0}, \varphi_{\xi_1}, \ldots, \varphi_{\xi_i})$. We derive this equation by analogy with (32) and using Eq. (37). As in (32) the Jacobian—a product of factors

$$d\varphi_j/d\varphi_{j+1}=1-2\gamma\sin(2\varphi_{j+1}+\delta)$$

occurs on the right-hand side. Retaining everywhere only terms of order γ^2 (the terms which are linear in γ contain fast-oscillating factors depending on one of the phases) and rewriting $w(\varphi_{\xi_0}, \ldots, \varphi_{\xi_i}) \equiv w(\theta_1, \ldots, \theta_i)$, where $\theta_i = \varphi_{\xi_i} - \varphi_{\xi_0}$ we get as a result of elementary calculations

$$-\sum_i \nu_i \frac{\partial w}{\partial\theta_i}+\sum_i \frac{\partial^2}{\partial\theta_i^2}[(1-\cos 2\theta_i)w]$$
$$+\sum_{i>k}\frac{\partial^2}{\partial\theta_i\partial\theta_k}[(1+\cos 2(\theta_i-\theta_k)-\cos 2\theta_i-\cos 2\theta_k)w]=0, \quad (38)$$

where, in turn, $\nu_i = (k_i - k_0)l$ and l is defined through (34).

We did not manage to find the general solution of (38). The limit $\nu_i \gg 1$ denotes the statistical independence and corresponds to a vanishing of the dependence on the angle θ_i of the function $w(\theta_1, \ldots, \theta_i)$. Equation (38) has asymptotic solutions for all $\nu_i \gg 1$.

Let us discuss in detail the case of two phases: $w(\varphi_1, \varphi_2) \equiv w(\theta)$. For the sake of argument let $\nu > 0$. In that case

$$-\nu\frac{\partial}{\partial\theta}w(\theta)+2\frac{\partial^2}{\partial\theta^2}(\sin^2\theta\, w(\theta))=0. \quad (39)$$

The first integral of Eq. (39) is

$$-\nu w(\theta)+2\frac{\partial}{\partial\theta}(\sin^2\theta\, w(\theta))=-\frac{\nu}{\pi^2} \quad (39')$$

(the probability distribution function of the two phases is normalized to unity). Equation (39) has a singularity at $\theta=0$.

The general form of the solution is:

$$w(\theta)=-\frac{\nu}{2\pi^2\sin^2\theta}\exp\left(-\frac{\nu}{2}\operatorname{ctg}\theta\right)\int^{\theta}\exp\left(\frac{\nu}{2}\operatorname{ctg}\theta'\right)d\theta'.$$

For the final definition of it we must require the finiteness (integrability) of $w(\theta)$ for $\theta=0$ and impose the periodicity condition $w(\theta+\pi)=w(\theta)$. [In other words, when $\theta<0\, w(\theta)\equiv w(\theta+\pi)$.] One checks easily that such a solution is

$$w(\theta)=\frac{\nu}{2\pi^2\sin^2\theta}\exp\left(-\frac{\nu}{2}\operatorname{ctg}\theta\right)\int_0^{\pi-0}\exp\left(\frac{\nu}{2}\operatorname{ctg}\theta'\right)d\theta'. \quad (40)$$

The asymptotic properties of the solution for $\nu \ll 1$ follow at once from (40). In particular, one checks easily that for small $\theta \sim \nu$ we can expand (40) as follows

$$w(\theta)=\frac{1}{\nu}w_0\left(\frac{\theta}{\nu}\right)+w_1\left(\frac{\theta}{\nu}\right)+\ldots \quad (41)$$

The main term $(1/\nu)w(\theta/\nu)$ is then non-vanishing only when $\theta>0$:

$$\frac{1}{\nu}w_0\left(\frac{\theta}{\nu}\right)=\frac{\nu}{2\pi\theta^2}\exp\left(-\frac{\nu}{2\theta}\right), \quad (42)$$

and thus when $\theta<0$ the first-order term is important:

$$w_1\left(\frac{\theta}{\nu}\right)=\frac{\nu^2}{2\pi^2\theta^2}\exp\left(\frac{\nu}{2|\theta|}\right)\int_{-\infty}^{-\nu/|\theta|}\frac{dx}{x^2}e^{x/2}, \quad (43)$$

with an asymptotic behavior when $|\theta|/\nu \gg 1$

$$w_1(\theta/\nu)\approx-\nu/2\pi^2\theta \quad (43')$$

In applications one must often evaluate rather complicated correlators of wave functions which occur in different matrix elements. When writing down and solving the appropriate equations it turns out to be more convenient to look not for transition probabilities of the kind $w_{jj'}(\varphi, \varphi')$ in Eq. (31) (see A.6) themselves, but to use a somewhat different method which we now explain, evaluating again by the proposed method the frequency dependence of the conductivity due to static defects.[3]

Substituting the expressions for the wave functions (14) into (11) and retaining under the integral sign of the correlator only slowly varying factors at the points x and x' we get

$$(\pi e^2v^2/2)\int dx'\left\langle\sum_{\nu\nu'}\delta(\xi-\varepsilon_\nu)\delta(\varepsilon_{\nu'}-\varepsilon_\nu-\omega)\right.$$
$$\left.\times\sin\theta(x)R_\xi(x)R_{\xi'}(x)\sin\theta(x')R_\xi(x')R_{\xi'}(x')\right\rangle,$$

where $\xi' = \xi+\omega$, $\theta(x)=\varphi_{\xi'}(x)-\varphi_\xi(x)$. As the average depends only on $|x-x'|$ let $x<x'$ and take the matching point in x'. We rewrite the conductivity in the form

$$\sigma(\omega)=4\pi^2e^2\int_0^\infty dR\int_{-\pi/2}^{\pi/2}d\theta g(R,\theta)w(-\theta)\sin\theta \quad (44)$$

(the extra π is due to the integral over φ). The function $g(R, \theta)$ is thus

$$g(R,\theta)=\langle\sin\theta(x)\alpha_\xi(x|x')\alpha_{\xi+\omega}(x|x')\delta(\theta-\theta^>(x'))$$
$$\times\delta(\varphi-\varphi_\xi^>(x'))\rangle_\sigma, \quad (45)$$

and we can write down an equation for it without introducing a transition probability, but connecting instead of that directly its values in the points x_j and x_{j+1} similarly to (32):

$$g(R+a,\theta_{j+1})=(1-p)g\left(R,\theta_{j+1}-\frac{\omega a}{v}\right)+pg(R,\theta_j(\theta_{j+1}))$$
$$\times\left(\frac{R_{\xi j}}{R_{\xi j+1}}\right)\left(\frac{R_{\xi' j}}{R_{\xi' j+1}}\right)\frac{d\theta_j}{d\theta_{j+1}}\frac{d\varphi_{\xi j}}{d\varphi_{\xi j+1}}.$$

Expanding this relation we find the equation ($\nu = l\omega/v = \omega\tau$)

$$l\frac{\partial g}{\partial R} = -\nu\frac{\partial g}{\partial \theta} + (1-\cos 2\theta)\frac{\partial^2 g}{\partial \theta^2} + 2\sin 2\theta\frac{\partial g}{\partial \theta} - (1-\cos 2\theta)g. \quad (46)$$

We take its Laplace transform

$$g_\lambda(\theta) = \int_0^\infty dR\, g(R,\theta)e^{-\lambda R}.$$

We have

$$-l\sin\theta w(\theta) + \lambda\, l g_\lambda(\theta) = -\nu g_\lambda'(\theta) + (1-\cos 2\theta)g_\lambda''(\theta) + 2\sin 2\theta g_\lambda'(\theta) - (1-\cos 2\theta)g_\lambda(\theta). \quad (46')$$

We have used the boundary condition $g(R=0,\theta) \equiv w(\theta)\sin\theta$ which is obvious from (45). We have not solved Eq. (46′) in its general form. However, for small ν small angles are important:

$$-\theta l w(\theta) = -\nu g_\lambda' + 2\theta^2 g_\lambda'' + 4\theta g_\lambda' - \lambda l g_\lambda. \quad (46'')$$

The solution of (46″) can be expressed in terms of Bessel functions.

To evaluate the conductivity (44) one requires only $g_{\lambda=0}(\theta)$. We see that the main contribution to (44) comes from the region $\nu \ll |\theta| \ll 1$. To find the solution we proceed as follows. On the one hand, in the small angle region we easily solve Eq. (46″) ($\lambda = 0$):

$$g(\theta) = -\frac{l}{\nu}\left\{\int_0^\theta \theta' w(\theta')\left[1-\exp\frac{\nu}{2}\left(\frac{1}{\theta'}-\frac{1}{\theta}\right)\right]d\theta' + \alpha\exp\left(-\frac{\nu}{2\theta}\right) + \varkappa\right\}, \quad \theta > 0, \quad (47)$$

$$g(\theta) = -\frac{l}{\nu}\left\{\int_0^{|\theta|} x w(-x)\left[1-\exp\frac{\nu}{2}\left(\frac{1}{|\theta|}-\frac{1}{x}\right)\right]dx + \varkappa\right\}, \quad \theta < 0.$$

(We have taken solutions finite at $\theta = 0$; α and $\varkappa$ are integration constants.)

For large angles we can neglect the term with ν. As a result (46′) becomes

$$-\frac{l\nu}{4\pi^2}\frac{\pi-\theta}{\sin^2\theta} = \frac{\partial^2}{\partial\theta^2}(g\sin\theta)$$

and can be integrated:

$$g\sin\theta = \frac{l\nu}{4\pi^2}\left\{(\pi-\theta)\ln\sin\theta + 2\int_0^\theta \ln\sin\theta' d\theta' + c_1\theta + c_2\right\}.$$

One must match the expansion of this expression in the region $\theta \ll 1$ and $\pi - \theta \ll \pi$ with the asymptotic solution (47) in the region $1 \gg |\theta| \gg \nu$ using the fact that $g(\theta) = -g(\pi+\theta)$ when $\theta < 0$. This procedure enables us to determine the constants c_1, c_2, α, and $\varkappa$. We write down the main terms of the function $g(\theta)$ in the region $\nu \ll |\theta| \ll 1$:

$$g(\theta) = \frac{l\nu}{4\pi^2}\begin{cases}(\pi/\theta)\ln(\theta/\nu); & \theta > 0\\ \ln|\theta| + \ln\nu; & \theta < 0\end{cases}.$$

Using them in (44) we find[3]

$$\sigma(\omega) = (l\nu^2/\pi)\ln^2\nu.$$

The answer is the same as the equations in Refs. 3, 5 if we substitute for the localization length (34) $l = 2l_-$ (see Ref. 7) where l_- is the kinetic mean free path for backward scattering.

Finally, Eq. (6) allows us to point out those changes caused by a case important for localized states, when there are more than two wave functions with the same energy in the averaged quantity:

$$F(x-x') = \langle\sum_\nu \delta(\xi-\varepsilon_\nu)\psi_\xi^2(x)\psi_\xi^2(x')\rangle_v. \quad (48)$$

Let the point of joining be x' and let $x' > x$. In agreement with (26) and (27) we have for (48)

$$F(x-x') = (1/v)\langle\delta(\varphi^>(x') - \varphi^<(x'))\alpha_\xi^2(x|x')[\zeta^>(x')-\zeta^<(x')]^{-1}\rangle_v. \quad (48')$$

We can proceed here in two ways. Firstly, we can introduce a probability distribution function for the phases and their derivatives simultaneously. For instance, the stationary function corresponding to the interval $(-\infty, x)$ is

$$w_\xi(\varphi,\zeta) = \langle\delta(\varphi-\varphi_\xi^>(x))\delta(\zeta_\xi^>(x)-\zeta)\rangle. \quad (49)$$

We can obtain an equation for it independently, but we can also use Eq. (38) together with the definition

$$\zeta_\xi = \lim_{k\to 0}(\varphi_{\xi+\varepsilon k} - \varphi_\xi)/k.$$

For (49) we have

$$-l\frac{\partial}{\partial\zeta}w_\xi(\varphi,\zeta) + 2\frac{\partial^2}{\partial\zeta^2}[\zeta^2 w_\xi(\varphi,\zeta)] = 0. \quad (50)$$

It is, however, convenient to use in (48′) the representation

$$R^2(x') = [\zeta^>(x')-\zeta^<(x')]^{-1} = \int_0^\infty ds\exp(-s[\zeta^>(x')-\zeta^<(x')]). \quad (51)$$

There occur then under the integral sign in (48′) two factors each of which can be averaged independently

$$F(x-x') = \frac{1}{v}\int_0^\infty ds\int_{-\pi/2}^{\pi/2} d\varphi\,\bar{g}(R,s,\varphi)w_\xi(-\varphi,s). \quad (52)$$

One obtains the equation for the function $w_\xi(\varphi, s) \equiv w(s)$ by taking the Laplace transform of (50)

$$-lsw(s) + 2s^2(\partial^2 w(s)/\partial s^2) = 0, \quad (53)$$

and the equation determining $\bar{g}(R,s,\varphi) \equiv \bar{g}(R,s)$ is derived, like (46), by using (18) in the Born approximation

$$\zeta_{j+1} = (\zeta_j + a)\{1 + {}^1/{}_2\gamma^2 + 2\gamma\sin(2\varphi_j + 2ka - \delta)\}. \quad (54)$$

After simple calculations we find

$$l(\partial\bar{g}/\partial R) = -ls\bar{g} + 2s^2(\partial^2\bar{g}/\partial s^2) + 4s(\partial\bar{g}/\partial s). \quad (55)$$

The boundary condition for (55) is $\bar{g}(R=0,s) = w(s)$. The solutions of both (53) and (55) can be expressed in terms of Bessel functions, but we shall not obtain these formulae here (see Refs. 3, 4).

In concluding this section we give a brief derivation of the main equations for the case when the potential $U(x)$ represents Gaussian white noise with δ-function pair correlations.

For the energy range $E = E_0 + \varepsilon$, where $\varepsilon = v(k-k_0) \ll E_0$ it is convenient to introduce $\psi = \psi^+ + \psi^-$:

$$\psi^\pm(x) = a_\pm\exp(\pm ik_0 x),$$

where we can neglect the ε-dependence of v. The approximate Schrödinger equation has the form

$$v(da_\pm/dx)=\pm i\varepsilon a_\pm\mp i[a_\pm+a_\mp\exp(\mp 2ik_0x)]U(x).$$

We introduce the amplitude $a(x)$ and the phase $\varphi(x)$, for instance, by the relation

$$a_\pm(x)=a(x)\exp(\pm i\varphi(x)).$$

One then gets easily equations for the new quantities:

$$v(da/dx)=-U(x)\sin(2k_0x+2\varphi)a, \tag{56}$$

$$v(d\varphi/dx)=\varepsilon-[1+\cos(2k_0x+2\varphi)]U(x). \tag{57}$$

Assuming $a(x)$ and $\varphi(x)$ to be slowly changing functions it is convenient to write:

$$v\frac{da}{dx}=-\frac{a}{2i}[U_+(x)e^{2i\varphi}-U_-(x)e^{-2i\varphi}],$$

$$v\frac{d\varphi}{dx}=\varepsilon-U_0(x)-\frac{1}{2}U_+(x)e^{2i\varphi}-\frac{1}{2}U_-(x)e^{-2i\varphi},$$

where we have introduced the notation

$$U_0(x)=U(x),\ U_\pm(x)=U(x)\exp(\pm 2ik_0x). \tag{58}$$

As before, we denote the solutions of the equations for the phase with the boundary condition $\varphi(x_0)=\varphi_0$ for $x>x_0$ and $x<x_0$ by $\varphi_\varepsilon^>(x|x_0,\varphi_0)$ and $\varphi_\varepsilon^<(x|x_0,\varphi_0)$, respectively. We introduce again the factor

$$\alpha_\varepsilon(x|x')=\frac{a(x)}{a(x')}=\exp\left\{-\frac{1}{v}\int_{x'}^{x}U(\xi)\sin(2k_0\xi+2\varphi(\xi))d\xi\right\},$$

which describes the change in the amplitude when we go from x' to x. We have exactly:

$$\alpha_\varepsilon(x|x'')\alpha_\varepsilon(x''|x')=\alpha_\varepsilon(x|x'),$$

$$\frac{vd}{dx}\alpha_\varepsilon(x|x')=-\frac{1}{2i}[U_+(x)e^{2i\varphi(x)}-U_-(x)e^{-2i\varphi(x)}]\alpha_\varepsilon(x|x'),$$

$$\frac{vd}{dx'}\alpha_\varepsilon(x|x')=\frac{\alpha_\varepsilon(x|x')}{2i}[U_+(x')e^{2i\varphi(x')}-U_-(x')e^{-2i\varphi(x')}].$$

To derive (26) we vary Eq. (57):

$$v(d\delta\varphi/dx)=\delta\varepsilon+2U(x)\sin(2k_0x+2\varphi)\delta\varphi.$$

Comparing this with the equation for $\alpha_\varepsilon(x|x')$ we find for the two forms of the variation ($\delta\varepsilon$ and $\delta\varphi_0$):

$$(\partial/\partial k)\varphi_\varepsilon(x|x_0,\varphi_0)=\int_{x_0}^{x}\alpha_\varepsilon^2(x'|x)dx,$$

$$(\partial/\partial\varphi_0)\varphi_\varepsilon(x|x_0,\varphi_0)=\alpha_\varepsilon^{-2}(x|x_0).$$

In order to obtain the analog of Eqs. (32) we must write down the Fokker–Planck equations for the interval

$$l\gg\Delta x>k_0^{-1}$$

assuming that the correlation radius for the average $\langle U(x)U(x')\rangle$ is small compared to the mean free path l. In the approximation $k_0l\gg 1$ the following moments of (58) are nonvanishing:

$$\begin{aligned}1/l&=(\Delta xv^2)^{-1}\iint_{\Delta x>\xi'>\xi>0}\langle U_+(x+\xi')U_-(x+\xi)\rangle d\xi\,d\xi'\\&=v^{-2}\int_0^\infty\langle U(\xi)U(0)\rangle\exp(2ik_0\xi)d\xi,\\1/l^*&=(\Delta xv^2)^{-1}\iint_{\Delta x>\xi'>\xi>0}\langle U_-(x+\xi')U_+(x+\xi)\rangle d\xi\,d\xi'\\&=v^{-2}\int_0^\infty\langle U(\xi)U(0)\rangle\exp(-2ik_0\xi)d\xi,\end{aligned} \tag{59}$$

$$1/l_0=(\Delta xv^2)^{-1}\iint_{\Delta x>\xi'>\xi>0}\langle U_0(x+\xi')U_0(x+\xi)\rangle d\xi\,d\xi'=v^{-2}\int_0^\infty\langle U(\xi)U(0)\rangle d\xi.$$

If we assume that $U(x)=U(-x)$ (in the sense that both have the same probability distribution) then $1/l$ is real and $l=l^*$.

To derive the Fokker-Planck equations we must accurately write down expressions for the increases up to terms of second order in $U(x)$. For the increase in the phase $\Delta\varphi(\xi)=\varphi(\Delta x+\xi)-\varphi(\xi)(\Delta\varphi=\Delta_0\varphi+\Delta_1\varphi+\Delta_2\varphi)$ we have

$$\Delta_0\varphi=(k-k_0)\Delta x,$$

$$\Delta_1\varphi(\xi)=-\frac{1}{v}\int_0^\xi d\xi'\left[U_0(x+\xi')+\frac{1}{2}U_+(x+\xi')\exp[2i\varphi(\xi')]+\frac{1}{2}U_-(x+\xi')\exp[-2i\varphi(\xi')]\right],$$

$$\Delta_2\varphi(\Delta x)=-\frac{i}{v}\int_0^{\Delta x}d\xi'[U_+(x+\xi')e^{2i\varphi(x)}-U_-(x+\xi')e^{-2i\varphi(x)}]\Delta_1\varphi(\xi').$$

Using (59) we get

$$\left\langle\frac{\Delta_0\varphi}{\Delta x}\right\rangle=k-k_0,\quad\left\langle\frac{\Delta_2\varphi}{\Delta x}\right\rangle=\frac{2}{l''}=i\left(\frac{1}{l}-\frac{1}{l^*}\right),$$

$$\frac{1}{\Delta x}\langle\Delta\varphi_\varepsilon\Delta\varphi_{\varepsilon'}\rangle=\frac{2}{l_0}+\left(\frac{1}{l}+\frac{1}{l^*}\right)\cos(2\varphi_\varepsilon-2\varphi_{\varepsilon'}).$$

In the general case we must add to these formulae

$$\left\langle\frac{\Delta\alpha_\varepsilon}{\Delta x'}\right\rangle=-\frac{1}{4}\left(\frac{1}{l}+\frac{1}{l^*}\right)\alpha_\varepsilon,$$

$$\left\langle\frac{\Delta\alpha_\varepsilon\Delta\alpha_{\varepsilon'}}{\Delta x}\right\rangle=-\frac{1}{2}\left(\frac{1}{l}+\frac{1}{l^*}\right)\cos 2(\varphi_\varepsilon-\varphi_{\varepsilon'})\alpha_\varepsilon\alpha_{\varepsilon'},$$

$$\left\langle\frac{\Delta\alpha_\varepsilon\Delta\varphi_{\varepsilon'}}{\Delta x}\right\rangle=\frac{1}{2}\left(\frac{1}{l}+\frac{1}{l^*}\right)\sin 2(\varphi_\varepsilon-\varphi_{\varepsilon'})\alpha_\varepsilon.$$

We take an average function $F(\ldots,\varphi_{\varepsilon_i}(x),\ldots)$ and define

$$\langle F(\ldots,\varphi_{\varepsilon_i}(x),\ldots)\rangle=\int\prod d\varphi_{\varepsilon_i}w(\ldots,\varphi_{\varepsilon_i},\ldots;x)F(\ldots,\varphi_{\varepsilon_i},\ldots), \tag{60}$$

i.e., we introduce a phase probability distribution function $w(\ldots,\varphi_{\varepsilon_i},\ldots;x)$. Considering the change of that average over the interval Δx we get by means of the formulae given above

$$\begin{aligned}\frac{\partial}{\partial x}\langle F\rangle&=\sum_i\left\langle\frac{\partial F}{\partial\varphi_{\varepsilon_i}}\frac{\Delta\varphi_{\varepsilon_i}}{\Delta x}\right\rangle+\frac{1}{2}\sum_{i,k}\left\langle\frac{\partial^2F}{\partial\varphi_{\varepsilon_i}\partial\varphi_{\varepsilon_k}}\frac{\Delta\varphi_{\varepsilon_i}\Delta\varphi_{\varepsilon_k}}{\Delta x}\right\rangle\\&=\sum_i\left[(k_i-k_0)-\frac{2}{l''}\right]\left\langle\frac{\partial F}{\partial\varphi_{\varepsilon_i}}\right\rangle\\&\quad+\frac{1}{2}\sum_{i,k}\left\langle\left[\frac{2}{l_0}+\left(\frac{1}{l}+\frac{1}{l^*}\right)\cos 2(\varphi_{\varepsilon_i}-\varphi_{\varepsilon_k})\right]\frac{\partial^2F}{\partial\varphi_{\varepsilon_i}\partial\varphi_{\varepsilon_k}}\right\rangle\end{aligned}$$

and substituting Eq. (60) into the definitions of the averages and integrating by parts we find immediately an equation for the distribution function $w(\ldots,\varphi_{\varepsilon_i},\ldots;x)$:

$$\frac{\partial w}{\partial x}=-\sum_i\left(k-k_0-\frac{2}{l''}\right)\frac{\partial w}{\partial\varphi_{\varepsilon_i}}+\frac{1}{2}\sum_{i,k}\frac{\partial^2}{\partial\varphi_{\varepsilon_i}\partial\varphi_{\varepsilon_k}}\left(\left[\frac{2}{l_0}+\left(\frac{1}{l}+\frac{1}{l^*}\right)\cos 2(\varphi_{\varepsilon_i}-\varphi_{\varepsilon_k})\right]w\right).$$

To obtain equations for more general transition functions, in particular, ones which contain factors $\alpha_\varepsilon(y|x)$ we can use the same method, substituting instead of an arbitrary function $F(\ldots,\varphi_{\varepsilon_i}(x),\ldots)$ into (60)

$$F\to\prod\alpha_{\varepsilon_i}\tilde{F}(\ldots,\varphi_{\varepsilon_i},\ldots),$$

where $\tilde{F}$ is a new arbitrary function.

APPENDIX

Let there be an average of a product of several $\rho_{\xi_l}(x_i, x_j)$ which appears when we average a product of several Green functions:

$$\prod(\xi_1,\ldots,\xi_n|x_1,\ldots,x_{2n})=\left\langle\prod_{l=1}^{n}\rho_{\xi_l}(x_i,x_j)\right\rangle_U, \qquad (A.1)$$

where the coordinates of the wave functions may be placed in order of increasing values: $x_1 < x_2 < \cdots < x_{2n}$. We assume that all ξ_l are different. This product then has the form

$$\prod(\xi_1,\ldots,\xi_n|x_1,\ldots,x_{2n})$$
$$=v^{-n}\left\langle\prod_l \delta(\varphi_{\xi_l}^{>}(j_0)-\varphi_{\xi_l}^{<}(j_0))\alpha_{\xi_l}(x|j_0)\sin\varphi_{\xi_l}(x_i)\right\rangle_U. \qquad (A.2)$$

We have chosen the joining point $x_0 \equiv j_0$ arbitrarily so that

$$-L=-\infty<x_1<x_2\ldots,<x_m<x_0<x_{m+1}<\ldots,<x_{2n}<L=\infty. \qquad (A.3)$$

In (A.2) $\alpha(x_i|j_0)$ can be written as

$$\alpha(x_i|j_0)=\alpha(x_i|x_{i+1})\alpha(x_{i+1}|x_{i+2})\ldots\alpha(x_m|j_0) \qquad (A.4)$$

in agreement with the sequence (A.3), if $x_i < j_0$ and

$$\alpha(x_i|j_0)=\alpha(x_i|x_{i-1})\alpha(x_{i-1}|x_{i-2})\ldots\alpha(x_{m+1}|j_0), \qquad (A.4')$$

if $x_i > j_0$.

We introduce the stationary probability distributions $w^>$ and $w^<$:

$$w^{>}(\varphi_1,\ldots,\varphi_n)=\left\langle\prod_{l=1}^{n}\delta(\varphi_{\xi_l}^{>}-\varphi_l)\right\rangle_U,$$
$$w^{<}(\varphi_1,\ldots,\varphi_n)=\left\langle\prod_{l=1}^{n}\delta(\varphi_l-\varphi_{\xi_l}^{<})\right\rangle_U \qquad (A.5)$$

and the generalized transition functions going from i to $i+1$:

$$w_{i,i+1}^{>(s)}(\varphi_1^{i},\ldots,\varphi_n^{i}|\varphi_1^{i+1},\ldots,\varphi_n^{i+1})$$
$$=\left\langle\prod_{l=1}^{n}\delta(\varphi_l^{i+1}-\varphi_{\xi_l}^{>}(x_{i+1}|\varphi_l^{i},x_i))\prod\alpha_{\xi_p}^{s_p}(x_{i+1}|x_i)\right\rangle, \qquad (A.6)$$

and similarly for the functions

$$w_{i,i+1}^{<(s)}(\varphi_1^{i},\ldots,\varphi_n^{i}|\varphi_1^{i+1},\ldots,\varphi_n^{i+1}),$$

which determine the set of quantities α and phases in the point $x_i < x_{i+1}$ if the set of phases is given in the point x_{i+1}. The $\varphi_{\xi_l}^{>}(x_{i+1}|\varphi_l^{i}, x_i)$ and $\varphi_{\xi_l}^{<}(x_i|\varphi_l^{i+1}, x_{i+1})$ which occur in these definitions are, respectively, the solutions of Eqs. (17) and (17′) in the points $x_{i+1}(x_i)$ under the condition that in the point $x_i(x_{i+1})$ the phase equals $\varphi_l^{i}(\varphi_l^{i+1})$ for an energy ξ_l. In the vector $\mathbf{s}=\{s_1,\ldots,s_n\}$ each component s_l indicates the power of the factor $\alpha_{\xi_l}(x_{i+1}|x_i)$ in the product $\Pi\,\alpha_{\xi_p}^{s_p}(x_i|x_{i+1})$. The vector $\mathbf{s}$ is determined by the coordinate distribution in (A.3) and by the writing down of the factors $\alpha_{\xi_l}(x_i|j_0)$ in the ordered form (A.4) or (A.4′). The separate component s_l can thus have the values 0, 1, and 2. The rules for the construction of the average are simple. It is clear that the intervals $(-\infty, x_1)$ and (x_{2n}, ∞) are associated, respectively, with the stationary probability distributions $w^>(\varphi_1,\ldots,\varphi_n)$ and $w^<(\varphi_1,\ldots,\varphi_n)$. Each interval (x_i, x_{i+1}) in (A.3) is associated with a transition function of the form (A.6), $w^{>(s)}$, if $x_i \leq j_0$ and $w^{<(s)}$, if $x_i \geq j_0$. Finally, each of the ψ-functions in (A.1) gives not only $\alpha_{\xi_l}(x_i|j_0)$, but, according to (14), also a factor $\sin\varphi_{\xi_l}(x_i)$. The expression obtained must be integrated over all phases occurring in it in the points x_i and j_0. The meaning and advantage of such a representation of (A.1) lies in the fact that it is divided into a product of factors, each of which can, in agreement with the assumption about the Markovian nature of the random process, be averaged independently in the intervals (A.3). The functions (A.5) and (A.6) satisfy, of course, equations with partial derivatives.

In actual problems one does normally not use the expressions for the transition probabilities as it is easier to write down recurrence relations relating the point j_0 to one of the coordinates occurring in (A.1).

For a proof that nothing depends on the choice of the point j_0 we note that as long as the point j_0 lies in one of the intervals (A.3), (A.4), the independence of the results with respect to its position is determined exclusively by the properties of the probability distribution:

$$w_{x_m,x_{m+1}}^{>(s)}(\varphi_1^{m},\ldots,\varphi_n^{m}|\varphi_1^{m+1},\ldots,\varphi_n^{m+1})=\int w_{x_m x_0}^{>(s)}(\varphi_1^{m},\ldots,\varphi_n^{m}|\varphi_1',\ldots,\varphi_n')$$
$$\times w_{x_0,x_{m+1}}^{>(s)}(\varphi_1',\ldots,\varphi_n'|\varphi_1^{m+1},\ldots,\varphi_n^{m+1})\prod_{c=1}^{n}d\varphi_l'.$$

If the point j_0 changes an interval of (A.3) we must use for a proof of the independence the relations

$$w_{x,x'}^{>(s)}(\varphi_1,\ldots,\varphi_n|\varphi_1',\ldots,\varphi_n')=w_{x,x'}^{<(s-2)}(\varphi_1,\ldots,\varphi_n|\varphi_1',\ldots,\varphi_n'). \qquad (A.7)$$

The verification of the latter statement consists in the observation that δ-functions such as $\delta(\varphi'-\varphi^{>}(x'|x,\varphi))$ express the fact that $\varphi=\varphi^{<}(x|x',\varphi')$ and therefore

$$\delta(\varphi'-\varphi^{>}(x'|x,\varphi))=\delta(\varphi-\varphi^{<}(x|x',\varphi'))[\partial\varphi^{>}(x'|x,\varphi)/\partial\varphi]^{-1}.$$

Differentiating (17) and using the boundary condition we find easily

$$\partial\varphi_\xi(i'|i,\varphi)/\partial\varphi=\alpha_\xi^{-2}(i'|i),$$

which proves (A.7).

We show in conclusion that the results are independent of the choice of the intervals $(0,\pi)$ or $(0,2\pi)$. The possibility to change from one interval to the other is connected with the fact that when $k_F l \gg 1$ all stochastic equations are invariant with respect to the transformation $\varphi_{\xi_l} \to \varphi_{\xi_l} \pm \pi$ for each variable φ_{ξ_l} independently. We show how one can use that fact through the example of averaging a two-point expression of the kind

$$\sum_{\nu} f_1(\varphi_i)f_2(\varphi_{i'})\delta(\xi-\xi_\nu),$$

i.e., we shall consider Eq. (28), dropping the unimportant factor $\alpha(j|j')$. The total expression for this average is

$$A_{jj'}=\frac{2}{v}\int_0^{2\pi} d\varphi\int_0^{2\pi} d\varphi'\langle[\delta(\varphi_j^{>}-\varphi)+\delta(\varphi_j^{>}-\varphi-\pi)]f_1(\varphi)\cdot\delta(\varphi-\varphi^{<}(j|j',\varphi'))f_2(\varphi')\delta(\varphi'-\varphi_{j'}^{<})\rangle_v.$$

Here $\langle\delta(\varphi^{\gtrless}_{j(j')}-\varphi)\rangle$ and $\langle\delta(\varphi-\varphi^{<}(j|j',\varphi'))\rangle$ are, respectively, the probability distribution function and the transition probability referring to the interval $(0, 2\pi)$. We have

$$A_{jj'}=\frac{2}{v}\int_0^{2\pi} d\varphi\int_0^{2\pi} d\varphi'\,\langle\delta(\varphi_j^{>}-\varphi)[f_1(\varphi)\delta(\varphi-\varphi^{<}(j|j',\varphi'))+f_1(\varphi+\pi)\delta(\varphi+\pi-\varphi^{<}(j|j',\varphi'))]f_2(\varphi')\delta(\varphi'-\varphi_{j'}^{<})\rangle_v$$

$$=\frac{2}{v}\int_0^{2\pi} d\varphi\int_0^{2\pi} d\varphi'\,\langle\delta(\varphi_j^{>}-\varphi)[f_1(\varphi)\delta(\varphi-\varphi^{<}(j|j',\varphi'))f_2(\varphi')\delta(\varphi'-\varphi_{j'}^{<})+f_1(\varphi+\pi)\delta(\varphi-\varphi^{<}(j|j',\varphi'))f_2(\varphi'+\pi)\delta(\varphi'+\pi-\varphi_{j'}^{<})]\rangle_v.$$

As was explained above

$$\langle\delta(\varphi^{\gtrless}_{j(j')}-\varphi)\rangle=\langle\delta(\varphi^{\gtrless}_{j(j')}-\varphi-\pi)\rangle={}^1/_2\,w^{\gtrless}(\varphi),$$

where $w^{\gtrless}(\varphi)$ is the probability distribution (29) referring to the interval $(0, \pi)$. If the functions f_1 and f_2 are both even or odd with respect to the substitution $\varphi \to \varphi \pm \pi$ we get again Eq. (31).

When there are several energies we can perform the indicated transformations independently for each variable φ_{ε_i}.

[1)] Hence it follows that in the averaged single-particle Green functions there is no information about localization or delocalization.

[2)] Writing it down in terms of the spectral density means that among all integrations the integration over E must be performed last. In accordance with §37 of Ref. 8 we must then drop the term with $e^2\bar{p}\mathbf{A}/m$.

[3)] See the Appendix about the choice of interval.

[1] R. E. Borland, Proc. Roy. Soc. (London) **A274**, 529 (1963).

[2] N. F. Mott and W. D. Twose, Adv. Phys. **10**, 107 (1961).

[3] V. L. Berezinskiĭ, Zh. Eksp. Teor. Fiz. **65**, 1251 (1973) [Sov. Phys. JETP **38**, 620 (1974)],

[4] A. A. Gogolin, V. I. Mel'nikov, and É. I. Rashba, Zh. Eksp. Teor. Fiz. **69**, 328 (1975) [Sov. Phys. JETP **42**, 168 (1975)].

[5] A. A. Abrikosov and I. A. Ryzhkin, Adv. Phys. **27**, 147 (1978).

[6] H. Schmidt, Phys. Rev. **105**, 425 (1957).

[7] L. P. Gor'kov and O. N. Dorokhov, Fiz. Nizk. Temp. **4**, 332 (1978) [Sov. J. Low Temp. Phys. **4**, 160 (1978)]; Solid State Comm. **20**, 789 (1976).

[8] A. A. Abrikosov, L. P. Gor'kov, and I. E. Dzyaloshinskii, Metody kvantovoĭ teorii polya v statisticheskoĭ fizike (Quantum Field Theoretical Methods in Statistical Physics) Fizmatgiz, 1962 [English translation published by Pergamon Press, Oxford].

Translated by D. ter Haar

Energy level correlations in a disordered one-dimensional chain

L. P. Gor'kov, O. N. Dorokhov, and F. V. Prigara

L. D. Landau Institute of Theoretical Physics, Academy of Sciences of the USSR

(Submitted 24 August 1983)
Zh. Eksp. Teor. Fiz. **84**, 1440–1452 (April 1983)

The spatial dependence of the density-of-states correlator for energetically close states in a linear chain containing defects is found. The curve obtained can be interpreted as indicating the existence of attraction between close states over distances of the order of the mean free path l. Next, there is a dip—an interlevel repulsion region extending to distances of the order of $z_0 = 2l\ln(8/\nu)$ (where ν is a dimensionless energy difference). The pattern of independent levels is restored at large distances. It is shown that in the case of a finite chain of great length (greater than z_0) the levels of the system as a whole are statistically independent.

PACS numbers: 71.20. + c

1. INTRODUCTION

Back in 1961 Mott and Twose[1] adduced convincing arguments to show that, in a one-dimensional system, the electron wave functions for all energies are localized in the presence of impurities or defects. A mathematically rigorous proof of this fact was given by Berezinskiĭ,[2] who, in particular, found a frequency dependence of the conductivity, that basically agrees with the predictions made in Ref. 1.

The concept of centers on which the electron is localized in the first approximation is used in the argumentation in Ref. 1. The tunneling overlap of wave functions on different centers splits the levels of the centers, and guarantees transitions between them. In applications (to quasi-one-dimensional organic metals, for example), the first approximation corresponds rather to a one-dimensional metallic band, and the localization is a result of the quantum interference effects due to the multiple scattering on the impurities.[2] The electron wave functions (at least in the quasiclassical region, which will be discussed below) oscillate rapidly over atomic distances, i.e., over distances $\sim\hbar/p_F$, and only their envelopes decay over distances of the order of the localization length, which is of the order of the mean free path.[2,3]

Thus, contrary to the relatively simple picture painted in Ref. 1, the exact electron wave functions are a rather complicated object, and in Refs. 2 and 3 quite a complicated diagrammatic technique is used to compute the corresponding averages from them. That simplifications connected with the specific nature of the one-dimensional problem are possible had been pointed out earlier by Schmidt[4] (see also Halperin's paper[5]). Subsequently, Schmidt's method[4] was successfully applied in Ref. 6 to the problem of computing density of states. Finally, in Ref. 7 a method is developed which allows us, at least in principle, to compute mean values of entirely general form, from an arbitrary combination of the exact wave functions of an electron localized in the field of defects.

The purpose of the present paper is to determine more accurately the scope of the ideas about electron localization, or, more strictly speaking, to determine the correlations between levels corresponding to close energy values. The greater part of the paper is devoted to the computation of the correlator for densities of states at different points:

$$F_{E,E+\omega}(x-x')=\left\langle \sum_{\rho\eta}\delta(E-\varepsilon_\rho)\,\delta(E+\omega-\varepsilon_\eta)\,\psi_\rho^2(x)\,\psi_\eta^2(x')\right\rangle, \tag{1}$$

where $\varepsilon_\rho,\varepsilon_\eta$ and ψ_ρ,ψ_η are respectively the exact energy levels and exact wave functions of states localized in the field of the defects and the brackets denote averaging over the disposition of the defects.

The correlator in (1) is defined for an arbitrary segment of an infinitely long chain. For a finite segment we can formulate a different problem concerning the correlation and repulsion of the exact energy levels of the whole system of fixed dimension L. The solution to the last problem is known,[8,9] and the answer is that there are no correlations. We shall briefly discuss this question in the last section.

2. THE DENSITY-OF-STATES CORRELATOR

The method developed in Ref. 7 is used below to compute (1). Let us, following (7), write the wave function of, say, the state ε_η in the form

$$\psi_\eta(x)=R_\eta(x)\sin\varphi_\eta(x). \tag{2}$$

The main variation of the phase in (2) (the rapid oscillations) is connected with the free motion: $\nabla\varphi\approx p_F$. At distances large compared to the atomic distances, the square of the wave function should be averaged in order to obtain a smoothed out density of states at the point x', i.e., we should make the substitution

$$\psi_\eta^2(x')\to {}^1/_2 R_\eta^2(x') \tag{3}$$

and similarly for $\psi_\rho(x)$. Therefore, for $|x-x'|\gg 1/p_F$, we shall everywhere replace (1) by

$$F_\omega(x-x')\equiv F_{E,E+\omega}(x-x')=\frac{1}{4}\left\langle\sum_{\rho\eta}\delta(E-\varepsilon_\rho)\,\delta(E+\omega-\varepsilon_\eta)\,R_\rho^2(x)\,R_\eta^2(x')\right\rangle. \tag{1'}$$

In the case of identically coincident x and x' the phases of the wave functions cannot be averaged independently:

$$F_\omega(x\equiv x')=\frac{1}{4}\left\langle\sum_{\rho\eta}\delta(E-\varepsilon_\rho)\,\delta(E+\omega-\varepsilon_\eta)\,R_\rho^2(x)\,R_\eta^2(x')\left(\frac{1}{2}+\cos^2\theta_\omega\right)\right\rangle. \tag{4}$$

Here

$$\theta_\omega=\varphi_\eta-\varphi_\rho \tag{5}$$

is the phase difference for the two states, which, generally speaking, changes over distances greater than, or of the order of, the mean free path, and carries the main information about the correlation of the states.

According to (7), if the potential of the defects has a sufficiently short range, then the problem of averaging expressions of the type (1), (1′) reduces to Markov processes. For $x<x'$, it is convenient to take the point x' to be a matching point, and perform two independent averagings: one over the defects located to the right, the other over the defects located to the left, of x'. With (1) expressed in terms of the notation used in (7), let us, by transforming the energy δ function, rewrite it in the form

$$F_\omega(x-x')=\frac{1}{v_F^2}\int_0^\pi\int_0^\pi\langle\delta(\varphi_E^{>}(x')-\varphi_1)\delta(\varphi_{E+\omega}^{>}(x')-\varphi_2)$$

$$\times\delta(\varphi_1-\varphi_E^{<}(x'))\delta(\varphi_2-\varphi_{E+\omega}^{<}(x'))\alpha_E^2(x|x')\rangle d\varphi_1\,d\varphi_2,$$

where $\alpha(x|x')=R(x)/R(x')$.

For $p_Fl\gg1$, only the dependence on the relative phase difference (5) is important in the last expression, since the energy ε_ρ itself is not distinguished by anything in the quasi-classical limit. The steady-state probability distribution $w(\varphi,\theta)$ for the phases $\varphi\equiv\varphi_\rho$ and $\theta=\varphi_\eta-\varphi_\rho$ depends only on the phase difference θ:

$$w(\varphi,\theta)=w(\theta), \tag{6}$$

where the function $w(\theta)$ is found in Ref. 7. Finally, let us rewrite (1′) in the form

$$F_\omega(x-x')=\frac{\pi}{v_F^2}\int_0^\pi g(x-x',\theta)\,w(\pi-\theta)\,d\theta, \tag{7}$$

where the equation for the function

$$g(x-x',\theta)=\langle\delta(\theta_\omega^{>}(x')-\theta)\,\alpha^2(x|x')\rangle \tag{8}$$

can easily be derived in accordance with Ref. 7:

$$\frac{\partial g}{\partial z}=\frac{\partial}{\partial\theta}\left(\sin^2\theta\frac{\partial g}{\partial\theta}\right)-\nu\frac{\partial g}{\partial\theta}. \tag{9}$$

Everywhere below we use the dimensionless notation

$$\nu=\omega\tau,\quad z=(x-x')/l,\quad x\to x/l,$$

where the quantities $l=v_F\tau$ and τ are the corresponding kinetic quantities–the mean free path and mean free time.

The boundary condition for (9) at $x=x'$ follows from the definition

$$g_\omega(x=x',\theta)=w(\theta). \tag{9'}$$

For coincident pints $x\equiv x'$ the expression (4) has the form

$$F_\omega(x\equiv x')=\frac{\pi}{v_F^2}\int_0^\pi w(\theta)\,w(-\theta)\left(\frac{1}{2}+\cos^2\theta\right)d\theta, \tag{4'}$$

and can be determined directly.

Before proceeding to carry out the computations, let us say something about Fig. 1, which schematically shows the spatial behavior of the correlator (1), whose general form follows from the foregoing and obvious physical considerations. In the figure, the point $z=x'-x=0$ corresponds to the expression (4′); the hatched region, to atomic distances, i.e., to distances for which $p_Fz\sim1$, where the answer depends on the specific model. Further, the behavior of the correlator (7) smoothed out over atomic distances is depicted. At small distances (as compared to l, i.e., for $z\ll1$) the behavior of the curve is determined by its value at zero value of its argument,

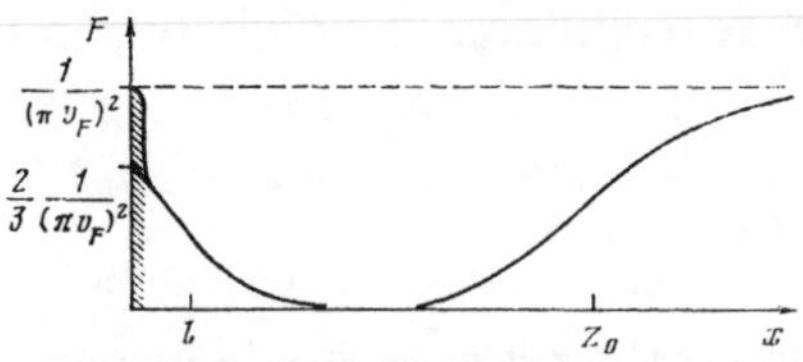

FIG. 1. Diagrammatic representation of the spatial dependence of the density-of-states correlator (1) for energetically close states. The region $z\sim l$ corresponds to the overlap of the corresponding wave functions. The hatched region corresponds to atomic distances ($zp_F\sim1$). The right-hand side of the curve is a plateau corresponding to independent, spatially-separated states. The characteristic scale $z_0=2l\ln(8/\nu)$.

$$F_\omega(0)=\frac{\pi}{v_F^2}\int_0^\pi w(\theta)\,w(\pi-\theta)\,d\theta, \tag{10}$$

and the slope (first derivative)

$$\frac{dF_\omega}{dz}(0)=-\frac{\pi}{v_F^2}\int_0^\pi w(\pi-\theta)\frac{\partial}{\partial\theta}(w(\theta)\sin2\theta)\,d\theta. \tag{10'}$$

The equality (10) follows from (9′) and (7), while (10′) is a consequence of Eq. (9) above and Eq. (39) in Ref. 7 for the steady-state phase distribution function $w(\theta)$. (Notice that the symbol ν used in Ref. 7 differs by a factor of 2 from the one used in the present paper.)

As z increases, the function $F_\omega(z)$ falls off (for sufficiently small ν). This decrease reflects the repulsion of levels with sufficiently close energies. At large distances the correlator (1) must go over into a product of densities of states:

$$F_\omega(z\to\infty)\approx(\pi v_F)^{-2}, \tag{11}$$

which corresponds to independent spatially-distant states of the one-dimensional chain.

Of primary physical interest is the correlation of two energetically close levels, i.e., levels for which $\nu\ll1$, which we shall assume below to be fulfilled. In this case we have

$$F_\omega(0)={}^2/_3(\pi v_F)^{-2}, \tag{12}$$

$$F_\omega'(0)=-{}^2/_3(\pi v_F)^{-2}. \tag{12'}$$

To find the general dependence on z, let us go over in (9) to the Laplace transforms in the coordinate z:

$$g_\omega(z,\theta)=\frac{1}{2\pi i}\int_{\delta-i\infty}^{\delta+i\infty}e^{\varkappa z}\tilde{g}_\omega(\varkappa,\theta)\,d\varkappa. \tag{13}$$

With allowance for (9′) we obtain in place of (9) an equation for the transform:

$$\frac{\partial}{\partial\theta}\left(\sin^2\theta\frac{\partial}{\partial\theta}\bar{g}_\omega\right)-\nu\frac{\partial}{\partial\theta}\bar{g}_\omega-\varkappa\bar{g}_\omega=-w(\theta). \tag{14}$$

This equation cannot be solved in closed form. It is well known,[7] however, that, for small ν, the function $w(\theta)$ is large in the domain of variation of the phase close to zero, i.e., for $\theta\sim\nu$:

$$w(\theta)\approx(1/\nu)\,w_0(\theta/\nu),\quad \theta\sim\nu,$$

and in the region close to π:

$$w(\theta)\approx w_1((\pi-\theta)/\nu),\quad \pi-\theta\sim\nu$$

(see the formulas (41)-(43) in Ref. 7). Therefore, it follows from the representation of F_ω in the form (7) that we shall also need for what follows the expression for $\bar{g}_\omega$ in the same regions. Let us recall that the function $\bar{g}_\omega$ is a periodic function with period π:

$$\bar{g}_\omega(\varkappa,0)=\bar{g}_\omega(\varkappa,\pi). \tag{15}$$

Thus, for small ν, the mathematical aspect of the problem consists in the solution of Eq. (14) in the regions of θ values close to zero and π and the matching of the solutions with the aid of the periodicity condition (15). In the region of small θ we have in place of (14) the equation

$$\frac{\partial}{\partial\theta}\left(\theta^2\frac{\partial\bar{g}_\omega}{\partial\theta}\right)-\nu\frac{\partial\bar{g}_\omega}{\partial\theta}-\varkappa\bar{g}_\omega=-w(\theta) \tag{16}$$

(and similarly for $\pi-\theta\ll\pi$). The solution to Eq. (16) in this region can be expressed in terms of the modified Bessel functions:

$$\xi^{1/2}e^{-\xi}\{K_\mu(\xi),\ I_\mu(\xi)\}, \tag{16'}$$

where

$$\mu=(1/4+\varkappa)^{1/2},\quad \xi=\nu/2\theta.$$

In the main region $\theta\sim1$ we can neglect the term $\nu\partial\bar{g}_\omega/\partial\theta$ in (16), which yields

$$\frac{\partial}{\partial\theta}\left(\sin^2\theta\frac{\partial}{\partial\theta}\bar{g}_\omega\right)-\varkappa\bar{g}_\omega=-w(\theta). \tag{17}$$

The solution to Eq. (17) can be expressed in terms of the Legendre polynomials[10]:

$$(\sin\theta)^{-1/2}P_{1/2}^{\mu}(\cos\theta);\quad(\sin\theta)^{-1/2}P_{1/2}^{-\bar{\mu}}(-\cos\theta). \tag{18}$$

The solutions should be matched in the region

$$\nu\ll\theta\ll1\quad(\nu\ll\pi-\theta\ll1).$$

In this region the asymptotic forms of the functions (16') and (18) coincide, and have the form

$$\theta^{-1/2+\mu},\quad\theta^{-1/2-\mu}. \tag{19}$$

For correct matching to be possible, i.e., in order that the coefficients can be determined, the quantity $\mathrm{Re}\,\mu$ should, as we shall see, satisfy the condition

$$\mathrm{Re}\,\mu\leqslant1. \tag{20}$$

Only in this case can both terms in (19) be separated against the background of the neglected term of the order θ^2 in (16) and the neglected terms with ν in (17). It is clear that the two terms in (19) are of the same order of magnitude for negative $\varkappa<-\frac{1}{4}$, i.e., for imaginary $\mu=(\frac{1}{4}+\varkappa)^{1/2}$.

We shall first investigate the solution to Eq. (14) in the general form, and then reduce the computation of the required quantities in the asymptotic regions to their determination under the condition (20).

It is in certain aspects convenient to rewrite Eq. (14), which is naturally solved as a standard second-order inhomogeneous equation if both solutions of the homogeneous equation are known, in terms of new variables in the infinite range $(-\infty,\infty)$, introducing for this purpose the variable $u=\cot\theta$ and a new function $\psi(u)$ defined by the relation

$$g_\omega(\varkappa,\theta)=e^{-\nu u/2}\psi(u). \tag{21}$$

For the function $\psi(u)$ we now have

$$-\psi''+\frac{\varkappa}{1+u^2}\psi+\frac{\nu^2}{4}\psi=\frac{w(u)}{1+u^2}e^{\nu u/2}\equiv f(u). \tag{22}$$

The left-hand side of (22) has the form of the Schrödinger equation (this transformation is noted also in Ref. 9) with potential

$$V(u)=\varkappa/(1+u^2) \tag{23}$$

and negative energy $-\nu^2/4$, which is small when $\nu\ll1$. A characteristic of the potential (23) is its decrease at infinity according to the law u^{-2}, from which it follows, according to Ref. 11, that a system with such a potential will have an infinite number of levels when $\varkappa<-\frac{1}{4}$. Other boundary conditions for our problem (periodicity, Eq. (15)) do not allow us to express the solutions (22) in terms of the eigenfunctions of the quantum-mechanical problem, but the indicated circumstance manifests itself in rapid oscillations of the solutions for $\varkappa<-\frac{1}{4}$, which, by the way, is already evident from (19).

It is convenient to define the solutions to the homogeneous equation (14) in such a way that the first of them $g_1(\varkappa,\theta)$ is finite at the point π. It is easy to verify that the second linearly independent solution is

$$g_2(\varkappa,\theta)=e^{-\nu\,\mathrm{ctg}\,\theta}g_1(\varkappa,\pi-\theta). \tag{24}$$

The pair of independent functions $\psi_1(\varkappa,u)$ and $\psi_2(\varkappa,u)$ figuring in (22), and corresponding to this definition are uniquely determined by the requirement that ψ_1 decrease exponentially as $u\to-\infty$:

$$\begin{aligned}&\psi_1(\varkappa,u)\approx e^{\nu u/2}\quad(u\to-\infty),\\&\psi_1(\varkappa,u)\approx C_1(\varkappa)e^{\nu u/2}+C_2(\varkappa)e^{-\nu u/2}\quad(u\to+\infty),\end{aligned} \tag{24'}$$

while the relation (24) corresponds to $\psi_2(u)=\psi_1(-u)$, which is a consequence of (21) and the evenness of the homogeneous part of Eq. (22) with respect to the substitution $u\to-u$. The Wronskian is equal to

$$\psi_1\psi_2'-\psi_1'\psi_2=-\nu C_1(\varkappa). \tag{25}$$

That solution $\psi(u)$ to the inhomogeneous equation (22) which satisfies the condition that $g_\omega(\varkappa,\theta)$ be finite at $\theta=0,\pi$ and also guarantees the fulfillment of the periodicity condition (15) has the form

$$\psi(u)=\frac{1}{\nu C_1(\varkappa)}\Big\{\psi_1(u)\int_u^\infty\psi_1(-u')f(u')du'+\psi_1(-u)\int_{-\infty}^{u}\psi_1(u')f(u')du'+\frac{1}{C_1(\varkappa)-1}\psi_1(u)\int_{-\infty}^{\infty}\psi_1(-u')f(u')du'\Big\}. \tag{26}$$

To the points where $C_1(\varkappa) = 0$ correspond, according to (24), solutions of the Schrödinger equation (22) that decrease on both sides at $u = \pm \infty$. To such solutions would correspond a function $g(\varkappa,\theta)$ that does not satisfy the periodicity condition. It is easy to verify that, since $\psi_2(\varkappa,u) = \psi_1(\varkappa,u)$ for such $\varkappa$ values, the expression (26) as a function of $\varkappa$ actually does not have poles wherever $C_1(\varkappa) = 0$. The singularities in the expression (26) as a function of $\varkappa$ are connected with the last term in it, or, more precisely, with the zeros of the denominator

$$C_1(\varkappa)=1, \tag{27}$$

which determine the eigenvalues and eigenfunctions of the non-self-adjoint boundary-value problem (14). The fact that the boundary points of the interval $(0,\pi)$ are both singular points imposes, on the face of it, mathematical limitations on the applicability of the results of the general theory (see, for example, Ref. 12). An analysis performed by us with a quasiclassical representation of the function $\psi_1(\varkappa,u)$ for large $|\varkappa|$ shows, however, that it is admissible to deform the contour for the $\varkappa$ integration in (23) into the left half-plane Re $\varkappa < 0$.

The next step consists in the use of the condition $\nu \ll 1$ to find in the region (20) an approximate expression for the function $g_1(\varkappa,\theta)$ (or $\psi_1(u)$) in the entire interval by matching the solutions to the homogeneous equation (14) or (22) near $\theta = 0$ and $\theta = \pi$ (in the region $\nu \ll \theta \ll 1$). Such an operation can easily be performed, since the asymptotic forms of the solutions (16′) and (18) are well known.[10] Here we give only the final results. In the region $u \gg 1$ the function $\psi_1(\varkappa,u)$ has the form $(u' = \nu u/2)$

$$\psi_1(u)=C_1(\varkappa)(\pi/2)^{1/2}(u')^{1/2}[I_\mu(u')+I_{-\mu}(u')] + (2/\pi)^{1/2}\,\mathrm{ctg}\,\mu\pi(\gamma_\mu-\gamma_{-\mu})(u')^{1/2}K_\mu(u'), \tag{28a}$$

in the region $u < 0, |u| \gg 1$ $(u' = -\nu u/2)$

$$\psi_1(u)=(2/\pi)^{1/2}(u')^{1/2}K_\mu(u'), \tag{28b}$$

and in the intermediate region $\nu|u| \ll 1$

$$\psi_1(u)=(\nu/\pi)^{1/2}(\sin\theta)^{-1/2}[A_\mu P^{\mu}_{1/2}(\cos\theta)+A_\mu' P^{-\mu}_{1/2}(-\cos\theta)]. \tag{28c}$$

Above we have used the notation:

$$\gamma_\mu=\frac{\Gamma(3/2-\mu)}{\Gamma(3/2+\mu)}\left[\frac{\Gamma(1+\mu)}{\Gamma(1-\mu)}\right]^2\left(\frac{8}{\nu}\right)^{2\mu}, \tag{29}$$

$$C_1(\varkappa)=[1-{}^1/_2(\gamma_\mu+\gamma_{-\mu})\cos\mu\pi]\sin^{-2}\mu\pi. \tag{30}$$

The constants A_μ and A_μ' are determined from the system

$$A_\mu\frac{\Gamma(\mu)}{\pi}=\frac{\pi}{2\sin\mu\pi}\frac{1}{\Gamma(1+\mu)}\left(\frac{\nu}{8}\right)^\mu,$$

$$\frac{\pi}{2\sin\mu\pi}\frac{1}{\Gamma(1-\mu)}\left(\frac{8}{\nu}\right)^\mu \tag{31}$$

$$=A_\mu\frac{\Gamma(-\mu)}{\Gamma(3/2-\mu)\Gamma(-1/2-\mu)}+\frac{A_\mu'}{\Gamma(1+\mu)}. \tag{31′}$$

The expressions for A_μ and A_μ' will not be needed below, and their definitions (31) and (31′) are given in the form in which they figure in the matching conditions for the asymptotic forms for $\nu \ll \theta \ll 1$. Allowance for the higher-order terms from the expansion of $\sin^2\theta$ in powers of θ in (16) will give rise to corrections of relative magnitude ν^2 in (31) and (31′), whence we obtain the condition, (20), limiting the applicability of the matching procedure used.

We have already indicated above that the θ-integration domains near zero and π play the major role in (7). In fact, the expression for the relative order of magnitude of the quantity $w(\theta)$ in the region $\theta \sim 1$ contains an additional factor that is of the order of ν in smallness. Retaining in (7), (21), and (26) only the contribution from the principal regions, where the function $\psi_1(\varkappa,u)$ should, in its turn, be replaced by the expressions (28a) and (28b), rearranging the terms in (26) so as to get rid of spurious poles determined by the condition $C_1(\varkappa) = 0$, and, finally, discarding all the terms that are analytic in the left half-plane Re $\varkappa < 0$, we obtain the following representation:

$$F_\omega(\varkappa)=\frac{\pi}{v_F^2}\frac{1}{\nu[C_1(\varkappa)-1]}\left(\int_{-\infty}^{\infty}f(u)\psi_1(-u)\,du\right)^2. \tag{32}$$

The integral in (32) is a sum of two terms $J^{(0)}$ and $J^{(\pi)}$, which contain contributions from the θ integration near zero $(u \gg 1)$ and near π (u is negative and large), respectively.

Let us transform the denominator in (32) in accordance with (30) to the form

$$C_1(\varkappa)-1=\cos\mu\pi\sin^{-2}\mu\pi[\cos\mu\pi-\mathrm{ch}\,y(\mu)], \quad e^{y(\mu)}=\gamma_\mu. \tag{33}$$

So long as μ is small, we can, according to (29), use the expression

$$y(\mu)=2\mu L^*\approx 2\mu\ln(8/\nu). \tag{34}$$

The poles determined by the zeros of (33) correspond to a simple pole at $\varkappa = 0$ $(\mu = \frac{1}{2})$ and two sets of complex-conjugate poles, which, for small μ (i.e., for $\varkappa$ close to $-\frac{1}{4}$), have the form

$$\mu_n^{(1)}=in\pi/(L^*+i\pi/2), \quad \mu_n^{(2)}=-in\pi/(L^*-i\pi/2). \tag{35}$$

The poles (35) consequently lie on both sides of the negative axis $\varkappa < -\frac{1}{4}$. The use of the set of eigenfunctions determined in accordance with (35) does not, however, offer any special advantage, and is useful only in the computation of certain asymptotic forms of little interest. The numerical computation of the behavior of the correlator (1) for, say, arbitrary $z = x - x' \sim 1$ requires the ability to evaluate the integral in (32) with rapidly oscillating eigenfunctions, the oscillations of these functions being connected with the specific nature of the potential $V(u)$, (23).

Having noted these mathematical characteristics, we proceed to directly compute (1) in the region $z \sim z_0$ (see Fig. 1), where strong repulsion occurs between two energetically close states in the case of small ν. We shall presently see that $z_0 \sim \ln(1/\nu) \gg 1$ and, consequently, small $\varkappa$ values play a role in the representation (13). In (32) the dominant term is connected with $[J^{(0)}]^2$:

$$J^{(0)}=({}^1/_4-\mu^2)/\cos\mu\pi\approx 1/\pi, \quad \mu\approx{}^1/_2+\varkappa. \tag{36}$$

The expression (36) is the result of the integration of the function $w(\theta) \approx v^{-1}w_0(\theta/\nu)$, taken from (7), with (28b). We used the relation 6.621(3) from Ref. 13.

We should expand $\cos\mu\pi$ in the denominator of (32)

and take in (33) the dominant terms for small $\varkappa$:

$$F_\omega(\varkappa)=\left(\frac{1}{\pi v_F}\right)^2\frac{1}{\varkappa}\exp\left[-(2\mu-1)\ln\frac{8}{\nu}\right]$$
$$\approx\left(\frac{1}{\pi v_F}\right)^2\frac{1}{\varkappa}\exp\left[(-\varkappa+\varkappa^2)2\ln\frac{8}{\nu}\right].$$

The last expression should be substituted into the integral

$$F_\omega(z)=\frac{1}{2\pi i}\int_{\delta-i\infty}^{\delta+i\infty}e^{\varkappa z}F_\omega(\varkappa)\,d\varkappa,$$

in order to obtain the coordinate dependence of $F_\omega(z)$. Let us draw attention to the fact that the exponential function contains the combination $z-z_0$, where $z_0=2\ln(8/\nu)$, in view of which we have retained the terms quadratic in $\varkappa$ in the exponent in the preceding expression. The integration over $\varkappa$ is easy to perform. We see that, after a deep dip, the function $F_\omega(z)$, (1), begins to increase again in the vicinity of z_0:

$$F_\omega(z)=\frac{1}{(\pi v_F)^2}\left\{\frac{1}{2}+\frac{1}{2}\operatorname{erf}\left(\frac{z-z_0}{2z_0^{1/2}}\right)\right\}, \quad (37$$

going over at $z<z_0$ into a product of two independent densities of states. The width of the transition region is

$$\Delta z\sim[\ln(8/\nu)]^{1/2}. \quad (37')$$

It is of interest to compute the asymptotic law of decrease of the correlator (1) for small z in the left part of Fig. 1, i.e., for $1\ll z\ll z_0$. The term with $2J^{(0)}J^{(\pi)}$ in (32) is responsible for the corresponding variation, since the contribution from $[J^{(\pi)}]^2$ is everywhere small. In evaluating the integrals in $J^{(\pi)}$, which contain the expression

$$W_1(\theta)=W_1(u)=\frac{\nu|u|}{\pi^2}\int_0\frac{dt}{(1+t)^2}\exp(-\nu|u|t)$$

from Ref. (7) and Bessel functions, we found the formulas 6.622(3) and 7.512(10) in Ref. 13 useful. Let us, without going into the details of the computations performed, note that the answer can be obtained either by taking the residues (in the integration over $\varkappa$) at the poles (35) and then going over from summation to integration over $|\mu_n|\approx n\pi/L^*$, or by directly continuing the terms dominant with respect to ν in the region (20) with $\operatorname{Re}\mu>0$ to the $\varkappa<-\frac{1}{4}$ axis. Let us give the answer obtained in the indicated region:

$$F_\omega(z)=\frac{1}{(\pi v_F)^2}\frac{\pi^{7/2}e^{-|z|/4}}{16\,|z|^{3/2}}. \quad (38)$$

This result differs by a numerical factor from the decay law found in Ref. 3 for the correlation of the moduli of the wave function of a single localized state at distances $|z|\gg1$. Thus, the wave functions of energetically close states almost coincide with each other at small distances. This does not contradict the ideas expressed in Ref. 1, according to which the wave functions of energetially close states are a result of the tunneling splitting of a pair of wave functions localized at two separate centers, with the important proviso, however, that the radius of localization in the problem in question is determined by the kinetic mean free path. Finally, let us, for completeness, give the correlator value, determined in accordance with (4'), at coincident points:

$$F_\omega(x'\equiv x)=(\pi v_F)^{-2}.$$

This value is exact, and does not depend on ν.

Let us say a few more words about the shape of the curve in the case when ν is not small. As ν increases, the dip gradually decreases, disappearing at $\nu\sim1$. At large ν the major part of the curve would correspond to a plateau (independent states). It is not difficult to obtain the corresponding small corrections in the correlator (1) at $\nu=\omega\tau\gg1$:

$$F_\omega(z)=\frac{1}{(\pi v_F)^2}\left[1-e^{-2z}\frac{\cos 2\nu z}{2\nu^2}\right].$$

These corrections, however, have over small distances an oscillating character, which reflects the beating of two energetically close wave functions.

3. ENERGY-LEVEL CORRELATION IN A FINITE CHAIN

The problem of the relative disposition of the energy levels in a fairly complex finite system was first formulated by Wigner[14] for nuclear levels. Its solution was subsequently offered by Dyson[15] in some general mathematical formulation. One of the formulations of the problem consists in the following. Let it be known that the mean level spacing is Δ. What is the probability of finding two levels a distance $E_2-E_1=\Delta$ apart if the interactions or the other parameters of the system can be considered to be random? It is suggested in Ref. 16 that it is precisely such a situation that, possibly, is realized for the electron levels in a small metallic particle containing scattering defects. Recently, Efetov[17] successfully demonstrated that the Dyson formulas are indeed reproduced in such objects.

In the one-dimensional case the density of states of a finite section (of length L) of a chain containing defects is

$$1/\Delta=L/\pi v_F. \quad (39)$$

A characteristic feature of the results obtained in Refs. 15 and 17 is the prediction of a significant correlation between levels over the energy range Δ. In the one-dimensional case we should rather expect the opposite assertion, since the energy levels are now connected with states localized at different points of the chain over a distance of the order of the mean free path $l\ll L$. Indeed, according to Refs. 8 and 9, the level distribution in this case is a Poisson distribution. We shall presently prove this assertion, using the formulas of the preceding section. Certain results obtained below supplement the results presented in Refs. 8 and 9.

Thus, we shall denote the correlator defined above by

$$S_\omega(x)=\left\langle\sum_{\mu\nu}\delta(E-\varepsilon_\mu)\delta(E+\omega-\varepsilon_\nu)\right\rangle, \quad (40)$$

where $\omega=E_2-E_1$. Rewriting the energy δ functions in terms of the phase δ functions and $d\phi/dE$, as is done in, say, Ref. 7, we introduce the function

$$S_\omega(x,\varphi,\theta)=\langle\delta(\varphi-\varphi_1^>)\delta(\theta-(\varphi_2^>-\varphi_1^>))\zeta_1^>\zeta_2^>\rangle. \quad (41)$$

The phases $\phi=\phi_1$ and $\theta=\phi_2-\phi_1$ pertain to the corresponding wave functions and for the present we regard the chain length x as a variable. The phases are equal to zero at $x = 0$, and equal to integral multiples of π at $x = L$ if the states with energies ε_μ and ε_ν are eigenstates. The notation $\zeta = \partial\phi/\partial k$ was taken from Ref. 7. Similarly, $\phi_1^>$ and $\phi_2^>$ are that solution for the phases of the wave functions with energies E_1 and E_2 which satisfies the condition $\phi = 0$ at $x = 0$. The functions $\phi_1^>$ and $\phi_2^>$ increase monotonically with x and E ($\zeta^> > 0$ (Ref. 7)). In contrast to the preceding section where we considered an infinite chain, here it is expedient to refer the phase θ to the interval $(0,\pi)$. Therefore, the integral part of $[\theta^>(x)/\pi]$, i.e., the number of zeros of the wave function, directly indicates the number of levels lying in the interval between E_2 and E_1 along a section of the chain with length x. As to the phase ϕ, in the quasiclassical problem the phase with energy E_1 rapidly settles by itself into a steady-state distribution (over distances of the order of l). The brackets in (40) or (41) denote averaging over the various defect realizations in a finite segment and, hence, over the quasiclassical part of the energies E_1.

Besides the function $S_\omega(x,\phi,\theta)$, we must determine two other quantities:

$$\tilde{S}_\omega(x,\varphi,\theta)=\langle\zeta_1^>\delta(\varphi-\varphi_1^>)\delta(\theta-\theta^>)\rangle, \tag{42}$$

$$W_\omega(x,\varphi,\theta)=\langle\delta(\varphi-\varphi_1)\delta(\theta-\theta^>)\rangle. \tag{43}$$

Using the method of Ref. 7, we can easily derive the corresponding equation for each of these functions. Thus, we have

$$\frac{\partial w}{\partial x}=-\nu\frac{\partial w}{\partial\theta}+\frac{\partial^2}{\partial\theta^2}(\sin^2\theta w), \tag{44}$$

where, regarding the dependence on ϕ, we assume that the function $w_\omega(\phi,\theta) = w_\omega(x,\theta)$ has already settled into a steady-state distribution with respect to ϕ, and that it does not depend on this variable. Similar equations follow for $\tilde{S}_\omega$:

$$\frac{\partial\tilde{S}}{\partial x}=lw-\nu\frac{\partial\tilde{S}}{\partial\theta}+\frac{\partial}{\partial\theta}\left(\sin^2\theta\frac{\partial\tilde{S}}{\partial\theta}\right) \tag{45}$$

and $S_\omega(x,\theta)$ itself:

$$\frac{\partial S}{\partial x}=2l\tilde{S}-\nu\frac{\partial S}{\partial\theta}+\sin^2\theta\frac{\partial^2 S}{\partial\theta^2}. \tag{46}$$

The boundary conditions for these equations are obvious:

$$w(x=0,\theta)=\delta(\theta),\quad \int_0^\infty w(\theta,x)d\theta=1,\quad w(x,\theta\to\infty)=0. \tag{44'}$$

According to Ref. 7, $\zeta^>(x = 0) = 0$. It follows from the definitions (41) and (42) that

$$S_\omega(x=0,\theta)=\tilde{S}_\omega(x=0,\theta)=0,\quad \tilde{S}_\omega(\theta\to\infty,x)=0. \tag{45'}$$

With the aid of Eqs. (44)-(46) and these boundary conditions, we can easily verify the identities

$$\tilde{S}_\omega=-l\int_0^\theta\frac{\partial w}{\partial\nu}d\theta', \tag{47}$$

$$S_\omega=-l\int_0^\theta\frac{\partial\tilde{S}_\omega}{\partial\nu}d\theta'=l^2\frac{\partial^2}{\partial\nu^2}\int_0^\theta d\theta'\int_0^{\theta'}wd\theta''. \tag{48}$$

Thus, the problem has been reduced to the problem of determining the function

$$\Phi(x,\theta)=\int_0^\theta w_\omega(x,\theta')d\theta', \tag{49}$$

which, in its turn, satisfies the equation

$$\frac{\partial\Phi}{\partial x}=-\nu\frac{\partial\Phi}{\partial\theta}+\frac{\partial}{\partial\theta}\left(\sin^2\theta\frac{\partial\Phi}{\partial\theta}\right).$$

On account of (44′), the boundary condition for this equation is

$$\Phi(x=0,\theta)=1. \tag{44''}$$

Thus, the last equation differs from (9) only in that it has a different boundary condition, and that the domain of variation of θ is not limited to the interval $(0,\pi)$, but extends over the infinite interval $(0,\infty)$, reflecting the fact that, for a sufficiently long chain of length $L = x$, the number of levels in the energy range (E_1,E_2) is arbitrary. The physical meaning of the function $\Phi(x,\theta)$ consists, according to (49), in precisely the fact that the values of the function at the points $\theta = n\pi$ determine the probability of the system's having n levels in the chosen energy range. The function $\Phi(x,\theta)$ is a fundamental quantity for what follows. In the Laplace representation, the equation for its transform $\Phi_\varkappa(\theta)$ has the form

$$\frac{\partial}{\partial\theta}\left(\sin^2\theta\frac{\partial\Phi_\varkappa}{\partial\theta}\right)-\nu\frac{\partial\Phi_\varkappa}{\partial\theta}-\varkappa\Phi_\varkappa=-1. \tag{50}$$

The solution of (50) is

$$\Phi_\varkappa(\theta)=\varkappa^{-1}+\bar{\Phi}_\varkappa(\theta), \tag{51}$$

where in (51) $\bar{\Phi}_\varkappa(\theta)$ denotes that solution to the homogeneous equation in (50) which decreases as $\theta\to\infty$. In each interval $\theta_n = (n\pi,(n+1)\pi)$ the function $\bar{\Phi}_\varkappa$ is, as can easily be verified, directly expressible in terms of the function $g_1(\varkappa,\theta)$ (or $\psi_1(u)$) introduced in the preceding section by the formulas (24). In this case we have

$$\bar{\Phi}_\varkappa((n+1)\pi)=C_1^{-1}(\varkappa)\bar{\Phi}_\varkappa(n\pi), \tag{52}$$

and from the definition of $\Phi(x,\theta)$ for $x\neq 0$ it follows that $\Phi_\varkappa(\theta\to 0)\to 0$, i.e.,

$$\bar{\Phi}_\varkappa(\theta=0)=-\varkappa^{-1}. \tag{52'}$$

The formulas (52) and (52′) constitute, in principle, the exact solution to the problem. But since we posed at the beginning of this section the question of level correlation under conditions when the characteristic energy scale $\omega = E_2 - E_1$ is comparable to the mean level spacing Δ, (39), of a large system, it is sufficient for this purpose to know all the quantities in the region $|\varkappa|\sim\nu\ll 1$. Carrying out the corresponding expansion in powers of $\varkappa$ in (29) and (30), we obtain

$$C_1^{-1}(\varkappa)\approx\frac{\nu}{\nu+\pi\varkappa},\quad \Phi_n(\varkappa)=\bar{\Phi}_\varkappa(n\pi)=-\frac{1}{\varkappa}\left(\frac{\nu}{\nu+\pi\varkappa}\right)^n. \tag{53}$$

Returning to the formulas (47)-(49), we find that the probability for a finite system of length $x = L$ to have n levels in the energy range ω is

$$w_n=(\nu x/\pi)^n e^{-\nu x/\pi}/n!. \tag{54}$$

In its turn, the condition for the absence of levels in this range is

$$w_0 = e^{-\nu\varkappa/\pi}. \tag{55}$$

Summation over all n gives the statistical number of levels in the range ω:

$$N_\omega = (L/\pi v_F)\,\omega. \tag{56}$$

The results (54)-(56) naturally coincide with the results obtained in Refs. 8 and 9, thus indicating the independence of the individual localized levels in the one-dimensional system. The expression for the correlator $S_\omega(x = L)$ has the form

$$S_\omega(L) = (\pi v_F)^{-2}[L - 2l \ln (8/\nu)]^2; \qquad L > 2l \ln (8/\nu).$$

We should have more complete expressions if we left under the $\varkappa$-integration sign (in the inversion of the Laplace transformation) terms of the form

$$S_\omega(\varkappa) = \frac{l^2}{\pi\varkappa^2}\frac{\partial^2}{\partial\nu^2}\frac{\nu}{C_1(\varkappa)-1}$$
$$\approx \left(\frac{l}{\pi}\right)^2 \frac{2}{\varkappa^3} \exp\left[2(-\varkappa+\varkappa^2)\ln\frac{8}{\nu}\right]. \tag{57}$$

Going over from $\varkappa$ to L (i.e., inverting the Laplace transformation), we find

$$S_\omega(L) = \frac{1}{(\pi v_F)^2}\int_0^L dx \int_0^x dx' \left[1 + \operatorname{erf}\left(\frac{x'-x_0}{2(x_0 l)^{1/2}}\right)\right]; \; x_0 = 2l \ln\frac{8}{\nu}.$$

These formulas are not given in Refs. 8 and 9. Notice that we already know from the results of the preceding section the estimate for the magnitude of the energy interval $\omega_0(L)$ in which there is, for a fixed chain length, appreciable level respulsion:

$$\omega_0 = (8/\tau)\, e^{-L/2l}$$

(here we have gone over to dimensional units, using the gas-kinetic definitions of the time interval between collisions and the mean free path). This estimate differs somewhat from the one given in Ref. 9 by numerical factors.

4. CONCLUSION

We have obtained the density-of-states correlation function for energetically close states in a one-dimensional chain which defects at different points. The result is represented in Fig. 1 by the curve, whose asymptotic behaviors are given by the formulas (37) and (38). This result is somewhat unusual in respect of that part of it (in the region $z \sim l$) which shows that two wave functions with close energies always overlap appreciably at distances $z \sim l$. (Attraction of levels!) And what is more, the expression (38) essentially shows that the two wave functions are virtually identical in the region $z \sim l$ prior to the onset of an irregularity or a dip, which reflects the fact that two electrons with nearly equal energies will, with high probability, occupy independent levels located far from each other. This circumstance is due to the fact that localization on defects is a complicated quantum-mechanical interference phenomenon, and that the localized state is centered on many defects at the same time. Nevertheless, the main results do not contradict the intuitive ideas. Moreover, the characteristic scale $z_0 = 2\ln(8/\nu)$ of the distances over which the interlevel interaction occurs is in accord with the tunneling estimates given in Ref. 1.

Our results also confirm the results obtained in Refs. 8 and 9 in respect of the statistical independence of the levels in a long segment of a linear chain containing defects. The general expression (57) allows us to find the level statistics for either small energy differences or shorter segments (but $l \ll L$).

The authors are grateful to K. B. Efetov, A. I. Larkin, and V. L. Pokrovskiĭ for useful discussions and to S. A. Molchanov for kindly making reprints of Refs. 8 and 9 available to him.

[1]N. F. Mott and W. D. Twose, Adv. Phys. **10**, 107 (1961).
[2]V. L. Berezinskiĭ, Zh. Eksp. Teor. Fiz. **65**, 1251 (1973) [Sov. Phys. JETP **38**, 620 (1974)].
[3]A. A. Gogolin, V. I. Mel'nikov, and É. I. Rashba, Zh. Eksp. Teor. Fiz. **69**, 327 (1975) [Sov. Phys. JETP **42**, 168 (1975)].
[4]H. Schmidt, Phys. Rev. **105**, 425 (1957).
[5]B. I. Halperin, Phys. Rev. **139**, A104 (1965).
[6]L. P. Gor'kov and O. N. Dorokhov, Solid State Commun. **20**, 789 (1976); L. P. Gor'kov and O. N. Dorokhov, Fiz. Nizk. Temp. **4**, 332 (1978) [Sov. J. Low Temp. Phys. **4**, 160 (1978)].
[7]V. L. Berezinskiĭ and L. P. Gor'kov, Zh. Eksp. Teor. Fiz. **77**, 2498 (1979) [Sov. Phys. JETP **50**, 1209 (1979)].
[8]S. A. Molčanov, Commun. Math. Phys. **78**, 429 (1981).
[9]V. M. Malkin, Preprint, Institute of Nuclear Physics, Siberian Branch of the USSR Academy of Sciences, 1981.
[10]A. Erdelyi (ed.), Higher Transcendental Functions (California Institute of Technology H. Bateman MS Project), Vol. 1, McGraw-Hill, New York, 1953 (Russ. Transl., Nauka, Moscow, 1973).
[11]L. D. Landau and E. M. Lifshitz, Kvantovaya mekhanika (Quantum Mechanics), Nauka, Moscow, 1974 (Eng. Transl., Pergamon, Oxford, 1977).
[12]M. A. Naĭmark, Lineĭnye differentsial'nye operatory (Linear Differential Operators), Nauka, Moscow, 1969.
[13]I. S. Gradshteĭn and I. M. Ryzhik, Tablitsy integralov, summ, ryadov i proizvedeniĭ (Tables of Integrals, Series, and Products), Fizmatgiz, Moscow, 1962 (Eng. Transl., Academic Press, New York, 1965).
[14]E. P. Wigner, Ann. Math. **53**, 36 (1951); **62**, 548 (1955).
[15]F. Dyson, J. Math. Phys. **3**, 140,157,166 (1962).
[16]L. P. Gor'kov and G. M. Éliashberg, Zh. Eksp. Teor. Fiz. **48**, 1407 (1965) [Sov. Phys. JETP **21**, 940 (1965)].
[17]K. B. Efetov, Zh. Eksp. Teor. Fiz. **83**, 833 (1982) [Sov. Phys. JETP **56**, 467 (1982)].

Translated by A. K. Agyei

On the ac conductivity of slightly disordered one-dimensional conductors

L. P. Gor'kov, O. N. Dorokhov, and F. V. Prigara
L. D. Landau Institute of Theoretical Physics, Academy of Sciences of the USSR

(Submitted 2 March 1983)
Pis'ma Zh. Eksp. Teor. Fiz. **37**, No. 8, 377–380 (20 April 1983)

The Austin-Mott formula for the frequency dependence of the conductivity is derived for the case of one-dimensional conductors containing disorder. It is shown under what circumstances the "pair approximation" is valid in these systems. The conditions, for which the two states making up a pair are statistically independent, are considered.

PACS numbers: 72.10.Bg

The calculation of the conductivity for localized electrons (say, in semiconductors) is greatly simplified by model representations.[1] An electron is localized at separate centers, and far from the centers its wave function decays exponentially with some atomic constant a. The overlap is ignored—at low concentrations of centers. At low temperatures only the transitions between pairs of nearest centers are important, since the probability of finding two levels (μ,ν) with nearby energies at small spatial separations is generally small. If the transitions between states are due to phonons, the corresponding relaxation time is[2]

$$1/\tau_{\mu\nu} = \nu_{ph} \exp(-2z/a). \tag{1}$$

At zero temperature and for $\omega \to 0$, the absorption of energy from the external field is due to transitions between the two quantum states formed by the tunneling splitting of the levels of two "identical" wells[3]:

$$\hbar\omega = I_0 \exp(-r/a). \tag{2}$$

This expression implies the well-known Mott law for the phononless conductivity, $\sigma(\omega) \propto \omega^2 \ln^2 I_0/\omega$ (in the $1D$ case).

According to Refs. 3 and 4, a one-dimensional system containing disorder has only localized states even in the limit $k_F l \gg 1$, which would otherwise correspond to a regime of metallic conductivity. As an initial approximation here one takes the rapidly oscillating band wave functions. The density of states $\nu(E_F) = (\hbar\pi v_F)^{-1}$ is not small, and the energy scale is determined by the reciprocal $1/\tau$ of the characteristic time for (back)scattering by defects (the range is $l = v_F\tau$).

Our goal here is to study the applicability of the phenomenological representations (1) and (2) to this case. In particular, we shall obtain the Austin-Mott formula for the conductivity.[2]

The spectrum of the Schrödinger equation in a $1D$ system containing disorder consists of localized states with "unit" probability (see Ref. 5). The (Lyapunov) growth exponent of the amplitude $R(x)$ of the wave function $\psi(x) = R(x)\sin\phi(x)$ is a self-averaging quantity which is given by[6]

$$\lim_{z \to \infty} \frac{1}{z} \ln \frac{R(z)}{R(0)} = \frac{1}{2l} \cdot \tag{3}$$

Consequently, we have $a = 2l$ in (1).

The phononless contribution to the conductivity $\sigma(\omega)$ is determined by the average of the square of the dipole-moment matrix element $x_{\mu\nu}$ between two states (μ,ν):

$$\sigma(\omega) \propto < \sum_{\mu\nu} \delta(E-\epsilon_\mu)\delta(E+\omega-\epsilon_\nu)\, x^2_{\mu\nu} > .$$

The latter is expressed in terms of a density-density correlator of the two eigenfunctions:

$$F^1_\omega(z) = < \sum_{\mu\nu} \delta(E-\epsilon_\mu)\,\delta(E+\omega-\epsilon_\nu)\,\psi_\mu(0)\,\psi_\nu(0)\,\psi_\mu(z)\,\psi_\nu(z) > . \tag{4}$$

We are interested in the case $\omega\tau \ll 1$. If all the localized wave functions were independent, then, according to (3), the function $F^{(1)}_\omega(z)$ would have a peak at $z \sim l$ and be small for $z \gg l$. A calculation (which we shall not reproduce here) shows that this correlator has another peak as well:

$$F^{(1)}_\omega(z) = -\nu^2(E_F)\,(l/4\pi z_0)^{1/2} \exp[-(z-z_0)^2/4 l z_0] . \tag{5}$$

at

$$z_0 = 2l \ln(8/\omega\tau) . \tag{6}$$

The properties of $F^{(1)}_\omega(z)$ should be discussed in conjunction with the properties of the correlator of the densities of states at two different points[7]:

$$F_\omega(z) = < \sum_{\mu\nu} \delta(E-\epsilon_\mu)\,\delta(E+\omega-\epsilon_\nu)\,\psi^2_\mu(0)\,\psi^2_\nu(z) > . \tag{7}$$

The behavior of the correlators $F_\omega(z)$ and $F^{(1)}_\omega(z)$ is shown in Fig. 1 (for $\omega\tau \ll 1$) by the solid and dashed curves, respectively. For $z \to \infty$ we have $F_\omega(z) = \nu^2(E_F)$, which corresponds to statistically independent states at large distances. For $z < z_0$ a "dip" appears, demonstrating a "repulsion" of the levels. However, there is no "repulsion" at small distances $z \sim l$, where the two functions coincide (the hatched region in Fig. 1 corre-

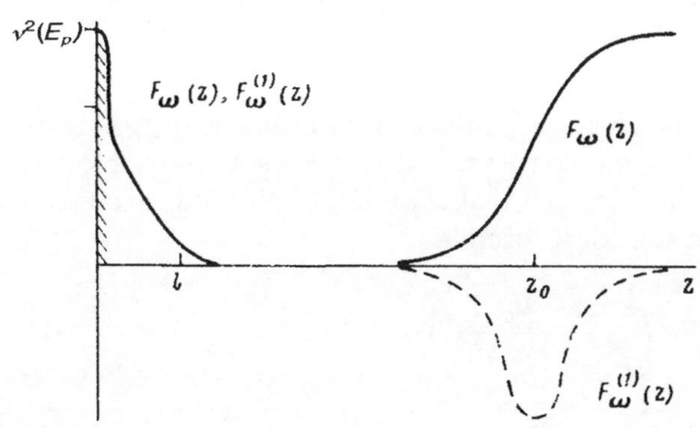

FIG. 1.

sponds to the closest distances $z \sim k_F^{-1}$). The figure indicates the presence of a strong correlation of two nearby levels for $z \lesssim z_0$.

Molcanov[8] has proved that for a very long chain ($L \to \infty$), "all" of the energy levels are statistically independent. We have evaluated the probability $W_N(L,\omega)$ that N levels appear in the energy interval ω for a chain of *finite* length L ($L < [\omega\nu(E_F)]^{-1}$):

$$W_N(L,\omega) = (N!)^{-1} [\omega \nu(E_F)(L - N z_0)]^N. \tag{8}$$

This expression contains the so-called "exceptional volume," which indicates once again that the levels do not approach each other closer than z_0. At these distances the two levels cease to be independent and form a correlated pair analogous to the two coherent states which arise on account of tunneling in the two identical potential wells.[3,9] [Ternary correlations are already missing in (8)]. The coincidence of the two correlators for $z \sim l$ indicates the possibility of a limiting transition $\epsilon_\mu - \epsilon_\nu \to 0$, i.e., the phases of these states are nearly the same. Analysis of the behavior of the phase difference from $z \sim l$ to $z \sim z_0$ shows that if the two wave functions are correlated for a given state of disorder, this corresponds to the rare case in which the phase advance at the distance z_0 is equal to π. The large value of z_0 [Eq. (6)] permits interpretation of the shape and width of the distribution of $F_\omega^{(1)}(z)$ in terms of Gaussian fluctuations of the defect concentration. These results establish the limits of applicability of the model representations of localized and independent states.

The frequency dependence of the conductivity for $T \neq 0$ is due to the Debye loss,[1,10] i.e., to the relaxation ($1/\tau_{\mu\nu}$) due to phonons of the populations of two neighboring (close in both energy and space) levels. The contribution to the dissipation from an individual pair (μ,ν), as we know,[10,11] is proportional to the factor

$$\frac{\omega^2}{\tau_{\mu\nu}} \left[\omega^2 + \frac{1}{\tau_{\mu\nu}^2} \right]^{-1}; \tag{9}$$

it follows that the important pairs are those for which $1/\tau_{\mu\nu} \sim \omega$. If the one-dimensional filaments make up a crystal, the transitions between states at low temperatures are governed by three-dimensional acoustic phonons. An estimate of the transition matrix elements by the method of Ref. 12 gives

$$\frac{1}{\tau_{\mu\nu}} = \frac{\alpha g_{ph}^2}{n_\perp l^2} T (\tau T)^2 \frac{v_F}{s} \frac{\Delta\epsilon}{T} \mathrm{cth} \frac{\Delta\epsilon}{T} \exp\left(-\frac{z - z_0}{l}\right), \tag{10}$$

where $\Delta\epsilon = \epsilon_\mu - \epsilon_\nu$, s is the speed of sound, $n_\perp$ is the density of filaments, $g_{\rm ph}^2$ is the dimensionless electron-phonon interaction constant, and α is a numerical factor. In expression (10) it is explicitly taken into account that $z > z_0$ in (6). With this expression we obtain, by averaging (9), the Austin-Mott formula:

$$\sigma(\omega) = \frac{\pi^3}{4} e^2 \nu^2 (E_F) n_\perp l^3 \omega T \ln^2 \left(\frac{T}{\hbar\omega} \frac{v_F}{s n_\perp l^2} \right). \tag{11}$$

At high frequencies there is a plateau region, for which the following microscopic expression has been found[12,13]:

$$\sigma = g_{ph}^2 \frac{\pi^2}{45} \frac{e^2}{\hbar v_F} \frac{1}{\tau} \left(\frac{v_F}{s}\right) \left[\frac{2T\tau}{\hbar} \ln\left(\frac{2T\tau}{\hbar}\right)\right]^4 . \tag{12}$$

In (10) and (11), $z - z_0$ is determined by the condition $\omega \sim 1/\tau_{\mu\nu}$. It is seen that (11) joins with (12) (i.e., $z \simeq z_0$) at frequencies

$$\omega \sim \frac{\alpha g_{ph}^2}{n_\perp l^2} \frac{T}{\hbar} \left(\frac{T\tau}{\hbar}\right)^2 \frac{v_F}{s} .$$

In conclusion, we note that in $1D$ conductors the long-range Coulomb interaction is screened: At distances shorter than l the screening is due to the displacement of the charges on neighboring filaments, while at larger distances the Coulomb term is simply small on account of the anomalously large dielectric constant (the localization effect[4]). The Coulomb effects proposed by Éfros and Shklovskiĭ[14,15] are therefore small, in contrast to the case of semiconductors. The short-range (screened) interactions, although altering the coefficient in (11), do not otherwise introduce any new frequency or temperature dependence in the conductivity.

[1]N. F. Mott and E. A. Davis, Electronic Processes in non-Crystalline Materials, Clarendon Press, Oxford, 1971.
[2]I. G. Austin and N. F. Mott, Adv. Phys. **18**, 41 (1969).
[3]N. F. Mott and W. D. Twose, Adv. Phys. **10**, 107 (1961).
[4]V. L. Berezinskiĭ, Zh. Eksp. Teor. Fiz. **65**, 1251 (1973) [Sov. Phys. JETP **38**, 620 (1974)].
[5]I. M. Lifshitz, S. A. Gredeskul, and L. A. Pastur, Vvedenie v Teoriyu Neuporyadochennykh Sistem [Introduction to the Theory of Disordered Systems], Nauka, Moscow, 1982.
[6]R. Borland, Proc. R. Soc. London Ser. A **274**, 809 (1963).
[7]L. P. Gor'kov, O. N. Dorokhov, and F. V. Prigara, Zh. Eksp. Teor. Fiz. **84**, 1440 (1983) [Sov. Phys. JETP **57** (1983)].
[8]S. A. Molcanov, Commun. Math. Phys. **78**, 429 (1981).
[9]I. M. Lifshitz, Usp. Fiz. Nauk **83**, 617 (1964) [Sov. Phys. Usp. **7**, 549 (1965)].
[10]M. Pollak and T. H. Geballe, Phys. Rev. **122**, 1742 (1961).
[11]H. Böttger and V. V. Bryksin, Phys. Status Solidi B **78**, 9, 415 (1976).
[12]V. L. Berezinskiĭ, L. P. Gor'kov, and O. N. Dorokhov, Solid State Commun. **37**, 61 (1981).
[13]O. N. Dorokhov, S. I. Matveenko, and F. V. Prigara, Fiz. Nizk. Temp. **7**, 738 (1981) [Sov. J. Low. Temp. Phys. **7**, 360 (1981)].
[14]A. L. Efros, Philos. Mag. Ser. B **43**, 829 (1981).
[15]B. I. Shklovskiĭ and A. L. Éfros, Zh. Eksp. Teor. Fiz. **81**, 406 (1981) [Sov. Phys. JETP **54**, 218 (1981)].

Translated by Steve Torstveit
Edited by S. J. Amoretty

Particle conductivity in a two-dimensional random potential

L. P. Gor'kov, A. I. Larkin, and D. E. Khmel'nitskiĭ
L.D. Landau Institute of Theoretical Physics, USSR Academy of Sciences

(Submitted 16 July 1979)
Pis'ma Zh. Eksp. Teor. Fiz. **30**, No. 4, 248–252 (20 August 1979)

It is shown that the conductivity in a two-dimensional system with disorder at low frequencies depends logarithmically on the frequency. The connection between this result and the conclusions of Ref. 1 about the total localization in the two-dimensional case is discussed.

PACS numbers: 72.10.Bg

This work was stimulated by a paper[1] in which it was asserted that all single-particle states in the defect field in the two-dimensional case are localized. We shall investigate the quantum corrections to the diffusion law or to the conductivity of an infinite system as a function of the frequency ω.

The conductivity σ can be expressed in terms of the Green's functions of the Schrödinger equation with a random potential[2]:

$$\sigma = e^2 \int dE \nu(E) \frac{n_E - n_{E+\omega}}{\omega} D(\omega) , \tag{1}$$

$$D(\omega) = \frac{1}{d} \int dr < \mathbf{v}(0) G^R_{E+\omega}(0, r) \mathbf{v}(\mathbf{r}) G^A_E(\mathbf{r}, 0) >, \tag{2}$$

where $D_E(\omega) \equiv D(\omega)$ is the diffusion coefficient of a particle with an energy E, d is the dimension of the space, G^R and G^A are the lagging and leading Green's functions, respectively, and the angle brackets $\langle \quad \rangle$ denote averaging over the possible locations of the impurities. The energy E is assumed to be large compared with the reciprocal time of the mean free path, $\hbar/\tau$. We shall use the "cross technique" of averaging[2] to calculate the average values. As is well known, for $E\tau/\hbar \gg 1$ the diagrams with crossed dashed lines (for example, the plot in Fig. 1a) are small. Disregarding these diagrams, we obtain $D = D_0 = v^2\tau/d$ (for simplicity, the scattering is assumed to be isotropic $\tau = \tau_{tr}$).

To calculate the quantum corrections, we must examine the diagrams in Fig. 1. Such diagrams were earlier examined by Langer and Neal,[3] who showed that each

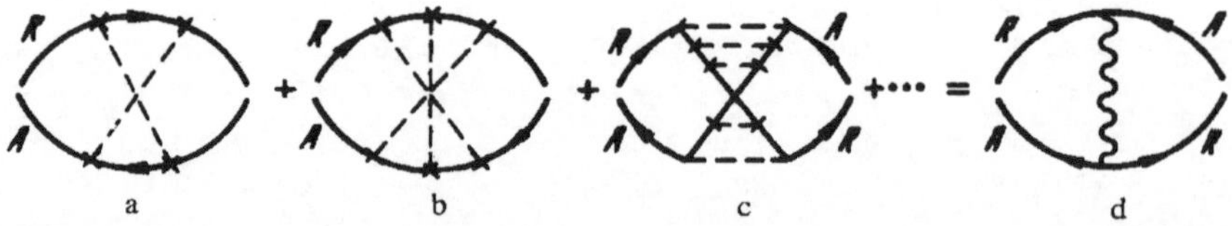

FIG. 1.

individual diagram gives a small correction to the diffusion coefficient. It is important that the *sum* of all the ladder-type diagrams, which is illustrated in Fig. 1d by a wavy line, has a characteristic diffusion pole

$$C(\mathbf{q},\omega) = \frac{1}{\tau^2}\frac{1}{Dq^2 - i\omega} - \frac{1}{\tau}; \quad \mathbf{q} = \mathbf{p} + \mathbf{p}_1, \quad qv\tau \ll 1, \quad \omega\tau \ll 1. \tag{3}$$

This peculiarity of the "fan-shaped" ladder coincides in total momentum with the well-known property of the ordinary diffusion ladder

$$\langle \rho, \rho \rangle_{\omega,\mathbf{q}} \sim \frac{1}{Dq^2 - i\omega} \tag{4}$$

as a function of the transferred momentum **q**. Such a coincidence is a consequence of the invariance with respect to time reversal as a result of transformation of the wave functions $\psi(p) \to \psi^*(-p)$. Henceforth, the wavy line in the diagrams will denote both the ordinary diffusion ladder and the fan-shaped ladder.

In diagram 1*d* we examine integration over the momenta **q** of the wavy line. The singular frequency contribution comes from the region of small **q**. Therefore, $\mathbf{p}_1 \approx -\mathbf{p}$ and the singular quantum correction to the conductivity is negative. After calculation we obtain[1)]

$$D = D_0\left\{1 - \frac{3\hbar^2\pi}{2SmE\sqrt{6\omega\tau}}\right\}, \qquad d = 1 \tag{5}$$

$$D = D_0\left\{1 - \frac{\hbar}{2\pi E\tau}\ln\frac{1}{\omega\tau}\right\}, \qquad d = 2 \tag{6}$$

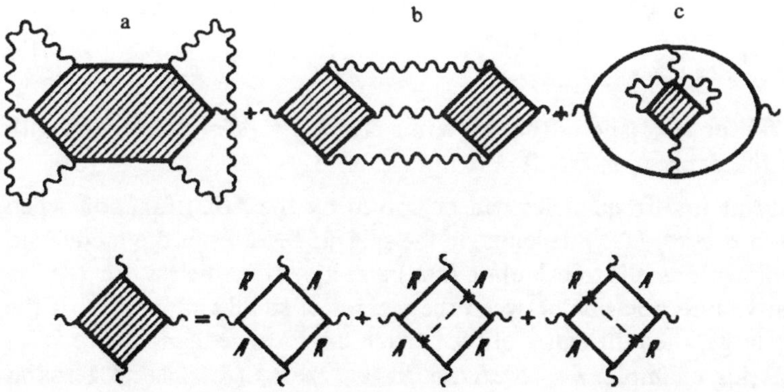

FIG. 2.

$$D = D_0 \left\{ 1 + \frac{3\hbar^2\sqrt{6\omega\tau}}{16E^2\tau^2} \right\}, \qquad d = 3 \tag{7}$$

For the one-dimensional case the localization length is of the order of the mean free path $l \sim v\tau$ and the diffusion region $ql \ll 1$ is missing. Therefore, Eq. (5) will apply to a wire of cross section $S \gg \hbar^2 p^{-2}$, for which a diffusion region exists,[6] and the second term in the braces is a small correction.

For simplicity, we have explicitly calculated the conductivity. The quantum correction for the correlator $\langle \rho\rho \rangle$ can also be determined. We find that the diffusion equation (4) remains true and the diffusion coefficient is in fact given by Eqs. (5)–(7).

Of particular interest is the two-dimensional case, in which the correction to the conductivity increases logarithmically with decreasing frequency, and hence a question arises about the summation of the set of principal logarithms of the type $[(\hbar/E\tau)\ln(1/\omega\tau)]^n$. The diagrams giving contributions of order $[(\hbar/E\tau)\ln(1/\omega\tau)]^n$. are illustrated in Fig. 2. The number of different diagrams and the complexity of their structure increase dramatically with increasing order. In each order and for any block diagram $[(\hbar/E\tau)\ln(1/\omega\tau)]^n$. coincides with the number of integrations over the momenta of the wavy lines. It was found, moreover, that in second order the diagrams in Fig. 2b cancel each other and the contribution such as the diagram in Fig. 2a is cancelled by the contribution from the diagram in Fig. 2c. In other words, the term $[(\hbar/E\tau)\ln(1/\omega\tau)]^2$. in Eq. (6) is missing.

We can assume that this result, rather than being accidental, is a consequence of a renormalized invariance.[6] In fact, for $\omega\tau \ll 1$ and $q = 0$ there are no parameters with dimensions of length or frequency, and there is only one dimensionless parameter $R_0 = 1/E\tau$, which plays the role of a bare interaction constant and coincides with the resistance of a square of the film for $\omega\tau \sim 1$, measured in units of $\pi e^{-2}\hbar \sim 13$ kΩ. Therefore, if the frequency scale is changed $\omega\tau \to \omega/\omega_1$, it can be compensated for by a corresponding change of concentration of the impurities according to the law $R_0 \to R(\omega_1)$, so that $R(\omega\tau, R_0) = R[\omega/\omega_1, R(\omega_1)]$. Differentiating this equation with respect to $\omega_1 = \omega$, we obtain an equation for the renormalized group in the form

$$\frac{dR(\omega)}{d\ln\omega} = f(R) = -\frac{R^2}{2\pi} + 0(R^3). \tag{8}$$

The solution of Eq. (8) for $R \ll 1$ shows that all terms of order $(1/E\tau)^n \ln^n(1/\omega\tau)$ with $n > 1$ are missing in the expression for $D = \hbar/mR$.

The conductivity at low frequencies can be solved by the $f(R)$ function when $R \approx 1$. In this region, however, $f(R)$ depends on the specific features of the model and to determine it is as difficult as it is to calculate $R(\omega)$ directly. Nonetheless, Eq. (8) can be used for a qualitative interpolation between the regions of small and large ω. If the disorder is extremely large, then the states of the particle are localized, and hence $f(R)$ increases when $R \gg 1$ (for example, $f = -2R$ for $R \sim \omega^{-2}$). If $f(R)$ has no zeros at finite R, then $R(\omega)$ decreases to zero as $\omega \to 0$. Thus, in the two-dimensional case all the states are localized at arbitrarily low concentrations of the impurities. If $f(R)$ vanishes at a certain point R_1, then $R(\omega) \to R_1$ as $\omega \to 0$. Generally, the vanishing of the static conductivity is not equivalent to localization, which corresponds[7] to a density correlator of the form

$$< \rho\rho >_{\omega,\mathbf{q}} = \frac{A(\mathbf{q})}{\omega} . \tag{9}$$

If, however, a single invariant charge R exists in the scale transformations, then the small R's correspond to a large disorder and hence to localization.[8]

Thus our results are very similar to those of Abrahams *et al.*,[1] although the details of their calculations have not been published. The difference is that the variable in the renormalized group of Abrahams *et al.*[1] is the sample size L rather than the frequency ω. For a sample of finite size it is important to know the method used to determine the conductivity and the frequencies in question. In a finite-size sample absorption occurs at frequencies $\omega > \hbar/mL^2$. On the other hand, if $\omega \gg DL^{-2}$, then the logarithms in the quantum corrections are independent of L. In other words, if $R \ll 1$, then there exists a frequency region $DL^{-2} \gg \omega \gg \hbar/mL^2$ for which the quantum corrections are given by Eq. (6) in with $(L/l)^2$ is substituted for $1/\omega\tau$ in the argument of the logarithm. Therefore, we can assume[1] that

$$\frac{dR}{d \ln L^{-2}} = \beta(R) = \frac{1}{2\pi} R^2 + 0(R^3). \tag{10}$$

If the sample is connected in a circuit, then the discrete levels will have a width of the order of $\gamma \sim \hbar DL^{-2}$. Therefore, the frequency can be set equal to zero. The divergence in this case is cut by γ and we can again attempt to write Eq. (10). It is important to note that the function f coincides with the function β only when $R \ll 1$, i.e., in the approximation of the principal logarithms, and when $R \gg 1$ f is different from β. Everything said about the behavior of the function f at small R evidently applies to the function β.

The conductivity in a two-dimensional disordered system was numerically modeled in Refs. 9 and 10, in which the level shift in a finite sample due to a change of the boundary conditions was calculated. This quantity, which is useful for determining the onset of localization in numerical experiments, is equal in order of magnitude to the conductivity. However, the exact relation between these quantities is not known. Our conclusions coincide better with those of Ref. 9.

In conclusion, we note that the energy-relaxation processes cut off the logarithms in Eq. (6) at an appropriate parameter $1/\tau_\epsilon$ if $\omega\tau_\epsilon \ll 1$.

[1]The importance of taking into account the diffusion poles in the calculation of the quantum corrections for the conductivity was pointed out for the first time by Maleev and Totrverg[4]; however, they did not examine the fan-shaped diagrams and hence the corrections calculated by them have a smallness ω/E that complements Eqs. (5)–(7).

[1]E. Abrahams, P.W. Anderson, D.C. Licciardello, and T.V. Ramakrishnan, Phys. Rev. Lett. **42**, 673 (1979).
[2]A.A. Abriksov, L.P. Gor'kov, and I.E. Dzyaloshinskiĭ, Metody kvantovoĭ teorii polya v statisticheskoĭ fizike (Methods of Quantum Field Theory in Statistical Physics), Fizmatgiz, 1962.
[3]J.S. Langer and T. Neal, Phys. Rev. Lett. **16**, 984 (1966).
[4]S.V. Maleev and B.P. Totrverg, Zh. Eksp. Teor. Fiz. **69**, 1440 (1975) [Sov. Phys. JETP **42**, 734 (1975)].
[5]D.J. Thouless, Phys. Rev. Lett. **39**, 1167 (1977).
[6]M. Gell-Mann and F. Low, Phys. Rev. **95**, 1300 (1954).
[7]V.L. Berezinskiĭ, Zh. Eksp. Teor. Fiz. **65**, 1251 (1973) [Sov. Phys. JETP **38**, 620 (1973)].
[8]P.W. Anderson, Phys. Rev. **109**, 1492 (1958).
[9]D.C. Licciardello and D.J. Thouless, J. Phys. **C8**, 4157 (1975); **C11**, 925 (1978).
[10]P.A. Lee, Phys. Rev. Lett. **42**, 1492 (1979).

Chapter 8

Hydrodynamics

Landau considered hydrodynamics as the fascinating branch of theoretical physics. Not everyone among his disciples would agree with him probably, but howbeit, everyone was obliged to pass hydrodynamics as one of the exams of the Landau "Theoretical Minimum".

8.1 The Bernar's convection

Paper [8a] was the first paper that I have finished after being admitted to the Landau group in 1956. I took an interest in the formal analogy between the Bernar convection and the Landau theory of the second order phase transitions. In fact, in Landau theory the free energy is expanded in powers of the order parameter near the critical temperature. Similarly, it seemed possible to expand the hydrodynamic equations near the threshold of stability of the homogenous heat flow between two parallel plates. From the physical point of view, the two problems, of course, have nothing in common because heat transfer means the dissipative process.

I have had no full access to the hydrodynamic literature of those days, but from several references to my paper, I gather that I probably was among the first ones who applied such an approach to studying weekly non-linear hydrodynamic regimes.

To simplify calculations, I modeled the onset of a Bernar convection in the heat flow between two *free* surfaces, the lower one being at a temperature higher than the upper; for such geometry I arrived to the hexagonal convection cells. The stability of the solution was not studied, although it was known that depending on the experimental conditions, the cells may have other symmetries (cells of the tetragonal symmetry, or the parallel rolls).

I had no time to think of more, for soon, Landau, Abrikosov and I were bogged down in analysis of singularities at the second order phase transitions. We have even elaborated the diagrammatic technique reinvented later by many who used it in the *scaling* approach to the problem, but, instead of calculating the critical indexes, we were trying to single out the "leading" contributions into the specific heat from fluctuations and determine the *exact form* of the singularity at the transition; understandably, our attempts led us nowhere.

8.2 Acoustic coagulation

At the end of the 50's, one group at the Acoustic Institute of the Soviet Academy of Sciences in Moscow was studying the behavior of small particles suspended in a field of standing ultrasound waves. The group probably worked on the problem of air purification for some industries plagued by the formation of dangerous combustive suspensions, such as the sugar powder of sugar-mills, or the precipitation of harmful aerosols. What they have been observing is the clustering of suspended particles into fanciful complexes of various symmetries. The head of the group, N. N. Andreev, an academician, and one from the old school, asked Landau for advice and help. As I have had the interests in hydrodynamics, Landau told me to go and see what could be done.

I derived for them the general expression for the average forces acting on a particle suspended in the acoustic field of the *arbitrary* profile [8b]. The approach could be directly extended to the forces between the particles forming their clusters. I do not know whether my paper was of consequence in the Soviet Union back in the day, for, most probably, the project was classified; recently I have discovered that the results of [8b] became popular in different biological and medical applications in which acoustic forces can be employed for manipulations by small objects, e.g. as non-invasive "acoustic micro-tweezers", or for coagulation of gas bubbles in a liquid (say, air bubbles in blood).

Few omissions and typos made in the English version in translation must be corrected as follows:

$$\text{Eq. (4): } \varphi_{sc} = \frac{a(t-r/c)}{r} - \frac{(\vec{A}(t-r/c)\cdot n)}{r^2} + \cdots \quad \text{Eq. (6): } a(t) = -\frac{R^3}{3\rho}\dot{\rho}_{in}(t);$$

$$\text{Eq. (7): } \varphi_{sc} = -\frac{R^3}{3\rho}\dot{\rho}_{in} f_1 - \frac{R^3}{2} f_2 div\left(\vec{v}_{in}\frac{1}{r}\right);$$

In Eqs. (8, 9) (under the integral):

$$v_{in}v_{in} \rightarrow v_{in}v_{sc};$$

Unnumbered, below Eq. (11):

$$\Delta\varphi_{sc} - \frac{1}{c^2}\frac{\partial^2\varphi_{sc}}{\partial t^2} = \frac{4\pi}{3\rho} f_1 R^3 \dot{\rho}_{in}\delta(\vec{r}) + 2\pi f_2 R^3 div(\vec{v}_{in}\delta(\vec{r})).$$

8.3 Running shock wave at acoustic resonance

To the two papers above I would be glad to add one more. Unfortunately, it was published only in Russian in a rather specialized engineering journal.

The work was commissioned by Kapiza. Apart from the discovery of superfluidity of *liquid He II* Petr Kapiza made many other important contributions in the field of the low temperature physics. Recall also that he was a brilliant engineer famous for the legendary Kapiza's helium liquefiers; he conducted experiments and has published several papers on hydrodynamic problems.

One day Kapiza came forward with an idea of the cooling making use of non-linear standing acoustic waves in a gas. That is, in a closed cylinder in the regime close to the acoustic resonance when the amplitude of the ultrasound wave becomes large; inside the cylinder forms a shock wave running back and forth between the piston and closed end of the cylinder. With negligible viscosity, the shock wave becomes the only mechanism of dissipation. I don't think that Kapiza had thought it over in all the technical details, but somehow his idea was to take away the heated air behind the passing shock wave with a few synchronized valves (at that, the interior of the cylinder should, of course, be thermally isolated).

I solved the hydrodynamic equations and obtained the time and the spatial profile of the non-linear oscillations. From the formulas the engineers from the Kapitza group could calculate the amplitude of the shock wave and, hence, the losses at the given amplitude of the piston oscillations. To my thinking, the solution was quite elegant. I regret not knowing what happened to Kapiza's idea, for I left the institute too soon. Howbeit, it was worth seeing that huge steel tube of ten meters long and hearing the rumble!

List of original papers

[8a] L. P. Gor'kov, *Stationary convection in a plane liquid layer near the critical heat transfer point*, Sov. Phys. JETP **6**, 31 (1958) {96}

[8b] L. P. Gor' kov, *On the force acting on a small particle in an acoustic field in an ideal liquid*, Sov. Phys. Doklady **6**, 773 (1962) {431}

SOVIET PHYSICS JETP VOLUME 6(33) NUMBER 2 FEBRUARY, 1958

STATIONARY CONVECTION IN A PLANE LIQUID LAYER NEAR THE CRITICAL HEAT TRANSFER POINT

L. P. GOR'KOV

Institute of Physical Problems, Academy of Sciences, U.S.S.R.

Submitted to JETP editor February 5, 1957

J. Exptl. Theoret. Phys. (U.S.S.R.) 33, 402-407 (August, 1957)

Thermal convection which arises above the limit of stability is investigated for a plane liquid layer. A dependence of the amplitude of motion on a parameter which characterizes the departure from critical conditions has been obtained. Possible symmetry of the flow is discussed.

---

*An analogous situation can in principle arise for the scattering of electrons by heavy atoms. Here an $\sim E^{2/3}$ dependence for the current is also obtained, but with different numerical coefficients; this is understandable, since A_2 is different for ions. In this case, however, the electrons have sufficient energy for ionization, which must be taken into account.

HEAT transfer in a plane liquid layer with constant temperature gradient was studied experimentally by Benard,[1] who showed that for certain values of a downward-directed temperature gradient pure thermal conduction is replaced by convection with certain characteristics in a stationary mode; the flow is divided among regular vertical hexagonal prisms and thus possesses hexagonal symmetry. The critical gradient at which convection occurs has been calculated in several papers (see Ref. 2), but the utilization of linearized hydrodynamic equations permitted the determination of only the magnitude of the periods. The determination of the symmetry and amplitude of the motion for supercritical gradients requires study of the nonlinear equations. The present paper presents the results of such an investigation for modes close to the critical value.

A plane horizontal liquid layer of thickness h is bounded above and below by two planes between which a constant temperature difference is maintained. The boundary conditions can be taken in three forms: (a) two rigid planes, (b) one rigid plane and one free surface, and (c) two free surfaces. The last case is distinguished by the simplicity of its equations. Benard's experiments were performed under conditions (b). We shall write the basic equations of stationary convection as follows (see Ref. 3):

$$(\mathbf{v}\nabla)\,\mathbf{v} = -\frac{1}{\rho}\frac{\partial p}{\partial x} - \beta \mathbf{g}\theta + \nu\Delta\mathbf{v}, \quad \operatorname{div}\mathbf{v} = 0, \quad (\mathbf{v}\nabla\theta) - Av_z = \chi\Delta\theta. \tag{1}$$

Here p and θ are small deviations, due to convection, from the equilibrium pressure P and temperature T:

$$P = p_0 - \rho g z + p; \quad T = -Az + T_0 + \theta,$$

v is the velocity of the motion; $\beta = -\rho^{-1}(\partial\rho/\partial T)_p$. The boundary conditions at the rigid surface are:

$$\theta = 0, \quad v_z = 0; \quad \partial v_z/\partial z = 0 \quad (v_x = v_y = 0). \tag{1a}$$

and at the free surface:

$$\theta = 0; \quad v_z = 0; \quad \partial^2 v_z/\partial z^2 = 0 \quad (\sigma_{xz} = \sigma_{yz} = 0). \tag{1b}$$

For convenience we shall retain the vector notation only for variables in the horizontal plane $\mathbf{x} = (x, y)$, separating the z coordinate; for all quantities we shall employ a Fourier expansion in these variables:

$$\mathbf{v}(x, y, z) = \sum_k \mathbf{v}_k(z)\, e^{i\mathbf{k}\mathbf{x}} \tag{2}$$

and similarly for v_z, p and θ. By virtue of the boundary conditions $v_{0z} = 0$. The velocity $\mathbf{v}_k(z)$ is resolved into components which are parallel and perpendicular to the wave vector **k**:

$$\mathbf{v}_k = \tilde{\mathbf{v}}_k + \mathbf{k}\,(\mathbf{k}\mathbf{v}_k)/k^2.$$

The second and third equations of (1) become

$$i\,(\mathbf{k}\mathbf{v}_k) + dv_{kz}/dz = 0, \tag{3}$$

$$\chi\left(\frac{d^2}{dz^2} - k^2\right)\theta_k + Av_{kz} = \sum_{k'+k''=k}\left\{i\,(\mathbf{k}\mathbf{v}_{k'})\,\theta_{k''} + \frac{d}{dz}(v_{k'z}\theta_{k''})\right\}. \tag{4}$$

Eliminating the pressure in the first equation,

$$\frac{1}{\rho}p_k = \frac{\nu}{k^2}\left(\frac{d^2}{dz^2} - k^2\right)\frac{dv_{kz}}{dz} + \frac{i}{k^2}\sum_{k'+k''=k}\left\{i\,(\mathbf{k}\mathbf{v}_{k'})(\mathbf{k}\mathbf{v}_{k''}) + \frac{d}{dz}(v_{k'z}(\mathbf{k}\mathbf{v}_{k''}))\right\},$$

we also obtain

$$-\nu\left(\frac{d^2}{dz^2} - k^2\right)^2 v_{kz} + k^2\beta g\theta_k = \sum_{k'+k''=k}\left\{k^2\left[i\,(\mathbf{k}\mathbf{v}_{k'})\,v_{k''z} + \frac{d}{dz}(v_{k'z}v_{k''z})\right] + i\frac{d}{dz}\left[i\,(\mathbf{k}\mathbf{v}_{k'})(\mathbf{k}\mathbf{v}_{k''}) + \frac{d}{dz}(v_{k'z}(\mathbf{k}\mathbf{v}_{k''}))\right]\right\}, \tag{5}$$

$$\nu\left(\frac{d^2}{dz^2} - k^2\right)\tilde{\mathbf{v}}_k = \sum_{k'+k''=k}\left\{i\,(\mathbf{k}\mathbf{v}_{k'})\left[\mathbf{v}_{k''} - \frac{\mathbf{k}\,(\mathbf{k}\mathbf{v}_{k''})}{k^2}\right] + \frac{d}{dz}\left(v_{k'z}\left[\mathbf{v}_{k''} - \frac{\mathbf{k}\,(\mathbf{k}\mathbf{v}_{k''})}{k^2}\right]\right)\right\}. \tag{6}$$

Finally, we give the equations for the Fourier components with $k = 0$:

$$\chi \frac{d^2\theta_0}{dz^2} = \frac{d}{dz}\sum_{k'}(v_{k'z}\theta_{-k'});\ \frac{d}{dz}\left\{\nu\frac{d\mathbf{v}_0}{dz} - \sum_{k'} v_{k'z}\mathbf{v}_{-k'}\right\} = 0;\quad \frac{1}{\rho}\frac{dp_0}{dz} = \beta g\theta_0 - \frac{d}{dz}\sum_{k'}(v_{k'z}v_{-k'z}).$$

Before proceeding to study the equations which have been derived we shall determine the vectors $\mathbf{k}$ of the Fourier series (2). As was shown by the investigation of stability in Ref. 2, when the critical temperature gradient is attained there arises convective motion of the form $f(z)e^{i\mathbf{kx}}$ with a certain value $k = \kappa_0$ which depends on the experimental conditions (a, b or c). In (2), and similarly in the expansions for v_z and θ, we shall call $v_{\mathbf{k}z}$ with $k = \kappa_0$ the principal term. It is easy to determine all such vectors of the flow. Indeed, as can be seen from (3)–(6), corrections to the principal terms with $k = \kappa_0$ will consist of terms with wave vectors which are again equal in magnitude to κ_0. Therefore, if (2) contains a term with the wave vector $\kappa(|\kappa| = \kappa_0)$ there must also be terms with the vectors $\mathbf{a}$ and $\mathbf{b}$ ($a = b = \kappa_0$), thus forming the equilateral triangle of Fig. a. It is clear that there are altogether six such vectors $\kappa, \mathbf{a}, \mathbf{b}, \tilde{\kappa}, \tilde{\mathbf{a}}, \tilde{\mathbf{b}}$, which are represented together in Fig. b. The Fourier components which are thus designated, for all quantities, comprise the set of principal terms of the solution; the remaining terms correspond to wave vectors representing various sums of the six principal vectors and will be small for temperature gradients near the critical value.

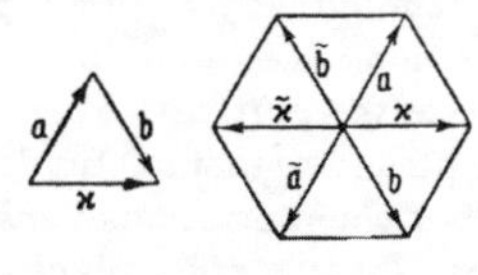

a b

We shall write the principal terms in the form

$$v_{iz} = X_i f(z);\quad \theta_i = X_i\varphi(z),$$

where[2]

$$\{(d^2/dz^2 - \varkappa_0^2)^3 + (A\beta g\varkappa_0^2/\chi\nu)\}f(z) = 0;\ \varphi(z) = (\nu/\varkappa_0^2\beta g)(d^2/dz^2 - \varkappa_0^2)^2 f(z). \tag{7}$$

It will be sufficient hereinafter to confine ourselves to terms of the third order in X, i.e., to take into account all triple combinations of the principal vectors. This means that all quantities $v_{\mathbf{k}z}$ and $\theta_{\mathbf{k}}$ can be obtained up to terms of the second order in X; in other words, in addition to the corrections $\delta v_{\kappa z}$ and $\delta\theta_\kappa$ (and similarly for $\mathbf{a}$, $\mathbf{b}$ etc.) to the expressions in (7), it is necessary to obtain the Fourier components for the velocities and temperature corresponding to wave vectors each of which is a sum of two principal vectors; for example $\mathbf{f} = \boldsymbol{\kappa} + \mathbf{b}$, $2\boldsymbol{\kappa}$, o etc. We note that by virtue of the equation of continuity (3) expressions of the form $(\mathbf{qv}_l)$ in (4) and (5) can be rewritten as

$$(\mathbf{qv}_l) = (\mathbf{q}\tilde{\mathbf{v}}_l) + i(\mathbf{ql})\,l^{-2}\,dv_{lz}/dz.$$

From (6) it follows, to terms of the second order, that

$$\nu\left(\frac{d^2}{dz^2} - l^2\right)(\mathbf{q}\tilde{\mathbf{v}}_l) = i\sum_{k'+k''=l}\left\{\left[\frac{(\mathbf{qk''})}{k''^2} - \frac{(\mathbf{ql})(\mathbf{lk''})}{l^2k''^2}\right]\left[\frac{d}{dz}\left(v_{k'z}\frac{dv_{k''z}}{dz}\right) - \frac{(\mathbf{lk'})}{k'^2}\frac{dv_{k'z}}{dz}\frac{dv_{k''z}}{dz}\right]\right\}.$$

Substituting the expressions in (7) we see that for all required $\mathbf{l}$ and $\mathbf{q}$ the right-hand side vanishes. Second-order terms in $(\mathbf{q}\tilde{\mathbf{v}}_l)$ vanish so that in (4) and (5) we can write

$$(\mathbf{qv}_l) \to i(\mathbf{ql})\,l^{-2}\,dv_{lz}/dz.$$

For brevity we introduce the notation

$$D_\varkappa = (d^2/dz^2 - \varkappa_0^2).$$

From (4) and (5), including third-order terms in X, we obtain

$$D_k^3\theta_k + \frac{k^2 A\beta g}{\chi\nu}\theta_k = \frac{1}{\chi}\sum_{k'+k''=k} D_k^2\left\{\frac{d}{dz}(v_{k'z}\theta_{k''}) - \frac{(\mathbf{kk'})}{k'^2}\frac{dv_{k'z}}{dz}\theta_{k''}\right\} + \frac{A}{\chi\nu}\sum_{k'+k''=k}\left\{k^2\left[\frac{d}{dz}(v_{k'z}v_{k''z}) - \frac{(\mathbf{kk'})}{k'^2}\frac{dv_{k'z}}{dz}v_{k''z}\right]\right.$$
$$\left. + \frac{(\mathbf{kk''})}{k''^2}\frac{d}{dz}\left[\frac{(\mathbf{kk'})}{k'^2}\frac{dv_{k'z}}{dz}\frac{dv_{k''z}}{dz} - \frac{d}{dz}\left(v_{k'z}\frac{dv_{k''z}}{dz}\right)\right]\right\}. \tag{8}$$

The magnitude of the dimensionless parameter $\gamma = A\beta gh^4/\chi\nu$ determines the possibility of the existence of stationary convective motion. The critical value γ_0 at which instability arises[2] is for two rigid surfaces ≈ 1708, for one free surface ≈ 1100 and for two free surfaces ≈ 657. We shall write (8) for the κ component as follows:

$$D_\varkappa^3\theta_\varkappa + \gamma_0\theta_\varkappa = -\Delta\gamma\theta_\varkappa + R_\varkappa(\theta^3)$$

where $R_\kappa(\theta^3)$ is the right-hand side of (8) and $\Delta\gamma = \gamma - \gamma_0$. The homogeneous equation has the solution $\varphi(z)$, with γ_0 possessing the same meaning. The condition

$$\int f\{R_\varkappa(\theta^3) - \Delta\gamma\theta_\varkappa\}\,dz = 0, \tag{9}$$

which is required for the solution of (8), leads to the determination of the amplitude of motion X. This condition has the form

$$\frac{\varkappa_0^2\Delta\gamma}{h^4}X_\varkappa\int f\varphi dz = \frac{1}{\chi}\int(D_\varkappa^2 f)\sum_{k'+k''=\varkappa}\left\{\frac{d}{dz}(v_{k'z}\theta_{k''}) - \frac{(\varkappa\mathbf{k}')}{k'^2}\frac{dv_{k'z}}{dz}\theta_{k''}\right\}dz + \frac{A\varkappa_0^2}{\chi\nu}\int f\sum_{k'+k''=\varkappa}\left\{\frac{d}{dz}(v_{k'z}v_{k''z}) - \frac{(\varkappa\mathbf{k}')}{k'^2}\frac{dv_{k'z}}{dz}v_{k''z}\right\}dz$$

$$+\frac{A}{\chi\nu}\int\frac{df}{dz}\sum_{k'+k''=\varkappa}\left\{\frac{(\mathbf{kk}'')}{k''^2}\left[\frac{d}{dz}\left(v_{k'z}\frac{dv_{k''z}}{dz}\right) - \frac{(\mathbf{kk}')}{k'^2}\frac{dv_{k'z}}{dz}\frac{dv_{k''z}}{dz}\right]\right\}dz.$$

Second-order terms in X drop out, as is shown by substitution of (7) into the right-hand side;* for the purpose of calculating third-order terms it is necessary to obtain the corrections to v_z and θ. Such corrections appear for the following quantities (see Fig. b):

$$\delta\theta_a = X_{\tilde{b}}X_\varkappa\xi_1(z);\ \delta\theta_b = X_{\tilde{a}}X_\varkappa\xi_1(z);\ \delta\theta_f = X_\varkappa X_b\xi_2(z);\quad \delta\theta_g = X_aX_\varkappa\xi_2(z);\ \delta\theta_{2\varkappa} = X_\varkappa^2\xi_3(z);$$

$$\delta\theta_0 = (|X_a|^2 + |X_a|^2 + |X_\varkappa|^2)\,\xi_4(z).\ \ \delta v_{az} = X_{\tilde{b}}X_\varkappa\eta_1(z);\ \delta v_{bz} = X_{\tilde{a}}X_\varkappa\eta_1(z);\ \delta v_{fz} = X_\varkappa X_b\eta_2(z);$$

$$\delta v_{gz} = X_\varkappa X_a\eta_2(z);\ \delta v_{2\varkappa} = X_\varkappa^2\eta_3(z).$$

Using these designations (9) becomes

$$\frac{\chi\nu\Delta\gamma}{\beta g\hbar^4}\int(fD^2f)\,dz = (|X_a|^2 + |X_b|^2)\int\left\{\frac{A}{2\nu}[Df'f(\eta_1+\eta_2) + 2f'Df\eta_1] - fD^2f'(\xi_1+\xi_2+\xi_4) - \frac{D^2f}{2}f'(\xi_1-\xi_2+2\xi_4)\right\}dz$$

$$+|X_\varkappa|^2\int\left\{\frac{A}{2\nu}[f'''f - f''f']\,\eta_3 + \left[\xi_3(D^2ff' - D^2f'f) - \xi_4\frac{d}{dz}(fD^2f)\right]\right\}dz$$

where primes denote differentiation with respect to z. Two other equations are then obtained directly by the substitutions: $\kappa\to a$, $a\to b$, $b\to\kappa$, or $\kappa\to b$, $a\to\kappa$, $b\to a$.

The solution of these equations is obviously $|X_\kappa| = |X_a| = |X_b|$; the amplitude of convective motion is then proportional to the square root of the parameter Δ_γ which characterizes the supercritical heat transfer mode. But our equations contain only the absolute value of X so that the phase relation cannot be determined. It is clear how this has come about: when we write (9) for θ_κ, say, we must select in R all combinations corresponding to any three principal vectors whose vector sum is x. It is easily seen from the figure that any such combination must be of the form $X_\kappa|X_i|^2$ so that X_κ is cancelled and (10) contains only the square of the modulus of X. Fourth-order terms in X in (9) obviously drop out. In the fifth order it is possible to obtain a combination which is not proportional to X_κ: the sum $2\mathbf{a} + 2\mathbf{b} + \boldsymbol{\kappa}$ in the right-hand side gives a term which is proportional to $X_\kappa \sim X_a^2X_b^2$. Writing $X = |X|e^{i\delta}$, we obtain because of the reality of the coefficients the following phase relation:

$$\delta_a + \delta_b - \delta_\varkappa = 0,\ \pi/2.$$

In the first case it is possible to make all phases equal to zero by shifting the coordinate origin. The principal term in the solution becomes

$$v_z = B\left(\frac{\chi}{\nu}\right)f(z)\sqrt{\frac{\Delta\gamma}{\gamma}}\left\{\cos\varkappa_0 y + \cos\frac{\varkappa_0}{2}(y+\sqrt{3}x) + \cos\frac{\varkappa_0}{2}(y-\sqrt{3}x)\right\} \tag{11a}$$

where the y axis is parallel to κ. This solution possesses hexagonal symmetry; in the (x, y) plane it represents a periodic structure of regular hexagonal prisms of the length $4\pi/\kappa_0$ along each side of the base; the liquid flows upward at the center of each prism and downward along the walls.[2]

In the second case, with a suitable choice of the coordinate origin, the solution can be written as follows:

$$v_z = B\left(\frac{\chi}{\nu}\right)f(z)\sqrt{\frac{\Delta\gamma}{\gamma}}\left\{\sin\frac{\varkappa_0}{2}(y+\sqrt{3}x) + \sin\frac{\varkappa_0}{2}(y-\sqrt{3}x) - \sin\varkappa_0 y\right\}. \tag{11b}$$

This flow permits three-fold rotations around the vertical axis (the z axis) and is symmetric with respect to the x axis, but changes sign with the substitution $y\to -y$. In the horizontal plane it breaks up

*For symmetrical boundary conditions this is clear from simple considerations, since the solution of f(z) is symmetrical with respect to z = 0.

into alternate equilateral triangles of side length $4\pi/\sqrt{3}\kappa_0$, in each of which the liquid flows upward or downward, respectively.

A decision as to which of the two types of flow will occur in reality could be reached theoretically only by investigating fifth-order terms in X in the basic equations of convection, but in actuality it would be difficult to do this. For a liquid layer on a rigid plate experiment[1] favors the first type of flow.

The equations for the ξ functions are

$$\left[D^3+\frac{\varkappa_0^2\gamma}{h^4}\right]\xi_1=\frac{2\nu}{\varkappa_0^2\beta g\chi}\left\{D^2\left(fD^2f'+\frac{f'}{2}D^2f\right)-\frac{A\beta g\varkappa_0^2}{\nu^2}\left(fDf'+\frac{f'}{2}Df\right)\right\};$$

$$\left[(D-2\varkappa_0^2)^3+\frac{3\varkappa_0^2\gamma}{h^4}\right]\xi_2=\frac{2\nu}{\varkappa_0^2\beta g\chi}\left\{(D-2\varkappa_0^2)^2\left(fD^2f'-\frac{f'}{2}D^2f\right)-\frac{3A\beta g\varkappa_0^2}{2\nu^2}fDf'\right\}; \tag{12}$$

$$\left[(D-3\varkappa_0^2)^3+\frac{4\varkappa_0^2\gamma}{h^4}\right]\xi_3=\frac{\nu}{\varkappa_0^2\beta g\chi}\left\{(D-3\varkappa_0^2)^2(fD^2f'-f'D^2f)+\frac{2A\beta g\varkappa_0^2}{\nu}(f'f''-ff''')\right\};\quad d^2\xi_4/dz^2=(2\nu/\varkappa_0^2\beta g\chi)\,d(fD^2f)/dz,$$

and the expressions for the η functions are

$$A\eta_1=-\chi D\xi_1+2(f\varphi'+{}^1/_2\,\varphi f');\quad A\eta_2=-\chi(D-2\varkappa_0^2)\xi_2+2(f\varphi'-{}^1/_2\,\varphi f');\quad A\eta_3=-\chi(D-3\varkappa_0^2)\xi_3+(f\varphi'-f'\varphi). \tag{13}$$

The calculation of these quantities subject to the boundary conditions 1a and 1b would be very laborious for the general case. The problem becomes simpler only in the somewhat hypothetical case of a liquid with two free surfaces. We present the results in so far as they are of a general character.

In this case the critical value is $\gamma_0 = 27\pi^4/4$, $\kappa_0^2 = \pi^2/2h^2$ and $f(z)$, which determines the dependence of the vertical velocity component on height, can be taken in the form

$$f(z)=\cos\frac{\pi z}{h}\left(\text{or}\quad \varphi(z)=\frac{9\nu\pi^2}{2\beta gh^2}\cos\frac{\pi z}{h}\right).$$

Equations (12) and (13) can be solved simply and give for ξ and η

$$\left.\begin{aligned}\xi_1&=\frac{9\nu\pi[9+\chi/\nu]}{52\chi\beta gh}\\ \xi_2&=\frac{9\nu\pi[121+27\chi/\nu]}{2500\chi\beta gh}\\ \xi_4&=\frac{9\nu\pi}{4\chi\beta gh}\end{aligned}\right\}\times\sin\frac{2\pi z}{h},\quad \left.\begin{aligned}\eta_1&=\frac{h[1+3\chi/\nu]}{26\pi\chi}\\ \eta_2&=\frac{9h[3+11\chi/\nu]}{1250\pi\chi}\end{aligned}\right\}\times\sin\frac{2\pi z}{h},$$

$$\xi_3=0.$$

By substituting these expressions in (10) we can determine X. For example, the temperature distribution for the first type of flow is given by

$$\theta(xyz)=\frac{4Ah}{3\pi}\sqrt{\frac{\Delta\gamma}{\gamma}}\frac{\cos(\pi z/h)}{\sqrt{1.35+0.08\chi/\nu+0.14\chi^2/\nu^2}}\left\{\cos\frac{\pi y}{2h}+\cos\frac{\pi}{4h}(y+\sqrt{3}x)+\cos\frac{\pi}{4h}(y-\sqrt{3}x)\right\} \tag{14}$$

The development of convection is conveniently characterized by the ratio of the maximum temperature change in the (x, y) plane, $T_{max} - T_{min}$, to the vertical temperature difference $T_1 - T_2 = Ah$:

$$\frac{T_{max}-T_{min}}{T_1-T_2}=\frac{4}{\pi}\sqrt{\frac{\Delta\gamma}{\gamma}}\left(1.35+0.08\,\frac{\chi}{\nu}+0.14\,\frac{\chi^2}{\nu^2}\right)^{-1/2}.$$

We note in conclusion that the form of the dependence of the coefficient in (14) on the Prandtl number is the same for any boundary conditions. This results from the fact that the solution of the linear equation $f(z) \equiv f(z/h)$ depends only on the critical value γ_0, and also on how (12), (13) and (10) contain the different dimensional quantities. Thus the strength of stationary thermal convection in a plane liquid layer for modes near the critical point is proportional to the square root of the parameter which characterizes supercriticality, whereas dependence on the Prandtl number appears in a denominator as the square root of a polynomial of not higher than the second degree. The flow belongs to one of the two types of symmetry represented by (11a) and (11b).

The author is indebted to Academician L. D. Landau for valuable suggestions and assistance.

[1]H. Benard, Ann. chim. et phys. **23**, 62 (1901)

[2]A. Pellew and R. Southwell, Proc. Roy. Soc. (London) **A176**, 312 (1940)

[3]L. D. Landau and E. M. Lifshitz, Механика сплошных сред (The Mechanics of Continuous Media) Gostekhizdat, 1954.

Translated by I. Emin

SOVIET PHYSICS – DOKLADY VOLUME 6, NUMBER 9 MARCH, 1962

PHYSICS

ON THE FORCES ACTING ON A SMALL PARTICLE IN AN ACOUSTICAL FIELD IN AN IDEAL FLUID

L. P. Gor'kov

S. I. Vavilov Institute of Physical Problems, Academy of Sciences, USSR

(Presented by Academician L. D. Landau, April 28, 1961)
Translated from Doklady Akademii Nauk SSSR, Vol. 140, No. 1, pp. 88-91, September, 1961
Original article submitted April 22, 1961

When a particle is suspended in the field of a sound wave, the fluid exerts hydrodynamical forces on it. In the linear approximation, these forces are proportional to the velocity of the fluid [1] and, on the average, do not lead to a displacement of the particle. In problems of acoustical coagulation, an important part is played by average forces acting on the particle which arise as the result of second order effects. These forces are of a nature related to that of the radiation-pressure forces in a sound wave [2, 3]. The magnitude of the forces has been found only for particles which are in an ideal fluid, in a paper by King [2]. The method of [2] was to solve exactly the problem of the flow around a small sphere in the field of a sound wave. So far as is known to us, no account was taken of effects associated with viscosity and thermal conductivity.

We present here a simple method which allows us to determine the magnitude of the average forces that act on the particle in an arbitrary acoustical field when the size of the particle is much smaller than the wavelength of the sound. In this paper we shall treat the case of an ideal fluid.

The magnitude of the force is equal to the average flux of momentum through any closed surface in the fluid which encloses the particle:

$$F_i = -\oint \overline{\Pi_{ik}} df_k.$$

Here Π_{ik} is the momentum flux density tensor in the ideal fluid: $\Pi_{ik} = p\delta_{ik} + \rho v_i v_k$. We shall take as the surface a sphere with a radius much larger than λ, the wavelength of the sound. In the linear approximation the velocity potential φ is the sum of waves incident on the particle and scattered by it:

$$\varphi(\mathbf{r}, t) = \varphi_{in}(\mathbf{r}, t) + \varphi_{sc}(\mathbf{r}, t). \qquad (1)$$

At the same time, we have from Euler's equations ($p = p_0 + p'$)

$$p' = -\rho\frac{\partial\varphi}{\partial t} - \rho\frac{v^2}{2} + \frac{\rho}{2c^2}\left(\frac{\partial\varphi}{\partial t}\right)^2$$

Since $\overline{(\partial\varphi/\partial t)} \equiv 0$, we get for the average force

$$F_i = -\oint\left\{\left[-\rho\frac{\overline{v^2}}{2} + \frac{\rho}{2c^2}\overline{\left(\frac{\partial\varphi}{\partial t}\right)^2}\right]\delta_{ik} + \rho\overline{v_i v_k}\right\} df_k.$$

Thus, for the calculation of the average force correct up to terms of the second order in the velocity, it is sufficient to find the solution of the linear scattering problem.

In the wave zone the potential of the scattered wave can be expanded in terms of multipoles:

$$\varphi_P = \frac{a(t - r/c)}{r} - \frac{(\dot{\mathbf{A}}(t - r/c), \mathbf{n})}{rc} + \ldots \qquad (4)$$

For particles of size $R \ll \lambda$, the coefficients in Eq. (4) are in turn determined from the solution of the problems of the flow of an incompressible fluid around the body (cf. [1]). In fact, for $r \ll \lambda$ the wave equation goes over into the equation of an incompressible liquid, $\Delta\varphi_{sc} = 0$, and the decreasing solutions of this equation are:

$$\varphi_{sc} = \frac{a(t)}{r} - \frac{(\mathbf{A}(t), \mathbf{n})}{r^2} + \ldots \qquad (4')$$

For the potential flow of an ideal fluid around a stationary sphere the second term is of the form:

$$-(\mathbf{v}_{in}(t), \mathbf{n})\frac{R^3}{2r^2}. \qquad (5)$$

where $\mathbf{v}_{in}(t)$ is the velocity in the incident wave at the place where the sphere is located. The first term corresponds to the "ejection" of mass owing to the compression of the gas in the incident wave, and is given by

$$a(t) = -\frac{R^3}{3\rho}\rho_{in}(t). \qquad (6)$$

The expressions (5) and (6) relate only to a stationary rigid sphere. In the general case, the sphere is partially entrained by the fluid and acquires a velocity $\mathbf{u}(t)$:

$$\mathbf{u}(t) = \frac{3\rho}{\rho + 2\rho_0}\mathbf{v}_{in}(t),$$

where ρ_0 is the density of the sphere. Therefore, if ρ_0 is comparable with the density of the fluid we must insert in Eq. (5) the relative velocity $\mathbf{v}_{in} - \mathbf{u}$.

If one takes into account compressibility of the sphere itself, one gets instead of Eq. (6),

$$a(t) = -\frac{R^3}{3\rho}\dot{\rho}_{in}(t)\left(1-\frac{c^2\rho}{c_0^2\rho_0}\right)$$

Collecting the results, we find for the potential (4) of the waves scattered by the sphere the expression

$$\varphi_p = -\frac{R^3}{3\rho r}\dot{\rho}_{in} f_1 - \frac{R^3}{2} f_2 \operatorname{div}\left(\mathbf{v}_{in}\frac{1}{r}\right) \qquad (7)$$

where

$$f_1 = 1 - c^2\rho/c_0^2\rho_0, \quad f_2 = 2(\rho_0 - \rho)/(2\rho_0 + \rho).$$

As was shown in [2], the magnitude of the average force in a standing wave is larger than in a plane running wave. In the former case, in the quadratic expression (3) for the force there are also important contributions from the interference terms between the incident and scattered waves, whereas for a running wave the magnitude of the momentum imparted to the particle by the waves is determined only by the momentum carried away by the scattered wave. In fact, by using the relations between the quantities in a plane running wave, we can rewrite Eq. (3) in the following form:

$$F_x = -\rho \oint [\overline{v_{in} v_{in}}(1+\cos\theta) + \overline{v_{sc}^2}\cos\theta]\, df \qquad (8)$$

(the $\underline{x}$ axis is taken in the direction of propagation of the incident wave, and v_{sc} is the projection of the velocity of the scattered waves in the direction of the radius of the sphere). On the other hand, in the absence of dissipation, the average flux of energy through the closed surface is zero. Expanding the expression for the energy flux $\mathbf{Q} = \rho\mathbf{v}(v^2/2 + w)$ [1] to terms of second order, we find

$$\oint \mathbf{Q}\, d\mathbf{f} = \rho c \oint [\overline{v_{in} v_{in}}(1+\cos\theta) + \overline{v_{sc}^2}]\, df \equiv 0. \qquad (9)$$

From this we have

$$F_x = \rho \oint \overline{v_{sc}^2}(1-\cos\theta)\, df. \qquad (8')$$

From a monochromatic wave we get

$$F = \frac{4\pi I}{9c} R^2 (kR)^4 \left[f_1^2 + f_1 f_2 + \frac{3}{4} f_2^2\right], \qquad (10)$$

where $I = \rho c u_0^2/2$ is the average energy flux density in the incident waves.

The formula (8') is valid only in a plane running wave. In an arbitrary acoustical field the main contribution to the expression (3) for the force comes from interference terms between the incident and scattered radiation, which lead to forces much larger than that given by Eq. (10). Confining ourselves to these terms in Eq. (3), we find

$$\mathbf{F}_i = -\oint \left\{\left[-\rho\,(\overline{\mathbf{v}_{sc}\mathbf{v}_{in}}) + \frac{c^2}{\rho}\overline{\rho'_{in}\rho'_{sc}}\right]\delta_{ik} + \rho\,(\overline{v_{isc} v_{kin}} + \overline{v_{in} v_{ksc}})\right\} df_k.$$

Changing to a volume integral, and using the Euler equations and the fact that the flow is a potential flow, we get

$$\mathbf{F} = -\int_V \overline{\mathbf{v}_{in}\left(\rho \operatorname{div}\mathbf{v}_{sc} + \frac{\partial\rho_{sc}}{\partial t}\right)}\, dV$$

$$= -\rho\int_V \overline{\mathbf{v}_{in}\left(\Delta\varphi_{sc} - \frac{1}{c^2}\frac{\partial^2\varphi_{sc}}{\partial t^2}\right)}\, dV; \qquad (11)$$

but, according to Eq. (7):

$$\Delta\varphi_{sc} - \frac{1}{c^2}\frac{\partial^2\varphi_p}{\partial t^2} = \frac{4\pi}{3\rho} f_1 R^3 \dot{\rho}_{in}\,\delta(\mathbf{r}) + 2\pi f_2 R^3 \operatorname{div}(\mathbf{v}_{in}\delta(\mathbf{r})).$$

Substituting this in Eq. (11) and introducing the quantity U (r), the potential of the forces $\mathbf{F} = -\nabla U$ we get

$$U = 2\pi R^3\rho\left\{\frac{\overline{p'^2_{in}}}{3\rho^2c^2} f_1 - \frac{\overline{v^2_{in}}}{2} f_2\right\} \qquad (12)$$

($\overline{p'^2_{in}}$ and $\overline{v^2_{in}}$ are the mean square fluctuations of the pressure and velocity in the wave at the point where the particle is located). The force in a standing wave $\varphi_{in} = -(u_0/k)\cos\omega t\cos kx$:

$$F = 4\pi\bar{E}R^2(kR)\sin 2kx\left\{\frac{\rho_0 + {}^2/_3(\rho_0 - \rho)}{2\rho_0 + \rho} - \frac{1}{3}\frac{c^2\rho}{c_0^2\rho_0}\right\} \qquad (13)$$

(E is the average density of the acoustical energy). The formulas (10) and (11) agree with the results of [2, 3].

In the field U (r), particles collect near the minimum of the potential energy (12). The equilibrium density distribution of the suspended particles n (r) follows the Boltzmann formula

$$n(\mathbf{r}) \sim \exp\{-U(\mathbf{r})/kT\}. \qquad (14)$$

At the places where the density is greatest the further coagulation of the particles is facilitated. At ordinary sound intensities $I \sim 0.1$ w/cm^2 and $\omega \sim 10^4$ sec^{-1} the accumulation of particles at the nodes or antinodes of the wave plays an important role. For example, the width Δ of the region in which the particles are concentrated in a standing wave, Eq. (13), is given in order of magnitude by*

*For very small distances between the particles, or near a wall, forces of interaction between the particles and between the particles and the wall become important.

$$(k\Delta)^2 = \frac{3\kappa T}{10\pi R^2 \bar{E}}.$$

(For $E \approx 4 \cdot 10^{10}$ erg /cm^3, $\nu \approx 1$ kc, $\Delta \approx 1.5 \cdot 10^{-6} R^{-3/2}$) The time after which the distribution (4) is established is $\tau \sim \eta\ (kR)^{-2}\ (E)^{-1}$, where η is the viscosity of the medium.

We emphasize that Eq. (12) holds for an arbitrary field, with the exception of fields closely similar to that of a plane running wave. In particular, in a spherically converging or diverging wave

$$U(r) = \frac{\bar{Q}R^3}{2c}\left\{\frac{f_1}{3r^2} - \frac{f_2}{2}\left(\frac{1}{r^2} + \frac{1}{k^2r^4}\right)\right\} \qquad (15)$$

(Q is the power of the source).

When $f_2 > 0$, $f_1 > {}^3/_2 f_2$, the particles either collect at the center or move out to infinity, depending on the distance from the center; for $f_2 > 0$, $f_1 < {}^3/_2 f_2$ there is a collapse to the center; if $f_1 < {}^3/_2 f_2$, $f_2 < 0$ Eq. (15) has a minimum at distances of the order of a wavelength; finally, if $f_1 > {}^3/_2 f_2 > 0$, the particles are pushed out to infinity. Of course, at very large distances from the center the quantity (15) becomes small, the spherical wave goes over into a plane wave, and the main part is played by the forces (10):

$$U(r) = \pm\frac{\bar{Q}R^2}{9cr}(kR)^4\left\{f_1^2 + f_1 f_2 + \frac{3}{4}f_2^2\right\} \qquad (16)$$

(the plus sign is for a diverging wave and the minus sign for a converging wave).

The values of (16) and (15) are comparable for $r \sim R\,(kR)^{-4}$. For the results given here to be valid it is necessary for the condition $\sqrt{\eta/\rho\omega} \ll R$. to hold.

In conclusion the author expresses his gratitude to Academician L. D.Landau for a discussion of this work and for valuable advice.

LITERATURE CITED

1. L. D. Landau and E. M. Lifshits, Mechanics of Continuous Media [in Russian] (1954).
2. L. V. King, Proc. Roy. Soc., A147, 212 (1954).
3. R. Yosioka and G. Kawasima, Acoustica, 5, 167 (1955).

All abbreviations of periodicals in the above bibliography are letter-by-letter transliterations of the abbreviations as given in the original Russian journal. *Some or all of this periodical literature may well be available in English translation.* A complete list of the cover-to-cover English translations appears at the back of this issue.

Chapter 9

Quantum Electrodynamics

The war-time progress in microwave technology made possible the detection in 1947 of a minor difference between two energy levels of the hydrogen atom, $2^2S_{\frac{1}{2}}$ and $2^2P_{\frac{1}{2}}$ that otherwise should be degenerate in the Dirac theory. Hans Bethe was first to ascribe the experimental effect (known as the "Lamb shift") to perturbations coming from Quantum Electrodynamics.

It is known that quantum-mechanical calculations in Electrodynamics suffered from the appearance of ultraviolet divergences in expressions for the vacuum corrections. Discoveries in 1947 stimulated development of new theoretical approaches. In brief, the newly elaborated methods allowed reducing all expressions to an invariant relativistic form, so that the divergences could be unambiguously singled out, "regularized" and included into the definitions either as the physical charge, or the mass (the procedure termed the "renormalization"). The diagrammatic technique elaborated by R. Feynman turned out one of the most convenient for the purpose.

9.1 "Moscow nil"

By 1953, the time when I started as a postgraduate at the Institute for Physical Problems, the main activity seemed to be over and Landau took disapprovingly my intention to work in Quantum Electrodynamics. He said nothing to me, but to others he told that unless you have an idea of your own, a novice should never start with the formal subject; he was, of course, quite right.

Landau was not involved in the activity before, but in 1954 he took interest in several attempts in literature to go beyond the few first perturbation terms in the quantum electrodynamics. As a way to treat the ultraviolet divergences, he put forward the idea of summing the "main" terms in powers of $e_1^2 \ln(\Lambda^2/-p^2)$, where e_1 is the *small* "bare" charge and Λ is a *large* ultraviolet cut-off. In such a product, smallness of the charge can be compensated by the large logarithm: $e_1^2 \ln(\Lambda^2/-p^2) \sim 1$ ($-p^2$ is square of the *4*-vector).

With Abrikosov and Khalatnikov, they found, for the renormalized (i.e., the observable!) charge the expression $e^2 = e_1^2[1 - (e_1^2/3\pi)\ln(\Lambda^2/m^2)]^{-1}$. If their result was correct, its interpretation could be that the pole in the denominator defines the scale of a momentum (or of the spatial distance $\sim 1/\Lambda$) at which interactions would become *strong*.

My adviser I. M. Khalathikov suggested that I repeat their calculations but for the Electrodynamics of scalar particles with zero spin. In such field theories, both the charge and the mass of the meson need be renormalized. The solution was first obtained in the representation of the so-called "Kemmer matrices" that formally resemble the algebra of the Dirac matrices. Solving the problem turned out to be not straightforward. Actually, I could have done it to the very end only using the renormalization group method by M. Gellman and F. Low. Anyways, the details are in the paper [9a].

Important at the moment was that for the renormalized charge, I found $e^2 = e_1^2[1 + e_1^2 \ln(\Lambda^2/m^2)]^{-1}$, i.e. the *plus* sign in the denominator, instead of their *minus*. I was cursed on the spot, but M. D. Galanin from the group of I. Ya. Pomeranchuk soon restored the same plus sign in their solution as well. With the *plus* sign the interpretation should be different, for the expression seemed to indicate on a *screening* of the electronic charge at short distances.

Indeed, in few papers Landau and Pomeranchuk suggested strong arguments that the result is actually quite general and correct *beyond* the logarithmic accuracy. So, at $\Lambda \to \infty$ the renormalized charge is screened to zero $e^2 = 1/\ln(\Lambda^2/m^2) \to 0$. This is the famous "zero of the charge," known in days of old also as the "Moscow nil".

The paper [9b] discusses the problem of the strong interactions in the very same light; it argued that any quantum field theory based on the idea of *local* interactions is self-contradictory.

9.2 "Infra-red catastrophe"

In Quantum Electrodynamics, each electron is "dressed" in a cloud of vacuum fluctuations, thus carrying along a virtual electromagnetic field. When electrons interact or an electron just scatters, say on an external potential, the cloud changes, so the process of scattering is accompanied by the emission of an *arbitrary* number of soft photons; this problem is known as the *infrared catastrophe*. In [9c] I have shown that because many photons are involved in the effect, fields in the cloud surrounding the electron are classical fields and the infrared catastrophe can be reduced to the problem found in classical electrodynamics.

List of original papers

[9a] L. P. Gorkov, I. M. Khalatnikov, *Electrodynamics of charged scalar particles*, Sov. Phys. JETP **4**, 777 (1957)

[9b] A. A. Abrikosov, A. D. Galanin, L. P. Gorkov, L. D. Landau, I. Ya. Pomeranchuk, and K. A. Ter-Martirosyan, *Possibility of formulation of a theory of strongly interacting fermions*, Phys. Rev., II Ser. 111, 321 (1958)

[9c] L. P. Gor'kov, *Charged particle Green function in the infrared catastrophe region*, Sov. Phys. JETP **3**, 762 (1956)

SOVIET PHYSICS

JETP

A translation of the Journal of Experimental and Theoretical Physics of the USSR.

SOVIET PHYSICS—JETP VOL. 4, NO. 6, PP 777-979 JULY, 1957

Electrodynamics of Charged Scalar Particles

L. P. GOR'KOV AND I. M. KHALATNIKOV

Institute for Physical Problems, Academy of Sciences, USSR

(Submitted to JETP editor November 23, 1955)

J. Exptl. Theoret. Phys. (U.S.S.R.) 31, 1062-1078 (December, 1956)

The behavior of the Green's function in the region of high momenta for the electrodynamics of spin-zero charged particles is investigated on the basis of the Kemmer formalism.

THE asymptotic behavior of the Green's function for high momenta has been found for spin-1/2 electrodynamics by Landau, Abrikosov, and Khalatnikov[1] by means of the direct solution of the integral equations. The same problem has been treated by Gell-Mann and Low[2], who made use of the renormalizability of the theory. The present work is a similar attempt for the electrodynamics of spin-zero particles in the Kemmer formalism.

1. β-FORMALISM

Two equivalent formulations can be used for the description of particles with spin zero and one: the second order Klein-Gordon equation and the matrix formalism of Kemmer. The latter is characterized by a deep analogy with the Dirac equation for the electron. In the β-formalism, the wave function of the spin-zero or spin-one particles satisfies a first order equation

$$(\hat{p} - m_1)\Psi(x) = 0,$$

where* $p = p_k \beta_k$, p is the momentum operator, and the β-matrices satisfy the following commutation relations

$$\beta_\mu \beta_\sigma \beta_\nu + \beta_\nu \beta_\sigma \beta_\mu = \beta_\nu \delta_{\mu\sigma} + \beta_\mu \delta_{\sigma\nu}. \qquad (1)$$

The algebra determined by Eq. (1) has two nontrivial irreducible representations, a five-dimensional and a ten-dimensional one; the first corresponds to spin-zero particles, and the second to spin-one particles. From the point of view of the Klein-Gordon equation, the five-component function for the spin-zero particles consists of one scalar component and its four derivatives with respect to space and time. The components of the wave function are therefore not dynamically independent. This is related to the fact that the β-matrices determined by relations (1) have no inverses.

The interaction of mesons with an electromagnetic field can be described both in the Klein Gordon formalism and in the β-formalism. A very significant disadvantage of the scalar formulation is the extremely complicated structure of the perturbation theory series, due to diagrams with vertices connected by two photon lines. The Kemmer formalism is more convenient, since the class of diagrams possible in its perturbation theory is the same as that in the electrodynamics of electrons, so that the rules for constructing the matrix elements from the diagrams, as has been shown by Peasly,[4] are similar to Feynman's rules for electrodynamics[3] with the difference that the γ-matrices are everywhere replaced by the β-matrices, and the Feynman propagation factor

* In the following we shall use the notation of Feynman[3].

$(p - m_1)^{-1}$ is equal to

$$(\hat{p} - m_1)^{-1} = \{\hat{p} + m_1 + (\hat{p}^2 - p^2)/m_1\}(p^2 - m_1^2)^{-1}. \quad (2)$$

In addition, a meson closed loop enters the matrix element with the sign opposite to that in ordinary electrodynamics. The whole formalism is equally applicable to spin zero and to spin one. The difference between the two spins occurs in considering the convergence of the perturbation theory integrals.

It is well known[5] that all existing field theories may be divided into the renormalizable and nonrenormalizable ones. In the renormalizable theories there exist several types of irreducible divergent diagrams, and the infinite expressions due to these can be included in the physical constants of the theory in a unique way. In a nonrenormalizable theory the number of different kinds of divergent diagrams is infinite. The electrodynamics of vector mesons belongs to this class, which can be seen immediately from expression (2) for $G_0(p)$, since for large values of the momentum p this factor is approximately unity. We will therefore consider only spin-zero mesons, using the five-dimensional representation of the β-matrices. For spin zero the situation is simplified, since a large number of the matrix expressions vanish due to the properties of the five-dimensional representation of the β-matrices.

We shall define two matrices[6]

$$X = (\beta_\lambda\beta_\lambda - 1)/3; \quad (3)$$

$$Y = (4 - \beta_\lambda\beta\ )/3; \quad X + Y = 1.$$

For spin zero

$$XY = 0, \quad (4)$$

$$X\beta_\nu\beta_\mu = \delta_{\mu\nu}X. \quad (5)$$

We present several useful expressions which follow immediately from (1), (3)-(5):

$$X\beta_\mu = \beta_\mu Y; \quad (6)$$

$$\beta_\mu(\hat{p}^2 - p^2)\beta_\nu = X(p_\mu p_\nu - \delta_{\mu\nu}p^2);$$

$$X(\hat{p}^2 - p^2) = 0; \quad \beta_\mu\beta_\mu = 3X + 1; \quad \beta_\mu\beta_\sigma\beta_\mu = \beta_\sigma;$$

$$\beta_\mu\beta_\sigma\beta_\tau\beta_\mu = \delta_{\sigma\tau}; \quad \beta_\mu\beta_\sigma\beta_\tau\beta_\rho\beta_\mu = X\beta_\sigma\delta_{\rho\tau} + \beta_\rho X\delta_{\sigma\tau}.$$

In calculating traces, we must bear in mind that the trace of a product of an odd number of β-matrices vanishes. In addition,

$$\mathrm{Sp}\,1 = 5; \quad \mathrm{Sp}\,X = 1; \quad \mathrm{Sp}\,\beta_\lambda\beta_\mu = 2\delta_{\lambda\mu};$$
$$\mathrm{Sp}\,\beta_\lambda\beta_\mu\beta_\nu\beta_\rho = \delta_{\lambda\mu}\delta_{\nu\rho} + \delta_{\lambda\rho}\delta_{\mu\nu}. \quad (7)$$

Consideration of the perturbation theory expressions shows[6] that in the electrodynamics of a spin-zero meson there exist in general five irreducibly divergent diagrams: 1) and 2) the self-energy diagrams of the meson and photon, which are quadratically divergent; 3) the vertex part, which is logarithmically divergent; 4) a Compton type diagram, which is logarithmically divergent; 5) the diagram corresponding to scattering of a meson by a meson, which is logarithmically divergent.

Scattering of light by light gives a finite expression when summed over all permutations of the emitted quanta.

2. FUNDAMENTAL EQUATIONS

The character of the divergence which appears in the theory is determined by the asymptotic behavior of the Green's function for high momenta. Two of the articles already cited[1,2] are devoted to the problem of finding the asymptotic behavior of the Green's functions in electrodynamics and in pseudoscalar theory. In solving the corresponding problem in the electrodynamics of spin-zero particles, we shall henceforth base our considerations on the idea of a smeared out interaction[1], according to which the interaction which is usually described by a δ-function is assumed smeared out and nonvanishing in some finite region whose linear dimensions are about a. In calculating radiation effects, integration over virtual quanta is cut off at momenta $\sim \Lambda$ where $(\Lambda \sim 1/a)$, and the integrals are then finite. The transition to the exact interaction corresponds to $\Lambda \to \infty$.

Let us write the integral equations which are satisfied by the quantities entering into the theory. The equation for the meson Green's function $G(p)$ can be written, in complete analogy with the usual electrodynamics, in the following way[7]

$$G(p)\Big\{\hat{p} - m_1 - (e_1^2/\pi i) \times \int B_\mu(p, p-k, k)\,G(p-k)\,\beta_\nu D_{\mu\nu}(k)\,d^4k\Big\} = 1. \quad (8)$$

Here m_1 is the "bare" mass, e_1 is the "bare" meson charge, and $B_\mu(p, p-k, k)$ is the operator of the vertex part. The photon propagation factor $D_{\mu\nu}(k)$ as has previously been noted[1], can be written as the sum of the transverse and longitudinal parts

$$D_{\mu\nu}(k) = \left(\delta_{\mu\nu} - \frac{k_\mu k_\nu}{k^2}\right)\frac{d_t(k^2)}{k^2} + \frac{k_\mu k_\nu}{k^2}\frac{d_l(k^2)}{k^2}, \quad (9)$$

of which only the transverse part has physical meaning for the interaction; it must therefore be determined from the equations, whereas the longitudinal part may be chosen arbitrarily. The equation for the transverse part of the photon Green's function is of the form

$$D^t_{\mu\sigma}(k)\left\{k^2\delta_{\sigma\nu} - \frac{e_1^2}{\pi i}\int \mathrm{Sp}\,[G(p)B_\sigma(p,p-k,k) \quad (10)$$

$$\times\, G(p-k)\,\beta_\nu]\,d^4p\right\} = \delta_{\mu\nu} - \frac{k_\mu k_\nu}{k^2}.$$

The gauge invariance of the theory should follow directly from the transverse nature of the polarization tensor in this equation. Let us also note that the different sign for the integral term in Equation (10), as compared with the electro-dynamics of an electron, is related to the Bose statistics for spin-zero particles. Equations (8) and (10) for the Green's functions of an interacting meson and photon are exact and contain no approximations. The situation with respect to the equations for the operators of the vertex parts and the Compton diagrams is much more complicated. It is in general impossible to obtain exact integral equations for them in closed form. It turns out, however, that for certain conditions it is possible to write an approximate integral equation for the operator of the vertex part. In order to do this, let us go over to a consideration of the expressions furnished by perturbation theory. We shall write the expression for the meson Green's function obtained from perturbation theory to first order in e_1^2 in the high momentum region (for the space-components, $-p^2 \gg m^2$) up to terms containing a large logarithm

$$G^{-1}(p) = \hat{p} - m_1 + \left(\frac{3X}{m_1}\right)\left(\frac{e_1^2}{\pi i}\right)\int\frac{d^4k}{k^2} \quad (11)$$

$$-\left(\frac{e_1^2}{8\pi}\right)(\hat{p}+2m_1)\ln\left(\frac{\Lambda^2}{-p^2}\right)$$

$$+\left(\frac{e_1^2}{16\pi}\right)\Big[\hat{p} - m_1$$

$$+\left(\frac{3X}{m_1}\right)(p^2-m_1^2)\Big]\ln\left(\frac{\Lambda^2}{-p^2}\right)(d_l-1).$$

Calculation of the matrix elements for the first order correction to the vertex part leads to the following asymptotic expression for $B_\sigma(p, p-l, l)$ for large space components, the largest of which we shall call f:

$$B_\sigma(p,p-l,l) = \beta_\sigma - \frac{e_1^2}{8\pi}\beta_\sigma\ln\left(\frac{\Lambda^2}{-f^2}\right) \quad (12)$$

$$+\frac{e_1^2}{16\pi}\left[\beta_\sigma + \frac{3X}{m_1}(2p-l)_\sigma\right]\ln\left(\frac{\Lambda^2}{-f^2}\right)(d_l-1).$$

As can be seen from these expressions, not only the longitudinal, as for ordinary electrodynamics, but also the transverse part of the photon Green's function leads to terms logarithmic in the meson self-energy and vertex parts. The last (third) term in Eq. (12) shows, in view of the Ward identity, that the logarithmic divergence in the radiation corrections to the Compton effect occur only in combination with the factor $d_l - 1$. This result can also be obtained by considering the matrix element of the first correction to the Compton effect (Appendix I). Thus with the choice $d_l = d_t$ the Compton diagrams are finite in first order perturbation theory, and give no logarithmic contribution. It is interesting to note that this situation follows from the use of the β-formalism; to one diagram for the correction to the Compton effect in the Kemmer formalism corresponded several diagrams in the second order equations, each of which diverges separately. Only for the special choice $d_l = d_t$ is their β-matrix sum finite.

Except for the quadratically diverging integrals in the meson and photon mass, the perturbation theory series thus contains only logarithmic terms in the limiting momentum. Use of perturbation theory effectively implies a condition $e_1^2\ln(\Lambda^2/m^2) \ll 1$. We shall, following an earlier work[1], use the much weaker condition $e_1^2 \ll 1$. Subjecting the theory to such a limitation means that in the asymptotic region we must sum only over those diagrams whose contribution to the matrix elements contains the large logarithm raised to a power equal to the perturbation theory order. In this approximation the Compton diagrams in all perturbation theory orders are the same as in the zeroth approximation (for $d_l = d_t$; this assumption, based at this point on calculations to the first approximation in e_1^2 is later verified by the expression obtained for the operator of the vertex part). For the vertex part $B_\sigma(p, p-l, l)$ we can write the approximate integral (so-called three-vertex) equation

$$B_\sigma(p,p-l,l) = \beta_\sigma + (e_1^2/\pi i) \quad (13)$$

$$\times\int B_\mu(p,p-k,k)G(p-k)B_\sigma(p-k,p-k-l,l)$$

$$\times\, G(p-k-l)\,B_\mu(p-k-l,$$

$$p-l,-k)\,D(k)\,d^4k.$$

This equation expresses the fact that in summing

the radiation corrections to the vertex part, only diagrams with nonintersecting photon lines are accounted for, and is derived on the assumption that the longitudinal part of the photon Green's function is chosen equal to its transverse part ($d_l = d_t$). In other words

$$D_{\mu\nu}(k) = \delta_{\mu\nu} D(k) = \delta_{\mu\nu} d_t(k^2) / k^2.$$

Direct calculations show that the second order diagram in e_1^2 with two intersecting photon lines gives no contribution in this approximation (Appendix I). Let us note also that the matrix element for scattering of a meson by a meson is not "linked" with the fundamental equations and can be considered separately.

In the following sections Equations (8), (10), and (13) will be used to find the asymptotic forms of the photon and meson Green's functions.

3. PHOTON GREEN'S FUNCTION

Let us now go on to a solution of the above equations. We should note that the matrix composition of the exact Green's functions for the meson and vertex parts will not in general, as can be seen, for instance, from expression (11), be the same as the corresponding expressions for these quantities in the zeroth approximation for noninteracting fields. Considerations of relativistic invariance, although they narrow the possible matrix combinations, still leave a sufficient number of different invariant combinations both in the meson Green's function and, especially, in the vertex part. The reason for this is the triple (and not double, as in ordinary electrodynamics) permutation rules, so that a large number of invariants are independent. For $d_l = d_t$, however, a consideration of the perturbation theory formulas (11) and (12) allow us to assume that in the high-momentum region the meson Green's function $G(p)$ and the vertex part $B_\sigma(p, p-l, l)$ can be written in the following form:

$$G^{-1}(p) = \left\{\hat{p} - m_1(p^2) - \frac{X\varepsilon}{m_1}\right\}\left(\frac{1}{\beta(p^2)}\right); \tag{14}$$

$$B_\sigma(p, p-l, l) = \beta_\sigma \alpha(f^2).$$

Here $\beta(p^2)$, $\alpha(f^2)$, and $m_1(p^2)$ are slowly (logarithmically) varying functions of their arguments, $m_1(p^2)$ is proportional to the "bare" mass m_1 of the meson. and f is the largest of the three arguments in $B_\sigma(p, p-l, l)$. The quantity ϵ in the expression for $G(p)$ is of order $\sim e_1^2 \Lambda^2$, where Λ is the upper limit of the interaction, and is the quadratic correction to the mass. We note that the operator which is the inverse of $(\hat{p} - m_1 - XA/m_1)$ is equal to

$$\left(\hat{p} - m_1 - \frac{XA}{m_1}\right)^{-1} \tag{14a}$$

$$= \left\{\hat{p} + m_1 + \frac{1}{m_1}(\hat{p}^2 - p^2) + \frac{YA}{m_1}\right\}\frac{1}{p^2 - m_1^2 - A}.$$

We have already seen in Eq. (11) that ϵ comes from the transverse part of the photon Green's function. It is significant that the quantity ϵ drops out of all further calculations, so that it effects only a change in mass. In our approximation ϵ may be considered constant. We make the following additional remark: in integral equations (8), (10), and (13) the high momentum region is of fundamental importance. Therefore in order to calculate these integrals we must know the behavior of all the functions in the high momentum region. This makes it possible to determine the variation of all quantities in the asymptotic region, for this approximation, without considering regions of momentum small compared to the meson mass. In addition, after all intermediate operations the integrands can be written as the sum of various combinations of matrices multiplied by a scalar function whose argument is of the form $(k-a)^2$, where a may be any of the vectors in Eqs. (8), (10), and (13). These functions have no singularities in the upper half plane. If a is a spacelike vector, then integration over k may be replaced by integration over four-dimensional Euclidean space by replacing k_0 by ik_0. Therefore in solving the equations which determine $G(p)$, $B_\sigma(p, p-l, l)$, $D(k)$, we may limit ourselves to the region of spacelike vectors. For timelike arguments, these functions may be obtained by analytic continuation through the upper half plane. In those regions of integration over k which give the main contribution (logarithmic) to arguments of the form $(k-a)^2$, we may neglect a in comparison with k, and therefore all the functions will depend on k^2; the transition to integration over Euclidean space is accomplished particularly simply:

$$d^4k \to i/4\,(-k^2)\,d(-k^2). \tag{15}$$

The importance of this transformation was first noted by Landau and co-workers[1]. Turning to Equation (13), let us replace $G(p)$ and $B_\sigma(p, p-l, l)$ by the corresponding expressions from (14). It is easy to see that the integral diverges logarithmically in the region $-f^2 \ll -k^2 \ll \Lambda^2$, where f is the largest of the arguments of the vertex part.

Bringing similar terms into the numerator, going over to integration over Euclidean space, and averaging over angles, the integral on the right side becomes

$$-\beta_\sigma \frac{e_1^2}{8\pi} \int_{-k^2 \gg -f^2}^{-k^2 \ll \Lambda^2} \alpha^3(-k^2)\,\beta^2(-k^2)\,d(-k^2)\frac{d-k^2}{-k^2}$$

in agreement with the assumption that the right side of Eq. (13) is again proportional to B_σ. This also verifies the validity of both expressions (14), since $G(p)$ and $B_\sigma(p, p-l, l)$ are related by the Ward identity

$$\partial G^{-1}(p)/\partial p_\sigma = B_\sigma(p, p, 0).$$

The function α satisfies the following equation:

$$\alpha(\xi) = 1 - \frac{e_1^2}{8\pi}\int_\xi^L \alpha^3(z)\,\beta^2(z)\,d(z)\,dz. \tag{16}$$

Here we have introduced the new variable $z = \ln(-k^2/m^2)$ so that $\epsilon = \ln(-f^2/m^2)$ and $L = \ln(\Lambda^2/m^2)$.

The equation which determines the function $\beta(p)$ should be obtained from (8); however, since the integral contains a quadratically diverging part, in order to obtain the logarithmic terms in it, it is necessary to have not only expressions (14) for $B_\mu(p, p-k, k)$, but the contributions to it up to terms of the order of about p^2/k^2 inclusive. For simplicity, we shall use the Ward identity, which we write in the form

$$\alpha(\xi)\,\beta(\xi) = 1. \tag{17}$$

With the aid of this expression, Eq. (16) can be written

$$\alpha(\xi) = 1 - \frac{e_1^2}{8\pi}\int_\xi^L \alpha(z)\,d(z)\,dz. \tag{18}$$

Eq. (10) for the photon Green's function contains a quadratically diverging integral. The quadratically diverging part can be separated, and represents the photon mass. Because of the gauge invariance of the theory, the interaction should be smeared out in such a way that this quantity remain zero. The remaining part again contains logarithmic integrals, and in order to calculate these it is necessary to know the logarithmic corrections to the vertex parts $B_\sigma(p, p-l, l)$ in the region $p \gg l$ accurately to terms of second order in l/p inclusive. To obtain this information we again refer to Eq. (3), replacing the variable $p-k$ by k:

$$B_\sigma(p, p-l, l) \tag{19}$$

$$= \beta_\sigma + \frac{e_1^2}{\pi i}\int B_\mu(p, k, p-k)\,G(k)\,B_\sigma(k, k-l, l)$$

$$\times\, G(k-l)\,B_\mu(k-l, p-l, k-p)\,D(p-k)\,d^4k.$$

Inserting Eq. (14) for the appropriate factors, and multiplying out the numerator of the integrand, we see that the appropriate terms occur on the right side for the region of integration $p \gg k \gg l$ in expanding $B_\sigma(k, k-l, l)$ and the denominators of $G(k)$ and $G(k-l)$. Expanding the rest of the factors, as can be easily seen, leads to higher order terms. In order to clarify the form of the corrections in this region, let us consider the additions which come from expanding the denominators in $G(k)$ and $G(k-l)$ in powers of l/k, $(m_1^2+\epsilon)/k^2$. It turns out that the corresponding term is of the form

$$\frac{e_1^2}{24\pi}(\hat{l}l_\sigma - \beta_\sigma l^2)\,(p^{-2})\,\alpha^2(\xi)\,d(\xi)\int_\eta^\xi \beta^2(z)\,\alpha(z)\,dz,$$

$$\xi = \ln(-p^2/m^2), \quad \eta = \ln(-l^2/m)^2.$$

We shall try to obtain $B_\sigma(p, p-l, l)$ in the region $p \gg l$ in the form

$$B_\sigma(p, p-l, l) = \beta_\sigma\alpha(p^2) \tag{20}$$

$$+\,(l\hat{l}_\sigma - \beta_\sigma l^2)\,(p^{-2})\,\alpha^2(p^2)\,d(p^2)\,S_0(p^2, l^2).$$

We then obtain the following equation (in closed form) for the function $S_0(\xi, \eta)$:

$$S_0(\xi,\eta) = -\frac{e_1^2}{8\pi}\int_\eta^\xi \alpha^2(z)\,\beta^2(z)\,d(z)\,S_0(z,\eta)\,dz \tag{21}$$

$$+\,\frac{e_1^2}{24\pi}\int_\eta^\xi \alpha(z)\,\beta^2(z)\,dz.$$

Corrections of other forms do not occur.

If we insert (20) into expression (10) for the polarization tensor, we obtain an equation which determines the function $d_t = d_l$

$$d_t^{-1}(\eta) = 1 + 2S_0(L, \eta), \tag{22}$$

where L is the logarithm of the cut-off parameter.

Equations (18), (21) and (22) must be solved simultaneously. With the aid of the Ward identity, Eq. (21) can be written

$$S_0(\xi,\eta) = -\frac{e_1^2}{8\pi}\int_\eta^\xi d(z)\,S_0(z,\eta)\,dz + \frac{e_1^2}{24\pi}\int_\eta^\xi \beta(z)\,dz.$$

Differentiating this by ξ, we obtain

$$\partial S_0(\xi,\eta)/\partial\xi = -(e_1^2/8\pi)\,d(\xi)\,S_0(\xi,\eta) + (e_1^2/24\pi)\,\beta(\xi).$$

But

$$d\beta(\xi)/d\xi = -(e_1^2/8\pi)\,d(\xi)\,\beta(\xi), \tag{23}$$

and therefore

$$\partial[S_0(\xi,\eta)\beta^{-1}(\xi)]/\partial\xi = e_1^2/24\pi$$

and with the conditions $S_0(\xi,\xi) = 0$, we obtain

$$S_0(\xi,\eta) = (e_1^2/24\pi)(\xi-\eta)\,\beta(\xi).$$

Since $\beta(L) = 1$, we have $S_0(1,\eta) = (e_1^2/24\pi) \times (L - n)$. It is interesting to note that $S_0(L,\eta)$ is exactly identical with the same function in spinor electrodynamics. We note in this respect that in addition to the polarization of spin-zero particles by the photon, there actually occurs also a polarization of the spin 1/2 particles. Therefore, instead of (10), we should have written the photon Green's function equation in the form

$$D^t_{\mu\sigma}(k)\Big\{k^2\delta_{\sigma\nu} - \Big(\frac{e_1^2}{\pi i}\Big) \times \int \mathrm{Sp}\,[G^0(p)\,B_\sigma(p,p-k,k)\,G^0(p-k)\,\beta_\nu]\,d^4p + \Big(\frac{e_1^2}{\pi i}\Big)\int \mathrm{Sp}[G^{1/2}(p)\,\Gamma_\sigma(p,p-k,k) \times G^{1/2}(p-k)\,\gamma_\nu]\,d^4p\Big\} = \delta_{\mu\nu} - \frac{k_\mu k_\nu}{k^2}.$$

Making use of previously derived results [1], we can write down immediately

$$d_t^{-1}(\eta) = 1 + 8\nu S(L,\eta).$$

The quantity ν, here introduced, plays the role of the effective number of different particles. Spin-1/2 gives 1 for ν, and spin zero, 1/4. Thus

$$d_t(\eta) = [1 + (e_1^2/3\pi)\,\nu\,(L-\eta)]^{-1}$$

or

$$d_t(k^2) = [1 + (e_1^2/3\pi)\,\nu\ln(\Lambda^2/-k^2)]^{-1}. \tag{24}$$

We note that this formula is valid not only for $k \gg m$, but also for $k \gtrsim m$, since in the calculation both m_1 and ϵ drop out. For $-k^2 \ll m^2$

$$d_t(k^2) = [1 + (e_1^2/3\pi)\,\nu\ln(\Lambda^2/m^2)]^{-1}$$

which corresponds to the fact that the logarithmic region of integration in Eq. (10) lies* above m.

The physical charge is given by

$$e^2 = e_1^2[1 + (e_1^2/3\pi)\,\nu\ln(\Lambda^2/m^2)]^{-1}. \tag{25}$$

The expression we obtain for the "bare" charge e_1^2 is then

$$e_1^2 = e^2[1 - (e^2/3\pi)\,\nu\ln(\Lambda^2/m^2)]^{-1}. \tag{26}$$

For a sufficiently large radius of interaction $(e^2/3\pi)\nu\ln\Lambda^2/m^2 \sim 1$, and the bare charge e_1^2 becomes of the order of unity in contradiction to the original assumption of weak interaction $(e_1^2 \ll 1)$. As has been shown by Pomeranchuk[8], this restriction is not of importance, as Eq. (25) is the exact result for large Λ^2. From this we obtain the fundamental conclusion that in the point-interaction limit ($\Lambda \to \infty$) the physical charge vanishes (for more details see Pomeranchuk[8]).

The interaction $e_1^2 d_t$ can be written in a form containing only renormalized quantities

$$e_1^2 d_t(k^2) \equiv e^2 d_{tc}(k^2) = e^2[1 - (e^2/3\pi)\,\nu\ln(-k^2/m^2)]^{-1}. \tag{27}$$

4. MESON GREEN'S FUNCTION

With the function $d(p^2)$ obtained, it is easy to derive the formula for $\beta(p^2)$ from Eq. (18):

$$\beta(p^2) = [1 + (e_1^2/3\pi)\,\nu\ln(\Lambda^2/-p^2)]^{3/8\nu}. \tag{28}$$

The function $\alpha(p^2) = 1/\beta(p^2)$. We note that the function $\beta(p^2)$ can be written as the product of a constant and a function independent of the cut-off parameter,

$$\beta(p^2) = (e_1/e)^{3/4\nu} \times [1 - (e^2/3\pi)\,\nu\ln(-p^2/m^2)]^{3/8\nu} = [e_1^2 d_t(p^2)/e_1^2]^{-3/8\nu}. \tag{29}$$

* We are here neglecting the difference between the meson and electron masses.

In this form $\beta(p^2)$ decreases as $-p^2$ increases, and becomes small at the same point where the denominator of expression (27) vanishes. Equation (29) expresses the so-called renormalizability of the function $\beta(p^2)$.

Let us write down the meson Green's function $G(p)$ for large momenta

$$G(p) = \left\{\hat{p} + m_1(p^2) + \frac{1}{m_1(p^2)}(\hat{p}^2 - p^2) - \frac{Y\varepsilon}{m_1}\right\}\frac{\beta(p^2)}{p^2}.$$

In the β-formalism $G(p)$ consists of a scalar, a vector, and a tensor component. The dependence of each of these components on the radius of cut-off Λ is different. Using a special and simply derivable representation of the β-matrices, which establishes the relation between the Klein-Gordon equation and the Kemmer equation*, we can separate out the scalar component

$$G_{55}(p) \approx \beta(p^2)\, m_1(p^2)/p^2, \tag{30a}$$

The vector component

$$G_{i5}(p) \approx \beta(p^2)(p_i/p^2) \tag{30b}$$

and the tensor component

$$G_{ik}(p) \approx \frac{\beta(p^2)}{m_1(p^2)}\left(\frac{p_i p_k}{p^2} - \delta_{ik}\right) \tag{30c}$$

in the region $-p^2 >> m^2$. In these expressions, in addition to $\beta(p^2)$, the function $m_1(p^2)$ also enters. We shall not try to find an equation for $m_1(p^2)$, but shall make use of the fact that, as can be seen from the expression for the meson Green's function in coordinate space[9],

$$G(x, x') = \langle T(\Psi(x), \overline{\Psi}(x'))\rangle_0 - (i/m_1)(1 - \beta_0^2)\,\delta(x - x')$$

the component $G_{55}(p)/m_1$ for our representation of the β-matrices is the meson propagation kernel in the scalar formulation of the theory. The asymptotic behavior of this quantity can be determined by the method of Gell-Mann and Low[2]. The applicability of this method to one or another theory of interacting fields depends on whether or not this theory refers to quantities renormalized in the sense of Z-factors. Let us bear in mind that the requirement of the renormalizability means the following. Assume that certain functions, say $f_i(p^2)$, referring to the particle Green's function and the vertex parts appear in the theory. We construct some combination $H(p^2)$ of these functions, such that $g_1 H(p^2)$ represents an interaction, where g_1 is the bare coupling constant. Let Λ be the interaction cut-off radius, chosen so that $f_i(\Lambda^2) = 1$. The requirement of renormalizability means that the function $f_i(p^2)$ differs only by some factor from a function independent of the cut-off radius

$$f_i(p^2/\Lambda^2, g_1) = f_{ic}(p^2/m^2, g)/f_{ic}(\Lambda^2/m^2, g) \tag{31}$$

(for large values of $-p^2$, the function f_i depends only on the ratio $-p^2/\Lambda^2$). The renormalized interaction constant is determined by the condition

$$g_1 = g\dot{H}_c(\Lambda^2/m^2, g).$$

It can then be shown[2] that the following relation holds:

$$g_1 H(-p^2/\Lambda^2, g_1) = F(-\varphi(g_1)\, p^2/\Lambda^2),$$

where F and φ are inverse functions: $F[\varphi(x)] = x$.

Equation (31) is not trivial. We shall see later that in the Kemmer formalism for $d_l \neq d_t$ by no means do all functions satisfy Eq. (31). In the scalar formulation, however, this relation holds; Eq. (29) shows that this is valid both with relation to the function $\beta(p^2)$ in our approximation and for the condition $d_l = d_t$. We shall present all the calculations anew in this case, using only (31). We then have

$$e_1^2 d_t(k^2) = F(-\varphi(e_1^2)\, k^2/\Lambda^2). \tag{32}$$

Inverting this expression and taking logarithms, we obtain

$$\ln\varphi(e_1^2 d_t) - \ln\varphi(e_1^2) = \ln(-k^2/\Lambda^2).$$

Expanding the left side of this equation in powers of $e_1^2(d_t - 1)$ and taking only the first term, we obtain

$$\frac{1}{\varphi(e_1^2)}\frac{d\varphi(e_1^2)}{de_1^2}\, e_1^2(d_t - 1) = \ln\frac{-k^2}{\Lambda^2}. \tag{33}$$

The quantity $d_t - 1$ may be found by perturbation theory. Vacuum polarization for spin 1/2 particles gives the following value for $d_t - 1$ in the first approximation: $(e_1^2/3\pi)\ln(-k^2/\Lambda^2)$,

* This representation is $(\beta_\mu)_{\alpha\tau} = \delta_{\mu\alpha}\delta_{5\tau} + \delta_{\mu\tau}\delta_{5\alpha}$; $(\mu = 1\ldots 4;\ \alpha\tau = 1\ldots 5)$.

vacuum polarization for spin zero particles:

$$(e_1^2/12\pi)\ln(-k^2/\Lambda^2);$$

introducing again the quantity ν, we obtain $d_t - 1 = (e_1^2/3\pi)\nu\ln(-k^2/\Lambda^2)$. . Inserting this expression in (32) gives a differential equation for the function $\varphi(x)$

$$\varphi^{-1}(x)\,d\varphi(x)/dx = (3\pi/\nu)\,x^{-2},$$

whose solution is $\varphi(x) = \exp(-3\pi/\nu x)$, and for the inverse function $F(x)$

$$F(x) = -(3\pi/\nu\ln x),$$

which leads again to Eq. (24), which was obtained by solving the integral equations.

If the functions $\beta(p^2)$ and $m_1(p^2)/m_1$ are renormalizable, then it follows from Eq. (32) that they can be written in the following form:

$$\beta(p^2) = H[e_1^2 d_t]/H[e_1^2]; \tag{34}$$

$$m_1(p^2)\beta(p^2)/m_1 = Q[e_1^2 d_t]/Q[e_1^2].$$

At the same time, perturbation theory, according to (11), leads to the following expression:

$$\beta(p^2) \approx 1 - (e_1^2/8\pi)\ln(-p^2/\Lambda^2); \tag{35}$$

$$m_1(p^2)\beta(p^2)/m_1 \approx 1 - (e_1^2/2\pi)\ln(-p^2/\Lambda^2).$$

Taking logarithms in expression (34), expanding in powers of $e_1^2(d_t - 1)$ and comparing with (35), we obtain the following equations for the functions $H(x)$ and $Q(x)$, $H^{-1}(x)\,dH(x)/dx = -(3/8\nu)\,x^{-1}$; $Q^{-1}(x)\,dQ(x)/dx = -(3/2\nu)\,x^{-1}$, whose solutions are

$$H(x) = x^{-3/8\nu}; \quad Q(x) = x^{-3/2\nu},$$

which gives the previous expression (28) for the function $\beta(p^2)$, and for $m_1(p^2)$ gives the following result:

$$m_1(p^2) = m_1[1 + (e_1^2/3\pi)\nu\ln(\Lambda^2/-p^2)]^{9/8\nu} \tag{36}$$

or

$$m_1(p^2) = [e_1^2 d_t/e_1^2]^{-9/8\nu}$$

$$= (e_1/e)^{9/4\nu}[1 - (e_1^2/3\pi)\nu\ln(-p^2/m^2)]^{9/8\nu}.$$

Knowing the functions $m_1(p)$ and $\beta(p^2)$ we can write an expression for all components of the meson Green's function for $-p^2 \gg m^2$.

For the scalar component

$$G_{55}(p) \approx m_1[1 + (e_1^2/3\pi)\nu\ln\Lambda^2/-p^2]^{3/2\nu}(1/p^2), \tag{37a}$$

For the vector component

$$G_{i5}(p) \approx [1 + (e_1^2/3\pi\nu)\ln\Lambda^2/-p^2]^{3/8\nu}(p_i/p^2) \tag{37b}$$

and, finally, for the tensor component

$$G_{ik}(p) \approx \frac{1}{m_1}[1 + (e_1^2/3\pi)\nu\ln\Lambda^2/-p^2]^{-3/4\nu}\left(\frac{p_i p_k}{p^2} - \delta_{ik}\right). \tag{37c}$$

Thus when used for all the components of the meson Green's function, method (2) leads to the same results as the direct solution of the integral equations. We emphasize, however, that in the β-formalism, the use of this method requires preliminary proof of the renormalizability by solving the integral equations. Equation (31) is satisfied for all the functions only in the approximation we are considering and for the special choice $d_l = d_t$. For arbitrary d_l, Eq. (31) is not satisfied even in the first approximation. We are not, of course, referring to the function $m_1(p^2)$.

5. GAUGE TRANSFORMATION FOR THE MESON GREEN'S FUNCTION

The equations of electrodynamics remain invariant under the following transformation of the four-dimension potential

$$A_\mu(x) \to A_\mu(x) + \partial\varphi(x)/\partial x_\mu$$

with the simultaneous transformation of the particle wave function according to

$$\psi(x) \to e^{ie\varphi(x)}\psi(x),$$

and therefore an electromagnetic field operator can be separated into its transverse and longitudinal parts. The longitudinal part of the electromagnetic field operator may be arbitrary in view of gauge invariance, and does not interact (commutes) with all other quantities. For the photon Green's function this separation leads to expression (9). Gauge transformation of charged particle Green's functions has been studied by Landau and Khalatnikov[10] and is given by the following formula

$$G(x, x') = G_0(x, x')\exp\{ie_1^2(\Delta_F(0) - \Delta_F(x - x'))\},$$

where $G_0(x, x')$ is the corresponding Green's function in the absence of scalar quanta, and

$$\Delta_F(x-x') = i\langle T(\varphi(x), \varphi(x')\rangle_0.$$

In momentum space it is more convenient to make use of another result of the same work[10], which was obtained for the infinitesimal gauge transformation

$$\delta G(p) = \frac{ie_1^2}{\pi}\int \{G(p) - G(p-k)\}\frac{\delta d_l(k^2)\, d^4k}{(k^2)^2}.$$

Restricting ourselves to slowly varying functions $d_l(k^2)$, we see that this integral diverges logarithmically in the region $-k^2 \gg -p^2$. Let us write this relation for the scalar component $G_{55}(p)$:

$$\delta G_{55}(p) = \frac{ie_1^2}{\pi}\int_{\gg p^2} G_{55}(p)\frac{\delta d_l(k^2)\, d^4k}{(k^2)^2}.$$

Transforming this to integration over Euclidean space and writing $Z = \ln(-k^2/m^2)$, we obtain

$$\delta G_{55}(p) = -\frac{e_1^2}{4\pi} G_{55}(p)\int_{\ln(-p^2/m^2)}^{\ln(\Lambda^2/m^2)} \delta d_l(z)\, dz.$$

This relation should be integrated with the boundary condition that for $d_l = d_t\, G_{55}(p)$ it gives Eq. (37a). The result is then

$$G_{55}(p) \approx m_1[d_t(-p^2)]^{-9/8\nu} \tag{38a}$$

$$\times \exp\left\{-\frac{e_1^2}{4\pi}\int_{\ln(-p^2/m^2)}^{\ln(\Lambda^2/m^2)} d_l(z)\, dz\right\}(1/p^2).$$

In exactly the same way we can obtain expressions for the vector components $G_{i5}(p)$ for $-p^2 \gg m^2$

$$G_{i5}(p) \approx [d_t(-p^2)]^{-9/8\nu} \tag{38b}$$

$$\times \exp\left\{-\frac{e_1^2}{4\pi}\int_{\ln(-p^2/m^2)}^{\ln(\Lambda^2/m^2)} d_l(z)\, dz\right\}(p_i/p^2).$$

This shows that the difference between the exact functions and the interacting field functions does not arise only from the gauge transformation, which is not the case in electron electrodynamics. We note that according to (30), (36), and (38) the function $m_1(p^2)$ does not depend on d_l, since the vector and scalar components of $G(p)$ transform in the same way under a gauge transformation. This, furthermore should have been expected, since the function $m_1(p^2)$ gives a contribution to the experimental meson mass m

$$m^2 = m_1^2(m^2) + \varepsilon.$$

The tensor component of the meson Green's function for arbitrary d_l cannot now be written in the form of Eq. (30c). For $-p^2 \gg m^2$, however, it can always be represented in the form

$$G_{ik}(p) \approx \frac{1}{m_1}\left[\left(\frac{p_i p_k}{p^2} - \frac{1}{4}\delta_{ik}\right)\lambda(p^2) - \frac{3}{4}\delta_{ik}\mu(p^2)\right]. \tag{38c}$$

Let us write the equation for an infinitesimal gauge transformation of the tensor component

$$\left[\left(\frac{p_i p_k}{p^2} - \frac{1}{4}\delta_{ik}\right)\delta\lambda(p^2) - \frac{3}{4}\delta_{ik}\delta\mu(p^2)\right]$$

$$= \frac{ie_1^2}{\pi}\int\left\{\left(\frac{p_i p_k}{p^2} - \frac{1}{4}\delta_{ik}\right)\lambda(p^2) - \frac{3}{4}\delta_{ik}\mu(p^2) - \left(\frac{k_i k_k}{k^2} - \frac{1}{4}\delta_{ik}\right)\lambda(k^2) + \frac{3}{4}\delta_{ik}\mu(k^2)\right\}\frac{\delta d_l(k^2)\, d^4k}{(k^2)^2},$$

from which we obtain, after averaging over angles

$$\delta\lambda(\xi) = -\frac{e_1^2}{4\pi}\lambda(\xi)\int_\xi^L \delta d_l(z)\, dz, \tag{39a}$$

$$\delta\mu(\xi) = -\frac{e_1^2}{4\pi}\int_\xi^L [\mu(\xi) - \mu(z)]\,\delta d_l(z)\, dz, \tag{39b}$$

$$\xi = \ln(-p^2/m^2),\quad L = \ln(\Lambda^2/m^2).$$

With the boundary condition (37c) we obtain the following expression for the function $\lambda(p)$:

$$\lambda(p^2) = \exp\left\{-\frac{e_1^2}{4\pi}\int_{\ln(-p^2/m^2)}^{\ln(\Lambda^2/m^2)} d_l(z)\, dz\right\}. \tag{40}$$

Differentiating Eq. (39) by ξ

$$\delta\frac{d\mu(\xi)}{d\xi} = -\frac{e_1^2}{4\pi}\frac{d\mu(\xi)}{d\xi}\int_\xi^L \delta d_l(z)\, dz$$

and making use of boundary condition (37c) for $d_l = d_t$, we obtain

$$\mu(\xi) = 1 - \frac{e_1^2}{4\pi}\int_\xi^L \lambda(z)\, d_t(z)\, dz. \tag{41}$$

For $d_l = d_t$ as follows from Eq. (41), $\mu(\xi) = \lambda(\xi)$; for $d_l = 0$, the function $\lambda(\xi)$ becomes unity, and $\mu(\xi)$ is given by

$$\mu(\xi) = 1 - (3/4\nu)\ln\{1 + (e_1^2/3\pi)\nu(L-\xi)\}. \tag{42}$$

Thus for $d_l \neq d_t$, the function $\mu(\xi)$ no longer satisfies the renormalizability relation

6. CONCLUSION

The above considerations show that spin-zero electrodynamics in the β-formalism exhibit certain formal similarities with spin 1/2 electrodynamics. Beyond this, the general situation is more complicated. In order to derive the integral equations, it is necessary to make wide use of the results of perturbation theory. The problem is simplified with the choice $d_l = d_t$, in which case it is possible to write the three-vertex equation for the vertex parts. For this choice of d_l in the approximation we are considering it becomes possible to restrict ourselves to finding only the photon and meson Green's functions, and the vertex part; the Compton diagrams give contributions equal to their zeroth approximation. In this case, use of the three-vertex equation leads to the correct expression for the photon Green's function, which follows from a comparison with a result obtained by a separate method using only perturbation theory and the requirement of renormalizability. Both methods give the same results for the meson Green's function, since expression (28), which is found directly from the equations, satisfies the renormalizability relation (31).

For $d_l \neq d_t$ Eq. (13) for the vertex part is not applicable; the expression for the meson Green's function in this case can be obtained by a gauge transformation of the expression obtained for the condition $d_l = d_t$ to arbitrary d_l. It is interesting that the difference between the exact meson Green's function and its zero-approximation form does not arise only from the guage transformation, as it does in ordinary electrodynamics.

The authors are deeply grateful to Academician L. D. Landau for many valuable suggestions.

APPENDIX I

a) Correction to the Compton effect in the first approximation in e_1^2. Let us consider the matrix element for the Compton type diagram (Fig. 1)

$$C_{\sigma\tau}(p,\ p-q_1-q_2;\ q_1,\ q_2)$$
$$= -\frac{e_1^2}{\pi i}\int \beta_\mu G(p-k)\,\beta_\sigma G(p-k-q_1)$$
$$\times\ \beta_\tau G(p-k-q_1-q_2)\,\beta_\nu D_{\mu\nu}(k)\,d^4k.$$

Here $(q_1;\sigma)$ and $(q_2;\tau)$ are the momenta and polarizations of the two omitted quantities. Writing out the products of the various factors according to Eq. (6), we obtain

$$C_{\sigma\tau}(p,\ p-q_1-q_2;\ q_1,\ q_2)$$
$$= -\frac{e_1^2}{\pi i}\int \Big\{\beta_\mu(\hat{p}_1+m_1)\beta_\sigma(\hat{p}_2+m_1)\beta_\tau(\hat{p}_3+m_1)\beta_\nu\cdot$$
$$+(p_{1\mu}p_{1\sigma}-\delta_{\mu\sigma}p_1^2)\,X\beta_\tau(\hat{p}_3+m_1)\,\beta_\nu$$
$$+\beta_\mu X(p_{2\sigma}p_{2\tau}-\delta_{\sigma\tau}p_2^2)(\hat{p}_3+m_1)\,\beta_\nu$$
$$+\frac{X}{m_1}(p_{1\mu}p_{1\sigma}-\delta_{\mu\sigma}p_1^2)\,p_{2\tau}(\hat{p}_3+m_1)\,\beta_\nu$$
$$+\frac{X}{m_1}p_{1\mu}(p_{2\sigma}p_{2\tau}-\delta_{\sigma\tau}p_2^2)(\hat{p}_3+m_1)\,\beta_\nu$$
$$+\beta_\mu(\hat{p}_1+m_1)\,\beta_\sigma(\hat{p}_2+m_1)\,\frac{X}{m_1}(p_{3\tau}p_{3\nu}-\delta_{\tau\nu}p_3^2)$$
$$+\frac{X}{m_1}(p_{1\mu}p_{1\sigma}-\delta_{\mu\sigma}p_1^2)(p_{3\tau}p_{3\nu}-\delta_{\nu\tau}p_3^2)\Big\}$$
$$\times\frac{[\delta_{\mu\nu}+(k_\mu k_\nu/k^2)(d_l-1)]\,d^4k}{(p_1^2-m_1^2)(p_2^2-m_1^2)(p_3^2-m_1^2)\,k^2}$$

(for simplicity we have written $p_1 = p - k$; $p_2 = p - k - q_1$; $p_3 = p - k - q_1 - q_2$).

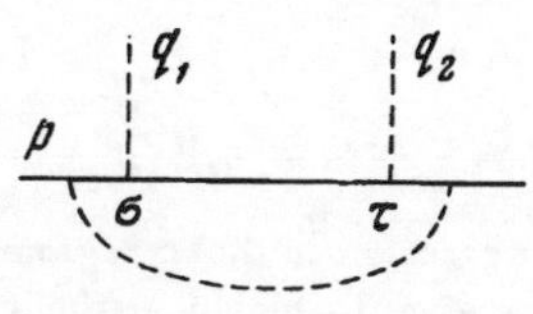

FIG. 1

The highest power of the virtual momentum k in the numerator is equal to 4, and therefore the integral diverges logarithmically in the region of large k. In this region $p_1 = p_2 = p_3 = k \gg m$:

$$-\frac{e_1^2}{\pi i}\frac{X}{m_1}\int\limits_{-k^2\gg -f^2}[(k_\mu k_\sigma-\delta_{\mu\sigma}k^2)\,k_\tau k_\nu$$
$$+k_\mu k_\nu(k_\sigma k_\tau-\delta_{\sigma\tau}k^2)+k_\mu k_\sigma(k_\tau k_\nu-\delta_{\tau\nu}k^2)$$
$$+(k_\mu k_\sigma-\delta_{\mu\sigma}k^2)(k_\tau k_\nu-\delta_{\tau\nu}k^2)]$$
$$\times[\delta_{\mu\nu}+(k_\mu k_\nu/k^2)(d_l-1)]\,d^4k/(k^2)^4.$$

The nonzero result gives, in the numerator, the product of the longitudinal part of the photon function by the second term in the square brackets. Averaging over angles and integrating, we find

$$\frac{e_1^2}{16\pi}\frac{3X}{m_1}\delta_{\sigma\tau}\int\limits_{\ln(-f^2/m^2)}^{\ln(\Lambda^2/m^2)}(d_l(z)-1)\,dz$$

in complete agreement with (12) (the expression obtained should be doubled to take account of exchanging the positions of quanta (q_1 and q_2). Therefore for $d_l = 1$ the Compton diagrams are finite.

b) Vertex part with two intersecting photon lines. The vertex diagram with two intersecting photon lines (Fig. 2) can be treated similarly. The matrix element for this diagram is of order e_1^4 and is proportional to the integral

$$\Lambda_\sigma(p, p-l, l) \tag{A1}$$

$$\sim \int \{\beta_\mu G(p-k)\beta_\rho G(p-k_1-k_2)\beta_\sigma$$

$$\times G(p-l-k_1-k_2)\beta_\nu\, G(p-l-k_2)\beta_\tau\} D_{\mu\nu}(k_1)$$

$$\times D_{\rho\tau}(k_2)\, d^4k_1 d^4k_2.$$

We shall not write out the whole expression in the brackets in the integrand, which is extremely complicated, and shall consider only the denominators

$$(p-k_1)^2(p-k_1-k_2)^2 \tag{A2}$$

$$\times\ (p-l-k_1-k_2)^2(p-l-k_2)^2k_1^2k_2^2.$$

If we exclude integration in those regions of k_1 and k_2 which can give double logarithmic integrals, then it can be seen that there are two possibilities.

1. $k_1 \gg k_2$ (or $k_2 \gg k_1$). Then the numerator of the integrand should contain terms in the fourth power of k_1. With the choice $d_l = 1$, no such terms appear, as has been shown above.

2. The case in which $k_1 + k_2$ is small in comparison with k_1 (although large in comparison with p and l), which corresponds to a change of variables $k_2 = q - k_1$ or $k_1 = q - k_2$). The denominator of Eq. (A2) is about equal to

$$(k_1^2)^2[(k_1-q)^2]^2(p-l-q)^2(p-q)^2.$$

We must therefore look for terms in the expression for $\Lambda_\sigma(p, p-l, l)$ which can give the power k_1^4 in the numerator. Such terms are easily found. The corresponding double logarithmic integral (for $d_l = d_t$) will be

$$\frac{X}{m_1}\int \{(k_\mu k_\rho - \delta_{\mu\rho}k^2)(k_\mu k_\rho - \delta_{\mu\rho}k^2)(2p-l-2q)_\sigma\}$$

$$\times \frac{d^4k\, d^4q}{(k^2)^4 q^4 (1-2(p,q)/q^2)(1-2(p-l,q)/q^2)}.$$

It thus follows that the logarithmic term gives only integration over k, and integration over q after averaging over angles vanishes.

With respect to the diagram being considered, we must bear in mind the following. In calculating the polarization tensor

$$P_{\mu\nu}(l) = (e_1^2/\pi i)$$

$$\times \int \mathrm{Sp}\,[B_\mu(p-l, p, l)\,G(p)\,\beta_\nu G(p-l)]\, d^4p$$

it turns out that the combination of factors

$G(p)\beta_\nu G(p-l)$ contains the following powers of p in the numerator: p^2, p^3/m_1. Therefore, in general, in $B_\sigma(p-l, p, l)$ in the region $p \gg l$ we must know, in addition to the term proportional to β_σ, also the corrections of order l^2/p^2 and m_1l^2/p^3. Terms proportional to β_σ, as we have seen above, do not occur. Terms $\sim l^2/p^2$ and ml^2/p^3 occur in (A1) from the integration region

$$p \gg k_1' \gg k_1' + k_2 \gg l\ (k_1' = p - k_1).$$

Their total contribution, however, to the polarization tensor vanishes. The same is true about the diagram of Fig. 1, for all corrections having a matrix structure other than expression (20). Consideration of higher order diagrams in the vertex part is complicated by the increasing complexity of the matrix expressions and the integrals. The above consideration of the matrix element of the vertex diagram with two intersecting photon lines allows us, however, to assume that for $d_l = d_t$ the higher approximations also give no contribution to the equation for the vertex part. The validity of this assumption is demonstrated by the final expression for the photon Green's function (24).

APPENDIX II

Gauge transformation of the vertex part. We shall find the vertex part $B_\mu(p, p-l, l)$ for arbitrary d_l on the basis of the relation [10]

$$G(p)\,\delta B_\mu(p, p-l, l)\,G(p-l) \tag{A3}$$

$$= -\frac{ie_1^2}{\pi}\int \{G(p)\,B_\mu(p, p-l, l)\,[G(p-l)$$

$$-G(p-l-k)] + G(p-k)$$

$$\times\ [B_\mu(p-k, p-l-k, l)\,G(p-l-k)$$

$$-B_\mu(p, p-l, l)\,G(p-l)]\}\,\delta d_l(k^2)\, d^4k/(k^2)^2,$$

which gives the form of the infinitesimal gauge transformation of $B_\mu(p, p-l, l)$. Considering only slowly varying $d_l(k^2)$, we see that the integral on the right side of (A3) is logarithmic in the region $|-k^2| \gg |(p-l)^2|$.

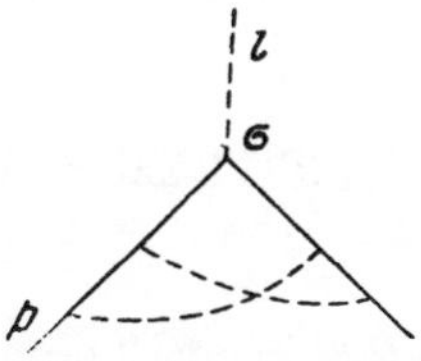

FIG. 2

We note, first of all, that the expressions (38) for the meson Green's function can be written in the matrix form

$$G(p) \approx \beta(p^2)\Big\{\hat{p} + m_1(p^2) + \frac{\hat{p}^2 - p^2}{m_1(p^2)} - \frac{Yp^2}{m_1(p^2)}[\varphi(p^2) - 1]\Big\}\frac{1}{p^2}, \tag{A4}$$

where $m_1(p^2)$ is given by (36),

$$\beta(p^2)/m_1(p^2) = \lambda(p^2)/m_1, \; \varphi(p^2) = \frac{1}{4\lambda(p^2)}[\lambda(p^2) + 3\mu(p^2)].$$

From (14a) and (A4) it follows that

$$G^{-1}(p) = \Big\{\hat{p} - m_1(p^2) + \frac{Xp^2}{m_1(p^2)}[\varphi(p^2) - 1]\Big\}\frac{1}{\beta(p^2)\varphi(p^2)}. \tag{A5}$$

In the region $p \gg l$ or $p \sim l$, we shall attempt to find the vertex part in the form

$$B_\mu(p, p-l, l) = \beta_\mu \alpha[(p-l)^2] + (2p-l)_\mu X\rho[(p-l)^2]. \tag{A6}$$

Turning to (A3), we see that the first term in the second square bracket can be dropped, and Eq. (A3) can be written in the form

$$\delta B_\mu(p, p-l, l) = -\frac{ie_1^2}{\pi}\Big\{B_\mu(p, p-l, l)\int\frac{\delta d_l(k^2)\,d^4k}{(k^2)^2} - B_\mu(p, p-l, l)\int G(p-l-k) \times G^{-1}(p-l)\frac{\delta d_l(k^2)\,d^4k}{(k^2)^2} - G^{-1}(p)\int G(p-k) \times\frac{\delta d_l(k^2)\,d^4k}{(k^2)^2}B_\mu(p, p-l, l)\Big\}. \tag{A3a}$$

In order to obtain the logarithmic expressions in the second and third terms, we need consider only the quadratic terms in the numerator

$$(\beta(k^2)/m_1(k^2))\{\hat{k}^2 - k^2 - Yk^2[\varphi(k^2) - 1]\}.$$

After averaging over angles this expression becomes

$$-(3/4m_1)\mu(k^2)Yk^2.$$

Going over to Euclidean space $k_0 \to ik_0$, introducing the new variable $\xi = \ln[(p-l)^2/m^2]$, and comparing corresponding terms in both parts of Equation (A3a), we obtain the following equations for the functions $\alpha(\xi)$ and $\rho(\xi)$:

$$\delta\alpha(\xi) = \frac{e_1^2}{4\pi}\alpha(\xi)\Big\{\int_\xi^L \delta d_l(z)\,dz - \frac{3}{\lambda(\xi) + 3\mu(\xi)}\int_\xi^L \mu(z)\,\delta d_l(z)\,dz\Big\}; \tag{A7a}$$

$$\delta\rho(\xi) = \frac{e_1^2}{4\pi}\Big\{\rho(\xi)\int_\xi^L \delta d_l(z)\,dz + \frac{3\alpha(\xi)}{m_1(\xi)(\lambda(\xi) + 3\mu(\xi))}\int_\xi^L \mu(z)\,\delta d_l(z)\,dz\Big\}, \tag{A7b}$$

where $L = \ln(\Lambda^2/m^2)$.

By comparing (A7a) with Eqs. (39a) and (39b), we obtain

$$\delta\alpha(\xi)/\alpha(\xi) = -\delta[\lambda(\xi) + 3\mu(\xi)]/[\lambda(\xi) + 3\mu(\xi)].$$

Using the boundary condition

$$\alpha(\xi)|_{d_l = d_t} = \beta^{-1}(\xi)|_{d_l = d_t}$$

and the fact that the function $m_1(\xi)$ does not depend on d_l, we obtain

$$\alpha(\xi) = [\varphi(\xi)\beta(\xi)]^{-1},$$

which is in agreement with (A5) and the Ward identity. Dividing (A7a) by $m_1(\xi)$ and adding it to (A7b), we find

$$\delta\Big[\rho(\xi) + \frac{\alpha(\xi)}{m_1(\xi)}\Big] = \frac{e_1^2}{4\pi}\Big[\rho(\xi) + \frac{\alpha(\xi)}{m_1(\xi)}\Big]\int_\xi^L \delta d_l(z)\,dz.$$

We have $\rho(\xi)|d_l = d_t = 0$. Integrating the previous equation, we obtain the following expression for $\rho(\xi)$:

$$\rho(\xi) = [1 - \alpha(\xi)\beta(\xi)]/m_1(\xi)\beta(\xi), \quad \text{(A9)}$$

which again is in agreement with (A5).

In the other case, when $l \gg p$, expression (A6) for the vertex part becomes inexact. Instead of this, $B_\mu(p, p-l, l)$ is represented in the following form:

$$B_\mu(p, p-l, l) = \beta_\mu \alpha[(p-l)^2] + (2p-l)_\mu X\rho[(p-l)^2] + p_\mu X\chi_1(p-l, p) + \beta_\mu X\chi_2(p-l, p). \quad \text{(A10)}$$

The logarithmic functions $\chi_1(\xi, \eta)$ and $\chi_2(\xi, \eta)$ vanish for $d_l = 0$ and when $\xi = \eta$. Insertion of (A10) into Eq. (A3a) leads to equations for the functions χ_1 and χ_2, which can be transformed to the following expressions:

$$\delta[\chi_1(\xi, \eta) + \rho(\xi)] = \frac{e_1^2}{4\pi}\left\{[\chi_1(\xi, \eta) + \rho(\xi)]\int_\xi^L \delta d_l(z)\,dz + \frac{3[\alpha(\xi) + \chi_2(\xi, \eta)]}{[\lambda(\eta) + 3\mu(\eta)]\,m_1(\eta)}\int_\xi^L \mu(z)\,\delta d_l(z)\,dz\right\};$$

$$\delta[\chi_2(\xi, \eta) + \alpha(\xi)] = \frac{e_1^2}{4\pi}[\chi_2(\xi, \eta) + \alpha(\xi)]\left\{\int_\xi^L \delta d_l(z)\,dz - \frac{3}{\lambda(\eta) + 3\mu(\eta)}\int_\xi^L \mu(z)\,\delta d_l(z)\,dz\right\}.$$

The solutions of these equations cannot be simply expressed in terms of the functions $\lambda(\xi)$, $\mu(\xi)$, and $m_1(\xi)$. We note only that the following relation holds:

$$m_1(\eta)\chi_1(\xi, \eta) + \chi_2(\xi, \eta) = [m_1(\xi) - m_1(\eta)]\,\rho(\xi).$$

The expressions we have obtained for the vertex part show again that for arbitrary d_l the functions entering into this expression do not satisfy the renormalizability condition.

1 Landau, Abrikosov and Khalatnikov, Dokl. Akad. Nauk SSSR **95**, 497, 773, 1177 (1954).

2 M. Gell-Mann and F. Low, Phys. Rev. **95**, 1300 (1954)

3 R. Feynman, Phys. Rev. **76**, 749, 769 (1949).

4 D. Peasly, Phys. Rev. **81**, 94 (1951)

5 P. Matthews and A. Salam, Rev. Mod. Phys. **23**, 311 (1951).

6 A. Salam, Proc. Roy. Soc. (London) **211A**, 276 (1952).

7 F. Dyson, Phys. Rev. **75**, 1736 (1949).

8 I. Ia. Pomeranchuk, Dokl. Akad. Nauk SSSR **103**, 1005 (1955).

9 J. Schwinger, Proc. Nat. Acad. Sci. **37**, 452 (1951).

10 L. D. Landau and I. M. Khalatnikov, J. Exptl. Theoret. Phys. (U.S.S.R.) **29**, 89 (1955); Soviet Phys. JETP **2**, 69, (1956).

Translated by E. J. Saletan
221

PHYSICAL REVIEW VOLUME 111, NUMBER 1 JULY 1, 1958

Possibility of Formulation of a Theory of Strongly Interacting Fermions

A. A. ABRIKOSOV, A. D. GALANIN, L. P. GORKOV, L. D. LANDAU, I. YA. POMERANCHUK, AND K. A. TER-MARTIROSYAN
Academy of Sciences of U.S.S.R., Moscow, U.S.S.R.

(Received December 26, 1957)

A fermion field is investigated with the interaction Lagrangian density equal to $g(\bar{\psi}O_j\psi)\cdot(\bar{\psi}O_j\psi)$. This point interaction is considered as a limit of an extended one, where it is supposed that the interaction vanishes if the momentum of a particle exceeds Λ and/or if the momentum transferred in a collision of two particles exceeds λ. The relation between Λ and λ is such as to make the quantity $\lambda^2/\ln(\Lambda/\lambda)$ arbitrarily small as $\lambda\to\infty$ and $\Lambda\to\infty$. This choice of the limiting procedure considerably simplifies the investigation of the theory. It is shown that in the limit $\lambda\to\infty$, $\Lambda\to\infty$, the physical interaction between particles vanishes in all types of four-fermion interactions. The case of two interacting fields ψ and χ with different isotopic spin is also considered. Going over to the local theory, the physical interaction vanishes in this case as well.

This result shows that in the cases considered no strongly interacting fermion theory can be constructed. In the case of the weak interaction, although no logically consistent theory can be built up, there does exist the perturbation theory, as in electrodynamics, which is valid for sufficiently small energies.

1. INTRODUCTION

THE transition to the limit of a point interaction in electrodynamics[1] or meson theories[2] entails a difficulty which is connected with vanishing of the renormalized charge and the disappearance of physical interactions between particles. It would seem to be of interest to ascertain whether this difficulty can be overcome by replacing the Yukawa-type interaction $g_0(\bar{\psi}O_j\psi)\varphi_j$ between fermions and bosons by other types of interactions, such as that between bosons alone or that between fermions alone. The difficulty created by the vanishing of the renormalized charge remains in the case of a system of bosons whose interaction is determined by the operator

$$\frac{g_0}{4!}\int\varphi^4(x)dv.$$

In this case the dependence of g_c on g_0 and Λ (Λ is the cutoff momentum) has the form[3]

$$g_c=\frac{g_0}{1+\frac{3}{2}g_0\ln(\Lambda^2/\mu^2)},\tag{1}$$

and g_c vanishes for $\Lambda\to\infty$ and any arbitrary dependence of g_0 on Λ, providing that g_0 is positive. Negative values of g_0 (for which, in the limit $\Lambda\to\infty$, g_c may not vanish) are in general inadmissible because no stationary states of a boson system exist for $g_0<0$. Indeed, for boson fields a classical limiting case exists in which each state may contain many particles. For $g_0<0$ the energy of the classical field φ,

$$\tfrac{1}{2}\left[\left(\frac{\partial\varphi}{\partial x_\mu}\right)^2+\mu^2\varphi^2\right]+\frac{g_0}{4!}\varphi^4(x),$$

is not positive definite and can decrease indefinitely with increase of the field amplitude φ. Physically this means that it should be energetically possible for an infinite number of particles to be created from vacuum. Thus the vacuum cannot exist for $g_0<0$.

The constant g_0 for the Fermi interaction

$$V=2\pi^2g_0\int(\bar{\psi}O_j\psi)(\bar{\psi}O_j\psi)dv\tag{2}$$

(where the O_j are the ordinary spin and isotopic spin operators for fermions) can have any sign, since the occupation numbers cannot exceed unity and stationary states exist for any sign of g_0. In the given case, the turning on of the interaction simply leads to a redistribution of the levels of negative and positive energy. The new stationary state with minimal energy which arises after the interaction is "turned on" is the one of a physical vacuum.

Hence if the relation between g_c and g_0 in this case were also determined by a formula analogous to (1), the renormalized charge g_c would not vanish for $g_0<0$. Meson theories could be based on interaction (2) and the mesons from the very start would be similar (in the sense of the Fermi-Yang concept[4]) to nonlocal formations of fermions.

In the following we shall consider the possibility of setting up a theory of this type.[5]

2. EQUATION FOR THE VERTEX OPERATOR

If the interaction has the same form as (2), the matrix elements will contain quadratic and logarithmic divergences. We shall cut off the diverging integrals by assuming that interaction (2) is somewhat "smeared out": it will be assumed that the interaction vanishes in the momentum representation if momenta P and

[1] L. D. Landau and I. Ya. Pomeranchuk, Doklady Akad. Nauk S.S.S.R. **102**, 489 (1955).
[2] I. Ya. Pomeranchuk, Doklady Akad. Nauk S.S.S.R. **103**, 1005 (1955); **104**, 51 (1955); **105**, 461 (1955).
[3] Pomeranchuk, Sudakov, and Ter-Martirosyan, Phys. Rev. **103**, 784 (1956). This formula was derived for $g_0<1$; if, however, two cutoff momenta are introduced it can be shown that it will be valid for an *arbitrary* positive value of g_0.
[4] E. Fermi and C. Yang, Phys. Rev. **76**, 1739 (1949).
[5] An interaction of the same type as (2) has been discussed in a number of papers. See for example, B. Jouvet, Nuovo cimento **5**, 1 (1957).

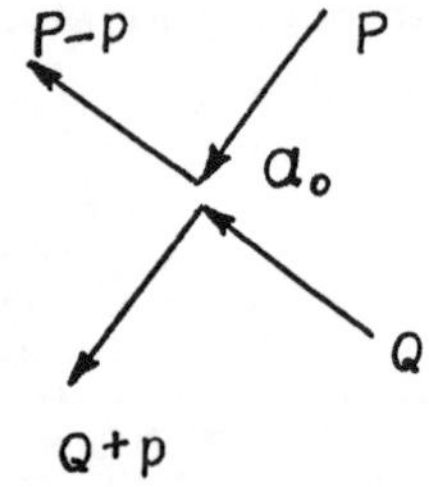

FIG. 1. Simplest four-fermion vertex.

Q, directed along the fermion lines according to Fig. 1,[6] exceed some limiting momentum Λ or if the momentum p transferred from one line to another exceeds λ, where $\lambda \ll \Lambda$.[7] In order to go over to the limit of a point interaction one should put $\Lambda \to \infty$, $\lambda \to \infty$, the relation between Λ and λ being arbitrary.

If the theory is internally consistent, one may expect the result of the limiting process to be independent of the nature of the transition. The latter, however, should not violate the general conditions required by any physically reasonable theory, such as the general theorems regarding the behavior of Green's functions,[8] gauge invariance, etc. We shall restrict ourselves to the case of a limiting transition in which λ/Λ remains arbitrarily small, since in this case the analysis is considerably simplified.

It will be shown that if g_0 is assumed in general to depend on the cutoff limits, one finds that for any form of this dependence, the exact solution yields in the limit of a point interaction the result that no physical interaction exists between fermions.

Of those graphs which define the vertex operator $\Gamma(P,Q,p)$ (at which the momentum arrangement corresponds to that drawn in Fig. 1), the largest contribution comes from those in which integration over virtual momenta is performed along closed loops (that is, up to Λ^2) and in which the degree of divergence of the integrals is maximal. This is the loop chain shown in Fig. 2.

We shall now find the total contribution (Γ_1) of all of these graphs to the vertex operator and show, by using the value of the vertex operator thus obtained, that the contribution of the remaining graphs (Fig. 4) to Γ is, for sufficiently small λ/Λ, arbitrarily small, and that the Green's function of a fermion is identical with the function $G_0 = (-i\hat{p}-m)^{-1} \cong i/\hat{p}$ for a free fermion.[9]

[6] In the figures, line discontinuities signify that a large momentum (exceeding λ but smaller than Λ) cannot be transferred at the point of discontinuity.

[7] This corresponds to writing (2) with a form factor $F_{\Lambda,\lambda}$:
$$\int_{-\infty}^{+\infty} V(t)dt = 2\pi^2 g_0 \int (\bar{\psi}(x)O_j\psi(x'))(\bar{\psi}(y)O_j\psi(y')) \times F_{\Lambda,\lambda}(x-x', y-y', x-y)dxdx'dydy',$$
in which the width of the distributions with respect to $x-x'$ and $y-y'$ are identical and equal $1/\Lambda$, whereas the width of the distribution with respect to $x-y$ is $1/\lambda$.

[8] H. Lehmann, Nuovo cimento **11**, 342 (1954).

[9] Or is equal to βG_0 where β is a constant, the value of which can be determined from the equation for G (see below). We have employed the following notation here: $\hat{p} = \gamma_\mu p_\mu$; $\gamma_\mu^\dagger = \gamma_\mu$.

The infinite sum of quantities corresponding to the graphs in Fig. 2 satisfies the equation (see Fig. 3)

$$\Gamma_1(P,Q,p) = a_0 - i\int^{\Lambda^2} a_0 G(l-p)G(l)\Gamma_1(l,Q,p)d^4l, \quad (3)$$

where

$$a_0 = g_0(O_j \times O_j) \quad (3a)$$

is a quantity which corresponds[10] to the simplest graph in Fig. 1; $d^4l = (2\pi)^{-2}dl_1dl_2dl_3dl_4$. The spinor indices are arranged in the integral term in accord with the closed loop in Fig. 3; that is, the integral contains a trace. It is easy to verify that a consequence of this is that no interference occurs between the various interactions (3a) in (3).

3. SCALAR INTERACTION

Equation (3) possesses a solution which depends only on p; in the simplest case of scalar ($O_j=1$) or pseudoscalar ($O_j=i\gamma_5$) theory the solution is $\Gamma_1 = \alpha(p) \times (O_j \times O_j)$, where

$$\alpha(p) = g_0\left\{1 + g_0 i\int^{\Lambda^2} \mathrm{Sp}[O_jG(l-p)O_jG(l)d^4l]\right\}^{-1}.$$

Inserting $G(l) = G_0(l) = i/\hat{l}$, we evaluate the quadratically divergent integral involved:

$$i\int^{\Lambda^2} \mathrm{Sp}[O_jG(l-p)O_j'G(l)]d^4l = \tfrac{1}{4}\,\mathrm{Sp}(O_j\gamma_\nu O_j'\gamma_\mu)\left(\tfrac{1}{2}J_0\delta_{\mu\nu} + \frac{p_\mu p_\nu}{p^2}J_1\right), \quad (4)$$

where[11]

$$J_0 = \frac{8}{3i}\int^{\Lambda^2} \frac{k^2-(pk)^2/p^2}{(k-\frac{1}{2}p)^2(k+\frac{1}{2}p)^2}d^4k = \frac{8}{3\pi}\int^{\Lambda^2}(k^2)^2dk^2\int_0^1 \frac{(1-x^2)^{\frac{3}{2}}dx}{(k^2+\frac{1}{4}p^2)^2 - p^2k^2x^2} = \tfrac{1}{2}\Lambda^2 - \tfrac{1}{6}p^2\ln(\Lambda^2/\zeta_0^2p^2),$$

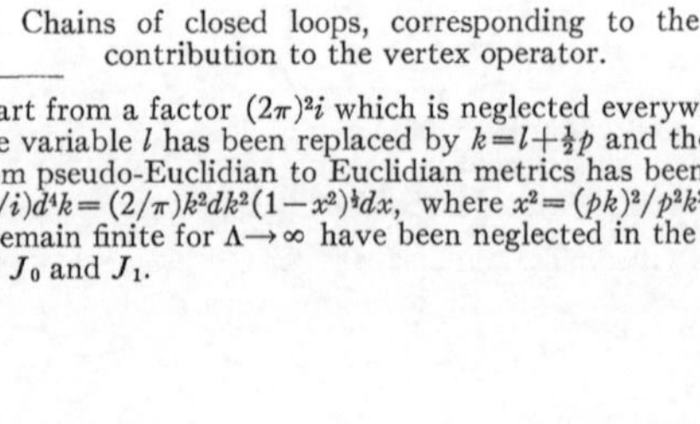

FIG. 2. Chains of closed loops, corresponding to the largest contribution to the vertex operator.

[10] Apart from a factor $(2\pi)^2 i$ which is neglected everywhere.

[11] The variable l has been replaced by $k = l + \frac{1}{2}p$ and the transition from pseudo-Euclidian to Euclidian metrics has been carried out: $(4/i)d^4k = (2/\pi)k^2dk^2(1-x^2)^{\frac{1}{2}}dx$, where $x^2 = (pk)^2/p^2k^2$. Terms which remain finite for $\Lambda \to \infty$ have been neglected in the calculations of J_0 and J_1.

$$J_1=\frac{4}{i}\int^{\Lambda^2}\frac{\frac{4}{3}\{[(pk)^2/p^2]-\frac{1}{4}k^2\}-\frac{1}{4}p^2}{(k-\frac{1}{2}p)^2(k+\frac{1}{2}p)^2}d^4k$$

$$=\frac{4}{\pi}\int^{\Lambda^2}k^2dk^2\int_0^1\frac{\frac{4}{3}k^2(x^2-\frac{1}{4})-\frac{1}{4}p^2}{(k^2+\frac{1}{4}p^2)^2-p^2k^2x^2}(1-x^2)^{\frac{1}{2}}dx$$

$$=-\tfrac{1}{6}p^2\ln(\Lambda^2/\zeta_1^2p^2),$$

and ζ_0 and ζ_1 are constants whose exact values are of no significance for the following. Substituting these values of J_0 and J_1 in (4), we get

$$\alpha(p)=g_0\{1+g_0[\Lambda^2-\tfrac{1}{2}p^2\ln(\Lambda^2/\zeta p^2)]\}^{-1}, \qquad (5)$$

where $\zeta=(\zeta_0^2\zeta_1)^{\frac{1}{3}}$. The quadratic divergence in the denominator cancels out if $g_0=-1/\Lambda^2$ (or if $1+g_0\Lambda^2$ is a quantity which decreases with increase of Λ^2); if, however, $g_0\neq-1/\Lambda^2$, $\alpha(p)$ will be practically independent of p^2 (it may be recalled that $p^2\leqslant\lambda^2\ll\Lambda^2$) and for $\Lambda\rightarrow\infty$ it vanishes as $1/\Lambda^2$. Postponing the analysis of this case, we shall first assume that

$$1+g_0\Lambda^2=\mu^2/\Lambda^2,$$

where μ is a quantity of the order of the lower cutoff limit. Neglecting μ^2/Λ^2 compared with unity, we get from (5)

$$\alpha(p)=-1/(p^2L+\mu^2), \qquad (5a)$$

where $L=\ln(\Lambda/\lambda)$ [ζp^2 has been replaced by λ^2 as a result of which $\alpha(p)$ simply increases]. With the aid of this value of $\alpha(p)$, we shall estimate the magnitude of the contribution from the graph in Fig. 4(a) which determines the difference $G-G_0$, where $G\equiv G_0\beta$; that is, the difference between β and unity.

The spinor indices can be neglected in estimating the integral corresponding to Fig. 4(a), since it is only the order of magnitude that matters. The integral for Fig. 4(a) has the following form (α being a function of f^2 in the given case):

$$g_0\int^{\Lambda^2}\frac{\alpha(f^2)d^4f}{p-f}\int^{\Lambda^2}\frac{d^4l}{l(l+f)}. \qquad (6)$$

Integration over l yields Λ^2 (inasmuch as $f^2\leqslant\lambda^2\ll\Lambda^2$ and the integral over l diverges and therefore $l+f\approx l$). For integration over f, the factor $(p-f)^{-1}$ should be expanded into a series in p:

$$(p-f)^{-1}=-\frac{1}{f}-\frac{1}{f}\frac{1}{f}p-.$$

Integration of the first term yields zero; if (5a) is taken into account, the second term yields $(1/L)\times\ln(\lambda^2/p^2)$ after integration. Taking into account that $g_0=-1/\Lambda^2$, we get

$$G^{-1}-G_0^{-1}=-ip\frac{\ln(\lambda^2/p^2)}{L},$$

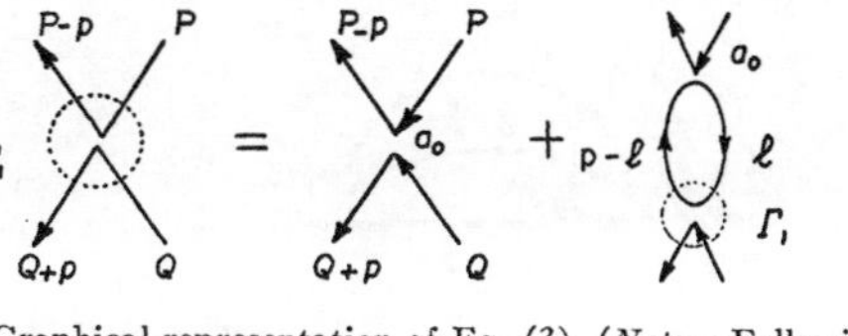

FIG. 3. Graphical representation of Eq. (3). (*Note.*—Following the plus sign, "$p-l$" should read "$l-p$.")

or, for $L\gg\ln(\lambda^2/m^2)$,

$$\beta-1=-\frac{\ln(\lambda^2/p^2)}{L}.$$

Hence for sufficiently large values of L [and for $L\gg\ln(\lambda^2/m^2)$], the difference between β and unity is arbitrarily small.

In a similar manner, we consider the ratio of the integrals corresponding to Figs. 4(b), (c), (d), and (e) [not included in Eq. (3)] to $\alpha(p)$. For Fig. 4(b) this ratio is

$$\frac{1}{\alpha(p)}\int^{\Lambda^2}\frac{\alpha(p)\alpha(f)}{(P-f)(Q+f)}d^4f.$$

Depending on the magnitude of the momenta P and Q, *viz.*, $\lambda\ll P\sim Q\ll\Lambda$, $Q\ll\lambda\ll P\ll\Lambda$, or $p\sim P\sim Q\ll\lambda\ll\Lambda$, we obtain, respectively, taking into account (5a),

$$\frac{\lambda^2}{PQL},\quad \frac{P}{Q}\frac{\ln(\lambda^2/p^2)}{L},\quad \text{or}\quad \frac{\ln(\lambda^2/p^2)}{L}.$$

In any of the cases considered the ratio can be made arbitrarily small.

For the diagrams in Fig. 4(c), we get (see reference 6)

$$\frac{1}{\alpha(p)}\int^{\Lambda^2}d^4l\int^{\Lambda^2}d^4l'\int^{\lambda^2}d^4k\frac{1}{l(l+p)(l+k)}$$

$$\times\frac{1}{l'(l'+k)(l'-p)}\alpha^2(p)\alpha(k)\alpha(p-k).$$

To calculate the integral over l (and l') it should be expanded in k, and p should be set equal to zero. The result is

$$\alpha(p)\int^{\lambda^2}k^2\alpha(k)\alpha(p-k)[\ln(\Lambda^2/k^2)]^2d^4k$$

$$\cong\frac{1}{p^2L}\int^{\lambda^2}\frac{d^4k}{k^2}\cong\frac{\lambda^2}{p^2L}.$$

Figure 4(d) leads to the following integral:

$$\frac{1}{\alpha(p)}\int^{\Lambda^2}d^4l\int^{\lambda^2}d^4k\frac{1}{l(l+p)(l+k)(l+p+k)}\alpha^2(p)\alpha(k).$$

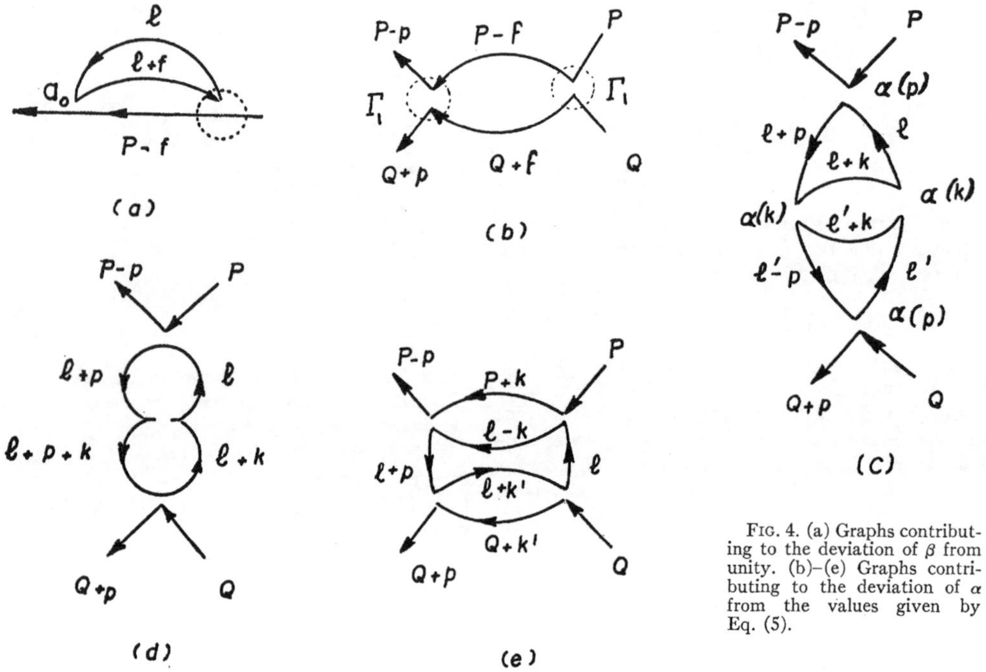

FIG. 4. (a) Graphs contributing to the deviation of β from unity. (b)–(e) Graphs contributing to the deviation of α from the values given by Eq. (5).

Integration over l yields $\ln(\Lambda^2/k^2)$, and we obtain

$$\frac{1}{p^2L}\int^{\lambda^2}\alpha(k)\ln\left(\frac{\Lambda^2}{k^2}\right)d^4k \cong \frac{\lambda^2}{p^2L}.$$

Finally, we shall consider Fig. 4(e). The integration up to a small limit performed in Fig. 4(b) is repeated twice in this diagram. One may therefore expect that it will be smaller than Fig. 4(b). The corresponding ratio is

$$\frac{1}{\alpha(p)}\int^{\Lambda^2}d^4l\int^{\lambda^2}d^4k\int^{\lambda^2}d^4k' \times\frac{\alpha(k)\alpha(k')\alpha(k-p)\alpha(k'-p)}{l(l+p)(l-k)(l+k')(P+k)(Q+k')}.$$

Depending on the magnitudes of P and Q, namely, $\lambda \ll P \sim Q \ll \Lambda$, $p \sim Q \ll \lambda \ll P$, or $p \sim P \sim Q \ll \lambda$, we correspondingly get

$$\frac{p^2}{PQ}\left(\frac{\ln\lambda^2/p^2}{L}\right)^2,\quad \frac{p}{P}\frac{\ln\lambda^2/p^2}{L^2},\quad \text{or}\quad \frac{1}{L^2}.$$

These quantities can be made arbitrarily small if the limiting procedure is carried out in such a manner that λ^2/m^2L remains sufficiently small for $\Lambda \to \infty$. More complex graphs of the vertex operator not indicated in Fig. 4 are proportional to higher powers of the same ratio λ^2/m^2L (or of a smaller quantity) and hence are certainly small.

The expression (5a) obtained for $\alpha(p)$ is thus an exact solution.

The physical interaction between two fermions is determined by the product $\alpha\beta^2$. The results obtained above indicate that $\alpha\beta^2$ vanishes for $\Lambda \to \infty$; that is, physical interaction between point fermions is absent.

The case when $1+g_0\Lambda^2$ is negative should be excluded. Indeed, if $1+g_0\lambda^2$ is negative and equal to -1, by order of magnitude, the logarithmic term in the expression for Γ can be neglected. Evaluating the additional term in the Green's function G with a negative Γ, we see that β would exceed unity, and this is inconsistent with the general theorems of field theory.[8]

If the absolute value of $1+g_0\Lambda^2$ is less than $g_0\lambda^2$, a pole in the space values of p^2 will arise in the formula for Γ. Since, in treating the interaction between fermions as an interaction between bosons, Γ has the meaning of a boson Green's function, it is evident that this pole corresponds to an imaginary boson mass. Such bosons would yield a term $-|m|^2\varphi^2$ in the Hamiltonian and this would point to nonstability of the vacuum in this case.

Returning to the case $g_0 \neq -1/\Lambda^2$, we note that Eq. (3) should be solved simultaneously with the equation for the Green's function $G(p)$,

$$\left[-ip-\frac{g_0}{2}\int a_0G(p-f)\Gamma_1(f)G(l)G(l+f)d^4l d^4f\right]G(p)=1,$$

[the order of the spinor indices corresponds to Fig. 4(a); that is, the integral term contains an integral over l of the same type as (4)].

Neglecting terms of the form $p^2 \ln(\Lambda^2/p^2)$ compared with Λ^2 and taking into account that in this case the integral (4) is equal to Λ^2, we obtain the following solution:

$$\Gamma_1(p) = (O_j \times O_j)\alpha; \quad G(p) = i\beta/p,$$

where $\alpha = g_0[1 + g_0\beta^2\Lambda^2]^{-1}$ and $\beta = [1 + \frac{1}{8}g_0\lambda^2\Lambda^2\alpha\beta^3]^{-1}$ are quantities which do not depend on p. Inserting α in the formula for β, we obtain an equation of the fourth degree with respect to β. Its solution can easily be found in the cases when $g_0\Lambda^2 \ll 1$ or $g_0\Lambda^2 \sim 1$, $g_0\lambda^2 \ll 1$. In the first case we get $\beta = (4/g_0\lambda^2)[(1+\frac{1}{2}g_0\lambda^2)^{\frac{1}{2}} - 1]$, and in the second, $\beta \cong 1$. In both cases (which include the complete range of values of g_0 which are of any interest), the quantity $\xi = \alpha\beta^2\Lambda^2$, in the limit for $\Lambda \to \infty$, remains of the order of unity. Now it is easy to demonstrate that the unaccounted graphs of the type depicted in Fig. 4 are much less than α. For example, the ratio of the quantities corresponding to the graphs in Figs. 4(b) and (d) to α is $\alpha\beta^2(\lambda^4/PQ) = \xi(\lambda^4/PQ\Lambda^2)$, and $\alpha^2\beta^4\Lambda^2\lambda^2 = \xi^2(\lambda^2/\Lambda^2)$. These quantities are arbitrarily small when $\lambda^2/\Lambda^2 \to 0$. It follows that the solutions obtained for Γ and G are exact ones, and since $\alpha\beta^2\Lambda^2 \sim 1$, the physical interaction $\alpha\beta^2$ vanishes as $1/\Lambda^2$ for $\Lambda^2 \to \infty$.

A point which was important for the foregoing analysis was that the integral J_0 contains Λ^2. The magnitude of the quadratically divergent integral, however, may significantly depend on the form factor employed in computing this quantity. Suppose, for example, that the quadratically divergent integral is made to vanish by using an oscillating form factor. Then instead of (5) we get

$$\alpha(p) = \frac{g_0}{1 - \frac{1}{2}g_0 p^2 \ln(\Lambda^2/p^2)}. \tag{7}$$

If g_0 behaves as $1/m^2L$ for $\Lambda \to \infty$, the difference between β and 1 will be of the order of λ^2/m^2L. Indeed, in formula (6) the integral over l now does not contain a quadratically divergent part and hence is equal to $\sim f^2 \ln(\Lambda^2/f^2)$. The remaining integral over f yields $g_0\lambda^2 p \sim (\lambda^2/m^2L)p$, if one takes into account that $\alpha \sim 1/f^2L$ and expands in a series in p.

If g_0 is not small, the equations for α and β should be solved simultaneously. The equation for β can be satisfied in this case by assuming that β is independent of p. Then, instead of (7), we get

$$\alpha(p) = \frac{g_0}{1 - \frac{1}{2}g_0\beta^2 p^2 \ln(\Lambda^2/p^2)}. \tag{7a}$$

The integral in the equation for β has the following form:

$$g_0\beta^3 \int \frac{\alpha(f)}{p-f} d^4f \int \frac{d^4l}{l(l-f)} \cong g_0\beta^3 p \int \frac{\alpha(f)}{f^2} d^4f \cdot f^2 \ln\frac{\Lambda^2}{f^2}.$$

For a sufficiently large L, when $g_0\beta^2\lambda^2 L \gg 1$, this integral equals $g_0\beta\lambda^2 p$ and the following equation is obtained for β:

$$\beta[1 + \tfrac{1}{16}g_0\beta\lambda^2] = 1.$$

Hence

$$\beta = \frac{8}{g_0\lambda^2}[(1 + \tfrac{1}{4}g_0{}^2\lambda^4)^{\frac{1}{2}} - 1];$$

that is, β is independent of L. It therefore follows from (7a) that for a sufficiently large value of L,

$$\alpha\beta^2 \sim 1/p^2L$$

[if $p^2 \to 0$, formula (7a) should be refined by taking into account the finiteness of the mass; then $\alpha\beta^2 \sim 1/m^2L$].

It can readily be verified that in this case the diagrams of Fig. 4 and the more complicated ones are small compared with α in (7a).

Thus the conclusion drawn above regarding the vanishing of the interaction of point fermions does not depend on whether the quadratically divergent integral is considered equal to Λ^2 or to zero. This confirms our viewpoint that the *physical* results of the theory (in the given case, the vanishing of the interaction) should not depend on the form of the cutoff factor.

4. VECTOR AND TENSOR INTERACTION

Consider now the vector interaction theories ($O_j = i\gamma_\mu$, or $O_j = i\gamma_5\gamma_\mu$). We start with the pseudovector theory. According to (4), insertion of Γ_1 in the form $\alpha(p) \times (i\gamma_5\gamma_\mu \times i\gamma_5\gamma_\mu)$ into (3) leads to the appearance of a new spinor form, $(1/p^2)(i\gamma_5 p \times i\gamma_5 p)$, in the right-hand part of the equation. Thus the solution of (3) should be sought in the form

$$\Gamma_1 = \alpha(p)(i\gamma_5\gamma_\mu \times i\gamma_5\gamma_\mu) + \alpha_1(p)(i\gamma_5 p \times i\gamma_5 p)\frac{1}{p^2}. \tag{8}$$

Inserting (8) into Eq. (3) and separately equating the coefficients in each spinor form, we get

$$\begin{aligned} \alpha(p) &= g_0 - g_0(J_0 + J_1), \\ \alpha_1(p) &= 2g_0 J_1 \alpha(p) - g_0(J_0 - J_1)\alpha_1(p), \end{aligned} \tag{9}$$

or

$$\begin{aligned} \alpha(p) &= \frac{g_0}{1 + g_0(J_0 + J_1)}, \\ \alpha(p) + \alpha_1(p) &= \frac{g_0}{1 + g_0(J_0 - J_1)}. \end{aligned} \tag{10}$$

Taking into account the values of the integrals J_0 and J_1, we get

$$\begin{aligned} \alpha(p) &= \frac{g_0}{1 + g_0[\frac{1}{2}\Lambda^2 - \frac{1}{3}p^2 \ln(\Lambda^2/p^2)]}, \\ \alpha(p) + \alpha_1(p) &= \frac{g_0}{1 + \frac{1}{2}g_0\Lambda^2}; \end{aligned} \tag{11}$$

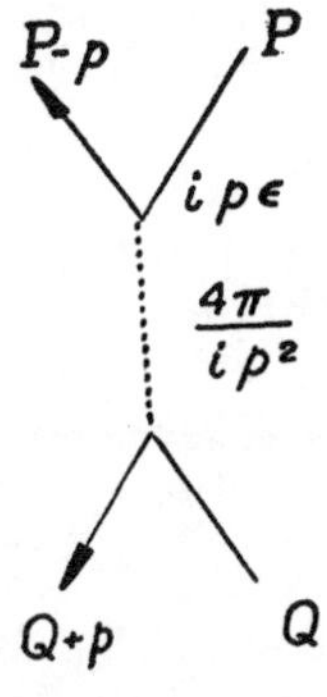

FIG. 5. Simplest graph corresponding to the second-order interaction arising from the term in $\mathfrak{L}_f^{(0)}$.

that is $\alpha(p)$ has the same form as in (5) and for decreases with increasing Λ^2 as $1/L$ for $g_0\sim 1/\Lambda^2$, whereas $\alpha_1(p)$ does not contain $\ln(\Lambda^2/p^2)$ in the denominator and does not tend to zero for $\Lambda^2\rightarrow\infty$.

The term containing $\alpha_1(p)$, however, does not correspond to any real interaction, as it can be excluded with help of a transformation similar to a guage transformation. Let us examine, for example, a system consisting of fermions and of scalar bosons not interacting with the fermions. The Lagrangian of the system has the form

$$\mathfrak{L}=\mathfrak{L}_b^{(0)}+\mathfrak{L}_f^{(0)}+\mathfrak{L}_f^{(1)},$$

where $\int\mathfrak{L}_f^{(1)}dv=-V$, $\mathfrak{L}_b^{(0)}$ is the Lagrangian of the boson free field, and $\mathfrak{L}_f^{(0)}=-\bar{\psi}(\gamma_\mu\partial/\partial x_\mu+m)\psi$ is the Lagrangian for free fermions.

The fermion system can be characterized by a wave field ψ_1 related to ψ by the relation

$$\psi_1=\exp[i\gamma_5\bar{\epsilon}\varphi(x)]\psi;\quad \bar{\psi}_1=\bar{\psi}\exp[i\gamma_5\bar{\epsilon}\varphi(x)],\quad (12)$$

where $\varphi(x)$ is a quantized boson field and $\bar{\epsilon}$ is an operator defined in the momentum representation by the function $\epsilon(p^2)$ [i.e., $\bar{\epsilon}=\epsilon(-\partial^2/\partial x_\mu^2)$]. Under this transformation of the field $\psi(x)$, the Lagrangian $\mathfrak{L}_f^{(1)}$ [i.e., interaction (2)] remains constant, whereas $\mathfrak{L}_f^{(0)}$ changes:

$$\mathfrak{L}_f^{(0)}=\mathfrak{L}_{f1}^{(0)}+\mathfrak{L}_1+\mathfrak{L}_2,$$

where

$$\mathfrak{L}_1=-\bar{\psi}_1 i\gamma_\mu\gamma_5\bar{\epsilon}\frac{\partial\varphi}{\partial x_\mu}\psi_1,$$

and

$$\mathfrak{L}_2=m\bar{\psi}_1\{\exp[-2i\gamma_5\bar{\epsilon}\varphi(x)]-1\}\psi_1.$$

The Lagrangian $\mathfrak{L}_1$ corresponds to the addition to (2) of a fictitious interaction

$$V'=\int\bar{\psi}_1 i\gamma_\mu\gamma_5\bar{\epsilon}\frac{\partial\varphi}{\partial x_\mu}\psi_1 dv.$$

In second order, it leads to scattering of fermions by fermions, $(f_0/p_2)(i\gamma_5 p\times i\gamma_5 p)$ corresponding to the simplest diagram in Fig. 5. Here $f_0=\epsilon^2(p^2)/\pi\geqslant 0$. The presence of this interaction (that is, of all the diagrams created from the simplest one in Fig. 5) can be taken into account in calculating Γ_1 if instead of (3a) we insert in (3) the expression

$$a_0'=g_0(i\gamma_5\gamma_\mu\times i\gamma_5\gamma_\mu)+(f_0/p_2)(i\gamma_5 p\times i\gamma_5 p).$$

After simple transformations, the following results can be derived from Eq. (3):

$$\alpha(p)=\frac{g_0}{1+g_0(J_0+J_1)};\qquad \alpha(p)+\alpha_1(p)=\frac{g_0+f_0}{1+(g_0+f_0)(J_0-J_1)}.\qquad (11a)$$

Inserting the upper equation in the lower, we note that $\alpha_1(p)$ vanishes if

$$f_0=-2g_0^2J_1/(1+2g_0J_1).\qquad (13)$$

The expression for $\alpha(p)$ is identical with (11); it yields the result that the interaction is maximal if $g_0=-2/\Lambda^2$. From (13) we then obtain

$$f_0=(p^2/3\Lambda^4)L.$$

Thus, in order to exclude the term $\alpha_1(p)$ in the expression for Γ_1, it is sufficient to choose $\epsilon(p^2)$ equal to $[(p^2/3\Lambda^4)L]^{\frac{1}{2}}$; it is arbitrarily small for $\Lambda\rightarrow\infty$. It was thus sufficient to consider the fictitious interaction V' in the lower approximation, the Lagrangian $\mathfrak{L}_2$ disappearing for $\Lambda\rightarrow\infty$.

Formula (11a) for $\alpha(p)$ is in all respects similar to the expression (5) obtained in scalar theories and therefore, as in the latter cases, it leads to the disappearance of physical interaction for $\Lambda\rightarrow\infty$.

We shall now consider the vector interaction theory. In this case the quadratically divergent integral can be considered to equal zero. Indeed, in vector theory the integral (4) can be written in the following form:

$$\int T\langle j_\nu(x)j_\mu(y)\rangle_0 e^{ip(x-y)}d^4x d^4y,\qquad (14)$$

where $j_\nu(x)=\bar{\psi}(x)\gamma_\nu\psi(x)$ and[12] $\partial j_\nu(x)/\partial x_\nu=0$. If we put $p=0$ in (14), we formally obtain a quadratically divergent integral. Its structure is the same as that of the integral which in electrodynamics determines the photon mass (since in electrodynamics j_ν satisfies the continuity equation), and it therefore must vanish.[13]

[12] This equation is valid not only for free operators but for coupled operators as well.

[13] A formal proof, not based on the analogy with electrodynamics, can be given, for example, in the following way. Consider the integral

$$\int T\langle j_\nu(x)j_\mu(y)\rangle\frac{\partial\varphi(x)}{\partial x_\nu}d^4xd^4y=\int\frac{\partial}{\partial x_\nu}T\langle j_\nu(x)j_\mu(y)\rangle\varphi(x)d^4xd^4y-\int T\left\langle\frac{\partial j_\nu(x)}{\partial x_\nu}j_\mu(y)\right\rangle\varphi(x)d^4xd^4y.$$

The first integral on the right-hand side is zero if $j_\nu(x)$ vanishes at infinity; the second integral is zero because $\partial j_\nu(x)/\partial x_\nu=0$. If $\partial\varphi(x)/\partial x_\nu$ is a sufficiently slowly varying function of the

In vector theory the term $\alpha_1(p)(ip\times ip/p^2)$ appears in Γ_1, and formula (10) is valid as before; however, instead of (11), we have

$$\alpha(p)=g_0[1-\tfrac{1}{3}g_0p^2\ln(\Lambda^2/p^2)]^{-1};$$
$$\alpha(p)+\alpha_1(p)=g_0. \quad \text{(11b)}$$

We now take notice of the fact that for real particles the interaction $\alpha_1(p)(ip\times ip/p^2)$ is identically equal to zero. Indeed, p is the difference between the final and initial particle momenta. Since the wave functions of the initial and final states obey the free Dirac equation, the matrix element of ip will vanish.[14] As $\alpha(p)$ in (11b) is practically identical with (7), the reasoning and also the results obtained in Sec. 3 for scalar theory (in which the quadratic integral vanishes) are also valid in the given case.

It remains for us to consider tensor theory. Replacing O_j in (4) by $\sigma_{\mu\nu}=\frac{1}{2}i(\gamma_\mu\gamma_\nu-\gamma_\nu\gamma_\mu)$, we see that

$$\mathrm{Sp}(\sigma_{\alpha\beta}\gamma_\nu\sigma_{\lambda\rho}\gamma_\nu)=0.$$

Thus no quadratic divergence is involved in tensor theory. The new spinor forms for Γ_1 arise, but in tensor theory $\alpha_1(p)$ contain $\ln\Lambda^2/p^2$ in the denominator. The considerations do not differ from those applied in the case of scalar theory (in which the quadratic integral is assumed to be zero).

Consequently, if only a single fermion field is considered, physical interaction will be absent in all types of four-fermion interaction.

5. INTERACTION OF SEVERAL FIELDS

The interaction between several fields will now be considered. If two types of neutral particles exist (the respective fields being designated by ψ and χ), we arrive at three types of interaction:

$$V=2\pi^2\int[g_1(\bar\psi O_j\psi)(\bar\psi O_j\psi)+g_2(\bar\chi O_j\chi)(\bar\chi O_j\chi)$$
$$+2g_3(\bar\psi O_j\psi)(\bar\chi O_j\chi)]dv, \quad (15)$$

the respective constants being g_1, g_2, g_3. One may inquire whether these constants can be chosen in such a way so as to cancel out the logarithm in the denominator of the expression for $\alpha(p)$, which is of the same type as (5).

For convenience, the field of the two particles will be characterized by the column matrix

$$\phi=\begin{pmatrix}\psi\\ \chi\end{pmatrix}.$$

Then in interaction (15),

$$a_0=\sum_{i=1}^{3} g_iV_i',$$

where

$$V_1'=Q_1'\times Q_1';\quad V_2'=Q_2'\times Q_2';$$
$$V_3'=Q_1'\times Q_2'+Q_2'\times Q_1';$$
$$Q_1'=O_j\begin{pmatrix}1&0\\0&0\end{pmatrix};\quad Q_2'=O_j\begin{pmatrix}0&0\\0&1\end{pmatrix}$$

will correspond to the simplest diagram in Fig. 1 and Eq. (3) will have a solution of the type

$$\Gamma_1=\sum_{i=1}^{3}\alpha_i(p)V_i'.$$

We shall consider scalar theory ($O_j=1$, $O_j=i\gamma_5$), similar results obtaining in the other theories.

Inserting the foregoing expressions for a_0 and Γ_1 in (3), we obtain, after equating the coefficients before V_i', the following set of equations for α_i:

$$(1+g_1I)\alpha_1+g_3\alpha_3I=g_1,$$
$$(1+g_2I)\alpha_2+g_3\alpha_3I=g_2, \quad (16)$$
$$\tfrac{1}{2}g_3\alpha_1I+\tfrac{1}{2}g_3\alpha_2I+[1+\tfrac{1}{2}(g_1+g_2)I]\alpha_3=g_3,$$

where $I=2J_0+J_1\cong\Lambda^2-p^2L$. The solutions of these equations are

$$\alpha_1=\frac{g_1+f^2I}{\Delta},\quad \alpha_2=\frac{g_2+f^2I}{\Delta},\quad \alpha_3=\frac{g_3}{\Delta}, \quad (17)$$

where $\Delta=1+2gI+f^2I^2$, $g=\frac{1}{2}(g_1+g_2)$, and $f^2=g_1g_2-g_3^2$. Taking into account only terms linear in p^2L (it can easily be seen that these are the leading terms if the $g_i\Lambda^2$ are of the order of unity), we obtain

$$\Lambda^2\Delta=(1+F_0)\Lambda^2-F_1p^2L,$$

where $F_0=2\tilde g+\tilde f^2$ and $F_1=2(\tilde g+\tilde f^2)$. The dimensionless constants $\tilde g=\Lambda^2g$ and $\tilde f=\Lambda^2f$ have been introduced; in the case of interest these quantities are of the order of unity. In the expression $\Lambda^2\Delta$ [that is, in the denominator of the $\alpha_i(p)$] there will be no term proportional to Λ^2, and the quantity L will cancel out if $1+F_0=0$ and $F_1=0$. Inserting the values of F_0 and F_1, we get $\tilde g=-1$ and $\tilde f^2=1$; that is, $\tilde g_1+\tilde g_2=-2$ and $\tilde g_1\tilde g_2-\tilde g_3^2=1$, where $\tilde g_i=\Lambda^2g_i$. Hence $\tilde g_1=-2-\tilde g_2$ and $-\tilde g_3^2=(1+\tilde g_2)^2$.

Thus quadratic and logarithmic infinities can be removed from the denominator of the expressions for α_i only when the constant g_3 is imaginary (if the numerators of the expressions for α_i are multiplied by Λ^2 they will be of the order of unity, providing $g_i\sim 1$).

---

coordinates, it can be written before the integral; and since the vector $\partial\varphi/\partial x_\nu$ is arbitrary, we obtain

$$\int T\langle j_\nu(x)j_\mu(y)\rangle d^4x d^4y=0.$$

[14] See the similar arguments in R. Feynman's paper [Phys. Rev. **76**, 769 (1949)]. At small p^2 the denominator of the expression $(ip\times ip)/p^2$ will not vanish if finiteness of the mass is taken into account. Vanishing of the interaction α_1 can also be proved by employing a transformation similar to (12).

Therefore, for Hermitian Lagrangians, physical interaction will vanish in this case also if $\Lambda^2 \rightarrow \infty$.

If the quadratic integral vanishes (vector and tensor theories), the same type of reasoning as that used in Sec. 3 for a single field should be applied. No physical interaction will arise in this case either.

We now turn to consideration of charged fields. Care should be exercised in using form factors to analyze the charged field. Thus if the form factor is introduced simply as the integration limit in momentum space, a contradiction with Ward's identity will arise when $\lambda^2 \ll \Lambda^2$ (in particular, because of the interaction, a neutral particle acquires a charge). In order to avoid this inconsistency, a form factor should be introduced which has, for example, the form

$$\left(\frac{\Lambda^2}{(p-A)^2-\Lambda^2}\right)^2,$$

where A is the electromagnetic potential.[15] Gauge invariance will apply to the relation between λ^2 and Λ^2, and hence the methods employed above may be used.

Consider now the interaction between two fields with isotopic spin $\frac{1}{2}$ (field ψ) and isotopic spin 1 (field Σ). Instead of (15) we obtain

$$V=2\pi^2\int\left[g_1(\bar{\psi}\psi)(\bar{\psi}\psi)+g_2(\bar{\Sigma}\Sigma)(\bar{\Sigma}\Sigma)+2g_3(\bar{\psi}\psi)(\bar{\Sigma}\Sigma)\right.$$
$$+g_4(\bar{\psi}\tau_\alpha\psi)(\bar{\psi}\tau_\alpha\psi)+g_5(\bar{\Sigma}T_\alpha\Sigma)(\bar{\Sigma}T_\alpha\Sigma)$$
$$\left.+2g_6(\bar{\psi}\tau_\alpha\psi)(\bar{\Sigma}T_\alpha\Sigma)+g_7(\bar{\Sigma}S_{\alpha\beta}\Sigma)(\bar{\Sigma}S_{\alpha\beta}\Sigma)\right]dv, \quad (18)$$

where the τ_α are operators of isotopic spin $\frac{1}{2}$ and T_α of spin 1, and $S_{\alpha\beta}=T_\alpha T_\beta+T_\beta T_\alpha-\frac{4}{3}\delta_{\alpha\beta}$; for the sake of brevity, the operators of ordinary spin are not explicitly included. To compute the integral (4), the trace over the isotopic spin variables should also be taken. It can readily be verified in this case that interaction (18) splits into three independent sets: (1) g_1, g_2, g_3; (2) g_4, g_5, g_6; and (3) g_7. The first two can be reduced to the case of interaction of two neutral fields, discussed above, by changing the notation. Thus, in the first case, the following substitution should be made: $2g_1=g_1'$; $3g_2=g_2'$; $6^{\frac{1}{2}}g_3=g_3'$; $2\alpha_1=\alpha_1'$; $3\alpha_2=\alpha_2'$; and $6^{\frac{1}{2}}\alpha_3=\alpha_3'$. In the second case, we obtain the previous equation (16) by performing the substitution $2g_i=g_i'$. In the third case, only one field is involved.

More complicated cases of interaction may be considered, such as the interaction between three different fields. Then, besides interactions of type (18), an interaction of the β-decay type $(\bar{\psi}\psi)(\bar{\varphi}\chi)$ can be set up. These cases may be handled by the same methods, but the problem becomes more unwieldy. There are no apparent physical reasons to expect other results to be obtained in these more complex cases, although a study of them would seem to be of interest.

Thus the expectation expressed in the introduction, that choice of the sign of the interaction constant might lead to a result differing from that in other cases of local interaction, is not confined.

In all of the simplest cases of local interaction considered up to the present, physical interaction between the fields is absent.

[15] The authors are thankful to B. L. Ioffe for pointing out that a form factor of the indicated type removes the contradiction with Ward's identity.

Charged Particle Green's Function in the "Infrared Catastrophe" Region

L. P. GOR'KOV
Institute for Physical Problems
Academy of Sciences, USSR

(Submitted to JETP editor January 3, 1956)
J. Exptl. Theoret. Phys. (U.S.S.R.) **30**, 790-791
(April, 1956)

IN the electrodynamics of the electron Abrikosov[1] has shown that the interaction with the electric field leads to the appearance in the Green's function of the electron in the infrared region ($|p^2 - m^2| \ll m^2$) of the additional singularity

$$\left(\frac{m^2}{p^2 - m^2}\right)^{(e^2/2\pi)\,[3-d_l\,(0)]} \tag{1}$$

as compared with the simple pole for the Green's function of the free electron. An analogous investigation in the electrodynamics of spin zero[2] shows that the same singularity arises in the Green's function of a charged scalar particle, where the result obtained coincides exactly with (1). Hence, it seems worthwhile to present a result from which this effect would follow independently of the nature (spin) of the charged particle. The present letter is devoted to this question.

The Green's function of the particle is defined in the usual way:

$$G(x, x') = \langle (\psi(x), \overline{\psi}(x'))_+ \rangle_0$$

The brackets indicate a time-ordered average over the vacuum, $\psi(x)$ is the particle field operator, $\overline{\psi}(x)$ is the adjoint operator. The operations of conversion between ψ and $\overline{\psi}$ are evidently Hermitian conjugates.

The Fourier components of $G(x, x')$ in the region $p^2 \sim m^2$ are determined by the matrix element $\langle 0 | \psi(x) | p \rangle$, where $p^2 \sim m^2$. In other words, in the Fourier expansion of the operator $\psi(x)$, it is sufficient to determine only the part of the spectrum of $\psi(p)$ where $p^2 \sim m^2$. In the state in which $p^2 \sim m^2$, there is a single particle interacting with the electric field, and the magnitude $\Delta = |p^2 - m^2| m^{-2}$ is a measure of the energy of the photons which this particle may emit or absorb. We make the assumption, justified below, that for the effect under consideration only a connection with the low frequency part of the electromagnetic field is necessary. Choosing a system of reference in which the motion of the particle is nonrelativistic, we may write the nonrelativistic Schrödinger equation

$$i\,\frac{\partial \psi(x)}{\partial t} = \left\{ m + eA_0(x) + \frac{1}{2m}(\hat{\mathbf{p}} - e\mathbf{A}(x))^2 \right\} \psi(x), \tag{2}$$

where $\hat{\mathbf{p}} = -i\nabla$. This equation is correct for the description of the interaction of the "free" part of the Fourier expansion of $\psi(x)$ ($p^2 \sim m^2$) with the low frequency part of the electromagnetic field. Neglecting terms quadratic in the field, since, as may be shown, they give rise to a higher order contribution in the final result, we substitute into Eq. (2)

$$\psi(\mathbf{x}, t) = P_\tau \exp\left\{ -i \int^t j_\mu A_\mu(\mathbf{x}, \tau)\, d\tau \right\} \psi_0(\mathbf{x}, t), \tag{3}$$

where P_τ is the time-ordering operator and $j_\mu = \{e, e\mathbf{v}\}^*$. The operator $\psi_0(\mathbf{x}, t)$ satisfies the free particle equation

$$i\partial\psi_0 / \partial t = (m + \mathbf{p}^2/2m)\psi_0.$$

In the approximation $e^2 \ll 1$ we may consider $A_\mu(\mathbf{x}, \tau)$ in Eq. (3) as the free-field operator; hence, ψ_0 and A_μ commute. For the Green's function $G(x, x')$ (more exactly, for the "free" part $p^2 \sim m^2$ in the Fourier expansion) we obtain

$$G(x, x') = G_0(x, x') \left\langle P_\tau P_{\tau'} \left(\exp\left\{ -i \int^t j_\mu A_\mu(\mathbf{x}, \tau)\, d\tau \right\} \times \exp\left\{ i \int^{t'} j_\nu A_\nu(\mathbf{x}', \tau')\, d\tau' \right\} \right)_+ \right\rangle_0,$$

where $G_0(x, x')$ is the free particle Green's function.

Using the formula

$$\exp(A + B) = \exp(A) \cdot \exp(B) \cdot \exp\left(-\frac{1}{2}[A, B] \right),$$

which is correct when there exists a number $[A, B]$, and taking an average over the photonic vacuum, we obtain the following simple transformations:

$$G(x, x') = G_0(x, x') \tag{4}$$

$$\times \exp\Big\{ -\frac{1}{2}\Big[\int^t\!\!\int^t \langle (A_\mu(\mathbf{x}, \tau), A_\nu(\mathbf{x}, \tau'))_+ \rangle_0\, j_\mu j_\nu\, d\tau\, d\tau' + \int^{t'}\!\!\int^{t'} \langle (A_\mu(\mathbf{x}', \tau), A_\nu(\mathbf{x}', \tau'))_+ \rangle_0\, j_\mu j_\nu\, d\tau\, d\tau' \Big]$$

$$+ \int^t\!\!\int^{t'} \langle (A_\mu(\mathbf{x}, \tau), A_\nu(\mathbf{x}', \tau'))_+ \rangle_0\, j_\mu j_\nu\, d\tau\, d\tau' + I(t, t') \Big\},$$

where

$$I(t, t') = \begin{cases} \displaystyle\int^t\!\!\int^{t'} \theta(\tau' - \tau)[A_\mu(\mathbf{x}, \tau), A_\nu(\mathbf{x}', \tau')]\, j_\nu j_\mu\, d\tau\, d\tau'; & t > t', \\ \displaystyle -\int^t\!\!\int^{t'} \theta(\tau - \tau')[A_\mu(\mathbf{x}, \tau), A_\nu(\mathbf{x}', \tau')]\, j_\nu j_\mu\, d\tau\, d\tau'; & t < t'. \end{cases} \tag{5}$$

$$[A_\mu(x), A_\nu(x')] = -4\pi i \delta_{\mu\nu} D(x - x'); \qquad \langle (A_\mu(x), A_\nu(x'))_+ \rangle_0 = 4\pi D_{F\mu\nu}(x - x').$$

The assumption of adiabatic cessation of the interaction

$$e = e(t) = \begin{cases} e, & t > T_0 \\ e \exp \alpha t, & t < T_0 \end{cases} \quad (T_0 \to -\infty)$$

allows us to neglect the values of all the integrals taken at the lower limit. In particular, Eq. (5) depends only on the lower limit, and hence is equal to zero. Carrying out the integration with respect to $d\tau$ and $d\tau'$ in Eq. (4) in the indicated manner, and varying Eq. (4) with respect to δe^2, we take the Fourier component of the resulting relation in the momentum region $|p^2 - m^2| \ll m^2$ which is of interest to us. We find

$$\delta G(p) = \frac{i\delta e^2}{\pi}\left\{\int [G(p) - G(p-k)]\left[\left(v^2 - \frac{(vk)^2}{k^2}\right) + \frac{(vk)^2}{k^2} d_l(k^2)\right]\frac{d^4k}{\omega^2(k^2+i\varepsilon)}\right\}. \quad (6)$$

The calculation is most simply carried out in the system of reference in which the velocity of the particle $\mathbf{v} = \mathbf{p}/m$ is equal to zero. Then

$$\delta G(p) = \frac{i\delta e^2}{\pi}\left\{\int [G(p) - G(p-k)]\left[\left(1 - \frac{\omega^2}{k^2}\right) + \frac{\omega^2}{k^2} d_l(k^2)\right]\frac{d^4k}{\omega^2(k^2+i\varepsilon)}\right\}.$$

Carrying out the integration with respect to $d\omega$ according to the usual rules of contouring[3], we find that the terms sought are obtained by taking a calculation at the point $k^2 = 0$:

$$\delta G(p) = \frac{\delta e^2}{2\pi}\left\{\int [G(p) - G(p-k)]\frac{d|\mathbf{k}|}{|\mathbf{k}|}\right\}[3 - d_l(0)]; \quad (\omega = +|k|). \quad (7)$$

The essential logarithmic region of interaction in this integral is $|p^2 - m^2|/m \ll k \ll m$. The region of high frequencies leads to renormalization effects, which, naturally, cannot be correctly taken into account in this technique. Carrying out the integration in (7) for low frequencies, we obtain

$$\delta G(p) = G(p)\left(\frac{\delta e^2}{2\pi}\right)[3 - d_l(0)]\ln\left(\frac{m^2}{p^2 - m^2}\right)$$

or

$$G(p) = G_0(p)\left(\frac{m^2}{p^2 - m^2}\right)^{(e^2/2\pi)[3 - d_l(0)]}$$

where $G_0(p)$ differs from the Green's function of a free particle by the renormalization factors. Thus, the appearance of the additional singularity (1) in the Green's function of a particle interacting with an electromagnetic field is connected only with the classical properties of the electric current being produced by the particle in its uniform motion.

In conclusion, I wish to express my deep gratitude to A. A. Abrikosov and I. M. Khalatnikov for discussions of this work.

* Designations are those used in Ref. 3.

1 A. A. Abrikosov, Dissertation, Institute for Physical Problems, Academy of Sciences, USSR, 1955.

2 L. P. Gor'kov, Dissertation, Institute for Physical Problems, Academy of Sciences, USSR, 1955.

3 R. Feynman, Phys. Rev. 76, 749, 769 (1949).

Translated by M. G. Gibbons
159

Chapter 10

Materials: HTSC Cuprates, Manganites, Fe-based Superconductors

The following three main events have changed the face of contemporary condensed matter physics: the discovery of high temperature superconductivity (HTSC) in cuprates in 1986, the renaissance of the 90's in connection with the discovery of the so-called colossal magnetoresistance (CMR) in doped manganites, and, more recently, the 2006 discovery of HTSC in the broad class of iron-based superconductors (FeCS). The emerging new physics in all three cases promised great opportunities for the development of new technologies. Today, neither the nature of new phenomena is fully understood, nor have all the opportunities yet come to fruition.

As I see this, slow progress is mainly due to difficulties of experimentation. Without addressing this point, it is of no surprise, at least to me, that input of theory in understanding of the new physics is modest. As active studies continue, I place here only several papers with some ideas of a qualitative character related to interpretation of a few experimental findings.

10.1 Manganites

V. Kresin and I were the first to suggest that the transition from the *insulating* antiferromagnetic state of $LaMnO_3$ to *band*s of itinerant electrons in the ferromagnetic phase of $La_{1-x}Sr_xMnO_3$ takes place in the *percolation* regime with the critical value of strontium concentration $x_{cr} \approx 0.16$. Our conclusion was based on the analysis of data of the 1995 paper by the Tokura group (Ref. 1). Paper [10a] was fiercely rejected by the referees at Physical Review Letters; we had it first published in the Russian JETP Letters.

10.2 Cuprates

Concerning cuprates, I dare say that for the first time the idea of the phase separation controlled by the Coulomb forces (i.e., by the electro-neutrality condition) in doped cuprates was put forward in a paper [10b] for LSCO: the strontium-doped La_2CuO_4 $(La_{1-x}Sr_x)_2CuO_4$.

Later, the analysis of the experimental data on ARPES and on the Hall Effect in [10c] has led G. Teitel'baum and I to the conclusion that besides the band-like states, at least in LSCO, there are bound electron-hole complexes responsible for the activated temperature dependence in the Hall coefficient $R_H(T)$ present.

Much attention has been drawn to the recent de Haas–van Alphen and the angle resolved photoemission experiments (ARPES) in underdoped cuprates. The experiments suggest that the energy spectrum consists of small Fermi pocket(s) and of the coherent excitations on the so-called "Fermi arcs". In [10d], we corroborated this suggestion theoretically. In particular, we have proven that the experimental data for resistivity and the Hall coefficient are consistent with such model spectrum. The pocket size is too small for electrons to contribute to the Umklapp scattering. Electrons must follow motion of holes on the Fermi arcs. We show that a strong enough magnetic field can decouple the two sub-systems.

10.3 Fe-based superconductors

In [10e] we suggested that temperatures of the charge density wave (CDW) and of antiferromagnetic (SDW) transitions in iron-pnictides are so close together because both transitions in reality are one and the same SDW transition split by weak striction effects.

G. Teitel'baum and I argue in [10f] that in iron-pnictides the local moments and electrons in the conduction d-bands are not decoupled even in the paramagnetic phase, but interact forming a coherent whole.

Reference

1. A. Urusibara *et al.*, *Phys. Rev. B* **51**, 14103 (1995)

List of original papers

[10a] L. P. Gor'kov and V. Z. Kresin, *Manganites at low temperatures and light doping: band approach and percolation*, JETP Lett. 67, 934 (1998)

[10b] L.P. Gor'kov, A.V. Sokol, *Phase stratification of an electron liquid in the new superconductors*, Sov. Phys. JETP Lett., **46**, 420 (1987)

[10c] Gor'kov, L. P. and Teitel'baum, G. B., *The two-component physics in the real space and in the momentum representation,* J. Phys.: Conf. Series, **108**, 012009 (2008)

[10d] Lev P. Gor'kov and Gregory B. Teitel'baum, *Two regimes in conductivity and the Hall coefficient of underdoped cuprates in strong magnetic fields*, J. Phys.: Condens. Matter **26**, 042202 (2014)

[10e] Barzykin, V. V. and Gor'kov, L.P., *Role of striction at magnetic and structural transitions in iron pnictides,* Phys. Rev. B, **79**, 134510 (2009)

[10f] Gor'kov, L.P. and Teitel'baum G.B., *Dual role of d-electrons in iron pnictides*, Phys. Rev. B **87**, 024504 (2013)

Pis'ma v ZhETF, vol.67, iss.11, pp.934 - 939

MANGANITES AT LOW TEMPERATURES AND LIGHT DOPING: BAND APPROACH AND PERCOLATION

*L.P.Gor'kov, V.Z.Kresin**

National High Magnetic Field Laboratory, Florida State University
Tallahassee, FL 32310, U.S.A.
and
L.D.Landau Institute for Theoretical Physics,
117334 Moscow, Russia

**Lawrence Berkeley Laboratory, University of California*
Berkeley, CA 94720

Submitted 30 April 1998

A tight-band model is employed for the e_{2g}-orbitals in manganites. It is shown that a large intraatomic Hund's coupling, J_H and the resulting double-exchange mechanism lead to antiferromagnetic ordering along one of the cubic axis stabilized by the cooperative JT effect which further decreases the band energy of electrons. As a result, $LaMnO_3$ is a band insulator built of 2D ferromagnetic layers. The critical concentration $x_c \simeq 0.16$, for onset of ferromagnetic and metallic behavior at low temperatures in $La_{1-x}Sr_xMnO_3$ and the phase transition are treated in the percolation approach.

PACS: 75.10.Lp, 75.30.-m, 75.50.-y

Current efforts in studies of the "colossal magnetoresistance" (CMR) in the manganites, $R_{1-x}B_xMnO_3$ (usually, R=La, B=Ca, Sr; see e.g. [1–3] for a review and references) are, naturally, focused on phenomena occurring in the vicinity of the metal-insulator transition temperature, T^*, which for $x = 0.2 - 0.3$, reaches the room temperature range. There are no doubts that both the Jahn - Teller (JT) distortions and Zener's double exchange (DE) mechanism [4,5] are two key ingredients of CMR.

The transition between the paramagnetic and conducting ferromagnetic phases is treated in [6] as a localization-delocalization "crossover" driven by thermal fluctuations of the local JT-modes. In this paper we focus on the analysis of phenomena occurring at low T. In this region the approach should be different. Indeed, the two end members of the series, $LaMnO_3$ and $CaMnO_3$, are both antiferromagnetic (AF) insulators. "Doping" of $LaMnO_3$ by divalent atoms results in metallic conductivity only above some critical concentration, $x_c \simeq 0.16$.

Low temperature phases of manganites are habitually interpreted in terms of localized orbitals [7, 8], more often that not in the modified Hubbard model (see. e.g. [9]). Although it is widely accepted that electron-electron correlations play the key role in low temperature behavior of manganites, we show that their major properties can be understood within the framework of the band approach, as well, or in percolation terms.

The following facts have to be explained: 1) the insulating state of the parent $LaMnO_3$; 2) the A-type of antiferromagnetic ordering (alternating ferromagnetic planes); 3) the small value of the Néel temperature, $T_N \sim 150\,K$ in the low x-range, while the structural changes occur at about 900 K; 4) the localization of "holes" introduced into the Mn^{3+}-subsystem at $x < x_c$; 5) the threshold concentration, $x_c \simeq 0.16$, above which metallic conductivity sets in at low T.

In the band model the properties from 1) to 4) are brought about by the DE- and JT-mechanisms. We interpret the value $x_c \simeq 0.16$ in terms of percolation theory: both the cluster approach and the phase separation picture, as suggested first for the cuprates [10], seem to give the same criterion [11].

Let us start from the A-type of the ground state of $LaMnO_3$. According to our picture, appearance of such a magnetic state is *not* caused by any exchange interaction between localized spins on different sites. The mutual arrangement of the distorted octahedra and of the core $^3t_{2g}$-spins, $\mathbf{S}_i$ (S=3/2), is stabilized only by gains in the kinetic *band* energy of the "conduction" e_{2g}-electrons.

The single electron Hamiltonian is then of the form [6]

$$\hat{H} = \sum_{i,\delta} \left(\hat{t}_{i,i+\delta} - J_H \mathbf{S}_i \cdot \hat{\sigma} + g\hat{\tau}_i \cdot \mathbf{Q}_i + J_{el}\mathbf{Q}_i^2\right). \tag{1}$$

Here $\hat{t}_{i,i+\delta}$ is the nearest neighbor tunneling matrix, defined on some basis of the two-dimensional cubic representation, e_{2g}; $\mathbf{Q}_i$ are the active local JT-modes with the matrix elements on that electronic basis in terms of a "pseudospin" matrix, $\hat{\tau}$ (see e.g., in [7]; $\mathbf{Q}_i$ are defined as dimensionless parameters). The static JT deformations and the staggered magnetization, $\langle \mathbf{S}_i \rangle = (-1)^i \langle S \rangle$ of the A-type phase are treated below in the mean field approximation.

To avoid cumbersome expressions, we first discuss the competition between ferro- and antiferromagnetic order for the *one* band model:

$$\varepsilon(\mathbf{p}) = t\cos(ap_z) + \bar{t}(\mathbf{p}_\perp). \tag{2}$$

Assume first that the local ($^3t_{2g}$) spins are ordered ferromagnetically. At large $J_H \gg t$ the low energy band is shifted down, by $-J_H\langle S \rangle$. The main contribution to the total energy is $E_H = -J_H\langle S \rangle n$ (n is the number of electrons per unit cell), while the kinetic energy contribution due to the polarized electrons is linear in $t, \bar{t}$.

Consider the same problem for AF-ordering along the z-axis (with the period $2a$). It is convenient to discuss a more general case:

$$\mathbf{S}_i = (\pm\langle S_z \rangle, M_x) \quad (S_z^2 + M_x^2 = S^2), \tag{3}$$

i.e. the canted AF-structure [8] for the $^3t_{2g}$- spins. Solving the periodic electron spectrum problem in a manner similar to the solution for the two local sites [5], one obtains the following four branches:

$$\varepsilon_l(\mathbf{p}) - \bar{t}(\mathbf{p}_\perp) = \pm\sqrt{J_H^2 S^2 + t^2(p_z) \pm 2J_H \mid t(p_z) \mid\mid M_x \mid}. \tag{4}$$

At $J_H \gg t, \bar{t}$, the low energy spectrum reduces to:

$$\varepsilon(\mathbf{p})_{1,2} - \bar{t}(\mathbf{p}_\perp) \simeq -J_H S \pm \mid t(\mathbf{p}_z) \mid (\mid M_x \mid /S). \tag{5}$$

The second term restores the familiar expression for transport [5]: $M_x = 2S \mid \cos\theta/2 \mid$ (θ is the angle between the adjacent local spins).

For the AF-case ($M_x \equiv 0$), the Brillouin zone is reduced by half, but there are now two branches in (5). As the result, the main term, $-J_H S$, does not change. For a single band, the antiferromagnetic order may even be energetically favorable, although the energy gain would be small, of the order of (see (4)):

$$t^2/J_H \ll t. \tag{6}$$

For two bands there are terms linear in t. The JT-band splitting *is* necessary if we wish to decrease the ground state energy below that one for the ferromagnetic state.

We obtained the "bare" band spectrum of the cubic material with the e_{2g}-electrons. For calculating $\hat{t}_{ik}$, and hence, the electronic spectrum, $\varepsilon(\mathbf{p})$, use of the normalized basis functions of the form:

$$\psi_1 \propto z^2 + \epsilon x^2 + \epsilon^2 y^2, \ \psi_2 \equiv \psi_1^* \tag{7}$$

($\epsilon = \exp(2\pi i/3)$), proves to be more convenient to account for the cubic symmetry of the initial lattice. The functions (7) are connected with the real basis, $\varphi_1 \propto d_{z^2}$ and $\varphi_2 \propto d_{x^2-y^2}$:

$$\psi_1 = (\varphi_1 + i\varphi_2)/\sqrt{2}, \ \ \psi_2 = (\varphi_1 - i\varphi_2)/\sqrt{2}. \tag{8}$$

The tight-binding spectrum of the e_{2g}-electrons consists of two branches, which are discussed below.

Before proceeding further, we write down explicitly the local, JT-part of the Hamiltonian (1) in the basis (7):

$$-\frac{g}{2}Q_0 \begin{pmatrix} 0 & \exp(i\theta) \\ \exp(-i\theta) & 0 \end{pmatrix}. \tag{9}$$

In the standard notations [7]: $Q_2 = Q_0 \sin\theta$; $Q_3 = Q_0 \cos\theta$. Here Q_0 is the magnitude of the JT distortions, while the "angles", $\theta = 0, 2\pi/3, -2\pi/3$ correspond to elongations of the octahedron along the z, x, and y-axes, respectively.

With the above in mind, consider the options for possible ground states in manganites. In the case of stoichiometric $LaMnO_3$ there is exactly one e_{2g}-electron (Mn^{3+}) per unit cell. The ferromagnetic filling-up for two bands will produce a metallic spectrum. AF-ordering along one axis results in the same main contribution, $-J_H S$ (at large J_H), while the corresponding kinetic energy contribution becomes of the order of $t^2/J_H \ll t$. Second order terms in t in Eq. (4) imply that the spectrum is much less dispersive along the z-axis ($M_z \equiv 0$). Planes with the antiparallel core spins become almost isolated.

AF-ordering along any other direction would lack electronic transport along this direction, again shrinking bands. Thus, it is enough to have the AF-order along one axis to access the major energy gain, $-J_H S$. However, to stabilize the AF-order along *one* axis, one needs to further reduce the total energy by the antiferroelastic JT-distortions along the remaining directions.

Note again that after the AF-ordering sets in, say, along the z-axis, the communication between adjacent layers in real space becomes very weak. Therefore, at large J_H it is possible to reduce the problem, to main approximation, to the one of two-dimensional (2D) electrons. For decoupled layers, electrons inside the layer are ferromagnetically polarized along the ${}^3t_{2g}$-(core) spins. Again, in this 2D problem there is one electron per (2D) cell. The metallic spectrum simplifies to the form:

$$\begin{aligned} \epsilon_{1,2}^{2D}(\mathbf{p}) &= (A+B)(\cos ap_x + \cos ap_y) \pm \\ &\pm (A-B)\left\{(\cos ap_x + \cos ap_y)^2 + 3(\cos ap_x - \cos ap_y)^2\right\}^{1/2}, \end{aligned} \tag{10}$$

where $i, k = (x, y, z)$;

$$A \propto \overline{\varphi_1(z; x, y)\varphi_1(z + a; x, y)}\ , \quad B \propto \overline{\varphi_2(z; x, y)\varphi_2(z + a; x, y)}$$

are the two overlap integrals between the two sites separated by the lattice constant, a. Introduce now JT-distortions in the layer with the structure vector (π, π) in a plane. If the JT-splitting is strong enough, then instead of (9), the new spectrum will consist of two pairs of bands separated by energy gaps over the entire (new) Brillouin zone, with the low-energy bands fully occupied. This completes the band picture for the insulating A-type ground state in $LaMnO_3$.

For this interpretation one needs:

$$J_H \gg t,\ gQ_0\ ; \quad gQ_0 \stackrel{\sim}{>} t. \tag{11}$$

Eq.(10) does not impose any severe limitations on the model parameters. The above physics should be present in any realistic band structure. Usually the band calculations use the experimental lattice parameters, and it is not straightforward to single out (e.g. in [3]) the competition of the effects considered above.

Finding the single electron spectrum in presence of the JT-distortions from (8), (9), writing down the total energy of the filled-up bands and subsequent minimization of the total energy, is a straightforward but rather tedious task. In the general case, there is no small parameter. We will consider numerical aspects elsewhere.

Below we take advantage of the transparent physics of the independent 2D-layer approximation. First of all, the 2D ferromagnetic ground state in each layer which was obtained in the mean field approximation, will be smeared out by fluctuations: ferromagnetism does not exist as a stable thermodynamical phase in 2D, unless there is a weak interlayer coupling, t^2/J_H, of eq. (6). This results in a rather low Néel temperature, $T_N \simeq 150\,\mathrm{K}$, in $LaMnO_3$, at $t \sim 0.1\,\mathrm{eV}$, $J_H \sim 1\,\mathrm{eV}$.

To illustrate the above, consider the "symmetric" model [9], $A = B = t/2$ in (9):

$$\varepsilon(\mathbf{p}) = t(\cos ap_x + \cos ap_y) \equiv \tilde{t}(\mathbf{p}). \tag{12}$$

Eq. (11) gives the two degenerate bands. For one polarized 2D electron per unit cell, the two Fermi surfaces run along the lines $p_x \pm p_y = (\pm\pi/a)$. In the presence of the JT-distortion (8) the new spectrum (in the reduced Brillouin zone, with the structural vector $(\pi/a, \pi/a)$) consists of the four branches: $\varepsilon_l(\mathbf{p}) = \pm||t(\mathbf{p})| \pm \Delta|$, where $\Delta = |gQ_0/2|$. At $\Delta > 2t$ there are two pairs of energy bands, and the insulating gap sets in at $p_x = p_y = 0$, equal to $\Delta - 2t$. The distortion $gQ_0 = g^2/J_{el}$ is defined from the on-site problem.

Turning back to the parameters (J_H, J_{el}, g, A, B) of our general model, connections between gQ_0, J_{el} and A and B are established by minimization of the total (electronic and elastic) energies with respect to Q and M_x (see eq. (5)). The deformations of octahedra (Q_0) are known from the structural data (see e.g., in [3]). One may also expect that $B \ll A$: tunneling between Mn-sites takes places through the shared oxygen ions, and the notion that the d-shell size, d, is small compared to the lattice constant, $a(d \ll a)$, is quite helpful, as oxides of transition metals are concerned [7, 12, 13]. Currently the spin wave spectrum in $LaMnO_3$ is available [14]. Although we postpone the study of the spin wave spectrum for the future, it is worth mentioning that coexistence of the localized ${}^3t_{2g}$-spins and itinerant e_{2g}-electronic states (which in the AF-environment are not characterized by the spin-projection quantum number), may result in deviations from an anisotropic Heisenberg model.

Finally, we briefly discuss the concentration dependence. At small x the two mechanisms lead to localization of a doped hole. The first one is directly related to the 2D physics. According to (1), interaction of a hole with the JT-modes may decrease the hole's energy by an amount of order of g^2/J_{el}. If its energy is now below the band's bottom, the hole becomes self-trapped. Two-dimensionality plays the crucial role in that there are no energy barriers for the process [15]. If a dimensionless parameter $C \sim g^2/J_{el}W$ (W is a band-width) exceeds a critical value of order of unity, no itinerant states are possible in the 2D system. The second mechanism is the Coulomb binding of holes near dopants [8].

At larger x the band approach becomes useless. Instead, we adopt the point of view of the percolation theory. Let us start first with the *cluster* description [11]. In the so-called site problem, consider formation of an infinite cluster of the neighboring divalent ions. For the simple cubic lattice the critical $x_{cr}(s) = 0.31$. This value is not universal.

It is well known [11] that correlations between sites rapidly reduce $x_c(s)$ to the value $x_c \simeq 0.16$, which is also a threshold concentration for a continuous percolation problem. Although holes are located near divalent ions (forming the cluster's skeleton), the hole's hopping takes place along the adjacent Mn-sites correlated by the fact that the wave function of a single hole is spread over a few interatomic distances tending to suppress JT-distortions and align spins ferromagnetically in its vicinity. The cluster's structure is complicated, however, at $x > x_c$ one may expect that locally its properties are close to the ones of the homogeneous metallic ferromagnetic state. The rest of the sample does not conduct.

Another plausible view is that the material may separate into coexisting insulator and metallic phases [10]. In the cluster approach the Coulomb energy is playing the crucial role confining holes in the skeleton's vicinity. In the continuous description the Coulomb energy would limit sizes of the phases' domains. In both approaches the Coulomb energy may be strongly reduced by sufficiently large dielectric constant of the insulating phase and due to metallic screening in the ferromagnetic phase.

With the temperature increase, concentration of the metallic phase decreases: temperature-induced JT-distortions [6] provide stronger localization of "doped" holes. From such a point of view the metal-insulator transition temperature, T^*, may also be treated as a percolation point (at fixed x), and this agrees well with the extreme sensitivity of CMR to applied magnetic field.

To summarize, the view of $LaMnO_3$ as a band insulator, is consistent with the main experimental facts. Large intraatomic Hunds' exchange, J_H, forces the cubic system to order antiferromagnetically along one direction because the band energy of electrons can then be effectively reduced by the JT-instability resulting in the new (tetragonal) lattice ($\sqrt{2}a \times \sqrt{2}a \times 2a$). At $J_H \gg t, gQ_0$, the e_{2g}-electrons form almost disconnected 2D layers. This manifests itself in the low Néel temperature. The percolation analysis can explain the value of the critical concentration, $x_c \simeq 0.16$ and CMR itself suggesting new interesting physics.

The work was supported (L.P.G.) by the National High Magnetic Field Laboratory through NSF Cooperative Agreement # DMR-9016241 and the State of Florida, and (V.Z.K.) by the U.S. Office of Naval research under contract # N00014-97-F0006.

1. H.Kawano, R.Kajimoto, M.Kubota, and H.Yoshizawa, Phys. Rev. **B53**, 14709 (1996).
2. J.Coey, M.Vivet, and S. von Molnar, (unpublished, 1997).
3. W.Pickett and S.Singh, Phys. Rev. **B53**, 1146 (1996).
4. G.Zener, Phys. Rev. **82**, 403 (1951).
5. P.W. Anderson and H. Hasegawa, Phys. Rev. **100**, 675 (1955).
6. A.Millis, B.Shraiman, and R.Mueller, Phys. Rev. Lett. **77**, 175 (1996).
7. J.Kanamori, J. Appl. Phys. (Suppl.) **31**, 145 (1960).
8. P.G.de Gennes, Phys. Rev. **118**, 141 (1960).
9. For review, see K. Kugel and D. Khomskii, Sov. Phys. - Usp. **25**, 231 (1982).
10. L.P. Gor'kov and A. Sokol, JETP Lett. **46**, 420 (1987).
11. H.Scher and R. Zallen, J. Chem. Phys. **53**, 3759 (1970); for review see B.Shklovskii and A.Efros, *Electronic Properties of Doped Semiconductors*, Springer – Verlag, Berlin, 1984; G.Deutcher, in: *Chance and Matter*, p. 1, Eds. by J.Souletie, J. Vannimenus, and R. Stora (Elsevier, Amsterdam, 1987).
12. J.Coodenough, in: *Progress in Solid State Chemistry*, v.5, Ed. H.Reiss, Pergamon, NY, 1971, p.145.
13. P.W.Anderson, in: *Magnetism*, v.1, Eds. G.Rado and H.Suhl, Academic, NY, 1963, p. 75.
14. K.Hirota, N.Kaneko, A.Nishizwa, and Y.Endoh, J. Phys. Soc. Jpn. **65**, 3763 (1996).
15. E. Rashba, in: *Excitons*, Eds. E.Rashba and M.Sturge, North-Holland, Amsterdam, 1982, p.473; Y.Toyozawa and Y.Shinozuka, J. Phys. Soc. Jpn. **52**, 1446 (1983).

Phase stratification of an electron liquid in the new superconductors

L. P. Gor'kov and A. V. Sokol
L. D. Landau Institute of Theoretical Physics, Academy of Sciences of the USSR

(Submitted 17 September 1987)
Pis'ma Zh. Eksp. Teor. Fiz. **46**, No. 8, 333–336 (25 October 1987)

Analysis of the experimental situation in lanthanum cuprate suggests the coexistence of two phases: one conducting and one insulating. This situation would explain certain features of the normal state.

Over the time which has elapsed since the discovery of so-called high-temperature superconductivity,[1] many important pieces of experimental evidence have been accumulated. Nevertheless, the question of whether the high values of T_c can be explained by an ordinary model remains open. We accordingly consider the problem from a different viewpoint in the present letter. Specifically, if the new superconductivity is caused by certain unusual properties of these materials, we would naturally expect to see manifestations of these properties even in the normal state. In other words, we concentrate exclusively on the properties of the normal phase, particularly those of lanthanum compounds. "Pure" La_2CuO_4 is a special case in that it has a body-centered tetragonal lattice in its tetragonal phase. The primitive cell therefore contains precisely one structural unit of La_2CuO_4, corresponding to a half-filled Brillouin zone. [We assume below that since the structural transition from $T_{OT} \approx 500$ K is not of an electronic nature (in contrast with Ref. 2, for example), it is a small effect against the background of a broad zone, $\sim$1–2 eV.] However, the resistance of "pure" La_2CuO_4 is very large ($\sim 0.1\ \Omega\cdot$cm), depends weakly on the temperature above, say, 150 K, and exhibits a localization behavior at lower temperatures. On the other hand, doping with Sr to a level of a few percent increases the conductivity by two orders of magnitude, giving rise to a pronounced metallic behavior. A Mott-insulator state arising from strong Coulomb correlations would be the only presently known alternative to a Fermi-liquid state (Ref. 3, for example). It would then seem that the resistance of "pure" La_2CuO_4 should obey an activation law with a large energy gap. Since this is

 0021-3640/87/200420-04$01.00

not what we see experimentally, the finite conductivity is linked with an uncontrollable doping as a result of oxygen vacancies or a deviation from stoichiometry. Even at a lower doping level, however, the conductivity should vary with the temperature in accordance with the temperature dependence of the mobility.

We will not gloss over these contradictions, but rather treat them as important facts. We will attempt to propose yet another alternative explanation, specifically, the coexistence of two phases, conducting and insulating, in La_2CuO_4.

The La_2CuO_4 lattice is characterized by an octahedral coordination of CuO_6. With a cubic symmetry, the Cu^{2+} ion (d^9 electron configuration) corresponds to a degenerate Jahn-Teller term E_g. The presence of La atoms lowers the symmetry to tetragonal and in principle lifts the degeneracy of the E term. Let us assume, however, that the Jahn-Teller Hamiltonian is still of the form

$$\begin{pmatrix} \lambda Q_3 + E_3' & \lambda Q_2 \\ \lambda Q_2 & -\lambda Q_3 + E_2' \end{pmatrix} , \tag{1}$$

where the matrix acts on the coordinates (η_3, η_2) in the electron wave function $\psi = \eta_3 \psi_{3z^2 - r^2} + \eta_2 \psi_{x^2 - y^2}$ (the notation is that of Ref. 4). We are thereby assuming that the tetragonal splitting is smaller than the Jahn-Teller energy E_{JT}, so a strong polaron coupling is still characteristic of the copper ion in the d^9 configuration.

We first assume that there are no divalent states of copper at all in the system [for definiteness, we assume that the copper ions are monovalent (d^{10})]. We introduce a conduction band, constructed in a self-consistent way for this configuration, in terms of, say, oxygen p orbitals. Such a band will be half-filled. Now taking one electron (or, more precisely, one hole) and putting it in place of a copper, we form a local polaron level (d^9) with an energy E_0 (Fig. 1). If E_0 is above the Fermi level (μ_0) in the conduction band, a polaron state is not preferable. At E_0 below the bottom of the conduction band, all the carriers go into localized states (an analog of a Mott insulator[4]).

There is a third possibility, in which the polaron level occupies an intermediate position. In this case the chemical potential is pinned at E_0 (Refs. 5 and 6), and the

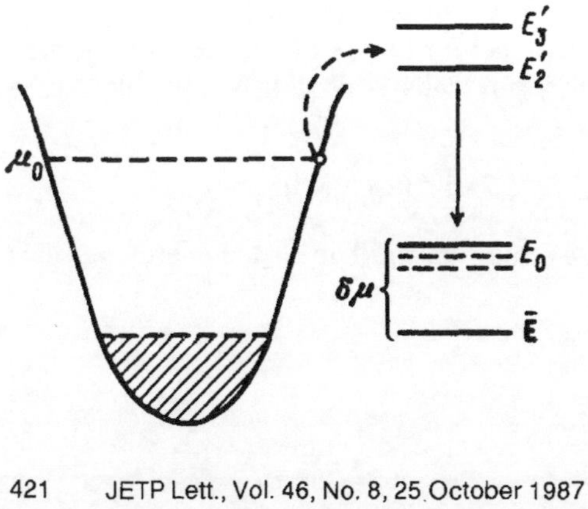

FIG. 1.

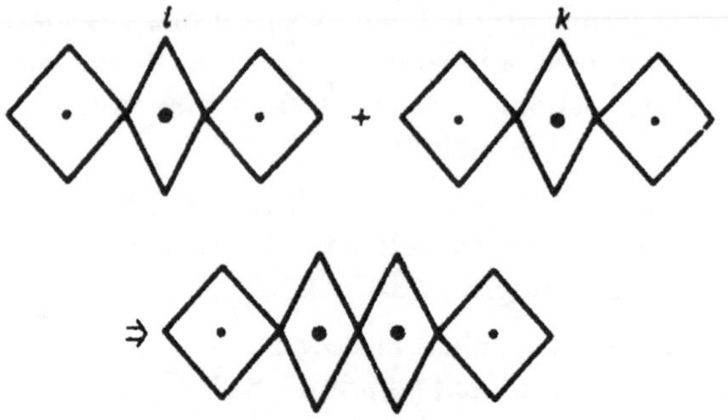

FIG. 2.

polaron levels are only partially filled. When we take the temperature into account, we conclude that upon hybridization each cell acquires an additional mean tetragonal distortion $\langle Q_3 \rangle \neq 0$. The hybridization of a local level with the conduction band is reduced by another factor because of the overlap of the lattice coordinates $\exp(-E_{JT}/\omega)$; this effect again corresponds to a slow restructuring of the lattice (slow at the scale of the phonon frequencies). If hybridization is ignored, the population of the local levels is determined by the conditions for a thermodynamic equilibrium.

The structure of La_2CuO_4 is such that an attraction between ordinary localized levels is essentially inescapable. As can be seen from Fig. 2, each of the two electrons (i and k) in the cells causes an elastic shear strain near its own cell. Ignoring Coulomb interactions for the moment, we see that there is a possible advantage in terms of the elastic energy if the two electrons are in adjacent cells. Such a formation would correspond to a bipolaron. Even larger clusters could then appear. The position of the chemical potential would correspondingly decrease by an amount $\delta\mu$, which is equal to the energy required for one localized level to split off from a cluster. The electron density in the band and that in the clusters satisfy the following respective proportionalities:

$$n_c = \int_0^{E_0 - \delta\mu} \nu(E) dE, \qquad n_i = \int_{E_0 - \delta\mu}^{\mu_0} \nu(E) dE.$$

Let us examine the structure of a cluster in La_2CuO_4 in more detail. Oxygen atoms in the ab plane are collectivized between two neighboring cells. For simplicity, we assume that the bond between the oxygens (O-La-La-O) along the c axis is also rigid. A Jahn-Teller strain in certain neighboring cells is obviously equivalent to a local shear:

$$Q_3 \Rightarrow 6^{-1/2}(2u_{zz} - u_{xx} - u_{yy}), \qquad Q_2 \Rightarrow 2^{-1/2}(u_{xx} - u_{yy}),$$

where $u_{ik} = 1/2(\partial u_i/\partial x_k + \partial u_k/\partial x_i)$ is the strain tensor. The elastic energy per unit volume is now

$$\frac{1}{2}(c_{11} - c_{12})(Q_3^2 + Q_2^2). \tag{2}$$

This energy should be considered along with Jahn-Teller Hamiltonian (1). According to this interpretation, the clusters mentioned above would be manifested as local shear in the La_2CuO_4 lattice.

What would the appearance of a cluster in real space imply? (If a cluster is large enough, it could be treated as a nucleating region of a new phase.) Since there must be one electron in each localized level (more precisely, one hole, for the d^9 configuration with respect to d^{10}), there are no carriers in the conduction band, because of electrical neutrality in this spatial region. In other words, a cluster corresponds to a "droplet" of an insulating phase. In the static picture of the coexistence of insulating and conducting phases, the conductivity would vanish if the ratio n_c/n_i were less than the percolation limit. In the three-dimensional case, this situation would correspond to $n_c/n_i \sim 0.15$, and in the two-dimensional case to ~ 0.5. We would like to link the sharp differences between the conductivity of "pure" La_2CuO_4 and that of the doped compound to specifically a proximity to the percolation limit. The coexistence of carriers of two types in La_2CuO_4 would be necessary in order to explain, for example, the presence of a normal Pauli contribution to the susceptibility in the case of a simultaneous localization nature of the resistance at low temperatures.

A thermodynamic analysis is capable of telling us only the relative concentrations of the insulating and conducting phases. It cannot tell us the extent to which the "droplets" are broken up. Coulomb effects must be taken into account in order to resolve this question. Since all the energies involved here are at the atomic order of magnitude, the dimensions of the droplets or clusters forming a "fog" in the La_2CuO_4 lattice would probably be small. The percolation approach would thus be of a limited nature, and even below the percolation threshold the conductivity might be low but nonzero. These questions are presently being studied. We wish to stress once more that the clusters are not static; i.e., the overall state is dynamic, although the characteristic frequency is associated with the lattice dynamics.

Doping changes the total number of carriers in the band in Fig. 1, lowers the local symmetry (by virtue of the Coulomb charge), and creates inhomogeneities, which may pin the clusters or droplets.

We do not have space here to do more than mention that a structural transition would not be difficult to link with a third-order anharmonicity in the product of representations $E_g \times F_{2g} \times F_{2g}$, where the three-dimensional representation F_{2g} contains a strain corresponding to rotations of octahedra.

We wish to thank G. M. Éliashberg fo- many discussions.

[1]J. G. Bednorz and K. A. Muller, Z. Phys. **B64**, 189 (1986).
[2]I. E. Dzyaloshinskiĭ, Pis'ma Zh. Eksp. Teor. Fiz. Supplement **46**, 110 (1987) [JETP Lett. Suppl. 46 (to be published)].
[3]P. W. Anderson, Science **235**, 1196 (1987).
[4]H. Kamimura, Jpn. J. Appl. Phys. **26**, L267 (1987); Proceedings of the Adriatico Research Conference on High Temperature Superconductivity, 1987 (in press).
[5]Yu. Kagan and K. A. Kikoin, Pis'ma Zh. Eksp. Teor. Fiz. **31**, 367 (1980) [JETP Lett. **31**, 335 (1980)].
[6]G. M. Éliashberg, Pis'ma Zh. Eksp. Teor. Fiz. Supplement **46**, 94 (1987) [JETP Lett. Suppl. 46 (to be published)].

Translated by Dave Parsons

International Symposium on Lattice Effects in Cuprate High Temperature Superconductors IOP Publishing
Journal of Physics: Conference Series **108** (2008) 012009 doi:10.1088/1742-6596/108/1/012009

The two-component physics in cuprates in the real space and in the momentum representation

Lev P Gor'kov[1,3] and G B Teitel'baum[2]

[1]NHMFL, Florida State University, 1800 E P. Dirac Dr., Tallahassee FL 32310, USA
[2]E.K.Zavoiskii Institute for Technical Physics of the RAS, Sibirskii Trakt 10/7, Kazan 420029, Russia

E-mail: gorkov@magnet.fsu.edu, grteit@yahoo.com

Abstract. Gradual evolution of two phase coexistence between dynamical and static regimes in cuprates is first investigated in the real space by making use of available neutron scattering, NMR and μSR data. Analyses of the Hall effect and the ARPES spectra reveals the presence of two groups of charge carriers in LSCO. The *T*-dependent component is due to the thermal activation of bound electron-hole structures seen near antinodal points in the Brillouin zone, thus introducing the two component physics also for the momentum representation. Interpretation of so-called "van Hove bands" undergoes drastic changes. Importance of the findings for pseudo-gap physics is stressed. Relation to some recent STM and STS results is discussed.

1. Introduction

Although the mechanism of high temperature superconductivity (HTSC) in cuprates remains unknown, it is worth mentioning that the superconductivity (SC) itself, beside its *d*-wave character, bears all the major features of SC in more common materials, including the double charge (2*e*) for the SC order parameter, as seen, e.g., from the vortex flux quantization. The most controversial issue concerns the normal state and stems from the fact that the very conductivity in cuprates is the result of an external doping, such, for instance, as doping the Mott insulator, the parent antiferromagnet (AFM) La_2CuO_4, with the divalent Sr. Numerous experimental facts demonstrate that the resulting system in no case behaves as an ordinary Fermi liquid (FL).

There currently exist compelling experimental evidences that the high T_c- cuprates in the part of the (T,x)-phase diagram known as the "pseudogap"(PG) region, are actually inhomogeneous in the real space. More recently, it has been shown [1] that the analysis of the ARPES and the Hall effect data together unequivocally indicates in favor of the two- component physics for these systems also in the momentum representation. The presentation below is the attempt to further comprehend these findings to unify them into a self-consistent picture.

We summarize first the NMR and μSR data from which it follows that the two-phase regions in the bulk, while displaying dynamical character at elevated temperatures, gradually freeze down to form heterogeneous static areas at lower temperatures.

Analysis of ARPES and the Hall coefficient data have led us to the conclusion that the so-called van Hove (vH) singularity at the anti-nodal points has nothing to do with any quasi-particle band structure.

[3] To whom any correspondence should be addressed.

International Symposium on Lattice Effects in Cuprate High Temperature Superconductors IOP Publishing
Journal of Physics: Conference Series **108** (2008) 012009 doi:10.1088/1742-6596/108/1/012009

Instead, in the vicinity of the anti-nodal "patches" a quasi- periodic (with doubled periodicity) structure appears representing an electron-hole bound state with the binding energy value gradually decreasing with doping, x. At x~0.2 the change of the regime takes place, corresponding probably to a quantum critical point (QCP)

We postulate that inhomogeneous areas seen in the new generation of the surface-sensitive experiments [2-4], where the heterogeneity itself is static due to low-T pinning at the sample's surface, have the same character as in the bulk. Correspondingly, the presence of such "exciton"-like states suggests the new interpretation for the two energy scales observed in these and other experiments. While the low temperature gap around the nodal regions at the Fermi surface (FS) is justly interpreted as the *d*-wave superconducting (SC) gap, the higher energy scale that persists above T_c and is seen near the $(0,\pi)$- points is related to the spatially inhomogeneous distribution of the bound *e-h*-pairs and may have nothing in common with a SC gap. We speculate that the aforementioned "excitonic" *e-h*-bound states come about from clusters with the predominant participation of the O-Cu pairs and involve the octahedral distortions.

In the discussion below we restrict ourselves mainly by the class of $La_{2-x}Sr_xCuO_4$ (LSCO) materials, for which the most extensive experimental results are available.

2. Evolution of the two phase regime with temperature

Experimentally the notion of two components should be dated back to the Johnston's experiments [5] where the decomposition of the magnetic susceptibility of LSCO into the Pauli- and a temperature dependent contributions had been performed for the first time.

Theoretically the tendency to the phase separation in cuprates into magnetic and conducting islands has been predicted in [6] with lattice interactions between neighboring Cu^{2+} ions as the driving mechanism, and, later, in [7] where the phase separation would be due to the exchange interactions between the copper spins. In both cases the electro-neutrality that otherwise would be violated by the presence of immobile Sr-charges was expected to severely restrict sizes of an AFM island, thus resulting in a dynamically frustrated first order transition. As shown in [8], it is energetically preferable for AFM phase, at least at small x, to appear in a form of so-called "stripes": charged incommensurate antiferromagnetic structures. The dynamical character of stripes' formation at temperatures above T_c has been demonstrated by observation of large peaks at the incommensurate wave vectors' values around the commensurate AFM (π,π)-point [9] only in the inelastic neutron scattering experiments.

The dynamical spatial separation into two components in *the real space* has been established in [10]. In [10] it was shown that for such local probe experiments as NMR and NQR, the nuclear magnetic relaxation time of the Cu-nuclei spins, $1/^{63}T_1(T,x)$, consists of the two contribution:

$$1/{}^{63}T_1(T,x) = 1/{}^{63}\overline{T}_1(x) + 1/{}^{63}T_1(T) \qquad (1)$$

The first component comes about when the participating nuclear spin belongs to a stripe area, while the second one is inherent to the stoichiometric CuO_2-plane. This last conclusion follows from the surprising fact that the concentration-independent term, $1/^{63}T_1(T)$, turned out to be same for a whole class of materials and, in particular, exactly coincides with the relaxation rate $1/^{63}T_1(T)$ measured for the two-chains' material, YBCO 1248 [11].

The standard analysis [10] with the use of the expression:

$$1/T_1 = \frac{k_B T}{2\mu_B^2\hbar^2\omega}\sum_i F(Q_i)\int\frac{d^2q}{(2\pi)^2}\cdot\chi''(q,\omega\to 0) \qquad (2)$$

and of the experimental inelastic neutron scattering data with $\chi''(q,\omega,T)$ from the incommensurate peaks [12] leads to the correct estimate of the first term value in Eq. (1). Even more important, this contribution turns out to be temperature-independent in the whole temperature interval up to 300 K. Both contributions are, of course, positive, confirming their origination from the two independent relaxation processes.

International Symposium on Lattice Effects in Cuprate High Temperature Superconductors IOP Publishing
Journal of Physics: Conference Series **108** (2008) 012009 doi:10.1088/1742-6596/108/1/012009

It is worth mentioning that $1/{}^{63}\overline{T}_1(x)$-value also provides an insight into the intensity of processes related to stripes for a whole class of cuprates. In Fig. 1 from [10] $1/{}^{63}\overline{T}_1(x)$ for LSCO is plotted as a function of the doping, x. Although its value rapidly decreases with the doping increase, it always

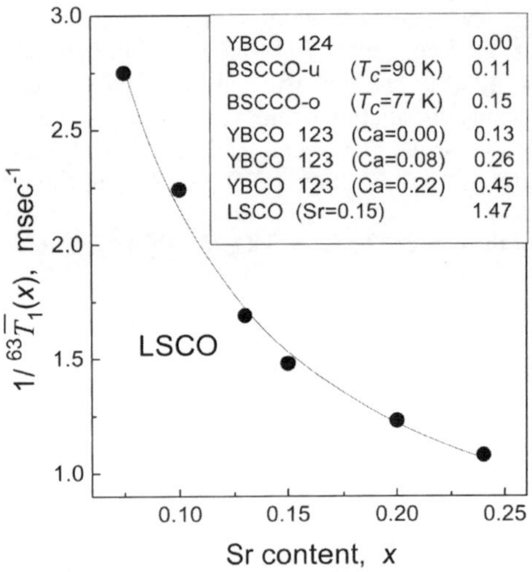

Figure 1. The $1/{}^{63}\overline{T}_1(x)$ vs Sr content x for LSCO (together with $1/x$ fit). Inset: $1/{}^{63}\overline{T}_1(x)$ for few other compounds, (u) and (o) indicate underdoped (u) and overdoped (o) BSCCO 2212. Details and references in [10].

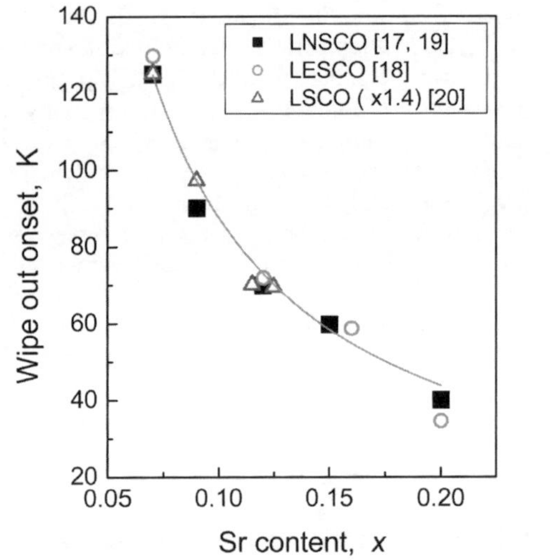

Figure 2. Doping dependence of the wipeout onsets for LSCO and rare earth co-doped materials LNSCO (Nd=0.4) and LESCO (Eu=0.2) together with $1/x$ fit. Note, that LSCO data are multiplied by a factor 1.4.

remains considerably larger (by one order of magnitude, approximately) than the same quantity in other cuprates, as shown in the Inset to Fig. 1. In other words, phenomena related to the dynamical stripes' formation are less pronounced in other compounds. Therefore, in what follows we mostly concentrate on the LSCO materials for which the interplay of SC with the magnetic (antiferromagnetic) phenomena seems to be strongest and has been extensively studied.

Note that at lower temperatures the coexistence of two static phases is a well documented experimental fact. Indeed, the static stripe phase has been directly seen in the elastic neutron experiments [13] for the Nd-and Eu- doped LSCO (LNSCO and LESCO below). The NQR copper spectra for LESCO at 1.3 K also reveal the in-plane distribution of the hole and spin densities typical of the static stripe phase [14]. For $La_{1.88}Sr_{0.12}CuO_4$ the presence of both SC and the magnetic phases (the volume fraction ~18% for the latter) has also been proven by μSR [15].

At temperatures below the LTT transition for the 1/8-Ba doped LCO (LBCO) the incommensurate antiferromagnetism (ICAFM) manifests itself as the thermodynamic equilibrium state through the magnetic susceptibility anisotropies and the observation of the spin-flop transition [16].

Whether the static magnetic phase occupies the whole sample's volume or only a part of it, can be also judged from measurements of the "wipe-out" fraction of the ^{63}Cu-signal. This fraction constitutes almost 100% in LNSCO and LESCO [17-19]. Meanwhile, the "wipe-out" is far from being complete

International Symposium on Lattice Effects in Cuprate High Temperature Superconductors IOP Publishing
Journal of Physics: Conference Series **108** (2008) 012009 doi:10.1088/1742-6596/108/1/012009

in LSCO, thus confirming again the static coexistence of the two different ground states at low temperatures in these materials [20].

On the other hand, as we have said above, the ICAFM-islands seem to manifest themselves at higher temperature only in a dynamical regime. Study of the temperature dependence for the "wipe-out" of the ^{63}Cu-signal allows to trace in details that the dynamical fluctuations gradually slow down to finally form static patterns at the lowest temperatures [14, 17-20]. In Fig. 2 the temperatures for onset of such a freezing are given for LNSCO, LESCO and LSCO materials. It seems that phases' pinning is governed by the charges of the dopants' ions.

One concludes from this Section that two phases may, indeed, coexist in the sample's volume statically at low temperatures, while at elevated temperatures heterogeneity in the real space manifests itself in a dynamical fashion.

3. Two components in the number of charge carriers

Doping of a Mott insulator is not the thermodynamic process, and so far is not understood well. It is commonly accepted that doped holes initially go onto the oxygens' sites and (at x>0.02 for LSCO) start to participate in conductivity and other transport phenomena. Naturally, for the above picture of the dynamical coexistence of two phases all transport properties are expected to behave different from predictions of any Boltzmann-like theory. No approach has been developed yet to address the issue theoretically, and we limit ourselves by the mere analysis of experimental data. Fortunately, the detailed data are now available on the temperature dependence both for resistivity and the Hall coefficient in LSCO in the whole range of concentrations and temperatures up to 1000 K [21].

In [1] it has been demonstrated that analyses of the data [21] for the Hall coefficient, $R_{\rm Hall}(T, x)$, again unambiguously leads to the notion of two components, now in the effective number of carriers:

$$n_{\rm eff} = n_0(x) + n_1 \cdot \exp(-\Delta(x)/T). \qquad (3)$$

The first component is temperature independent, while the second one bears the activation character. The presence of the activation contribution into Eq. (3) must be considered as the thermodynamic feature of the system.

In Fig. 3 we plot data [21] for $R_{\rm Hall}(T, x)$ against Eq. (3) to demonstrate the excellent fit for the whole concentration range (above temperatures where localization features begin to be seen in the resistivity behavior).

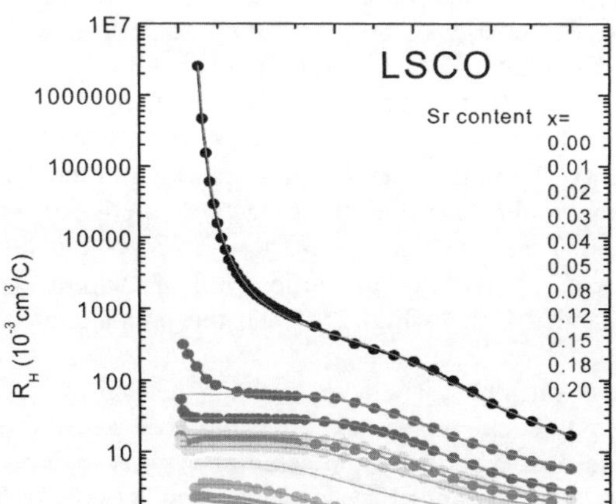

Figure 4. The doping dependent evolution of $R_{\rm H}$ [21] with fittings to Eq. (3).

International Symposium on Lattice Effects in Cuprate High Temperature Superconductors IOP Publishing
Journal of Physics: Conference Series **108** (2008) 012009 doi:10.1088/1742-6596/108/1/012009

Dependence of $n_0(x)$ presented in [1] does show that already at x=0.07 the number of carriers at low temperatures, $n_0(x)$, starts to exceed the number of the externally doped holes and rapidly increases with x to values close to one carrier per Cu-site. Similar low temperature doping behavior for the Hall coefficient has also been observed in the single–layer BSLCO [22]. We emphasize that our analysis in terms of Eq. (3) was restricted by the temperatures above 100 K. At low temperatures R_H is not monotonous for $0.15 \le x \le 0.17$. This may be due to interplay of holes with the charge carriers from the electron pockets whose mobility is increasing upon cooling [23]. The observed picture may be also affected by the stripes' pinning and by an increase in hybridization between the oxygen- and the copper-orbitals.

The coefficient $n_1(x)$ in front of the exponent is practically concentration-independent ($n_1(x)$ ~2.8), but rather abruptly drops to zero at x above ~0.2.

It is crucial to emphasize once more the activation character of the T-dependent contribution. While in [24] the decrease of $R_{Hall}(T, x)$ with temperature has been ascribed to a "spill-over" of electrons from the so-called vH "flat bands" near the antinodal points $(0,\pi)$, such an interpretation would be unable to explain the activation form of the T-dependence in Eq. (3). The vH-type singularity with its logarithmic character, it is too week to account for the exponential term in (3). For the activation gap in Eq.(3) to exist there must be a sharp feature in the spectrum of the system, with a large density of states that is well separated from the position of the chemical potential. (Note also the large value of the pre-factor, n_1 ~2.8 proportional to the density of states (DOS) per unit cell).

Eq. (3) also adequately describes the earlier data for $R_{Hall}(T, x)$ for temperatures and compositions available in [25] and [26]. (The x-dependence of the gap, $\Delta(x)$, itself is discussed in the next sub-Section).

4. Comparison with the ARPES results

The activation term itself in Eq. (3) of the previous Section could, in principle, be interpreted in sort of a band picture as if due to the thermal excitation of an electron from the chemical potential to the bottom of an unoccupied narrow band or an empty energy level. In [21] $R_{Hall}(T, x)$-data are decomposed into a "nodal" term, $R_{Hall}(T, x)^{arc}$, and $R_{Hall}(T, x)^{LFS}$, a contribution from "the large FS". For the latter the activation gap is taken as the x-independent "charge transfer" gap (from its value at small x), while additional T-dependences for the two components come about through T-dependence of their relative fractions. Both assumptions [21], however, are not consistent with the ARPES data [27-29].

In the ARPES experiments one measures the energy of out-coming photoelectron with a known momentum with respect to the chemical potential's energy position. (Theoretically, ARPES can "reconstruct" the electronic band spectra of a system in the whole Brillouin Zone (BZ) provided electrons were not interacting strongly). In [27-29] data are available, in particular, for the quasi momentum near the vicinity of the antinodal (vH) points, $(0,\pi)$. Ideally, according to the band picture, the electronic ARPES spectra at these points would determine the (negative) energy of the "vH flat band" with respect to the chemical potential. These data are now shown in Fig. 4 together with the values for $\Delta(x)$ extracted from the $R_{Hall}(T, x)$ [21]. The two quantities *coincide* within the limits of experimental errors. (In Fig. 4 new points added [29] to Fig. 2 of [1]; the points shown by green squares are discussed later).

This is the surprising result. Indeed, it is difficult to believe that the thermal excitation of an electron from the chemical potential to a hypothetical energy level that lies well *above* the latter (to create a hole, as needed for Eq. (3)), would cost practically the same energy as to transfer to the chemical potential an electron taken from the "vH flat band" that lies well *below* it (at least at small x). Although a numerical coincidence may occur at some doping, Fig. 4, as a whole, obviously, manifests the new physics.

In [1] we made a comment that the result is an indication in favor of localized states near the antinodal points. Here we elaborate further that the equal energy for creation of an electron (ARPES) and a hole (thermal activation) can be understood only in terms of *bound* states for electron-hole pairs.

International Symposium on Lattice Effects in Cuprate High Temperature Superconductors IOP Publishing
Journal of Physics: Conference Series **108** (2008) 012009 doi:10.1088/1742-6596/108/1/012009

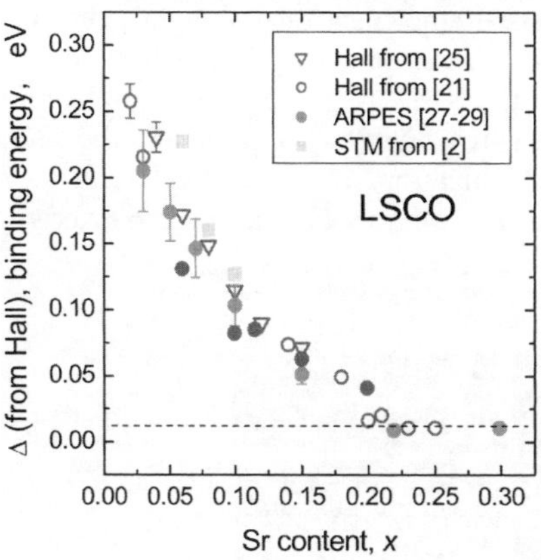

Figure 4. The ARPES binding energy vs the Hall activation energy. (See [1] for the details.)

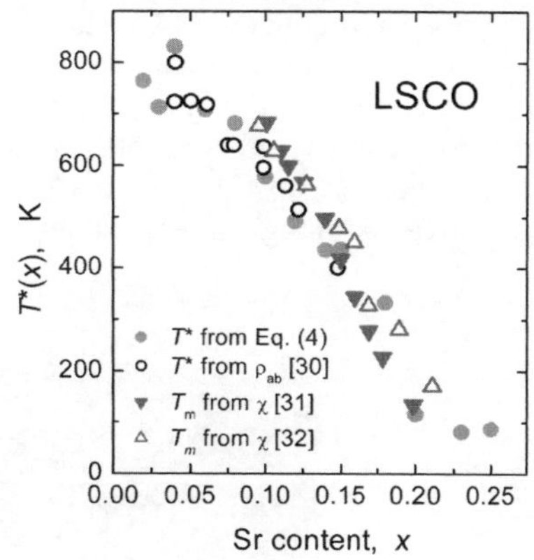

Figure 5. The doping dependence of $T^*(x)$ for $La_{2-x}Sr_xCuO_4$ [1].

These states, being positioned in the momentum representation near the vH points, according to ARPES, form quasi-periodic structures (close to the double periodicity along the bonds directions). Large DOS (large n_1~2.8 in Eq. (3)!) show that these formation densely populate the real space.

ARPES is the fast experiment. Each emitted photoelectron leaves behind the perturbed lattice. In other words, the energy cost, as seen by ARPES, for breaking the bound *e-h* pair should be less than the energy needed for breaking the same bound pair in the thermal activation process. This feature is indeed present in Fig. 4: all points from the ARPES data [27-29] lie below the corresponding $\Delta(x)$-values by some ~ 15 meV, a scale more typical for the lattice effects.

In Fig. 4 one can notice some sharp feature at x above ~0.2, the same concentration at which n_1 begins to drop to zero [1]. However, the binding energy does not go to zero: from more recent ARPES data [29] one concludes that the bound state persists above x~0.2, as shown by the dashed line. The energy gap still persists, but now is of order of ~ 15 meV. We speculate that the lattice effects are also involved into formation of such *e-h* structures.

Change with temperature of the balance between number of carriers introduced through external doping process, x, and the concentration of the thermally activated holes should strongly impact transport and other characteristics of cuprates. In Fig. 5 we plotted the temperature at which the number of carriers from the each two components becomes equal:

$$T^*(x) \approx -\Delta(x) / \ln x \qquad (4)$$

Other symbols in Fig. 5 show the marks derived from behavior of the magnetic susceptibility and resistivity data at their crossing into the pseudogap regime in LSCO. The result shows that Eq. (4) actually determines so-called pseudogap temperature, $T^*(x)$, quite well. With the increase of number of carriers at elevated temperatures to ~ one carrier per Cu one may expect the restoration of the FL features in cuprates, including a large Fermi surface.

5. Brief discussion

We do not attempt to provide any strict theoretical interpretation. The facts above are phenomenological ones, obtained from the analyses of experimental data, and below we shall suggest a few speculations only.

To start with, one may ask whether same energy scales, as shown in Fig. 4 for LSCO, have ever been seen in other materials, by other methods. APRES data, while abundant for two-layer systems

International Symposium on Lattice Effects in Cuprate High Temperature Superconductors IOP Publishing
Journal of Physics: Conference Series **108** (2008) 012009 doi:10.1088/1742-6596/108/1/012009

like YBCO or BSCO 2212, are unable to provide a reliable answer being masked by the inter-plane bonding / anti-bonding splitting of spectra near the "vH patches".

In Fig. 6 we reproduce STM results from [2]. Positions of the ticks on the right side of *dJ/dV* curves are plotted as the green squares in Fig. 4 for three concentrations. The agreement with the data for LSCO is excellent. More intriguing is that, according to the STM map in Fig. 6, these features are seen for *dJ/dV* –curves in the bright ("*metallic*") regions. It is reasonable to assume that the shoulders indicated by ticks correspond to the bias that matches the "ionization" energy of the bound *e-h* states discussed above. Although in [2] resolution is worse than the atomic scale, the pattern of bright and

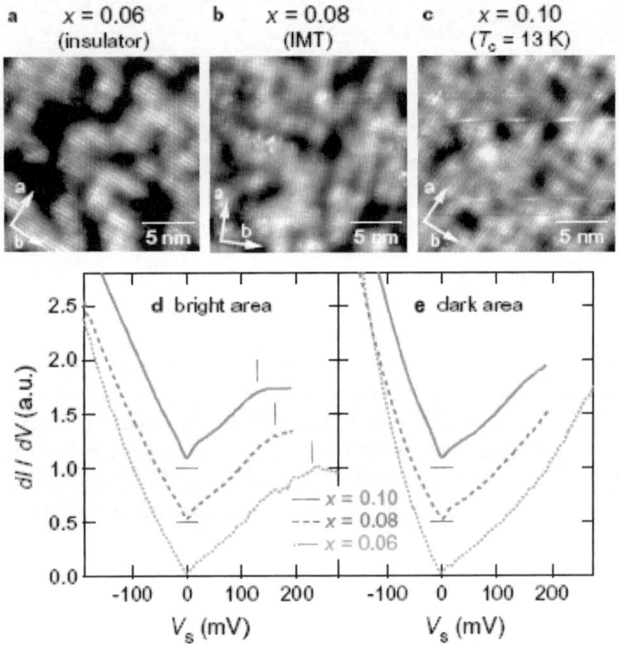

Figure 6. *dJ/dV* curves obtained by STM imaging/spectroscopy of $Ca_{2-x}Na_xCuO_2Cl_2$ [2]. Positions of the ticks on the right side of *dJ/dV* curves are plotted as the green squares in Fig. 4 for three concentrations.

dark pinned spots shown there, looks very similar to the one in Fig. 3 of [3] for the two materials, with the Dy-doped BSCO 2212 among them.

It should be supposed that the "vH" spots in the BZ observed within the ARPES accuracy are in some correspondence with the inhomogeneous distribution of pinned nanoscale clusters seen in STS by the Davis group and others (See [3] and references there in). From the results above it becomes clear that the energy values plotted in Fig. 4 are actually related to the inhomogeneous electronic structure. Therefore, it seems that the language of the two "gaps", often used in the literature (see, e.g., [33]), i.e., of the SC gap and so-called "pseudogap" that would persist above *Tc* near $(0,\pi)$ points, can only obscure the real physics. There is no "pseudogap": what one sees near the antinodal, or "vH", points corresponds to the internal scale of the bound electron-hole structures in Fig. 4. Their double-periodicity is heterogeneously distorted along the sample at low temperatures due to random pinning.

To emphasize again, no theoretical description exists for the *e-h* bound states that we hypothesized above. One can only speculate that adjacent pairs of oxygen and copper ions may be involved, and the hybridization between O- and Cu-orbitals results in an intermediate valence of the Cu-ions belonging to a cluster. Unlike the Zhang-Rice singlet, such a construction may possess a spin. One may

International Symposium on Lattice Effects in Cuprate High Temperature Superconductors IOP Publishing
Journal of Physics: Conference Series **108** (2008) 012009 doi:10.1088/1742-6596/108/1/012009

anticipate a local variation of distances between O and Cu. The current STM or STS experiments are not sensitive enough so far to verify such an assumption.

Acknowledgements
The work of L.P.G. was supported by the NHMFL through NSF cooperative agreement DMR-9527035 and the State of Florida, that of G.B.T. through the RFBR Grant N 07-02-17137.

References

[1] L.P. Gor'kov and G.B. Teitel'baum, Phys. Rev. Lett. **97**, 247003 (2006)
[2] Y. Kohsaka *et al.*, Phys. Rev. Lett. **93**,097004 (2004)
[3] Y. Kohsaka *et al.*, Science **315**, 1380 (2007)
[4] P. Abbamonte *et al.*, Nature Phys. **1**, 155 (2005)
[5] D.C. Johnston, Phys. Rev. Lett. **62**, 957 (1989)
[6] L.P. Gor'kov, A.V. Sokol, JETP Lett. **46** 420 (1987)
[7] V. Emery, S. Kivelson, and H. Lin, Phys.Rev.Lett. **64,** 475 (1990)
[8] J. Zaanen and O. Gunnarson, Phys. Rev., B **40** 7391 (1989)
[9] Cheong, S. W., Aeppli, G., Mason *et al.*, Phys. Rev. Lett. **67**, 1791 (1991)
[10] L.P. Gor'kov and G.B. Teitel'baum, JETP Letters **80**, 221 (2004)
[11] G.V.M. Williams *et al.*, Phys. Rev. B **58**, 15053 (1998)
[12] G. Aeppli et al., Science **278**, 1432 (1997)
[13] J.M. Tranquada *et al.*, Nature **375**, 561 (1995)
[14] G. B. Teitel'baum, B. Buechner, and H. de Gronckel, Phys.Rev.Lett. **84**, 2949 (2000)
[15] A.T. Savichi *et al.*, Phys.Rev. B **66**, 014524 (2002)
[16] M. Huecker *et al.*, cond-mat/0503417v2
[17] P. M. Singer, Phys. Rev. B **60**, 15345 (1999)
[18] A. W. Hunt *et al.*, Phys. Rev. B **64**, 134525 (2001)
[19] G. B. Teitel'baum *et al.*, Phys. Rev. B **63**, 020507 (2001)
[20] A. W. Hunt *et al.*, Phys. Rev. Lett. **82**, 4300 (1999)
[21] S. Ono *et al.*, Phys. Rev. B **75**, 024515 (2007)
[22] F. Balakirev *et al.*, Nature **424**, 912 (2003)
[23] F. Balakirev *et al.*, cond-mat /07104612v2
[24] Y. Ando *et al.*, Phys. Rev. Lett. **92**, 197001 (2004)
[25] T. Nishikawa *et al.*, J. Phys. Soc. Jpn. **63**, 1441 (1994)
[26] H.Y. Hwang *et al.*, Phys. Rev. Lett. **72**, 2636 (1994)
[27] T. Yoshida *et al.*, Phys.Rev.Lett. **91**, 027001 (2003)
[28] T. Yoshida *et al.*, cond-mat/0510608
[29] T. Yoshida *et al.*, J. Phys.: Condens. Matter **19**, 125209 (2007)
[30] T. Timusk and B. Statt, Rep. Prog. Phys. **62**, 61 (1999)
[31] T. Nakano *et al.*, Phys. Rev. B **49**, 16 000 (1994).
[32] R. Yoshizaki *et al.*, Physica C **166**, 417 (1990).
[33] M. Boyer *et al.*, Nature Physics **3**, 802 (2007)

Fast Track Communications

IOP Publishing | Journal of Physics: Condensed Matter

J. Phys.: Condens. Matter **26** (2014) 042202 (6pp) | doi:10.1088/0953-8984/26/4/042202

Fast Track Communication

Two regimes in conductivity and the Hall coefficient of underdoped cuprates in strong magnetic fields

L P Gor'kov[1,2] **and G B Teitel'baum**[3]

[1] NHMFL, Florida State University, 1800 East Paul Dirac Drive, Tallahassee, FL 32310, USA
[2] L D Landau Institute for Theoretical Physics of the RAS, Chernogolovka 142432, Russia
[3] E K Zavoiskii Institute for Technical Physics of the RAS, Kazan 420029, Russia

E-mail: grteit@kfti.knc.ru

Received 19 November 2013, revised 7 December 2013
Accepted for publication 16 December 2013
Published 6 January 2014

Abstract

We address recent experiments shedding light on the energy spectrum of under and optimally doped cuprates at temperatures above the superconducting transition. Angle resolved photoemission reveals coherent excitation only near nodal points on parts of the 'bare' Fermi surface known as the Fermi arcs. The question debated in the literature is whether the small normal pocket, seen via quantum oscillations, exists at higher temperatures or forms below a charge order transition in strong magnetic fields. Assuming the former case as a possibility, expressions are derived for the resistivity and the Hall coefficient (in weak and strong magnetic fields) with both types of carriers participating in the transport. There are two regimes. At higher temperatures (at a fixed field) electrons are dragged by the Fermi arcs' holes. The pocket being small, its contribution to conductivity and the Hall coefficient is negligible. At lower temperatures electrons decouple from holes behaving as a Fermi gas in the magnetic field. As the mobility of holes on the arcs decreases in strong fields with a decrease of temperature, below a crossover point the pocket electrons prevail, changing the sign of the Hall coefficient in the low temperature limit. Such behavior finds its confirmation in recent high-field experiments.

Keywords: superconductivity, cuprates, pseudogap phase, Hall effect

1. Introduction

An understanding of high temperature superconductivity (HTSC) in cuprates is a pressing issue both on the theoretical side of the problem and for practical implications. While in superconductors of the 'old' generation it was at least known that superconductivity emerges from the normal Fermi liquid phase, HTSC is preceded by the so-called pseudogap phase with many abnormal properties. The most unexpected feature revealed in angle resolved photoemission (ARPES) experiments is that coherent electronic excitations exist only on the Fermi arcs along the 'bare' Fermi surface, separated from each other by large energy gaps [1].

It was absolutely unclear how to interpret even such basic experimental data as those for resistivity or the Hall coefficient. The importance of the question was probably realized for the first time when it turned out that the Hall coefficient [2] in $La_{2-x}Sr_xCuO_4$ (LSCO) manifests an activation temperature dependence [3].

In 2007 the discovery of quantum oscillations (QOs) in underdoped (UD) *ortho II* $YBa_2Cu_3O_{7-\delta}$ (YBCO) [4] showed the presence of small Fermi liquid (FL) pocket(s).

Low frequency QOs in high magnetic fields were found in $HgBa_2CuO_{4+\delta}$ (Hg1201) as well [6].

So far, ARPES experiments have been performed on two-layer $Bi_2Sr_2CaCu_2O_8$ (Bi2212), single-layer $Bi_2Sr_2CuO_6$ (Bi2201) and LSCO (see the recent review [7]). As yet, ARPES is not available for Hg1201 and YBCO because of problems with the surface [8]. The consensus is that, except for minor details, ARPES findings bear the general character and reflect the basic physics.

It is all the more remarkable that the recent experiments [9] added new and convincing evidence in support of the Fermi arcs concept. In brief, it was pointed out in [9] that the characteristic for temperature dependence of resistivity of clean Hg1201 (and a few other cuprate compounds) is the contribution proportional to the square of the temperature, as if in a Fermi liquid. In [9] the latter was related to free carriers on the Fermi arcs.

In turn, non-monotonic temperature and field dependence of the Hall and Seebeck coefficients in YBCO and Hg1201 in experiments [11, 12] below $T \approx 100$–150 K was considered as due to the contribution from electrons on the pocket.

These results pose a question about the interrelation between pocket(s) and the Fermi arcs. Currently, one highly debated issue in the literature concerns the nature of the pocket, that is, whether the pocket is a band feature and is omnipresent in the spectrum, including at higher temperatures, or whether it emerges below a charge order (CDW) transition. As to the latter, according to [13, 14], a CDW ordering indeed occurs in the relevant intervals of temperature and magnetic field.

Let us emphasize that most QO experiments have been performed in high fields and at not too high temperatures ($T < 10$ K). We investigate conductivity and the Hall coefficient assuming that carriers of both types participate in the transport properties in the whole temperature range of the pseudogap phase. We adopt the view [5] that only one small pocket exists in UD YBCO that lies away from a Fermi arc.

The model of holes on the Fermi arcs interacting via short-range interactions has been examined in the framework of the kinetic equation in [10] for interacting holes on the Fermi arcs. Below, the model is generalized to include electrons on the pocket and extended to the case of strong magnetic fields. Looking ahead, our results do not contradict [11, 12]; more importantly, in our opinion, the results give support to the notion of such a pocket as a band feature.

In its most general form the kinetic equation is

$$\frac{dn}{dt} = I_{col}. \tag{1}$$

On the left, the total derivative (dn/dt) accounts for variations of the Fermi distribution of the quasiparticles in their motion in the real and the momentum space in the presence of external fields. The right-hand term, the collision integral I_{col}, is responsible for the relaxation processes.

2. Results from the model of Fermi arcs

Consider first carriers on the Fermi arcs. In weak constant electric and magnetic fields the kinetic equation (1) is presented in the commonly accepted form

$$e\left(\vec{E} + \frac{1}{c}[\vec{v} \times \vec{H}]\right)\vec{\nabla}_{\vec{p}}n(\vec{p}) = I_{coll}. \tag{2}$$

Changes n_1 of the Fermi distribution function $n(\vec{p})$ are also small. At a non-zero current the system of charged carriers moves as a whole with a drift velocity $\vec{u}$. Correspondingly, the left-hand side (LHS) in (2) reduces to

$$e\left\{(\vec{E}\cdot(\vec{v}-\vec{u})) - \frac{1}{c}([\vec{v}\times\vec{H}]\cdot\vec{u})\right\}\frac{\partial n_o}{\partial\varepsilon}. \tag{3}$$

(Here $(\partial n_o/\partial\varepsilon)$ is the derivative of the equilibrium Fermi function: $\vec{v} = d\varepsilon(\vec{p})/d\vec{p}$.) Similarly, the collision integral on the right-hand side (RHS) of equation (3) should be expanded over small n_1. The balance between LHS and RHS gives expressions for the quadratic resistivity term and for the Hall coefficient, correspondingly [10],

$$\rho_{2D}(T) = \left(\frac{\pi\hbar}{e^2}\right)\frac{4\pi^2|Z\tilde{V}(1;2)|^2}{3(\Delta\varphi)^2}\sqrt{\frac{\Delta}{\varepsilon_F^*}}\left(\frac{T}{\varepsilon_F^*}\right)^2 \tag{4}$$

and

$$R_H = \frac{1}{ecn_{eff}} > 0. \tag{5}$$

Here $Z\tilde{V}(1;2)$ is the renormalized matrix element of the electron–electron interaction, $m^* = Z^{-1}m$ is the renormalized mass. The combination $\varepsilon_F^* = p_F^2/2m^*$ is defined as the renormalized Fermi energy.

In equation (4), $\Delta = v_F p_F[(K/4p_F) - 1]$, $K^{\|}$ is the projection of the Umklapp vector $\vec{K} = (2\pi/a, 2\pi/a)$ on the diagonal in the Brillouin zone (BZ). Processes with Umklapp scattering are critical for relaxation of the momentum and, hence, for non-zero resistivity. Unlike ordinary metals with large Fermi surfaces, for the system of narrow Fermi arcs the resistivity is obtained in the explicit form of equation (4); $\Delta\varphi$ is the arc width. For the isotropic energy spectrum the effective number of carriers is $n_{eff} = \frac{\Delta\varphi p_F^2}{\pi^2}(\frac{s}{c_z})$. Here c_z is the lattice constant in the perpendicular-to-plane direction, s is the number of CuO_2-planes per unit cell. Note in passing that each parameter in (4) and (5) can be found experimentally. In fact, the position of the nodal points p_F is known from ARPES; the Fermi arc width $\Delta\varphi(x)$ follows directly from the experimental value of R_H. (The nodal points usually lie rather close to the $(\pm\pi/2, \pm\pi/2)$-points [7].)

Rigorously speaking, expression (5) for the Hall coefficient applies only in the weak-field regime $\omega_c\tau_{col} \ll 1$ ($\omega_c = (eH/m^*c)$) [10]. (The notation τ_{col} stands for the mean scattering time.) Meanwhile, the order of magnitude τ_{col} in (4) is $\tau_{col} \propto \varepsilon_F/T^2$, i.e., τ_{col} is large in clean samples. For instance, at $T = 100$ K the magnetic field should be much smaller than 3–5 T for (5) to apply. Data for R_H in Hg1201, YBCO and LSCO in the whole temperature range have been obtained only in higher fields (see [2, 11] and [12]).

3. The role of the electronic pocket

The Hall coefficient equation (5) is temperature independent. Meanwhile, the notable feature in the Hall data [2, 11, 12] is the temperature dependence, $R_H(T)$ (hence, the same for $\Delta\varphi(T)$).

Two mechanisms come to mind in regard to the T-dependence of the Hall effect. The first one is a temperature dependence of the Fermi arcs' width that is seen in ARPES experiments [15–18]. For the model this means an increase in the number of carriers with temperature increase.

Recall that the notion of two scattering times [19] was introduced to reconcile the resistivity linear in temperature and the T^2-dependence of the Hall angle: $\cot(\theta_H) \propto T^2$ *above* $T^*(x)$. Here we concentrate on the universal experimental regime of the T^2-resistivity [9]. Increase of the arc width with temperature (i.e. of the number of carriers) causes deviations from quadratic dependence in the resistivity without changing the relaxation mechanisms. Indeed, for LSCO it was already pointed out in [20] that the quadratic temperature dependence in the product $n_{\rm Hall}(T) \times \rho(T) \to 1/\tau(T)$ means nothing but the T^2-dependence of the relaxation rate $1/\tau(T) = \text{const} \times (T^2/E_F)$ for the carriers.

The second mechanism of the Hall effect's temperature dependence concerns the role of the pocket. The latter was detected by high-field quantum oscillations [4]. It is known that in the superconducting phase and *in zero field* the pocket carriers are independent of holes paired on the arcs and contribute to the specific heat the residual *normal metal* linear term [5].

Experimentally, the temperature dependence of the Hall coefficient has a maximum. In *high field* experiments [11, 12] $R_H(T)$ even changes sign at lower temperatures. This was ascribed to *electrons on a pocket* for which the mobility is increasing as $T \to 0$. Correspondingly, we consider the resistivity and the Hall coefficient for the two-component system of the Fermi arcs' holes and the pocket's electrons, first from the side of *higher* temperatures where the classic kinetic equation approach is applicable.

The general form of a current in dimension d ($d = 2, 3$) is

$$\vec{j}_d = 2e \int \vec{v} \frac{d^d\vec{p}}{(2\pi)^d} (\vec{p} \cdot \vec{v}_{\rm drift}) \left(\frac{1}{4T}\right) ch^{-2}\left(\frac{\varepsilon_p}{2T}\right) \qquad (6)$$

with $\vec{v}_{\rm drift}$ a drift velocity. We introduce the notations $\vec{v}_p$ and $\vec{u}_{FA}$ for the drift velocities of electrons and holes, respectively. Since the size of the pocket is small [4, 5] scattering events do not involve electrons directly in the Umklapp processes, so that holes on the Fermi arcs tend to 'drag' electrons. An electric field pulls electrons in the opposite direction from holes. Solving the linear kinetic equation allows one to write the relation between $\vec{v}_p$ and $\vec{u}_{FA}$ as follows:

$$(\vec{v}_p - \vec{u}_{FA})\frac{1}{\tau_{eh}} + \vec{v}_p \frac{1}{\tau_{imp}} = \frac{|e|\vec{E}}{m_e}. \qquad (7)$$

Here $1/\tau_{eh}$ is the inverse time for relaxation of the electrons *relative to* holes ($1/\tau_{imp}$ for scattering of electrons on defects is small in YBCO and Hg1201 and is neglected below).

An estimate of the Fermi energy of electrons at low temperatures E_{Fe} from the data [4, 5] gives $E_{Fe} \approx 500$ K (below E_{Fe} and p_{Fe} stand for the Fermi energy and Fermi momentum of electrons; the notations ε_F and p_F, as before, are reserved for holes). One sees that already at temperatures of 100–150 K the electrons may be treated as a degenerate Fermi gas.

The expression for u_{FA} [10] rewritten as

$$u_{FA} = -|e|E \frac{3\Delta\varphi\varepsilon_F}{8\pi m^* |ZV(1,2)|^2 T^2} \sqrt{\frac{\varepsilon_F}{\Delta}} = -|e|E\frac{\tau_{FA}}{m*} \qquad (8)$$

defines the relaxation time of holes $\tau_{FA}(T)$:

$$\tau_{FA}(T) = \frac{3\Delta\varphi\varepsilon_F}{8\pi |ZV(1;2)|^2 T^2}\sqrt{\frac{\varepsilon_F}{\Delta}}. \qquad (9)$$

Interactions between electrons and holes are due to the same short-range Coulomb interactions as the interactions between holes on different Fermi arcs. Consequently, the matrix elements that determine $1/\tau_{eh}$ have the same order of magnitude. This makes it possible from the analysis of the kinetic equation to derive the following estimate for $1/\tau_{eh}$:

$$\frac{1}{\tau_{eh}} \approx \text{const} \times \left(\frac{m_e}{m^*}\right)\left(\frac{p_F}{p_{Fe}}\right)\frac{T^2}{\varepsilon_F}, \qquad (10)$$

with a const ~ 1. The factor $p_F/p_{Fe} \gg 1$ in (10) shows that $1/\tau_{eh} \gg 1/\tau_{FA}$. From (7) the conductivity follows:

$$\sigma_{3D}(T) = e^2 \left\{\frac{n_{\rm eff}}{m^*} - \frac{n_e}{m_e}\right\}\tau_{FA} \qquad (11)$$

where the second term is the contribution from *electrons dragged* by holes (hence the negative sign). In (11) n_e is the electronic density. Comparison between $n_{\rm eff} = \Delta\varphi s p_F^2/c\pi^2$, the effective density of holes on the Fermi arcs, and $n_e = (p_{Fe}^3/2\pi^2)$ shows that $n_{\rm eff} \gg n_e$ ($m^* \sim m_e$; $p_F/p_{Fe} \gg 1$), i.e. the contribution of electrons to the conductivity in this regime is very small.

In vector notation the Hall coefficient $R_H(T)$ is defined by the relation

$$\vec{E} = \rho\vec{J} + R_H[\vec{H} \times \vec{J}]. \qquad (12)$$

Here ρ is the resistivity and $\vec{J}$ the longitudinal current. The two drift velocities $v_{p,H}$ and $u_{FA,H}$ for electrons and holes in the presence of the magnetic field are found from the kinetic equations after substituting into equation (3) for the Lorentz forces:

$$-|e|E \Rightarrow \frac{|e|}{c}Hu_{FA} \qquad \text{(for holes)}; \qquad (13a)$$

$$|e|E \Rightarrow -\frac{|e|}{c}Hv_p \qquad \text{(for electrons)}. \qquad (13b)$$

Calculations of the Hall currents are similar to those for the two electrical currents. The Hall coefficient of the two-component system is

$$R_H(T) = \frac{1}{c|e|n_{\rm eff}} \times \left\{1 + \left(\frac{n_e}{n_{\rm eff}}\right)\left(\frac{m^*}{m_e}\right)\right\}. \qquad (14)$$

(Note that the small contributions from electrons in (11) and (14) have different signs.)

Pockets have been seen in e-doped $Nd_{2-x}Ce_xCuO_4$ [21]. With the estimate $E_F \sim 50$ meV in hole-doped cuprates and the ARPES resolution of typically 10 meV, observation of the pocket in photoemission experiments looks, in principle, feasible, although difficult. We recall again that it is the surface quality that plagues ARPES in YBCO and Hg1201. Doping in LSCO produces more disorder. (Two-layer Bi2212 has other problems related to super-structure.) Currently, ARPES experiments are mostly focused on the vicinity of the 'bare' Fermi surface. In our model the pocket lies away from the Fermi arcs. To repeat, the main experimental difficulties lie in the preparation of good crystalline surfaces.

4. Interplay between holes and electrons in strong magnetic fields

As mentioned above, formally, there are restrictions on the strength of the magnetic field. For holes it reads $\omega_c\tau_{FA} \propto (\omega_c/T)(\varepsilon_F/T) \ll 1$.

The argument is given below that expression (5) applies for the Hall coefficient of holes even if that restriction is not fulfilled. At the same time, for electrons the restriction $\omega_{ce}\tau_{eh} \ll 1$ ($\omega_{ce} \equiv (eH/m_e c)$) at a *fixed* value of the magnetic field is violated at temperatures considerably *lower* than for carriers on the Fermi arcs due to the factor $p_{Fe}/p_F \ll 1$ in $\omega_{ce}\tau_{eh} \approx \omega_{ce}(p_{Fe}/p_{Fh})\tau_{FA}$. Thus, at high enough temperatures the contribution from holes on the Fermi arcs prevails over that from electrons.

Thereby, while holes on the Fermi arcs can already be in the strong field regime for *electrons*, the new physics sets in only at $\omega_{ce}\tau \approx \omega_{ce}(p_{Fe}/p_{Fh})\tau_{FA} \approx 1$. Below, the pocket *decouples* from holes on the Fermi arcs and its contribution to $R_H(T)$ finally becomes that of *free* electrons in a magnetic field and scattering on impurities finally becomes the only mechanism of relaxation. To the best of the authors' knowledge no rigorous theoretical expression for $R_H(T)$ is available in general except in the limit of *strong* magnetic field.

For the diagonal (symmetric) component of conductivity in high fields estimates give [22] $\sigma^S \sim \sigma_0(\omega_c\tau)^{-2}$, where σ_0 is the conductivity in the absence of the field. For holes on the Fermi arcs $\sigma^s_{FA} \sim (n_{eff}e^2/\omega_h^2 m^*)(T^2/\varepsilon_F)$, while for electrons $\sigma^s_p \sim (n_e e^2/\omega_{cp}^2 m_p)\tau_{imp}^{-1}$. One sees directly that at low temperatures the electronic mobility will exceed the mobility of carriers on the Fermi arcs. Comparison of the two expressions defines the temperature scale T_0: $T_0^2 \approx (n_e/n_{eff})(\varepsilon_F/\tau_{imp})$ (by the order of magnitude it can be rewritten as $T_0^2 \approx (p_{Fe}/p_{Fh})(E_{Fe}/\tau_{imp})$).

At lower temperatures the Hall coefficient is

$$R_H(T) = -\frac{1}{c|e|n_e}. \tag{15}$$

Rewriting $\omega_c\tau \approx \omega_c(p_{Fe}/p_{Fh})\tau_{FA} \approx 1$ as $T^2 \approx (p_{Fe}/p_{Fh})\omega_c\varepsilon_F$ defines the approximate position of the Hall coefficient maximum:

$$T_{max} \approx \sqrt{(p_{Fe}/p_F)\omega_{ce}\varepsilon_F}. \tag{16}$$

5. Hall coefficient for holes on Fermi arcs

We return to holes on the Fermi arcs in the opposite limit $\omega_c\tau_{col} \gg 1$ and show that the expression (5) for $R_H(T)$ is correct in the limits of both weak and strong magnetic fields.

Assuming tetragonal symmetry for the CuO_2-plane in cuprates, the Hall coefficient in strong fields can be defined as $R_H H \equiv 1/\sigma_{xy}$ (below we use the system of coordinates with the x, y-axes along the two diagonals of the BZ). For normal metals in the strong field limit the value of the non-diagonal conductivity component σ_{xy} is known to depend critically on whether electrons at the Fermi surfaces in the magnetic field move along closed or open trajectories (see e.g. [22]). In particular, in the former case,

$$\sigma_{xy} = -\frac{|e|c}{H}n. \tag{17}$$

(Here n is the number of carriers for a closed Fermi surface.)

Returning to equation (1), in strong magnetic fields $\omega_c\tau_{col} \gg 1$ the collision integrals obviously should be omitted. Finding the distribution functions for carriers placed in the strong magnetic field and a weak electric field reduces to solving the equations of motion, as given by Liouville's theorem,

$$\frac{dn}{dt} = 0. \tag{18}$$

The approach to solving (18) was elaborated in [23]. (See the summary of the method in [22].)

Unlike weak fields, as in equation (3), the charges now move along trajectories bent by the strong magnetic field. Because of this, presenting the distribution functions in terms of the Cartesian momentum components, p_x, p_y is not convenient. Instead, in [23] it was suggested to go over to new variables: energy ε and time variable τ defined via the relation

$$d\tau = \frac{ds}{v_\perp}(c/eH). \tag{19}$$

(Here ds is the element of 'length' along the Fermi surface: $ds^2 \equiv dp_x^2 + dp_y^2$; $v_\perp^2 = v_x^2 + v_y^2$.) Equation (8) means that in a strong field, H carriers move along the Fermi surface with high speed.

For a weak *electrical* field equation (18) takes the form [22, 23]

$$\frac{dn}{dt} \approx -\frac{dn_0}{d\varepsilon}e(\vec{v}\cdot\vec{E}) + \frac{\partial n_1}{\partial\tau} \approx 0. \tag{20}$$

(The collision integrals were discarded.) As for n_1, the latter is presented [22, 23] in the form

$$n_1 = (\partial n_0/\partial\varepsilon)e(\vec{E}\cdot\vec{g}). \tag{21}$$

The equation for $\vec{g}$ is

$$\partial\vec{g}/\partial\tau = \vec{v}. \tag{22}$$

It is important that the solution of (22) cannot contain a constant vector $\vec{g}_0$. (Such a term would result in a meaningless constant shift of the distribution function (21)). The further calculations for the Fermi arcs model are, in some sense, even simpler than in [23].

In fact, there is no dilemma now concerning motion along either open or closed trajectories in the momentum space: carriers cannot go away from arcs because on both sides of the arc the distribution function is restricted by large energy gaps. In [22] the 'time' τ for a carrier was formally defined by its initial position on the Fermi surface. The actual physical meaning of such a procedure [23] is that it indirectly characterizes carriers by their velocities $\mathrm{d}s/\mathrm{d}\tau$ along the Fermi surface, as follows from equation (19). As the Fermi arcs are separated from each other by large energy gaps where the distribution function is zero, the time variables τ of equation (19) can be introduced for each Fermi arc *independently*.

One can now refer to equations (84) and (11) in [22]. The latter reads now as the sum over the four arcs:

$$\sigma_{\alpha\beta} = \frac{2e^3H}{(2\pi)^2c}\sum\int_{\Delta\varphi_i} v_\alpha g_\beta \,\mathrm{d}\tau\,\mathrm{d}p_z \equiv \frac{2e^3H}{\pi^2c}\int_{\Delta\varphi} v_\alpha g_\beta\,\mathrm{d}\tau\,\mathrm{d}p_z. \tag{23}$$

(Sgn $\Delta\varphi$ in the integral reminds one that the integration in (23) is limited by one Fermi arc.) For definiteness, σ_{xy} is chosen as follows:

$$\sigma_{xy} = \frac{2e^3H}{\pi^2c}\int_{\Delta\varphi} v_x g_y\,\mathrm{d}\tau\,\mathrm{d}p_z. \tag{24}$$

After writing out the equation

$$\frac{\mathrm{d}\vec{p}}{\mathrm{d}\tau} = \frac{e}{c}[\vec{H}\times\vec{v}] \tag{25}$$

in the x, y-components, solving equation (11) gives $g_y = -(c/eH)p_x$ while $v_x = (c/eH)(\mathrm{d}p_y/\mathrm{d}\tau)$; one finds that integration over the τ-variable in (24) is actually integration over the momentum component p_y:

$$\sigma_{xy} = \frac{2e^3H}{\pi^2c}\int_{\Delta\varphi} v_x g_y\,\mathrm{d}\tau\,\mathrm{d}p_z \equiv -\frac{2ec}{\pi^2H}\left(\frac{s}{c_z}\right)\int_{\Delta\varphi} p_x\,\mathrm{d}p_y, \tag{26}$$

where the last integral is over the area δS limited by the single arc:

$$\sigma_{xy} = -\frac{ec}{H}n_{\mathrm{eff}}. \tag{27}$$

In the isotropic case $\delta S = (p_F^2/2)\Delta\varphi$ and $n_{\mathrm{eff}} = (\Delta\varphi)p_F^2 s/c_z\pi^2$ is defined as in equation (5). At $\Delta\varphi \Rightarrow \pi/2$ one gradually returns to the expression of the Hall constant for the closed Fermi surface [22]. In the anisotropic case the number of carriers is expressed in terms of the area δS limited by the single arc. (For holes, $(-e) \Rightarrow |e|$. One sees that introduction of the time variable τ for the *finite* Fermi arc was nothing more than a convenient mathematical device.) Equations (26) and (27) conclude the proof.

6. Conclusions

In summary, the expressions for resistivity and the Hall coefficient were derived for two types of carriers participating in transport. There are two regimes depending on the strength of the magnetic field. At a fixed magnetic field and at higher temperatures electrons on the pocket are bound to holes on the Fermi arcs. Since the size of the pocket is small, its contribution to the conductivity and the Hall coefficient is negligible.

On lowering the temperature electrons decouple from holes and behave as a Fermi gas in the magnetic field. In the limit of low temperatures electrons contribute most to the Hall effect, changing the sign of the Hall coefficient. The result seems to agree with recent high-field experiments.

Contributions from holes on the Fermi arcs at higher temperatures prevail and the Hall coefficient $R_H(T)$ determines the actual number of carriers on the arcs. Its decrease with increase of temperature reflects the increase of the widths of the Fermi arcs $\Delta\varphi(T)$.

At higher temperatures the contribution to $R_H(T)$ from holes on the Fermi arcs is proven to be rigorous in the limits of both weak and strong magnetic field. In the expression for the Hall coefficient the width of the Fermi arcs can be not small.

Regarding the nature of the electronic pocket, the results are consistent with the notion of its being a particular feature in the spectrum.

Acknowledgments

The work of LPG was supported by the NHMFL through NSF Grant No. DMR-1157490, the State of Florida and the US Department of Energy; that of GBT was supported by the Russian Academy of Sciences through Grants No. P19 and OFN 03.

References

[1] Damascelli A, Hussain Z and Shen Z X 2003 *Rev. Mod. Phys.* **75** 473
[2] Ono S, Komiya S and Ando Y 2007 *Phys. Rev.* B **75** 024515
[3] Gor'kov L P and Teitel'baum G B 2006 *Phys. Rev. Lett.* **97** 247003
[4] Doiron-Leyraud N, Proust C, LeBoeuf D, Levallois J, Bonnemaison J B, Liang R, Bonn D A, Hardy W N and Taillefer L 2007 *Nature* **447** 565
[5] Riggs S C, Vafek O, Kemper J B, Betts J B, Migliori A, Balakirev F F, Hardy W N, Liang R, Bonn D A and Boebinger G S 2011 *Nature Phys.* **7** 332
[6] Barišić N *et al* 2013 *Nature Phys.* **9** 761
[7] Yoshida T, Hashimoto M, Vishik I M, Shen Z X and Fujimori A 2012 *J. Phys. Soc. Japan* **81** 011006
[8] Fournier D *et al* 2010 *Nature Phys.* **6** 905
[9] Barišić N, Chan M K, Li Y, Yu G, Zhao X, Dressel M, Smontara A and Greven M 2013 *Proc. Natl Acad. Sci.* **110** 12235
[10] Gor'kov L P 2013 *Phys. Rev.* B **88** 041104
[11] LeBoeuf D *et al* 2011 *Phys. Rev.* B **83** 054506
[12] Doiron-Leyraud N *et al* 2013 *Phys. Rev.* X **3** 021019
[13] Wu T, Mayaffre H, Krämer S, Horvatić M, Berthier C, Hardy W N, Liang R, Bonn D A and Julien M-H 2011 *Nature* **477** 191
[14] LeBoeuf D, Krämer S, Hardy W N, Liang R, Bonn D A and Proust C 2013 *Nature Phys.* **9** 79

[15] Kanigel A *et al* 2006 *Nature Phys.* **2** 447

[16] Storey J G, Tallon J L and Williams G V M 2008 *Phys. Rev.* B **78** 140506

[17] Lee W S, Vishik I M, Tanaka K, Lu D H, Sasagawa T, Nagaosa N, Devereaux T P, Hussain Z and Shen Z X 2007 *Nature* **450** 81

[18] Kanigel A, Chatterjee U, Randeria M, Norman M R, Souma S, Shi M, Li Z Z, Raffy H and Campuzano J C 2007 *Phys. Rev. Lett.* **99** 157001

[19] Anderson P W 1991 *Phys. Rev. Lett.* **67** 2092

[20] Gor'kov L P and Teitel'baum G B 2008 *Phys. Rev.* B **77** 180511

[21] Armitage N P *et al* 2002 *Phys. Rev. Lett.* **88** 257001

[22] Lifshitz E M and Pitaevskii L P 1997 *Physical Kinetics* (*Landau and Lifshitz Course of Theoretical Physics* vol 10) (Oxford: Butterworth–Heineman)

[23] Lifshitz I M, Azbel' M Ya and Kaganov M I 1957 *Sov. Phys.—JETP* **4** 41

PHYSICAL REVIEW B **79**, 134510 (2009)

Role of striction at magnetic and structural transitions in iron pnictides

Victor Barzykin and Lev P. Gor'kov*
National High Magnetic Field Laboratory, Florida State University, 1800 East Paul Dirac Drive, Tallahassee, Florida 32310, USA
(Received 22 December 2008; revised manuscript received 4 March 2009; published 9 April 2009)

We discuss the role of striction in the intertwined magnetic and structural phase transitions in the underdoped iron pnictides. The magnetoelastic coupling to acoustic modes is then derived and estimated in framework of the multiband spectrum for itinerant electrons with nesting features. We argue that the first-order character of the magnetoelastic phase transition originates from the lattice instabilities near the onset of spin-density wave order introducing, thus, a shear acoustic mode as another order parameter. Taking nonharmonic terms in the lattice energy into account may explain the splitting of the structural and magnetic transitions in some oxypnictides. Fluctuations of the magnetic order parameter show up in the precursory temperature dependence of the elastic moduli.

DOI: 10.1103/PhysRevB.79.134510 PACS number(s): 74.70.−b, 74.20.Rp, 74.20.Mn

Considerable efforts have been concentrated recently on studies of the interdependence for the antiferromagnetic (AFM) and the superconducting (SC) transitions in the new iron-based compounds.[1] Doping is known to destroy AFM and increase T_C—the temperature for the onset of SC.[2]

In this presentation, we address the peculiarities and interrelations for the magnetic (T_m) and structural (T_{str}) transitions in the parent and underdoped iron pnictides. For concreteness, we discuss below the properties of two systems only: the quarternary REFeAsO ("1111;" RE stands for a rare earth) and the bilayered $A\mathrm{Fe_2As_2}$ ("A22," where A = Sr, Ca, Ba) materials. In both classes, the intertwined magnetic and the structural transitions of the weak first order are observed in a temperature interval of ~100–200 K. In 1111's, the structural change precedes the magnetic ordering,[2,3] while in A22's magnetic order and change in the lattice symmetry occur simultaneously in the single first-order transition.[4]

It is known that coupling to the lattice may transform a magnetically driven transition into the weak first-order transition accompanied by structural changes.[5] In Ref. 6 the problem has been rigorously solved for the elastically isotropic solid, taking fluctuations into account. Unfortunately, the method of Ref. 6 does not apply to anisotropic materials, such as iron pnictides, with the tetragonal symmetry and layered structure. The solution of Ref. 6, however, undoubtedly contains the main physics for these phenomena. Correspondingly, below we simplify the approach of Ref. 6 to model the magnetoelastic interactions (striction) and phase transitions in the parent and underdoped FeAs systems.

The magnetic order in oxypnictides is built of the alternating (AFM) spins running along one of the lattice axis, while in the other direction the spins are ordered ferromagnetically.[3] In the tetragonal notations, the structure vectors are commensurate: $\boldsymbol{Q}_1=(0,\pi)$ and $\boldsymbol{Q}_2=(\pi,0)$. With the local picture of interacting Fe spins and the two Heisenberg exchange interactions constants J_1 (nn) and J_2 (nnn), such a ground state realizes itself at the inequality $J_2 > J_1/2$. Accompanying orthogonal lattice distortions can be understood, as in Ref. 5, in terms of a "spin-Peierls" effect, i.e., the variation in the exchange integrals at the lattice deformation.[7]

The present materials, however, are semimetals. Pnictides are better described in terms of an itinerant scheme. The consensus is that the energy spectrum obtained in the "first-principles" calculations[8] presents a good starting point. According to Ref. 8, the electron spectrum bears the multiband character. There are Fermi surfaces (FSs) for the two hole (h) pockets at the Γ point (0, 0) and two electronic (e) pockets located in the tetragonal (unfolded)[9] reciprocal lattice at $(0,\pi)$ and $(\pi,0)$. An approximate nesting between the e and h pockets is believed to be responsible for a spin-density wave (SDW) instability with the two vectors $\boldsymbol{Q}_1$ and $\boldsymbol{Q}_2$ mentioned above.[9] The main features of this spectrum have been reproduced in many numerical calculations for all classes of the new Fe-based materials. In Ref. 10 this spectrum has been directly observed for LaFePO in the de Haas–van Alphen (dHvA) experiments. Below we accept this model. With $T_c \sim 50$ K, $T_m \sim T_{str} \sim 100$–$200$ K, and the Fermi energy $E_F \sim 0.1$–0.2 eV for the pockets' sizes, the model is expected to allow a mean-field treatment both for SC and magnetic phenomena.[11]

"First-principles" calculations provide a reasonable description of the ground-state properties, albeit some disagreements are not uncommon in the literature (see Ref. 12). However, subtleties, especially in a vicinity of phase transitions, remain beyond the reach of numerical analysis.

Among advantages of the "nesting" model[9,13,14] is that its formalism is practically identical to the well-studied BCS scheme for SC. It is a weak-coupling mean-field scheme if $T_{SDW} \ll E_F$. (Fluctuations become important only in a narrow temperature interval $|\Delta T|/T_0 = \mathrm{Gi} \ll 1$, as given by sort of a Levanyuk-Ginzburg criteria Gi.) The Landau functional near the transition can be derived exactly as it has been done for superconductivity near T_c.[15]

Nevertheless, we prefer not to write equations explicitly. Numerous unknown parameters that include parameters of the e and h pockets, interactions, and six elastic moduli for the tetragonal symmetry of pnictides should make it useless. For the sake of transparency we hold, where possible, the discussion on the qualitative level.

The mathematical analogy of the model[13,14,16] with the Cooper pairing in a BCS-type scheme stems from a logarithmic divergence for the scattering in the e-h channel in Fig. 1(a), which has the following form:

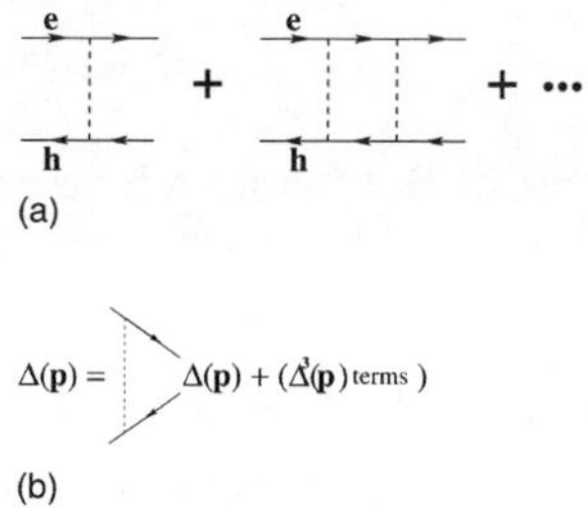

FIG. 1. (a) The logarithmic nesting model. (b) Mean-field equations near T_c.

$$V(\boldsymbol{p}-\boldsymbol{p}')T\sum_{\omega_n}\int\frac{\nu(E_F)d\epsilon' d\Omega_{p'}}{(i\omega_n-\epsilon_e')(i\omega_n-\epsilon_h')}=V\nu(E_F)\ln(\bar{\omega}/T,\delta), \tag{1}$$

where

$$\epsilon_h=-\epsilon_e+\delta \tag{2}$$

are e and h energies, while δ accounts for deviations from the ideal nesting. With the notations $\boldsymbol{\Delta}_1$ and $\boldsymbol{\Delta}_2$ for the triplet (SDW) order parameters corresponding to the structure vectors $\boldsymbol{Q}_1=(0,\pi)$ and $\boldsymbol{Q}_2=(\pi,0)$, the equations for Δ's have the same structure as in SC theory,[15] as demonstrated in Fig. 1(b). Let T_0 be the "bare" transition temperature in the absence of coupling to the lattice. The Landau functional[17] near T_0 is of the form,

$$\Omega=\int\frac{A\nu}{2}\left[\tau\Delta^2+b_0\frac{\Delta^4}{T_0^2}+c_0\xi_0^2(\boldsymbol{\nabla}\boldsymbol{\Delta})^2\right]d\boldsymbol{r}, \tag{3}$$

where $\nu\equiv\nu(E_F)$ is the characteristic density of states (DOS). A, c_0, and b_0 are constants and $\tau=(T-T_0)/T_0$; $\xi_0=v_F/2\pi T_0$. The functional (3) is the starting point in the general theory of the second-order phase transitions.[17,18] To account for interactions with the lattice, one adds to Eq. (3) a coupling to the lattice of the form

$$H_{\text{str}}=-q\int\hat{u}(\boldsymbol{r})\hat{\Delta}(\boldsymbol{r})^2d\boldsymbol{r}, \tag{4}$$

and the elastic energy itself that we also write schematically as

$$H_{\text{el}}=\int K\hat{u}^2d\boldsymbol{r}, \tag{5}$$

where K is an elastic modulus (such as the bulk modulus, for instance); the notation $\hat{u}$ stands for components of the strain tensor,

$$u_{ik}=\frac{1}{2}\left(\frac{\partial u_i}{\partial x_k}+\frac{\partial u_k}{\partial x_i}\right). \tag{6}$$

For magnetic phenomena, Eq. (4) describes magnetoelastic interactions and q is called the striction constant. The total Gibbs energy near the phase transition is

$$\Phi=\Omega(\tau)+H_{\text{el}}+H_{\text{str}}. \tag{7}$$

In the itinerant model of Eq. (1), the parameters Δ_1 and Δ_2 are built on the Bloch wave functions. These spatial features present in the SDW parameters immediately lead to the quadratic coupling between Δ's and acoustic degrees of freedom ($\boldsymbol{Q}=0$) given by Eq. (4). An estimate of the striction constants in Eq. (4) follows directly from the logarithmic contribution of Eq. (1). Indeed, elastic deformations change the parameter δ in Eq. (3) that controls the degree of the nesting,

$$\delta\equiv\delta(u)=\delta_0+\lambda\hat{u}. \tag{8}$$

From Eq. (1) and Fig. 1(b) it is seen that

$$\tau\rightarrow\tau+\lambda\frac{\hat{u}}{T_0}. \tag{9}$$

Making use of Eq. (3), we estimate

$$q\sim\frac{\nu\lambda}{T_0}\sim\nu\left(\frac{E_F}{T_0}\right) \tag{10}$$

(taking λ on order of characteristic atomic energy E_F).

Return to Eqs. (4)–(7) and assume homogeneous $\hat{u}$. Minimizing Φ over $\hat{u}$, we obtain

$$K\hat{u}=q\left(\frac{\int\Delta^2 d\boldsymbol{r}}{V}\right) \tag{11}$$

(with V for the volume). The effective Landau functional $\tilde{\Omega}$ is

$$\tilde{\Omega}_{\min}=\Omega(\tau)-\left(\frac{1}{2K}\right)\left[q\frac{\int\Delta^2(\boldsymbol{r})d\boldsymbol{r}}{V}\right]^2. \tag{12}$$

In the mean field $\Delta^2(\boldsymbol{r})=$const, and Eq. (12) reproduces the result of Ref. 6: the functional Ω acquires a negative contribution to the biquadratic term in Eq. (3). Should the total be negative, the transition becomes first order. [Higher-order terms in the expansion (3) become necessary.]

In the presence of an external homogeneous deformation, $\hat{u}_{\text{ext}}$ substitution into Eq. (9), $\tau\rightarrow\tau+\lambda(\hat{u}_{\text{ext}}/T_0)$, leads to

$$\Omega_{\text{eff}}=\Omega\left(\tau+\lambda\frac{\hat{u}_{\text{ext}}}{T_0}\right). \tag{13}$$

Differentiating, we find

$$\frac{\partial^2\Omega_{\text{eff}}}{\partial\hat{u}_{\text{ext}}^2}=\frac{\lambda^2}{T^2}\frac{\partial^2\Omega}{\partial\tau^2}=-\frac{\lambda^2C(\tau)}{T_0}, \tag{14}$$

where $C(\tau)$ is the specific heat. According to Eq. (14), already in the Landau mean-field approach the jump ΔC in the specific heat at the transition is accompanied by a negative steplike change in elastic moduli. The variation is of the atomic order. Therefore, the lattice would become unstable, and the second-order character of a transition changes to the first-order one.

In Refs. 6 and 19, the attention has been drawn to the fact that the specific heat $C(T)$ in Eq. (14) actually has a singularity at T_0,

$$C(T) \propto |\tau|^{-\alpha}. \tag{15}$$

[α is the so-called scaling index for the specific heat $C(T)$.] As to the order parameter, it appears below T_0 (see Ref. 17; §148, p. 484),

$$\Delta(T) \propto (-\tau)^{\beta}(\tau < 0). \tag{16}$$

Therefore, an elastic instability is ubiquitous for any magnetic transition. In the model chosen above, the instability occurs inside a narrow temperature interval controlled by the Levanyuk-Ginzburg parameter. Provided that layered pnictides can be treated, in the first approximation, as two-dimensional (2D), we write

$$\left|\frac{\Delta T}{T_0}\right| \equiv G_i^{2D} \sim \left(\frac{T_{SDW}}{E_F}\right) \ll 1. \tag{17}$$

By the order of magnitude, the same parameter also controls the width of the hysteresis.

Note that even for homogeneous lattice deformations, the magnetoelastic coupling [Eq. (4)] leads to Eq. (12), with the present biquadratic term that describes nonlocal interactions for the order parameter Δ. In Ref. 6 inhomogeneous fluctuations result in the cancellation of the nonlocal term (12) for the isotropic (liquid) media. However, the nonlocal coupling terms are always present for any anisotropic solid. As it was mentioned above, the problem could be rigorously solved in Ref. 6 only for the elastically isotropic solid (i.e., for a solid characterized by the bulk and shear moduli). In all other cases, the nonlocal term (12) that comes about due to exchange by acoustic phonons becomes strongly anisotropic. We now simplify the model[6] by restricting our consideration to homogeneous lattice fluctuations. Inhomogeneous fluctuations should not qualitatively change the physics of the problem.

With this in mind, one can rewrite Eq. (3) in the form

$$\Phi = \frac{K}{2}\hat{u}^2 + \Omega\left(\tau + \frac{\lambda}{T_0}\hat{u}\right) - \hat{\sigma}\hat{u}, \tag{18}$$

where $\Omega(\tau + \frac{\lambda}{T_0}\hat{u})$ is the exact functional (3), i.e., the functional that would describe the second-order phase transition driven by the order parameter Δ neglecting striction effects. In particular, it has the contribution Ω_{sing} responsible for the singular behavior of the specific heat $C(T)$ [Eq. (15)]. The term $-\hat{\sigma}\hat{u}$ stands for the applied external stress ($\hat{\sigma}$ is a proper component of the stress tensor; this term is needed at calculations of the temperature behavior of elastic moduli across the transition.)

In the presence of nonzero stress $\hat{\sigma} \neq 0$, fluctuations take place around the present equilibrium point $\hat{u}_{\text{ext}}$,

$$\hat{u}_{\text{ext}} = \frac{\hat{\sigma}}{K}. \tag{19}$$

Rewriting in Eq. (18) $\hat{u} \to \hat{u} + \hat{u}_{\text{ext}}$, we obtain

$$\Phi = -\frac{\hat{\sigma}^2}{2K} + \frac{K}{2}\hat{u}^2 + \Omega\left(\tau + \frac{\lambda}{T_0}\hat{u} + \frac{\lambda\hat{\sigma}}{T_0 K}\right). \tag{20}$$

The contribution $\lambda\hat{\sigma}/(T_0 K)$ in the argument of $\Omega(x)$ [Eq. (20)] shifts the transition temperature T_0 under the applied stress,

$$T_0(\hat{\sigma}) = T_0 - \frac{\lambda\sigma}{K}. \tag{21}$$

The Gibbs energy (20) must be minimized over fluctuations of (homogeneous) deformations $\hat{u}$,

$$\frac{\delta\Phi}{\delta\hat{u}} = K\hat{u} + \frac{\lambda}{T_0}\Omega'(x) = 0, \tag{22}$$

where

$$x = \tau + \frac{\lambda}{T_0}u + \frac{\lambda\sigma}{T_0 K}. \tag{23}$$

One may easily recognize in Eqs. (20) and (22) the analogy to equations in Ref. 6 describing the transitions in the elastically isotropic solid body.

To be now more specific, note that the elastic energy of homogeneous deformations for the tetragonal lattice can be rewritten in terms of the even irreducible representation of the D_{4h} group,

$$A_{1g}{:}u_{xx} + u_{yy}, u_{zz}; B_{1g}{:}u_{xx} - u_{yy}; B_{2g}{:}u_{xy}; E_g{:}u_{zx}, u_{zy}. \tag{24}$$

In principle, any distortion in Eq. (24) perturbs the electronic spectrum and, hence, affects the nesting features. Therefore, the number of the independent striction constants q's in Eq. (4) [or λ's in Eqs. (8) and (9)] may be large. In view that the layered character of pnictides makes them close to two-dimensional systems, the nesting parameter δ in Eq. (8) is controlled mainly by the strain components in Eq. (24) that do not have the z indices. Rewriting the 2D part of the elastic energy (5) in the form

$$H_{\text{el}} = \frac{K}{2}(u_{xx} + u_{yy})^2 + \frac{\mu_1}{2}(u_{xx} - u_{yy})^2 + 2\mu_2 u_{xy}^2, \tag{25}$$

with three independent moduli according to Eq. (24) and three magnetoelastic constants q_i [or λ_i in Eqs. (8) and (9)] and taking into account the uniaxial symmetry along $(0,\pi)$ or $(\pi,0)$ directions for the parameters Δ_1 and Δ_2, we simplify the problem further and rewrite Eq. (9) as

$$\tau \to \tau + \frac{\lambda_+}{T_0}u_+ + \frac{\lambda_-}{T_0}u_-, \tag{26}$$

where $u_+ = u_{xx} + u_{yy}$ and $u_- = u_{xx} - u_{yy}$. The second term is responsible for the orthorhombic deformation of the lattice.

For the actual calculations, one would need the expression for $\Omega(\tau)$. In our case the fluctuations are strong only in a narrow vicinity of the phase transition, and we expect that the first-order transition occurs inside the same interval. Therefore, only the *singular* part $\delta\Omega_{\text{sing}}(\tau)$ is of importance. We limit ourselves by the first fluctuation correction to the mean field Ω, which we calculate in exactly the same manner as in Ref. 17 (see §147, problem, p. 482). The minor difference is that for the strongly anisotropic pnictides, one needs to introduce in Eq. (3) the in-plane and out-of-plane coherence lengths $\xi_{0\parallel} \gg \xi_{0\perp}$. After an elementary calculation it follows:

$$\delta\Omega_{\rm sing}(\tau) = -\,{\rm const}(T_0\xi_{0\parallel}^{-2}\xi_{0\perp}^{-1})|\tau|^{3/2} \equiv -B|\tau|^{3/2}, \quad (27)$$

where const is a numeric factor that depends on the model details. [Strictly speaking, the values of B in Eq. (27) differ by a factor of $2^{3/2}$ on the two sides of T_0 (Ref. 17) (see §146, footnote on p. 475).] For the singularity in the specific heat, Eq. (27) gives

$$\delta C(\tau) = \frac{3}{4T_0}B|\tau|^{-1/2}. \quad (28)$$

Comparing Eq. (28) with the normal specific heat $C_n \sim \nu T_0$ leads to the criterion,

$$\frac{B}{\nu T_0^2}|\tau|^{-1/2} \equiv \alpha|\tau|^{-1/2} \ll 1. \quad (29)$$

[In the strictly 2D limit ($\xi_{\perp 0} \to \infty$) one would obtain the criteria (17)]. With Eqs. (14) and (23) one finds that corrections to the elastic moduli of Eq. (25) become strong in the same order of magnitude temperature interval Eq. (17) [or Eq. (29)] if both K and μ are of the atomic scale.

To demonstrate the emergence of the first-order transition from Eqs. (20) and (22), assume—for simplicity sake—$\alpha = 1/2$ in Eq. (15), i.e., extrapolate Eq. (27) over the whole fluctuation interval. From Eq. (22) (for each $u_{+,-}$) it follows:

$$Ku_+ = \frac{3}{2}B\frac{\lambda_+}{T_0}\frac{x}{|x|}|x|^{1/2}; \quad u_-\mu_1 = \frac{3}{2}B\frac{\lambda_-}{T_0}\frac{x}{|x|}|x|^{1/2}. \quad (30)$$

With the notations

$$\alpha_+ = \frac{3}{2}B\frac{\lambda_+^2}{KT_0^2}, \quad \alpha_- = \frac{3}{2}B\frac{\lambda_-^2}{\mu_1 T_0^2}, \quad (31)$$

$$\frac{\lambda_+ u_+}{T_0} = \rho, \quad \frac{\lambda_- u_-}{T_0} = \mu, \quad x \equiv \tau + \rho + \mu, \quad (32)$$

we rewrite Eq. (30) as

$$\frac{1}{\alpha_+}\rho = \frac{x}{|x|}|x|^{1/2}, \quad \frac{1}{\alpha_-}\mu = \frac{x}{|x|}|x|^{1/2} \quad (33)$$

or

$$\alpha_+^{-1}\rho = \alpha_-^{-1}\mu, \quad x = \tau + \rho\left(1 + \frac{\alpha_-}{\alpha_+}\right) = \tau + v. \quad (34)$$

Finally, with

$$v = (\alpha_+ + \alpha_-)\frac{x}{|x|}x^{1/2}, \quad (35)$$

we arrive to the single equation

$$x = \tau + (\alpha_+ + \alpha_-)\frac{x}{|x|}x^{1/2}. \quad (36)$$

Equation (36), such as Eq. (13) of Ref. 6, reveals the typical features of a first-order transition. At small enough τ, there are three solutions for x. If coefficients B in Eq. (27) are equal on both sides of T_0, the transition takes place at $\tau=0$, where

$$x_+ = -x_- = (\alpha_+ + \alpha_-)^2 \quad (37)$$

determines the jumps of u_+, u_- at the transition (note that at negative $x_- \neq 0$, the driving parameter, according to Eq. (16), is finite). The area of hysteresis is determined by the equation

$$2 = (\alpha_+ + \alpha_-)|x|^{-1/2}. \quad (38)$$

These results qualitatively agree with the observed simultaneous onset both of orthorhombic distortions and the "stripe" SDW order in the iron system $A22$.[4] The distortion (32) just accompany the magnetic transition. In case of the "1111" class, the temperature $T_{\rm str}$ for the onset of the orthorhombic deformation precedes the onset of the magnetic order at T_m.[3] The two temperatures are rather close: $\Delta T = T_{\rm str} - T_m > 0$ is on order of 10–20 K, so that by order of magnitude $\Delta T/T_{\rm str} \sim 0.1$ falls into the range of Eq. (17) with $E_F \sim 0.1-0.2$ eV taken for the pockets' depths obtained in "first-principle" calculations.[8,20] Such closeness seen among most of REFeAsO [e.g., compare La (Ref. 3) and Nd (Ref. 21)] is strongly in favor that the structural transition is directly related to the magnetic instability. In the language of local Fe spins, the attempt was made in Ref. 22 to ascribe the temperature interval separating $T_{\rm str}$ and T_m to the appearance of a "nematic phase" that comes about due to strong spin fluctuations above T_m. We suggest that both transitions have the common origin and come about as the result of the lattice instabilities caused by the striction. Indeed, according to Eq. (14), the striction triggers softening of elastic moduli as temperature approaches the transition interval (17) or Eq. (29). Assume that it takes place more strongly for modulus μ_1 in Eq. (25). Recall that the orthorhombic distortion is the symmetry change by itself and for the tetragonal lattice is characterized by the symmetry parameter $u_- = u_{xx} - u_{yy}$. So far as the dependence $\delta\Omega_{\rm sing}(\tau + (\lambda_-/T_0)u_-)$ on u_- is the only form of the nonlinear elastic energy, the above analysis applies. However, when the renormalized modulus $\mu_{1\rm eff}$ becomes small, other nonlinear terms ever present in the lattice must be also taken into account. As the result, the Landau functional for the parameter $u_- = u_{xx} - u_{yy}$ also depends on those contributions. The higher-order terms in u_- becoming important when renormalized $\mu_{1\rm eff}$ is small nearby magnetic T_0. The mean-field treatment of the present symmetry parameter u_- could then be applied in the usual way for the second-order tetraorthotransition at $T_{\rm str}$.

Experimentally, the structural distortions at $T_{\rm str}$ appear in the weak first-order transition.[3] That last result immediately follows from the fact that by symmetry, the orthorhombic transition in the tetragonal lattice infers its *own quadratic striction*. Indeed, the following cubic terms are allowed in the tetragonal lattice by the symmetry reasons:

$$q_\parallel u_+ u_-^2 + q_\perp u_{zz} u_-^2, \quad (39)$$

in addition to the elastic terms

$$\frac{K}{2}u_+^2 + \frac{K_\perp}{2}u_{zz}^2. \tag{40}$$

In terms of our models (1), (3), and (7), the fact that $T_m < T_{str}$ should mean that the nonzero distortion u_- below T_{str} makes the nesting condition (1) worse.

Let us add a few final comments to the analyses above. The order of magnitude estimate of the in-plane lattice distortions at low enough temperatures follows from Eqs. (10) and (11):

$$u \sim \frac{q}{K} \sim \frac{\nu E_F}{KT_0} T_0^2 \sim \frac{T_0}{E_F} \sim 10^{-2} - 10^{-3} \tag{41}$$

that agrees well with the experimental data of Refs. 23–25. The hysteresis ΔT from Eq. (38) agrees by the order of magnitude with data.[4]

There are no experimental data for the volume change $\Delta V/V$. The lattice deformation along the c axis, in principle, could be obtained from Eqs. (39) and (40) together

$$u_{zz} \sim \frac{q_\perp}{K_\perp} u_-^2. \tag{42}$$

In Eq. (42) both $q_\perp$ and $K_\perp$ are expected to be small in a layered material. These parameters remain unknown in the oxypnictides.

Equation (14) together with Eq. (27) provides another observable feature. At temperatures above T_0 the fluctuation corrections to the elastic moduli should behave as

$$\sim (T - T_0)^{-1/2}, \tag{43}$$

while in the 2D limit of isolated planes

$$\sim (T - T_0)^{-1}. \tag{44}$$

The data[25] do not allow to distinguish between the exponents in Eqs. (43) and (44).

To summarize, in the frameworks of multiband electronic spectrum with nesting features, the theoretical scheme is elaborated to treat striction in iron pnictides. Magnetoelastic coupling changes the second-order character of magnetic transition. The transition becomes of the first order. Whether the transition bears strong or weak first-order character may depend on details. Provided that $T_0 \ll E_F$, the weak first-order transition is predicted. The model, when applied to the layered FeAs systems, leads to estimates of the correct order of magnitude. Discontinuities of all parameters at the transition are due to lattice instabilities. Magnetoelastic interactions may split the magnetic (SDW) transition at T_m and the orthorhombic deformations at $T_{str} > T_m$. The model predicts a noticeable precursory temperature dependence above the transition temperature in the elastic moduli.

The obtained results are in good qualitative agreement with peculiarities of the phase diagram of present parent or underdoped FeAs materials well above the temperature of superconducting transition.

The authors are thankful to Z. Fisk, A. Gurevich, J. Lynn, A. Migliori, V. Pokrovsky, I. Liuksyutov, D. Khmelnitsky, and M. Sadovskii for helpful discussions. This work was supported by NHFML through the NSF Cooperative Agreement No. DMR-008473 and the State of Florida.

*Also at L.D. Landau Institute for Theoretical Physics, Chernogolovka, 142432, Russia.

[1] Y. Kamihara, T. Watanabe, M. Hirano, and H. Hosono, J. Am. Chem. Soc. **130**, 3296 (2008).

[2] Q. Huang, J. Zhao, J. W. Lynn, G. F. Chen, J. L. Luo, N. L. Wang, and P.-C. Dai, Phys. Rev. B **78**, 054529 (2008).

[3] C. de la Cruz, Q. Huang, J. W. Lynn, J. Li, W. Ratcliff, J. L. Zaretsky, H. A. Mook, G. F. Chen, J. L. Luo, N. L. Wang, and P. Dai, Nature (London) **453**, 899 (2008).

[4] A. I. Goldman, D. N. Argyriou, B. Ouladdiaf, T. Chatterji, A. Kreyssig, S. Nandi, N. Ni, S. L. Bud'ko, P. C. Canfield, and R. J. McQueeney, Phys. Rev. B **78**, 100506(R) (2008).

[5] C. P. Bean and R. D. Rodbell, Phys. Rev. **126**, 104 (1962); S.-H. Lee, C. Broholm, T. H. Kim, W. Ratcliff, and S.-W. Cheong, Phys. Rev. Lett. **84**, 3718 (2000).

[6] A. I. Larkin and S. A. Pikin, Zh. Eksp. Teor. Fiz. **56**, 1664 (1969) [Sov. Phys. JETP **29**, 891 (1969)].

[7] T. Yildirim, Phys. Rev. Lett. **101**, 057010 (2008).

[8] D. J. Singh and M.-H. Du, Phys. Rev. Lett. **100**, 237003 (2008).

[9] I. I. Mazin, D. J. Singh, M. D. Johannes, and M. H. Du, Phys. Rev. Lett. **101**, 057003 (2008).

[10] A. I. Coldea, J. D. Fletcher, A. Carrington, J. G. Analytis, A. F. Bangura, J.-H. Chu, A. S. Erickson, I. R. Fisher, N. E. Hussey, and R. D. McDonald, Phys. Rev. Lett. **101**, 216402 (2008).

[11] V. Barzykin and L. P. Gor'kov, JETP Lett. **88**, 131 (2008).

[12] I. I. Mazin, M. D. Johannes, L. Boeri, K. Koepernik, and D. J. Singh, Phys. Rev. B **78**, 085104 (2008).

[13] L. V. Keldysh and Yu. V. Kopaev, Fiz. Tverd. Tela (Leningrad) **6**, 2791 (1964) [Sov. Phys. Solid State **6**, 2219 (1965)].

[14] A. N. Kozlov and L. A. Maksimov, Zh. Eksp. Teor. Fiz. **49**, 1284 (1965) [Sov. Phys. JETP **22**, 889 (1966)].

[15] L. P. Gor'kov, Zh. Eksp. Teor. Fiz. **36**, 1918 (1959) [Sov. Phys. JETP **9**, 1364 (1959)].

[16] V. Cvetkovic and Z. Tesanovic, arXiv:0808.3742 (unpublished).

[17] L. D. Landau and E. M. Lifshits, *Statistical Physics: Part 1* (Butterworth-Heinemann, Oxford, 1980).

[18] A. Z. Patashinskii and V. L. Pokrovskii, *Fluctuation Theory of Phase Transitions* (Pergamon, Oxford, 1979).

[19] O. K. Rice, J. Chem. Phys. **22**, 1535 (1954).

[20] I. A. Nekrasov, Z. V. Pchelkina, and M. V. Sadovskii, JETP Lett. **87**, 560 (2008); M. V. Sadovskii, arXiv:0812.0302 (unpublished).

[21] Y. Chen, J. W. Lynn, J. Li, G. Li, G. F. Chen, J. L. Luo, N. L. Wang, P. Dai, C. dela Cruz, and H. A. Mook, Phys. Rev. B **78**, 064515 (2008).

[22] C. Fang, H. Yao, W.-F. Tsai, J. P. Hu, and S. A. Kivelson, Phys. Rev. B **77**, 224509 (2008).

[23] M. Fratini, R. Caivano, A. Puri, A. Ricci, Z.-Z. Ren, X.-L. Dong, J. Yang, W. Lu, Z.-X. Zhao, L. Barba, G. Arrighetti,

M. Polentarutti, and A. Bianconi, Supercond. Sci. Technol. **21**, 092002 (2008).

[24] M. Rotter, M. Tegel, I. Schellenberg, F. M. Schappacher, R. Poettgen, J. Deisenhofer, A. Guenther, F. Schrettle, A. Loidl, and D. Johrendt, arXiv:0812.2827 (unpublished).

[25] M. A. McGuire, A. D. Christianson, A. S. Sefat, B. C. Sales, M. D. Lumsden, R. Jin, E. A. Payzant, D. Mandrus, Y. Luan, V. Keppens, V. Varadarajan, J. W. Brill, R. P. Hermann, M. T. Sougrati, F. Grandjean, and G. J. Long, Phys. Rev. B **78**, 094517 (2008).

PHYSICAL REVIEW B **87**, 024504 (2013)

Dual role of *d* electrons in iron pnictides

Lev P. Gor'kov
NHMFL, Florida State University, 1800 East Paul Dirac Drive, Tallahassee, Florida 32310, USA and L.D. Landau Institute for Theoretical Physics of the RAS, Chernogolovka 142432, Russia

Gregory B. Teitel'baum
E.K. Zavoiskii Institute for Technical Physics of the RAS, Kazan 420029, Russia
(Received 18 August 2012; revised manuscript received 26 December 2012; published 8 January 2013)

Recent Fe x-ray emission spectroscopy experiments [H. Gretarsson *et al.*, Phys. Rev. B **84**, 100509 (2011)] unveiled sizable local moments in iron pnictides in the room-temperature paramagnetic state. In an effort to further clarify the notion of coexisting magnetic moments and itinerant carriers in iron pnictides we focus on the interactions between the two subsystems. At a moderate on-site Coulomb repulsion the intra-atomic Hund's interaction leads to the formation of nonzero ("bare") local moments on the Fe sites. We show that the Kondo-like exchange with the itinerant electrons may significantly renormalize the "bare" value of the moments manifested in different experiments. In turn, the itinerant carriers scatter on the renormalized moments that remain disordered in the paramagnetic phase. On the one hand, the scattering mechanism is responsible for high values of resistivity of the stoichiometric pnictides at the temperature of their transition into the antiferromagnetic phase, and on the other hand, it washes out the fine details of the Fermi surfaces. The results are rigorous and were obtained without use of any Born-type approximation. It also turned out that the value of the local moment and the inverse free time for scattering of carriers on the moments tend in the limit of the strong Kondo exchange to the finite universal values. The independence of the results on the on-site Coulomb repulsion is then illustrated in the frameworks of a simplistic model. It is shown that the spin density wave transition is driven by the Ruderman-Kittel-Kasuya-Yosida interactions between the renormalized moments via exchange by the electron-hole pairs. The applicability of the Boltzmann approach to transport in multiband pnictides is discussed briefly.

DOI: 10.1103/PhysRevB.87.024504 PACS number(s): 74.70.Xa, 74.25.F−, 75.30.Fv

I. INTRODUCTION

The discovery in 2008 of superconductivity in doped ferropnictides[1] with transition temperatures T_c up to 55 K (Ref. 2) has opened new perspectives in condensed matter physics. The avalanche of subsequent publications confirmed the emergence of the whole broad class of Fe-based materials possessing many interesting properties, both in the normal and in the superconducting state. Encouraged by some similarities with the high-temperature superconductivity (HTSC) in cuprates, experimental efforts were initially concentrated on the search for materials with even higher temperatures of the superconducting transition.

These new materials are remarkable for the apparent competition between superconductivity and magnetism. They also reveal a rather uncommon behavior already at elevated temperatures. In what follows, we address some of these properties. To be more specific, we focus on the basic properties of local moments on the Fe sites interacting with itinerant electrons in the paramagnetic phase. In addition, we study mechanisms that govern the structural and magnetic instability in the stoichiometric materials. We also discuss the electrical transport in the same temperature interval.

The structural and antiferromagnetic transitions in iron pnictides are close to each other and take place at temperatures T_{SDW} that lie in the interval between 100 and 200 K.[2–4] It is commonly accepted that the transition is driven by the magnetic instability, but is split into two by the magnetostriction effects with the structural transition preceding.[5]

In the literature there is no consensus yet concerning mechanisms that govern the phase transitions. Theoretical interpretations of the magnetic transition vary with the assumption about the nature of the paramagnetic phase. In the scenario of the itinerant electrons the magnetic transition could be driven by the spin density wave (SDW) instability coming from the so-called "nesting" of the Fermi surfaces. "Nesting" was very popular in the past at describing the phase transitions in organic conductors. The nesting mechanism is the analog to the Peierls transition in one dimension (1D) and proceeds from the instability of the supposedly congruent holelike and electronlike Fermi surfaces (FSs) at the interactions.[6] Technically, the nesting model is rather flexible and is convenient for systematizing and interpreting even such delicate features in iron pnictides as the competition between the SDW and superconductivity at doping.[7,8] The fine details of the electronic spectrum in Fe pnictides are usually studied in the band structure calculations and may depend on the assumptions concerning the strength of the competing on-site interactions.

An alternative scenario of the antiferromagnetic transition is known as the J_1-J_2 model.[9] The antiferromagnetic ordering in Ref. 9 is ascribed to the superexchange mechanism characterized by competition between the two integrals, J_1 and J_2, for interactions between the local Fe moments on the nearest-neighbor (n.n.) and the next-nearest-neighbor (n.n.n.) sites, correspondingly.

Although FS nesting is compatible with a number of experimental observations for the evolution of spin excitations in electron- or hole-doped iron-based superconductors, there are several aspects of the problem where such a scenario cannot be reconciled with experiments (see Ref. 10 for the review).

Thus, the neutron scattering experiments reveal that fluctuating moments for $BaFe_2As_2$ and $BaFe_{1.9}Ni_{0.1}As_2$ are $m^2 = 3.17 \pm 0.16$ and $3.2 \pm 0.16\,\mu_B^2$ per Fe(Ni), respectively. While these values are considerably smaller than those of the fully localized case, they are much larger than expected for the value of the staggered magnetization in the nesting SDW scenario.

Recent Fe x-ray emission spectroscopy experiments unveiled the existence of sizable local moments in the room-temperature paramagnetic state[11] for almost all of the pnictides studied. The observed local moments are similar in magnitude to those reported in the low-temperature neutron scattering experiments.

The vulnerability of the nesting scenario was emphasized already in an early review.[12] Angle-resolved photoemission spectroscopy (ARPES) data[13] and a more recent ARPES experiment,[14] for instance, seem to show that the electronic spectrum of the antiferromagnetic phase of $BaFe_2As_2$ disagrees with the "nesting model."

The main unresolved issue concerns the role of electron-electron (*e-e*) interactions. The authors[15,16] argued that electrons in the Fe-based materials are on the verge of localization in the Mott-Hubbard-type transition. Consequently, the electronic excitations cannot manifest the same coherent character as in ordinary metals.

The view that neither the localized nor itinerant approach can be fully correct for iron pnictides was expressed by many other authors. However, as noticed, for instance, in Ref. 17, the direct Coulomb interactions seem to be moderate and, unlike Refs. 15 and 16, correlations are driven by the Hund's rule coupling rather than by the on-site Hubbard repulsion.

These basic questions cannot be resolved without a better understanding of the nature of the paramagnetic phase. In the SDW phase the values of the staggered magnetization turned out to be surprisingly small.[18] Elastic neutron scattering experiments are in favor of larger values of the local magnetic moments as well.[10,19]

Theoretically the latter problem was addressed within the numerical approach that combines the density functional and the dynamical mean-field theory (DFT + DMFT).[20,21] It was demonstrated that, although the *d* electrons possess an itinerant character, nevertheless, due to the strong on-site Hund's interactions, they simultaneously contribute to the fluctuating magnetic moments on the Fe sites.

As a whole, these results[20,21] draw a picture of the "mixed valence" for the Fe ions. The concept has found further development in recent photoemission experiments[22] that can be summarized as follows. In the classical language, the *d* electrons always enter and leave the local Fe sites. For a short time, while electrons are on the *d* orbital, the strong local Hund's interaction tends to align their spins. These instantaneous values of the moment can be measured only by the "fast" experiments with a time scale $\sim 10^{-16}$ s.[22] (Other evidences in favor of such a view were already known in the experimental literature. For the list of references, see Ref. 22.)

Neither the direction nor even the amplitude of the fluctuating moment is then fixed: Locally, the spin is not a good quantum number. In the language of quantum mechanics, the wave function of a single *d* electron has both local and itinerant components. The interrelation between the components' amplitudes determines the average local moment. Thus, experiments[22] measure the probabilities of finding the electron in one state or another.

Accounting for such behavior of the *d* electrons in all generality would present a formidable task for the theory. Most often than not, the studied phenomena are characterized by much longer time scales and the experiments measure some average spin values on the Fe sites. The thermodynamics of the SDW transition gives one such example. The averaged local spins appear also in such kinetic characteristics as resistivity. The sizes of the electron-hole pockets in the Brillouin zone (BZ) are small, and the Fermi velocities on single pockets are lower than the atomic value. This feature finds its reflection in the enlarged effective masses. In fact, the calculated band masses m_b are usually equal to $\sim$2–3 m_e.[18,23] The Hund's interaction increases these values further to $m^* \sim (3\text{–}4)\, m_b$.[21] Thereby, one may expect the effective masses to be as large as $m^* \sim 6\text{–}8\, m_e$.

To the best of our knowledge, so far there has only been one attempt to theoretically account for the dual character of the *d* electrons. The authors[24,25] considered the model in which, among all the *d* bands, some have energy below the Fermi level; these electrons create the local magnetic moments on the Fe sites. The rest form the bands of the itinerant electrons. The coexistence of the preformed moments and the itinerant electrons is postulated; the value of the local moments is fixed. The authors[24,25] were interested in properties of the SDW phase and in the possibility of superconducting pairing on the SDW background.

We consider the paramagnetic phase mainly of the so-called "1111" and "122" Fe pnictides. For the latter the staggered magnetic moments on the Fe sites in the SDW state are low and may lie in the range of 0.2–1 μ_B (see Ref. 18).

In all theoretical references cited before not much attention has been paid to the properties of the itinerant electrons in the systems. Therefore, besides discussing the instability of the paramagnetic phase itself, we concentrate on the resistivity behavior at temperatures above the transitions. The resistivity above T_{SDW} is controlled by scattering of the electrons on the randomly oriented magnetic moments. Although the resistivity is metallic, $\rho(T)$ is large even in clean single crystals. Thus, in $BaFe_2As_2$, $\rho(T_{SDW}) \approx 0.25$ mΩ cm;[26] it is even higher (at 200 K), $\sim$0.5 mΩ cm, in Sr122.[26,27] The authors[15,16] argued that such high values signify a violation of the Ioffe-Regel criterion. As distinct from Refs. 15 and 16, we argue that to address this controversy one has to take into account the multiband character of the energy spectrum of the Fe pnictides. In addition, we compare our theoretical results with the resistivity data for LaOFeP.[28,29] This "1111" compound is remarkable for the absence of the SDW transition.

II. INTERACTIONS BETWEEN LOCAL SPINS AND ITINERANT ELECTRONS

The purpose of this section is to study the renormalization of the local moments in the presence of Kondo-like exchange interactions with the itinerant subsystem. As a point of departure, we assume a single itinerant band of *d* electrons with one electronlike and one holelike pocket.

The interaction between the local sites and the band electrons is chosen in the form

$$\hat{H}_{\text{int}} = -\sum_i J S_i \hat{s}(R_i). \tag{1}$$

In Eq. (1), $\hat{s}(R_i)$ is the operator for the electronic spin of the unit cell located at the point R_i,

$$\hat{s}(R_i) = A_\square \hat{\psi}_\sigma^+(R_i) s_\sigma \hat{\psi}_\sigma(R_i), \tag{2}$$

where $A_\square \equiv a^2$ is the unit cell area. Each operator $\hat{\psi}$ (and $\hat{\psi}^+$) consists of the two components for the electronic and the hole pockets, $\hat{\psi}(R_i) = \hat{\psi}_e(R_i) + \hat{\psi}_h(R_i)$ correspondingly. For better transparency of arguments, we consider first the interactions in the Hamiltonian (1) for only one itinerant pocket.

At the ferromagnetic sign $J > 0$, Eq. (1) is the version of the Hund's interaction. Positive $J > 0$ is therefore the natural choice in applications to the Fe pnictides. [For a quantum spin $\hat{S}_i$ and the antiferromagnetic sign of $J < 0$, the Hamiltonian Eq. (1) describes the Kondo problem.] Following Ref. 22, the spin S is a *certain average* of the fluctuating nonconserving spin on Fe sites. Correspondingly, we assume below that neither S nor s_σ are quantum operators. (The sign of J is of no importance below.) The sum in Eq. (2) is over the repeating spin index σ runs over the two directions of the electronic spin, $s_\sigma = \pm 1$.

To start with, calculate the correction second order in J, $\delta E_i^{(2)}$, to the energy of the center. It can be written as

$$\delta E_i^{(2)} = J^2 S_i^2 \int dt' G(0; t - t') G(0; t' - t). \tag{3}$$

In Eq. (3), $G(0; t - t')$ stands for the Green's functions, and $G(R_i - R_i'; t - t')$ at the coinciding spatial arguments: $R_i' \equiv R$. As in the Kondo problem, the physics below is related to the properties of the Green's functions at large time arguments. By definition,[30]

$$G_0(0; t) = -i \int [1 - n(\varepsilon)] \nu(E_F) \exp(-i\varepsilon t) d\varepsilon; \quad (t > 0;\ \varepsilon > 0) \tag{4a}$$

and

$$G_0(0; t) = i \int n(\varepsilon) \nu(E_F) \exp(-i\varepsilon t) d\varepsilon; \quad (t < 0;\ \varepsilon < 0). \tag{4b}$$

Here $n(\varepsilon)$ and $\nu(E_F)$ are the electrons' occupation number and the density of states at the Fermi level, correspondingly. One easily finds from Eqs. (4a) and (4b) the asymptotic behavior of the Green's function at $|tE_F| \gg 1$:

$$G_0(0; t) \cong -\nu(E_F)[|t| - i\delta \text{sign}(t)]^{-1}. \tag{5}$$

For classical spins S_i the calculations below become simpler in the frequency representation. For instance, the second-order correction, Eq. (3), takes the form

$$\delta E_i^{(2)} = J^2 S_i^2 \int \frac{d\omega}{2\pi} G_0^2(0; \omega). \tag{6}$$

In notations, $\xi = \varepsilon(p) - \mu$ $(\mu = p_F^2/2m \equiv E_F)$, the Green's function at the equal spatial arguments is[30]

$$G_0(R_i = R_i; \omega) = \int \frac{A_\square d^2\vec{p}}{(2\pi)^2} G_0(p; \omega) \equiv \int \frac{A_\square d^2\vec{p}}{(2\pi)^2} \left\{ \frac{1}{\omega - \xi + i\pi\delta \text{sign}(\omega)} \right\}. \tag{7}$$

[Notice that from now on we include the factor $A_\square$ in the definition of the Green's function, Eq. (7). In other words, the Green's function at $R_i = R_i'$ is an "average" over the unit cell. Correspondingly, everywhere below $\nu(E_F)$ has the meaning of the density of states (DOS) per one spin direction and per one unit cell.]

When substituting the identity $\{\frac{1}{\omega - \xi + i\pi\delta \text{sign}(\omega)}\} = \text{P}\{\frac{1}{\omega - \xi}\} - i\pi \text{sign}(\omega)\delta(\omega - \xi)$ into Eq. (7) for the Green's function, the first term can be omitted, because it produces no contribution from the vicinity of the chemical potential. Indeed, P stands here for the principal value of the integral. For the band energy, $\varepsilon(p)$ were chosen to be symmetric respecting the chemical potential, E_F as such a term would give zero *exactly*. Otherwise, the P term leads to an insignificant spatial dependence in the Green's function on the atomic scale. For this reason it is enough to leave it under the integral sign in Eq. (7) only:

$$G_0(p; \omega) = -i\pi \nu(E_F) \text{sign}(\omega) \delta(\omega - \xi). \tag{8}$$

We study the polaronlike effects in the system of spins on the local centers strongly coupled to spins of the itinerant carriers. Such physics has many formal similarities to the polaronic effect in a system with strong electron-phonon interactions. The latter has been studied recently in Ref. 31. (The polaronic self-localization was discussed earlier for the A15 compounds in the Holstein model.[32]) The mathematics is basically the same in both problems and below we merely list the main steps of the calculations.

For classical spins one can calculate all energy contributions from the interaction of Eq. (1) with the itinerant electrons by using the identity $\partial E(S_1 \ldots S_N)/\partial J = \sum_{i,\alpha} S_i \hat{s}_\alpha(R_i)$. As in Ref. 31, equations for the Green's function presented in the real space have the following form:

$$\begin{aligned} G_\sigma(R, R'; \omega) &= G_{0,\sigma}(R - R'; \omega) + \sum_i G_{0,\sigma}(R - R_i; \omega) J \bar{S}_{i,\sigma}^L(\omega) G_{0,\sigma} \\ &\times (R_i - R'; \omega) + \sum_{i \neq k} G_{0,\sigma}(R - R_i; \omega) J \bar{S}_{i,\sigma}^L(\omega) G_{0,\sigma} \\ &\times (R_i - R_k; \omega) J \bar{S}_{k,\sigma}^L(\omega) G_{0,\sigma}(R_k - R'; \omega) + \cdots. \end{aligned} \tag{9}$$

[In accordance with the notations in Eq. (2), in Eq. (9) we restored the dependence on the electronic spin indices.]

The total *local* contribution into the center's energy from Eq. (9) has the following form:

$$\delta E_i = J S_i \int \frac{d\omega}{2\pi} G_0^2(0; \omega) J \bar{S}_i^L(\omega). \tag{10}$$

In Eqs. (9) and (10) at the expansion for spin $S_i(t)$ the summation over J may be taken from *the left side*; with the help of Eq. (9) this will give the following self-evident equation for $\bar{S}_i^L(\omega)$,

$$\bar{S}^L_{\sigma,i}(\omega) = S_i s_\sigma + S_i s_\sigma G_{0,\sigma}(R_i = R_i;\omega) J \bar{S}^L_{\sigma,i}(\omega). \quad (11a)$$

The same procedure defines the variable $\bar{S}_j^R(\omega)$,

$$\bar{S}^R_{\sigma,j}(\omega) = S_j s_\sigma + J\bar{S}^R_{\sigma,j}(\omega) G_{0,\sigma}(R_j = R_j; -\omega) S_j s_\sigma, \quad (11b)$$

if the summation over J were performed from *the right side.* $\bar{S}_j^R(\omega)$ and $\bar{S}_j^L(\omega)$ are complex conjugates.

It must be emphasized that at the derivation of Eqs. (11a) and (11b) no smallness of the interactions J in the Hamiltonian (1) was assumed. Omitting details, the total contribution into the local energy of one center is

$$\delta E_i = -(W/2\pi)\ln\{1 + [\pi \nu(E_F) J S_i]^2\}. \quad (12)$$

(W is the bandwidth.) Other terms in Eq. (9) lead to contributions into the total energy from the interactions between different centers. As it will be shown, relevant to the physics of Fe pnictides are the Ruderman-Kittel-Kasuya-Yosida (RKKY)-type interactions via exchange by the electron-hole pairs. The latter are investigated in the two-pocket model below.

Consider now the itinerant electrons. The scattering of electrons on the local moments above $T_{\rm SDW}$ is given by the average of $J\bar{S}^L_{\sigma,i}$ that stands in the denominator of the Green's function, $G(p;\omega) = [\omega - \xi + \langle J\bar{S}^L\rangle]^{-1}$. From Eq. (11a),

$$J\bar{S}^L(\omega) = JS[1 - i\,\mathrm{sign}(\omega)\pi\nu(E_F)JS]^{-1}. \quad (11')$$

Above the SDW transition the on-site Fe spins S_i are not oriented and for $\langle J\bar{S}_i^L\rangle$ it follows

$$\langle J\bar{S}_i^L\rangle = i\,\mathrm{sign}(\omega)\pi\nu(E_F)(JS)^2/\{1 + [\pi\nu(E_F)JS]^2\}. \quad (13)$$

Using again Eq. (11′) in the expression for the *correlator* $\langle S_i(0)S_i(t)\rangle$ one arrives back to the notion of the randomly oriented *renormalized* spins:

$$\langle S_i(0)S_i(t)\rangle = \int \frac{d\omega}{2\pi}\langle S\bar{S}_i^L(\omega)\rangle e^{i\omega t} \equiv \delta(t)\overline{S_i^2}, \quad (14)$$

where the average square of renormalized moment is

$$\overline{S_i^2} = \frac{\pi\nu(E_F)JS^2}{1 + [\pi\nu(E_F)JS]^2}. \quad (14')$$

[In the paramagnetic phase at averaging in Eqs. (13) and (14) the odd terms in S have dropped out.]

At large J one has $\overline{S_i^2} = 1/[\pi\nu(E_F)J]$. The itinerant carriers form a "cloud" at a site (i). The average value of its spin is

$$\langle s_i\rangle = \int \frac{d\omega}{\pi} G_0^2(0,\omega) J \bar{S}_i^L(\omega)$$
$$= JSW\pi\nu^2(E_F)/\{1 + [\pi\nu(E_F)JS]^2\}. \quad (14'')$$

III. MULTIBAND CASE

An extension of the above results in the case of one electronic and one hole pocket is trivial. The Hamiltonian of Eq. (1) would now include interactions with carriers from both pockets, including the nondiagonal terms. Correspondingly, in its general form the Hamiltonian is

$$\hat{H}_{\rm int} = -\lambda \sum_{i;l,t} S_i J_{lt}\hat{s}_{lt}(R_i), \quad (1')$$

where

$$\hat{s}_{lt}(R_i) = A_\square \hat{\psi}^+_{l,\sigma}(R_i) s_\sigma \hat{\psi}_{t,\sigma}(R_i). \quad (2')$$

The indices (l,t) stand for the electron and the hole pockets. (As usual, the summation runs over the repeated Latin indices.)

It is convenient to introduce one common parameter λ in $\partial E(S_1\ldots S_N)/\partial\lambda = \sum_{i,lt} S_i J_{lt}\hat{s}_{lt}(R_i)$ $(\lambda \in \{0,1\})$. Equation (9) for the Green's functions preserves the same structure, except that the Green's function now depends on λ and has the matrix form $G^{(\lambda)}_{lt;\sigma}(R,R';\omega)$. The free Green's function $G_{0,\sigma;lt}(R,R';\omega)$ is the diagonal matrix

$$G_{0,\sigma;lt}(R,R';\omega) \Rightarrow \begin{pmatrix} G^{(e)}_{0,\sigma}(\cdots) & 0 \\ 0 & G^{(h)}_{0,\sigma}(\cdots)\end{pmatrix}. \quad (15)$$

Instead of Eq. (11) for the single center, one has to calculate $J_{lt}G_{lt}(0;\omega)$, where $G_{lt}(0;\omega)$ obeys the equation

$$G_{\sigma;lt}(0;\omega) = G_{0,\sigma;lt}(0;\omega) + G_{0,\sigma;lm}(0;\omega) S_i J_{mp}(\omega) G_{pt}(0;\omega). \quad (16)$$

While the matrix form of Eqs. (15) and (16) significantly complicates the calculations of the total energy, the physics eventually suffers only minor changes.

IV. SDW INSTABILITY

Turn now to the antiferromagnetic (SDW) transition that is driven by the RKKY interaction. In the case of the stoichiometric Fe pnictides the structural vector $\vec{Q}$ of the SDW phase is commensurate and connects centers of the holelike pocket at the Γ point and the electron pocket at the X points [in the unfolded Brillouin zone (BZ) it is commensurate]. Thereby, the transition is controlled by the interaction J_{eh} between carriers from the two pockets. To demonstrate the basic physics, we restrict ourselves by the model of Sec. III with the two bands only and assume that the nondiagonal exchange J_{eh} is relatively small, $J_{eh} \ll J_e, J_h$. The RKKY interaction acquires the form

$$E_{i,k} \equiv E(R_i,R_k)$$
$$= -i2S_iS_kJ^2_{eh}\Pi_{eh}\int (d\omega/2\pi) G_{0,\sigma;e}$$
$$\times (R_i - R_k;\omega) G_{0,\sigma;h}(R_i - R_k;\omega). \quad (17)$$

The factor Π_{eh} stands for the renormalized value of the spins S_i, S_j. Making use of Eqs. (11a) and (11b), one finds that the product

$$\Pi_{eh} \equiv |\bar{S}_i^L(\omega)/S_i|^2 |\bar{S}_j^L(\omega)/S_j|^2$$
$$= \{1 + [\pi\nu_e(E_F)J_eS]^2\}^{-1}\{1 + [\pi\nu_h(E_F)J_hS]^2\}^{-1} \quad (17')$$

is frequency independent. A comparison between Eq. (17′) and Eqs. (13) and (14′) shows that the renormalization factors may differ depending on the effect under consideration and on the structure of the electronic spectrum.

In the diagrammatic representation at finite temperatures,[30] Eq. (17) is

$$E_{i,k} = -4S_iS_kJ^2_{eh}\Pi_{eh}T\pi\sum_{\omega_n}\int \frac{A_\square d\vec{p}}{(2\pi)^2} G_{0,h}(\varepsilon(\vec{p});\omega_n)$$
$$\times G_{0,e}(\varepsilon(\vec{p} - \vec{Q});\omega_n). \quad (18)$$

(An additional factor of 2 comes from the two spin directions.) It is convenient to introduce the notation

$$E_{i,k} \equiv S_i S_k J_{eh}^2 \Pi_{eh} \pi \nu(E_F) F(T,\eta). \quad (19)$$

After summation over the thermodynamic frequencies, $\omega_n = (2n+1)\pi T$ in Eq. (18), one arrives at the following well-known expression:

$$F(T,\eta) = \int \frac{\Omega d\vec{p}}{(2\pi)^2 \nu(E_F)} \left(\frac{1}{\varepsilon_h(\vec{p}) - \varepsilon_e(\vec{p}-\vec{Q})} \right) \times \left\{ \tanh\left(\frac{\varepsilon_h(\vec{p})}{2T}\right) - \tanh\left(\frac{\varepsilon_e(\vec{p}-\vec{Q})}{2T}\right) \right\}. \quad (20)$$

[Below the parameter η will characterize the dependence of $F(T,\eta)$ in Eq. (20) on such factors as the anisotropy or doping. The integration in Eq. (20) extends over the energy interval $\tilde{W}_{eh}$ that corresponds to the overlap of the electronlike and holelike bands.]

No rigorous solutions for the antiferromagnetic transition are known and one usually resorts to the mean-field approach. The method consists in analyzing the linear response of the system to the external "staggered" field, $B_{\text{ext}}^{\text{stag}} \Rightarrow B_{\text{ext}}(Q)$.

In the mean-field approach the system is unstable with respect to the magnetic transition when the following criterion is fulfilled:

$$1 = \frac{J_{eh}^2 S^2 \Pi_{eh} \pi \nu(E_F)}{T_{\text{SDW}}} F(T_{\text{SDW}},\eta). \quad (21)$$

Here $F(T_{\text{SDW}},\eta)$ plays a secondary role as a coefficient in front of the $1/T$ prefactor. Its magnitude depends on the fine details of the e-h spectrum, i.e., on the anisotropy or doping. Generally, we do not specify the band spectrum, but such a factor would increase due to the logarithmic singularity at the wave vectors coinciding with the nesting vector Q, if any. Experimentally, in the stoichiometric Fe pnictides, Q is commensurate, but may deviate from the commensurability at a large enough doping level.

The choice of the overlap $\tilde{W}_{eh}$ as the parameter is not convenient in what follows. A more familiar energy scale can be defined by considering first $F(T,\eta)$, Eq. (19), in the approximation of the *exact nesting*:

$$F(T,0) = \int \frac{A_{\square} d\vec{p}}{(2\pi)^2 \nu(E_F)} \left(\frac{\tanh(\xi/2T)}{\xi} \right) = 2 \ln\left(\frac{\gamma \tilde{W}_{eh}}{\pi T} \right). \quad (22)$$

Let $T_{\text{SDW}}^{(0)}$ be the temperature of the SDW transition for this case:

$$T_{\text{SDW}}^{(0)} = 2J_{eh}^2 S^2 \pi \nu(E_F) \ln\left(\frac{\gamma \tilde{W}_{eh}}{\pi T_{\text{SDW}}^{(0)}} \right). \quad (23)$$

Considering doping or the anisotropy recall that, realistically, the latter is always large enough to cut the logarithmic singularity in Eq. (22); then $\ln(\gamma \tilde{W}_{eh}/\pi T) \Rightarrow \ln(\gamma \tilde{W}_{eh}/\eta)$ and from Eq. (21) for the dependence $T_{\text{SDW}}(\eta)$ it follows

$$T_{\text{SDW}} = T_{\text{SDW}}^{(0)} - 2J_{eh}^2 S^2 \pi \nu(E_F) \int \ln\left[\frac{\eta(\vec{k})}{\pi T_{\text{SDW}}^{(0)}} \right] \frac{d\Omega_{\vec{k}}}{2\pi}. \quad (24)$$

[Integration in Eq. (24) averages over the Fermi surface the anisotropy in the parameter η, if any.]

So far, in the discussion above we fully ignored the electron-electron interactions. The latter can be of interest, especially at calculations of $F(T,\eta)$ in the logarithmic approximation. The corresponding analysis would lead to the expressions for $F(T,\eta)$ that are familiar from numerous publications on the "nesting" scenario. (See, e.g., in Ref. 33.) It does not change the general structure of Eq. (21) and we do not discuss further details here.

Equation (21) reveals the main difference from the "nesting" scenario. In our case the interaction takes place between the nonzero (renormalized) moments on different sites that already exist in the paramagnetic phase, while in the nesting scenario the finite magnetization appears only below T_{SDW}. We repeat once again that the electronic spectrum in Eq. (20) has a general character. Even more, the method of Eq. (21) and the analytic $T_{\text{SDW}}(\eta)$ dependence in Eq. (24) differ from the one in the "nesting" scenario. The "nesting" scenario relies on the "congruency" of the electron and hole pockets that results in the logarithmic singularity in the $F(T_{\text{SDW}},\eta)$ function that is compensated then by the interaction constant.[34] [Besides, note that Eq. (24) involves the independent parameter J_{eh}.]

V. LIMIT OF STRONG INTERACTIONS

From Eq. (11′) we derived the expression for the imaginary part of the self-energy Eq. (13) that determines the inverse mean time for scattering of electrons on the disordered moments in the paramagnetic phase:

$$\frac{\hbar}{\tau} = \frac{\pi \nu(E_F) J^2 S^2}{1 + [\pi \nu(E_F) J S]^2}. \quad (25)$$

The combination JS is the product of the coupling strength and the "bare" spin. It is remarkable that in the limit of large J this expression acquires a very simple form:

$$\hbar/\tau = 1/\pi \nu(E_F). \quad (25')$$

Thereby, any information about the local spins on the Fe site drops out from Eq. (25′).

This prompts the question as to whether these results could be correct even in the extreme limit of the Coulomb repulsion (or the Hubbard's U) on a Fe site being much larger than the *intra-atomic* Hund's exchange. In other words, the competition between the *on-site* Coulomb and Hund's interactions becomes replaced by the competition between a strong local Hubbard repulsion U and the strong Kondo exchange with the itinerant carriers, Eq. (1). For brevity, consider the simplistic model of a single itinerant d band with one electronlike and one holelike pocket centered at different points in the BZ and a Hubbard U strong enough to forbid the double on-site occupancy (the "U model").

In the U model E_i to build up the on-site magnetic moment S_i would cost energy[35]

$$E_i = U \frac{S_i^2}{2}. \quad (26)$$

[In the above S could be considered either as the spin ($S > 1$) or as the magnetic moment measured in units of the Bohr magneton (see also Ref. 35).]

Return now to the calculation of the second-order correction $\delta E_i^{(2)}$, Eq. (6). Substituting Eq. (8) for the Green's functions and performing the trivial integration, one arrives at

$$\delta E_i^{(2)} = -(W/2\pi)J^2 S_i^2[\pi\nu(E_F)]^2. \tag{27}$$

[Here again W is the bandwidth, and $\nu(E_F)$ stands for the density of states (DOS) per unit cell per single spin. For a parabolic two-dimensional (2D) spectrum DOS does not depend on E_F.]

The negative sign in Eq. (27) shows that the nonmagnetic state of the center at a large enough coupling constant may become unstable. In that case the electron-ion interactions, Eq. (1), will build up the local moments on the each site R_i. The threshold value J_0 for the onset of the instability is defined by equating (27) to the energy of the "bare" center, Eq. (26):

$$U\frac{S_i^2}{2} = (W/2\pi)J^2 S_i^2[\pi\nu(E_F)]^2. \tag{28}$$

Introduce the notation $J^2 = J_0^2\Lambda^2$. From Eq. (28) for J_0 one finds

$$J_0^2 = U\frac{1}{\pi W\nu(E_F)^2}. \tag{28a}$$

The total energy is the sum of Eqs. (12) and (26):

$$E_{i,\text{tot}} = U\frac{S_i^2}{2} - \frac{W}{2\pi}\ln\left[1+\pi^2\nu(E_F)^2 J_0^2\Lambda^2 S_i^2\right]. \tag{29}$$

Denote also

$$\tilde{S}_i^2 = [\pi J_0\nu(E_F)]^2 S_i^2. \tag{28b}$$

In the new notations, Eq. (29),

$$E_{i,\text{tot}} = \left(\frac{W}{2\pi}\right)\left\{\tilde{S}_i^2 - \ln\left[1+\Lambda^2\tilde{S}_i^2\right]\right\}. \tag{30}$$

At $\Lambda^2 > 1$, $E_{i,\text{tot}}$ in Eq. (30) has minima at $\tilde{S}_\pm = \pm\sqrt{1-\Lambda^{-2}}$.

Thereby, the Kondo exchange induces on site (i) the nonzero spin:

$$S_i^2 = \left(\frac{W}{U\pi}\right)(1-\Lambda^{-2}). \tag{31}$$

The energy at the minimum

$$E_{i,\text{tot}}^{\min} = (W/2\pi)[1-\Lambda^{-2}-\ln\Lambda^2]. \tag{32}$$

At small $\Lambda^2 \ll 1$, Eq. (30) is the potential energy of an anharmonic oscillator:

$$E_{i,\text{tot}} = \left(\frac{W}{2\pi}\right)\left\{\tilde{S}_i^2(1-\Lambda^2) + \left(\frac{1}{2}\right)\Lambda^4\tilde{S}_i^4\right\}. \tag{33}$$

All the results of Sec. II apply once the Ising spins are formed at $\Lambda^2 > 1$. In particular, in the paramagnetic phase the inverse mean free time for scattering of an electron on disordered spins and the average square of the local moment are given by the same Eqs. (13) and (14′). Thus

$$\overline{S_i^2} = \frac{\pi\nu(E_F)JS^2}{1+[\pi\nu(E_F)JS]^2}. \tag{34}$$

Hence, again $\overline{S_i^2} = 1/[\pi\nu(E_F)J]$ at large J. Actually, according to Eq. (28b), the limit of $[\pi\nu(E_F)JS]^2 \gg 1$ is the limit of large $\Lambda^2 \gg 1$: $[\pi\nu(E_F)JS]^2 \equiv \Lambda^2\tilde{S}^2 \gg 1$.

Returning to the case of the two pockets, the stability condition [see Eq. (27)] acquires the form

$$U\frac{S_i^2}{2} = S_i^2\{(W_{ee}/2\pi)J_{ee}^2[\pi\nu_e(E_F)]^2 + 2(W_{eh}/2\pi)J_{eh}^2 \times[\pi^2\nu_e(E_F)\nu_h(E_F)] + (W_{hh}/2\pi)J_{hh}^2[\pi\nu_h(E_F)]^2\}. \tag{35}$$

In Eq. (35) W_{ee} and W_{hh} are the bandwidths for the electronlike and the holelike bands, correspondingly. As to W_{eh}, it was already defined above as the energy difference between the top and the bottom of the h and the e bands. (The chemical potential must be inside each of the two bands.)

Again, Eq. (35) demonstrates that, provided the strength of interactions exceeds some threshold, the system becomes unstable for the multiband case as well. At nonzero J_{eh} the subsequent calculation, however, become cumbersome; we will not dwell further upon such a case.

The same calculations as above show that at $\Lambda^2 > 1$ the SDW instability corresponds to the transition in the system of the interacting Ising spins. Then $\tilde{S}^2 = (1-\Lambda^{-2})$ and Eq. (21) becomes

$$T_{\text{SDW}} = \left(\frac{\Pi_{eh}}{\pi\nu(E_F)}\right)\left(\frac{\Lambda^2-1}{\Lambda^2}\right)\left(\frac{J_{eh}^2}{J_0^2}\right)F(T_{\text{SDW}},\eta). \tag{21′}$$

For completeness, consider also the case of small $\Lambda^2 < 1$. Above T_{SDW} local moments appear only due to the thermal fluctuations:

$$\langle\langle\tilde{S}^2\rangle\rangle = \left\{\int\exp(-E_{\text{tot}}/T)\tilde{S}^2 d\tilde{S}\right\}\Big/\int\exp(-E_{\text{tot}}/T)d\tilde{S}. \tag{36}$$

Analytically, the average (36) can be calculated only at $\Lambda^2 \ll 1$. Substitution of Eq. (33) for E_{tot} gives

$$\langle\langle\tilde{S}^2\rangle\rangle \approx \left(\frac{2\pi T}{W}\right)\left\{1-\frac{9}{16}\Lambda^4\left(\frac{2\pi T}{W}\right)\right\}. \tag{36′}$$

The second term in the brackets at $T = T_{\text{SDW}}$ is small and instead, Eq. (21′), one has

$$\frac{1}{\pi\nu(E_F)}\left(\frac{J_{eh}^2}{J_0^2}\right)F(T_{\text{SDW}},\eta) = \frac{W}{2\pi}. \tag{37}$$

In Eq. (37), $W \sim 1$ eV, and one may immediately realize that for a solution for T_{SDW} to exist on the same scale of 100–200 K an enhancement should occur in the left side of Eq. (37) at $T \sim T_{\text{SDW}}$. "Nesting" features in the electronic spectrum under the integral in Eq. (20) for $F(T,\eta)$ could help.

However, since experimentally in stoichiometric pnictides the SDW transition always takes place via ordering of the local moments, we do not dwell any further upon details of the SDW transition described by Eq. (37).

We conclude the discussion of the simple model Eq. (26) by calculating the inverse mean free time for scattering of electrons on the thermally fluctuating local moments at $\Lambda^2 < 1$:

$$\hbar/\tau = [2\Lambda^2/W\nu(E_F)]T. \tag{38}$$

[Actually, as in Eq. (36′), $\hbar/\tau$ is calculated with $\Lambda^2 \ll 1$, with the small terms omitted.]

VI. RESISTIVITY

Rough estimates of the mean free path ℓ in Refs. 15 and 16 from data on the resistivity of polycrystalline samples of LaFeAsO led the authors to the conclusion that conductivity in the material cannot be described within the standard Boltzmann equation: According to their analysis, $k_F\ell \sim 1$, where k_F stands for the Fermi momentum. First, the measurements on the polycrystalline samples do not provide the intrinsic value of resistivity. In addition, the multiband character of the electronic spectrum in iron pnictides was not taken into account. Below we repeat the analysis[15,16] for a few iron pnictides.

Data for the resistivity $\rho(T)$ at elevated temperatures from single crystals are scarce. For $BaFe_2As_2$ we use the single-crystal data.[26] Unlike Refs. 15 and 16, whenever it is possible, we analyze the contributions into conductivity from each single pocket separately:

$$\sigma_i = n_i e^2 \tau_i / m_{\mathrm{eff},i}. \qquad (39)$$

It is instructive to start with the universal form of the mean free time, Eqs. (14) and (14′), for each pocket:

$$\hbar/\tau_i = 1/\pi \nu_i(E_F). \qquad (40)$$

[Recall that in the limit of strong interaction Eq. (40) is the same in both models.]

Then

$$\tau_i = \pi \nu_i(E_F)\hbar = \frac{2\pi^2\hbar(1/2)m_{\mathrm{eff},i}a^2}{(2\pi\hbar)^2} = \frac{a^2 m_{\mathrm{eff},i}}{2\hbar}. \qquad (41)$$

Correspondingly,

$$\sigma_i = (e^2\pi/4c\hbar)(ap_{F,i}/\pi)^2 \approx (ap_{F,i}/\pi)^2 1.2 \times 10^{15}\ \mathrm{s}^{-1}. \qquad (39')$$

(Note that in this limit the masses drop out from the expression for conductivity.)

$BaFe_2As_2$: There are two FeAs layers in "122" per unit cell. Taking the sum of the values of $(ap_{F,i}/\pi)^2$ (Ref. 23) over all four 2D pockets, for resistivity in $BaFe_2As_2$ at $T = T_{\mathrm{SDW}}$, one obtains $\rho(T_{\mathrm{SDW}}) \approx 1$ mΩ cm. Experimentally,[26,36] $\rho(T_{\mathrm{SDW}}) \approx 0.14$ mΩ cm. In the U model this difference could be compensated, in principle, by an appropriate choice of Λ^2 in Eq. (31). Still, the difference is too big and the attempt to compensate it by a proper value of Λ^2 would pose $BaFe_2As_2$ too close to the threshold, Eq. (27). No matter how, it is obvious that the "universal" expression Eq. (40) significantly underestimates the conductivity in "122." Hence, in the scenario of the preformed local moments the Hund's interaction, Eq. (1) in the system must not be too strong.

"1111": Repeating the same analysis for a few "1111" compounds[23] in different compounds gave values of $\rho(T_{\mathrm{SDW}})$ that usually lie between 1 and 1.16 mΩ cm. This is in a rather good agreement with the experiments as discussed in Refs. 18, 37, and 38.

With the notion[22] of rapidly fluctuating moments on the Fe sites, the question arises as to whether or not the kinetic characteristics, such as conductivity, can be analyzed in terms of the same "average" spins, as it was done above for the SDW transition.

In the limit of strong interaction, Eq. (41), the characteristic time is proportional to the effective mass on the pocket. Even with the mass $m_{\mathrm{eff},i}$ enhanced up to 6 m_e (Refs. 21 and 23) and $a \approx 4 \times 10^{-8}$ cm, one obtains for $\tau_i \sim 0.5 \times 10^{-14}$ s $\ll 10^{-16}$ s. Depending on the mass, the energy scale is $\hbar/\tau \approx$ 0.13–0.4 eV and is comparable to the corresponding Fermi energies [typically of the order of approximately a few tenths of eV (Refs. 18 and 23)].

One concludes that the analysis with the use of Eqs. (40) and (41) for τ_i will give for iron pnictides the results that seem to be close to the border of applicability of the standard Boltzmann expression. Fortunately, as it was noticed above, for $BaFe_2As_2$, Eqs. (40) and (41) overestimate its resistivity at and above T_{SDW}.

Taking seriously the factor of 10 differences between the calculated value, $\rho(T_{\mathrm{SDW}}) \approx 1$ mΩ cm, and the experimental value, $\rho(T_{\mathrm{SDW}}) \approx 0.14$ mΩ cm,[36] one may try to estimate the Hund's interaction coupling. A comparison of expressions (40) and (14′) leads to $[\pi\nu(E_F)JS]^2 \approx 0.1$. With $1/\pi\nu(E_F) \approx 0.13$–$0.4$ eV it gives surprisingly low values for JS to be between 0.04 and 0.13 eV. With $S \sim 1$–2 from the photoemission experiment[22] such an estimate gives for the Hund's constants too low a value, which seems to be in conflict with the commonly accepted values of the Hund's integral of order of 0.2–0.35 eV.[20,39] It is the time to remember that the measured staggered SDW magnetic moments are between 0.2 and 1 μ_B.[18] Therefore, it seems that the rapidly fluctuating moments on the Fe sites contribute into the transport characteristics only through the time averages that are similar in value to those ones observed in the SDW phase.

We conclude this discussion by a comment on LaOFeP. The SDW transition is absent in the latter. Whether large moments on the Fe sites are present in this material is not known. Our model, however, provides, assuming $\Lambda^2 \ll 1$, a tentative explanation for the large coefficient in front of the linear in T dependence of resistivity of this material above ~20 K.[28,29] With Eq. (38), $\hbar/\tau_i = [2\Lambda^2/W_i\nu_i(E_F)]T$, one can make a rough estimate by choosing for each pocket the same bandwidth, $W_i \sim$ 1 eV, and taking for simplicity $\Lambda^2 = 1/2$; this would give for the resistivity $\rho \sim 1.6 \times 10^{-19}\ T$ s $\sim 1.6 \times 10^{-4}\ T$ mΩ cm, which is rather close to the experimental data.[28,29]

VII. CONCLUSION

In summary, we studied the dual role of the d electrons in stoichiometric Fe pnictides in the paramagnetic phase in a model of the local moments and the itinerant electrons interacting via a Kondo-type Hamiltonian. At a moderate on-site Coulomb interaction the "bare" local moments become renormalized via the Kondo-like exchange interactions with itinerant electrons. In turn, the itinerant carriers scatter off on the renormalized moments that are disordered in the paramagnetic state. The latter mechanism explains the surprisingly high values of resistivity of the stoichiometric pnictides at the temperature of their transition into the antiferromagnetic phase.

It was shown that instability at the onset of the SDW order is caused by the RKKY interactions between the *renormalized* local moments via exchange by the electron-hole pairs; in the case of stoichiometric Fe pnictides the structural vector of the SDW phase is predetermined by the positions of the electron and the hole pockets in the BZ. The dependence of the transition temperature T_{SDW} on anisotropy and doping is found to be different from the one in the "nesting" scenario.

Generally, the assumed *intra-atomic* Hund's exchange is comparable or can even prevail over the Coulomb on-site repulsion. However, in the limit of a strong Kondo-like exchange between the local moment and itinerant carriers the value of the local moment itself and the effective mean free time for carriers scattered on the moments tend to finite values. The independence of the result on the strength of the on-site Coulomb repulsion was then justified by solving a simplistic model.

At calculations of conductivity of iron pnictides in the paramagnetic phase, the multiband character of the electronic spectrum in iron pnictides had to be taking into account. Contributions into the total conductivity from several electronlike and holelike pockets were then considered one by one.

For the "1111" group even a simplistic model leads to experimental values of resistivity at T_{SDW}. In case of the "122" iron pnictides the limit of strong interactions overestimates the resistivity value at T_{SDW}. Unlike the "1111" pnictides, it is found that the Born approximation already provides satisfactory results for resistivity of $BaFe_2As_2$, thus confirming the applicability of the standard kinetic analysis.

The linear in T dependence of resistivity in LaOFeP was explained in terms of the thermal fluctuations of local spins. Unlike other "1111" iron pnictides, the SDW transition is not observed in LaOFeP. All the results in the paper were obtained owing to the use of exact equations for the electronic Green's functions in the real space.

ACKNOWLEDGMENTS

The authors are grateful to I. A. Nekrasov and M. V. Sadovskii for sharing their unpublished data on the band parameters for many iron pnictides compounds. The work of L.P.G. was supported by the NHMFL through NSF Grant No. DMR-0654118 and the State of Florida, and that of G.B.T. through the RFBR under Grant No. 10-02-01056.

[1] Y. Kamihara, T. Watanabe, M. Hirano, and H. Hosono, J. Am. Chem. Soc. **130**, 3296 (2008).

[2] See Physica C **469** (9–12) (2009), special issue on superconductivity in iron pnictides.

[3] C. de la Cruz, Q. Huang, J. W. Lynn, J. Li, W. Ratcliff, J. L. Zarestzky, H. A. Mook, G. F. Chen, J. L. Luo, N. L. Wang, and P. Dai, Nature (London) **453**, 899 (2008).

[4] R. R. Urbano, E. L. Green, W. G. Moulton, A. P. Reyes, P. L. Kuhns, E. M. Bittar, C. Adriano, T. M. Garitezi, L. Bufaiçal, and P. G. Pagliuso, Phys. Rev. Lett. **105**, 107001 (2010).

[5] V. Barzykin and L. P. Gor'kov, Phys. Rev. B **79**, 134510 (2009).

[6] I. I. Mazin, D. J. Singh, M. D. Johannes, and M. H. Du, Phys. Rev. Lett. **101**, 057003 (2008).

[7] A. B. Vorontsov, M. G. Vavilov, and A. V. Chubukov, Phys. Rev. B **79**, 060508(R) (2009).

[8] R. M. Fernandes and J. Schmalian, Phys. Rev. B **82**, 014521 (2010).

[9] Q. Si and E. Abrahams, Phys. Rev. Lett. **101**, 076401 (2008).

[10] P. Dai, J. Hu, and E. Dagotto, Nat. Phys. **8**, 709 (2012).

[11] H. Gretarsson, A. Lupascu, J. Kim, D. Casa, T. Gog, W. Wu, S. R. Julian, Z. J. Xu, J. S. Wen, G. D. Gu, R. H. Yuan, Z. G. Chen, N.-L. Wang, S. Khim, K. H. Kim, M. Ishikado, I. Jarrige, S. Shamoto, J.-H. Chu, I. R. Fisher, and Y.-J. Kim, Phys. Rev. B **84**, 100509 (2011).

[12] I. I. Mazin and J. Schmalian, Physica C **469**, 614 (2009).

[13] G. Liu, H. Liu, L. Zhao, W. Zhang, X. Jia, J. Meng, X. Dong, J. Zhang, G. F. Chen, G. Wang, Y. Zhou, Y. Zhu, X. Wang, Z. Xu, C. Chen, and X. J. Zhou, Phys. Rev. B **80**, 134519 (2009).

[14] D. V. Evtushinsky, V. B. Zabolotnyy, T. K. Kim, A. A. Kordyuk, A. N. Yaresko, J. Maletz, S. Aswartham, S. Wurmehl, A. V. Boris, D. L. Sun, C. T. Lin, B. Shen, H. H. Wen, A. Varykhalov, R. Follath, B. Büchner, and S. V. Borisenko, arXiv:1204.2432.

[15] Q. Si, E. Abrahams, J. Dai, and J.-X. Zhu, New J. Phys. **11**, 045001 (2009).

[16] Q. Si and E. Abrahams, J. Phys.: Condens. Matter **23**, 223201 (2011).

[17] M. D. Johannes and I. I. Mazin, Phys. Rev. B **79**, 220510(R) (2009).

[18] D. C. Johnston, Adv. Phys. **59**, 803 (2010).

[19] H.-F. Li, W. Tian, J.-Q. Yan, J. L. Zarestky, R. W. McCallum, T. A. Lograsso, and D. Vaknin, Phys. Rev. B **82**, 064409 (2010).

[20] K. Haule and G. Kotliar, New J. Phys. **11**, 025021 (2009).

[21] Z. P. Yin, K. Haule, and G. Kotliar, Nat. Mater. **10**, 932 (2011).

[22] P. Vilmercati, A. Fedorov, F. Bondino, F. Offi, G. Panaccione, P. Lacovig, L. Simonelli, M. A. McGuire, A. S. M. Sefat, D. Mandrus, B. C. Sales, T. Egami, W. Ku, and N. Mannella, Phys. Rev. B **85**, 220503(R) (2012).

[23] I. A. Nekrasov, Z. V. Pchelkina, and M. V. Sadovskii, JETP Lett. **88**, 679 (2008); **88**, 144 (2008); I. A. Nekrasov and M. V. Sadovskii (private communication).

[24] S.-P. Kou, T. Li, and Z.-Y. Weng, Europhys. Lett. **88**, 17010 (2009).

[25] Y.-Z. You, F. Yang, S.-P. Kou, and Z.-Y. Weng, Phys. Rev. B **84**, 054527 (2011).

[26] M. Nakajima, T. Lianga, S. Ishida, Y. Tomioka, K. Kihou, C. H. Lee, A. Iyo, H. Eisaki, T. Kakeshita, T. Ito, and S. Uchida, Proc. Natl. Acad. Sci. USA **108**, 12238 (2011).

[27] P. L. Bach, S. R. Saha, K. Kirshenbaum, J. Paglione, and R. L. Greene, Phys. Rev. B **83**, 212506 (2011).

[28] T. M. McQueen, M. Regulacio, A. J. Williams, Q. Huang, J. W. Lynn, Y. S. Hor, D. V. West, M. A. Green, and R. J. Cava, Phys. Rev. B **78**, 024521 (2008).

[29] R. E. Baumbach, J. J. Hamlin, L. Shu, D. A. Zocco, N. M. Crisosto, and M. B. Maple, New J. Phys. **11**, 025018 (2009).

[30] A. A. Abrikosov, L. P. Gor'kov, and I. E. Dzyaloshinskii, *Methods of Quantum Field Theory in Statistical Physics* (Dover, New York, 1977).

[31] L. P. Gor'kov, Phys. Rev. B **85**, 165142 (2012).

[32] C. C. Yu and P. W. Anderson, Phys. Rev. B **29**, 6165 (1984).

[33] A. V. Chubukov, Physica C **469**, 640 (2009).

[34] Note also that, according to the experimental data (see Refs. 2 and 18) for stoichiometric Fe pnictides, the structural vector

$\vec{Q}$ of the SDW phase is commensurate; deviations from the commensurability were reported at the doping level in the vicinity of SC compositions.

[35]Expression (26) actually has a broader meaning. Indeed, even at $U = 0$, consider the situation when *all d* electrons spill over into itinerant bands, so that the ionic configuration at the center becomes diamagnetic. The energy of the diamagnetic "center" at the point R_i is again of the same form as in Eq. (26), $E_i = \mu_B^2 S_i^2/2|\chi|$, where the local diamagnetic susceptibility $\chi < 0$ is the phenomenological parameter.

[36]M. A. Tanatar, N. Ni, G. D. Samolyuk, S. L. Bud'ko, P. C. Canfield, and R. Prozorov, Phys. Rev. B **79**, 134528 (2009).

[37]P. M. Aswathy, J. B. Anooja, P. M. Sarun and U. Syamaprasad, Supercond. Sci. Technol. **23**, 073001 (2010).

[38]Note that according to Ref. 18 the values of the in-plane resistivity of polycrystalline samples *Ln*FeAsOF (*Ln* = Sm, Nd, Gd, Pr, Ce, La) (Ref. 36) are up to one order of magnitude too large in comparison with that for single crystals.

[39]Y. X. Yao, J. Schmalian, C. Z. Wang, K. M. Ho, and G. Kotliar, Phys. Rev. B **84**, 245112 (2011).

Summary List of Publications

1. L.P. Gor'kov, *Charged particle Green function in the infrared catastrophe region*, Sov. Phys. JETP **3**, 762 (1956)
2. L.P. Gorkov, I.M. Khalatnikov, *Electrodynamics of charged scalar particles*, Sov. Phys. JETP **4**,777 (1957)
3. L. P. Gor'kov, *Steady convection in a plane liquid layer near the critical point*, Sov. Phys. JETP **6**, 311(1957)
4. A. A. Abrikosov, A. D. Galanin, L. P. Gorkov, L. D. Landau, I. Ya. Pomeranchuk, and K. A. Ter-Martirosyan, *Possibility of Formulation of a Theory of Strongly Interacting Fermions*, Phys. Rev., II Ser. 111, 321 (1958)
5. L.P. Gor'kov, *On the energy spectrum of superconductors*, Sov. Phys.-JETP 8, 505 (1958)
6. A. A. Abrikosov, L. P. Gor'kov and I. M.Khalatnikov, *Superconductor in a high frequency field*, Sov. Phys.-JETP **8**, 182 (1959)
7. A. A. Abrikosov and L. P. Gor'kov, *On the theory of superconducting alloys I. The electrodynamics of alloys at absolute zero*, Sov. Phys. JETP, **8**, 1090 (1959)
8. A. A. Abrikosov and L. P. Gor'kov, *Superconducting alloys at finite temperatures*, Sov. Phys. JETP, **9**, 220 (1959)
9. A. A. Abrikosov, L.P. Gor'kov and I. E. Dzyaloshinski, *On the application of quantum field theory methods to problems of quantum statistics at finite temperatures*, Sov. Phys. JETP, **9**, 636 (1959)
10. L. P. Gor'kov, *Microscopic derivation of the Ginsburg-Landau equations in the theory of superconductivity*, Sov. Phys. JETP, **9**,1364 (1959)
11. L. P. Gor'kov, *The critical supercooling field in superconductivity theory*, Sov. Phys. JETP, **10**, 593 (1960)
12. L. P. Gor'kov, *Theory of superconducting alloys in strong magnetic field near critical temperature*, Sov. Phys. JETP, **10**, 998, (1960)
13. A. A. Abrikosov and L. P. Gor'kov, *Contribution to the theory of superconducting alloys with paramagnetic impurities*, Sov. Phys. JETP, **12**, 1243 (1961)
14. L. P. Gor'kov and T.K. Melik-Barkhudarov, *Contribution to the theory of superfluidity of an imperfect Fermi gas*, Sov. Phys. JETP, **13**, 1018 (1961)
15. L. P. Gor'kov, *On the forces acting on a small particle placed int an acoustic field in an ideal liquid*, Sov. Phys. Doklady, **6**, 773 (1962)
16. Yu. A. Bychkov and L. P. Gor'kov, *Quantum oscillations of the thermodynamic values of metals in magnetic fields in the Fermi liquid model*, Sov. Phys. JETP, **14**, 1132 (1962)
17. A. A. Abrikosov and L. P. Gor'kov, *Spin-orbit interaction and the Knight shift in superconductors*, Sov. Phys. JETP, **15**, 752, (1962)
18. L. P. Gor'kov and L. P. Pitaevskii, *Development of a shock wave at reflection of a weak discontinuity from the sonic line*, Sov. Phys. Doklady **7**, 377(1962)

19. L. P. Gor'kov and L. P. Pitaevskii, *On the transition of liquid ^{3}He into superfluid state*, Sov. Phys. JETP, **15**, 417 (1962)
20. A. A. Abrikosov and L. P. Gor'kov, *On nature of the impurity ferromagnetism*, Sov. Phys. JETP, **16**, 1575 (1963)
21. L. P. Gor'kov, *An estimate of the limiting values o the critical fields for hard superconductors*, Sov. Phys. JETP, **17**, 578 (1963)
22. L. P. Gor'kov and I. E. Dzyaloshinskii, *Possibility of zero-sound oscillations in metals*, Sov. Phys. JETP, **17**, 1111 (1963)
23. L. P. Gor'kov, *Non-linear acoustic oscillations of a gas column in a closed pipe*, Inzh. Zh. (Engineer Journal) **3**, 236 (1963)
24. L. P. Gor'kov and L. P. Pitaevskii, *Energy terms splitting of the hydrogen molecule*, Sov. Phys. Doklady, **8**, 788 (1964)
25. L. P. Gor'kov and T. K. Melik-Barkhudarov, *Microscopic derivation of the Ginsburg-Landau equations in anisotropic superconductors*, Sov. Phys. JETP, **18**, 1031(1964)
26. L. P. Gor'kov and A. I. Rusinov, *Ferromagnetism in superconducting alloys*, Sov. Phys. JETP, **19**, 922 (1964)
27. L.P. Gor'kov., G.M. Eliashberg, *Minute Metallic Particles in an Electromagnetic Field*, Sov. Phys. JETP, 21, 940 (1965)
28. Yu.A. Bychkov, L.P. Gor'kov and I.E. Dzyaloshinskii, *Concerning One-dimensional Superconductivity*, Sov. Phys. JETP Lett., **2**, 92 (1965)
29. Yu.A. Bychkov, L.P. Gor'kov and I.E. Dzyaloshinskii, *Possibility of Superconductivity Type Phenomena in a One-Dimensional System*, Sov. Phys. JETP **23**, 489 (1966)
30. L.P. Gor'kov, I.E. Dzyaloshinskii, *Contribution to the Theory of the Mott Exciton in a Strong Magnetic Field*, Sov. Phys. JETP **26**, 449 (1968)
31. L.P. Gor'kov, G.M. Eliashberg, *Generalization of the Ginzburg-Landau Equations for Non-Stationary Problems in the Case of Alloys with Paramagnetic Impurities*, Sov. Phys. JETP **27**, 328 (1968)
32. L.P. Gor'kov, G. M. Éliashberg, *The Behavior of a Superconductor in a Variable Field*, Sov. Phys. JETP, **28**, 1291 (1969)
33. L.P. Gor'kov, G.M. Éliashberg, *Superconducting Alloys in a Strong Alternating Field*, Sov. Phys. JETP **29**, 698 (1969)
34. L.P. Gor'kov, G.M. Éliashberg, *Dynamical properties of gapless superconductors*, J. Low Temp. Phys., **2**, 161 (1970)
35. L.P. Gor'kov, N.B. Kopnin, *Viscous vortex flow in superconductors with paramagnetic impurities*, Sov. Phys. JETP **33**, 1251 (1971)
36. L.P. Gor'kov, M.A. Fedorov, *Transparency of Thick Superconducting Films*, Sov. Phys. JETP Lett., 14, 393 (1971)
37. L.P. Gor'kov, T.T. Mnatsakanov, *Role of anisotropy in exciton transitions*, Sov. Phys. JETP **36**, 361 (1973).
38. L. P. Gor'kov, N.B. Kopnin, *Some features of viscous flow of vortexes in superconducting alloys near critical-temperature*, Sov. Phys. JETP, **37**, 183 (1973)
39. L.P. Gor'kov, D.M. Chernikova, *Concerning the structure of a charged surface of liquid helium*, JETP Lett., **18**, 68 (1973)
40. L.P. Gor'kov, I.E. Dzyaloshinskii, *Conductivity of quasi-one-dimensional crystals in the metallic phase*, JETP Lett., **18**, 401 (1973)
41. L.P. Gor'kov, *Contribution to the theory of the properties of the superconductors with β-W structure*, Sov. Phys. JETP, **38**, 830 (1974)

42. L.P. Gor'kov, I.E. Dzyaloshinskii, *Fluctuation increment to the conductivity in quasi-one-dimensional system near a structural transition*, Sov. Phys. JETP Lett. **20**, 16 (1974)
43. L.P. Gor'kov, I.E. Dzyaloshinskii, *Possible phase transitions in systems of interacting metallic filaments (quasi-one-dimensional metals)*, Sov. Phys. JETP **40**, 198 (1975)
44. L. P. Gor'kov, O. N. Dorokhov, *Singularities in electronic density of states of one-dimensional conductors with disorder*, Solid State Comm. **20**, 789 (1976)
45. L.P. Gor'kov, *Remark on the Frohlich conductivity in one-dimensional systems*, Sov. Phys. JETP Lett., **25** , 358 (1977)
46. L.P. Gor'kov, O.I. Dorokhov, *Disorder and one-electron spectrum in a linear chain*, Sov. J. Low Temp. Phys. **4**, 160 (1978)
47. L.P. Gor'kov, E.N. Dolgov, *Motion of the Fröhlich mode and the conductivity of* $NbSe_3$, Sov. Phys. JETP **50**, 203 (1979)
48. V.L. Berezinskii, L.P. Gor'kov, *On the theory of electrons localized in the field of defects*, Sov. Phys. JETP **50**, 1209 (1979)
49. L.P. Gor'kov, A.I. Larkin, D.E. Khmel'nitskii, *Particle conductivity in a two-dimensional random potential*, Sov Phys. JETP Lett., **30**, 228–232 (1979)
50. V.L. Berezinskii, L.P. Gor'kov, *Thermodynamics of interacting electrons in disordered one-dimensional conductors*, Sov. Phys. JETP **51**, 409 (1980)
51. V.L. Berezinskii, L.P. Gor'kov, O.N. Dorokhov, *The plateau regime of the AC conductivity in the disordered 1d conductors*, Solid State Communs., **37**, (1981)
52. L.P. Gor'kov, O.N. Dorokhov, F.V. Prigara, *Localization and the origin of paramagnetic centers in 1D conductors*, Solid State Communs., **37**, 869 (1981)
53. L.P. Gor'kov, D.M. Chernikova, *On the kinetics of the development of an instability on a charged liquid surface*, Sov. Phys. JETP **53**, 1008 (1981)
54. S.A. Brazovskii, L.P. Gor'kov, J.R. Schrieffer, *Influence of Interchain Electron Hopping on Commensurate One-Dimensional Conductors*, Physica Scripta, **25**, 423 (1982)
55. L.P. Gor'kov, E.N. Dolgov, A.G. Lebed, *Phonon drag and the characteristics of the conductivity in one-dimensional metals*, Sov. Phys. JETP, **55**, 365 (1982)
56. S.A. Brazovskii, L.P. Gor'kov, A.G. Lebed', *Incommensurate superstructures in organic conductors with three-dimensional electronic spectra*, Sov. Phys. JETP, **56**, 683 (1982)
57. L.P. Gor'kov, O.N. Dorokhov, F.V. Prigara, *Energy level correlations in a disordered one-dimensional chain*, Sov. Phys. JETP, **57**, 838 (1983)
58. L.P. Gor'kov, O.N. Dorokhov, F.V. Prigara, *Structure of wave functions and ac conductivity in disordered one-dimensional conductors*, Sov. Phys. JETP, **58**, 852 (1983)
59. L.P. Gor'kov, O.N. Dorokhov, F.V. Prigara, *On the ac conductivity of slightly disordered one-dimensional conductors*, Sov. Phys. JETP Lett., **37**, 447 (1983)
60. L.P. Gor'kov, *Boundary conditions and generation of periodic noise by a space-charge wave*, Sov. Phys. JETP Lett., **38**, 87 (1983)
61. L.P. Gor'kov, A.G. Lebed', *On the stability of the quasi-onedimensional metallic phase in magnetic fields against the spin density wave formation*, J. Physique Lett., **45**, 433 (1984)
62. L.P. Gor'kov, *Generation of oscillations by a running charge density wave*, Sov. Phys. JETP, **59**, 1057(1984)
63. I. Batistic, A. Bjelis, L.P. Gor'kov, *Generation of the coherent pulses by the CDW-motion. Solutions of the microscopic model equations*, J. Phys. France, **45**, 1049 (1984)
64. G.E. Volovik, L.P. Gor'kov, *An unusual superconductivity in* UBe_{13}, Sov. Phys. JETP Lett., **39**, 674 (1984)

65. L.P. Gor'kov, *Anisotropy of the upper critical field in exotic superconductors*, Sov. Phys. JETP Lett., **40**, 1155 (1984)
66. L.P. Gor'kov, D. Jerome, *Back to the problem of the upper critical fields in organic superconductors*, J.Physique Lett. **46**, 643 (1985)
67. G.E. Volovik, L.P. Gor'kov, *Superconducting classes in heavy-fermions systems*, Sov. Phys. JETP **61**, 843 (1985)
68. L.P. Gor'kov, P.A. Kalugin, *Defects and an unusual superconductivity*, Sov. Phys. JETP Lett., **41**, 253 (1985)
69. A.V. Balatskii, L.I. Burlachkov, L.P. Gor'kov, *Magnetic properties of anisotropic type-II superconductors*, Sov. Phys. JETP **63**, 866 (1986)
70. L.P. Gor'kov, *Unusual superconductivity and its physical properties*, Pramana — J. Phys., **28**, 597 (1987)
71. L.P. Gor'kov, A.V. Sokol, *Phase stratification of an electron liquid in the new superconductors*, Sov. Phys. JETP Lett., **46**, 420 (1987)
72. V.A. Borodin, L.P. Gor'kov, F.F. Igoshin, P.A. Kononovich, S.F. Kondakov, V.A. Merzhanov, L.I. Chernyshova, I.F. Shchegolev, *Nature of the electron state of* La_2CuO_4, Sov. Phys. JETP Lett., **46(S1)**, S181 (1987)
73. L.P. Gor'kov, A.V. Sokol, *On the problem of the phase diagram of new superconductors*, J. Phys. France, 50, 2823 (1989)
74. L.P. Gor'kov and A.V. Sokol, *Nematic state in an exchange Heisenberg Hamiltonian*, Sov. Phys. JETP Lett., **52**, 505 (1990)
75. V. Barzykin, L.P. Gor'kov, A.V. Sokol, *On Symmetry Classification of Spin Nematics*, Europhys. Lett., **15**, 869 (1991)
76. L.P. Gor'kov, *Spin Correlators and Weak Antiferromagnetism in Heavy Fermions*, Europhys. Lett., **16**, 301(1991)
77. B. Arfi, L.P. Gor'kov, *Circular dichroism in high-temperature superconductors: A characteristic of broken space-inversion symmetry*, Phys. Rev. B **46**, 9163 (1992).
78. L.P. Gor'kov, A. Sokol, *Nontrivial magnetic order: Localized versus itinerant systems*, Phys. Rev. Lett., **69**, 2586 (1992)
79. S.E. Esipov, L.P. Gor'kov, T.J. Newman, *Fluctuations in fragmentation processes*, J. Phys. A **26**, 787 (1993)
80. V. Barzykin, L.P. Gor'kov, A. Sokol, *Additional quadrupole phases for S=1 ions*, Phys. Rev. B **48**, 593 (1993)
81. V. Barzykin, L.P. Gor'kov, *Possibility of observation of nontrivial magnetic order by elastic neutron scattering in magnetic field*, Phys. Rev. Lett. **70**, 2479 (1993)
82. L.P. Gor'kov, A.G. Lebed', *Cyclotron resonance on open orbits in organic materials*, Phys. Rev. Lett. **71**, 3874 (1993)
83. L.P. Gor'kov, A.G. Lebed, *Fast oscillations in the surface impedance of* $(TMTSF)_2ClO_4$ *in a magnetic field*, Phys. Rev. B **51**, 1362 (1995)
84. L.P. Gor'kov, *Precursor Effects and Anomalous Magnetoresistance in the Bechgaard Salts*, Europhys. Lett., **31**, 49 (1995)
85. V. Barzykin, L.P. Gor'kov, *Singlet magnetism in heavy fermions*, Phys. Rev. Lett. **74**, 4301 (1995)
86. L. P. Gor'kov, Ju H. Kim, *Effect of finite concentration in Kondo alloys: Approach to the heavy-fermion problem*, Phys. Rev. B **51**, 3970 (1995)
87. L.P. Gor'kov, *Quasi-one-dimensional organic metals: Theory and experiment*, J. Phys. I France, **6**, 1697 (1996)

88. L.P. Gor'kov, V.Z. Kresin, *Manganites at low temperatures and light doping: band approach and percolation*, JETP Lett., **67**, 985 (1998)
89. D.F. Agterberg, V. Barzykin, L.P. Gor'kov, *Exotic ground states and impurities in multiband superconductors*, Europhys. Lett., **48**, 449 (1999)
90. D.F. Agterberg, V. Barzykin, L.P. Gor'kov, *Conventional mechanisms for exotic superconductivity*, Phys. Rev. B **60**, 14868 (1999)
91. M.O. Dzero, L.P. Gor'kov, A.K. Zvezdin, *First-order valence transition in $YbInCu_4$in the (B, T)-plane*, J. Phys.: Cond. Mat., **12**, L711-L717 (2000)
92. V. Barzykin, L.P. Gor'kov, *Ferromagnetism and superstructure in $Ca_{1-x}La_xB_6$*, Phys. Rev. Lett. **84**, 2207 (2000)
93. L.P. Gor'kov, E.I. Rashba, *Superconducting 2D system with lifted spin degeneracy: Mixed single-triplet state*, Phys. Rev. Lett. **87**, 037004 (2001)
94. L. Balicas, J.S. Brooks, K. Storr, S. Uji, M. Tokumoto, H. Tanaka, H. Kobayashi, A. Kobayashi, V. Barzykin, L.P. Gor'kov, *Superconductivity in an Organic Insulator at Very High Magnetic Fields*, Phys. Rev. Lett. **87**, 067002 (2001)
95. L. Balicas, J.S. Brooks, K. Storr, S. Uji, M. Tokumoto, H. Tanaka, H. Kobayashi, A. Kobayashi, V. Barzykin, and L.P. Gor'kov, High Field Phase Diagram of the Field-Induced Superconducting State of l-(BETS)2FeCl4, Int. J. Mod. Phys. B **16**, 3101 (2002)
96. M. Dzero, L.P. Gor'kov, A.K. Zvezdin, K.A. Zvezdin, *Even-odd effects in magnetoresistance of ferromagnetic domain walls*, Phys. Rev. B 67, 100402(R) (2003)
97. V. Barzykin, L.P. Gor'kov, *Inhomogeneous stripe phase revisited for surface superconductivity*, Phys. Rev. Lett. **89**, 227002 (2002)
98. L.P. Gor'kov, P.L. Krotkov, *Spin relaxation and antisymmetric exchange in n-doped III-V semiconductors*, Phys. Rev. B **67**, 033203 (2003)
99. L.P. Gor'kov, P.L. Krotkov, *Exact asymptotic form of the exchange interactions between shallow centers in doped semiconductors*, Phys. Rev. B **68**, 155206 (2003)
100. L.P. Gor'kov, G.B. Teitel'baum, *Pseudogap behavior of nuclear spin relaxation in high Tc superconductors in terms of phase separation*, JETP Lett., **80**, 195 (2004)
101. L.P. Gor'kov, P.D. Grigoriev, *Soliton phase near antiferromagnetic quantum critical point in Q1D conductors*, Europhys. Lett., **71**, 425 (2005)
102. Gor'kov, L.P. and Teitel'baum, G.B., On electron-hole symmetry and phase separation in some electron doped cuprates, Solid State Commun., 136, 523 (2005)
103. Gor'kov, L.P. and Grigoriev, P.D., Antiferromagnetism and hot spots in CeIn3, Phys. Rev. **B** Rapid Commun**.**, **73**, 060401 (2006)
104. L.P. Gor'kov, G.B. Teitel'baum, *Interplay of externally doped and thermally activated holes in $La_{2-x}Sr_xCuO_4$ and their impact on the pseudogap crossover*, Phys. Rev. Lett. **97**, 247003 (2006)
105. L.P. Gor'kov, P.D. Grigoriev, *Nature of superconducting state in the new phase in $(TMTSF)_2PF_6$ under pressure*, Phys. Rev. B **75**, 020507(R) (2007)
106. V. Barzykin, L.P. Gor'kov, *Gapless Fermi surfaces of anisotropic multiband superconductors in a magnetic field*, Phys. Rev. Lett. **98**, 087004 (2007)
107. V. Barzykin, L.P. Gor'kov, *Gapless Fermi Surfaces in Superconducting $CeCoIn_5$*, Phys. Rev. B **76**, 014509 (2007)
108. Gor'kov, L. P. and Teitel'baum, G. B., *The two-component physics in the real space and in the momentum representation*, J. Phys.: Conf. Series, 108, 012009 (2008)
109. Chubukov, A.V. and Gor'kov, L.P., *Spin resonance in three-dimensional superconductors: the case of CeCoIn5*, Phys. Rev. Lett., **101**, 147004 (2008)

110. Barzykin, V. and Gor'kov, L.P., *On superconducting and magnetic properties of iron-oxypnictides*, JETP Lett., **88**, 131 (2008)
111. Barzykin, V. (Victor) and Gor'kov, L.P., Role *of striction at magnetic and structural transitions in iron pnictides*, Phys. Rev. B **79**, 134510 (2009)
112. Gor'kov, L.P. and Teitel'baum, G.B., *Spatial inhomogeneities in iron pnictide superconductors: The formation of charge stripes*, Phys. Rev. B **82**, (Rapid Comm.) 020510 (2010)
113. Stankiewicz, J.; Evangelisti M.; Fisk Z.; Schlottmann P. and Gor'kov, L., *Kondo physics in a rare earth ion with well localized 4f electrons*, Phys. Rev. Lett., **108**, 257201 (2012)
114. Gor'kov, L.P., *Strong electron-lattice coupling as the mechanism behind charge density wave transformations in transition-metal dichalcogenides*, Phys. Rev. B **85**, 165142 (2012)
115. Gor'kov, L.P., *Quantum oscillations in the vortex state of underdoped YBa2Cu3O6.5 and other multi-band superconductors*, Phys. Rev. B Rapid Commun., **86**, 060501 (2012)
116. Gor'kov, L.P., *Kinetics of excitations on the Fermi arcs in underdoped cuprates at low temperatures*, Phys. Rev. B Rapid Commun., **88**, 041104 (2013)
117. Gor'kov, L.P. and Teitel'baum G.B., *Dual role of d-electrons in iron pnictides*, Phys. Rev. B **87**, 024504 (2013)
118. Lev P. Gor'kov and Gregory B. Teitel'baum, *Two regimes in conductivity and the Hall coefficient of underdoped cuprates in strong magnetic fields*, J. Phys.: Condens. Matter **26**, 042202 (2014)

Books and Chapters

L.P. Gor'kov, Developing BCS Ideas in the Former Soviet Union, in BCS: 50 years, eds. L. Cooper and D. Feldman; World Scientific (2010).

L.P. Gor'kov, *Notes on microscopic theory of superconductivity*, Contemporary Concepts of Condensed Matter. Science, Vol. 4(C), 15-50 (2011) [Superconductivity in New Materials. Eds. Z. Fisk and H.R. Ott, Elsevier, ISBN: 978-0-444-53425-5].

L.P. Gor'kov, *Theory of Superconducting Alloys*, In: Superconductivity. Vol.1: Conventional and Unconventional Superconductors, Ch. 5, 201–224 (2008). ed. K.H. Bennemann, J.B. Ketterson. Springer, xxxii,1568 pp (in 2 vols.). ISBN 978-3-540-73252-5.

L.P. Gor'kov, *Superconductivity in heavy fermion systems*, Sov. Sci. Rev. Sect. A: Physics Reviews, Vol. 9, 1–116 (1987). ed. by I. M. Khalatnikov. London: Harwood Academic Publishers. 562 pp. ISBN 3718604566.

L.P. Gor'kov, *Disorder and interactions in the system of quasi one-dimensional electrons*, in: Electron-electron interactions in disordered systems, 1985, pp. 619–669. ed. by A.L. Efros, M. Pollak. North-Holland, 1985, xi, 690 pp.

L.P. Gor'kov, *Properties of the A-15 Compounds and One-Dimensionality*, Prog. Low Temperature Physics, 7, Part 2, 517–589 (1978) [ed. D.F. Brewer. Elsevier, 1978, xii, 400pp. ISBN 9780444852090].

A.A. Abrikosov, I.E. Dzyaloshinski, L.P. Gorkov, *Methods of Quantum Field Theory in Statistical Physics*, 2nd Edition, Prentice Hall, 1965, xii, 365 pp. [International series of monographs in natural philosophy, Vol. 4] (Selected Russian publications in the mathematical sciences).

A.A. Abrikosov, L.P. Gor'kov, I. Ye. Dzyaloshinskii, *Quantum Field Theoretical Methods in Statistical Physics*, 2nd Edition, Pergamon Press, Oxford, 1965, xii, 365 pp.